DUDLEY'S GEAR HANDBOOK

Other McGraw-Hill Books of Interest

Avallone, Baumeister • MARKS' STANDARD HANDBOOK FOR MECHANICAL ENGINEERS
Bhushan, Gupta • HANDBOOK OF TRIBOLOGY
Brady, Clauser • MATERIALS HANDBOOK
Bralla • HANDBOOK OF PRODUCT DESIGN FOR MANUFACTURING
Brunner • HANDBOOK OF INCINERATION SYSTEMS
Corbitt • STANDARD HANDBOOK OF ENVIRONMENTAL ENGINEERING
Ehrich • HANDBOOK OF ROTORDYNAMICS
Elliott • STANDARD HANDBOOK OF POWERPLANT ENGINEERING
Freeman • STANDARD HANDBOOK OF HAZARDOUS WASTE TREATMENT AND DISPOSAL
Ganic, Hicks • THE MCGRAW-HILL HANDBOOK OF ESSENTIAL ENGINEERING INFORMATION AND DATA
Gieck • ENGINEERING FORMULAS
Grimm, Rosaler • HANDBOOK OF HVAC DESIGN
Harris • HANDBOOK OF NOISE CONTROL
Harris • SHOCK AND VIBRATION HANDBOOK
Hicks • STANDARD HANDBOOK OF ENGINEERING CALCULATIONS
Jones • DIESEL PLANT OPERATIONS HANDBOOK
Juran, Gryna • JURAN'S QUALITY CONTROL HANDBOOK
Karassik et al. • PUMP HANDBOOK
Kurtz • HANDBOOK OF APPLIED MATHEMATICS FOR ENGINEERS AND SCIENTISTS
Maynard • INDUSTRIAL ENGINEERING HANDBOOK
Parmley • STANDARD HANDBOOK OF FASTENING AND JOINING
Rohsenow, Hartnett, Ganic • HANDBOOK OF HEAT TRANSFER APPLICATIONS
Rohsenow, Hartnett, Ganic • HANDBOOK OF HEAT TRANSFER FUNDAMENTALS
Rosaler, Rice • STANDARD HANDBOOK OF PLANT ENGINEERING
Rothbart • MECHANICAL DESIGN AND SYSTEMS HANDBOOK
Shigley, Mischke • STANDARD HANDBOOK OF MACHINE DESIGN
Tuma • HANDBOOK OF NUMERICAL CALCULATIONS IN ENGINEERING
Wadsworth • HANDBOOK OF STATISTICAL METHODS FOR ENGINEERS AND SCIENTISTS
Young • ROARK'S FORMULAS FOR STRESS AND STRAIN

DUDLEY'S GEAR HANDBOOK

Dennis P. Townsend Editor in Chief
Lewis Research Center, NASA

Second Edition

McGRAW-HILL, INC.
New York St. Louis San Francisco Auckland Bogotá
Caracas Lisbon London Madrid Mexico Milan
Montreal New Delhi Paris San Juan São Paulo
Singapore Sydney Tokyo Toronto

Library of Congress Cataloging-in-Publication Data

Dudley's gear handbook.—2nd ed. / Dennis P. Townsend, editor.
   p. cm.
   Rev. ed. of: Gear handbook. 1st ed. 1962.
   Includes index.
   ISBN 0-07-017903-4
   1. Gearing—Handbooks, manuals, etc. I. Dudley, Darle W.
II. Townsend, Dennis P. III. Gear handbook.
TJ184.D83 1992
621.8'3—dc20   91-1967
                                                                             CIP

Copyright © 1992, 1962 by McGraw-Hill, Inc. All rights reserved. Printed in the United States of America. Except as permitted under the United States Copyright Act of 1976, no part of this publication may be reproduced or distributed in any form or by any means, or stored in a data base or retrieval system, without the prior written permission of the publisher.

2 3 4 5 6 7 8 9 0   DOC/DOC   9 7 6 5 4 3 2

ISBN 0-07-017903-4

The sponsoring editor for this book was Robert W. Hauserman, the editing supervisor was Fred Dahl, and the production supervisor was Suzanne W. Babeuf. It was set in Times Roman by ComCom, Inc.

Printed and bound by R. R. Donnelley & Sons Company.

Information contained in this work has been obtained by McGraw-Hill, Inc., from sources believed to be reliable. However, neither McGraw-Hill nor its authors guarantees the accuracy or completeness of any information published herein, and neither McGraw-Hill nor its authors shall be responsible for any errors, omissions, or damages arising out of use of this information. This work is published with the understanding that McGraw-Hill and its authors are supplying information but are not attempting to render engineering or other professional services. If such services are required, the assistance of an appropriate professional should be sought.

CONTENTS

Contributors xi
Preface xiii

Chapter 1. Theory of Gearing and Application   1.1

1.1 Gear Ratio, Pitch Circles, Centrodes, Pitch Cylinders, Pitch Cones, Axodes, and Pitch Surfaces / *1.1*
1.2 Parametric Representation of Curves and Surfaces / *1.26*
1.3 Envelope of a Family of Surfaces (Curves), Basic Law of Meshing / Contact Lines on Mating Surfaces / *1.30*
1.4 Basic Kinematic Relations / *1.32*
1.5 Conditions of Nonundercuting / *1.33*
1.6 Envelope of Contact Lines on Generating Surface Σ_1 / *1.35*
1.7 Limit Contact Normal / *1.35*
1.8 Relations between Principal Curvatures and Directions for Mutually Enveloped Surfaces / *1.37*
1.9 Contact Ellipse and Bearing Contact / *1.38*
1.10 Modification of Gear Tooth Surfaces / *1.39*
References / *1.42*

Chapter 2. Gear Types and Nomenclature   2.1

2.1 Types of Gears / *2.1*
2.2 Nomenclature of Gears / *2.30*
References / *2.49*

Chapter 3. Gear Arrangements   3.1

3.1 Possibilities in Gear Arrangements / *3.1*
3.2 How the Arrangement May Drastically Affect Ratio, Power, Efficiency, and Gearbox Volume / *3.3*
3.3 Parallel Axis Arrangements / *3.6*
3.4 Parallel Axis Epicyclic Gearing / *3.14*
3.5 Right-Angle Gear Arrangements / *3.24*
3.6 Differential Gearing / *3.26*
3.7 Speed-Changing Transmissions Using Shifting / *3.28*
3.8 Speed-Changing Transmissions Using Clutches and Brakes / *3.30*
3.9 High Ratio Arrangements / *3.31*
References / *3.34*

Chapter 4. Gear Tooth Design — 4.1

4.1 Basic Requirements of Gear Teeth / *4.1*
4.2 Standard Systems of Gear Tooth Proportions / *4.28*
4.3 General Equations Relating to Center Distance / *4.72*
4.4 Elements of Center Distance / *4.83*
  References / *4.94*

Chapter 5. Special Calculations of Spur and Helical Gears — 5.1

5.1 Notation / *5.1*
5.2 Fundamental Gear Equation / *5.3*
5.3 Involute Function and Roll Angle / *5.3*
5.4 Line of Action and Contact Ratio / *5.4*
5.5 Tooth Thickness Calculation / *5.6*
5.6 Hertzian Stress Along Contact Path / *5.7*
5.7 Backlash and Center Distance / *5.8*
5.8 Tooth Thickness from Pin or Ball Measurement / *5.9*
5.9 Interference Limit / *5.13*
5.10 High Contact Ratio Gear / *5.15*
5.11 Speed Ratio of Epicyclic Gears / *5.16*
5.12 Compound Epicyclic Gear Index / *5.16*
5.13 Nonstandard-Addendum Gear / *5.16*
  References / *5.19*

Chapter 6. Gear Tooth Calculations — 6.1

6.1 Calculating Involute Tooth Profile Coordinates / *6.1*
6.2 Calculating Gear Tooth Fillet Coordinates / *6.3*
6.3 Calculating the Inverse of an Involute / *6.6*
6.4 Calculating Addendum Modification / *6.8*
6.5 Derivation of AGMA Hertzian Stress Equation / *6.15*
6.6 Gear Bending Stress Analysis / *6.19*
6.7 Interference of Internal Gears / *6.22*
  References / *6.27*

Chapter 7. Gear Tolerances — 7.1

7.1 Terminology and Definitions Pertaining to Gear Tolerances / *7.1*
7.2 Introduction of Gear Tolerances / *7.5*
7.3 Tolerance Effects / *7.8*
7.4 Types of Tolerances / *7.9*
7.5 Factors Affecting Choice of Tolerance / *7.11*
7.6 Drawings and Tolerances for Various Gear Types / *7.12*
7.7 Tolerance Standards / *7.15*
7.8 Special Considerations for Tolerance Values / *7.27*
7.9 Assembly and Application Considerations / *7.41*
  References / *7.44*

CONTENTS

Chapter 8. Gear Materials

8.1 Ferrous Gear Materials / *8.1*
8.2 Gear Materials / *8.2*
8.3 Carbon and Alloy Steels / *8.13*
8.4 Nonferrous Gear Material / *8.28*
References / *8.41*

Chapter 9. Gear Drawings — 9.1

9.1 General Comments Concerning Preparation of Gear Drawings / *9.1*
9.2 Gear Drawing Practice / *9.2*
9.3 Gear Drawing Specifics / *9.6*
9.4 Preparation of Detail Manufacturing Drawings Using Computer-Aided Design and Computer-Aided Drafting Equipment (CAD) / *9.22*
References / *9.28*

Chapter 10. Gear Reactions and Mountings — 10.1

10.1 Mechanics of Gear Reactions / *10.1*
10.2 Basic Gear Reactions, Bearing Loads, and Mounting Types / *10.4*
10.3 Basic Mounting Arrangements and Recommendations / *10.9*
10.4 Bearing Load Calculations for Spur Gears / *10.12*
10.5 Bearing Load Calculations for Helicals / *10.18*
10.6 Mounting Practice for Bevel and Hypoid Gears / *10.23*
10.7 Calculation of Bevel and Hypoid Bearing Loads / *10.34*
10.8 Bearing Load Calculations for Worms / *10.46*
10.9 Bearing Load Calculations for Spiroids / *10.52*
10.10 Bearing Load Calculations for Other Gear Types / *10.54*
10.11 Design of the Body of the Gear / *10.55*
References / *10.57*

Chapter 11. Load Rating of Gears — 11.1

11.1 Considerations / *11.1*
11.2 Main Nomenclature / *11.2*
11.3 Coplanar Gears (Involute Parallel Gears and Bevel Gears) / *11.4*
11.4 Coplanar Gears: Simplified Estimates and Design Criteria / *11.8*
11.5 Coplanar Gears: Detailed Analysis, Conventional Fatigue Limits, and Service Factors / *11.13*
11.6 RH—Conventional Fatigue Limit of Factor K / *11.21*
11.7 RF—Conventional Fatigue Limit of Factors U_L / *11.40*
11.8 Coplanar Gears: Detailed Life Curves and Yielding / *11.47*
11.9 Coplanar Gears: Prevention of Tooth Wear and Scoring / *11.54*
11.10 Crossed Helical Gears / *11.55*
11.11 Hypoid Gears / *11.56*
11.12 Worm Gearing / *11.57*
11.13 State of the Standards for Load Rating of Gears / *11.59*
References / *11.59*

Chapter 12. Loaded Gears in Action   12.1

12.1 Efficiency of Gears / *12.1*
12.2 Sliding Velocity / *12.2*
12.3 Efficiency of Coplanar Gears (Parallel Axes and Intersecting Axes) / *12.4*
12.4 Efficiency of Nonplanar Gears (Nonintersecting and Nonparallel Axes) / *12.11*
12.5 Gear Trains / *12.17*
12.6 Gear Performance / *12.36*
12.7 Factors That Affect the Performance of Gear Units / *12.37*
12.8 Gear Failures / *12.40*
12.9 Performance Testing of Gears / *12.60*
References / *12.69*

Chapter 13. Gear Vibration   13.1

13.1 Fundamentals of Vibration / *13.1*
13.2 Measurement of Vibration / *13.3*
13.3 Some Examples of Vibration in Geared Units / *13.7*
13.4 Approximate Vibration Limits / *13.10*
13.5 Control of Vibration in Manufacturing Gears and in the Field / *13.16*
13.6 Vibration Analysis Techniques / *13.19*
References / *13.20*

Chapter 14. Gear Noise   14.1

14.1 Gear Noise: The General Problem / *14.2*
14.2 Measurements and Standards / *14.7*
14.3 Power Considerations / *14.10*
14.4 Transmission Error in Gears / *14.11*
14.5 Gear Noise Reduction at the Design Stage / *14.19*
14.6 Gear Noise Modeling / *14.24*
14.7 Gear Rattle / *14.25*
14.8 Gear Noise Troubleshooting / *14.26*
14.9 Closing Comments / *14.28*
References / *14.29*

Chapter 15. Gear Lubrication   15.1

15.1 Power Losses / *15.2*
15.2 Lubrication and Cooling Methods / *15.4*
15.3 Lubrication of Loaded Tank Flank and Surface Distress / *15.8*
15.4 Estimation of the Influence of Lubrication on Surface Distress / *15.16*
15.5 Lubricant and Additive Effects on Gear Fatigue Life / *15.31*
References / *15.31*

Chapter 16. Gear Cutting   16.1

16.1 Gear Milling / *16.1*
16.2 Gear Hobbing / *16.6*
16.3 Gear Shaping / *16.26*
16.4 Gears Cut by Shear Speed / *16.38*
16.5 Gear Broaching / *16.45*

Chapter 17. Gears Made by Dies — 17.1

17.1 Die Processes / *17.1*
17.2 Injection Molding / *17.2*
17.3 Die-Cast Gears / *17.6*
17.4 Powdered Metal Gears / *17.7*
17.5 Forged Gears / *17.12*
17.6 Cold Drawing and Extrusion / *17.14*
17.7 Gear Teeth and Worm Thread Rolling / *17.16*
17.8 Gear Stamping / *17.17*
    References / *17.21*

Chapter 18. Gear Finishing by Shaving, Rolling, and Honing — 18.1

18.1 The Rotary Gear Shaving Process / *18.1*
18.2 Basic Principles / *18.4*
18.3 The Shaving Cutter / *18.12*
18.4 Shaving Machines / *18.14*
18.5 Gear Roll-Finishing / *18.18*
18.6 Rotary Gear Honing / *18.27*

Chapter 19. Grinding of Spur and Helical Gears — 19.1

19.1 Reasons for Grinding / *19.1*
19.2 Grinding Process Mechanics and Process Parameters / *19.2*
19.3 Abrasives / *19.4*
19.4 Grinding Wheels / *19.7*
19.5 Grinding Processes / *19.10*
19.6 Cycle Time Estimates / *19.19*
    References / *19.21*

Chapter 20. Bevel and Hyphoid Gear Manufacturing — 20.1

20.1 The Basic Process / *20.1*
20.2 The Basic Generator / *20.2*
20.3 The Basic Nongenerator / *20.7*
20.4 Gear Blank Requirements Prior to Tooth Processing / *20.9*
20.5 Bevel and Hypoid Gear Manufacturing Procedures / *20.11*
20.6 Bevel Gear Work-Holding Equipment / *20.12*
20.7 Straight Bevel Gear Machines / *20.12*
20.8 Face-Mill and Face-Hob Types of Cutting Machines and Grinders—Spiral Bevel, Zerol Bevel, and Hypoid Gears / *20.20*
20.9 Spiral Bevel and Hypoid Gear Lapping / *20.40*
20.10 Tooling System / *20.45*
    References / *20.49*

Chapter 21. High Ratio Right-Angle Gearing Manufacturing — 21.1

21.1 The Rearrangement of Material / *21.2*
21.2 The Thread-Milling Process / *21.2*

21.3 Thread-Hobbing Process / 21.4
21.4 Determination of Chip Load / 21.4
21.5 Speeds and Feeds / 21.5
21.6 Calculate Thread-Milling Time: Single Cutters / 21.6
21.7 Calculating Hobbing Time for Worms / 21.7
21.8 Hobbing Worm Gears / 21.7
21.9 Infeed Hobbing of Worm Gears / 21.10
21.10 Tangential Hobbing of Worm Gears / 21.12
21.11 Fly Cutting / 21.13
21.12 Feeds for Use in Cutting Worm Gears / 21.14
21.13 Time Required to Cut Worm Gears / 21.15
21.14 Double-Enveloping Worms and Worm Gears / 21.15
21.15 Cutting Time for Double-Enveloping Gear Sets / 21.18
21.16 Worm-Thread Grinding / 21.18
21.17 Worm Rolling / 21.21
21.18 Caburizing and Optional Finishing / 21.23
21.19 Powdered Metal Compacts / 21.23
      References / 21.30

Chapter 22. Gear Cutting Tools 22.1

22.1 Hobs for Spur and Helical Gears / 22.1
22.2 Gear Shaper Cutters / 22.11
22.3 Gear Milling Cutters / 22.21
22.4 Shaving Cutters / 22.25
22.5 Gear Tooth Broaches / 22.33
22.6 Hard Gear Finishing / 22.41
     References / 22.44

Chapter 23. Gear Inspection Devices and Procedures 23.1

23.1 Analytical Gear Checking / 23.2
23.2 Functional Gear Checking / 23.6
23.3 Checking Spur and Helical Gears / 23.13
23.4 Bevel and Hypoid Gear Inspection / 23.23
23.5 Worm Gear Inspection / 23.38
23.6 Checking for Gear Wear / 23.42
23.7 Checking Extra Large and Extra Small Gears / 23.45
     References / 23.51

Chapter 24. Numerical Data Tables 24.1

24.1 Gear Measuring Tables / 24.1
24.2 Trigonometric Functions / 24.20
24.3 Involute Functions / 24.20
24.4 Arc and Chord Data / 24.27
24.5 Hardness Testing Data / 24.27
     References / 24.30

Index follows Chapter 24

CONTRIBUTORS

Michael J. Broglie *The Dudley Technical Group, Inc., San Diego, Calif.* (CHAPS. 17, 23)

Prof. Ing. Giovanni Castellani *Studio Progettazione Macchine e Trasmissioni, Modena, Italy* (CHAP. 11)

Dr. Charles Chao *Superior Gear, Gardena, Calif.* (CHAP. 5)

Roy Cunningham *Boeing Vortel Co., Philadelphia, Pa.* (CHAP. 8)

Raymond Drago *Boeing Vertol Co., Philadelphia, Pa.* (CHAP. 2)

Darle Dudley *Dudley Engineering, San Diego, Calif.* (CHAPS. 2, 4, 10, 13, 24)

John Dugas *National Broach & Machine Co., Mt. Clemens, Mich.* (CHAP. 18)

Robert Errichello *Gear Tech, Albany, Calif.* (CHAP. 6)

Dezi J. Folenta *Pres., Transmission Technology Inc., Fairfield, N.J.* (CHAP. 9)

Paul F. Fox *Auburn Gear, Auburn, In.* (CHAP. 3)

Marvin Hartman *Gear Consultant, Thousand Oaks, Calif.* (CHAP. 10)

Robert Hotchkiss *The Gleason Works, Rochester, N.Y.* (CHAP. 20)

Prof. Donald Houser *Ohio State University, Department of Mechanical Engineering, Columbus, Ohio* (CHAP. 14)

Richard Kitchen *The Gleason Works, Rochester, N.Y.* (CHAP. 20)

Dr. Aizoh Kubo *Department of Precision Mechanics, Kyoto University, Japan* (CHAP. 15)

Prof. Faydor L. Litvin *The University of Illinois at Chicago Circle, Department of Mechanical Engineering, Chicago, Ill.* (CHAP. 1)

William R. McVea *The Gleason Works, Rochester, N.Y.* (CHAP. 20)

A. Donald Moncrieff *Gear Consultant, Vero Beach, Fl.* (CHAPS. 16, 22)

Raymond Paquet *Plastic Gearing Technology, Manchester, Conn.* (CHAP. 17)

Dr. J. W. Polder *Neunen 5671 ED5, The Netherlands* (CHAP. 6)

Dr. Suren B. Rao *National Broach & Machine Co., Mt. Clemens, Mich.* (CHAP. 19)

Eugene E. Shipley *Mechanical Technology Inc., Latham, N.Y.* (CHAP. 12)

Walter L. Shoulders *Reliance Electric Co., Columbus, In.* (CHAP. 21)

Danny F. Smith *The Dudley Technical Group, Inc., San Diego, Calif.* (CHAP. 23)

Robert E. Smith *RE Smith & Co., Inc., Rochester, N.Y.* (CHAP. 7)

Dennis P. Townsend *NASA, Lewis Research Center, Ohio* (CHAPS. 6, 15, 20, 23)

PREFACE

Considerable advances have been made in many areas of gear technology during the nearly 30 years since the publication of the first edition of the *Gear Handbook*. The most notable of these advances are the improved methods of gear manufacturing, gear inspection, gear vibration analysis, and computerized gear design, not to mention the developments in gear materials, gear noise analysis and evaluation, and gear lubrication methods.

In this new edition an attempt has been made to bring the information on these new gear technologies up to date, and to present them in a usable way for the practicing engineer working in the gearing discipline. Most of the chapters have been completely rewritten, and two new chapters have been added to provide the reader with recent developments in gear vibration and gear noise.

Chapter 1, "Theory of Gearing and Application," presents a new approach to gear geometry with the excellent work of Professor Faydor Litvin, which many readers will appreciate.

Chapter 7, "Gear Tolerances," has been completely revised and updated to include the new developments in transmission error measurements and a comparison of the various gear standards on gear tolerance.

"Gear Materials" (Chap. 8) includes the latest on advanced gear materials, including the new high hot hardness gear steels and the latest advances in plastic gear materials.

Chapter 11, "Load Rating of Gears," was written by Dr. Giovanni Castellani from Italy to give the reader a more international approach on this subject with a comparison of the latest rating methods used by the various international organizations, such as AGMA, DIN, and ISO.

"Gear Vibration" (Chap. 13) is completely new, providing the reader with practical information to deal with the complex problems of gear vibration and dynamics.

A new addition to the handbook, Chapter 14, "Gear Noise," provides the gear engineer with the very latest developments used to evaluate gear noise and the methods available to produce quieter gear boxes.

"Gear Lubrication" (Chap. 15) has been completely revised since many new and improved approaches have been developed in gear lubrication since the first edition. Methods are presented to determine the EHL film thickness, scoring predictions, and gear lubrication and cooling for various gear applications.

Donald Moncrieff's chapters on gear cutting and gear cutting tools (Chaps. 16 and 22) have benefited from the work of Professor Masato Ainoura from Kurume National Technical College in Japan who has conducted extensive research on gear cutting methods such as high-speed hobbing with TiN-coated hobs and finishing of hard gears by skiving.

I am very grateful to the 26 contributors, who were selected because of their expert knowledge and experience in the various gear technologies. The unselfish contribution of these authors has made it possible to complete this large task of putting together a complete revision of the *Gear Handbook*.

I am also grateful to the many companies and representatives who provided the photographs, drawings, and expertise on the various equipment for gear manufacturing, inspection, and so on, without which the coverage of these important areas of gear manufacturing would have been incomplete.

My special thanks to Darle Dudley who asked me to perform the task of revising the handbook and who has been a great help and encouragement in finding contributors—in addition to contributing some of the work himself during his "semiretirement."

I also would like to thank AGMA for the use of their many standards and their help in providing the necessary input when required.

Dennis P. Townsend
Cleveland, Ohio

CHAPTER 1
THEORY OF GEARING AND APPLICATION

Faydor L. Litvin
Professor of Mechanical Engineering
University of Illinois at Chicago, Illinois

The theory of gearing provides: (1) methods for generation of conjugate gear tooth surfaces, and (2) analytical and computational methods for analysis and synthesis of gearing.

Gears with conjugate tooth surfaces transform motion with constant angular velocity ratio. The modern tendency is to localize the bearing contact of gear tooth surfaces and reduce the sensitivity of gears to misalignment. Theoretically, gear tooth surfaces with a localized bearing contact are in tangency at a point at every instant but at a line. Such a contact for spur and helical gears with parallel axes is accomplished by crowning of the pinion surface. Due to the elasticity of gear tooth surfaces, their contact under a load is spread over an elliptical area. The center of symmetry of the instantaneous contact ellipse is the theoretical point of contact.

The main problems of analysis and synthesis of gearing are as follows:

1. Determination of tooth surfaces generated by cutting
2. Avoidance of tooth undercutting
3. Simulation of meshing and contact
4. Synthesis of gearing with improved criteria

The modern tendencies in the theory of gearing are directed at the skillful use of computers and display of the results by computer graphics, to improve the gear design and manufacture.

1.1 GEAR RATIO, PITCH CIRCLES, CENTRODES, PITCH CYLINDERS, PITCH CONES, AXODES, AND PITCH SURFACES

1.1.1 Gear Ratio

The instantaneous angular velocity ratio (or the instantaneous gear ratio) is represented by

$$m_{12} = \frac{\omega^{(1)}}{\omega^{(2)}} \tag{1.1}$$

The gear ratio is represented by

$$m_{12}^o = \frac{n_1}{n_2} = \frac{N_2}{N_1} \tag{1.2}$$

where n_1 and n_2 are the revolutions of gears, and N_1 and N_2 are the numbers of gear teeth. Both ratios are the same if the gears are provided with conjugate surfaces (tooth shapes for planar gears) and are free from errors of manufacturing and assembly. If not, the gears transform rotation with a varied instantaneous ratio m_{12} within any gear revolution. Here m_{12} may be represented by a function $m_{12}(\varphi_1)$ where φ_1 is the angle of rotation of the driving gear. Function $m_{12}(\varphi_1)$ is a discontinuous function, since the change of teeth in the process of meshing causes a jump in the ratio of angular velocities.

The gear ratio m_{12}^o is a constant value and may be applied for all types of gears, with conjugate and nonconjugate gear tooth surfaces. However, this gear characteristic can be used only as the ratio between the gear revolutions but not their angular velocities.

1.1.2 Pitch Circle of a Spur Gear

Consider a rack cutter is in mesh with the gear being generated (Fig. 1.1). Point I is the instantaneous center of rotation, since at this point the following equation is observed:

$$\mathbf{v}^{(2)} = \mathbf{v}^{(1)} = \omega \times \overline{OI} \tag{1.3}$$

Here $\mathbf{v}^{(2)}$ is the velocity of the rack cutter in translational motion; $\mathbf{v}^{(1)}$ is the velocity of gear point I in rotation about O; ω is the gear angular velocity.

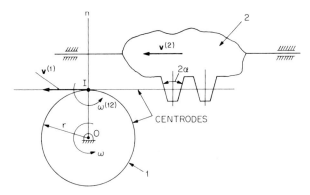

FIGURE 1.1 Centrodes of spur gear and rack cutter

Point I lies on line O_n which is perpendicular to the direction of the rack translation, and the distance OI is determined by

$$OI = \frac{v^{(2)}}{\omega} \quad (1.4)$$

Consider two coordinate systems S_1 and S_2 rigidly connected to the gear and rack cutter. Point I traces out a circle of radius $r = OI$ in system S_1 and a straight line which is tangent to the above-mentioned circle in system S_2. These circle and straight line are the loci of instantaneous center of rotation in S_1 and S_2, respectively, and are called the *centrodes*. The relative motion of the gear with respect to the rack cutter is pure rolling about I with the angular velocity $\omega^{(12)} = v^{(2)}/r$. The gear *pitch circle* is the centrode in meshing with the rack cutter. The radius r of the gear pitch circle depends on the ratio $v^{(2)}/\omega$ only and does not depend on the installment of the rack cutter with respect to the gear center. The diameter of the gear pitch circle is given by

$$d = \frac{pN}{\pi} \quad (1.5)$$

where N is the number of gear teeth, and p is the distance between two neighboring teeth measured: (1) along the rack centrode (for teeth taken at the rack), and (2) along the gear pitch circle (for teeth taken at the gear). Distance p between the gear neighboring teeth is called the *circular pitch*. The ratio

$$P = \frac{N}{d} \quad (1.6)$$

is called the *diametral pitch*. The magnitudes of P are standardized.

The circular and diametral pitch are related by the equation

$$Pp = \pi \quad (1.7)$$

1.1.3 Centrodes, Operating Pitch Circles, and Pitch Circles of Spur Gears

Consider that links 1 and 2 rotate about O_1 and O_2 (Fig. 1.2) with angular velocities $\omega^{(1)}$ and $\omega^{(2)}$, respectively; $C = O_1 O_2$ is the distance between the centers of rotation O_1 and O_2. Point I of line $O_1 O_2$, where $v^{(1)} = v^{(2)}$, is the instantaneous center of rotation. Here

$$v^{(1)} = \omega^{(1)} \times \overrightarrow{O_1 I} \qquad v^{(2)} = \omega^{(2)} \times \overrightarrow{O_2 I} \quad (1.8)$$

The relative velocity at point I, given by $v^{(12)} = v^{(1)} - v^{(2)}$ or $v^{(21)} = v^{(2)} - v^{(1)}$, is equal to zero.

The location of the *instantaneous center of rotation* (ICR) on line $O_1 O_2$ is represented by the equation

$$\frac{O_2 I}{O_1 I} = \frac{\omega^{(1)}}{\omega^{(2)}} = m_{12} \quad (1.9)$$

where m_{12} is the angular velocity ratio. ICR does not change its location on line $O_1 O_2$ in the process of meshing if the gear ratio is constant. In the case of noncircular gears

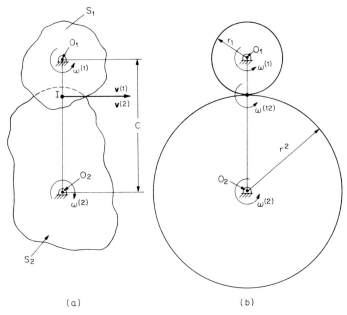

FIGURE 1.2 Centrodes of external spur gears

the gear ratio is represented by a function $m_{12}(\varphi_1)$ where φ_1 is the angle of rotation of the driving gear. For these gears ICR moves along line $O_1 O_2$ in the process of transformation of motion. Point I traces out circles of radii $r_1 = O_1 I$ and $r_2 = O_2 I$ (provided m_{12} is a constant) in coordinate systems S_1 and S_2, which are rigidly connected to gears 1 and 2. These circles, the loci of I in S_1 and S_2, are called the *gear centrodes*. The relative motion of gears may be represented as pure rolling of one centrode over the other, as rotation of one centrode with respect to the other one about their point of tangency. The angular velocity in rotation about I is

$$\omega^{(12)} = \omega^{(1)} - \omega^{(2)} \quad \text{or} \quad \omega^{(21)} = \omega^{(2)} - \omega^{(1)} \tag{1.10}$$

This equation yields

$$|\omega^{(12)}| = |\omega^{(21)}| = \omega^{(1)} + \omega^{(2)} \tag{1.11}$$

The radii of centrodes are represented by the equations

$$r_1 = \frac{C}{1 + m_{12}} \qquad r_2 = \frac{C}{1 + m_{21}} = \frac{C m_{12}}{1 + m_{12}} \tag{1.12}$$

where

$$m_{21} = \frac{\omega^{(2)}}{\omega^{(1)}} = \frac{N_1}{N_2} = \frac{1}{m_{12}}$$

Figure 1.3 shows the case where the gears are rotated in the same direction. The instantaneous center of rotation, I, lies on line O_1O_2, but it is offset with respect to the segment O_2O_1. The gear centrodes are in internal tangency at I. The relative motion is rotation about I with the angular velocity $\omega^{(12)}$ [or $\omega^{(21)}$] where

$$|\omega^{(12)}| = |\omega^{(21)}| = |\omega^{(1)} - \omega^{(2)}| \tag{1.13}$$

The angular velocity $|\omega^{(12)}| = |\omega^{(21)}|$ and, correspondingly, the sliding velocity are lessened for gears being in internal tangency. This is why such gears are used in planetary trains where a higher efficiency is desired. The radii of gear centrodes may be expressed in terms of C and m_{12} as follows:

$$r_1 = \frac{C}{|1 - m_{12}|} \qquad r_2 = \frac{C}{|1 - m_{21}|} = \frac{Cm_{12}}{|1 - m_{12}|} \tag{1.14}$$

It is necessary to emphasize that the gear centrodes coincide with the pitch circles if and only if the center distance C is chosen as follows:

1. In case of external tangency of centrodes (Fig. 1.2)

$$C = \frac{N_1 + N_2}{2P} \tag{1.15}$$

2. For internal tangency of centrodes (Fig. 1.3)

$$C = \frac{|N_2 - N_1|}{2P} \tag{1.16}$$

where N_1 and N_2 are the numbers of gear teeth.

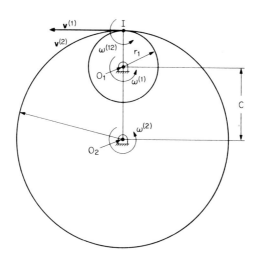

FIGURE 1.3 Centrodes of internal spur gears

1.6 CHAPTER ONE

The center distance for nonstandard involute gears may not satisfy Eqs. (1.15 and 1.16). In such a case we have to distinguish between the gear centrodes and gear pitch circles. The term *operating pitch circles* means gear centrodes and is used in the technical literature on gears instead of the term *centrodes*.

1.1.4 Spur Gears: Line of Action, Pressure Angle, and Center Distance

The line of action of meshing gears is the locus of contact points in the fixed coordinate system that is rigidly connected to the gear housing. The line of action of involute gears is the tangent KL to the base circles (Fig. 1.4). Point I of intersection of KL with the center distance $O_1 O_2$ is the point of tangency of the gear centrodes. The angle that is formed between KL and tangent t-t to the centrodes is called the *pressure angle*. For standard gears the pressure angle is equal to the shape angle α of the rack cutter (Fig. 1.1). The pressure angle for nonstandard gears is designated by ψ as shown in Fig. 1.4,

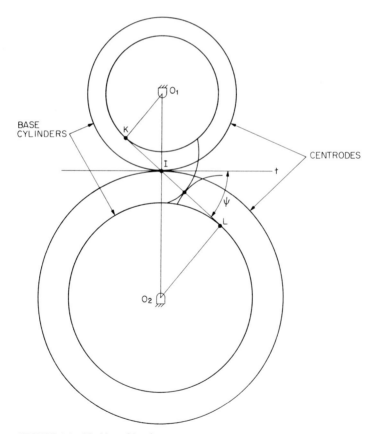

FIGURE 1.4 Meshing of involute spur gears

and, generally speaking, $\psi \neq \alpha$. The basic equation for the determination of the pressure angle ψ is as follows:

$$\text{inv } \psi = \text{inv } \alpha + \frac{2(e_P + e_G)P}{N_P + N_G} \tan \alpha \tag{1.17}$$

Parameter e (Fig. 1.5) determines the installment of the rack cutter by cutting; e is the distance between the middle line m-m of the rack cutter and the tangent a-a to the pitch circle; e is positive if m-m is located up to a-a (Fig. 1.5); e_P and e_G designate the installment of the rack cutter for the pinion and the gear, respectively. P is the diametral pitch. N_P and N_G are the pinion and gear tooth numbers. We recall that the designation inv x means: inv x = tan $(x) - x$ (x in radians).

Particular Cases

1. In the case of standard gears the rack cutter is installed in such a way that its middle line m-m coincides with tangent a-a to the pitch circle, and $e_P = 0$ and $e_G = 0$. Then $\psi = \alpha$ and the gear pressure angle is equal to the rack cutter shape angle.

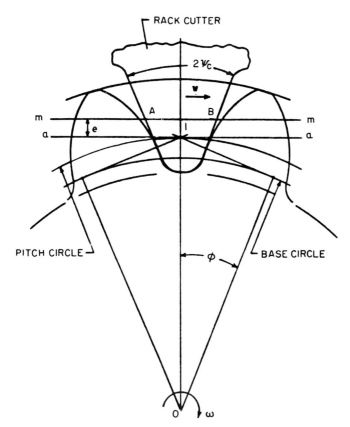

FIGURE 1.5 Standard and nonstandard settings of a rack cutter

2. The installments of the rack cutter, e_P and e_G, are related by $e_P + e_G = 0$. This means that e_P and e_G are of different signs but $|e_P| = |e_G|$. In this case we also have $\psi = \alpha$.

3. In the general case, $e_P + e_G \neq 0$ and $\psi \neq \alpha$.

The center distance of nonstandard gears is represented by the equation

$$C^* = \frac{(N_P + N_G) \cos \alpha}{2P \cos \psi} \tag{1.18}$$

In the case where $\psi = \alpha$ (because $e_P + e_G = 0$) we have

$$C^* = C = \frac{N_P + N_G}{2P}$$

where C is the standard center distance.

It is necessary to emphasize that the base circles of a standard and nonstandard spur gear are represented by the same equation as follows:

$$r_b = \frac{N}{2P} \cos \alpha \tag{1.19}$$

where α is the shape angle of the rack cutter.

The nominal value of the tooth thickness and the space width of the mating gears that are measured along their centrodes must be equal.

1.1.5 Helical Gears: Basic Concepts

The meshing of the rack cutter and a helical gear is to be considered in a three-dimensional space (Fig. 1.6). There is a straight line I-I where $\mathbf{v}^{(2)} = \mathbf{v}^{(1)}$ at any point of this line. Here $\mathbf{v}^{(2)}$ is the velocity of the rack cutter in translational motion; $\mathbf{v}^{(1)} = \boldsymbol{\omega} \times \overline{OI}$ is the rotational velocity of the gear at any point of line I-I. The straight line I-I is located in the fixed space; it is directed perpendicular to the direc-

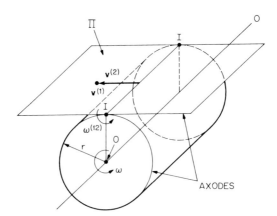

FIGURE 1.6 Axodes of a helical gear and a rack cutter

tion of the translational motion of the rack cutter, and, at the same time, it is parallel to the gear axis $O\text{-}O$. The distance between $I\text{-}I$ and the gear axis is determined by the equation

$$OI = \frac{v^{(2)}}{\omega}$$

The locus of straight lines $I\text{-}I$ in the coordinate system S_1, which is rigidly connected to the gear, represents the cylinder of radius $r = OI$ (Fig. 1.6). The locus of straight lines $I\text{-}I$ in coordinate S_2, which is rigidly connected to the rack cutter, is plane II, that is tangent to the cylinder of radius r. Plane II performs the translational motion with velocity $\mathbf{v}^{(2)}$, while the cylinder rotates about axis $O\text{-}O$ with angular velocity $\omega = v^{(2)}/r$. The relative motion of the cylinder and plane II is pure rolling. The cylinder rolls over plane II, which is fixed, or the plane rolls over the cylinder that is at rest. The relative motion may be represented at every instant as rotation about the line of tangency of the cylinder and the plane, the straight line $I\text{-}I$. $I\text{-}I$ is the instantaneous axis of rotation in relative motion, and the angular velocity in rotation about $I\text{-}I$, $\omega^{(12)}$, is equal to ω (Fig. 1.6).

To determine the basic parameters of a helical gear it is necessary to consider the transverse and normal section of the rack cutter that is in mesh with a helical gear.

Figure 1.7a shows a rack cutter for a spur gear; the direction of the rack cutter teeth is parallel to the gear axis. Figure 1.7b shows an imaginary rack cutter with skew teeth that will be in mesh with a helical gear. The direction of the teeth makes an angle $\beta = 90° - \lambda$ with the gear axis. It is evident that a rack cutter for spur gears may be "cut" out from the rack cutter that is designated for a helical gear, as shown in Fig. 1.7b. Figure 1.7c shows the transverse section $A\text{-}A$ and the normal $B\text{-}B$ section of the rack cutter for a helical gear. Considering the drawings of Fig. 1.7c we obtain

$$p_t = \frac{p_n}{\cos \beta} \qquad P_t = P_n \cos \beta \qquad p_t P_t = p_n P_n = \pi \qquad (1.20)$$

where p_t and p_n are the distances between the neighboring teeth that are measured in the transverse section and normal section, respectively. The shape angles in these sections are related as follows:

$$\tan \alpha_t = \frac{\tan \alpha_n}{\cos \beta} \qquad (1.21)$$

where α_n is the same shape angle that is standardized for spur gears. The radius of the pitch cylinder is represented as follows:

$$r = \frac{N}{2P_t} = \frac{N}{2P_n \cos \beta} \qquad (1.22)$$

The base cylinder is

$$r_b = \frac{N}{2P_t} \cos \alpha_t = \frac{N}{2P_n \cos \beta} \cos \alpha_t \qquad (1.23)$$

The helical gear tooth surface is a screw involute surface. The transverse section of the tooth surface is an involute curve (Fig. 1.8a). The intersection of the tooth surface

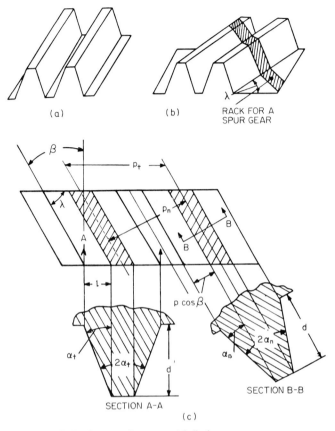

FIGURE 1.7 Rack cutters for spur and helical gears

by a cylinder, for instance with the pitch cylinder, represents a helix (Fig. 1.8b) The lead H represents the axial displacement for one revolution; β and λ represent the *helix* angle and the *lead* angle. The drawings of Figs. 1.8b and c yield

$$\tan \lambda = \cot \beta = \frac{H}{2\pi r} \tag{1.24}$$

The ratio

$$h = \frac{H}{2\pi} \tag{1.25}$$

is called the screw parameter and represents the axial displacement for the angle of one radian. The lead H and the screw parameter h are the same for helices that belong to various cylinders of the helical gear. This means that

$$\frac{\tan \lambda_x}{\tan \lambda} = \frac{r_x}{r} \tag{1.26}$$

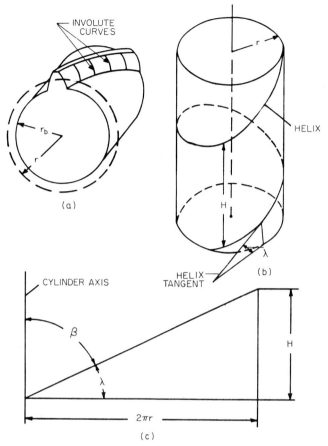

FIGURE 1.8 Screw involute surface

where r_x = radius of any cylinder of the helical gear
 λ_x = lead angle of the helix on this cylinder
 λ = lead angle of the pitch cylinder
 r = radius of the pitch cylinder

The screw involute surface may be generated by the screw motion of the transverse involute shape (Fig. 1.8a) or by a screw motion of a straight line that keeps its direction in this motion as a tangent to the helix on the base cylinder. Figure 1.9 shows the screw involute surface that is covered with straight lines—tangents to the base cylinder helix.

1.1.6 Helical Gears with Parallel Axes: Pitch Cylinders and Operating Pitch Cylinders

Henceforth, we consider two cases of meshing: (1) of a helical gear with rack cutter, and (2) of two helical gears in mesh with each other. When a single gear is in mesh

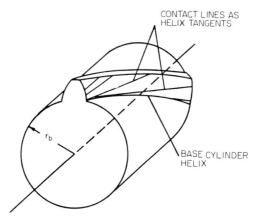

FIGURE 1.9 Screw involute surface and lines of contact

with the rack cutter, the gear axode is the pitch cylinder and the axode of the rack cutter is plane Π that is tangent to the gear pitch cylinder (Fig. 1.6). When two gears are in mesh with each other, their axodes are two cylinders—*operating pitch cylinders*—that contact each other along their common generatrix.

The determination of operating pitch cylinders is based on the following considerations:

Assume that two helical gears transform rotation between parallel axes (Fig. 1.10). The instantaneous axis of rotation, I-I, is parallel to the gear axes and perpendicular to the shortest distance between the gear axes. The gear axodes are two cylinders of radii r_1 and r_2 that are determined by Eqs. (1.12) and (1.14), represented for the cases where the axodes are in external and internal tangency. The relative motion of the cylinders is pure rolling-rotation about I-I with the angular velocity $\omega^{(12)} = \omega^{(21)} = \omega^{(1)} + \omega^{(2)}$ (external tangency) $\omega^{(12)} = \omega^{(21)} = |\omega^{(1)} - \omega^{(2)}|$ (internal tangency).

The operating pitch cylinders may coincide with the pitch cylinders if and only if the center distance C between the gears is chosen as follows:

1. In the case of external tangency we have

$$C = \frac{N_1 + N_2}{2P_t} = \frac{N_1 + N_2}{2P_n \cos \beta} \tag{1.27}$$

2. For internal tangency we have

$$C = \frac{|N_2 - N_1|}{2P_t} = \frac{|N_2 - N_1|}{2P_n \cos \beta} \tag{1.28}$$

Figure 1.10 shows that, due to the increase of center distance C, the radii of pitch cylinders, r_{p_1} and r_{p_2}, are less than the radii of gear axodes, cylinders of radii r_1 and r_2.

Consider that the center distance (C') of helical gears is not standard and r'_1 and r'_2 designate the radii of operating pitch cylinders. It is evident that

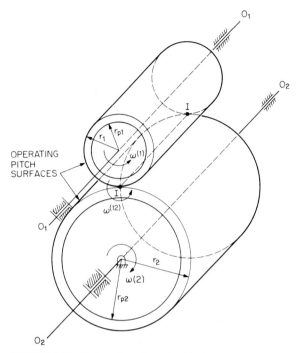

FIGURE 1.10 Pitch surfaces (axodes) of helical gears

$$r'_1 + r'_2 = C' \quad \text{and} \quad \frac{r'_1}{r'_2} = \frac{N_1}{N_2}$$

The gears have the same lead angle λ' on the operating pitch cylinders; the directions of the helices are opposite for the case of external tangency of the gears. The leads H_1 and H_2 (or the screw parameters h_1 and h_2) are not equal and are related as follows:

$$\frac{H_1}{H_2} = \frac{r'_1}{r'_2} \tag{1.29}$$

To determine the operating pressure angle α'_t we may use the following computational procedure. We assume that the input data are: α_n, P_n, β, N, and r' (the operating pitch cylinder radius). From this we obtain:

$$\tan \alpha_t = \frac{\tan \alpha_n}{\cos \beta} \qquad \cos \alpha'_t = \frac{N \cos \alpha_t}{2r' P_n \cos \beta} \tag{1.30}$$

1.1.7 Axodes and Pitch Cones of Bevel Gears

Bevel gears transform rotation between intersecting axes and their axodes are two cones whose axes coincide with the axes of gear rotation (Fig. 1.11). The common generatrix of the cones, OI, is the *instantaneous axis of rotation* (IAR).

Consider that the gears transform rotation between axes Oa and Ob with the

1.14 CHAPTER ONE

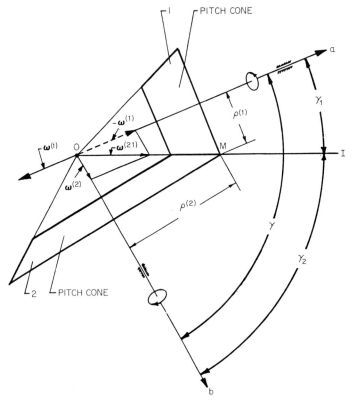

FIGURE 1.11 Pitch surfaces (axodes) of bevel gears

angular velocities, $\omega^{(1)}$ and $\omega^{(2)}$, respectively. Axes Oa and Ob form angle γ (Generally speaking γ is not equal to 90°). The orientation of OI is represented by the vector equation

$$\omega^{(21)} = \omega^{(2)} - \omega^{(1)} \tag{1.31}$$

where $\omega^{(21)}$ is the angular velocity in relative motion of gear 2 with respect to gear 1. Vector $\omega^{(21)}$ is directed along OI. The equation

$$\omega^{(12)} = \omega^{(1)} - \omega^{(2)} \tag{1.32}$$

where $\omega^{(12)} = -\omega^{(21)}$, may be used instead of Eq. (1.31).

Imagine that two coordinate systems, S_1 (x_1, y_1, z_1) and S_2 (x_2, y_2, z_2), are rigidly connected to gears 1 and 2. While the gears are rotated about $Oa = z_1$ and $Ob = z_2$, the instantaneous axis of rotation generates in coordinate systems S_1 and S_2 two mating cones with pitch angles γ_1 and γ_2. Thus the axodes of bevel gears are the loci of IAR in coordinate systems S_1 and S_2. The axodes of bevel gears are also called the pitch cones and the line of tangency of the pitch cones is called the pitch line. The linear velocities in rotation about axes Oa and Ob are equal and have the same direction for any point M of the pitch line OI. Thus

$$\mathbf{v}^{(1)} = \mathbf{v}^{(2)} \tag{1.33}$$

where $\quad |\mathbf{v}^{(1)}| = \omega^{(1)}\rho^{(1)} \quad |\mathbf{v}^{(2)}| = \omega^{(2)}\rho^{(2)}$

Vectors $\mathbf{v}^{(1)}$ and $\mathbf{v}^{(2)}$ pass through point M and are perpendicular to the plane of Fig. 1.11.
The pitch angles γ_1 and γ_2 are represented as follows:

$$\cot \gamma_1 = \frac{m_{12} + \cos \gamma}{\sin \gamma}$$
$$\cot \gamma_2 = \frac{m_{21} + \cos \gamma}{\sin \gamma} \tag{1.34}$$

where

$$m_{12} = \frac{\omega^{(1)}}{\omega^{(2)}} = \frac{N_2}{N_1} \quad m_{21} = \frac{\omega^{(2)}}{\omega^{(1)}} = \frac{N_1}{N_2} \tag{1.35}$$

The relative motion of the pitch cones is pure rolling about OI with the angular velocity

$$|\omega^{(12)}| = |\omega^{(21)}| = \{[\omega^{(1)}]^2 - 2\omega^{(1)}\omega^{(2)} \cos \gamma + [\omega^{(2)}]^2\}^{1/2} \tag{1.36}$$

For the case where the rotation is performed between two mutual perpendicular axes and $\gamma = 90°$, Eqs. (1.34) yield

$$\cot \gamma_1 = m_{12} \quad \cot \gamma_2 = m_{21} \tag{1.37}$$

There is a particular case where one of the pitch cones turns out into a plane. Figure 1.12 shows that the angle between axes Oa and Ob is $\gamma = 90° + \gamma_1$ and $\gamma_2 = 90°$. The gear ratio is

$$m_{12} = \frac{1}{\sin \gamma_1} \tag{1.38}$$

The axodes are the pitch cone of angle γ_1 and the pitch plane.

1.1.8 Axodes of Gears with Crossed Axes

Hypoid gears, worm gear drives and helical gears transform rotation between crossed axes. We have found for the previous cases where the gears transform rotation between parallel and intersected axes, that the relative motion of the axodes is pure rolling. For the case where the axes of gear rotation are crossed, the axodes are two hyperboloids of revolution, and the relative motion is not rolling but screw motion.

Consider that the gears transform rotation about axes z_f and z_2 with the angular velocities $\omega^{(1)}$ and $-\omega^{(2)}$, respectively (Fig. 1.13). The crossed axes form angle γ and the shortest distance between the axes is $O_1 O_2 = C$. The determination of location and orientation of the axis of screw motion is based on the following considerations:

1. Vectors of angular velocities $\omega^{(1)}$ and $-\omega^{(2)}$ which pass through points O_1 and O_2, respectively, may be substituted by equal vectors which pass through point B (Fig. 1.13a), and vector moments

$$\mathbf{m}_1 = \overline{BO_1} \times \omega^{(1)} \quad \text{and} \quad \mathbf{m}_2 = \overline{BO_2} \times [-\omega^{(2)}] \tag{1.39}$$

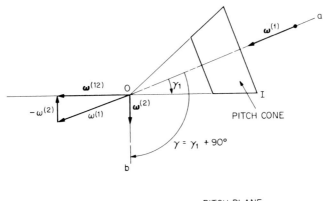

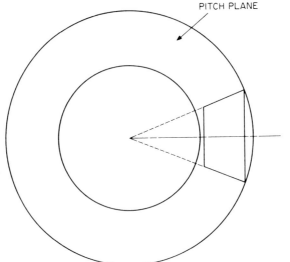

FIGURE 1.12 Pitch surfaces (axodes) of a bevel gear and a generation gear

2. Point B lies on the shortest distance and its location must satisfy the following requirements:

$$\omega^{(12)} = \omega^{(1)} - \omega^{(2)} \tag{1.40}$$

$$\mathbf{m} = \mathbf{m}_1 + \mathbf{m}_2 = (\overline{BO}_1 \times \omega^{(1)}) - (\overline{BO}_2 \times \omega^{(2)}) \tag{1.41}$$

$$\mathbf{m} = \lambda \omega^{(12)} \quad \lambda \neq 0 \tag{1.42}$$

Equations (1.40)–(1.42) represent that the relative motion of gear 1 with respect to gear 2 is a screw motion. The axis of screw motion (SS) (Fig. 1.13) passes through point B and the relative motion is rotation about SS with the angular velocity $\omega^{(12)}$ and translation along SS with the linear velocity $\mathbf{m}$. Vectors $\omega^{(12)}$ and $\mathbf{m}$ are collinear vectors that may be of the same or opposite directions. The location of point B in the

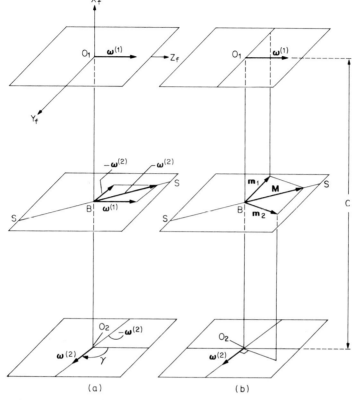

FIGURE 1.13 Axis of screw motion

coordinate system S_f (X_f, Y_f, Z_f) (shown in Fig. 1.13a) is given by the equation

$$X_f = \frac{-C\, m_{21}\, (m_{21} - \cos \gamma)}{1 - 2 m_{21} \cos \gamma + m_{21}^2} \qquad Y_f = Z_f = 0 \tag{1.43}$$

where

$$m_{21} = \frac{\omega^{(2)}}{\omega^{(1)}} \tag{1.44}$$

Equation (1.43) is derived for the direction of $\omega^{(2)}$ that is shown in Fig. 1.13. The sign of m_{21} in Eq. (1.43) must be changed to the opposite one for the case where $\omega^{(2)}$ has the opposite direction.

Consider now that two coordinate systems, S_1 (X_1, Y_1, Z_1) and S_2 (X_2, Y_2, Z_2), are rigidly connected to gears 1 and 2, respectively. When the gears rotate, the axis of screw motion, SS, generates in coordinate systems S_1 and S_2 two *hyperboloids of revolution*. One of these hyperboloids, generated in coordinate system S_1, is shown in Fig. 1.14. Thus, the axodes of gears with crossed axes are the loci of straight lines

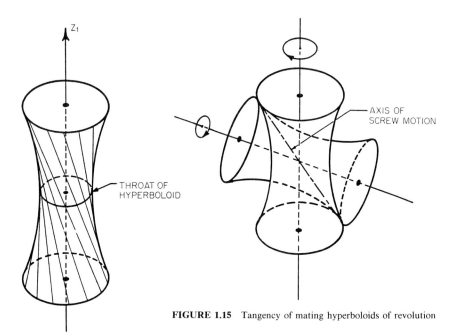

FIGURE 1.15 Tangency of mating hyperboloids of revolution

FIGURE 1.14 Hyperboloid of revolution

generated by the axis of screw motion in coordinate systems S_1 and S_2, rigidly connected to the gears. The axodes are two hyperboloids of revolution that are in tangency at every instant along a straight line (Fig. 1.15). The line of tangency of the hyperboloids is the axis of screw motion. The relative motion of one gear with respect to the other may be represented as a motion in two components: rotation *about* and translation *along* the line of tangency of the hyperboloids.

In the cases of spur gears, helical gears with parallel axes, and bevel gears, the axodes of gears coincide with their pitch surfaces. The pitch surfaces of gears with crossed axes differ from the axodes entirely. It is shown in the following that the pitch surfaces of crossed helical gears, a worm, and a worm gear are two cylinders with crossed axes being in contact at a point. The pitch surfaces of hypoid gears are two cones with crossed axes also being in contact at a point. Generally, the zone of meshing of tooth surfaces of gears with crossed axes is far away from the zone of contact of the hyperboloids.

1.1.9 Operating Pitch Cylinders of Helical Gears with Crossed Axes

The axes of operating pitch cylinders coincide with the gear axes (Fig. 1.16a) and form the same twist angle as the gear axes. The operating pitch cylinders contact each other at point P (called the pitch point) that is located on the line of shortest distance between the gear axes. The helices of the operating pitch cylinders have a common tangent (t-t) at the pitch point P. Tangent t-t lies in plane II that is tangent to the pitch cylinders at pitch point P. Figure 1.16b shows the helix angle β_2 on the gear pitch cylinder of radius $r_p^{(2)}$.

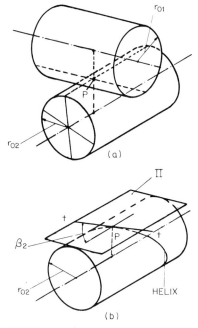

FIGURE 1.16 Pitch surfaces of crossed helical gears

It is necessary to differentiate between standard and nonstandard involute helical gears. The pitch cylinder of a standard helical gear coincides with the axoid of the gear in mesh with the rack cutter. The shortest distance for standard helical gears is represented by the following equation:

$$C = \frac{N_1}{2P_{t_1}} + \frac{N_2}{2P_{t_2}}$$

$$= \frac{1}{2P_n}\left(\frac{N_1}{\cos\beta_{p_1}} + \frac{N_2}{\cos\beta_{p_2}}\right) \quad (1.45)$$

where N_i = number of gear teeth
P_{ti} = diametral pitch (i = 1, 2)
$P_n = \pi/p_n$, where p_n is the distance between the teeth of the rack cutter that is measured in the normal section of the rack cutter
β_{p_1}, β_{p_2} = helix angles on the regular pitch cylinders.

The shortest distance for nonstandard involute helical gears, C', differs from C.

The gear ratio (angular velocity ratio) may be determined from the velocity polygon that is shown in Fig. 1.17. The sliding velocity $\mathbf{v}^{(12)}$ is represented by the equation $\mathbf{v}^{(12)} = \mathbf{v}^{(1)} - \mathbf{v}^{(2)}$. Here $\mathbf{v}^{(1)}$ is the velocity of pinion pitch point P; $\mathbf{v}^{(2)}$ is the velocity of gear pitch point P. Velocities $\mathbf{v}^{(1)}$, $\mathbf{v}^{(2)}$, and $\mathbf{v}^{(12)}$ belong to plane II that is tangent to the operating pitch cylinders at the pitch point P. The sliding velocity $\mathbf{v}^{(12)}$ is directed along the common tangent to the helices at P. Drawings of Fig. 1.17 yield the following expression for the gear ratio:

$$m_{12} = \frac{\omega^{(1)}}{\omega^{(2)}} = \frac{r_{O_2}\cos\beta_{O_2}}{r_{O_1}\cos\beta_{O_1}} = \frac{N_2}{N_1} \quad (1.46)$$

where r_{O_i} and β_{O_i} are the radius of the operating pitch cylinder and the helix angle on this cylinder. Unlike the case of spur gears or helical gears with parallel axes, the prescribed ratio m_{12} can be satisfied with various ratio r_{O_2}/r_{O_1} by choosing appropriate values for β_{O_1} and β_{O_2}.

Usually, mating helical gears are of the same helix hand as shown in Fig. 1.17. The crossing angle γ is represented in this case by

$$\gamma = \beta_{O_1} + \beta_{O_2} \quad (1.47)$$

It can be imagined that crossed helical gears with operating pitch cylinders are in mesh with a common rack. The normal section of such a rack is considered as a section by a plane that is perpendicular to plane II and the common tangent t-t (Fig. 1.17a). The shape angle of the rack in the normal section is called the *normal operating pressure*

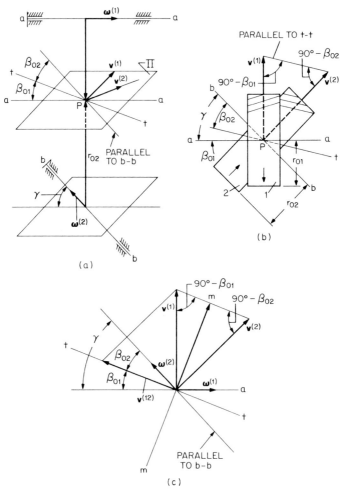

FIGURE 1.17 Velocity polygon for crossed helical gears

angle and is represented by the equation

$$\cos \alpha_{O_n} = \frac{\sin \beta_{b_i}}{\sin \beta_{O_i}} \qquad (i = 1,2) \qquad (1.48)$$

where β_{b_i} = the helix angle on the base cylinder, and β_{O_i} is the helix angle on the operating pitch cylinder. In the case of nonstandard gears α_{O_n} differs from the standard value of α_n — the standard pressure angle. However, α_{O_n} is equal to α_n in the particular case of standard crossed helical gears where the operating pitch cylinders coincide with the regular pitch cylinders. The normal tooth thickness and the normal space width on the operating pitch cylinders of mating crossed helical gears must be equal (assuming that the backlash is zero). It is necessary to emphasize that crossed helical gears are in point contact but not in line contact at every instant.

1.1.10 Worm Gear Drive: Operating Pitch Cylinders

A worm may be considered as a helical gear whose number of teeth is equal to the number of worm threads. However, the type of the worm thread surface depends on the method of manufacturing, and a screw involute surface is just a particular case of the worm thread surface. Unlike the case of crossed helical gears, the tooth surfaces of the worm and the gear are in line contact at every instant. However, the concept of operating pitch cylinders that has been discussed for crossed helical gears can be also applied for a worm gear drive.

Pitch point P is the point of tangency of operating pitch cylinders whose radii are designated by r_{O_1} and R_{O_2}, respectively (Fig. 1.18). The operating and regular worm pitch cylinders coincide in the case of a standard worm gear drive only. The radius r_{O_1} of the worm operating pitch cylinder may be represented by the equation

$$r_{O_1} = \frac{p_{O_1} N_1}{2\pi \tan \lambda_{O_1}} \tag{1.49}$$

where p_{O_1} = axial distance between two neighboring worm threads
λ_{O_1} = thread lead angle on the operating pitch cylinder
N_1 = number of threads

The axial thread distance p_{O_1} and the transverse gear tooth distance p_{O_2} are related by the following equation (Fig. 1.19):

$$p_{O_1} \cos \lambda_{O_1} = p_{O_2} \cos \beta_{O_2} = p_{O_n} \tag{1.50}$$

where β_{O_2} is the helix angle on the worm gear cylinder, and p_{O_n} is the normal tooth distance.

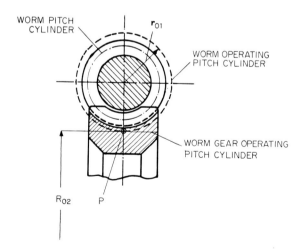

FIGURE 1.18 Operating pitch cylinders for a worm gear drive

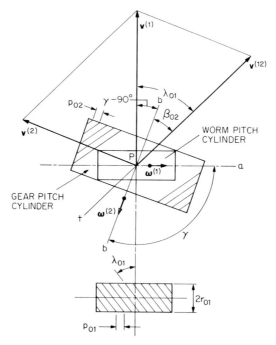

FIGURE 1.19 Velocity polygon for a worm gear drive

The radius of the worm gear operating pitch cylinder is

$$R_{O_2} = \frac{N_2 p_{O_2}}{2\pi} \tag{1.51}$$

The velocity polygon is shown in Fig. 1.19. The sliding velocity $\mathbf{v}^{(12)}$ is directed along the common tangent to the helices on the operating pitch cylinders. The worm and the worm gear are both right hand. The crossing angle is

$$\gamma = 90° - \lambda_{O_1} + \beta_{O_2} \tag{1.52}$$

The drawings of Fig. 1.19 yield the following expression for the gear ratio:

$$m_{12} = \frac{\omega^{(1)}}{\omega^{(2)}} = \frac{R_{O_2} \cos \beta_{O_2}}{r_{O_1} \sin \lambda_{O_1}} = \frac{N_2}{N_1} \tag{1.53}$$

In the particular case when $\gamma = 90°$ we have $\lambda_{O_1} = \beta_{O_2}$ and

$$m_{12} = \frac{R_{O_2}}{r_{O_1} \tan \lambda_{O_1}} = \frac{N_2}{N_1} \tag{1.54}$$

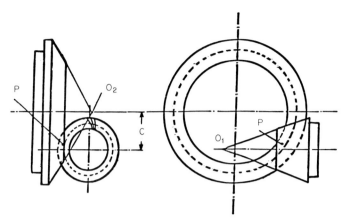

FIGURE 1.20 Pitch surfaces for hypoid gears

1.1.11 Operating Pitch Surfaces of a Hypoid Gear Drive*

The operating pitch surfaces are two cones that are in tangency at the pitch point P (Fig. 1.20). Figure 1.21 shows the pitch plane Π that is tangent to the pitch cones. Plane Π is drawn through the apexes O_1 and O_2 of the pinion and gear pitch cones and the pitch point P.

The relations between the pitch cone parameters may be derived using the equations of tangency of the pitch cones at the pitch point. These relations are represented as follows (Fig. 1.21):

$$\frac{r_1}{r_2} = \frac{\frac{C}{r_2}\cos\gamma_1}{\sqrt{\cos^2\gamma_1 - \sin^2\gamma_2}} - \frac{\cos\gamma_1}{\cos\gamma_2} \tag{1.55}$$

$$d_1 = \frac{C\cos\gamma_1\cot\gamma_1}{\sqrt{\cos^2\gamma_1 - \sin^2\gamma_2}} - \frac{r_2}{\sin\gamma_1\cos\gamma_2} \tag{1.56}$$

$$d_2 = \frac{2r_2}{\sin 2\gamma_2} - \frac{C\sin\gamma_2}{\sqrt{\cos^2\gamma_1 - \sin^2\gamma_2}} \tag{1.57}$$

where r_1, r_2 = radii of pitch cone circles that pass through the pitch point
  γ_1, γ_2 = pitch cone angles
  C = shortest distance between the pitch cone axes (Fig. 1.21)

d_1 and d_2 determine the location of the cone apexes with respect to the shortest distance (Fig. 1.21). The coordinates of pitch point P are represented in coordinate system S_f by the following equations:

---
*Trademark of Gleason Works.

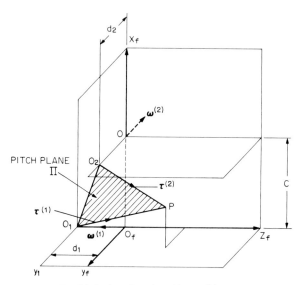

FIGURE 1.21 Pitch plane for a hypoid gear drive

$$Y_f^{(P)} = -r_2 \cot \gamma_2 + d_2 = -r_1 \frac{\sin \gamma_2}{\cos \gamma_1}$$

$$Z_f^{(P)} = r_1 \cot \gamma_1 - d_1 = r_2 \frac{\sin \gamma_1}{\cos \gamma_2} \qquad (1.58)$$

$$X_f^{(P)} = [r_1^2 - (Y_f^{(P)})^2]^{1/2}$$

Axis z_f coincides with the pinion axis; $\overline{O_2 O}$ is parallel to axis y_f and is directed along the gear axis; axis x_f coincides with the line of shortest distance $\overline{O_f O}$ between the pinion and gear axes.

The orientation of the pitch plane (Fig. 1.21) is represented in coordinate system S_f by the following column matrix:

$$[n_f^{(p)}] = \begin{bmatrix} \sqrt{\cos^2 \gamma_2 - \sin^2 \gamma_1} \\ -\sin \gamma_2 \\ -\sin \gamma_1 \end{bmatrix} \qquad (1.59)$$

The elements of column matrix (1.59) represent in coordinate system S_f the direction cosines of unit vector $\mathbf{n}_f^{(P)}$—the normal to the pitch plane.

The velocity polygon for pitch point P is shown in Fig. 1.22. The velocities are represented in the pitch plane that passes through points O_1, O_2, and P. Vectors $\mathbf{v}^{(1)}$ and $\mathbf{v}^{(2)}$ represent the velocities of point P in rotation about the axes of the pinion and the gear, respectively. The sliding velocity $\mathbf{v}^{(12)}$ is directed along the common tangent to tooth surfaces. Angles β_1 and β_2 represent the pinion and gear "spiral" angles.

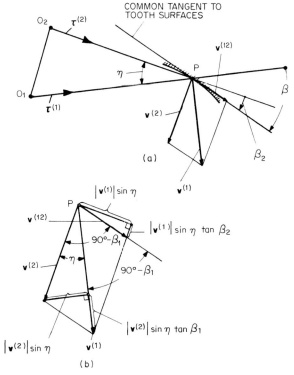

FIGURE 1.22 Velocity polygon for a hypoid gear drive

Vectors $\tau^{(1)}$ and $\tau^{(2)}$ represent the unit vectors of cone generatrices. Angle η that is formed by $\tau^{(1)}$ and $\tau^{(2)}$ is represented by the following equation:

$$\cos \eta = \cos(\beta_1 - \beta_2) = \tan \gamma_1 \tan \gamma_2 \tag{1.60}$$

The direct relation between γ_1, γ_2 and β_1 may be represented by the following equation:

$$\cos^2 \beta_1 = \frac{(1 - \tan^2 \gamma_1 \tan^2 \gamma_2) b^2}{1 + b^2 - 2b \tan \gamma_1 \tan \gamma_2} \tag{1.61}$$

where

$$b = \frac{N_1 \cos \gamma_2 (\cos^2 \gamma_1 - \sin^2 \gamma_2)^{1/2}}{N_2 \cos \gamma_1 \left[\dfrac{C}{r_2} \cos \gamma_2 - (\cos^2 \gamma_1 - \sin^2 \gamma_2)^{1/2}\right]} \tag{1.62}$$

The derivation of Eqs. (1.58) to (1.62) are presented in Litvin.[16]

Equations (1.60) and (1.61) represent a system of two nonlinear equations in three unknowns: β_2, γ_1, and γ_2 (β_1 is considered as given; usually, $\beta_1 \approx 45°$). The derivation of the third equation is based on: (1) the Wildhaber's concept of limiting normal for face-milled hypoid gears with tapered teeth,[16] and (2) special orientation of the head cutter axis for face-hobbed hypoid gears with uniform depth.[18]

1.2 PARAMETRIC REPRESENTATION OF CURVES AND SURFACES

A position vector in a three-dimensional space represented by

$$r(\theta) = x(\theta)\mathbf{i} + y(\theta)\mathbf{j} + z(\theta)\mathbf{k} \qquad (1.63)$$

associates the location of a point in the space with parameter θ. The equations $x = x(\theta)$, $y = y(\theta)$, and $z = z(\theta)$ determine the cartesian coordinates of the point; $\mathbf{i}$, $\mathbf{j}$ and $\mathbf{k}$ are the unit vectors of coordinate axes. Equation (1.63) represents a spatial curve if it associates the location of a point in the three-dimensional space with the *continuously* varied parameter θ. A *simple* curve means that there is one-to-one correspondence between the point in the three-dimensional space and θ. A simple curve does not have points of intersection. A curve is a *regular* one if the following inequality is observed:

$$x_\theta^2 + y_\theta^2 + z_\theta^2 > 0 \qquad (1.64)$$

where $$x_\theta = \frac{dx}{d\theta} \qquad y_\theta = \frac{dy}{d\theta} \qquad z_\theta = \frac{dz}{d\theta}$$

Inequality (1.64) means that the three derivatives are not equal to zero simultaneously.

Example 1.1. Equations

$$x = \rho \cos\theta \qquad y = \rho \sin\theta \qquad z = h\theta \qquad (1.65)$$

represent a *helix* on the cylinder of radius ρ, and $h = \dfrac{H}{2\pi}$ is the screw parameter (H is the *lead* of the helix—the axial displacement for the change of parameter θ of 2π).

A planar curve is represented in a two-dimensional space. For instance, a curve

$$r(\theta) = x(\theta)\mathbf{i} + y(\theta)\mathbf{j} \qquad z = c \qquad (1.66)$$

is represented in a plane that is perpendicular to axis z (c is a constant).

Consider that a planar curve and a point D that does not belong to the curve are given (Fig. 1.23*a*); tangent MD to the curve passes through point D.

The equation

$$\frac{X - x(\theta)}{x_\theta} - \frac{Y - y(\theta)}{y_\theta} = 0 \qquad (1.67)$$

where $$x_\theta = \frac{dx}{d\theta} \qquad y_\theta = \frac{dy}{d\theta}$$

relates coordinates (X, Y) of point D with the coordinates of point M, $x(\theta)$ and $y(\theta)$.

A normal to a planar curve is perpendicular to its tangent and may be represented as follows:

$$\mathbf{N} = \mathbf{T} \times \mathbf{k} = y_\theta \mathbf{i} - x_\theta \mathbf{j} \qquad (1.68)$$

where $\mathbf{T} = x_\theta \mathbf{i} + y_\theta \mathbf{j}$

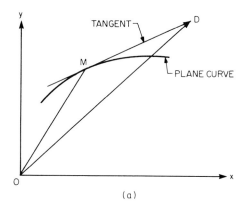

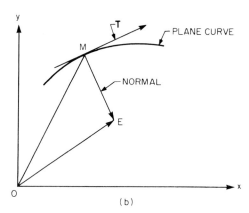

FIGURE 1.23 Tangent and normal for a planar curve

A unit normal is represented by

$$\mathbf{n} = \frac{\mathbf{N}}{|\mathbf{N}|} = \frac{1}{\sqrt{x_\theta^2 + y_\theta^2}} (y_\theta \mathbf{i} - x_\theta \mathbf{j}) \quad \text{(provided } |\mathbf{N}| \neq 0\text{)} \quad (1.69)$$

Consider that the normal to the planar curve is drawn from point E whose coordinates are X and Y. The equation

$$\frac{X - x(\theta)}{y_\theta} + \frac{Y - y(\theta)}{x_\theta} = 0 \quad (1.70)$$

relates coordinates of points E and M [Fig. 1.23(b)].

Example 1.2. The vector equation

$$\mathbf{R} = r_b[(\sin u - u \cos u)\mathbf{i} + (\cos u + u \sin u)\mathbf{j}] \quad 0 < u < \infty \quad (1.71)$$

where r_b is the radius of the base circle and $u = BM/r_b = M_0B/r_b$ is the variable, represents an involute curve (Fig. 1.24). At a point with $u = 0$ the curve is not a regular

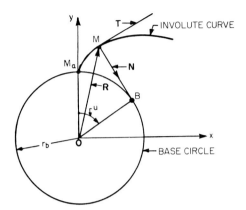

FIGURE 1.24 Involute curve

one; the tangent and the normal to the curve are indefinite since

$$x_\theta^2 + y_\theta^2 = 0$$

The vector equation

$$\mathbf{r}(u,\theta) = x(u,\theta)\mathbf{i} + y(u,\theta)\mathbf{j} + z(u,\theta)\mathbf{k} \qquad (1.72)$$

associates the cartesian coordinates of a point in a three-dimensional space with parameters (u, θ). Vector Eq. (1.72) represents a surface if there is a *continuous* correspondence between x, y, and z and the varied parameters, u and θ. A *simple* surface exists (it does not have points of self-intersection) if there is one-to-one correspondence between (x, y, z) and (u, θ). The vector equation

$$\mathbf{r}(u, \theta)\epsilon C^1 \qquad \mathbf{r}_u \times \mathbf{r}_\theta \neq 0 \qquad (u, \theta)\epsilon E \qquad (1.73)$$

represents a *regular* surface that does not have *singular* points. Symbol C^1 means that all partial derivatives of the first order at least are continuous functions, symbol E indicates the area of variables u and θ.

A surface may be covered with *coordinate* lines, the u-lines and the θ-lines. A position vector of a current point of a coordinate line is represented by $\mathbf{r}(u, \theta_k)$ for a u-line and $\mathbf{r}(u_k, \theta)$ for a θ-line, where θ_k and u_k are the fixed parameters. Figure 1.25 shows: (1) the position vector $\mathbf{r}(u_O, \theta_O)$ of a current point u of the surface, and (2) coordinate lines of the surface.

The normal $\mathbf{N}$ to the surface is represented by

$$\mathbf{N} = \mathbf{r}_u \times \mathbf{r}_\theta \qquad (1.74)$$

and the unit normal is

$$\mathbf{n} = \frac{\mathbf{r}_u \times \mathbf{r}_\theta}{|\mathbf{r}_u \times \mathbf{r}_\theta|} \qquad \text{provided } \mathbf{N} \neq 0 \qquad (1.75)$$

By changing the order of factors in the cross product $(\mathbf{r}_u \times \mathbf{r}_\theta)$, it is possible to change the direction of $\mathbf{N}$ and $\mathbf{n}$ for the opposite one. Singular points on the surface appear if $\mathbf{N} = 0$. At singular points the surface normal does not exist or its direction is indefinite.

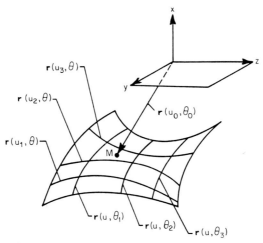

FIGURE 1.25 Surface representation in parametric form

Example 1.3. A cone (Fig. 1.26) is represented by the equations

$$x = \rho \cot \psi_c - u \cos \psi_c \quad y = u \sin \psi_c \sin \theta \quad z = u \sin \psi_c \cos \theta \quad (1.76)$$

where ψ_c is the apex angle, and u and θ are the surface coordinates. The u-line (θ_k is constant) is the cone generatrix and the θ-line (u_k is constant) is a circle on the cone

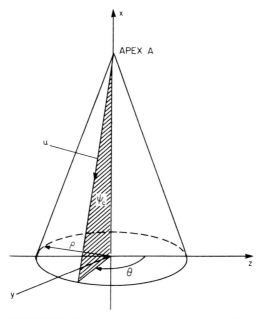

FIGURE 1.26 Cone surface in parametric representation

surface of the radius $u_k \sin \psi_c$. The normal to the surface is

$$\mathbf{N} = \mathbf{r}_\theta \times \mathbf{r}_u = u \sin \psi_c (\sin \psi_c \mathbf{i} + \cos \psi_c \sin \theta \mathbf{j} + \cos \psi_c \cos \theta \mathbf{k}) \quad (1.77)$$

The apex point A is the singular point, and at this point $\mathbf{N} = 0$. The unit normal is

$$\mathbf{n} = \frac{\mathbf{N}}{|\mathbf{N}|} = \sin \psi_c \mathbf{i} + \cos \psi_c \sin \theta \mathbf{j} + \cos \psi_c \cos \theta \mathbf{k} \quad \text{provided } u \sin \psi_c \neq 0 \quad (1.78)$$

1.3 ENVELOPE OF A FAMILY OF SURFACES (CURVES), BASIC LAW OF MESHING, CONTACT LINES ON MATING SURFACES

1.3.1 Envelope and Equation of Meshing

Consider coordinate systems S_1, S_2, and S_f that are rigidly connected to the driving gear, to the driven gear and to the frame, respectively. Gear 1 is provided with a regular surface Σ_1 that is represented in S_1 as follows:

$$\mathbf{r}_1(u_1, \theta_1) \qquad \frac{\partial \mathbf{r}_1}{\partial u_1} \times \frac{\partial \mathbf{r}_1}{\partial \theta_1} \neq 0 \qquad (u_1, \theta_1) \epsilon E \quad (1.79)$$

The gear tooth surfaces are in line contact at every instant and, being in continuous contact, transform rotation with the prescribed function $\phi_2(\phi_1)$. Usually, $\phi_2(\phi_1)$ is a linear function.

Consider as given: Σ_1, $\phi_2(\phi_1)$, the shortest distance C, and the crossing angle γ between the axes of rotation. It is necessary to determine surface Σ_2 of the driven gear teeth. Surface Σ_1 generates in coordinate system S_2 a *family* of surfaces Σ_ϕ that can be represented by the following matrix equation:

$$[r_2(u_1, \theta_1, \phi_1)] = [M_{21}] [r_1(u_1, \theta_1)] \quad (1.80)$$

Surface Σ_2 is the *envelope* to Σ_ϕ. We have to distinguish the *necessary* and *sufficient* conditions of the existence of an envelope. The *necessary* conditions provide the *possibility* of tangency of the envelope with any single surface of the family of surfaces Σ_ϕ. The *sufficient* conditions provide that the above-mentioned tangency exists indeed, and that the envelope is a *regular* surface. The avoidance of singular points on the envelope, the generated surface Σ_2, is discussed in the following as the problem of nonundercutting.

The necessary condition of envelope existence is represented in the book *Theory of Gearing*[16] by the equation

$$\mathbf{N}^{(1)} \cdot \mathbf{v}^{(12)} = 0 \quad (1.81)$$

where $\mathbf{N}^{(1)}$ is the normal to the generating surface, Σ_1 and $\mathbf{v}^{(12)}$ is the sliding velocity. For the case of transformation of rotation, vector $\mathbf{v}^{(12)}$ may be represented as follows

$$\mathbf{v}^{(12)} = [(\omega^{(1)} - \omega^{(2)}) \times \mathbf{r}^{(1)}] - (\mathbf{R} \times \omega^{(2)}) \quad (1.82)$$

where $\omega^{(1)}$, $\omega^{(2)}$ = angular velocities of gears 1 and 2 (the line of action of $\omega^{(1)}$ passing through the origin O_1 of coordinate system S_1)
$\mathbf{r}^{(1)}$ = position vector of a current point of Σ_1
$\mathbf{R}$ = position vector that is drawn from O_1 to an arbitrary chosen point of the line of action of $\omega^{(2)}$

The scalar product in Eq. (1.81) is invariant to the used coordinate systems. Thus, vectors $\mathbf{N}^{(1)}$ and $\mathbf{v}^{(12)}$ may be represented in any of the coordinate systems S_1, S_f, and S_2, particularly in coordinate system S_1. The equation

$$\mathbf{N}^{(1)} \cdot \mathbf{v}^{(12)} = f(u_1, \theta_1, \phi_1) = 0 \qquad (1.83)$$

relates the surface coordinates of Σ_1 and the parameter of motion, ϕ. Equation (1.83) is called the *equation of meshing*.

1.3.2 Contact Lines on Σ_1

The generating surface Σ_1 may be covered with lines that will become in turn lines of contact of Σ_1 and Σ_2 in the process of meshing. These lines are represented by the equations

$$\mathbf{r}_1 = \mathbf{r}_1(u_1, \theta_1), f(u_1, \theta_1, \phi_1^*) = 0 \qquad (1.84)$$

where ϕ_1^* is the fixed-in parameter.

1.3.3 Generated Surface Σ_2

This surface is represented in coordinate system S_2 by

$$[r_2(u_1, \theta_1, \phi_1)] = [M_{21}][r_1(u_1, \theta_1)], f(u_1, \theta_1, \phi_1) = 0 \qquad (1.85)$$

where $[M_{21}]$ describes the coordinate transformation from S_1 to S_2.

1.3.4 Surface of Action

This surface is considered as the set of lines of contact between Σ_1 and Σ_2 that are represented in the fixed coordinate system S_f. The surface of action is represented in S_f by the following equations:

$$[r_f(u_1, \theta_1, \phi_1)] = [M_{f_1}][r_1(u_1, \theta_1)], f(u_1, \theta_1, \phi_1) = 0 \qquad (1.86)$$

Particular Case. Consider that the gears transform rotation between parallel or intersected axes. In such a case the normal to the contacting surfaces intersects the instantaneous axis of rotation and the equation of meshing may be derived on the basis of the following equation:

$$\frac{X_1(\phi_1) - x_1(u_1, \theta_1)}{N_{x_1}(u_1, \theta_1)} = \frac{Y_1(\phi_1) - y_1(u_1, \theta_1)}{N_{y_1}(u_1, \theta_1)} = \frac{Z_1(\phi_1) - z_1(u_1, \theta_1)}{N_{z_1}(u_1, \theta_1)} \qquad (1.87)$$

where $X_1(\phi_1), Y_1(\phi_1), Z_1(\phi_1)$ = coordinates of a point of the instantaneous axis of rotation in coordinate system S_1
$x_1(u_1, \theta_1), y_1(u_1, \theta_1), z_1(u_1, \theta_1)$ = coordinates of a point of surface Σ_1
$N_{x_1}, N_{y_1}, N_{z_1}$ = projections on the coordinate axes of the normal $\mathbf{N}_1$ to Σ_1

1.3.5 Planar Gearing

The generating curve is a regular curve that is represented in S_1 by the vector equation

$$\mathbf{r}_1(\theta_1) \qquad \frac{\delta r_1}{\delta \theta_1} \neq 0 \tag{1.88}$$

The equation of meshing is

$$f(\theta_1, \phi_1) = 0 \tag{1.89}$$

and it can be derived on the basis of Eq. (1.81) or on the basis of equation

$$\frac{X_1(\phi_1) - x_1(\theta_1)}{N_{x_1}(\theta_1)} = \frac{Y_1(\phi_1) - y_1(\theta_1)}{N_{y_1}(\theta_1)} \tag{1.90}$$

Equation (1.90) is derived for a generating curve that is represented in a plane that is perpendicular to the z_1-axis. Equation (1.90) may be interpreted on the basis of the famous theorem of Willis[16] that states: *The common normal to the contacting shapes of gear teeth, Σ_1 and Σ_2, must pass through the instantaneous center of rotation.*

1.4 BASIC KINEMATIC RELATIONS

The following equations[5,16] relate the infinitesimal displacements (velocities) of the contact point and contact normal for two surfaces being in mesh. The velocity of a contact point (contact normal) with respect to the fixed coordinate system S_f is the same for both contacting surfaces, and it can be represented as a sum of two components: the velocity in *transfer* motion *with* the gear tooth surface, and the velocity in *relative* motion *over* the surface. The equations are as follows:

$$\mathbf{v}_r^{(2)} = \mathbf{v}_r^{(1)} + \mathbf{v}^{(12)} \qquad \dot{\mathbf{n}}_r^{(2)} = \dot{\mathbf{n}}_r^{(1)} + (\boldsymbol{\omega}^{(1)} - \boldsymbol{\omega}^{(2)}) \times \mathbf{n}^{(1)} \tag{1.91}$$

where $\mathbf{v}_r^{(i)}$ is the velocity of the contact point and $\dot{\mathbf{n}}_r^{(i)}$ is the velocity of the tip of the unit surface normal in the motion over the surface Σ_i ($i = 1, 2$) that is in addition to the translational velocity of the normal. The advantage of Eqs. (1.91) is that $\mathbf{v}_r^{(2)}$ and $\dot{\mathbf{n}}_r^{(2)}$ can be determined even if equations of Σ_2 have not been derived yet. Equations (1.91) are the basis for relationship between the principal directions and curvatures for contacting surfaces and determination of conditions of nonundercutting.[5,6,16]

1.5 CONDITIONS OF NONUNDERCUTTING

Conditions of nonundercutting of the generated surface Σ_2 may be discussed as conditions of avoidance of singular points on Σ_2. Singular points on Σ_2 occur if $\mathbf{v}_r^{(2)}$, the velocity of contact point on surface Σ_2, becomes equal to zero. The equations

$$\mathbf{r}_1 = \mathbf{r}_1(u_1, \theta_1) \qquad \mathbf{v}_r^{(1)} + \mathbf{v}^{(12)} = \mathbf{v}_r^{(2)} = 0 \qquad (1.92)$$

can be used for the determination of a line L on surface Σ_1. L is a set of points on surface Σ_1 that generate singular points on Σ_2. Singular points on Σ_2 will not appear if line L is out of the working part of Σ_1. This requirement can be satisfied with certain parameters and settings of the generating machine tool.

The computational procedure for the determination of L is based on the following equations:

$$\mathbf{r}_1 = \mathbf{r}_1(u_1, \theta_1) \qquad \mathbf{N}_1^{(1)} \cdot \mathbf{v}_1^{(12)} = f(u_1, \theta_1, \phi_1) = 0 \qquad (1.93)$$

$$\begin{vmatrix} \dfrac{\partial x_1}{\partial u_1} & \dfrac{\partial x_1}{\partial \theta_1} & -v_{x_1}^{(12)} \\ \dfrac{\partial y_1}{\partial u_1} & \dfrac{\partial y_1}{\partial \theta_1} & -v_{y_1}^{(12)} \\ \dfrac{\partial f}{\partial u_1} & \dfrac{\partial f}{\partial \theta_1} & -\dfrac{\partial f}{\partial \phi_1}\dfrac{d\phi_1}{dt} \end{vmatrix} = \begin{vmatrix} \dfrac{\partial x_1}{\partial u_1} & \dfrac{\partial x_1}{\partial \theta_1} & -v_{x_1}^{(12)} \\ \dfrac{\partial z_1}{\partial u_1} & \dfrac{\partial z_1}{\partial \theta_1} & -v_{z_1}^{(12)} \\ \dfrac{\partial f}{\partial u_1} & \dfrac{\partial f}{\partial \theta_1} & -\dfrac{\partial f}{\partial \phi_1}\dfrac{d\phi_1}{dt} \end{vmatrix} = \begin{vmatrix} \dfrac{\partial y_1}{\partial u_1} & \dfrac{\partial y_1}{\partial \theta_1} & -v_{y_1}^{(12)} \\ \dfrac{\partial z_1}{\partial u_1} & \dfrac{\partial z_1}{\partial \theta_1} & -v_{z_1}^{(12)} \\ \dfrac{\partial f}{\partial u_1} & \dfrac{\partial f}{\partial \theta_1} & -\dfrac{\partial f}{\partial \phi_1}\dfrac{d\phi_1}{dt} \end{vmatrix} = 0 \qquad (1.94)$$

Equations (1.94) yield the equation

$$F(u_1, \theta_1, \phi_1) = 0 \qquad (1.95)$$

The sought-for limiting line L is determined with Eqs. (1.93) and (1.95). For the case of a planar gearing considering that the generating shape is represented by the vector function $\mathbf{r}_1(\theta_1)$, it is necessary to use the following equations instead of Eqs. (1.94):

$$\begin{vmatrix} \dfrac{\partial x_1}{\partial \theta_1} & -v_{x_1}^{(12)} \\ \dfrac{\partial f}{\partial \theta_1} & -\dfrac{\partial f}{\partial \phi_1}\dfrac{d\phi_1}{dt} \end{vmatrix} = \begin{vmatrix} \dfrac{\partial y_1}{\partial \theta_1} & -v_{y_1}^{(12)} \\ \dfrac{\partial f}{\partial \theta_1} & -\dfrac{\partial f}{\partial \phi_1}\dfrac{d\phi_1}{dt} \end{vmatrix} = 0 \qquad (1.96)$$

This approach is now illustrated with an example using planar gearing.

Sample Problem 1.1. The shape Σ_1 of a rack cutter is an arc of a circle (Fig. 1.27) represented by the equations

$$\mathbf{r}_1(\theta_1) = (a + \rho \cos \theta_1)\mathbf{i}_1 + (b + \rho \sin \theta_1)\mathbf{\gamma}_1 \qquad (1.97)$$

where a and b are coordinates of point K, the center of the circular arc. Parameters a and b can be positive or negative depending on the location of K. While the rack cutter translates with velocity v, the gear being generated rotates with the angular

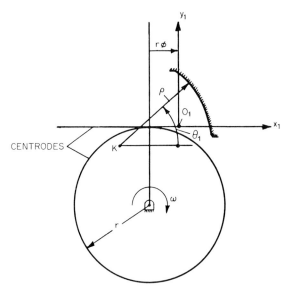

FIGURE 1.27 Rack cutter with a circular arc

velocity $\omega = v/r$, where r is the radius of the gear centrode by cutting (usually, this centrode coincides with the gear pitch circle). It is necessary to determine the limiting point of the generating shape Σ_1 to avoid the gear undercutting.

Step 1. Determination of velocity $\mathbf{v}^{(12)}$:

$$\mathbf{v}_1^{(12)} = \mathbf{v}_1^{(1)} - \mathbf{v}_1^{(2)} \tag{1.98}$$

where

$$\mathbf{v}_1^{(1)} = \omega r \mathbf{i}_1 \qquad \mathbf{v}_1^{(2)} = (\omega \times \mathbf{r}_1) + (\overrightarrow{O_1 O_2} \times \omega) = \omega[(y_1 + r)\mathbf{i}_1 - (x_1 - r\phi)\mathbf{j}_1] \tag{1.99}$$

Equations (1.96) to (1.98) yield

$$\mathbf{v}_1^{(12)} = \omega \begin{bmatrix} -y_1 \\ x_1 - r\phi \end{bmatrix} = \omega \begin{bmatrix} -(b + \rho \sin \theta_1) \\ a + \rho \cos \theta_1 - r\phi \end{bmatrix} \tag{1.100}$$

Step 2. Determination of the shape normal $\mathbf{N}_1$:

$$\mathbf{N}_1 = \mathbf{T}_1 \times \mathbf{k} = \rho(\cos \theta_1 \mathbf{i}_1 + \sin \theta_1 \mathbf{\gamma}_1) \tag{1.101}$$

where $\mathbf{T}_1 = \dfrac{\partial \mathbf{r}_1}{\partial \theta_1}$ is the shape tangent.

Step 3. Derivation of equation of meshing: The equation

$$\mathbf{N}_1 \cdot \mathbf{v}_1^{(12)} = 0$$

yields

$$f(\theta_1, \phi) = b \cos \theta_1 - a \sin \theta_1 - r\phi \sin \theta_1 = 0 \tag{1.102}$$

Step 4. Derivation of equation $F(\theta_1, \phi) = 0$: The equation

$$\begin{vmatrix} \dfrac{\partial x_1}{\partial \theta_1} & -v^{(12)}_{x_1} \\[6pt] \dfrac{\partial f}{\partial \theta_1} & -\dfrac{\partial f}{\partial \phi}\dfrac{d\phi}{dt} \end{vmatrix} = 0$$

yields

$$F(\theta_1, \phi) = -\rho r \sin^2 \theta_1 + (b + \rho \sin \theta_1)$$
$$(b \sin \theta_1 + a \cos \theta_1 + r\phi \cos \theta_1) = 0 \qquad (1.103)$$

Step 5. Derivation of final equation for the limiting value of θ_1: Equations (1.103) and (1.102), after elimination of ϕ, yield

$$\sin^3 \theta_1 - \frac{b}{r}\sin \theta_1 - \frac{b^2}{\rho r} = 0 \qquad (1.104)$$

1.6 ENVELOPE OF CONTACT LINES ON GENERATING SURFACE Σ_1

Consider contact lines on surface Σ_1 that is the generating surface if Σ_1 is the tool surface or is the surface of the input gear. Contact lines on Σ_1 may have an envelope as shown in Fig. 1.28 that represents the contact lines on the worm surface Σ_1 for a worm gear drive. The existence of an envelope on Σ_1 is accompanied with deteriorated conditions for lubrication and heat transfer. This also means that the part of tool surface (if Σ_1 is a tool) that is free from contact lines does not participate in the generation of Σ_2. Then a larger portion of the gear tooth surface will be generated by the fillet surface of the tool. The envelope of contact lines on Σ_1, if it exists, may be determined with the following equations:

$$\mathbf{r}_1 = \mathbf{r}_1(u, \theta) \qquad \mathbf{N} \cdot \mathbf{v}^{(12)} = f(u, \theta, \phi) = 0 \qquad \frac{\partial f}{\partial \phi} = 0 \qquad (1.105)$$

The existence of an envelope may be avoided by choosing certain design parameters, such as $\gamma \neq 90°$ for worm gear drives (where γ is the twist angle) or by certain pressure angles for the hypoid gears. In the case of hypoid gears the pressure angles are different for the driving and coast sides of the gear tooth surfaces and the tooth shapes are asymmetric.

1.7 LIMIT CONTACT NORMAL

The concept of limit contact normal has been proposed by E. Wildhaber[22]. The investigation performed by Litvin et al.[18] shows that, if a limit contact normal exists, either an envelope of contact lines appears on Σ_1 or singular points appear on Σ_2. To derive

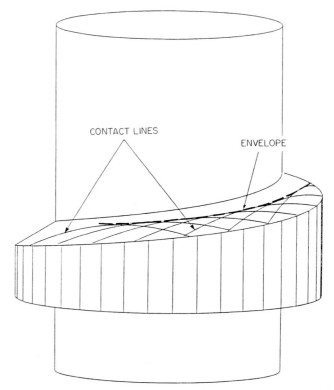

FIGURE 1.28 Contact lines and envelope of contact lines on the surface of an involute worm

equations of a limit contact normal that satisfy both above-mentioned conditions, we have to consider two equations:

$$\mathbf{n} \cdot \mathbf{v}^{(12)} = 0 \qquad \frac{d}{dt}\left[\mathbf{n} \cdot \mathbf{v}^{(12)}\right] = 0 \qquad (1.106)$$

where **n** is the surface unit normal. While differentiating the equation of meshing we have to consider two cases:

1. An envelope of contact lines on Σ_1 may occur. Then

$$\dot{\mathbf{n}}_r^{(1)} = 0 \quad \text{and} \quad \mathbf{v}_r^{(1)} = 0 \qquad (1.107)$$

2. Singular points on Σ_2 may appear. Then

$$\dot{\mathbf{n}}_r^{(2)} = 0 \quad \text{and} \quad \mathbf{v}_r^{(2)} = 0 \qquad (1.108)$$

Both cases yield the following equations that determine the limit contact normal:

$$\mathbf{n} \cdot \mathbf{v}^{(12)} = 0 \quad \mathbf{n} \cdot [(\omega^{(1)} \times \mathbf{v}_{tr}^{(2)}) - (\omega^{(2)} \times \mathbf{v}_{tr}^{(1)})] = 0 \qquad (1.109)$$

where

$$\mathbf{v}_{tr}^{(1)} = \omega^{(1)} \times \mathbf{r} \qquad \mathbf{v}_{tr}^{(2)} = (\omega^{(2)} \times \mathbf{r}) + (\mathbf{C} \times \omega^{(2)}) \qquad (1.110)$$

1.8 RELATIONS BETWEEN PRINCIPAL CURVATURES AND DIRECTIONS FOR MUTUALLY ENVELOPED SURFACES

Usually, the generated surface can be represented in a three-parameter form only with an additional relation between the parameters. Thus, the surface is represented by the vector equation

$$\mathbf{r}_2(u, \theta, \phi) \qquad f(u, \theta, \phi,) = 0 \qquad (1.111)$$

where (u, θ) are the parameters of the tool surface "inherited" in the process of generation, and ϕ is the generalized parameter of motion of the tool. The direct determination of the principal curvatures and directions of the generated surface is a complicated problem. The solution to this problem may be substantially simplified if the principal curvatures and directions on Σ_2 are expressed in terms of the principal curvatures and directions on Σ_1, the tool surface, and the parameters of relative motion (proposed by Litvin[6,16]). The final equations that can be used for this purpose are as follows:

$$\tan 2\sigma = \frac{2a_{13}a_{23}}{a_{23}^2 - a_{13}^2 + (k_s - k_q)a_{33}} \qquad (1.112)$$

$$k_f - k_h = \frac{2a_{13}a_{23}}{a_{33}\sin 2\sigma} = \frac{a_{23}^2 - a_{13}^2 + (k_s - k_q)a_{33}}{a_{33}\cos 2\sigma} \qquad (1.113)$$

$$k_f + k_h = (k_s + k_q) - \frac{a_{13}^2 + a_{23}^2}{a_{33}} \qquad (1.114)$$

where k_f, k_h = principal curvatures of Σ_2
k_s, k_q = principal curvatures of the tool surface
σ = angle formed between the unit vectors $\mathbf{e}_f$ and $\mathbf{e}_s$ measured clockwise from $\mathbf{e}_f$ to $\mathbf{e}_s$ (Fig. 1.29)

The coefficients in Eqs. (1.112) to (1.114) are expressed as follows:

$$a_{13} = -k_s v_s^{(12)} - [\omega^{(12)}\mathbf{n}\mathbf{e}_s]$$
$$a_{23} = -k_q v_q^{(12)} - [\omega^{(12)}\mathbf{n}\mathbf{e}_q]$$
$$a_{33} = k_s(v_s^{(12)})^2 + k_q(v_q^{(12)})^2 - [\mathbf{n}\omega^{(12)}\mathbf{v}^{(12)}] \qquad (1.115)$$
$$- \mathbf{n} \cdot \{[\omega^{(1)} \times \mathbf{v}_{tr}^{(2)}] - [\omega^{(2)} \times \mathbf{v}_{tr}^{(1)}]\} + (\{[\omega^{(1)}]^2 m_{21}'(\mathbf{n} \times \mathbf{k}_2)\} \cdot (\mathbf{r} - \mathbf{R}))$$

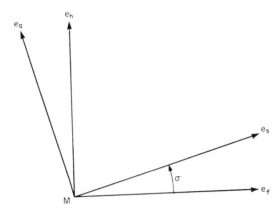

FIGURE 1.29 Surface principal directions

where
$\mathbf{v}^{(12)}$ = sliding velocity
$v_q = \mathbf{v}^{(12)} \cdot \mathbf{e}_q$, $v_s = \mathbf{v}^{(12)} \cdot \mathbf{e}_s$
$\boldsymbol{\omega}^{(12)} = \boldsymbol{\omega}^{(1)} - \boldsymbol{\omega}^{(2)}$ where $\boldsymbol{\omega}^{(i)}$ = the angular velocity (vector $\boldsymbol{\omega}^{(1)}$ passing through the origin O_f of fixed coordinate system S_f)
$\mathbf{r}$ = position vector of the point of contact of surfaces Σ_1 and Σ_2 that is drawn from O_f
$\mathbf{R}$ = position vector that is drawn from O_f to a point of the line of action of $\boldsymbol{\omega}^{(2)}$
$\mathbf{n}$ = surface unit normal
$\mathbf{v}_{tr}^{(i)}$ = transfer velocity of the point of surface Σ_i in motion *with* the surface
$\mathbf{v}_{tr}^{(1)} - \mathbf{v}_{tr}^{(2)} = \mathbf{v}^{(12)}$
$m_{21} = \omega^{(2)}/\omega^{(1)}$ = gear ratio
$m'_{21} = \dfrac{\partial f}{\partial \phi_1}(m_{21})$

For more information see Litvin (1989).[16]

1.9 CONTACT ELLIPSE AND BEARING CONTACT

The bearing contact is formed as a set of instantaneous contact ellipses. The center of symmetry of the instantaneous contact ellipse coincides with the respective contact point. To determine the orientation of the contact ellipse and its dimensions, it is necessary to know the principal curvatures and directions of the contacting surfaces and their elastic approach δ. It is assumed that the surfaces contact each other at every instant at a point. Figure 1.30a shows unit vectors $\mathbf{e}_f^{(1)}$ and $\mathbf{e}_f^{(2)}$ that represent the principal directions on surfaces Σ_1 and Σ_2; σ is the angle that is formed by $\mathbf{e}_f^{(1)}$ and $\mathbf{e}_f^{(2)}$; M is the point of contact of surfaces Σ_1 and Σ_2. It is assumed that the principal curvatures $k_I^{(1)}$, $k_{II}^{(1)}$, $k_I^{(2)}$, $k_{II}^{(2)}$, angle σ, and the elastic approach δ are known. The axes of the instantaneous contact ellipse are directed along the axes η and ξ. The orientation of the contact ellipse with respect to $\mathbf{e}_f^{(1)}$ is determined with $\alpha^{(1)}$, and the dimensions of the ellipse are designated by $2a$ and $2b$. To determine $\alpha^{(1)}$, $2a$, and $2b$

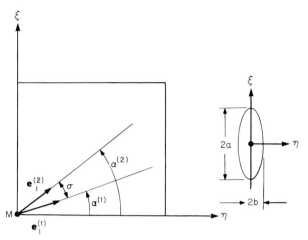

FIGURE 1.30 Orientation of contact ellipse

we may use the following equations[5,16]:

$$\cos 2\alpha^{(1)} = \frac{g_1 - g_2 \cos 2\sigma}{(g_1^2 - 2g_1g_2 \cos 2\sigma + g_2^2)^{1/2}} \qquad \sin 2\alpha^{(1)} = \frac{g_2 \sin 2\sigma}{(g_1^2 - 2g_1g_2 \cos 2\sigma + g_2^2)^{1/2}} \quad (1.116)$$

$$A = \frac{1}{4}[k_\Sigma^{(1)} - k_\Sigma^{(2)} - (g_1^2 - 2g_1g_2 \cos 2\sigma + g_2^2)^{1/2}] \quad (1.117)$$

$$B = \frac{1}{4}[k_\Sigma^{(1)} - k_\Sigma^{(2)} + (g_1^2 - 2g_1g_2 \cos 2\sigma + g_2^2)^{1/2}] \quad (1.118)$$

where $k_\Sigma^{(i)} = k_I^{(i)} + k_{II}^{(i)}$
$g_i = k_I^{(i)} - k_{II}^{(i)}$
$2a = 2\left|\frac{\delta}{A}\right|^{1/2}$
$$2b = 2\left|\frac{\delta}{B}\right|^{1/2} \quad (1.119)$$

1.10 MODIFICATION OF GEAR TOOTH SURFACES

Nonmisaligned gears transform rotation with a constant gear ratio m_{12}, and

$$\phi_2(\phi_1) = \frac{N_1}{N_2} \phi_1 \quad (1.120)$$

is a linear function that relates the angles of rotation of gears 1 and 2. N_1 and N_2 are the numbers of gear teeth.

An investigation of the influence of gear misalignment shows that $\phi_2(\phi_1)$ becomes a piecewise function which is almost linear for each cycle of meshing (Fig. 1.31a). The transmission errors are determined by

$$\Delta\phi_2(\phi_1) = \phi_2(\phi_1) - \phi_1 \frac{N_1}{N_2} \qquad (1.121)$$

and they are also represented by a piecewise linear function (Fig. 1.31b). Transmission errors of this type cause the jump of angular velocity at transfer points, and vibration becomes inevitable.

The gear misalignment also causes the bearing contact to shift to the edge of the tooth, and this is not acceptable. For these reasons it is necessary to modify the gear tooth surfaces to provide a localized bearing contact and absorb transmission errors. This goal can be achieved by the appropriate modification of gear tooth surfaces, for instance by crowning of the pinion tooth surface. The possibility to absorb linear functions of transmission errors is based on the following theorem (proposed by Litvin).[15]

Consider the interaction of a parabolic function given by (Fig. 1.32a)

$$\Delta\phi_2^{(1)} = -a\phi_1^2 \qquad (1.122)$$

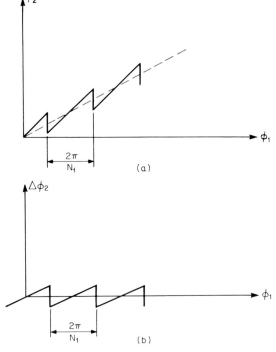

FIGURE 1.31 Transmission errors caused by gear misalignment

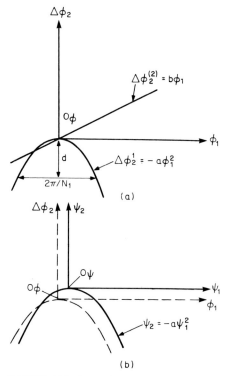

FIGURE 1.32 Interaction of parabolic and linear functions

with a linear function represented by

$$\Delta\phi_2^{(2)} = b\phi_1 \tag{1.123}$$

The resulting function

$$\Delta\phi_2 = b\phi_1 - a\phi_1^2 \tag{1.124}$$

represents again a parabolic function with the same parameter a that may be represented in a new coordinate system ψ_2, ψ_1 (Fig. 1.32b) as follows:

$$\psi_2 = -a\psi_1^2 \tag{1.125}$$

where

$$\psi_2 = \Delta\phi_2 - \frac{b^2}{4} \qquad \psi_1 = \phi_1 - \frac{b}{2a} \tag{1.126}$$

The new origin O_ψ is displaced with respect to the previous origin O_ϕ, but the axes of both coordinate systems are parallel.

The absorption of linear function $\Delta\phi^{(2)} = -b\phi_1$ by the parabolic function

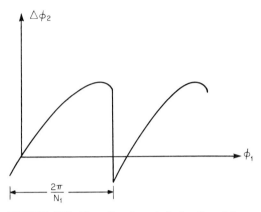

FIGURE 1.33 Discontinued parabolic function of transmission errors

$\Delta\phi_2^{(1)} = -a\phi_1^2$ means that gear misalignment will not change a specially predesigned parabolic function of transmission errors $\Delta\phi_2^{(1)} = -a\phi_1^2$. The predesigned parabolic function can be provided in the process of generation by special machine-tool settings or by controlling the motions of the tool and the gear in the process of generation.

It may happen that the absorption of a linear function by a predesigned parabolic function is accompanied with too large a change of transfer points, and the resulting parabolic function of transmission errors, $\psi_2(\psi_1)$, will be a discontinuous function for one cycle of meshing (Fig. 1.33). To avoid this it is necessary to limit the tolerances for gear misalignment.

REFERENCES

1. Baxter, M. L., "Basic Geometry and Tooth Contact of Hypoid Gears," *Industrial Mathematik*, No. 2, 1961.
2. Buckingham, E., *Analytical Mechanics of Gears*, 2nd ed., Dover, New York, 1963.
3. Dudley, D. W., *Gear Handbook, The Design, Manufacture, and Application of Gears*, McGraw-Hill, New York, 1962.
4. Gleason Works, *Understanding Tooth Contact Analysis*, Rochester, N.Y., 1970.
5. Litvin, F. L., *Theory of Gearing*, 2nd ed., Nauka, Moscow (in Russian), 1968.
6. Litvin, F. L., "Die Beziehungen zwishen den Krümmungen der Zahnoberflächen bei Räumlichen Verzahnungen," *Zeitschrift für Angewandte Mathematik und Mechanik*, Vol. 49, 1969, pp. 685–690.
7. Litvin, F. L., Petrov, K. M., and Ganshin, V. A., "The Effect of Geometrical Parameters of Hypoid and Spiroid Gears on Its Quality Characteristics," *Journal of Engineering for Industry*, Vol. 96, 1974, pp. 330–334.
8. Litvin, F. L., Coy, J. J., and Rahman, P., "Two Mathematical Models of Spiral Bevel Gears Applied to Lubrication and Fatigue Life," *Proceedings of the International Symposium on Gearing and Power Transmissions*, Japan Society of Mechanical Engineers, Tokyo, 1981, pp. 281–286.
9. Litvin, F. L., and Gutman, Y., "Methods of Synthesis and Analysis for Hypoid Gear-Drives of Formate and Helixform," *Journal of Mechanical Design*, Vol. 103, 1981a, pp. 83–113.

Part 1—"Calculations for Machine Setting for Member Gear Manufacture of the Formate and Helixform Hypoid Gears," pp. 83–88.
Part 2—"Machine Setting Calculations for the Pinions of Formate and Helixform Gears," pp. 89–101.
Part 3—"Analysis and Optimal Synthesis Methods for Mismatched Gearing and Its Application for Hypoid Gears of Formate and Helixform," pp. 102–113.

10. Litvin, F. L., and Gutman, Y., "A Method of Local Synthesis of Gears Based on the Connection Between the Principal and Geodetic Curvatures of Surfaces," *Journal of Mechanical Design*, Vol. 103, 1981*b*, pp. 114–125.

11. Litvin, F. L., Rahman, P., and Goldrich, N., "Mathematical Models for the Synthesis and Optimization of Spiral Bevel Gear Tooth Surfaces," NASA CR-3553, 1982.

12. Litvin, F. L.; and Tsay, C. B., "Helical Gears with Circular Arc Teeth: Simulation of Conditions of Meshing and Bearing Contact," *Journal of Mechanisms, Transmissions, and Automation in Design*, Vol. 107, No. 4, 1985, pp. 556–564.

13. Litvin, F. L., Tsung, W.-J., Coy, J. J., and Heine, C., "Method for Generation of Spiral Bevel Gears with Conjugate Gear Tooth Surfaces," *Journal of Mechanisms, Transmissions and Automation in Design*, Vol, 109, No. 2, 1987, pp. 163–170.

14. Litvin, F. L., Zhang, J., and Handschuh, R. F., "Crowned Spur Gears: Methods for Generation and Tooth Contact Analysis," *Journal of Mechanisms, Transmissions and Automation in Design*, Vol. 110, 1988, pp. 337–347.
Part 1—"Basic Concepts, Generation of the Pinion Tooth Surface by a Plane."
Part 2—"Generation of the Pinion Tooth Surface by a Surface of Revolution."

15. Litvin, F. L., Zhang, J., Handschuh, R. F., and Coy, J. J., "Topology of Modified Helical Gears," *Journal of Surface Topography*, Kogan press, Vol. 2, No. 1, 1989, pp. 41–38.

16. Litvin, F. L., 1989, *Theory of Gearing*, NASA Reference Publication 1212, AVSCOM Technical Report, 88–C–035.

17. Litvin, F. L., Kin, V., and Zhang, Y., "Limitations of Conjugate Gear Tooth Surfaces," *Journal of Mechanical Design*, Vol. 112, June 1990, pp. 230–236.

18. Litvin, F. L., Chaing, W. S., Lundy, M., and Tsung, W. J., "Design of Pitch Cones for Face-Hobbed Gears," *Journal of Mechanical Design*, Vol. 112, September 1990, pp. 413–418.

19. Stipelman, B. A., *Design and Manufacture of Hypoid Gears*, Wiley, New York, 1978.

20. Wildhaber, E., "Basic Relationships of Hypoid Gears," *American Machinist*, Vol. 90, No. 4, 1946*a*, pp. 108–111.

21. Wildhaber, E., "Conjugate Pitch Surfaces," *American Machinist*, Vol. 90, No. 13, 1946*b*, pp. 150–152.

22. Wildhaber, E., "Tooth Contact," *American Machinist*, Vol. 90, No. 12, 1946*c*, pp. 110–114.

23. Wildhaber, E., "Surface Curvature," *Product Engineering*, Vol. 27, No. 5, 1956, pp. 184–191.

24. Zalgaller, V. A., *Theory of Envelopes*, Nauka, Moscow (in Russian), 1975.

CHAPTER 2
GEAR TYPES AND NOMENCLATURE

Raymond J. Drago
Associate Technical Fellow, Boeing Helicopters
Philadelphia, Pennsylvania

Chief Engineer
Drive Systems Technology, Inc.
Glen Mills, Pennsylvania

Before we can discuss the different types of gears, we would do well to define just what we mean by a "gear." Perhaps the most succinct definition I have ever heard was provided by a loading dock worker who asked where he should deliver the "wheels with notches in them." In more technically correct terms, a *gear* is a toothed wheel that is usually, but not necessarily, round. The teeth may have any of an almost infinite variety of profiles. The purpose of gearing is to transmit motion and/or power from one shaft to another. This motion transfer may or may not be uniform, and may also be accompanied by changes in direction, speed, and shaft torque.

Unfortunately, even among gear specialists there is some ambiguity in the terms used to describe gears and gear-related parameters. In our treatment herein, to the extent possible, we follow the conventions recommended by the American Gear Manufacturers Association.[1]

2.1 TYPES OF GEARS

The first part of this chapter covers the types of gears in common use. The basic nomenclature and formulas for each type of gear are covered in the second part of this chapter.

The text material for each type of gear gives some comments on where the gears are used, how the gears may be made, the possible efficiency, and occasional comments on lubrication of the gears. These comments should all be considered introductory. Much more specific information is given in the following chapters of this book. Also, the references given in this book show the reader where more comprehensive data can be obtained.

2.1.1 Classifications

In general, gears may be divided into several broad classifications based on the arrangement of the axes of the gear pair. The most general type of gearing consists of a gear pair whose axes are neither perpendicular nor parallel and do not intersect (that is, they do not lie in the same plane). All other types are special cases of this basic form. In the ensuing sections we discuss most of the major gear types, generally in order of increasing complexity. We limit our discussion to specific gear types and do not include arrangements of these types.

Our discussions consider, qualitatively, the various aspects of many different types of gears. In order to provide an easy basis for comparing these many and varied gear types, Table 2.1 provides a broad-based comparison. It is by no means an exhaustive comparison, but it serves to provide a reasonable basis for preliminary design. In using this table, the reader must keep clearly in mind the fact that the data are "typical" or "nominal." Surely, any experienced gear technologist can point out specific cases that vary substantially in virtually every category.

Within each type of gear classification there exist many variations in actual tooth form. The involute or involute-based form is, at least for parallel axis gears, the most common tooth form; however, many other special forms exist. We do not discuss tooth forms, per se, in this section.

2.1.2 Parallel Axis Gears

The simplest types of gears are those that connect parallel shafts. They are generally relatively easy to manufacture and are capable of transmitting large amounts of power with high efficiency. Parallel axis gears transmit power with greater efficiency than any other type or form of gearing.

Spur Gears. The spur gear has teeth on the outside of a cylinder and the teeth are parallel to the axis of the cylinder. This simple type of gear is the most common type. Its volume of usage is the largest of all types.

The shape of the tooth is that of an involute form. There are, however, some notable exceptions. Precision mechanical clocks very often use cycloidal teeth since they have lower separating loads and generally operate more smoothly than involute gears and have less tendency to bind. The cycloidal form is not used for power gearing because such gears are difficult to manufacture, sensitive to small changes in center distance, and not as strong or as durable as their involute brothers.

Figure 2.1 shows a close-up view of the teeth on a set of spur gears having about a 5:1 ratio. The teeth are 20° involute tooth form. The pinion is made with more than a standard addendum and the gear addendum is shorter than standard. The whole depth is standard for high-strength gears. Note the large radius of curvature in the root fillet region. This reduces the bending stress due to a lower stress concentration factor. Also note that the pinion teeth look very sturdy. This is an effect of the long- and short-addendum design just mentioned. (Details of specific design strategy are covered

GEAR TYPES AND NOMENCLATURE

FIGURE 2.1 16-tooth spur pinion driving a gear with about 75 teeth. This is a special design of 20° involute teeth for high load-carrying capacity.

in later chapters of this book.)* Figure 2.2 shows a set of spur gears in an accessory gear box.

The most common pressure angles used for spur gears are 14½, 20, and 25°. In general, the 14½° pressure angle is not used for new designs (and has, in fact, been withdrawn as an AGMA standard tooth form); however, it is used for special designs and for some replacement gears. Lower pressure angles have the advantage of smoother and quieter tooth action because of the large profile contact ratio. In addition, lower loads are imposed on the support bearings because of a decreased radial load component; however, the tangential load component remains unchanged with pressure angle. The problem of undercutting associated with small numbers of pinion teeth is more severe with the lower pressure angle. Lower pressure angle gears also have lower bending strength and surface durability ratings and operate with higher sliding velocities (which contributes to their relatively poor scoring and wear performance characteristics) than their higher pressure angle counterparts.

Higher pressure angles have the advantage of better load-carrying capacity, with respect to both strength and durability, and lower sliding velocities (thus better scoring and wear performance characteristics). In some cases, very high pressure angles, 28, 30, and, in a few cases, as high as 45°, are employed in some special slow-speed gears for very high load capacity where noise is not the predominant consideration.

A minimum of equipment is required to produce this type of gear; thus it is usually the least expensive of all forms of gearing. While the most common tooth form for spur

---

*Drago[3] and Dudley[4] (specific books on gear design) give design strategy details beyond the scope of this book.

TABLE 2.1 General Comparison of Gear Types

Type of gear	Approx. range of efficiencies	Max.* widths generally used†	Type of load imposed on support bearings	Nominal range of reduction ratio‡	Nominal max. pitch line velocity, fpm		Member of pair	Methods of manufacture	Methods of refining
					Aircraft and high precision	Commercial			
Parallel Axes									
External spur gears	97–99.5	$F = d$	Radial	1:1–5:1	20,000	4,000	Both	Hob, shape, mill, broach, stamp, sinter	Grind, shave, lap (crossed axis only), hone
Internal spur gears	97–99.5	Dictated by mating gear	Radial	1:5–7:1	20,000	4,000	Both	Shape, mill, broach, stamp, sinter	Grind, shave, lap (crossed axis), hone
External helical gears	97–99.5	$F = d$	Radial and thrust	1:1–10:1	40,000	4,000	Both	Hob, shape, mill	Grind, shave, lap, hone
Internal helical gears	97–99.5	Dictated by mating gear	Radial and thrust	1:5:1–10:1	20,000	4,000		Shape, mill	Shave, grind, lap, hone
Internal herringbone or double helical	97–99.5	Dictated by mating gear	Radial	2:1–20:1	20,000	4,000		Shape	Lap, shave, hone
External herringbone or double-helical gear	97–99.5	$F = 2d$	Radial	1:1–20:1	40,000	4,000	Both	Hob, shape, mill	Grind, shave, lap, hone
Intersecting Axes									
Straight bevel gear	97–99.5	⅓ cone distance	Radial and thrust	1:1–8:1	10,000	1,000	Both	Generating, forming	Grind
Zerol bevel gear	97–99.5	28 % of cone distance	Radial and thrust	1:1–8:1	10,000	1,000	Both Gear	Generating Forming	Grind
Spiral bevel gear	97–99.5	⅓ cone distance	Radial and thrust	1:1–8:1	25,000	4,000	Both Gear	Generating Forming	Grind, lap
Face gear	95–99.5	From 0.2d at a low ratio to d at a high ratio	Radial and thrust	3:1–8:1	5,000	4,000	Pinion / Gear	Same as external spur / Shape	Lap, grind, shave, hone / Lap
Beveloid	95–99.5	$\dfrac{5}{p_n}$	Radial and thrust	1:1–8:1	5,000	4,000	Both	Hob, generating	Lap, grind

2.4

Nonintersecting Nonparallel Axes

Type	*(col2)*	†‡ Formula	Ratio range	Pitch line vel., ft/min	Efficiency, %	Loads	Driver	Method of cutting	Method of finishing
Crossed helical	50–95	$F_1 = 4p_n \sin \psi_1$ $F_2 = 4p_n \sin \psi_2$	1:1–100:1	10,000	4,000	Radial and thrust	Both	Hob, shape, mill	Grind, lap, shave
Cylindrical worm	50–90	$F_w = 5p_n \cos \lambda$ $F_g = 0.67d$	3:1–100:1	10,000	5,000	Radial and thrust	Pinion Gear	Mill, hob Hob	Grind Lap
Double-enveloping worm	50–98	$F_w = 0.9D$ $F_g = 0.9d$	3:1–100:1	10,000	4,000	Radial and thrust	Worm Wheel	Shape, hob Hob, mill	Lap, grind
Hypoid	90–98	$F_g = \frac{1}{3}$ cone distance	1:1–10:1	10,000	4,000	Radial and thrust	Both Gear	Generating Forming	Grind, lap
High-reduction hypoid	50–90	$F_g = 0.15D$	10:1–50:1	10,000	4,000	Radial and thrust	Pinion Gear	Generating Forming	Grind, lap Grind, lap
Spiroid	50–97	$F_p = 0.24D$ $F_g = 0.14D$	9:1–100:1	10,000	6,000	Radial and thrust	Pinion Gear	Mill, hob Hob, mold	Grind, chase
Planoid	90–98	$F = \frac{1}{3}$ cone distance	1.5:1–10:1	10,000	4,000	Radial and thrust	Pinion Gear	Hob Mill, broach	Lap, grind Grind
Helicon	50–98	$F_p = 0.21D$ $F_G = 0.12D$	3:1–100:1	10,000	6,000	Radial and thrust	Pinion Gear	Mill, hob Hob, mold	Grind, chase
Face gear	95–99.5	From $0.2D$ at a low ratio to D at a high ratio	3:1–8:1	10,000	4,000	Radial and thrust	Pinion Gear	Same as external spur Shape	Same as external spur Lap
Beveloid	50–95	$\dfrac{5}{p_n}$	1:1–100:1	10,000	4,000	Radial and thrust	Both	Generation, hob	Grind, lap

* Face width given is only a "nominal" maximum. Consult other parts of the Gear Handbook for detail limitations on face width.
† d = pinion pitch diameter
   D = gear pitch diameter
   λ = lead angle
   ψ = helix angle
   p_n = normal circular pitch
‡ The gear types showing an upper ratio of 100:1 can be built to 300:1 or higher. The ratio of 100:1 is shown as a normal maximum limit.

FIGURE 2.2 With the cover removed, a set of high-speed spur gears can be seen in an accessory box. Courtesy of the Gleason Works, Rochester, N.Y.

gears is the involute, other tooth forms are possible as long as they provide conjugate motion.

Helical Gears. When gear teeth are cut on a spiral that wraps around a cylinder, they are designated as helical. Helical teeth enter the meshing zone progressively and, therefore, have a smoother action than spur gear teeth and tend to be quieter. In addition, the load transmitted may be somewhat larger, or the life of the gears may be greater for the same loading, than with an equivalent pair of spur gears. Conversely, in some cases smaller-size helical gears (compared with spur gears) may be used to transmit the same loading. Helical gears produce an end thrust along the axis of the shafts in addition to separating and tangential (driving) loads of spur gears. Where suitable means can be provided to take this thrust, such as thrust collars or ball or tapered roller bearings, it is no great disadvantage. The efficiency of a helical gear set, which is dependent on the total normal tooth load (as well as the sliding velocity and friction coefficient, etc.), will usually be slightly lower than for an equivalent spur gear set.

Conceptually, helical gears may be thought of as stepped spur gears in which the size of the step becomes infinitely smaller. For external parallel axis helical gears to mesh, they must have the same helix angle but be of different hand. Just the opposite is true for an internal helical mesh; that is, the external pinion and the internal gear must have the same hand of helix. For crossed axis helical gears (discussed later in this section), the helix angles on the pinion and gear may or may not be of the same hand, depending on the direction of rotation desired and the magnitude of their individual helix angles.

Involute profiles are usually employed for helical gears and the same comments made earlier about spur gears hold true for helicals.

In order to provide significant improvement in noise over spur gears, the face overlap must be at least unity. That is, one end of the gear face must be advanced at least one circular pitch from the other. If the face overlap (also called face contact ratio) is less than unity, the helical gear, for most analytical purposes (that is, stress calculations) is treated as if it were a spur gear with no advantage taken of the fact that it is helical.

Helix angles from only a few degrees up to about 45° are practical. As the helix angle increases from zero, in general, the noise level is reduced and the load capacity is increased. At angles much above 15 to 20°, however, the tooth bending capacity generally begins to drop off. This is due to the fact that the transverse tooth thickness decreases rapidly.

Double-helical or herringbone gears are frequently used to obtain the noise benefits of single-helical gears without the disadvantage of thrust loading. The common types of helical gears are shown in Fig. 2.3, while Fig. 2.4 shows a typical single-helical set. The terms *double helical* and *herringbone* are sometimes used interchangeably. Actu-

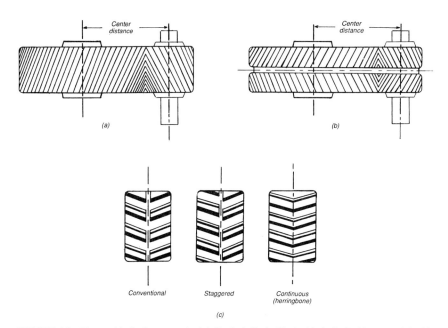

FIGURE 2.3 Types of helical-gear teeth. *(a)* Single helical; *(b)* double helical; *(c)* types of double helical.

FIGURE 2.4 Single-helical-gear set designed for high speed and high horsepower. Courtesy of Maag Gear Wheel Company, Zurich, Switzerland.

ally, herringbone is more correctly used to describe double-helical teeth that are cut continuously into a solid blank, as shown in Fig. 2.3c. Double-helical gears may be cut integral with the blank, either in line or staggered, or two separate parts may be assembled to form the double.

While double-helical or herringbone gears do eliminate the net thrust load on the shaft, it is important to note that the two halves of the gear must internally react to the full thrust load. This being the case, the design of the gear blank for one-piece gears, or the construction of the assembly, must be carefully evaluated to insure that it is strong enough to take the thrust loads and to provide enough rigidity so that the deflections are not excessive.

The efficiency of helical gears is quite good though not quite equal to that of simple spurs. This is due to the fact that all other things being equal, the normal (or total) tooth load on a helical gear is higher than that on a spur for an equivalent tangential load.

Internal Gears. The internal gear has teeth on the inside of a cylinder. The teeth may be made either spur or helical. The teeth of an involute form internal gear have a concave shape rather than a convex shape. Internal gears are generally more efficient since the sliding velocity along the profile is lower than for an equivalent external set. Because of the concave nature of the internal tooth profile, its base is thicker than an equivalent external gear tooth (either spur or helical). The tooth strength of an internal gear is greater than that of an equivalent external gear.

The internal gear has other advantages. It operates at a closer center distance with its mating pinion than do external gears of the same size. This permits a more compact design. The internal gear eliminates the use of an idler gear when it is necessary to have two parallel shafts rotate in the same direction. The internal gear forms its own guard over the meshing gear teeth. This is highly desirable for preventing accidents in some kinds of machines.

Internal gears cannot be used where the number of teeth in the pinion is almost the same as that of the gear. When this occurs, the tips of the pinion teeth interfere with the tips of the gear teeth. While a good guide is to maintain a ratio of 2:1 between the number of teeth on the internal gear and its mating pinion, lower ratios may be practical in some cases, particularly if the tooth utilizes shifted involute design characteristics. For full-depth 20°-pressure-angle teeth, for example, the internal gear pitch diameter must be at least one and one-half times that of the external gear; smaller ratios require considerable modification of the tooth shape to prevent interference as the teeth enter and leave the mating gear. Internal gears also have the disadvantage that few machine tools can produce them.

It may be more difficult to support a pinion in mesh with an internal gear because the pinion must usually be overhung mounted. Internal gears tend to find their greatest application in various forms of epicyclic gear systems and in large slewing-type drives, generally with integral bearings. In such configurations, the internal gear is frequently much larger than its mating external pinion, and thus it is usually possible to build a carrier structure inside the internal gear so that the pinion can be straddle mounted.

A typical internal gear arrangement is shown in Fig. 2.5, while Fig. 2.6 shows a large internal spur gear. In the case of an internal helical gear, its hand of helix must be the same as its mating pinion; that is, both the external helical pinion and its mating helical internal gear must be either right- or left-handed.

2.1.3 Nonparallel, Coplanar Gears (Intersecting Axes)

While the involute is the tooth form of almost universal choice for parallel axis gears, most gears that operate on nonparallel, coplanar axes do not employ involute profiles. A wide variety of gear types fall into this category. The most obvious use for gears of

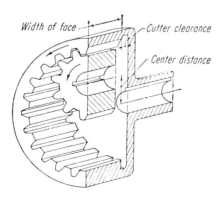

FIGURE 2.5 Typical internal-gear arrangement.

FIGURE 2.6 90-inch-diameter-internal-ring gear for an epicyclic speed reducer being measured on a 200-inch-diameter Maag gear checker. *(Courtesy of Philadelphia Gear Corp., King of Prussia, Pa.)*

this type is the redirection of power flow around a corner, such as might be required, for example, when connecting a horizontally mounted turbine engine to the vertically mounted rotor shaft on a helicopter. In fact, while power flow is a frequent reason for choosing one of these gear types, other features unique to a particular type of gear, such as the ability of straight bevel gears to be used in a differential arrangement for a rear-wheel-drive automobile, might well be the determining factor in making the decision to use one of these gear types. Descriptions of the major, and some interesting minor, ones follow.

Bevel Gears. There are four basic types of bevel gears: straight, Zerol,* spiral, and skew tooth. In addition, there are three different manufacturing methods (face milling, face hobbing, and tapered hobbing) that are employed to produce true curved tooth bevel gears, each of which produces a different tooth geometry. In our discussion herein we address the overall generic attributes of the different types of bevel gears without addressing the manufacturing variations.

All bevel gears impose both thrust and radial loads in addition to the transmitted tangential loads on their support bearings. The thrust is a result of the tapered nature of the gear blank, regardless of the tooth form employed (that is, even straight bevel gears produce an axial thrust). The diametral pitch for all bevel gears is conventionally measured at the heel of the tooth, while the pressure and spiral angles are conventionally measured at the mean section (tooth midface). In all cases, the pitch apex of the pinion and that of its mating gear must intersect at the point at which the axes of the

---

*Zerol is a registered trademark of the Gleason Works, Rochester, New York.

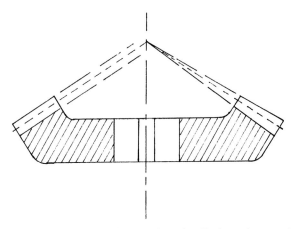

FIGURE 2.7 Basic bevel-gear configuration. Teeth may be tapered in depth or they may be made with a parallel depth, depending on manufacturing system used.

pinion and gear shafts intersect. Figure 2.7 shows the basic bevel gear configuration that is common to all types of bevel gears. By convention, the direction of rotation of a bevel gear is specified as clockwise or counterclockwise as viewed from the heel end of the gear looking toward the pitch apex.

Straight. The simplest form of the bevel gear is the straight bevel gear. Except for the thrust load produced by the blank taper, straight bevel gear teeth are a three-dimensional analogy to spur gear teeth. One very common use for straight bevel gears is in a bevel gear differential. Extensions of the straight teeth intersect at the axis of the gear. The tooth profile in a section normal to the tooth approximates that of an involute spur gear having a number of teeth equal to the actual number of teeth in the bevel gear divided by the cosine of the pitch angle (equivalent number of teeth). The teeth are always tapered in thickness and may have either constant or tapering height. The outer or heel part of the tooth is larger than the inner part called the toe. See Fig. 2.8. Although straight bevel gears have instantaneous line contact, they are often manufactured to have localized contact to permit more tolerance in mounting. One method of achieving this type of contact is the Coniflex* system, employed by Gleason Works.

Zerol. An improvement over straight bevel gears in terms of contact conditions, noise level, and power capacity is the Zerol bevel gear (Fig. 2.9). This gear is similar to a straight bevel except that the teeth are curved along their axis; however, the mean spiral angle is zero,† thus the bearing reaction loads (particularly the thrust loads) are the same as for straight bevels. Zerol gears are manufactured with the same type of cutters and on the same machines as spiral bevel gears. This is economically important as it eliminates the need for more than one type of bevel gear cutting equipment.

---

*Coniflex is a registered trademark of the Gleason Works, Rochester, New York.
†Actually, the term *Zerol* applies to any "spiral" bevel gear with a spiral angle of 10° or less, but most often Zerol bevels have a zero mean spiral angle.

FIGURE 2.8 Straight-bevel gears. *(Courtesy of the Gleason Works, Rochester, N.Y.)*

FIGURE 2.9 Zerol-bevel gears. *(Courtesy of the Gleason Works, Rochester, N.Y.)*

Localized tooth bearing of Zerol gears is accomplished by lengthwise mismatch (face curvature mismatch) and profile modification. The face width of a Zerol bevel gear is restricted to specific values due to cutter limitations. Zerol bevel gears are very frequently used in turbine engine, helicopter, and other high-speed devices such as accessory drives.

Zerol bevel gears may be compared to double helical or herringbone parallel axis gear teeth in that they have no more thrust load than their straight counterparts but provide advantages related to the improved contact ratio due to their lengthwise curvature. A pressure angle of 20° is the most common; however, Zerols may be manufactured with 14½, 22½, or 25° as well. Because Zerols are not generally used in very-high-load applications, the use of high pressure angles, which tend to reduce contact ratio and thus increase noise, should be carefully considered.

Spiral. The most complex form of bevel gear is the spiral bevel. It is commonly used in applications that require high load capacity at higher operating speeds than are typically possible with either straight or Zerol bevel gears. The relationship between spiral and straight bevel gears is comparable to that between helical and spur gears. By convention, the hand of spiral is defined as either left- or right-hand as determined by viewing the gear from the apex end looking toward the heel end.

The teeth of spiral bevel gears are curved and oblique (Fig. 2.10), and as a result they have a considerable amount of overlap. This insures more than one tooth in contact at all times and results in gradual engagement and disengagement with continuous multiple-tooth contact. This results in higher overall contact ratios than are possible with either straight or Zerol bevel gears. Because of this improved contact ratio and the resultant load sharing, spiral bevels have better load-carrying capacity and run more smoothly and quietly than either straight or Zerol bevel gears of the same size. The improved load capacity of spiral bevels also allows them to be smaller in size for a given load capacity than an equivalent straight or Zerol bevel gear.

FIGURE 2.10 Spiral-bevel gears. *(Courtesy of the Gleason Works, Rochester, N.Y.)*

Spiral bevel gears impose significantly more thrust load on their support bearings than do either Zerol or straight bevel gears. While plain thrust bearings have been used successfully, a rolling element thrust bearing is usually required with spiral bevel gears. By comparison, plain bearings are quite often used with straight or Zerol bevel gears. The more common pressure angles for spiral bevel gears are 16 and 20°, with the latter now being almost standard. The most common spiral angle in use is 35°. In more highly loaded applications, particularly those that operate at high speeds as well, such as aircraft or helicopter transmissions, a higher pressure angle (typically $20\frac{1}{2}$ to 25°) and lower spiral angles (typically 15 to 25°) are more often used.

The profile contact ratio on all bevel gears is usually substantially lower than equivalent spur gears; thus they tend to be somewhat noisier. In addition, due to their higher total load for a given tangential load and the higher sliding that accompanies the localized contact, they are less efficient than either spur or helical gears. Still, the efficiency of a well made set of bevel gears is generally better than 98 percent, and that of hardened and ground sets is usually above 99.2 percent.

Spiral bevel gears can be made by any of three generic methods. Each of these three methods produces a unique lengthwise tooth curvature and profile. In general, while gears within one system may be interchangeable, and gear sets made by each system are setwise interchangeable, pinions and gears are not interchangeable between systems, despite the fact that the basic configuration (that is, diametral pitch, spiral angle, pressure angle, etc.) may be identical.

1. *Circular lengthwise tooth curvature:* For this system, developed in the United States, face-milling cutters with multiple blades are used. While several methods of implementing this process exist, they are all somewhat similar in their basic motions. The blank executes a rolling motion relative to the cutter, which simulates one tooth of the imaginary crown gear. This is repeated for every tooth space, the cutter being withdrawn and turned back into starting position every time while the job is indexed for the next gap. This process is known as single indexing.

All tooth spaces have first to undergo a roughing operation and are finished by a subsequent operation. In general, the root lines of the teeth are not parallel to the pitch line, but set at a specific root angle. The cutters must be set parallel to the root lines while producing the gears; that is, they must be tilted relative to the pitch line. This causes the normal pressure angle to change, so that the pressure angles of the cutters must be corrected accordingly. This correction depends on the root angle and on the spiral angle and is different for either tooth side. In operation, the concave tooth surfaces of one member are always in mesh with the convex tooth side of the other. This system is used by Gleason machines and is the only one for which grinding machines have been developed.

2. *Involute lengthwise tooth curvature:* This system was originally developed in Germany. The tooth is a tapered hob, usually single-thread, with a constant pitch determined in accordance with the normal diametral pitch. The machining process is by continuous indexing: Tool and job rotate continuously and uniformly, the relationship between the two rotary motions being established by change gears as a function of the number of teeth being cut. The job is adjusted with its cone surface tangential to the surface of the imaginary flat gear.

This system is utilized by some Klingelnberg machines but is rapidly being replaced by the more versatile face-milling and face-hobbing processes.

3. *Epicycloidal lengthwise tooth curvature:* This system, originally evolved in Italy and England but still not free from some shortcomings, has been further developed in Switzerland and Germany. The machines available for producing this type of spiral bevel gear, also known as cyclo-palloid tooth form bevels, are similar to the one for the production of circular tooth length curves in that the tool is also a face-type milling cutter with inserted blades, which can be of the inside-cutting, outside-cutting, and gashing type, alternatively. The indexing is now, however, not intermittent but continuous. The cutter and the blank rotate continuously and uniformly, so that not only is one tooth space generated at a time, but all tooth spaces consecutively. Accordingly, the blades are not evenly distributed over the periphery of the cutter, but combined in groups (that is, the cutter has multiple starts). One group generally comprises one roughing, one outside-cutting, and one inside-cutting blade, and allows simultaneous machining of the tooth bottom and both tooth sides. The number of blade groups depends on the size of the cutter and usually varies from 3 to 11. This basic system is utilized by both Oerlikon and Klingelnberg, with some variation on each.

Skew. The final type of bevel gear is the skew tooth. This is similar to a spiral bevel gear and is quite often also referred to (incorrectly) as a spiral bevel gear. Actually, the skew tooth has no lengthwise curvature; rather the teeth are simply cut straight but at an angle to the shaft centerline (as shown in Fig. 2.11) so that some face overlap may occur. This provides an improvement in load capacity (Fig. 2.12) when compared with a straight bevel gear, but not to the level possible with a true spiral (curved tooth) bevel gear. Skew tooth gears are used primarily in large (over 30-in. gear pitch diameter) sizes only. They are produced on planing generator machines.

Until the late 1980s, because of limitations in the size of the face-milling machines available, most "spiral" bevel gears over 30 to 40 in. in diameter were planed; thus they were actually skew tooth, not true spiral bevel gears. Since pinion and gear must be conjugate, the pinions that mate with these large gears must also be planed. Relatively recent advances in spiral bevel gear manufacturing have led to the development of large machines of both the face-milling and face-hobbing type that can cut true spiral bevel gear teeth on gears over 100 in. in pitch diameter.

Face Gears (On-Center). A face gear set is actually composed of a spur or helical pinion that is in mesh with a "face" gear. The pinions are really not any different from their parallel axis counterparts except for the fact that they are in mesh with a face gear. Face gears have teeth cut into the blank such that the axis of the teeth lie in a plane

FIGURE 2.11 Skew-bevel gear.

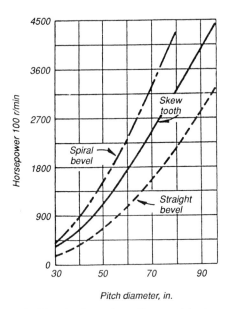

FIGURE 2.12 Comparison of large-straight skew and spiral-bevel-gear capacities. *(Data supplied by Philadelphia Gear Corp., King of Prussia, Pa.)*

that is perpendicular to the shaft axis. The mating pinion is either a spur or a helical gear. The pinion and gear axes are coplanar for on-center face gears. The pinion and face gear axes most often form a 90° shaft angle; however, small deviations from 90° are possible. In operation, this type of gear is similar to an equivalent set of straight bevel gears.

The face gear tooth changes shape from one end of the tooth to the other. The load capacity of face gears, compared with that of bevel gears, is rather small; thus they are used mostly for motion transmission rather than as power gears. Face gears are, relative to bevel gears, easy to make and somewhat less expensive as well. A typical arrangement is shown in Fig. 2.13. Toys and games make extensive use of face gears.

Conical Involute Gearing. Conical involute gearing is a completely generalized form of involute gear. It is an involute gear (as shown in Fig. 2.14) with tapered tooth thickness, root, and outside diameter. Commonly known as Beveloid* gears, they are useful primarily for precision instrument drives where the combination of high precision and limited load-carrying ability fits the application. Like other involute gears, Beveloids are relatively insensitive to positional errors although shaft angles must be quite accurate. They will mate conjugately with all other involute gears (spurs, worms, helicals, and racks). Like crossed axis helicals (see later discussion), however, the area

---

*Beveloid is a registered trademark of the Invincible Gear Co., Livonia, Michigan.

GEAR TYPES AND NOMENCLATURE

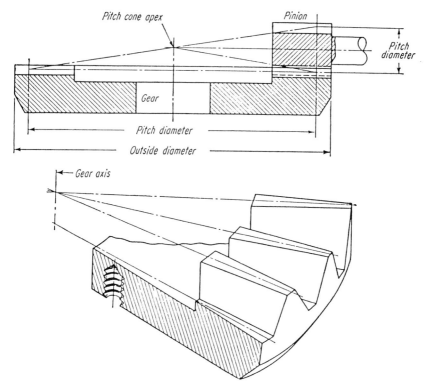

FIGURE 2.13 Face-gear arrangement.

of contact between these gears and Beveloids is limited; thus load capacity is low. For small tooth numbers and large cone angles, undercutting is quite severe.

This type of gear is not in widespread use at this time, but may be used to advantage in some circumstances, especially for precision motion-transmission applications in which the gear shafts must be located at varying orientations in space.

2.1.4 Nonparallel, Noncoplanar Gears (Nonintersecting Axes)

Gears in this classification are generally the most complex, both in terms of geom-

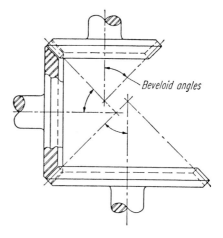

FIGURE 2.14 Beveloid-gear arrangement.

etry and manufacture. The simpler types, discussed first in the following discussion, are, however, quite easy to manufacture and are reasonably inexpensive, but they do not carry large loads. The more complex types are generally more expensive but provide better load capacity and other features that make them especially suited for a wide variety of special applications.

Crossed Axis Helicals. Crossed helical gears are satisfactory for the normal range of ratios used for single reduction helical gears. They provide both speed reduction and extreme versatility of shaft positioning at a relatively low initial cost. At higher ratios, or for anything above moderate loads, however, worm, Spiroid, or Helicon gears are generally preferable. (See Fig. 2.15 for typical crossed axis helical gears.)

Crossed axis helical gears are usually cut in the same manner as conventional helical gears using identical tooling, since they are, until mounted on their crossed axes, actually nothing more than conventional helical gears! Since crossed helical gears have a great deal of sliding action, special attention must be given to the selection of gear materials and their lubricants to reduce friction to a minimum and eliminate any possibility of seizing between mating gears. Experience has indicated that iron on iron functions satisfactorily. However, a hardened steel pinion driving a bronze gear will also work quite well.

Figure 2.16 shows some of the special relations that must exist for this type of gearing to function properly. It is interesting to note that, while mating parallel axis external helical gears must have opposite hands of helix, crossed axis helicals can have

FIGURE 2.15 Crossed-axis-helical-gear set, driving a pump and governor for a steam turbine application. *(Courtesy of the General Electric Co., Lynn, Mass.)*

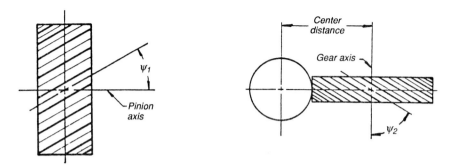

FIGURE 2.16 Crossed-axis-helical-gear arrangement.

either the same or opposite hands of helix, depending on the shaft angle and the relative directions of rotation of the driver and driven gears. Additionally, the reduction ratio between the pinion and the gear is a function of their relative numbers of teeth but not directly of their pitch diameters. This provides a great deal of flexibility in choosing the ratio, center distance, and diameters of the gears.

Finally, because the relative contact between the mating gears is the theoretical intersection of two cylinders, the load capacity of the gears is not directly influenced by the face width; that is, once the face width is extended such that it covers the full cylindrical intersection, extending either or both face widths farther is of no practical value since the extra face width will not be in contact.

Cylindrical Worm Gearing. The most basic form of worm gearing is a straight cylindrical worm in mesh with a simple helical gear. In reality, this is really an extreme case of crossed axis helical gearing. Such gears can provide considerably higher reduction ratios than simple crossed axis gear sets, but their load capacity is low, and the wear rate is high. For light loads, however, this configuration can be an economical alternative.

Single-Enveloping Worm Gearing. Better load capacity can be achieved if the simple helical gear is modified such that it is throated to allow the worm to fit down farther into the gear to achieve greater tooth contact area and thus smoother operation and improved load capacity. This common worm gear set is often referred to as single-enveloping since the gear envelops the worm but the worm remains straight. While many variations of this configuration are used, in the most common case the worm is essentially straight-sided while the gear or wheel is generally involute in form. The contact point is theoretically a line varying in length up to full face width of the gear with different tooth designs. Under load, this line becomes a thin elliptical band of contact. See Fig. 2.17 for a two-stage worm gear drive.

Basic arrangement views are shown in Fig. 2.18. The hand of helix for both members is the same. While the worm is generally made of steel, the wheel is usually made of either cast iron or, more frequently, one of several types of bronze. The worm teeth may be through hardened, but they are often, especially in higher-load and higher-speed applications, case hardened and ground. In some cases, however, to improve accuracy and efficiency, the worm teeth are ground even though they are not hardened after cutting. The wheel, due to the high sliding conditions that exist in this type of gearing, is usually made of cast iron or bronze, as noted in the foregoing; however, in some high-temperature environments it is necessary to make the wheel from steel as well.

FIGURE 2.17 Two stages of cylindrical-worm gears. *(Courtesy of Hamilton Gear and Machine Co., Ltd., Hamilton, Canada)*

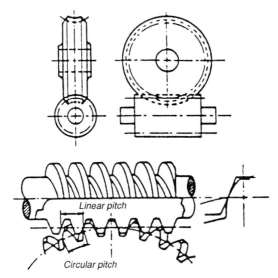

FIGURE 2.18 Schematic of cylindrical-worm gears.

By virtue of their inherent high contact ratio, the mechanical power rating of worm gears is quite high; however, in practice, their actual continuous-duty rating is substantially lower. This is due to high heat generation that can raise the lubricant temperature to unacceptable levels when the box is operated continuously. Fan-cooled worm boxes are quite common, and higher-power-capacity worm housings are almost always finned to aid heat dissipation. This large difference between the thermal and mechanical ratings of the typical worm gear drive gives rise to a peculiar property associated with worm drives: that is, their apparent ability to sustain relatively high short-term overloads without experiencing any damage. In reality, worm gears do not have a particularly good overload capability; rather, their thermal limitations cause them to be operated at loads below their mechanical limits. When operated for short periods of time at "overload," they are actually operating above their continuous-duty thermal limits but below their mechanical (stress-related) ratings; thus, since it takes an appreciable period of time for the temperature to rise, they sustain these short-term "overloads" quite well.

The advent of synthetic fluids has been a boon for worm drives of all types for two reasons. First, synthetic fluids have the ability to operate at higher bulk temperatures than the compounded mineral-based fluids that are commonly used for worm gears. Second, the friction coefficient associated with the use of synthetic fluids tends to be somewhat lower than that associated with compounded worm gear oils; thus less heating is produced. These factors combine to decrease the margin between the thermal and mechanical limits of newer work gear sets designed and rated to run with synthetic fluids; thus their apparent overload capability is reduced.

Worm gear efficiency is quite dependent on operating speed. The same set may show an efficiency of, say, 75 percent at a low speed and 85 percent at a higher speed. Ratio, material, accuracy, and geometric design all affect worm gear efficiency. Typical efficiencies run from 35 to 90 percent, with higher or lower values occurring in special cases.

A worm set can be used where irreversibility is desired, since, if the lead angle is less than the friction angle, the wheel cannot drive the worm. Usually, worms with lead angles less than 5° are self-locking. Care should be exercised when designing self-locking worms, since this feature is a static one. Vibration can cause the set to slip under dynamic conditions (that is, during a cutoff of power under load, the wheel, due to the inertia of the driving load, may overdrive the worm for a considerable time). Similarly, a worm set that, when stationary, cannot be driven through the gear shaft, may well begin to rotate if the unit is subjected to vibrations. The property of self-locking or irreversibility is better thought of as anti-back-driving rather than positive irreversibility. In critical applications, a brake on the worm should also be provided to insure that the system will positively not back-drive.

When the gears are assembled into their housing, the position of the gear along its axis must be adjusted such that an acceptable contact pattern is obtained on the bench. This contact pattern should favor the leaving side of the mesh so that a wedge is formed at the entering side of the mesh. This wedge at the entering side will cause lubricant to be drawn into the contact zone and thus minimize wear.

Double-Enveloping Worm Gearing. The capacity of a single-enveloping worm gear set as described in the foregoing discussion is improved by allowing the gear or wheel to envelop the worm. A further improvement in capacity may be achieved by allowing the worm to envelop the wheel as well. Such drives are known as double-enveloping. Double-enveloping worm gear drives, by getting more teeth into contact, tend to provide higher load capacity than do cylindrical or single-enveloping worm sets. This is accomplished by changing the shape of the worm (as shown in Fig. 2.19) from a cylinder to an hourglass.

All of the comments made earlier pertaining to cylindrical worms also apply to double-enveloping worms. Because of the shape of the worm, clearly shown in Fig. 2.20, this type of worm gearing is more expensive to produce; but where weight or size are considerations, the cost differential is relatively small.

The forms of the worm and the gear in the double-enveloping drive design are regenerative; that is, the worm and gear tend to reproduce each other in use. This is true of no other worm and gear design. This condition aids proper break-in and contact pattern development. Since the worm and wheel envelop each other, assembly is not as straightforward with double-enveloping gear sets as it is with a single-enveloping set. In general, the worm and wheel must be assembled obliquely, and thus provisions must be made in the design of the housing to accommodate this requirement. This can often be accomplished by splitting the housing such that the worm is one part and the wheel is the other half. In addition, the position of the worm along its axis and the position of the gear along its axis must be adjusted simultaneously such that an acceptable pattern is obtained. This bench patterning procedure is in contrast to the simpler procedure required of a single-enveloping set in which only the position of the gear along its axis need be adjusted at assembly to obtain an acceptable contact pattern at assembly.

Hypoid Gears. Hypoid gears (Fig. 2.21) resemble spiral bevel gears except that the teeth are asymmetrical; that is, the pressure angle on each side of the tooth is different. Many of the same machines used to manufacture spiral bevel gears can also be used to manufacture hypoids.

The pitch surfaces of hypoid gears are hyperboloids of revolution. The teeth in mesh have line contact; however, under load, these lines spread to become elliptical regions of contact inclined across the face width of the teeth. One condition that must exist if a hypoid gear set is to have conjugate action is that the normal pitch of both members must be the same. The number of teeth in a gear and pinion are not, however, directly proportional to the ratio of their pitch diameters. This makes it possible to make large pinions while minimizing the size of the driven gear. This is one of the most attractive features of hypoid gearing. High reduction ratios with small offsets must usually utilize an overhung-mounted pinion. Frequently, however, the pinion must be straddle mounted (Fig. 2.22) if sufficient offset exists. Aside from space considerations, the tradeoff usually centers on efficiency, since efficiency generally decreases with increasing offset.

In operation, hypoid gears are usually smoother and quieter than spiral bevel gears due to their inherent higher total contact ratio. However, as in all cases of nonintersecting gear sets, high sliding takes place across the face of the teeth. The efficiency of hypoid gears is thus much less than that of a similar set of spiral bevel gears, typically 90 to 95 percent as compared with over 99 percent for many spiral bevel gears. Hypoids do, however, generally have greater tolerance to shock loading and can frequently be used at much higher single-stage ratios than spiral bevel gears.

*Spiroid and Helicon Gearing.** Like most other skew axis concepts, Spiroid and Helicon are primarily screw action gears while, by comparison, spur, helical, and bevel gears are primarily rolling action gears.

---

*Spiroid and Helicon are registered trademarks of Illinois Tool Works, Chicago, Illinois.

GEAR TYPES AND NOMENCLATURE

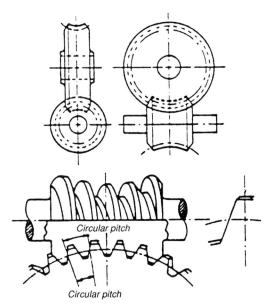

FIGURE 2.19 Schematic of double-enveloping-worm gears.

FIGURE 2.20 Double-enveloping-worm-gear set.

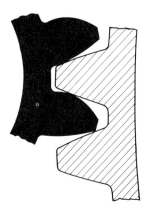

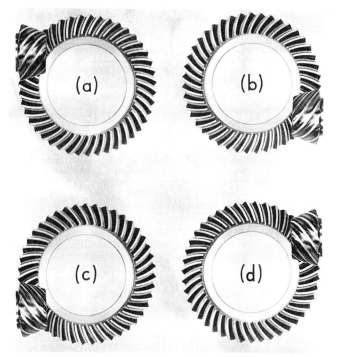

FIGURE 2.21 Hypoid-gear data. (1) Hypoid tooth profile showing unequal pressure angles and unequal profile curvatures on the two sides of the tooth. (2) Hypoid gears and pinions *(a)* and *(b)* are referenced to having an offset below center, while those in *(c)* and *(d)* have an offset above center. In determining the direction of offset, it is customary to look at the gear with the pinion at the right. For below-center offset, the pinion has a left-hand spiral, and for above-center offset the pinion has a right-hand spiral.

FIGURE 2.22 Hypoid-gear set. Both pinion and gear are straddle mounted. *(Courtesy of the Gleason Works, Rochester, N.Y.)*

The gear member of either a Spiroid (Fig. 2.23), or Helicon (Fig. 2.24), set resembles a high-pitch-angle bevel gear, while the pinions are more akin to worms. Both forms have more teeth in contact than an equivalent size worm set. The primary difference between Spiroid and Helicon gears is their minimum ratio capabilities [about 10:1 for Spiroid and 4:1 for Helicon (although lower ratios are practical if powdered metal fabrication is employed)] and the range of acceptable offsets. Spiroid pinions (Fig. 2.23) are typically tapered by about 5 or 10° on a side, while Helicon pinions (Fig. 2.24) are cylindrical. Pinions of both types, used in lower ratio designs (that is, less than 30:1), typically utilize multiple threads so that the number of teeth in the gear may be maintained at least at 30.

This type of gearing fits in between bevel and worm gears (Fig. 2.25) in terms of ratio, and generally provides better performance than do worms. They can, like worms, be designed to be anti-back-driving, but, again like worms, they cannot be designed to be entirely self-locking under all operating conditions, particularly where significant vibrations are present. They can also incorporate accurate control of backlash since the contact pattern is largely controlled by adjusting the pinion position, while the backlash is largely controlled by adjusting the position of the gear along its axis at assembly. These features make them a good choice for positioning drives such as radar or other antenna. These gear types are relatively insensitive to small position shifts; thus, in precision position-control applications, they are usually shimmed into intimate double flank contact to eliminate backlash.

The contact conditions are line, again like worm and hypoid gears, but the contact line is essentially radial; thus, the flow of oil into the contact zone is good, and efficiency

FIGURE 2.23 Spiroid-gear set. *(Courtesy of Illinois Tool Works, Spiroid Division, Chicago, Ill.)*

FIGURE 2.24 Helicon-gear set. *(Courtesy of Illinois Tool Works, Spiroid Division, Chicago, Ill.)*

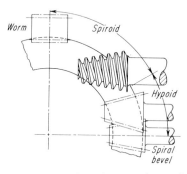

FIGURE 2.25 Schematic comparisons of worm gears, Spiroid gears, Helicon gears, hypoid gears, and spiral-bevel gears.

is improved over worms—usually about as good as hypoids. As with worms, the efficiency of a Spiroid or Helicon set varies with speed, but not to the large extent that occurs with worms.

One feature of Spiroids and Helicons that makes them both quiet and capable of great positional accuracy is the relatively high number of teeth in contact. The multiplicity of teeth in contact (for a typical design perhaps 10 percent of the gear teeth are in theoretical contact) provides an error averaging function so that individual tooth errors are not significant in terms of gear position.

Although generally less efficient than spur or helical gears, Spiroid or Helicon gears are usually superior to worm drives particularly at ratios less than 40:1, when compared on the basis of constant pinion size.

The offset required for a typical Spiroid gear set is about one-third the gear diameter, but small variations can also be accommodated. Similarly, the shaft angle is usually limited to 80 to 100°, with 90° being the standard in the vast majority of cases. In some special applications, shaft angles as low as 70° or as high as 120° are possible, but each case is a special design.

In general, the gears may be cut on standard hobbing machines. For a given configuration, a Spiroid gear set will have a greater load capacity than a Helicon set, but the Helicon position is somewhat easier to manufacture due to the cylindrical nature of its pinion as compared with the tapered shape of the Spiroid pinion.

Face Gears (Off-Center). If the pinion of a face gear set is offset such that its axis and that of the gear do not intersect, it is termed *off-center*. The discussion provided earlier for on-center face gears applies equally well here. Such gear sets are not ordinarily used to transmit significant amounts of power; rather they are generally used in motion-transmission applications where uniformity of motion is not a critical factor.

2.1.5 Special Gear Types

Our discussion to this point would lead the reader to believe that all gears are in the form of surfaces of revolution (that is, cylinders and cones). Such is not the case. There are many applications in which gears are used to transmit motions that are intentionally nonuniform. There are few areas in which man's mechanical ingenuity can be employed more effectively than in the design of special-motion gears. The variety is almost endless, limited only by application and imagination. In order to provide some insight into the subject, we discuss some representative cases.

Square or Rectangular Gears. Consider the case in which a constant input speed is to be converted into a varying output speed. Gears that are square or rectangular, depending on the actual output required, can be used to produce such motion (as shown in Fig. 2.26). While the gears are in the position shown, the greatest radius of the driver mates with the smallest radius of the driven gear; the speed of the driven gear is, therefore, then at its maximum. As the gears revolve in the directions indicated by the arrows, the radius of the driver gradually decreases, and that of the driven gear gradually increases, until the points b and b' are in contact. The speed of the driven gear, therefore, gradually decreases during this eighth of a revolution. From the moment of contact between the points b and b', the reverse action takes place, and the speed of the driven gear gradually increases until the points c and c' are in contact. Thus, during the entire revolution, the driven gear continues to alternate from a gradually decreasing to a gradually increasing speed, and vice versa.

For rectangular gears to work properly together, it is necessary first that the pitch peripheries of the two gears be equal in length, and, second, that the sum of the radii of each pair of points (points that come into contact with each other) on the two pitch peripheries be equal to the distance between the centers of the gears.

Triangular Gears. Figure 2.27 represents a pair of triangular gears, the object of which is to obtain an alternating, varying speed from the uniformly rotating driver, as in rectangular gears. Triangular gears give fewer changes of speed per revolution than do rectangular gears. In Fig. 2.27, the speed of the driven gear is at its minimum when

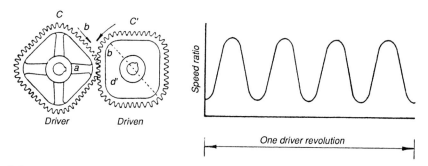

FIGURE 2.26 Square-gear characteristics.

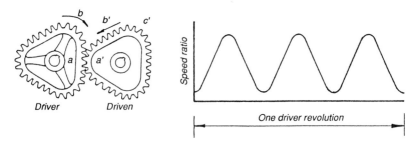

FIGURE 2.27 Triangular-gear characteristics.

the gears are in the positions shown in the figure. Since, from the positions, the radius of the driver gradually increases, and that of the driven gear decreases, as far as the points b and b', the speed of the driven gear will gradually increase until the points b and b' are in contact—or for one-sixth of an entire revolution. The reverse action will then take place until the points c and c' are in contact, and so on. Thus, while in rectangular gears each gradually increasing or decreasing period takes place during one-eighth of a revolution, in triangular gears each of these periods occupies one-sixth of a revolution; that is, in rectangular gears there are eight alternatively increasing and decreasing periods in one entire revolution of the driven gear, and in triangular gearing there are but six.

Elliptical Gears. With elliptical gears (Fig. 2.28), we have still another means of obtaining the same result, with the difference that in elliptical gears each period of gradually increasing and decreasing speed takes place during one-fourth of a revolution. In other words, there are but four periods of increasing and decreasing speed during an entire revolution of the driver.

Scroll Gears. Figure 2.29 shows a pair of scroll gears. From the positions shown in the figure (in which the greatest radius of the driver meshes with the smallest radius of the driven gear), as the gears revolve in the directions indicated by the arrows, the radius of the driver gradually and uniformly decreases, while that of the driven gear gradually and uniformly increases. The speed of the driven gear is, therefore, at its

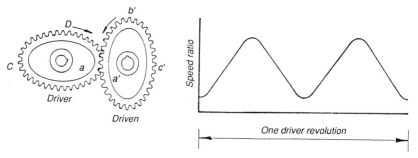

FIGURE 2.28 Elliptical-gear characteristics.

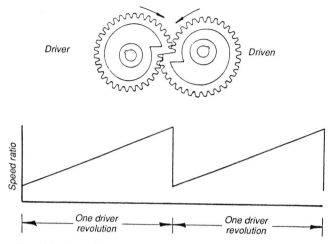

FIGURE 2.29 Scroll-gear characteristics.

maximum when the gears are in the positions shown, and gradually and uniformly decreases during the entire revolution. The moment before the positions shown in the figure are reached, the smallest radius of the driver gears with the greatest radius of the driven gear, the speed of the latter is at its minimum and suddenly (as the gears assume the positions in the figure) changes to its maximum.

Multiple-Sector Gears. The mechanism represented in Fig. 2.30 is a pair of multiple-sector gears, and the object is to obtain a series of discrete, different uniform speeds. In the figure, as long as the arcs ab and $a'b'$ are in mesh, the speed of the driven gear is the same. When the arcs cd and $c'd'$ come into mesh, the speed of the driven gear becomes slower, but remains the same throughout the meshing of these two arcs. Similarly, when the arcs ef and $e'f'$ come into mesh, the speed of the driven gear becomes still slower, but remains uniform during the meshing of these arcs. Thus, during each revolution, the driven gear has three periods of uniform speed, each differing from the others. For sector gears to work properly together, it is necessary that the arcs that mesh together be equal in length ($ab = a'b'$, $cd = c'd'$, etc.) and that the sum of the arc lengths on one gear be equal to the sum of the arc lengths on the other ($ab + cd + ef = a'b' + c'd' + e'f'$). Also, the sum of the radii of each two arcs together must be equal to the distance between the centers of the gears. Sector gears are somewhat difficult to construct because considerable care must be taken to insure that no two sectors of the driver gear mesh at the same time with the driven gear. To illustrate, suppose that the arcs ab and $a'b'$ mesh together at the same time as do the arcs ef and $e'f'$; that is, that the last few teeth of ab mesh with the driven gear at the moment when the first few teeth of cf do the same. The driven gear will then strive to drive the driven gear at its maximum and minimum speeds at the same time, an attempt that must obviously result in a fracture. In the figure, the arc cf ceases to mesh with the driven gear at the moment when the arc ab begins to gear. Thus, each arc of the driver must escape engagement just in time for its successor to begin engagement,

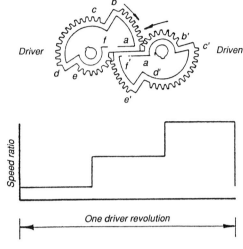

FIGURE 2.30 Multiple-sector-gear characteristics.

and yet leave between these events no appreciable interval to disturb the uniformity of motion.

In actuality, a discontinuous condition will exist for all sector-type gears at the time when different sectors exchange control of the speed of the driven member. These seemingly instantaneous periods of acceleration and deceleration limit the application of such gears to very low speeds lest the acceleration forces become exceedingly large.

2.2 NOMENCLATURE OF GEARS*

In this part of Chap. 2 the technical nomenclature of the various types of gears is given. Table 2.2 shows metric system symbols and English system symbols. The following tables (due to space limitations) show nomenclature tables in English symbols only.

2.2.1 Spur Gear Nomenclature and Basic Formulas

Figure 2.31 shows tooth elements that apply directly to spur gears. The same elements apply to other gear types.

Table 2.3 shows spur gear formulas. Note that the value of π may need to be used as 3.14159265 when very precise calculations are needed. Also, the determination of tooth thickness has complications of what backlash is needed and effects of long or short addendum. (These matters are covered later in this book.)

---

*Most of the material for this part of Chap. 2 was furnished by Darle W. Dudley, President of the Dudley Engineering Co., Inc., San Diego, California.

TABLE 2.2 Gear Terms, Symbols, and Units Used in the Calculation of Gear Dimensional Data

Term	Metric Symbol	Metric Units	English Symbol	English Units
Number of teeth, pinion	z_1	—	N_P or n	—
Number of teeth, gear	z_2	—	N_G or N	—
Number of threads, worm	z_1	—	N_W	—
Number of crown teeth	z	—	N_c	—
Tooth ratio	u	—	m_G	—
Addendum, pinion	h_{a1}	mm	a_P	in.
Addendum, gear	h_{a2}	mm	a_G	in.
Addendum, chordal	$\bar{h}_a$	mm	a_c	in.
Rise of arc	—	mm	—	in.
Dedendum	h_f	mm	b	in.
Working depth	h'	mm	h_k	in.
Whole depth	h	mm	h_t	in.
Clearance	c	mm	c	in.
Tooth thickness	s	mm	t	in.
Arc tooth thickness, pinion	s_1	mm	t_P	in.
Arc tooth thickness, gear	s_2	mm	t_G	in.
Tooth thickness, chordal	$\bar{s}$	mm	t_c	in.
Backlash, transverse	j	mm	B	in.
Backlash, normal	j_n	mm	B_n	in.
Pitch diameter, pinion	d_{p1}	mm	d	in.
Pitch diameter, gear	d_{p2}	mm	D	in.
Pitch diameter, cutter	d_{p0}	mm	d_c	in.
Base diameter, pinion	d_{b1}	mm	d_b	in.
Base diameter, gear	d_{b2}	mm	D_b	in.
Outside diameter, pinion	d_{a1}	mm	d_o	in.
Outside diameter, gear	d_{a2}	mm	D_o	in.
Inside diameter, face gear	d_{i2}	mm	D_i	in.
Root diameter, pinion or worm	d_{f1}	mm	d_R	in.
Root diameter, gear	d_{f2}	mm	D_R	in.
Form diameter	d'_f	mm	d_f	in.
Limit diameter	d_ℓ	mm	d_ℓ	in.
Excess involute allowance	Δd_ℓ	mm	Δd_ℓ	in.
Ratio of diameters	ϵ	—	m	—
Center distance	a	mm	C	in.
Face width	b	mm	F	in.
Net face width	b'	mm	F_e	in.
Module, transverse	m or m_t	mm	—	—
Module, normal	m_n	mm	—	—
Diametral pitch, transverse	—	—	P_d or P_t	in.$^{-1}$
Diametral pitch, normal	—	—	P_n	in.$^{-1}$
Circular pitch	p	mm	p	in.
Circular pitch, transverse	p_t	mm	p_t	in.
Circular pitch, normal	p_n	mm	p_n	in.
Base pitch	p_b	mm	p_b	in.
Axial pitch	p_x	mm	p_x	in.
Lead (length)	p_z	mm	L	in.

TABLE 2.2 Gear Terms, Symbols, and Units *(Continued)*

	Metric		English	
Term	Symbol	Units	Symbol	Units
Pressure angle	α or α_t	deg	ϕ or ϕ_t	deg
Pressure angle, normal	α_n	deg	ϕ_n	deg
Pressure angle, axial	α_x	deg	ϕ_x	deg
Pressure angle of cutter	α_0	deg	ϕ_c	deg
Helix angle	β	deg	ψ	deg
Lead angle	γ	deg	λ	deg
Shaft angle	Σ	deg	Σ	deg
Roll angle	θ_r	deg	ϵ_r	deg
Pitch angle, pinion	δ'_1	deg	γ	deg
Pitch angle, gear	δ'_2	deg	Γ	deg
Pi	π	—	π	—
Contact ratio	ϵ_a	—	m_p	—
Zone of action	g_a	mm	Z	in.
Edge radius, tool	r_{ao}	mm	r_T	in.
Radius of curvature, root fillet	ρ_f	mm	ρ_f	in.
Circular thickness factor	k	—	k	—
Cone distance	R	mm	A	in.
Outer cone distance	R_a	mm	A_o	in.
Mean cone distance	R_m	mm	A_m	in.
Inner cone distance	R_f	mm	A_i	in.

NOTE: Abbreviations for units: mm = millimeters, in. = inches, deg = degrees.

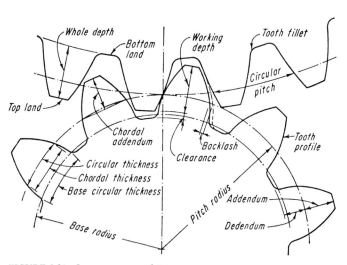

FIGURE 2.31 Spur-gear nomenclature.

GEAR TYPES AND NOMENCLATURE

TABLE 2.3 Spur Gear Formulas

To find	Having	Formula
Diametral pitch	Number of teeth and pitch diameter	$P = \dfrac{N}{D}$
Diametral pitch	Circular pitch	$P = \dfrac{3.1416^*}{p}$
Pitch diameter	Number of teeth and diametral pitch	$D = \dfrac{N}{P}$
Outside diameter	Pitch diameter and addendum	$D_o = D + 2a$
Root diameter	Outside diameter and whole depth	$D_R = D_o - 2h_t$
Root diameter	Pitch diameter and dedendum	$D_R = D - 2b$
Number of teeth	Pitch diameter and diametral pitch	$N = D \times P$
Base-circle diameter	Pitch diameter and pressure angle	$D_b = D \times \cos \phi$
Circular pitch	Pitch diameter and number of teeth	$p = \dfrac{3.1416 D}{N}$
Circular pitch	Diametral pitch	$p = \dfrac{3.1416}{P}$
Center distance	Number of gear teeth and number of pinion teeth and diametral pitch	$C = \dfrac{N_G + N_P}{2P} = \dfrac{D_G + D_P}{2}$
Approximate thickness of tooth	Diametral pitch	$t_t = \dfrac{1.5708}{P}$ (see design formulas, Chap. 5)

*Exact value of π is 3.14159265.

2.2.2 Helical Gear Nomenclature and Basic Formulas

Helical gear teeth spiral around a "base" cylinder. The angle of this spiral with respect to the axis is the helix angle. The advance of the spiral in going 360° around the base cylinder is a fixed quantity called the *lead*.

The spur gear tooth is parallel to the axis. This means that a spur gear can be thought of as a special case of a helical gear in which the numerical value of the lead is infinity!

Figure 2.32 shows the nomenclature of helical gear teeth. View *a* shows a section through the teeth that is perpendicular to the gear axis. This is known as the *transverse section*. View *b* shows how a "normal" section can be taken through a tooth perpendicular to a point on the pitch helix.

Both the transverse section and the normal section data must be used in helical gear calculations. Table 2.4 shows the formulas needed to calculate data for both of these sections.

A *rack* is a gear part in which the number of teeth in a 360° circle is infinite. Circular gears, of course, become straight when the radius of curvature is infinite.

Involute teeth become straight-sided for the rack tooth form. Figure 2.33 shows rack tooth parts for spur, helical, and straight bevel teeth. The helical rack shows very clearly the difference between the normal plane and the front plane (which corresponds to the transverse plane of circular gears).

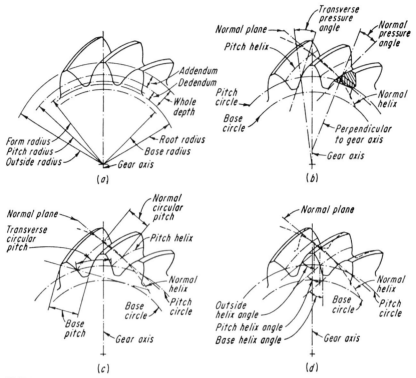

FIGURE 2.32 Helical-gear nomenclature.

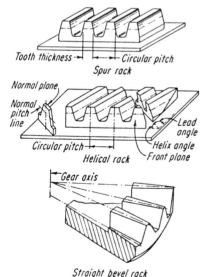

FIGURE 2.33 Spur-, helical-, and bevel-gear racks.

2.2.3 Internal Gear Nomenclature and Formulas

The internal gear has teeth on the *inside* of a ring rather than on the outside. An internal gear must mesh with a pinion having external teeth. (Two internal gears cannot mesh with each other.)

The nomenclature for the transverse section of an internal gear drive is shown in Fig. 2.34. Note that the internal tooth tends to have concave curves rather than convex curves. Also note that the tips of the teeth tend almost to hit each other as they enter and leave the meshing zone.

Formulas for internal gear dimensions are given in Table 2.5. Note that some of these formulas are similar to formulas in Table 2.3 except for a minus sign.

GEAR TYPES AND NOMENCLATURE

TABLE 2.4 Helical Gear Formulas

To find	Having	Formula
Normal diametral pitch	Number of teeth, pitch diameter, and helix angle	$P_n = \dfrac{N}{D \cos \psi}$
Normal diametral pitch	Normal circular pitch	$P_n = \dfrac{3.1416*}{p_n}$
Normal diametral pitch	Transverse diametral pitch and helix angle	$P_n = \dfrac{P}{\cos \psi}$
Normal circular pitch	Normal diametral pitch	$p_n = \dfrac{3.1416}{P_n}$
Normal circular pitch	Transverse circular pitch	$p_n = p \cos \psi$
Pitch diameter	Number of teeth, normal diametral pitch, and helix angle	$D = \dfrac{N}{P_n \cos \psi}$
Center distance	Pinion and gear pitch diameter	$C = \dfrac{D_P + D_G}{2}$
Outside diameter	Pitch diameter and addendum	$D_o = D + 2a$
Approximate normal tooth thickness	Normal diametral pitch	$t_n = \dfrac{1.571}{P_n}$ (see design formulas, Chap. 5)
Transverse tooth thickness	Normal tooth thickness and helix angle	$t_t = \dfrac{t_n}{\cos \psi}$
Normal pressure angle	Transverse pressure angle and helix angle	$\tan \phi_n = \tan \phi \cos \psi$

*Exact value of π is 3.14159265.

TABLE 2.5 Internal Spur Gear Formulas

To find	Having	Formula
Pitch diameter:		
Of gear	Number of teeth in gear and diametral pitch	$D = \dfrac{N_G}{P}$
Of pinion	Number of teeth in pinion and diametral pitch	$d = \dfrac{N_P}{P}$
Internal diameter (gear)	Pitch diameter and addendum	$D_i = D - 2a$
Center distance	Number of teeth in gear and pinion; diametral pitch	$C = \dfrac{N_G - N_P}{2P}$
	Pitch diameter of gear and pinion	$C = \dfrac{D - d}{2}$

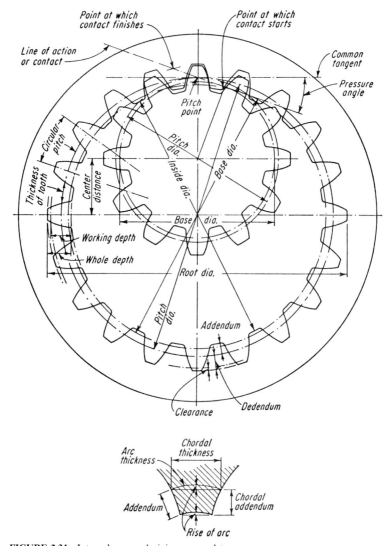

FIGURE 2.34 Internal-gear and pinion nomenclature.

Internal gears may be either spur or helical. If the teeth are helical, data for the teeth in the normal section may be determined by formulas in Table 2.4.

2.2.4 Crossed Helical Gear Nomenclature and Formulas

Since the axes of crossed helical gears are not parallel to each other and do not intersect, the two meshing gears tend to have different helix angles. (If the numerical amount of each helix angle is the same, then the two gears cannot have opposite helix angles—because this would make the axes parallel instead of crossed.) Figure 2.35 shows standard nomenclature for the crossed axis situation with helical gears. The basic formulas are given in Table 2.6.

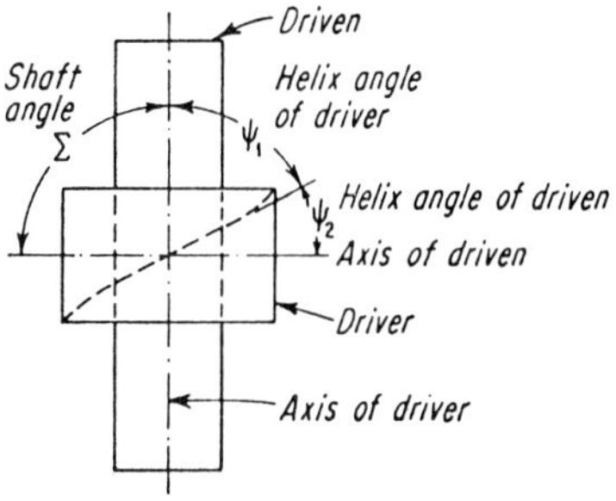

2.2.5 Bevel Gear Nomenclature and Formulas

Section 2.1.3 gave general information on gear types in the bevel gear family of gears. Nomenclature and formulas will now be given for three of these types that are in rather common usage.

FIGURE 2.35 Crossed-helical-shaft angle and helix.

Straight. Figure 2.36 shows section through of a pair of straight bevel gears in mesh. Note how the teeth are on the outside of cones. Dimensions are measured to a "crown point" that exists in space but may not exist in metal! (Usually this sharp corner is rounded off in the process of actual manufacture.)

The formulas for the straight bevel are given in Table 2.7. By tradition, the pitch diameters are taken at the large end of the tooth. This end also goes by names such as the *back face* or the *heel* end. (The other end of the tooth is termed the *front face* or *toe* end.)

Spiral. The spiral bevel gear has a curved shape (lengthwise). This curved shape is positioned at an angle to a pitch cone element. The angle is specified at the center of the face width (not the large end of the tooth where the pitch diameter is specified). Figure 2.37 shows the spiral angle.

The formulas for conventional spiral bevel gears are given in Table 2.8. Spiral bevel gears are often made to pressure angles other than 20°. They can also be made for shaft angles other than 90°. Beyond this, there is more than one system for the design and manufacture of bevel gears. Chapter 20 gives considerably more bevel gear data. The

TABLE 2.6 Crossed Helical Gear (Nonparallel Shaft, Helicals)

To find	Having	Formula
Shaft angle	Helix angle of pinion and helix angle of gear	$\Sigma = \psi_P + \psi_G$
Pitch diameter of pinion	Number of teeth in pinion, normal diametral pitch, and helix angle of pinion	$D_P = \dfrac{N_P}{P_n \times \cos \psi_P}$
Pitch diameter of gear	Number of teeth in gear, normal diametral pitch and helix angle of gear	$D_G = \dfrac{N_G}{P_n \times \cos \psi_G}$
Center distance	Pitch diameter of pinion, and pitch diameter of gear	$C = \dfrac{D_P + D_G}{2}$
Speed ratio	Number of teeth in gear and number of teeth in pinion	$m_G = \dfrac{N_G}{N_P}$

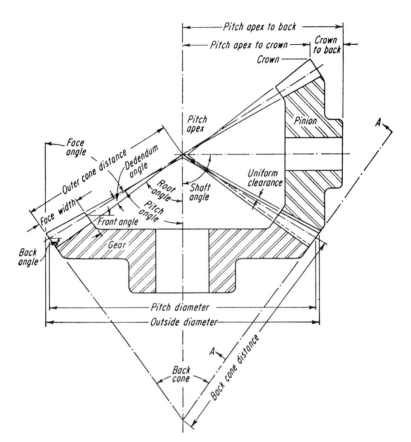

FIGURE 2.36 Bevel-gear nomenclature.

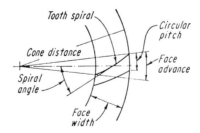

FIGURE 2.37 Spiral-bevel-gear nomenclature.

TABLE 2.7 Straight Bevel Gear Formulas (20° pressure angle, 90° shaft angle)

To find	Having	Formula
Pitch diameter of pinion	Number of pinion teeth and diametral pitch	$d = \dfrac{n}{P_d}$
Pitch diameter of gear	Number of gear teeth and diametral pitch	$D = \dfrac{N}{P_d}$
Pitch angle of pinion	Number of pinion teeth and number of gear teeth	$\gamma = \tan^{-1}\dfrac{n}{N}$
Pitch angle of gear	Pitch angle of pinion	$\Gamma = 90° - \gamma$
Outer cone distance of pinion and gear	Gear pitch diameter and pitch angle of gear	$A_o = \dfrac{D}{2 \sin \Gamma}$
Circular pitch of pinion and gear	Diametral pitch	$p = \dfrac{3.1416*}{P_d}$
Dedendum angle of pinion	Dedendum of pinion and outer cone distance	$\delta_P = \tan^{-1}\dfrac{b_{oP}}{A_o}$
Dedendum angle of gear	Dedendum of gear and outer cone distance	$\delta_G = \tan^{-1}\dfrac{b_{oG}}{A_o}$
Face angle of pinion blank	Pinion pitch angle and dedendum angle of gear	$\gamma_o = \gamma + \delta_G$
Face angle of gear blank	Gear pitch angle and dedendum angle of gear	$\Gamma_o = \Gamma + \delta_P$
Root angle of pinion	Pitch angle of pinion and dedendum angle of pinion	$\gamma_R = \gamma - \delta_P$
Root angle of gear	Pitch angle of gear and dedendum angle of gear	$\Gamma_R = \Gamma - \delta_G$
Outside diameter of pinion	Pinion pitch diameter, pinion addendum and pitch angle of pinion	$d_o = d + 2a_{oP} \cos \gamma$
Outside diameter of gear	Pitch diameter of gear, addendum, and pitch angle of gear	$D_o = D + 2a_{oG} \cos \Gamma$
Pitch apex to crown of pinion	Pitch diameter of gear, addendum, and pitch angle of pinion	$x_o = \dfrac{D}{2} - a_{oP} \sin \gamma$
Pitch apex to crown of gear	Pitch diameter of pinion, addendum, and pitch angle of gear	$X_o = \dfrac{d}{2} - a_{oG} \sin \Gamma$
Circular thickness of pinion	Circular pitch and gear circular tooth thickness	$t = p - T$
Chordal thickness of pinion	Circular tooth thickness, pitch diameter of pinion, and backlash	$t_c = t - \dfrac{t^3}{6d^2} - \dfrac{B}{2}$
Chordal thickness of gear	Gear circular tooth thickness, pitch diameter of gear, and backlash	$T_c = T - \dfrac{T^3}{6D^2} - \dfrac{B}{2}$
Chordal addendum of pinion	Addendum angle, circular tooth thickness, pitch diameter, and pitch angle of pinion	$a_{cP} = a_{oP} + \dfrac{t^2 \cos \gamma}{4d}$
Chordal addendum of gear	Addendum angle, circular tooth thickness, pitch angle, and pitch diameter of gear	$a_{cG} = a_{oG} + \dfrac{T^2 \cos \Gamma}{4D}$
Tooth angle of pinion	Outer cone distance, tooth thickness, dedendum of pinion, and pressure angle	$\dfrac{3{,}438}{A_o}\left(\dfrac{t}{2} + b_{oP} \tan \phi\right)$ minutes

2.40 CHAPTER TWO

TABLE 2.7 (*Continued*)

To find	Having	Formula
Tooth angle of gear..........	Outer cone distance, circular tooth thickness, dedendum of gear, and pressure angle	$\dfrac{3{,}438}{A_o}\left(\dfrac{T}{2} + b_{oG} \tan \phi\right)$ minutes

NOTE: $\tan^{-1}$ means "the angle whose tangent is."
*Exact value of π is 3.14159265.

references at the end of this chapter give further information—beyond the scope of this book.

Zerol. The Zerol bevel gear tooth has lengthwise curvature like the spiral bevel gear tooth. At the center of the face width, though, the spiral angle is 0°.

Figure 2.38 shows how the hand of a Zerol bevel is designated. Note that a right-hand spiral bevel pinion has to have left-hand spiral bevel gear to mesh with it, and vice versa. This same rule applies to spiral bevel gears.

Table 2.9 shows formulas for Zerol bevel gears. Both Zerol and spiral bevel gears have the pitch diameters specified at the large end of the tooth.

2.2.6 Worm Gear Nomenclature and Formulas

When the word *worm* is used in connection with gear types, the implication is that there is some "enveloping." A single-enveloping set of worm gears has a cylindrical worm

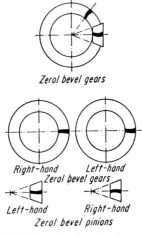

FIGURE 2.38 Zerol-bevel-gear nomenclature.

TABLE 2.8 Spiral Bevel Gear Formulas (20° pressure angle, 90° shaft angle)

To find	Having	Formula
Pitch diameter of pinion	Number of pinion teeth and diametral pitch	$d = \dfrac{n}{P_d}$
Pitch diameter of gear	Number of gear teeth and diametral pitch	$D = \dfrac{N}{P_d}$
Pitch angle of pinion	Number of pinion teeth and number of gear teeth	$\gamma = \tan^{-1} \dfrac{n}{N}$
Pitch angle of gear	Pitch angle of pinion	$\Gamma = 90° - \gamma$
Outer cone distance of pinion and gear	Pitch diameter of gear and pitch angle of gear	$A_o = \dfrac{D}{2 \sin \Gamma}$
Circular pitch of pinion and gear	Diametral pitch	$p = \dfrac{3.1416*}{P_d}$
Dedendum angle of pinion	Dedendum of pinion and outer cone distance	$\delta_P = \tan^{-1} \dfrac{b_{oP}}{A_o}$
Dedendum angle of gear	Dedendum of gear and outer cone distance	$\delta_G = \tan^{-1} \dfrac{b_{oG}}{A_o}$
Face angle of pinion blank	Pitch angle of pinion and dedendum angle of gear	$\gamma_o = \gamma + \delta_G$
Face angle of gear blank	Pitch angle of gear and dedendum angle of gear	$\Gamma_o = \Gamma + \delta_P$
Root angle of pinion	Pitch angle of pinion and dedendum angle of pinion	$\gamma_R = \gamma - \delta_P$
Root angle of gear	Pitch angle of gear and dedendum angle of gear	$\Gamma_R = \Gamma - \delta_G$
Outside diameter of pinion	Pitch diameter, addendum, and pitch angle of pinion	$d_o = d + 2a_{oP} \cos \gamma$
Outside diameter of gear	Pitch diameter, pitch angle, and addendum of gear	$D_o = D + 2a_{oG} \cos \Gamma$
Pitch apex to crown of pinion	Pitch diameter of gear, pitch angle, and addendum of pinion	$x_o = \dfrac{D}{2} - a_{oP} \sin \gamma$
Pitch apex to crown of gear	Pitch diameter of pinion, pitch angle, and addendum of gear	$X_o = \dfrac{d}{2} - a_{oG} \sin \Gamma$
Circular thickness of pinion	Circular pitch of pinion and circular thickness of gear	$t = p - T$

*Exact value of π is 3.14159265.

TABLE 2.9 Zerol Bevel Gear Formulas (20° pressure angle, 90° shaft angle)

To find	Having	Formula
Pitch diameter of pinion	Number of pinion teeth and diametral pitch	$d = \dfrac{n}{P_d}$
Pitch diameter of gear	Number of gear teeth and diametral pitch	$D = \dfrac{N}{P_d}$
Pitch angle of pinion	Number of pinion teeth and number of gear teeth	$\gamma = \tan^{-1} \dfrac{n}{N}$
Pitch angle of gear	Pitch angle of pinion	$\Gamma = 90° - \gamma$
Outer cone distance of pinion and gear	Pitch diameter of gear and pitch angle of gear	$A_o = \dfrac{D}{2 \sin \Gamma}$
Circular pitch of pinion and gear	Diametral pitch	$p = \dfrac{3.1416*}{P_d}$
Face angle of pinion blank	Pitch angle of pinion and dedendum angle of gear	$\gamma_o = \gamma + \delta_G$
Face angle of gear blank	Pitch angle of gear and dedendum angle of pinion	$\Gamma_o = \Gamma + \delta_P$
Root angle of pinion	Pitch angle of pinion and dedendum angle of pinion	$\gamma_R = \gamma - \delta_P$
Root angle of gear	Pitch angle of gear and dedendum angle of gear	$\Gamma_o = \Gamma - \delta_G$
Outside diameter of pinion	Pitch diameter, pitch angle, and addendum of pinion	$d_o = d + 2a_P \cos \gamma$
Outside diameter of gear	Pitch diameter, pitch angle, and addendum of gear	$D_o = D + 2a_G \cos \Gamma$
Pitch apex to crown of pinion	Pitch diameter of gear, pitch angle, and addendum of pinion	$x_o = \dfrac{D}{2} - a_P \sin \gamma$
Pitch apex to crown of gear	Pitch diameter of pinion, pitch angle, and addendum of gear	$X_o = \dfrac{d}{2} - a_G \sin \Gamma$
Circular thickness of pinion	Circular pitch of pinion and circular thickness of gear	$t = p - T$
Circular thickness of gear	Circular pitch, pressure angle, and addendum of pinion and gear	$T = \dfrac{p}{2} - (a_P - a_G) \tan \phi$

NOTE: $\tan^{-1}$ means "the angle whose tangent is."
*Exact value of π is 3.14159265.

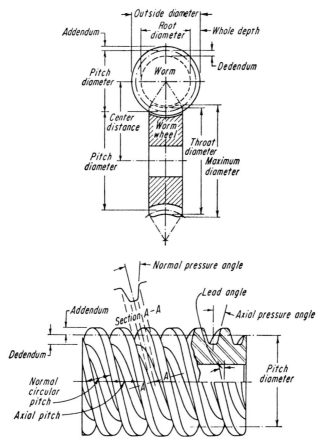

FIGURE 2.39 Cylindrical-worm-gear nomenclature.

in mesh with a gear that is throated to tend to wrap around the worm. Figure 2.39 shows a typical arrangement.

When both the worm and the gear wrap around each other, the combination is designated as double-enveloping. Figure 2.40 shows a typical arrangement of this kind.

Cylindrical Worm Gears. The worm thread (or tooth) may be dimensioned in both a normal section and an axial section. The transverse section is normally not used. Figure 2.39 shows the nomenclature that is used.

The special formulas for worm gearing are given in Table 2.10. The pitch diameter of the throated gear changes as you go across the face width. Standard practice is to take the pitch diameter at the center of the throat. Note the details of how this is done in Fig. 2.39.

Double-Enveloping Worm Gears. The typical style of design and the normal nomenclature for this kind of gearing is shown in Fig. 2.40. The pitch diameters are

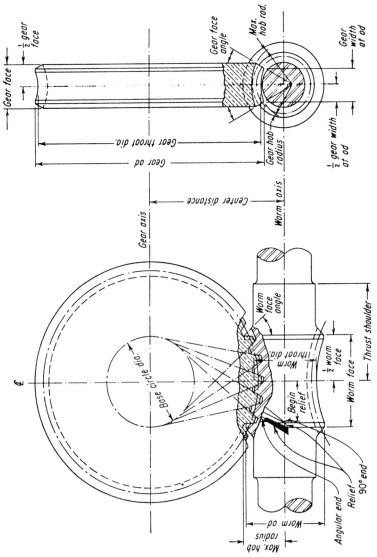

FIGURE 2.40 Double-enveloping-worm-gear nomenclature.

GEAR TYPES AND NOMENCLATURE

TABLE 2.10 Worm Gear Formulas

To find	Having	Formula
Worm:		
Lead	Number of threads in worm and axial pitch	$L = N_W \times p_x$
Pitch diameter	Center distance	$d = \dfrac{C^{0.875}}{1.7}$ to $\dfrac{C^{0.875}}{3}$
Root diameter	Outside diameter of pinion and whole depth of tooth	$d_R = d_o - 2h_t$
Outside diameter	Pitch diameter and addendum	$d_o = d + 2a$
Minimum face	Throat diameter, pitch diameter, and addendum	$f = 2\sqrt{\left(\dfrac{D_t}{2}\right)^2 - \left(\dfrac{D}{2} - a\right)^2}$
Lead angle	Lead and pitch diameter	$\tan \lambda = \dfrac{L}{3.1416 \times d}$
Normal pitch	Axial pitch and lead angle	$p_n = p_x \times \cos \lambda$
Gear:		
Nominal pitch diameter	Number of teeth in gear and axial pitch	$D = \dfrac{N_G \times p_x}{3.1416^*}$
Throat diameter	Nominal pitch diameter and addendum	$D_t = D + 2a$
Effective face	Pitch diameter and working depth	$F_e = \sqrt{(d + h_k)^2 - d^2}$
Center distance	Pitch diameter of gear and pitch diameter of worm	$C = \dfrac{D + d}{2}$

*Exact value of π is 3.14159265.

specified at the center of the throat on the worm and the center of the throat on the gear.

The formulas for double-enveloping worm gears are given in Table 2.11.

2.2.7 Face Gears

The gear member of a face gear set has teeth cut on the end of a cylindrical-shaped blank. Figure 2.41 shows a typical face gear arrangement and the nomenclature used.

The formulas for an on-center 90°-shaft-angle set of face gears are given in Table 2.12. Additional design data (for a standard design) are given in Tables 2.13 and 2.14.

There are other possible face gear designs with shaft angles that are not 90° or conditions of not being on-center. These can be thought of as special designs beyond the scope of this book.

TABLE 2.11 Double-Enveloping Worm Gear Formulas

To find	Having	Formula
Worm root diameter*	Pitch diameter of pinion and dedendum of gear; center distance	$d_R = d - 2b_G$ $d_R = \dfrac{C^{0.875}}{3}$ (approx.)
Worm pitch diameter*	Center distance	$d = \dfrac{C^{0.875}}{2.2}$ (approx.)
Gear pitch diameter	Center distance and worm pitch diameter	$D = 2C - d$
Axial circular pitch	Gear pitch diameter and number of teeth in gear	$p_x = \dfrac{\pi D}{N_G}$
Normal circular pitch	Axial circular pitch and pitch cone angle of pinion	$p_n = p_x \cos \lambda$
Whole depth of tooth*	Normal circular pitch	$h_t = \dfrac{p_n}{2}$
Working depth of tooth*	Whole depth of tooth	$h_k = 0.9 h_t$
Dedendum*	Working depth of tooth	$b_G = 0.611 h_k$
Normal pressure angle*		$\phi_n = 20°$
Axial pressure angle	Normal pressure angle and lead angle, deg	$\phi_x = \tan^{-1} \dfrac{\tan \phi_n}{\cos \lambda_c}$
Lead angle at center of worm	Pitch diameter, gear ratio, and worm pitch diameter	$\lambda_c = \tan^{-1} \dfrac{D}{m_G d}$
Lead angle average*	Pitch diameter, gear ratio, and worm pitch diameter	$\lambda = \tan^{-1} \dfrac{0.87 D}{m_G d}$

*Proportions given in these formulas represent recommendations for Cone-drive gears. Cone-drive is a registered trademark of Cone-Drive Textron, Traverse City, Michigan.

2.2.8 Spiroid Gear Nomenclature and Formulas

There is a family of gears that is generally described by the name *Spiroid*. The members of this family are Spiroid and Helicon. The gear member of each of these kinds of gear sets has teeth cut on the end of a blank. The pinion member does not mesh on center but instead meshes with a considerable offset from the central position. Figure 2.42 shows a standard Spiroid arrangement. The dimension A is the offset.

The Spiroid pinion is tapered. The teeth spiral around the pinion somewhat like threads on a worm, but they are on the surface of a cone rather than on a cylinder. The Helicon pinion differs from the Spiroid pinion by being cylindrical. This makes it quite comparable to a cylindrical worm.

Table 2.16 gives some formulas for Spiroid gear sets. Other dimensions needed to make a standard design are given as part of Fig. 2.42. Specific data for Helicon gear sets are not given. The Spiroid Division of the Illinois Tool Works in Chicago, Illinois should be consulted for additional data.

2.2.9 Beveloid Gears

This special type of gear is patented by the Vinco Corporation. The Vinco Corporation went out of business about twenty years ago, but this type of gear is still in use. The

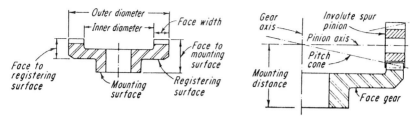

FIGURE 2.41 On-center-face-gear nomenclature.

TABLE 2.12 Face Gears—On Center 90° Shaft Angle*

To find	Having	Formula
Gear ratio	Number of pinion and gear teeth	$m_G = \dfrac{N_G}{N_P}$ (from 1.5 to 12.5)
Inside diameter	Diametral pitch, number of gear and pinion teeth, pressure angle	$D_{iG} = \dfrac{1}{P}\sqrt{8N_P + N_G \cos^2 \phi}$
Outside diameter (pointed teeth)	Diametral pitch, number of gear teeth, pressure angle, and tangency pressure angle	$D_{oG} = \dfrac{N_G \cos \phi \sec \phi_o}{P}$ (see Table 2.13)
Max. face width	Face-gear inside and outside diameters	$F_G = \dfrac{D_o - D_i}{2}$
Face to pinion axis	Pinion outside diameter and working depth of tooth	Face to pinion axis $= \dfrac{d_o}{2} - h_k$
Face to mounting surface	Mounting distance and face to pinion axis	Face to mounting surface = mounting distance − face to pinion axis
Min. numbers of teeth		See Table 2.14

*Based upon AGMA standard 203.03, "Fine-Pitch On-Center Face Gears for 20-degree Involute Spur Pinions."

TABLE 2.13 Tangency Pressure Angle at Face Gear Outside Diameter

N_P	ϕ_o	$\sec \phi_o$	N_P	ϕ_o	$\sec \phi_o$	N_P	ϕ_o	$\sec \phi_o$
12	40.765	1.32031	25	34.596	1.21481	38	32.837	1.19016
13	39.669	1.29913	26	34.429	1.21237	39	32.730	1.18874
14	38.692	1.28120	27	34.268	1.21005	40	32.627	1.18736
15	37.815	1.26583	28	34.113	1.20783	41	32.525	1.18602
16	37.022	1.25249	29	33.965	1.20572	42	32.427	1.18473
17	36.300	1.24081	30	33.821	1.20369	43	32.332	1.18352
18	36.012	1.23626	31	33.683	1.20176	44	32.240	1.18228
19	35.778	1.23261	32	33.550	1.19990	45	32.150	1.18111
20	35.557	1.22919	33	33.421	1.19812	46	32.061	1.17996
21	35.346	1.22598	34	33.297	1.19640	47	31.976	1.17887
22	35.145	1.22295	35	33.176	1.19475	48	31.893	1.17781
23	34.954	1.22009	36	33.059	1.19317	49	31.810	1.17675
24	34.771	1.21738	37	32.946	1.19164	50	31.734	1.17578

TABLE 2.14 Minimum Numbers of Teeth in Pinion and Face Gear

Diametral pitch range	Min. numbers of teeth		Diametral pitch range	Min. numbers of teeth	
	Pinion	Gear		Pinion	Gear
20–48	12	18	73–76	19	29
49–52	13	20	77–80	20	30
53–56	14	21	81–84	21	32
57–60	15	23	85–88	22	33
61–64	16	24	89–92	23	35
65–68	17	26	93–96	24	36
69–72	18	27	97–100	25	38

TABLE 2.15 Spiroid Gear Dimensions

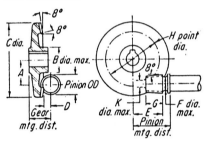

A	B	C	D	E	F	G	H	K
0.500	0.625	1.500	0.129	0.596	0.2969	0.365	1.101	0.2031
0.750	1.000	2.250	0.176	0.894	0.4219	0.548	1.651	0.2812
1.000	1.375	3.000	0.248	1.192	0.5938	0.731	2.202	0.3750
1.250	1.625	3.750	0.295	1.490	0.7031	0.914	2.752	0.4531
1.500	2.000	4.500	0.338	1.788	0.8281	1.096	3.303	0.5312
1.875	2.625	5.625	0.402	2.236	1.0156	1.370	4.129	0.6406
2.250	3.187	6.750	0.461	2.683	1.1562	1.644	4.954	0.7500
2.750	4.000	8.250	0.536	3.279	1.3594	2.010	6.055	0.8750
3.250	4.750	9.750	0.608	3.875	1.5625	2.375	7.156	1.0156
3.750	5.625	11.250	0.677	4.471	1.7656	2.741	8.257	1.1406
4.375	6.750	13.125	0.757	5.216	2.0000	3.197	9.633	1.2969
5.125	8.000	15.375	0.851	6.111	2.2812	3.745	11.285	1.4844

GEAR TYPES AND NOMENCLATURE

TABLE 2.16 Spiroid Gear Formulas

To find	Having	Formula
Ratio	Number of teeth in gear, number of threads in pinion	$m_G = \dfrac{N_G}{N_P}$
Pinion spiral angle	Theoretical lead, pinion OD cone angle, pinion pitch radius	$\tan \alpha_P = \dfrac{L \sec \tau}{2\pi r}$
Gear spiral angle	Ratio, pinion pitch radius, gear pitch radius, pinion spiral angle	$\sin \alpha_G = m_G \left(\dfrac{r}{R_G}\right) \sin \alpha_P$
Pinion pitch point	Length of pinion primary pitch cone	$= \frac{1}{3}$ axial length primary pitch cone from small end

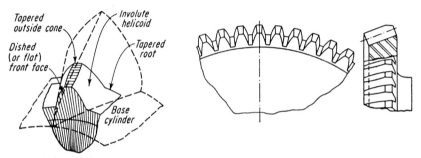

FIGURE 2.42 Beveloid-tooth and gear nomenclature.

Invincible Gear Company in Livonia, Michigan now holds the patent rights on this gear type and still manufactures Beveloid gears.

Figure 2.42 shows a Beveloid gear and its features. Beveloid gears can be engaged with spur gears, helical gears, cylindrical worms, racks, and other Beveloid gears. They can be used with intersecting, parallel, or skew shafts. They are relatively insensitive to mounting errors.

REFERENCES

1. ANSI/AGMA 1012-F90, "Gear Nomenclature, Definition of Terms with Symbols," American Gear Manufacturers Association, April 1990.
2. AGMA 904-B89, "Metric Usage," American Gear Manufacturers Association, May 1989.
3. Drago, R. J., *Fundamentals of Gear Design,* Butterworth, Stoneham, Massachusetts, 1988.
4. Dudley, D. W., *Handbook of Practical Gear Design,* McGraw-Hill, New York, 1984.
5. Klingelnberg Söhne, The Klingelnberg Spiral Bevel Gear Systems, Remscheid, Germany.
6. Machine Tool Works, Oerlikon-Bührle Ltd., Spiralmatic, Spiral Bevel Gear Cutting Machines, Zurich, Switzerland.
7. Smith, L. J., "The Involute Helicoid and The Universal Gear," American Gear Manufacturers Association, Fall Technical Meeting, 1989.

CHAPTER 3
GEAR ARRANGEMENTS

Paul F. Fox
*Product Engineer,
Auburn Gear
Auburn, Indiana*

Gears may be arranged in a great variety of ways. In complex pieces of machinery, the problem of fitting a gear arrangement into the available space is one of the major gear design problems. Even for relatively simple designs, the task of designing an optimum arrangement is very important. In these situations, factors such as the cost and manufacturability of a design must be optimized.

This chapter discusses various gear arrangements that can be found in industry today. It should be understood that there are an infinite number of arrangement possibilities for complex machinery drives. As a result, the arrangements shown in this chapter represent just a few possibilities for designs rather than a complete listing of all combinations.

Chapter 3 can be best understood if the reader is already familiar with the theory of gearing given in Chap. 1 and the kinds of gear teeth given in Chap. 2.

3.1 POSSIBILITIES IN GEAR ARRANGEMENTS

In general terms, a gear drive implies one toothed member that engages another toothed member and transmits motion and/or torque between the two. The second toothed member may be externally toothed or internally toothed, or it may be a section of a rack (a rack is a gear of infinite pitch diameter). A spur or helical pinion can be used to mesh with any of these three gear types. Many of the other gear types cannot mesh with the gear member if it is anything other than externally toothed. In some cases it is the geometry of the gear part that is limiting. In most cases, though, the gear part might be made in theory, but the available machinery may not have the capability to make the part.

Some types of gear teeth are interchangeable and some are not. In an interchangeable gear system, toothed gear wheels a, b, c, and d might be made. These gears may mesh in sequence with a meshing with b, b meshing with c, and c

meshing with d. If these gears were interchangeable, they could be taken apart and reassembled and run properly with a meshing with c or b meshing with d. Generally speaking, involute profile gears have the property of interchangeability, and other types of gears do not. Spur and helical gears are normally made with involute teeth and are interchangeable. An exception to this is the single-enveloping worm gear, where interchangeability does not exist even when the worm member is made to an involute profile.

In some gear types, it is geometrically impractical to make the pinion member with as little as one tooth. A spur pinion with only one tooth would not be capable of continuous, or conjugate, gear action. On the other hand, a worm with one thread is very practical and is often used. Some gear types can be made with the pinion member having as few as five teeth. Most gear types can have pinions with over 16 teeth. A single-enveloping worm, however, is impractical to manufacture at 16 threads, and there is seldom a good reason to use this many threads on a worm.

Table 3.1 shows in tabular form the meshing possibilities of all the major types of gear teeth. This table is based on normal practices and should not be considered an absolute rule. For instance, an internal bevel gear is theoretically possible and has been used in a few special cases. However, normal bevel gear machines will not make such a gear.

TABLE 3.1 Gear Meshing Possibilities

Type of Gear Teeth	Pinion and Gear	Pinion and Rack	Pinion and Internal Gear	Inter-change-ability	One Tooth Pinion	Pinion of 5 Teeth	Pinion of 16 or More Teeth
Spur	Yes	Yes	Yes	Yes	No	No*	Yes
Helical	Yes	Yes	Yes	No	No*	No*	Yes
Straight bevel	Yes	No*	No	No*	No	No*	Yes
Zerol bevel	Yes	No	No	No	No	No*	Yes
Spiral bevel	Yes	No	No	No	No*	No*	Yes
Hypoid	Yes	No	No	No	Yes	Yes	Yes
Face gear	Yes	No	No	No	No	No*	Yes
Crossed helical	Yes	Yes	No	Yes	Yes	Yes	Yes
Single-enveloping worm	Yes	No*	No*	No	Yes	Yes	No*
Double-enveloping worm	Yes	No	No	No	Yes	Yes	No*
Beveloid	Yes	Yes	No	Yes	No	No*	Yes
Spiroid	Yes	No	No	No	Yes	Yes	No*
Planoid	Yes	No	No	No	No*	Yes	Yes
Helicon	Yes	No	No	No	Yes	Yes	Yes

*Items with an asterisk indicate that it is mechanically possible to have this design but that the design is not normally used in the gear trade.

Trademarks: Zerol, registered trademark of Gleason Works, Rochester, N.Y.; Beveloid, registered trademark of Vinco Corp., Detroit, Mich.; Spiroid, Planoid, Helicon, registered trademarks of Illinois Tool Works, Chicago, Ill.

3.2 HOW THE ARRANGEMENT MAY DRASTICALLY AFFECT RATIO, POWER, EFFICIENCY, AND GEARBOX VOLUME

The arrangement of the gear drive can play a very important part in determining how easy or hard it is to achieve things like high ratio, high power capacity, high efficiency, or a very compact unit.

In ordinary spur or helical design practice, it is reasonable to handle ratios from 1:1 to about 8:1 in a single reduction. Ratios of even 10:1 are possible. Suppose a ratio of 120:1 was needed. This would require a triple-reduction unit of spur or helical gears. It would also take a triple-reduction bevel gear unit. A triple-reduction unit has three sets of pinions and gears, or a total of six toothed parts. With a worm gear set, it is possible to handle this much ratio in just a single reduction (only two toothed parts).

For each toothed part in a gear set, a shaft and generally two bearings are required. Where large ratios are needed, the choice of gear arrangement is often determined by finding the arrangement with the least number of parts that will do the job adequately.

Table 3.2 gives some general information about how different ratios can be achieved and the number of gear parts required.

The power-transmitting capacity of different gear arrangements is quite variable. For instance, there has never been a worm gear set that would handle 3750 kW (5000 hp) as a continuous-duty rating. On the other hand, many helical gear sets are in regular service carrying more than 7500 kW (10,000 hp). Defining precisely the upper limit of power for a particular gear arrangement is about as difficult as defining the altitude at

TABLE 3.2 How to Obtain Ratios

Kind of Arrangement	Minimum Number of Toothed Parts	Ratio Range		
		5:1	50:1	100:1
Single reduction:				
Spur	2	Yes	No	No
Helical	2	Yes	No	No
Bevel	2	Yes	No	No
Hypoid	2	Yes	Yes	Yes
Face	2	Yes	No	No
Worm	2	Yes	Yes	Yes
Spiroid	2	No	Yes	Yes
Planoid	2	Yes	No	No
Simple planetary	3	Yes	No	No
Fixed differential	5	No	Yes	Yes
Planocentric	2	No	Yes	Yes
Harmonic drive	2	No	Yes	Yes

which the earth's atmosphere ceases. The upper limit of a gear's capacity is determined by such factors as:

1. How large a gear can be made by available gear manufacturing equipment?
2. What is the upper limit on tooth surface loading and pitch line velocity for the best material available?
3. Are the job requirements of ratio, input speed, life expectancy, and lubrication favorable for a maximum power level?

Table 3.3 shows some nominal maximum power capacities for different gear arrangements. These should be used only as a general guide. In all cases there are a few gear sets in use (or that could be put in use) at considerably higher power ratings. Conversely, the job requirements of many applications make it completely impractical to achieve the ratings shown for the specific application. No power limits are shown for the fixed differential, the planocentric, or the Harmonic Drive. These kinds of units have been used mostly at low power levels (less than 40 kW), but they can probably be extended up into fairly high power ranges.

For specific data on load rating for a given gear design, Chap. 11 should be consulted.

TABLE 3.3 General Survey of Power and Efficiency

Kind of Arrangement	Nominal Maximum kW (hp)	Typical Efficiency, %		
		5:1 Ratio	50:1 Ratio	100:1 Ratio
Single reduction:				
Spur	2,240 (3,000)	98		
Helical	22,400 (30,000)	98		
Straight bevel	370 (500)	98		
Zerol bevel	745 (1,000)	98		
Spiral bevel	3,730 (5,000)	98		
Hypoid	745 (1,000)	95	80	60
Crossed helical	75 (100)	95	80	60
Cylindrical worm	560 (750)	95	80	60
Double-enveloping worm	745 (1,000)	95	80	60
Spiroid	370 (500)	95	80	60
Planoid	745 (1,000)	95		
Helicon	75 (100)	95	80	60
Double reduction:				
Spur	2,240 (3,000)	97	96	94
Helical	22,400 (30,000)	97	96	94
Spiral bevel	3,730 (5,000)	97	96	
Simple planetary	7,460 (10,000)	97		
Fixed differential			80	60
Planocentric			90	85
Harmonic drive			90	85

The efficiency that can be achieved in a gear set is quite variable. The kind of oil, the amount of pitch line velocity, and the kind of bearings enter quite directly into the efficiency. In a spur gear set, the best designs made under favorable conditions will achieve as much as 99-percent efficiency. Some spur sets under poor conditions may not achieve even 90-percent efficiency.

Table 3.3 shows the nominal order-of-magnitude efficiencies for different gear arrangements. These represent general trade experience with good designs but are not the best possible efficiency that can be achieved with an optimum design and a favorable application.

Table 3.3 also shows that the efficiency drops off rather quickly as the ratio increases. Generally this is caused by the fact that the sliding velocity increases as the ratio increases. A worm, for instance, at high ratio has a very large amount of sliding.

The fixed differential gear is capable of high ratio, but it too has a drop off in efficiency. This is caused by the fact that a large amount of tooth meshing occurs at high torque (see Sec. 3.6).

Suppose a 100:1 ratio is needed. So far as arrangement goes, the double-reduction approach using either spur, helical, or bevel gears is the easiest way to achieve high efficiency. With a favorable application and a good design, an efficiency of 96 percent can be achieved with helical gears. It is nearly impossible to obtain this kind of efficiency with any of the single-reduction gears that can reach a 100:1 ratio.

Chapter 12 covers the general subject of gear efficiency and shows curves for many gear types. This chapter should be consulted for more detailed information on gear efficiency.

The volume or weight of a gearbox is also considerably affected by the arrangement. The weight of a gearbox for a given job may vary as much as 50:1! For instance, a single-reduction helical gear unit made of low-hardness material in industrial practice might weigh 4500 kg (10,000 lb). A planetary gear set made of fully hardened gears for an aircraft application might handle the same power level, for the same design life, with the same degree of reliability as the industrial gear, and weigh only 90 kg (200 lb).

In general terms, a 50:1 spread in gear weight can be broken down as follows:

- Weight ratio for low tooth loads on low-hardness material vs. high loads on high-hardness material—7:1
- Weight ratio for heavy construction using thick walls and solid shafts vs. thin walls and hollow shafts—3:1
- Weight ratio of best arrangement to save weight vs. normal arrangement—2½:1

The opportunity to save weight in gear designs depends somewhat on the ratio required. For instance, a worm gear set is not particularly lightweight at a low ratio of 5:1. At a high ratio of 50:1, the worm gear becomes rather light because it can still do the job with just two toothed parts, whereas a spur or helical set would take at least twice as many parts to do the job.

Table 3.4 shows a general picture of what arrangements might be small for different gear ratios. In using this table, it should be kept in mind that intensity of tooth loading is often more important than arrangement in reducing weight. For instance, a case-hardened gear set used in a nominal arrangement might result in a lighter unit than a set of low-hardness cut gears used in the most weight-saving arrangement.

In using Table 3.4 it should also be kept in mind that the desired power rating must be favorable for the design. Some designs cannot handle high torque. These designs are small in weight only in comparison with other designs when their required torque levels are well within their capability.

TABLE 3.4 Gearbox Relative Size and Weight

Kind of Arrangement	Ratio Range			
	5:1	20:1	50:1	100:1
Single reduction:				
Spur, helical, bevel	Small			
Worm		Small	Small	Small
Hypoid	Small	Small	Small	Small
Spiroid		Small	Small	Small
Planoid	Small			
Double reduction:				
Single power path, helical gears		Medium size		
Multiple power path, helical gears		Small	Very small	
Epicyclic gears:				
Simple planetary	Very small			
Compound planetary		Very small		
Double-reduction planetary		Very small	Very small	
Fixed differential		Small	Very small	Very small

Generally speaking, the volume of the gearbox is proportional to the weight—provided that the type of construction is the same. Aircraft gearboxes with magnesium casings will have more volume for their weight than industrial boxes with cast-iron casings.

3.3 PARALLEL AXIS ARRANGEMENTS

Perhaps the most common gear arrangements are those consisting of a pinion and gear meshing on parallel axes. The teeth for these kinds of arrangements may be either spur, helical, or herringbone. (Chapter 2 defines the kinds of gear teeth and defines and explains gear nomenclature.)

Figure 3.1 shows the basic forms of the parallel axis arrangements discussed in this section.

3.3.1 Simple Gear Train

A single pinion and gear is an example of a simple gear train. The ratio for such a unit is figured quite simply:

$$u = \frac{z_2}{z_1} \qquad (3.1)$$

where z_2 is the number of gear teeth, and z_1 is the number of pinion teeth.

A simple gear train is one that consists of an input pinion and an output gear with

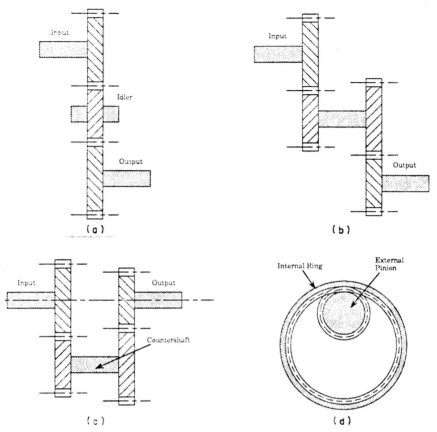

FIGURE 3.1 Parallel axis gear arrangements. *(a)* Simple gear train; *(b)* compound gear train; *(c)* reverted gear train; *(d)* internal gear train

any number of idlers in between. With a simple gear train the following relationships hold true:

1. Regardless of the number of idlers, the speed ratio is always the number of teeth in the output gear divided by the number of teeth in the input pinion.
2. An idler does not change the ratio of the gear train, but it does change the direction of rotation of the output gear.
3. The ratio of the output torque to the input torque is equal to the inverse of the speed ratio.

Figure 3.2 shows a simple gear train with four idlers. The ratio for this arrangement is calculated simply by dividing the number of teeth in the gear at the lower left corner of the picture by the number of teeth in the pinion at the upper right corner.

Figure 3.3 shows a helical pinion meshing with a gear of infinite pitch diameter, also known as a rack.

FIGURE 3.2 Simple gear train with three idlers

FIGURE 3.3 Helical pinion and rack

3.3.2 Compound Gear Train

A compound gear train is considered to be a set of simple gear trains mounted in series with each other. The ratio for this type of gear train is calculated by:

$$u = u_1 u_2 u_3 \text{ (etc.)}$$

or

$$u = \frac{z_{21}}{z_{11}} \frac{z_{22}}{z_{12}} \frac{z_{23}}{z_{13}} \text{ (etc.)} \quad (3.2)$$

where u_1 = ratio of the first reduction
u_2 = ratio of the second reduction

z_{11} = number of pinion teeth, first reduction
z_{12} = number of pinion teeth, second reduction
z_{21} = number of gear teeth, first reduction
z_{22} = number of gear teeth, second reduction

A wide variety of arrangements are possible with parallel axes using simple and compound gear trains. In addition to compound reductions, there may be multiple power inputs or multiple power outputs. There may also be multiple power paths through the unit. This would be analogous to parallel circuits in electrical wiring.

Figure 3.4 shows a compound gear train using helical gears.

Figure 3.5 shows a three-speed power shift transmission with multiple compound reductions.

FIGURE 3.4 Compound helical gear train

3.3.3 Articulated Gear Unit

In a double-reduction design, the first-reduction gear may be solidly mounted on the second-reduction pinion shaft with only two bearings supporting the assembly. Or the pinion and gear may each be mounted on a pair of bearings, and the two connected by a flexible quill shaft. The latter arrangement is called an *articulated design*. The articulated design is quite popular for the higher-speed, higher-horsepower gear units. It has advantages in that it provides flexibility between a first-reduction gear and a second reduction, which reduces dynamic tooth loading; and in the fact that both the gear and pinion have their own separate pair of bearings, making it possible to provide better control on the contact across the face width at each mesh.

Figure 3.6 shows a schematic arrangement for an articulated MDT unit. Further details of an articulated design are shown in Fig. 3.7. Figure 3.8 shows an MDT unit of articulated design that was used as a main drive on a super aircraft carrier.

3.3.4 Reverted Gear Train (Countershaft)

A special form of compound gear train is the reverted, or countershaft, arrangement. These gear trains enable multiple speed ratios to be achieved with a minimum number of shafts and bearings.

The significant characteristics of a reverted gear train are that the gear pairs have a common center distance with the input and output shafts being on the same center line, and both the input and output shafts rotate in the same direction. The ratio calculation for these arrangements is the same as for compound gear trains [Eq. (3.2)].

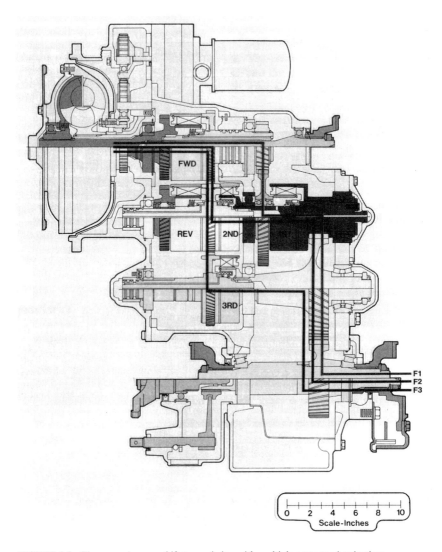

FIGURE 3.5 Three-speed power shift transmission with multiple compound reductions

GEAR ARRANGEMENTS

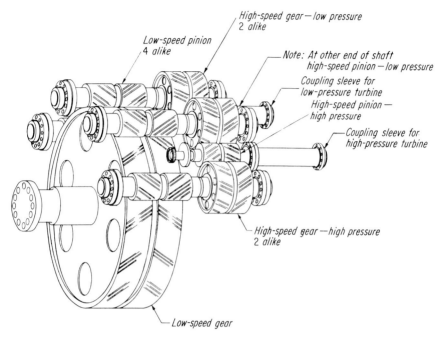

FIGURE 3.6 Schematic arrangement of articulated MDT unit

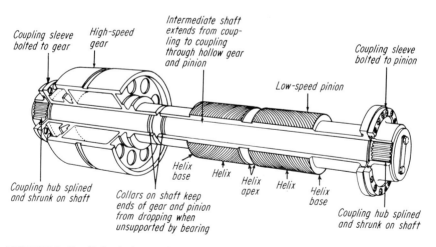

FIGURE 3.7 Detail of articulated design of high-speed and low-speed pinion. Note intermediate shaft passing through hollow pinion and gear, and the fact that both pinion and gear have their own separate bearings

FIGURE 3.8 Side view of MDT unit of type used on super aircraft carrier *Saratoga*. Note articulated design. Rating in excess of 50,000 hp. *(Courtesy of U.S. Navy and General Electric Co.)*

Figure 3.9 shows a six-speed transmission of the type used in trucks. Note the two countershafts in this design.

3.3.5 Internal Gear Train

The internal gear train is a special type of simple gear train. In an internal arrangement, the speed ratio is calculated in a manner similar to that for a simple gear train [Eq. (3.1)], with the exception that the rotation of the output member is in the same direction as the input.

With an internal gear train, a larger ratio reduction is possible when compared with an external mesh design.

However, because manufacturing difficulties are greater in producing internal gears than for external gears of similar size, internal mesh arrangements are seldom used for simple ratio reductions.

3.3.6 Parallel Axis Gears Combined with Right-Angle Gears

In many cases taking a basic parallel-axis gear unit and adding a first or last reduction of right-angle gearing works well. Doing so permits the drive to turn a right-angle corner, and it may make the drive more compact and result in the use of fewer bearings and other parts.

Figure 3.10 shows a multiple-reduction unit with a combination of spur and helical gear reductions and a bevel reduction.

GEAR ARRANGEMENTS

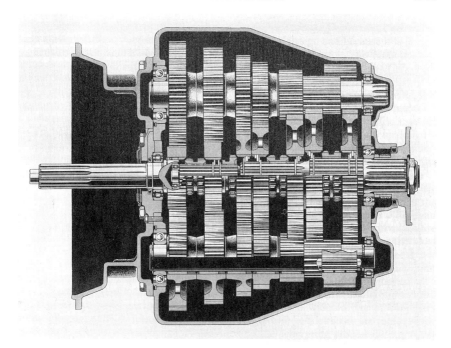

FIGURE 3.9 Six-speed truck transmission

FIGURE 3.10 Double-reduction differential drive and transmission gearing

3.3.7 Design Problems with Parallel Axis Gears

There are several problems in parallel axis gear design that should be addressed if a design is to be successful. In picking an arrangement, designers should be alert to the design problems and satisfy themselves that they can design the arrangement selected so as to obtain a practical design to manufacture and a reliable design in the field.

Some of the problems that may show up are how to:

1. Obtain uniform tooth loading across the face width
2. Compensate for excessive windup of long, skinny pinions or long shafts

3. Obtain equal torque through each power path in a multiple-power-path design
4. Choose numbers of teeth, gear locations, and gear-shaft connections so that multiple-power-path gearing will assemble
5. Position idlers so that the resultant bearing loads on the shafts are minimized
6. Select helix angles and hands-on helical gearing to minimize axial thrusts

The length-to-diameter ratios of helical and spur gears can be fairly large. In most other gear types, the pinion does not have a face width more than about one-third the pinion pitch diameter. In spur gearing the L/D ratio may be as great as 1, while in double helical gearing it may approach 2. Needless to say, high accuracy is needed in boring the casings and machining the teeth when a relatively large face width is used. Low-hardness gearing of about 200 BHN exhibits some tendency to wear in and improve contact. As much as 0.001 in. may be worn off the high ends of the teeth to equalize the contact without too serious a result. Medium- to high-hardness gears have but slight tendency to wear in. If the contact across the face width is not good at full load, the teeth may break before any equalizing wear occurs. (See Chap. 11 for data on how to handle misalignment when load-rating gears.)

Large pinions with an L/D ratio of 1 or more will, under load, twist elastically a significant amount. It may be necessary to provide helix modifications to compensate for this on large gears where the material is fairly hard and the horsepower is high.

In multiple-power-path units, each power path must be able to have the teeth engage simultaneously when a slight torque is applied. In addition, errors in tooth spacing and runout must be slight enough in the gears so that the flexibility of the connecting shafts can keep a uniform load on each path. If the connections are too stiff or the teeth not precise enough, the load will shift rapidly from one power path to another as the gear rotates.

3.4 PARALLEL AXIS EPICYCLIC GEARING

A large family of gear arrangements goes under the name of "epicyclic," or "planetary," gearing. Generally speaking, the epicyclic train has a central "sun" gear, several "planets" meshing with the sun and spaced uniformly around the sun, and an "annulus," or ring gear, meshing with the planets. The sun and planets are externally toothed while the ring is internally toothed. The name *epicyclic* comes from the fact that points on the planets trace epicycloidal curves in space.

The name *planetary* gear is often used interchangeably with epicyclic to denote the whole family of gears. In a stricter sense, the planetary is but one type of epicyclic gear arrangement.

There are a large number of possible epicyclic arrangements. These may be divided into three general groups:

1. Simple epicyclic trains
2. Compound epicyclic trains
3. Coupled epicyclic trains

Figure 3.11 shows schematic arrangements of the two most common types of epicyclic gear trains, the simple epicyclic train, and the compound epicyclic train.

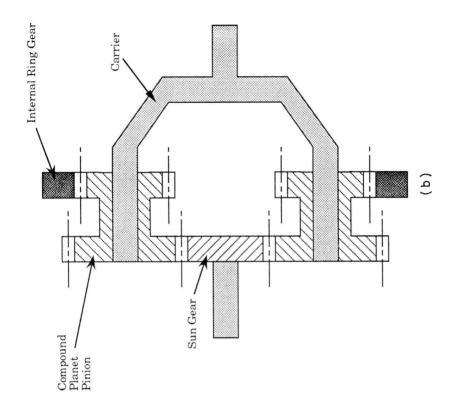
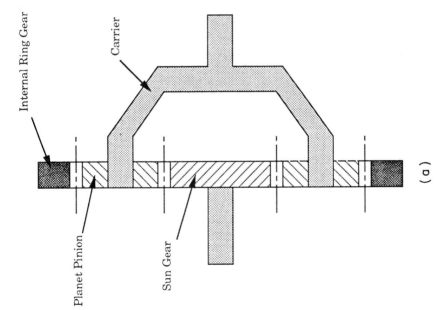

FIGURE 3.11 Epicyclic gear trains

FIGURE 3.12 Simple epicyclic gear arrangement

3.4.1 Simple Epicyclic Trains

The simple epicyclic train consists only of the elements of sun, planet, and ring gear. The arrangement is often called a "simple" epicyclic gear or "simple" planetary gear to distinguish it from the more complicated arrangements.

Figure 3.12 shows a simple epicyclic gear arrangement. The figure shows only three planets. The number of planets varies considerably in actual practice. With some ratios it has been possible to squeeze in as many as twenty planets. Sometimes as few as one is used for light load applications.

Figure 3.13 shows schematic arrangements of the three types of simple epicyclic gears. These are the planetary, the star, and the solar arrangements.

Table 3.5 shows the overall ratio for each of the three simple epicyclic trains. If one of these units were driven backward, the ratio would be the reciprocal of the ratio calculated from the table. A range of ratios is shown for each kind of epicyclic gear. Note that only the solar will result in small ratio reductions. The ratio range of 1.7:1 to 2:1 is a range that is difficult to handle with simple epicyclic gearing. The star will work in this range, but the design is impractical in most cases.

Epicyclic gears will not assemble unless the tooth numbers are properly chosen. In

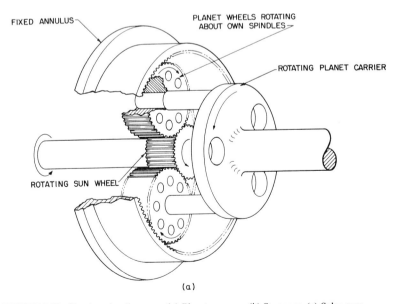

(a)

FIGURE 3.13 Simple epicyclic gears. *(a)* Planetary gear. *(b)* Star gear. (c) Solar gear

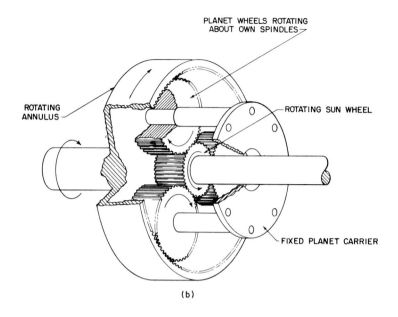

(b)

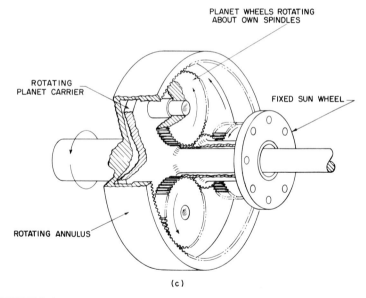

(c)

FIGURE 3.13 (*Continued*)

order to insure equal spacing of the planets around the sun, the following relationship must be met:

$$\frac{z_3 + z_2}{\text{number of planets}} = I \tag{3.3}$$

TABLE 3.5 Simple Epicyclic Gear Data

Kind of Arrangement	Fixed Member	Input Member	Output Member	Overall Ratio	Range of Ratios Normally Used
Planetary	Ring	Sun	Cage	$\dfrac{z_3}{z_2}+1$	3:1–12:1
Star	Cage	Sun	Ring	$\dfrac{z_3}{z_2}$	2:1–11:1
Solar	Sun	Ring	Cage	$\dfrac{z_2}{z_3}+1$	1.2:1–1.7:1

z_1 = number of planet teeth
z_2 = number of sun teeth
z_3 = number of ring teeth

where z_2 is the number of sun teeth and z_3 is the number of ring teeth, then I must equal an integer.

As an example, suppose you had a simple planetary arrangement with the following numbers of teeth: $z_2 = 21$, $z_1 = 15$, and $z_3 = 51$. Using Eq. (3.2), it can be determined that 3, 4, 6, or 8 planets are just a few of the possible numbers that can be equally spaced in this design. As a second example, let us look at a slightly different arrangement: $z_2 = 21$, $z_1 = 16$, and $z_3 = 53$. With this arrangement, it is possible to have only two equally spaced planets around the sun, which in most power transmission drives would be extremely impractical.

In cases where a specific planetary ratio is desired, but equal spacing of a given number of planets is impossible, the following equation can be used:

$$\frac{360}{z_3 + z_2} = x \qquad (3.4)$$

where x equals the angular spacing increment.

Using the example of the 21-tooth sun and 53-tooth ring, the angular spacing increment would be 4.8649°. As a result, the angular spacing between planets divided by 4.8649° must be an integer. Therefore, a four-planet design could be assembled with planets spaced 87.5676, 92.4324, 87.5676, and 92.4324° apart.

3.4.2 Compound Epicyclic Trains

The compound epicyclic train has two planet members attached together on a common shaft. Figure 3.11 shows a schematic arrangement of a compound planetary gear.

Table 3.6 shows the basic data for compound epicyclic gears. Note that the compound planetary and the compound star will handle much higher ratios than the simple epicyclic types.

Like the simple epicyclic gear arrangement, certain criteria must be met in order for the arrangement to assemble:

$$d_{p_3} = d_{p_3} + d_{p_{11}} + d_{p_{12}} \qquad (3.5)$$

TABLE 3.6 Compound Epicyclic Gear Data

Kind of Arrangement	Fixed Member	Input Member	Output Member	Overall Ratio	Range of Ratios Normally Used
Compound planetary	Ring	Sun	Cage	$\dfrac{z_3 z_{11}}{z_2 z_{12}} + 1$	6:1–25:1
Compound star	Cage	Sun	Ring	$\dfrac{z_{11} z_3}{z_2 z_{12}} + 1$	5:1–24:1
Compound solar	Sun	Ring	Cage	$\dfrac{z_2 z_{12}}{z_3 z_{11}} + 1$	1.05:1–2.2:1

z_2 = number of sun teeth
z_3 = number of ring teeth
z_{11} = number of first-reduction planet teeth
z_{12} = number of second-reduction planet teeth

where d_{p_3} = ring pitch diameter
    $d_{p_{11}}$ = first-planet pitch diameter
    $d_{p_{12}}$ = second-planet pitch diameter

Also

$$\frac{(z_{11} z_3 + z_{12} z_2)}{\text{number of planets}}$$

must equal an integer.

3.4.3 Coupled Epicyclic Trains

Coupled epicyclic trains consist of two or more simple epicyclic trains arranged so that two members in one train are common to the adjacent train. Figure 3.14 shows some schematic arrangements of coupled epicyclic trains. There are many more possible combinations.

As a rule, the equations for planet spacing for a simple epicyclic train can be applied to each simple train in the coupled arrangement. The calculation of the reduction ratio, however, is more complicated and cannot be solved using a general equation. Therefore a common method of ratio calculation for epicyclic arrangements is discussed.

3.4.4 Tabular Method

The tabular method of planetary ratio calculation involves tabulating the results of rotating the planetary train under two imaginary conditions and summing these results to determine the actual speed ratio of the train. This method can be used for all but the most difficult types of epicyclic arrangements (those with three or more coupled planetaries).

As an example, use arrangement A from Fig. 3.14. Construct a table as shown in the following, listing each member of the train.

Rotate the entire system one turn clockwise (+) and tabulate the results in the first row.

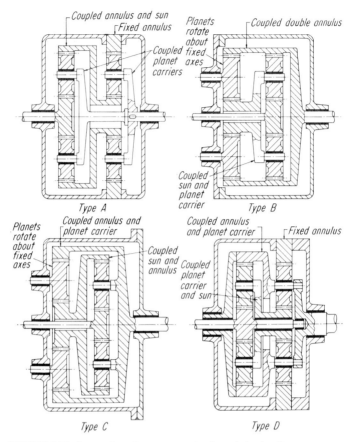

FIGURE 3.14 Some schematic arrangements of coupled epicyclic gears

Since the grounded member is the second ring gear, it must be rotated backward one turn (counterclockwise) in order to have a net rotation of zero. As the ring gear is rotated backward one turn, hold the carrier assembly stationary and calculate the resulting rotation of the other components, listing them in the second row.

The net rotation of each member can be found by summing the values in the first and second rows. From the totals in the third row of Table 3.7, it can be shown that the ratio of the input to the output of arrangement A is the following:

$$u = \frac{1 - \left(\dfrac{z_{32}}{z_{22}}\right)\left(\dfrac{z_{31}}{z_{21}}\right)}{1} \tag{3.6}$$

In general, the tabular method of ratio calculation will be successful if the following procedure is followed:

TABLE 3.7 Tabular Ratio Calculation Method

	Sun 1	Carrier 1	Ring 1	Sun 2	Carrier 2	Ring 2
Rotate system forward	$+1$	$+1$	$+1$	$+1$	$+1$	$+1$
Fix carriers Rotate ring backward	$-\dfrac{z_{32}z_{31}}{z_{22}z_{21}}$	0	$\dfrac{z_{32}}{z_{22}}$	$\dfrac{z_{32}}{z_{22}}$	0	-1
Net component Rotation	$1 - \dfrac{z_{32}z_{31}}{z_{22}z_{21}}$	$+1$	$1 + \dfrac{z_{32}}{z_{22}}$	$1 + \dfrac{z_{32}}{z_{22}}$	$+1$	0

z_{21} = number of teeth on first sun
z_{22} = number of teeth on second sun
z_{31} = number of teeth on first ring
z_{32} = number of teeth on second ring

1. Rotate the entire system clockwise one revolution and tabulate the results in the first row.
2. With the carrier fixed, rotate the grounded member counterclockwise one revolution. Calculate the rotation of the members and tabulate the results in the second row.
3. Add the values in the first and second rows for each component and record the results in the third row.
4. Calculate the ratio of any two components in the table.

3.4.5 Application of Epicyclic Gear Trains

Epicyclic gears have been used for many centuries in the gear trade. In 1878 James Watt patented a sun-and-planet gear that he used with one of his early engines. There have been many patents on epicyclic gears. Even in the recent period since World War II, several refinements in planetary gear design details have been patented.

The epicyclic gear is widely used in the aircraft industry. Almost all propeller-driven commercial aircraft use an epicyclic gear of some type to drive the propeller. Piston engines require a relatively low reduction ratio, therefore they often use a single-reduction planetary arrangement. Higher-speed gas turbine engines may use combinations of simple mesh reductions combined with epicyclic reductions, or compound epicyclic reductions. Figure 3.15 shows a turboprop reduction gearbox which combines a simple spur gear first reduction with a simple planetary second reduction. This figure also shows multiple simple and compound spur gear reductions for accessory drives.

There are some applications that require concentric output shafts rotating in opposite directions such as contrarotating turboprops. Figure 3.16 is a cross section of a contrarotating propfan gear arrangement. In this arrangement input power is supplied to the sun gear from the turbine while the planet carrier and ring gear rotate in opposite directions supplying power to the contrarotating propfan. The torque from the contrarotating propfan blades balances the speed of the system.

Epicyclic gears are widely used in tank transmissions, truck transmissions, and automatic transmissions for automobiles. In these applications the basic epicyclic gear may be combined with other gears, clutches, brakes, and pumps to make a total transmission package. Figure 3.17 shows a cross drive transmission for a track-laying vehicle like a tank. The transmission has the capability of controlling the speed ratio to each drive shaft. This transmission has almost all the gear arrangements discussed in this chapter.

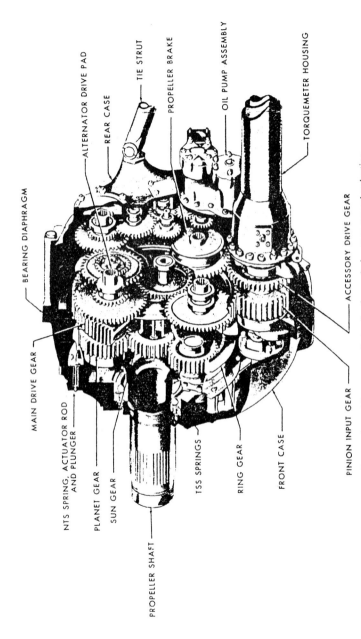

FIGURE 3.15 Turboprop reduction gearbox with spur gear first reduction and planetary second reduction

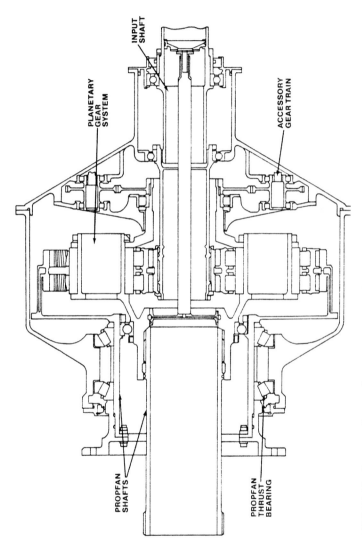

FIGURE 3.16 Contrarotating prop fan gear arrangement. *(Courtesy Allison Gear Turbine, Indianapolis, Indiana)*

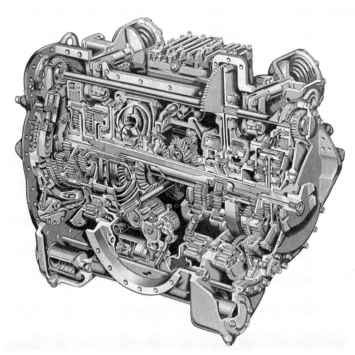

FIGURE 3.17 Tank transmission

3.4.6 Design Problems with Epicyclic Gears

Several of the problems in the design and use of epicyclic gears have already been discussed. A summary of the problems is given in the following:

1. Dividing load between planets
2. High bearing loads on planet pins
3. Balance and vibration of rotating cage
4. More complicated assembly
5. Hazard of jamming the whole drive if a piece of metal gets loose in the mesh

3.5 RIGHT-ANGLE GEAR ARRANGEMENTS

Right-angle gears are used in a wide variety of meshing arrangements. Table 3.8 shows a general survey of these gear types.

In theory all right-angle gear types could be used on either right angles or angles other than 90°. In practice, however, because of limitations in manufacturing equipment and other considerations, some of the gears are not ordinarily used on skew angles. Note the blank areas in the skew-angle column of Table 3.8.

Some types can be used in a planetary-style arrangement. When this is done, the ring member becomes externally toothed. The data given in Table 3.6 can be used with

GEAR ARRANGEMENTS

TABLE 3.8 Angular Gear Meshing Possibilities

Type of Gear Teeth	Pinion and Gear		Arrangements		
	Right Angle	Skew Angle	Planetary	Differential	Reversing
Straight bevel	Yes	Yes	Yes	Yes	Yes
Zerol bevel	Yes	Yes	Yes	Yes	Yes
Spiral bevel	Yes	Yes	Yes	Yes	Yes
Hypoid	Yes		No	Yes	Yes
Face gear	Yes	No		Yes	Yes
Crossed helical	Yes	Yes	No		No
Cylindrical worm	Yes		No	Yes	No
Double-enveloping worm	Yes		No	Yes	No
Beveloid	Yes	Yes	Yes	Yes	Yes
Spiroid	Yes		No	Yes	No
Planoid	Yes		No		Yes
Helicon	Yes		No	Yes	No

angular gears. Some angular gears are used in differentials. Even the worm gear may be used as the gear drive to rotate a differential package.

Some right-angle gears lend themselves to reversing drives. Two pinions may be shifted axially to make a quick reverse in a drive.

Figure 3.18 shows an axle design for an off-highway truck application. This design

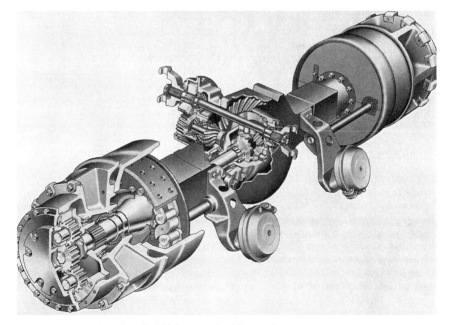

FIGURE 3.18 Axle design of off-highway truck with several gear arrangements

FIGURE 3.19 Bevel gear reverse unit

incorporates a spiral bevel gear set along with combinations of straight bevel gears in the differential, helical gears, and spur gears in the planetary final reduction.

One of the more important skew-angle drives is the V-drive for boats. This permits the engine to be horizontal and the propeller shaft to pass through the hull of the ship at an angle.

Figure 3.19 shows a bevel gear reverse unit. The bevel planets are attached to a fixed cage. Since the bevel sun and bevel ring are really the same size parts, this star type of arrangement has a 1:1 ratio. It is used only to reverse the direction of rotation.

3.5.1 Split Torque Transmission

The size of a gear mesh can be reduced by dividing the torque into two or more paths. There are several ways to accomplish a split torque arrangement such as an epicyclic drive or driving two gears with one pinion, etc. In split torque arrangements some method may have to be used to provide equal load sharing. Figure 3.20 is a schematic of a split torque arrangement for a 2700-kw (3600-hp) helicopter transmission. The input torque is split into two paths. The torque split is controlled by a planetary divider on the input shaft. Splitting the torque provides less power per gear mesh, giving the possibility of a lighter-weight transmission.

3.6 DIFFERENTIAL GEARING

In general, a *differential* gear is an arrangement in which the normal ratio of the unit can be changed by driving into the unit with a second input. Figure 3.21 shows a simple bevel gear differential of the type often used in instruments. Input rotations may be applied to the gears at either end of the bevel planetary set. The output speed is taken off the "spider," or cage shaft. Either input speed may be stopped, and the device will still transmit a ratio between the other input and output. By varying the speed of the second input, it is possible to get an infinite number of output ratios.

Figure 3.22 shows the elements of a spur gear differential. Again, there are two inputs and one output, or it is possible to have one input and two outputs.

Figure 3.23 shows a typical bevel gear differential. One of the most familiar uses of a differential is in the rear end of an automobile. When an automobile goes around a corner, one wheel must rotate faster than the other. A differential is needed with one input and two outputs. In Fig. 3.23 the large gears are hypoid, and the small gears are bevel. The hypoid type of car rear end allows the drive shaft to be lower than would be possible with an all-bevel differential. A low drive shaft permits the automobile to be built lower to the ground.

GEAR ARRANGEMENTS

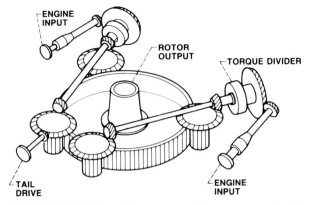

FIGURE 3.20 3600-hp split torque transmission arrangement based on combining gear and five pinions

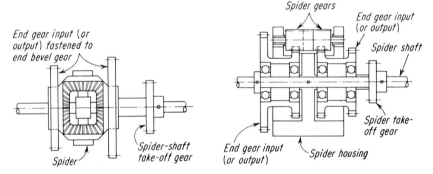

FIGURE 3.21 Simple bevel gear differential

FIGURE 3.22 Spur gear differential

3.6.1 Fixed Differential

The differentials just discussed, with two inputs and one output or two outputs and one input, are used to vary ratio. There is another kind of differential that has a fixed ratio and develops an unusually large ratio. This kind is called a differential because the output speed is the difference between the speeds of two parts of the drive that are running at almost the same speed.

There are dozens of arrangements of the (free) differential already discussed. Likewise, there are dozens of fixed differential arrangements. Figure 3.24 shows four typical arrangements of fixed differentials.

Table 3.9 shows some general data for the four fixed differentials shown in Fig. 3.24. Note that there is a subtractive item in the expression for overall ratio for each of these types. If the pitch were made infinitely fine, it would be possible to get the subtractive term almost to reach zero. This would tend to make the overall ratio infinitely large. Practical considerations on number of teeth, though, tend to limit the ratio that can

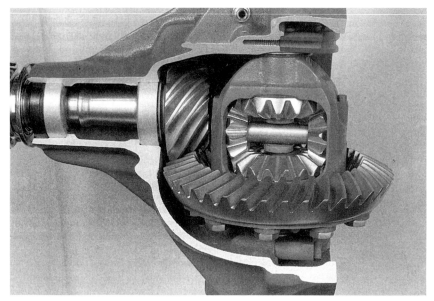

FIGURE 3.23 Typical spiral bevel or hypoid gear and differential

be obtained. The ratio range shown represents normal design practice rather than any theoretical limit on the ratio.

The fixed differential gears tend to be low in efficiency. The differential has a large amount of tooth meshing going on at very high tooth loads. Losses are much higher than for regular gearing. A fixed differential of 100:1 ratio might have an efficiency as low as 25 percent. With good design, a fairly good efficiency can be obtained. The chief value of the differential is that it permits a large ratio to be obtained in a very compact unit.

3.7 SPEED-CHANGING TRANSMISSIONS USING SHIFTING

Gear arrangements that change speed by shifting have been widely used in automobiles, trucks, tanks, and many other applications. The gear shift transmission may have a variety of parallel axis gears, right-angle gears, and clutch elements.

The normal mode of operation for these transmissions is to declutch the drive, shift a gear or coupling axially, and reengage the clutch.

Figure 3.25 shows a 20-speed manual transmission of the type used for heavy-duty on-highway hauling. Note the two countershafts in this design. Also note the clutches on the center shaft, which allow the center shaft to be coupled and decoupled in order to achieve a large number of ratios.

TABLE 3.9 Fixed Differential Gear Data

Kind of Differential	Fixed Member	Input Member	Output Member	Arrangement	Over-all Ratio	Range of Ratios Normally Used
A: All external mesh	No. 1 sun	Cage	No. 2 sun	No. 1 sun/No. 1 planet No. 2 planet/No. 2 sun	$\dfrac{z_{22}z_{11}}{z_{22}z_{11} - z_{21}z_{12}}$	10:1–50:1
B: Internal Gear	No. 1 ring	Cage	No. 2 ring	No. 1 planet/No. 1 ring No. 2 planet/No. 2 ring	$\dfrac{z_{32}z_{11}}{z_{32}z_{11} - z_{31}z_{12}}$	15:1–100:1
C: Planetary	No. 1 ring	Sun	No. 2 ring	No. 1 sun/No. 1 planet No. 1 planet/No. 1 ring No. 2 planet/No. 2 ring	$\dfrac{z_{31}}{z_{21}} + 1 \dfrac{z_{32}z_{11}}{z_{32}z_{11} - z_{31}z_{12}}$	20:1–500:1
D: Star-planetary	No. 1 cage	No. 1 sun	No. 2 ring	No. 1 sun/No. 1 planet No. 1 planet/No. 1 ring No. 2 sun/No. 2 planet No. 2 planet/No. 2 ring No. 1 sun attached to No. 2 sun No. 1 ring attached to No. 2 ring	$\dfrac{1}{\dfrac{z_{32}}{z_{22}} + 1} - \dfrac{1}{\dfrac{z_{31}}{z_{21}}\left(\dfrac{z_{22}}{z_{32}} + 1\right)}$	15:1–100:1

z_{11} = number of teeth, No. 1 planet
z_{21} = number of teeth, No. 1 sun
z_{12} = number of teeth, No. 2 planet
z_{22} = number of teeth, No. 2 sun
z_{31} = number of teeth, No. 1 ring
z_{32} = number of teeth, No. 2 ring
Follow rules of assembly for compound epicyclic for applicable parts of each differential (see Table 3.6). See Fig. 3.29 for schematic arrangement of each kind.

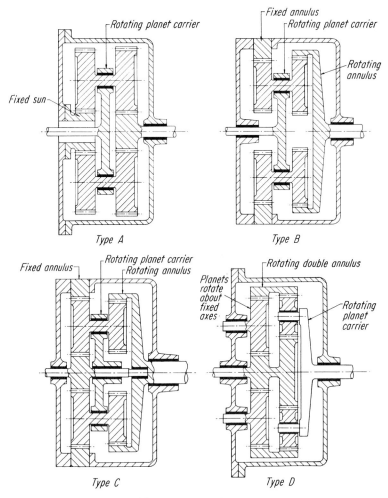

FIGURE 3.24 Fixed differential gear arrangements

3.8 SPEED-CHANGING TRANSMISSIONS USING CLUTCHES AND BRAKES

Many transmissions are made so that a gear shift can occur without disengaging the drive (or declutching). These transmissions use things like fluid couplings, brakes, or differentials to change speed while torque is being transmitted. The so-called "automatic" transmission used on most passenger automobiles is of this type. Some tanks and other military vehicles use the automatic transmission.

Figure 3.26 shows a power shift transmission for heavy-duty cycling applications. Note the fluid coupling and also the multiple shafts for obtaining large ratio reductions.

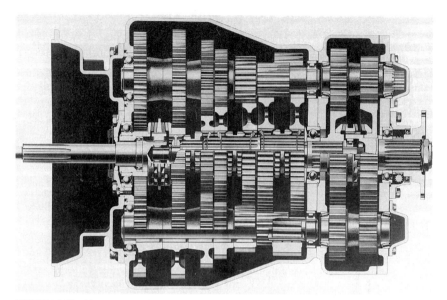

FIGURE 3.25 20-speed manual transmission for off-highway vehicle

3.9 HIGH RATIO ARRANGEMENTS

Besides the fixed differential gears discussed in Sec. 3.6, there are several other gear arrangements that have high ratio capabilities. These are used under varying trade names. Theoretically, they do not belong in any of the gear classifications already discussed in this chapter.

Planocentric Gear. An example of one of the high ratio types is the kind of gear set called the *planocentric*. This unit can have as few gear parts as a spur pinion and an internal gear and yet develop as much as 100:1 ratio! A ratio this great can be readily obtained with a single-thread worm and gear, but it is surprising that this much can be achieved with spur gearing or helical gearing. Figure 3.27 shows a schematic representation of a planocentric gear.

The pinion is mounted on an eccentric. A bearing separates the pinion from the eccentric and its drive shaft. The pinion is thus under no constraint to rotate at the speed of the input shaft.

The pinion wobbles instead of truly rotating. It does develop a rotation, superimposed on the wobble. Strangely, the rotation of the pinion is at output speed instead of input speed! The pinion transmits its rotation to the output shaft by a pin coupling (or other means). The internal gear is fixed to the casing.

The equation for ratio is

$$u = \frac{z_1}{z_2 - z_1} \qquad (3.7)$$

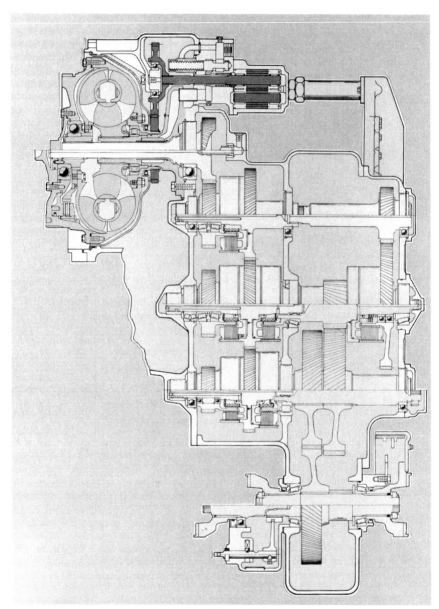

FIGURE 3.26 34000 series six-speed power shift transmission

GEAR ARRANGEMENTS

FIGURE 3.27 Planocentric gear, 64:1 ratio. High-speed end showing 64-tooth pinion mating with 65-tooth gear. Pinion drives cage pins. *(Courtesy of General Electric Co., Lynn, Massachusetts)*

This type of drive has been used for high-torque, high-reduction applications in aircraft, marine, and appliance fields. Under favorable conditions the efficiency may be as high as 90 percent or more.

3.9.1 Harmonic Drive

Harmonic Drive is the name given to a type of arrangement that used the controlled elastic deflection of one or more parts for the transmission, conversion, or change of mechanical motion. The basic mechanism of Harmonic Drive has the broad multipurpose capabilities of the simple lever and has been shown to be adaptable to such forms of mechanical systems as rotary-to-rotary motion transmissions, rotary-to-linear motion converters, linear-to-linear transmissions, and rotary pumps and valves.

A close look at the basic Harmonic Drive elements is provided in Fig. 3.28. As the wave generator is turned, the Flexspline is progressively deflected to follow the rotating elliptical shape. The Flexspline and circular spline are held in engagement at the major axis of the wave generator and are fully disengaged and clearing at the minor axis. At the major axis the teeth are in full spline contact and are rotationally stationary. At the minor axis where the teeth are disengaged, they are in angular motion because of deflection of the Flexspline. For a full rotation of the wave generator, the Flexspline counterrotates through an angle equivalent to 2 of its 198 teeth, resulting in a reduction ratio of 99:1 magnitude.

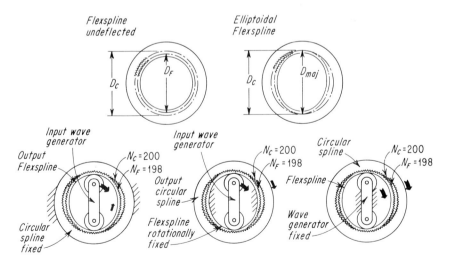

FIGURE 3.28 Design relationships and ratios of single-stage rotary Harmonic Drive

3.9.2 Special Properties of the Harmonic Drive

From the foregoing discussion of the basic rotary Harmonic Drive structure, the basis for several unique and inherent properties can be seen:

1. Tooth motion relationship between Flexspline and circular spline makes unusually high speed reduction or speed increase possible in a single stage.
2. Under load, many spline teeth are in simultaneous engagement, resulting in high torque capacity.
3. Spline teeth come into contact with an almost pure radial motion and have essentially zero sliding velocity, even at high input speeds. Tooth friction losses and tooth wear are thus very low.
4. Because of low friction losses, high mechanical efficiencies can be obtained, which are particularly outstanding at high ratios.
5. Spline teeth in contact and under load are practically stationary. Dynamic loading, under normal operating conditions, is very low.
6. Regions of tooth engagement and application of load torque are usually diametrically opposed and result in a force couple that is symmetrical and balanced.
7. Harmonic Drive elements are concentric and tend to be self-aligning.
8. Tubular construction of splined elements results in high torsional rigidity and thus less windup under load.
9. Backlash is normally low but can be completely eliminated between Flexspline and circular spline by spring-loading the two splines into mesh.
10. With Flexspline formed as an integral section of a flexible cylindrical wall, positive transmission of rotary mechanical notion through the wall can be achieved.

3.9.3 Harmonic Drive Applications

Industrial and military applications include speed reducers, combined motor reducers, linear and rotary actuators, servomechanisms, lifts and jacks, elevators and escalators, power takeoffs, rotary-to-linear motion converters, torque generators, instrument controls, radio and antenna drives, and special turret traversing mechanisms. Harmonic Drives are also well suited for hermetically sealed power transmissions, sealed and pressurized valve actuators, mixer and agitator drives, control rod drives for nuclear reactors, and special marine and medical uses.

REFERENCES

1. Coy, J. J., Townsend, D. P., and Zaretsky, E. V., "Gearing," NASA Reference Publication 1152, AVSCOM Technical Report 84–C–15, December 1985.
2. Bush, G. S., Osman, M. O. M., and Sankar, S., 1984, "On the Optimal Design of Multi-Speed Gear Trains," *Mechanism and Machine Theory,* Vol. 19, No. 2, pp. 183–195.
3. Savage, M., Coy, J. J., and Townsend, D. P., "Optimal Tooth Numbers for Compact Standard Spur Sets," *Journal of Mechanical Design,* Transactions of the ASME, Vol. 104, October 1982, pp. 749–757.

4. Anderson, N. E., and Loewenthal, S. H., "Design of Spur Gears for Improved Efficiency," *Journal of Mechanical Design,* Transactions of the ASME, Vol. 104, October 1982, pp. 767–774.
5. Sanger, D. J., and White, G., "A General Theory of Composite Gear Trains," *Proc. Instn. Mech. Engrs.,* 1969–70, Vol. 184, Pt. 1, No. 58, pp. 1063–1073.
6. Crawshaw, S. L., and Kron, H. O., "Gears," *Machine Design,* September 21, 1967, pp. 18–23.
7. Dudley, D. W., 1964, "How Increased Hardness Reduces the Size of Gear Sets," *Product Engineering,* November 6, pp. 92–102.
8. Shigley, J. E., *Mechanical Engineering Design,* McGraw-Hill, New York, 1963, pp. 398–503.
9. Bryant, R. C., and Dudley, D. W., "Which Right-Angle Gear System?" *Product Engineering,* November 7, 1960.
10. Eisbrenner, R. D., "Brake on Output Shaft Permits Rapid Forward-Reverse Shifting," *Design News,* March 28, 1960.
11. Wilkinson, W. H., "Four Ways to Calculate Planetary Gear Trains," *Machine Design,* January 7, 1960, pp. 155–159.
12. Tuplin, W. A., "Compound Epicyclic Gear Trains," *Machine Design,* April 4, 1957, pp. 100–104.
13. Michalec, G. W., "Gear Differentials," *Machine Design,* October–December 1955.
14. Sicklesteel, D. T., "Method for Design and Calculation of Planetary Gear Sets," *Design News,* June 15, 1952, pp. 45–49.
15. Poppinga, R., "The Efficiency of Planetary Gear Trains," *Engrs. Digest,* December, 1950, pp. 421–425.

CHAPTER 4
GEAR TOOTH DESIGN

Darle W. Dudley
President
Dudley Engineering Company, Inc.
San Diego, California

In this chapter the subject of the design of the gear tooth shape is covered.* To the casual observer, some gear teeth appear tall and slim while others appear short and fat. Gear specialists talk about things like the "pressure angle," long and short addendum, root fillet design, and the like. Obviously, there is some well-developed logic in the gear trade with regard to how to choose and specify factors related to the gear tooth shape. This chapter presents the basic data that are needed to exercise good judgment in gear tooth design.

4.1 BASIC REQUIREMENTS OF GEAR TEETH

Gear teeth mesh with each other and thereby transmit nonslip motion from one shaft to another. Those making and using gear teeth may expect the teeth to conform to some standard design system. If this is the case, the gear maker may be able to use some standard cutting tools that are already on hand. If a standard design is used that is already familiar to the gear user, the functional characteristics of the gears; such as relative load-carrying capacity, efficiency, quietness of operation, and the like, can be expected to be similar for new gear drives to those of gear drives already in service.

It should be kept in mind in the 1990s (and the years to follow) that the gear art has progressed to a point where there is much more versatility than previously. Machine tools are computer controlled and can be programmed to cover much more variations than were possible in the past. In the gear design office, computers can find what may be believed to be a truly optimum design for some important application.

---

*In the first edition of this book,[1] Chapter 5, "Gear-Tooth Proportions," and Chapter 6, "Center Distance," covered the material given in this chapter. The original material was written by Paul M. Dean, Jr. of General Electric Co. Since much of the gear tooth design logic does not tend to change with time, a considerable amount of Dean's original material is used again.

Frequently, the design chosen does not agree with a design that might have been considered standard in earlier years.

What this means is that a great variety of gear tooth designs are being used. This trend in gear engineering can be a good one—providing the gear designer has an in-depth knowledge of all the things that need to be considered. Serious mistakes, though, can be made when a decision is made by computer data and the computer program failed to consider some critical constraint in the application.

4.1.1 Definition of Gear Tooth Elements

Many features of gear teeth need to be recognized and specified with appropriate dimensions (either directly or indirectly).

Spur and helical gears are usually made with an involute tooth form. If a section through the gear tooth is taken perpendicular to the axis of the part, the features shown in Figs. 4.1 and 4.2 are revealed. Note the nomenclature used.

Outside diameter (dashed line in Fig. 4.2) is the maximum diameter of the gear blank for spur, helical, worms, or worm gears. All tooth elements lie inside this circle. A tolerance on this diameter should always be negative.

Modification diameter (dashed line in Fig. 4.2) is the diameter at which any tip modification is to begin. It is a reference dimension and may be given in terms of degrees of roll.

Pitch diameter (dashed line in Figs. 4.1 and 4.2) is the theoretical diameter established by dividing the number of teeth in the gear by the diametral pitch of the cutter to be used to produce the gear. This diameter can have no tolerance.

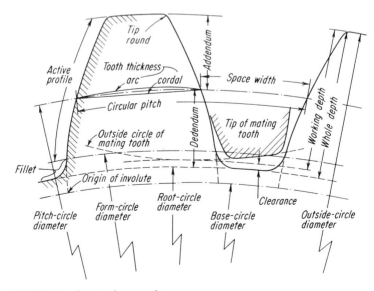

FIGURE 4.1 Gear tooth nomenclature

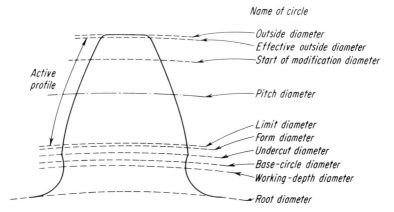

FIGURE 4.2 Nomenclature of gear circles

Limit diameter (dashed line in Fig. 4.2) is the lowest portion of a tooth that can actually come in contact with the teeth of a mating gear. It is a calculated value and is not to be confused with form diameter. It is the boundary between the active profile and the fillet area of the tooth.

Form diameter (dashed line in Fig. 4.2) is a specified diameter on the gear above which the transverse profile is to be in accordance with drawing specifications on profile. It is an inspection dimension and should be placed at a somewhat smaller radius than the limit diameter to allow for shop tolerances.

Undercut diameter (dashed line in Fig. 4.2) is the diameter at which the trochoid producing undercut in a gear tooth intersects the involute profile.

Base-circle diameter (dashed line in Figs. 4.1 and 4.2) is the diameter established by multiplying the pitch diameter (see above) by the cosine of the pressure angle of the cutter to be used to cut the gear. It is a basic dimension of a gear.

Root diameter (dashed line in Figs. 4.1 and 4.2) is the diameter of the circle that establishes the root lands of the teeth. All tooth elements should lie outside this circle. The tolerance should be negative.

Figure 4.3 shows a side view of a spur gear tooth to further depict the features and nomenclature of gear teeth.

Active profile (shaded area in Fig. 4.3a) is a *surface* and is that portion of the surface of the gear tooth which at some phase of the meshing cycle contacts the active profile of the mating gear tooth. It extends from the limit diameter (see Fig. 4.2) near the root of the tooth to the tip round (see Fig. 4.1) at the tip of the tooth and, unless the mating gear is narrower, extends from one side of the gear or edge round (see Fig. 4.3k) at one end of the tooth to the other side of the gear or edge round.

Top land (shaded area in Fig. 4.3b) is a *surface* bounded by the sides of the gear (see Fig. 4.3d) and active profiles; or if the tooth has been given end and tip rounds (see Fig. 4.3h and i), the top land is bounded by these curved surfaces. The top land forms the outside diameter of the gear.

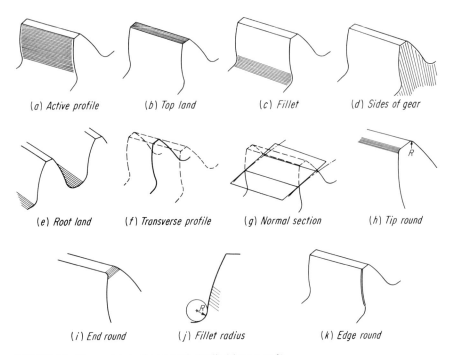

FIGURE 4.3 Nomenclature of gear tooth details (views a to k)

Fillet The fillet of a tooth (shaded area in Fig. 4.3c) is a surface that is bounded by the form diameter (see Fig. 4.2) and the root land (if present) (see Fig. 4.3e) and the ends of the teeth. In *full-fillet teeth,* the fillet of one tooth is considered to extend from the center line of the space to the form diameter.

Sides of gear The sides of gear (shaded area in Fig. 4.3d) are *surfaces* and are the ends of the teeth in spur and helical gears.

Root land (also known as bottom land) The root land (shaded area in Fig. 4.3e) is a *surface* bounded by fillets (see Fig. 4.3c) of the adjacent teeth and sides of the gear blank.

Transverse profile (heavy line in Fig. 4.3f) is the shape of the gear tooth as seen in a plane perpendicular to the axis of rotation of the gear.

Axial profile (heavy line in Fig. 4.3g) is the shape of the gear tooth as seen in a plane tangent to the pitch cylinder at the surface of the tooth. In the case of helical gears, it is the shape of a tooth as seen on a pitch cylinder and may be developed to be shown as a plane.

Tip round (shaded area in Fig. 4.3h) is a surface that separates the active profile and the top land. It is sometimes applied to gear teeth either to remove burrs or to lessen the chance of chipping, particularly in the case of hardened teeth. It may also be added as a very mild and crude form of profile modification.

End round (shaded area in Fig. 4.3i) is a surface that separates the sides and the top land of the tooth. It is sometimes applied to gear teeth to reduce the chance of chipping, particularly in the case of hardened teeth.

GEAR TOOTH DESIGN   4.5

Edge round (shaded area in Fig. 4.3k) is the surface that separates the active profiles of the teeth from the sides of the gear. These edges are of importance in the cases of helical gears, spiral bevel gears, and worms, since they become very sharp on the leading edge.

Fillet radius (dimension in Fig. 4.3j) is the minimum radius that a gear tooth may have.

4.1.2 Basic Considerations for Gear Tooth Design

Gear teeth are a series of cam surfaces that contact similar surfaces on a mating gear in an orderly fashion. In order to drive in a given direction and to transmit power or motion smoothly and with a minimum loss of energy, the contacting cam surfaces on the mating gears must have the following properties:

1. The height and the lengthwise shape of the active profiles of the teeth (cam surfaces) must be such that, before one pair of teeth goes out of contact during mesh, a second pair will have picked up its share of the load. This is called *continuity of action*.
2. The shape of the contacting surfaces of the teeth (active profiles) must be such that the angular velocity of the driving member of the pair is smoothly imparted to the driven member in the proper ratio. The most widely used shape for active profiles of spur and helical gears that meets these requirements is the involute curve. There are many other specialized curves, each with specific advantages in certain applications. This subject is developed further in the section, "Conjugate Action."
3. The spacing between the successive teeth must be such that a second pair of tooth-contacting surfaces (active profiles) are in the proper positions to receive the load before the first leave mesh.

Continuity of action and conjugate action are achieved by proper selection of the gear tooth proportions. Manufacturing tolerances on the gears govern the spacing accuracies of the teeth. Thus, to achieve a satisfactory design, it is necessary to specify correct tooth proportions, and, in addition, the tolerances on the tooth elements must be properly specified.

As a general rule, gearing designed in accordance with the standard systems will not have problems of continuity of action or conjugate action. In those cases where it is necessary to depart from the tooth proportions given in the standard systems, the designer should check both the continuity of action and the conjugate action of the resulting gear design.

Continuity of Action. As discussed in Chap. 1, all gear tooth contact must take place along the "line of action." The shape of this line of action is controlled by the shape of the active profile of the gear teeth, and the length of lines of action is controlled by the outside diameters of the gears (see Fig. 4.4). In order to provide a smooth continuous flow of power, at least one pair of teeth must be in contact at all times. This means that during a part of the meshing cycle, two pairs of teeth will be sharing the load. The second pair of teeth must be designed such that they will pick up their share of the load and be prepared to assume the full load before the first pair of teeth go out of action.

Control of the continuity of action is achieved in *spur* gears by varying

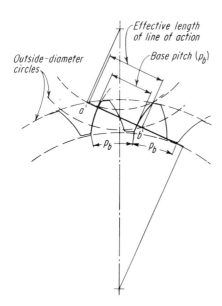

FIGURE 4.4 Zone of tooth action

1. The slope of the line of action (in the case of involute gears, the operating pressure angle).
2. The outside diameters of the pinion and the gear.
3. The shape of the active profile.
4. The relative sizes of the limit-diameter and undercut-diameter circles. (Limit circle must be larger.)

Control of the continuity of action in *internal*-type spur and helical gears is achieved by varying

1. The slope of the line of action (in the case of involute gears, the operating pressure angle).
2. The inside diameter of the internal gear and the outside diameter of the pinion.
3. The shape of the active profile.
4. The relative sizes of the limit-diameter and undercut-diameter circles. The internal gear will not be undercut, but the external member may. Thus the limit diameter on the pinion must be larger than the undercut diameter.

Control of the continuity of action in *helical* gears is achieved by varying

1. The slope of the line of action (in the case of involute gears, the operating pressure angle).
2. The outside diameters of the pinion and of the gear.
3. The shape of the active profile.
4. The relative sizes of the limit-diameter and the undercut-diameter circles. (Limit circles must be larger.) The foregoing elements (1 through 4) control the continuity of action in the transverse plane.
5. The lead of the tooth.
6. The length of the tooth. These two elements (5 and 6) control the continuity of action in the axial plane. In order to assure continuity of action, the portion of the line of action bounded by the outside-diameter circles $(a-b)$ must be somewhat longer than the base pitch (see Fig. 4.4).

The *base pitch* p_b is defined as follows:

$$p_b = p \cos \Phi = \frac{\pi \cos \Phi}{P_d} \qquad (4.1)^*$$

---

*Equations (4.1) through (4.6) give dimensionless answers. For English-system calculations, use inches for all dimensions. For metric-system calculations, use millimeters for all dimensions.

where p = circular pitch of gear
Φ = pressure angle of gear
P_d = diametral pitch

Thus, either the outside-diameter circles, the operating pressure angle, or the base pitch must be adjusted so that $(a-b)$ exceeds p_b by from 20 to 40 percent.

The most general way of checking continuity of action is by calculating the contact ratio.

A numerical index of the existence and degree of continuity of action is obtained by dividing the length of the line of action by the base pitch of the teeth (see Fig. 4.4). This is called contact ratio m_p.

The American Gear Manufacturers Association (AGMA) recommends that the contact ratio for spur gears not be less than 1.2.

$$m_p = \frac{L_a}{p_b} = 1.2 \text{ min} \tag{4.2}$$

A spur gear mesh has only a transverse contact ratio m_p, whereas a helical gear mesh has a transverse contact ratio m_p, an axial contact ratio m_f (face contact ratio), and a total contact ratio m_t.

Equations for *contact ratio* are

Spur gears and helical gears:

$$m_p = \frac{\left[\left(\frac{d'_O}{2}\right)^2 - \left(\frac{d_b}{2}\right)^2\right]^{1/2} + \left[\left(\frac{D'_O}{2}\right)^2 - \left(\frac{D_b}{2}\right)^2\right]^{1/2} - C' \sin \Phi'}{p_b} \tag{4.3}$$

Internal gears, spur, and helical gears:

$$m_p = \frac{\left[\left(\frac{d'_O}{2}\right)^2 - \left(\frac{d_b}{2}\right)^2\right]^{1/2} + \left[\left(\frac{D'_i}{2}\right)^2 - \left(\frac{D_{bi}}{2}\right)^2\right]^{1/2} - C' \sin \Phi'}{p_b} \tag{4.4}$$

Helical gears, axial contact ratio:

$$m_f = \frac{F \tan \psi}{p} \tag{4.5}$$

Helical gears, total contact ratio:

$$m_t = m_p + m_f \tag{4.6}$$

where d'_O = outside diameter (effective) of pinion
D'_O = effective outside diameter of gear (diameter to intersection of tip round and active profile); see the following discussion
D'_i = inside diameter (effective), internal gear
d_b = base diameter of pinion
D_b = base diameter of gear
D_{bi} = base diameter of internal gear
C' = operating center distance of pair

Φ' = operating pressure angle
p = circular pitch (in plane of rotation)
p_b = base pitch
F = length of tooth, axial, the face width of the gear
Ψ = helix angle of helical gears

Notes: To achieve correct answers on contact ratio, the following points should be observed.

1. The effective outside diameter d'_O or D'_O is actually the diameter to the beginning of the tip round, usually $D'_O = D_O - 2$ (edge round specification) (max). This value rather than the drawing outside diameter should be used since in many cases manufacturing practices for removing burrs produce a large radius, particularly in fine-pitch gears. Thus a considerable percent of the addendum may not be effective. If the teeth are given a very heavy profile modification, consideration should be given to performing the calculation under (a) full load, assuming contact to the tip of the active profile, and (b) light load, assuming contact near the start of modification. In this case d'_O or D'_O is selected to have a value close to the diameter at the start of modification. This will give an index to the smoothness of operation at these conditions.

2. This equation also assumes that the form diameter is a larger value than the undercut diameter. If not, use the value of undercut diameter.

3. On occasion, the outside diameter of one or both members is so large relative to the center distance and operating pressure angle that the tip extends below the base circle where it is tangent to the line of action. Since no involute action can take place below the base circle, the value $C \sin \Phi$ (the total length of the line of action) should be substituted in the equation in place of the value of $[(d'_O/2)^2 - (d'_b/2)^2]^{1/2}$ or $[(D'_O/2)^2 - (D'_b/2)^2]^{1/2}$ in the case that one or both become larger than the value of $C \sin \Phi$.

Conjugate Action.* Gear teeth are a series of cam surfaces that act on similar surfaces of the mating gear to impart a driving motion. Many different shapes of surfaces can be used on the teeth to produce uniform transmission of motion. Curves that act on each other with a resulting smooth driving action and with a constant driving ratio are called *conjugate curves*. The fundamental requirements governing the shapes that any pair of these curves must have is summarized in Buckingham's "basic law of gearing," which states, *"normals to the profiles of mating teeth must, at all points of contact, pass through a fixed point located on the line of centers."*[2]

In the case of spur- and helical-type gears, the curves used almost exclusively are those of the involute family. In this type of curve, the fixed point mentioned in the basic law is the pitch point. Since all contact takes place along the line of action, and since the line of action is normal to both the driving and driven involutes at all possible points of contact, and, lastly, since the line of action passes through the pitch point, it can be seen that the involute satisfies all requirements of the basic law of gearing.† Many years ago, the cycloidal family of curves was in common use; however, clockwork gears

---

*See Chap. 1 for a general discussion of gear theory.
†An exception to this rule is that of heavily loaded teeth that bend appreciably while under load. See Dudley,[1] for a discussion of how to modify the involute profile when loads are heavy. Note pages 8.12 and 8.13 in particular.

are about the only application of this type of tooth today. The involute curve has supplanted the cycloid because of its greater ease of design and because it is far less sensitive to manufacturing and mounting errors. In addition, the tools used to generate the cycloid are more difficult to make with the same degree of accuracy as those for the involute.

The use of the involute has become so universal that, with few exceptions, it is hardly necessary to specify on spur and helical gear drawings that the involute form is desired. It is only when a designer wishes to use a noninvolute tooth for a special application that it becomes apparent how much work must be done to design a satisfactory profile properly. The results of standardization and the deep understanding of gears on the part of competent manufacturers are appreciated most by those who require quality designs of a nonstandard nature.

The methods of determining conjugate profiles for various gear applications are beyond the scope of this chapter. The designer should be aware, however, that the basic law of gearing can be utilized to determine conjugate tooth profiles other than those shown in this chapter. Several of the unique properties of the involute are utilized by the data in this chapter to provide conjugate gears, made, for the most part, with standard tooling. Each gear will have special properties that make it best suited for its application. It will, however, not be the type of gear, often found, that would result from an indiscriminate application of cookbook gear formulas.

Bevel gear tooth forms covered by the standards noted in this chapter are *not* involute. In bevel gearing, it is practical to employ a wide variety of conjugate curves, since the problems of interchangeability and of tool costs are minor. This is due to the very great flexibility provided by the generating process used in bevel gear generators. The tools are usually single-edged, hence relatively easy to make, and are fundamental in nature. One of the most common forms for bevel gear teeth is called the *octoid* after the shape of its line of action.

Worm gearing, like bevel gearing, is noninvolute. The tooth form of worm gearing is usually based on the shape of the worm; that is, the teeth of the worm gear are made conjugate to the worm. In general, worms can be chased, as on a lathe, or cut by such processes as milling or hobbing, or can be ground. Each process, however, produces a different shape of worm thread and generally requires a different shape of worm gear tooth in order to run properly.

In the case of face gearing, the pinion member is a spur or helical gear of involute form, but the gear tooth is a special profile conjugate only to the specific pinion. Thus pinions having a number of teeth larger or smaller than the number for which the face gear was designed will not run properly with the face gear.

Pitch Diameter. Although pitch circles are not the fundamental circles on gears, they are traditionally the starting point on most tooth designs. The pitch circle is related to the base circle, which is the fundamental circle, by the relationships that follow. Some authorities list as many as nine distinct definitions of pitch circles. The following definitions cover the pitch circles considered in this chapter.

Standard Pitch Diameter. The diameter of the circle on a gear determined by dividing the number of teeth in the gear by the diametral pitch.

The diametral pitch is that of the basic rack defining the pitch and pressure angle of the gear.

The AGMA[5] gives the following equation for standard pitch diameter in the fine-pitch standards:

$$D = \frac{N}{P_d} \qquad (4.7)$$

where D = diameter of standard pitch circle
P_d = diametral pitch of basic rack
N = number of teeth in gear

Operating Pitch Diameter. The diameter of the circle on a gear which is proportional to the gear ratio and the actual center distance at which the gear pair will operate. A gear does not have an operating pitch diameter until it is meshed with a mating gear. The equations for operating pitch diameters are

External spur and helical gearing:

$$d' = \frac{2C'}{m_g + 1} \quad \text{pinion member} \tag{4.8}*$$

$$D' = \frac{2C'm_g}{m_g + 1} \quad \text{gear member} \tag{4.9}$$

Internal spur and helical gearing:

$$d' = \frac{2C'}{m_g - 1} \quad \text{pinion member} \tag{4.10}$$

$$D' = \frac{2C'm_g}{m_g - 1} \quad \text{gear member} \tag{4.11}$$

where d' = operating pitch diameter of pinion
D' = operating pitch diameter of gear
C' = operating center distance of mesh
m_g = gear ratio = $\dfrac{\text{number of teeth in gear}}{\text{number of teeth in pinion}}$

The operating and the standard pitch circles will be the same for gears operated on center distances that are *exactly* standard. The distinction to be made usually involves tolerances on the center distance. Most practical gear designs involve center-distance tolerances that are the accumulated effects of machining tolerances on the center bores and tolerances in bearings (clearances, runout of outer races, etc). Thus, gears all operate with maximum and minimum operating pitch diameters. The most important application occurs on gears operated on nonstandard center distances.

In Fig. 4.5, a pair of gears designed to operate on enlarged center distances is shown. Both the standard and operating pitch circles are shown. It will be noted that the pitch circles are related to the base circle as follows:

$$D = \frac{D_b}{\cos \Phi} \quad \text{standard pitch diameter} \tag{4.12}$$

$$D' = \frac{D_b}{\cos \Phi'} \quad \text{operating pitch diameter} \tag{4.13}$$

---

*Equations (4.8) through (4.13) should use inches for English calculations and millimeters for metric calculations.

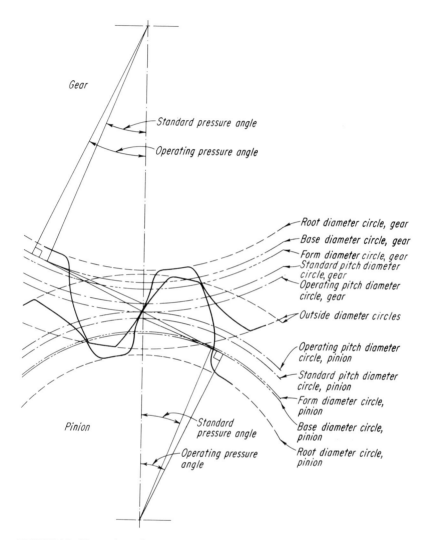

FIGURE 4.5 Nomenclature for gear cut to standard pressure angle and operated at an increased center distance

It will also be noted that the standard pitch circles do not contact each other by the amount that the center distance has been increased from standard.

In worm gearing, it is convenient to use pitch circles. In this case, however, it is common practice to make the pitch circle of the gear go through the teeth at a diameter at which the tooth thickness is equal to the space width. In the case of the worm, the pitch circle also defines a cylinder at which the width of the threads and spaces are equal. It is also good practice to modify the worm teeth slightly to achieve the required backlash. When this is done, the space widths are greater than the thread thicknesses, as measured on the "standard" pitch cylinder, by the amount of backlash introduced. The pitch cylinder also defines the diameter at which the lead angle, as well as the pressure angle, is to be measured.

Zones in Which Involute Gear Teeth Can Exist. Although many gear designs utilize "standard" or "equal-addendum" tooth proportions, it is not always necessary or even desirable to use these proportions. One of the outstanding features of the involute tooth profile is the opportunity it affords for the use of different amounts of addendum and tooth thicknesses on gears of any given pitch and numbers of teeth. These variations can be produced with standard gear tooth generating tools. It is not necessary to buy different cutting or checking tooling for each new value of tooth thickness or addendum if the proper tooth thickness–addendum relationship is maintained in the design.

As discussed in other chapters of this handbook, the limits of true involute action are established by the length of the line of action. It has been shown that the end limits defining the maximum usable portion of the line of action are fixed by the points at which this line becomes tangent to the base circles. These limits, a and b, are shown in Fig. 4.6a. The largest pinion or gear that will have correct gear tooth action is defined by circles that pass through points a and b (see Fig. 4.6b). Therefore, any gear teeth that lie fully within the crosshatched area of Fig. 4.6b will have correct involute action on any portions of the teeth that are not undercut. Undercut limitations are discussed more fully under "Undercut."* One should not infer that the largest usable outside diameters will be used simultaneously on both members on a given gear design.

The equations for maximum usable outside diameter are

$$D_{OG} = 2\left[(C' \sin \Phi')^2 + \left(\frac{D_G}{2} \cos \Phi\right)^2\right]^{1/2} \quad \text{max} \quad (4.14)\dagger$$

$$D_{OP} = 2\left[(C' \sin \Phi')^2 + \left(\frac{D_P}{2} \cos \Phi\right)^2\right]^{1/2} \quad \text{min} \quad (4.15)$$

where C' = operating center distance
 Φ = cutting or standard pressure angle
 Φ' = operating pressure angle
 D_G = pitch diameter of gear
 D_P = pitch diameter of pinion

The operating pitch circles for the gear ratio (2:1) and center distance chosen for this illustration are shown in place in Fig. 4.6c. If this center distance is "standard" for the number of teeth and pitch, the operating pitch circles shown would also be the "standard" pitch circles.

To transmit uniform angular motion, a series of equally spaced involute curves are arranged to act on each other. These are shown in place in Fig. 4.6d. On both the pinion and the gear member, the involute curves originate at the base circles and theoretically can go on forever. The more practical lengths are suggested by solid lines. The spacing of the involutes of both members measured on the base circles must be equal, and the interval chosen is called the base pitch. The base pitch of the gear or pinion times the number of teeth in the member must exactly equal the circumference of the base circles.

---
*See page 4.16.
†Equations (4.14) through (4.18) should use inches for English calculations and millimeters for metric calculations.

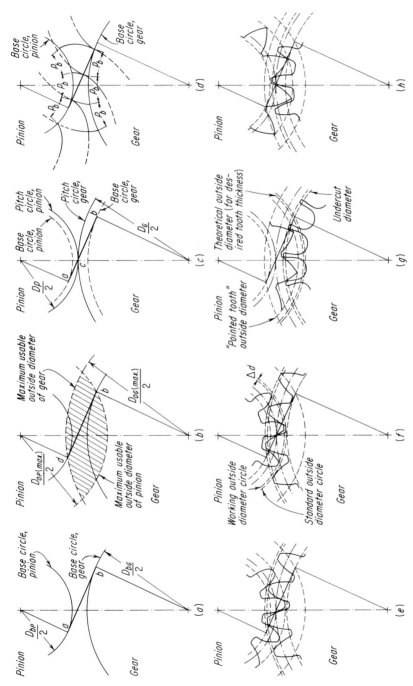

FIGURE 4.6 Study of involute tooth development. (a) Maximum usable length of line of action; (b) maximum zone in which conjugate action can take place; (c) operating pitch circles; (d) development of involute profiles; (e) standard addendum tooth proportions; (f) short- (pinion) addendum tooth proportions; (g) long- (pinion) addendum (abnormal) tooth proportions; (h) long- (pinion) addendum (normal) tooth proportions

If similar involute curves of opposite hand are drawn for both base circles, the familiar gear teeth are achieved. It is customary to measure the distance from one involute curve to the next along the standard pitch circle. This distance is called the *circular pitch*. It is also customary to make "standard" tooth proportions with tooth thicknesses equal to one-half the circular pitch. The standard addendum used for the gearing systems shown in this chapter is equal to the circular pitch divided by π.

Figure 4.6e shows "standard" teeth developed on the base circles of Figs. 4.6a through d. It will be noted that the addendum of the 12-tooth pinion is equal to that of the gear and that the outside diameters thus established do not reach out to the maximum values established by the line-of-action limits.

In modern gear-cutting practice, the tooth thickness of gears as measured on the "standard" pitch circle is established by the depth to which the generating-type cutter (usually hob or shaper cutter) is fed, relative to the "standard" pitch diameter. In order to obtain a correct whole depth of tooth, the outside diameters of the gear blanks are made larger or smaller than "standard" by twice the amount that the cutter will be fed in or held out relative to the normal or standard amount that would be used for standard gears. If the cutter is to be held out a distance Δc, the outside diameter of the blank is made $2\Delta c$ larger than standard and the resultant gear is said to be long addendum. Figure 4.6f shows an example of a gear in which the cutter was held out sufficiently so that the outside diameter of the gear was equal to the diameter of the maximum usable outside-diameter circle (see also Fig. 4.6b). This is called a long-addendum gear. The cutter was fed into the pinion an equal amount, making it short addendum. Reasons for making pinions short addendum are discussed under "Speed-Increasing Drive," and the effects of the undercutting produced are discussed under "Undercut."

In order to avoid undercut, or to achieve a more equal balance in tooth strength, it is customary to make the pinion addendum long and that of the gear short. If an attempt is made to design a pinion with the maximum usable outside diameter as established by the line of action (see Fig. 4.6b), difficulties may arise. In the example shown having 12 and 24 teeth, it is not possible to generate a pinion having such an outside diameter.

In Fig. 4.6g is shown the tooth form resulting from such an attempt. The pinion blank was turned to the maximum usable outside diameter, and the cutter was held out an equivalent distance Δc. As a result of the generating action, the sides of the teeth are involute curves "crossed over" at a diameter smaller than the desired outside diameter. Thus the teeth are pointed at the outside diameter and also the whole depth is less than anticipated. This effect is less serious in gears having larger numbers of teeth.

The gear tooth that results from this extreme modification is badly undercut. This effect would not have been so great if the gear had more teeth.

The amount of long and short addendum that may be applied to each member of a gear mesh is limited by the three following considerations:

1. The length of the usable portion of the line of action will form a maximum limit on the outside diameter of a gear. Diameters in excess of this will not provide additional tooth contact area, since there is no involute portion on the mating gear that can contact this area.

2. The diameter at which the teeth become pointed limits the actual or effective outside diameter.

3. Undercutting may limit the short-addendum gear. The undercut diameter should be always less than the form diameter.

GEAR TOOTH DESIGN

These three considerations show the extreme limits that bound gear tooth modifications. The designer should not infer that it is necessary to approach these limits in any given gear design. In the treatment of gear tooth modifications, some reasons for making given tooth modifications are considered. In each case, however, the designer must be sure that the amount actually used does not exceed the limits discussed here.

Pointed Teeth. The previous sections have shown how gear teeth generated by a specific basic rack can have different tooth thicknesses, and that the outside diameters of such gears are altered from standard as a function of the change in tooth thickness. In practice, these tooth profiles are achieved by "feeding in" or "holding out" the gear-generating tools. It is customary to feed a specific cutter to a definite depth into the gear blank but to achieve long-addendum designs, for instance, by starting with a blank which has a greater than standard outside diameter. The cutter, when working at its full cutting depth, will still be held out from the standard N/P_d pitch circle.

Teeth generated thicker than standard will have tips of a width less than standard, since the cutter must be "held out." For any given number of teeth, the tooth thickness can be increased such that the tip will become pointed at the outside-diameter circle. In Fig. 4.7 are shown four gears all having the same number of teeth, diametral pitch, and pressure angle. In Fig. 4.7a, the cutter shown by a rack has been fed to the standard depth. The outside diameter of this gear is standard, $(N + 2)P_d$. Note the tooth thickness at the tip. In Fig. 4.7b, the outside diameter was somewhat enlarged and the cutter fed in the standard whole-depth distance starting from the enlarged outside diameter. This results in the thicker tooth (which would operate correctly with a

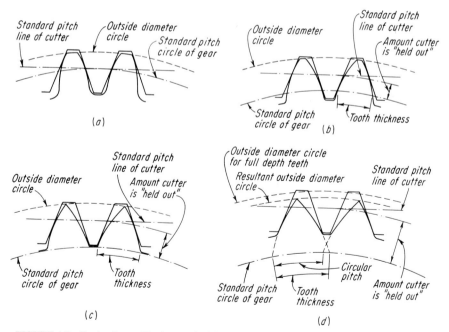

FIGURE 4.7 Cutting long-addendum teeth. (*a*) Standard tooth; (*b*) long-addendum tooth; (*c*) long-addendum (pointed) tooth; (*d*) long-addendum (overenlarged) tooth

standard gear on an enlarged center distance) and a somewhat thinner tip. In Fig. 4.7c, the outside diameter has been established at a maximum value at which it is still possible to achieve a standard whole-depth tooth; the tooth thus generated is pointed. Figure 4.7d shows what happens if the maximum outside diameter and tooth thickness are exceeded. The resulting tooth does not have the correct whole depth because the involute curves cross over below the expected outside diameter. This tooth is similar to the one shown on the pinion member in Fig. 4.6g.

The amount that the outside diameter of a gear is to be modified is usually a function of the tooth thickness desired. Equation (4.18) gives the relationship usually employed.

The maximum amount that the tooth thickness of a gear can be increased over standard to just achieve a pointed tooth can be calculated from the simultaneous* solution of Eqs. (4.16) and (4.17).

$$\Delta D_O \text{ max} = N \left(\frac{\cos \Phi}{\cos \Phi''} - 1 \right) - 2 \qquad (4.16)$$

$$\Delta D_O \text{ max} = \frac{N (\text{inv } \Phi'' - \text{inv } \Phi) - \frac{\pi}{2}}{\tan \Phi} \qquad (4.17)$$

The equivalent amount that the tooth thickness must be increased above the standard to achieve this increase in outside diameter is given by Eq. (4.18).

$$\Delta T = d(\text{inv } \Phi'' - \text{inv } \Phi) - T \qquad (4.18)$$

where D_o max is the maximum outside diameter at which a tooth having full working depth will come to a point, as follows:

$$D_o = \frac{N + 2}{P} + \Delta D_O$$

where T = tooth thickness at standard pitch diameter
 Φ = standard pressure angle of hob or rack
 Φ'' = pressure angle at tip of tooth

Undercut. An undercut tooth is one in which a portion of the profile in the active zone has been removed by secondary cutting action. Under certain conditions, the path swept out by the tip of a generating-type cutter will intersect the involute active profile at a diameter greater than the limit diameter. Such a tooth is said to be undercut since no contact with the mating gear can take place between the undercut and the limit circles.

The amount of undercut will depend on the type of tool used to cut the gears; in general, a hob will undercut a gear to a greater degree than will a circular, shaper-type cutter. Form-type cutters do not normally produce undercut. In calculating the undercut diameter, it is customary to use an equation based on the type of tooling that will produce the greatest undercut, usually a hob. This is done in order to give the shop

---

*Equations (4.16) and (4.17) are solved by assuming a series of values for Φ''. Curves are plotted for ΔD_O max vs. Φ''. The point where the curves cross indicate a *simultaneous* solution for the two equations. For instance, a 20-tooth pinion of 1 pitch having a 20° pressure angle and cut with a standard cutter has a crossing point for Φ'' of 39.75° and a ΔD_O max of 2.44 in. It should be kept in mind that these equations solve problems for one pitch only (divide by pitch to adjust answer for other pitches) and make no allowance for backlash. In the metric system the solution is in millimeters and for one module. (For other modules, multiply the answer by the module.)

the greatest freedom of choice in selecting tools. If the most adverse choice will produce a satisfactory tooth profile, then all other types of cutter that might be chosen by the shop should prove to be satisfactory, except in those cases where undercut is intended to provide clearance for the tips of the mating teeth.

In general, spur gears of 20° pressure angle having 18 or more teeth and made to standard- or long-addendum tooth proportions will not be undercut. In each of the "standard systems" for different types of gears, the minimum number of teeth recommended for each pressure angle is usually based on undercut.

Helical gears can usually be made with fewer teeth than can spur gears without getting into problems of undercut.

Figure 4.8 shows three gears having the same numbers of teeth, each produced by generating-type cutters. Figure 4.8a is the tooth form of a gear having a standard addendum. Figure 4.8b shows the shape of the tooth that results when the tip of the rack-type cutter is operating at a depth exactly passing through the point of intersection of the line of action and the base circle. Figure 4.8c shows the shape of a tooth which results when the cutter works at a depth somewhat below the intersection of the line of action and the base circle.

Undercutting is a product of generating-type cutters. Gears cut by forming-type cutters do not usually have undercut. In certain cases, undercut may ensure tip clearance in mating gears. Racks operating with gears having small numbers of teeth may show tip interference unless the pinions were generated with special shaper-type cutters or hobs.

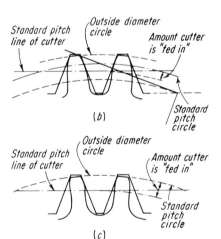

FIGURE 4.8 Cutting short-addendum teeth. (a) Standard tooth; (b) short-addendum tooth; (c) short-addendum (abnormal) tooth

4.1.3. Long- and Short-Addendum Gear Design

The previous article shows the geometric limitations on the amount that gears can be made long or short addendum. This section indicates cases in which long and short addendum should be considered.

Modification of the addendum of the pinion, and in most cases the gear member, is recommended for gears serving the following applications:

1. Meshes in which the pinion has a small number of teeth.
2. Meshes operating on nonstandard center distances because of limitations on ratio or center distances.
3. Meshes of speed-increasing drives.

4. Meshes designed to carry maximum power for the given weight allowance. (This type of gearing is usually designed to achieve the best balance in strength, wear, specific sliding, pitting, or scoring.)
5. Meshes in which an absolute minimum of energy loss through friction is to be achieved.

Addendum Modification for Gears Having Small Numbers of Teeth. Undercutting* is one of the most serious problems occurring in gearing having small numbers of teeth. The amount that gears with small numbers of teeth should be enlarged (made long addendum) to avoid undercut, has been standardized by the AGMA. The values of modification are based on the use of a hob or rack-type cutter and as a result are more than adequate for gears cut with circular, shaper-type cutters. The values of addendum modification recommended for each number of teeth are shown in Table 4.9.

When two gears, each containing a small number of teeth, must be operated together, it may be necessary to make both members long addendum to avoid undercut. Such teeth will have a tooth thickness which is larger than standard and which will necessitate the use of a greater-than-standard center distance for the pair.

Modifications to avoid undercut fall into three categories.

1. If both members have fewer teeth than the number critical to avoid undercut, increase the center distance so that the operating pressure angle is increased. Then get appropriate values of the addendum and whole depth for the pinion and for the gear.
2. If the pinion member has fewer teeth than the critical number, and the mating gear has considerably more, the usual practice is to decrease the addendum of the gear by the amount proportional to the amount that the pinion is increased. This results in a pair of gears free from undercut that will operate on a standard center distance.
3. If the pinion member has fewer teeth than the critical number, and the gear just slightly more, a combination of practices 1 and 2 above may be employed.

An alternate is to increase the pinion addendum by the required amount and increase the center distance by an equivalent amount to make it possible to use a standard gear.

Speed-Increasing Drives. Most gear trains are speed reducing (torque increasing), and most data on gear tooth proportions are based on the requirements of this type of gear application. The kinematics of speed-increasing drives is somewhat different, and as a result, special tooth proportions should be considered for this type of gear application. As in the case of conventional drives, the problems to be discussed here are most serious in meshes involving small numbers of teeth.

The first of these problems involves the tendency of the tip edge of the pinion tooth to gouge into the flank of the driving gear tooth. This gouging can come about as a result of spacing errors in the teeth of either member which allow the flank of the gear tooth to arrive at the theoretical contact point on the line of action before the pinion does. The pinion tooth has to deflect to get into the right position or it will gouge off a sliver of gear tooth side. If the gears are highly loaded, the unloaded pinion tooth entering the mesh will be out of position (lagging) since it is not deflected. The gear tooth is in effect slightly ahead of where it should be. The result is the same as if the gear tooth had an angular position error. The bearing and lubricating problems at the

---

*See Chap. 5, Sec. 5.9, for further information on undercutting.

beginning point of contact are particularly bad. The edge of the pinion tooth tends to act as a scraper and remove any lubricating film that may be present, for some distance along the flank of the tooth.

One possible solution to this problem is to give the tip of the pinion tooth a moderate amount of tip relief. This provides a sort of sled-runner condition which is easier to lubricate and which helps the pinion tooth find the proper position relative to the gear tooth with less impact.

A better solution, which may also be combined with the tip modification, is to modify the tooth thicknesses and addendums so as to get as much of the gear tooth contact zone in the arc of recess as possible.

The action of gear teeth in the arcs of approach and recess may be likened to a boy pushing or dragging a stick down the street. A gear tooth driving a pinion in the arc of approach is like the case where the boy pushes the stick along ahead of him. It tends to gouge into the ground. The gear tooth action in the arc of recess is like the case where the boy drags the stick along behind him. It has no tendency to gouge in; it rides up over bumps and is easier to pull. The relative gear efficiencies in each case are discussed under the topic "Low-Friction Gearing" later in this article.

Power Drives (Optimum Design). As shown in other chapters, gears fail in one or more of the following ways: actual breaking of the teeth, pitting, scoring, or by wear. In the case of drives made with gears of standard tooth proportions and similar metallurgy, the weakest member is the pinion, and if tooth breakage does occur, it is generally in the pinion. This is a result of the weaker shape of the pinion tooth, as well as the larger number of fatigue cycles that it accumulates. This problem can be relieved to a considerable degree by making the pinion somewhat longer addendum and, in so doing, increasing the thickness of the teeth and also improving their shapes. If a standard center distance is to be maintained, the gear addendum is reduced a proportional amount. If the proper values are chosen, the pinion tooth strength will be increased and the gear tooth strength somewhat reduced, which will result in almost equal gear and pinion tooth strengths. This will result in an overall increase in the strength of the gear pair.

Several authorities have suggested addendum modifications which will balance scoring, specific sliding, and tooth strength. Unfortunately, each balance results in different tooth proportions so that the designer has to use proportions that will balance only one feature or else use proportions that are a compromise. Table 4.10 gives values that are such a compromise.

Experimental data seem to indicate that a pair that is corrected to the degree that seems to be indicated by tooth layouts or by calculation for balanced tooth strength will usually result in an overcorrection to the pinion member. The gear is not so strong as form factors seem to indicate. Notch sensitivity in higher hardness ranges seems to be a problem, especially if the gear is to experience a great number of cycles of loading.

In small numbers of teeth, the correction required to avoid undercut on gears operated on standard center distances is excessive, in many cases, in respect to equal tooth strength. An overcorrection in pinion tooth thickness can lead to an excessive tendency to score. As a result, the values of addendum recommended in Table 4.10 represent a compromise among balanced strength, sliding, and scoring.

Gears with teeth finer than about 20 diametral pitch (generally) cannot score, since the tooth is not strong enough to support a scoring load; therefore, the values for addendum increase in fine-pitch gears are somewhat larger than the values for coarse-pitch power gearing.

Low-Friction Gearing. In cases where a speed-increasing gear train is to transmit power or motion with the least possible loss of energy, the selection of the tooth

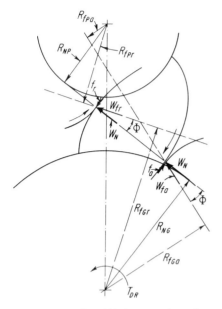

FIGURE 4.9 Effect of friction on tooth reactions

proportions is of considerable importance. The sliding should be kept as low as possible, and as much of the tooth action should be put into the arc of recess as possible.

Figure 4.9 shows two involute curves (tooth profiles) in contact at two different points along the line of action. The direction in which the driven pinion tooth slides along the driving gear tooth is shown by the arrows. This example is a speed-increasing drive which is the most sensitive to friction between the teeth. At the pitch point (where the line of centers crosses the line of action) there is no sliding, and it is at this point that the direction of relative sliding of one tooth on the other changes.

The forces shown in Fig. 4.9 are those acting on the driven pinion. The subscripts a are the values considered in the arc of approach and r are those considered in the arc of recess. The normal driving force W_N is the force that occurs at the pitch point and, if there were no friction at the point of gear tooth contact, would be the force at all other points of contact along the line of action. Since there is friction, the friction vectors f_a and f_r oppose the sliding of the gear teeth in the arcs of approach and recess. Note the change in direction due to the change in direction of sliding. The angle of friction is Φ and is assumed to be the same in both cases. The torque exerted by the shaft driving the driving gear T_{DR} manifests itself in arc of approach as

$$T_{DRa} = W_N \times R_{NG} \quad \text{if no friction}$$
$$T_{DRa} = W_{fa} \times R_{fGa} \quad \text{if friction is assumed}$$

and in the arc of recess

$$T_{DR_r} = W_N \times R_{NG} \quad \text{if no friction}$$
$$T_{DR_r} = W_{f_r} \times R_{fGr} \quad \text{if friction is assumed}$$

The resisting moments are, in the arc of approach,

$$T_{DN_a} = W_N \times R_{NP} \quad \text{if no friction}$$
$$T_{DN_a} = W_{fa} \times R_{fP_a} \quad \text{if friction is assumed}$$

and the corresponding moments are, in the arc of recess

$$T_{DN_r} = W_N \times R_{NP} \quad \text{if no friction}$$
$$T_{DN_r} = W_{f_r} \times R_{fP_r} \quad \text{if friction is assumed}$$

Note that, in all cases above, single tooth contact is assumed.

Efficiency is output divided by input and in this case is the torque that would appear

GEAR TOOTH DESIGN

on the driven shaft when friction losses are considered, compared with the torque that would result if no losses occurred.

Equation (4.19) shows the efficiency of the mesh (single tooth contact) for the contacts occurring in the arc of approach, and Eq. (4.20) shows the efficiency in the arc of recess.

$$E_{\text{approach}} = \frac{R_{fP_a}}{R_{fG_a}} \cdot \frac{R_{NG}}{R_{NP}} \qquad (4.19)*$$

$$E_{\text{recess}} = \frac{R_{fP_r}}{R_{fG_r}} \cdot \frac{R_{NG}}{R_{NP}} \qquad (4.20)$$

In the case of speed-increasing drives, the increased efficiency can have considerable significance; cases have occurred in which, for very high ratios and poor lubrication, the speed increaser actually became self-locking.

4.1.4. Special Design Considerations

Several special considerations should be kept in mind during the evolution of a gear design.

Interchangeability. Two types of interchangeability are related to gearing. The first, and most generally recognized, is part interchangeability. This means that, if a part made to a specific drawing is damaged, a similar part made to the same drawing can be put in its place and can be expected to perform exactly the same quality of service. In order to achieve this kind of interchangeability, all parts must be held to carefully selected tolerances during manufacture, and the geometry of the parts must be clearly defined. The second kind is engineering interchangeability. This type provides the means whereby a large variety of different sizes or number of teeth can be produced with a very limited number of standard tools. Parts so made may or may not be designed to have part interchangeability. The several systems of gear tooth proportions are based on engineering interchangeability. By careful design and skill in the application of each system, it is possible to design gears that will perform in almost any application and that can be made with standard tools. Only in the most exceptional cases will special tools be required. The advantages of gearing designed with engineering interchangeability are the lower tool costs and the ease with which replacement or alternate gears can be designed.

Gears procured through catalogue source are good examples of gears having both part and engineering interchangeability.

Tooth Thickness. The thickness of gear teeth determines the center distance at which they will operate, the backlash that they will have, and, as discussed in previous sections, their basic shape. One of the important calculations made during the design of gear teeth is establishment of tooth thickness.

It is essential to specify the distance from the gear axis at which the desired tooth thickness is to exist. The usual convention is to use the distance that is established by the theoretical pitch circle N/P. Thus, if no other distance is shown, the specified tooth thickness is assumed to lie on the standard pitch circle. In certain cases, the designer

---

*Equations (4.19) through (4.21) can be used in the metric system by using newtons for force and millimeters for distance. This makes the units of torque N·mm.

may wish to specify a thickness, such as the chordal tooth thickness, at a diameter other than the standard pitch diameter. This special diameter should be clearly defined on the gear drawings.

The actual calculation of tooth thickness is usually accomplished by the following procedure.

1. The theoretical tooth thickness is established.
   a. If the gears are of conventional design and are to be operated on standard center distances, the tooth thickness used is one-half the circular pitch.
   b. If special center distances are to be accommodated, Eqs. (4.21) and (4.22) may be used.

$$\cos \Phi_2 = \frac{C}{C_2} \cos \Phi \qquad (4.21)$$

where Φ_2 = operating pressure angle
C = standard center distance
Φ = standard pressure angle
C_2 = operating center distance

$$T_P + T_G = \pi + (\text{inv } \Phi_2 - \text{inv } \Phi)(N_P + N_G) \qquad (4.22)*$$

where T_P = tooth thickness of pinion member (at 1 diametral pitch)
T_G = tooth thickness of gear member (at 1 diametral pitch)
N_P = number of teeth in pinion
N_G = number of teeth in gear

   c. If special tooth thicknesses for tooth strength considerations are required, the total thickness (sum of pinion and gear) must satisfy Eq. (4.22). If standard centers are used the sum of the modified thickness must equal 1 circular pitch.

2. After the theoretical tooth thickness is established, the allowance for backlash is made.
   a. *Backlash* allowance may be shared equally by pinion and gear. In this case the theoretical tooth thickness of each member is reduced by one-half the backlash allowance.
   b. In case of pinions with small numbers of teeth which have been enlarged to avoid problems of undercut, it is customary to take all the backlash allowance on the gear member. This avoids the absurdity of increasing the tooth thickness to avoid undercut and then thinning the teeth to introduce backlash.

   In case *a*

$$T_{\text{actual}} = T_{\text{theoretical}} - \frac{B}{2} \qquad (4.23)$$

where B = backlash allowance

---

*The best way to use this equation is to start with C being the center distance for 1 diametral pitch (or for 1 module in metric dimensions). The C_2 is computed by Eq. (4.21). The tooth thicknesses determined are not at the operating pitch diameters but are at the 1 diametral pitch center distance of C.

In case *b*

$$T_{\text{actual, pinion}} = T_{\text{theoretical}} \qquad (4.24)$$

$$T_{\text{actual, gear}} = T_{\text{theoretical}} - B \qquad (4.25)$$

3. The teeth are lastly given an allowance for machining tolerance. This tolerance gives the machine operator a size or processing tolerance. Usually this is a unilateral tolerance. Chapter 7 deals more fully with the establishment of this value.

It is beyond the scope of this general-purpose book on gear technology to go deep into gear design details. Chapter 8 of Dudley[3] has the following sample problems worked out on pages 8.1 to 8.11.

Problem 1: Drop-Tooth Design, Fixed Center Distance
Problem 2: Fixed Center Distance, Standard Tools
Problem 3: Fixed Center Distance, Adjust Helix Angle Standard Tools
Problem 4: Fixed Design, Standard Tools, Adjust Center Distance for Tooth Proportions

Tooth Profile Modifications. Errors of manufacture, deflections of mountings under load, and deflections of the teeth under load all combine to prevent the attainment of true involute contact in gear meshes. As a result, the teeth do not perform as they are assumed to by theory. Premature contact at the tips or excessive contact pressures at the ends of the teeth give rise to noise and/or gear failures. In order to reduce these causes of excessive tooth loads, profile modification is a usual practice.

Transverse or Profile Modification. It is customary to consider the shape of the individual tooth profile. Actually, an operating gear presents to the mating gear tooth profiles whose shapes are distorted by errors in the profile shape (involute) and in tooth spacing, as well as by deflections due to contact loads acting at various places along the tooth. Ideally, teeth under load should appear to the mating gear to have perfect involute profiles and to have perfect spacing. This would require a tooth profile of such a shape when under no-load conditions as to deflect into a conjugate profile when subjected to a load. Since loads vary, this is not practical. Usual practice is to give tip or root profile modifications to otherwise correct involute profiles that will be distorted by contact loads.

Tip modification usually takes the form of a slight thinning of the tip of the tooth, starting at a point about halfway up the addendum. The amount of this modification is based on the probable accumulated effects of the following.

Allowances for Errors of Gear Manufacture. The correct involute profile may not be achieved as a result of manufacturing tolerances in the cutter used to cut the gear teeth or in the machine guiding the cutter. Errors in spacing may also be introduced by the cutter or machine. The result is that the tips of the teeth will attempt to contact the mating gear too soon or too late. Tip modification produces a sort of sled-runner shape to help guide the teeth into full contact with the least impact.

Allowances for Deflection under Load. Although the teeth of a gear may have a correct profile under static conditions, the loads imposed may deflect the teeth in engagement to such a degree that the teeth that are just entering mesh will not be in relative positions that are correct for smooth engagement. These tooth deflections cause two errors: The actual tooth profiles are not truly conjugate under load and therefore do not transmit uniform angular motion, and the spacing or relative angular placement of the driving teeth relative to the driven teeth is such that smooth tooth engagement cannot take place. High dynamic loads are the result.

Several methods may be used to compensate for these effects. The most usual is to provide tip relief as discussed above. The flanks of the teeth may also be relieved. Tip and flank relief assume that the normal tooth would tend to contact too soon. If the tips of the driven gear are made slightly thin, the tooth will be able to get into a position on the line of contact before contact between that tooth and its driving mate actually occurs. The sled-runner effect of the tip modification will allow the tooth to assume full contact load gradually.

In the case of spur gearing the amount of tip relief should be based on the sum of the allowance for probable tooth-to-tooth spacing errors and for assumed deflection of the teeth already in mesh. Equations (4.26) and (4.27) give good first approximations of the amount of tip relief required. The values obtained by these equations should be modified by experience for the best overall performance.

Modification at first point of contact

$$\text{Modification} = \frac{\text{driving load (lb)} \times 3.5 \times 10^{-7}}{\text{face width (in.)}} \qquad (4.26)$$

Remove stock from tip of driven gear.

Modification at last point of contact

$$\text{Modification} = \frac{\text{driving load (lb)} \times 2.0 \times 10^{-7}}{\text{face width (in.)}} \qquad (4.27)$$

Remove stock from tip of driven gear.

In general, gear teeth that carry a load in excess of 2000 lb per in. of face width for more than 1,000,000 cycles should have modification. Those under 1000 lb per in. of face do not generally require modification.

Bevel gear teeth are often modified to accommodate mounting misalignments, tooth errors, and deflections due to load. The geometry of bevel gear teeth allows profile modifications both along the length of the teeth and from root to tip. In the case of spiral and Zerol* bevel gears cut with face mill-type cutters, the contacting faces of the gear and pinion teeth can be easily made to a slightly different radius of curvature on each member. This is equivalent to crowning of spur gears and is called mismatch.

The tips of the teeth may be given relief, and in some cases the root of the flank is relieved (undercut). This is most commonly done on spiral bevel gears and hypoid gears that are to be lapped. It is done by cutters having a special protuberance called Toprem† cutters.

Mismatch is calculated into the machine settings and is therefore beyond the scope of this chapter.

---

*Zerol is a trademark registered by the Gleason Works, Rochester, NY.
†Toprem is a trademark registered by the Gleason Works, Rochester, NY.

GEAR TOOTH DESIGN

Axial Modifications. In general, it is expected that a gear tooth will carry its driving load across the full face width (if a spur gear).

Because of deflections in shafts, bearings, or mountings, when under load, or because of errors in the manufacturing of these parts or in the gears, the teeth may not be quite parallel, and end loading may result. Since a heavy load on the end of a gear tooth will often cause it to break off, attempts to avoid loading by relieving the ends of the teeth (crowning) are often made. A gear tooth that has been crowned is slightly thicker at the center section of the tooth than at the ends when measured on the pitch cylinder.

In effect, crowning allows a rocking-chair-like action between the teeth when the shafts deflect into increasingly nonparallel positions. Heavy concentrations of load at the ends of the teeth are avoided. The fact that the whole face cannot act when the shafts are parallel requires that the load imposed upon a set of crowned gears be less than could be carried on a similar noncrowned pair if parallel teeth could be maintained.

In general, the ends of crowned gears are made from 0.0005 to 0.002 in. thinner at the ends as compared with the middle, when measured on the circular-arc tooth thickness.

Spur and helical gearing is crowned by means of special attachments on the gear tooth finishing machines.

Bevel gearing is crowned, in effect, by the shape of the cutting tools and the way in which they are driven in relation to the teeth. Coniflex* teeth are straight bevel teeth cut on special generating machines which produce lengthwise crowning. Spiral and Zerol bevel gears and hypoid gears are given the effect of crowning by the use of a different radius of lengthwise curvature on the convex and the concave sides of the teeth.

Throated worm gears can be given the effect of crowning by the use of slightly oversize hobs.

Face gears are given the effect of crowning by the selection of a cutter having one or more teeth than the mating gear.

Root Fillets. The shape and minimum radius of curvature that the root fillet of a tooth will have depends on the type and design of the cutting tool used to produce the gear tooth. The shape of the root fillet, as well as its radius, and the smoothness with which it blends into the root land and active profile of a gear tooth can have a profound effect on the fatigue strength of the finished gear. The radius of the fillet at the critical cross section of the tooth is controlled on drawings by specifying the minimum acceptable fillet radius. The point of minimum radius occurs almost adjacent to the root land in the case of gears cut by hobs. An equation giving a reasonable evaluation of the minimum radius for teeth cut by hobs is

$$r_f = 0.7 \left[r_T + \frac{(h_t - a - r_T)^2}{\dfrac{d}{2\cos 2} + h_t - (a + r_T)} \right] \quad (4.28)\dagger$$

---

*Coniflex is a trademark registered by the Gleason Works, Rochester, N.Y.
†Equation (4.28) is valid for the metric system by using millimeters instead of inches.

where r_f = minimum calculated fillet radius produced by hobbing or generating grinding
r_T = edge radius of generating rack, hob, or grinding wheel
a = addendum of gear
h_t = whole depth of gear
d = pitch diameter of gear
ψ = helix angle (use 0° for spur gears)

The edge radius specified for the generating tool will, in general, depend on the service the gear is to perform or on special manufacturing considerations. Table 4.16 shows suggested values of edge radius for various gear applications.

Other types of manufacturing tools can be designed to produce the minimum fillet radius as obtained from Eq. (4.28). The shape of the fillet will be somewhat different, however.

A common method of checking the minimum radius of the coarser-pitch gears is to lay a pin in the fillet zone and note that the contact is along a single line.

The constant 0.7 is to allow a reasonable working tolerance to the manufacturer of the tools. Edge breakdown of hobs and cutters will tend to increase the radius produced. This is particularly true of hobs and cutters for very-fine-pitch gears.

Effective Outside Diameter. It is customary to consider the outside diameter of a gear as the outer boundary of the active profile of the tooth. In several cases this approximation is not good enough.

1. In very-fine-pitch gears that have been burr brushed, the tip round may be quite large in proportion to the size of the teeth even though it is only a few thousandths in radius by actual measurement. Since no part of this radius can properly contact the mating tooth, the outside diameter is, from the standpoint of conjugate action, limited to the diameter where the tip radius starts. *Effective outside diameter* should be used instead of *outside diameter* in calculations of contact ratio.

2. In some gear meshes in which the pinion member contains a small number of teeth, the tips of gear teeth may be found to be extending into the pinion spaces to a depth greater than that bounded by the line of action. In Fig. 4.6b the dimension $D_{oP}/2$ is the maximum effective diameter of the pinion. The actual outside diameter of the pinion may exceed this value, however. In calculating contact ratio, the effective diameter as limited by the pinion base circle and center distance should be used.

Width of Tip of Tooth. The tooth thickness at the tip of the tooth is a convenient index of the quality of a gear design. For most power gearing applications, the thickness of the gear tooth should not be more than 1½ to 2 times that of the pinion tooth at the tip.

If the tooth thickness (arc), at the pitch diameter (standard) is known, the following equation will give its thickness at the tip:

$$t_o = D_o \left(\frac{t}{D} + \text{inv } \Phi - \text{inv } \Phi_o \right) \tag{4.29}$$

$$\cos \Phi_o = \frac{D \cos \Phi}{D_o} \tag{4.30}$$

where t_o = thickness at outside diameter D_o
 D_o = outside diameter of gear, or diameter where tooth thickness is wanted
 t = arc tooth thickness at D, or at known diameter
 D = standard pitch diameter, or diameter where t is known
 Φ = standard pressure angle, or pressure angle where t is known
 Φ_o = pressure angle at outside diameter or at diameter where tooth thickness is wanted

Pointed Tooth Diameter. An independent method of checking the quality of a long- and short-addendum gear design is to calculate the diameter at which the teeth would come to a point. If this value is smaller than the value of outside diameter chosen by other means, the design should be recalculated. Equations (4.31) and (4.32) provide one method of calculating pointed tooth diameters.

$$\operatorname{inv} \Phi_{o(P)} = \frac{t}{D} + \operatorname{inv} \Phi \qquad (4.31)$$

$$D_{o(P)} = \frac{D \cos \Phi}{\cos \Phi_{o(P)}} \qquad (4.32)$$

where $\Phi_o(P)$ equals the pressure angle at pointed tooth diameter, and $D_{o(P)}$ is the pointed tooth diameter. See Eq. (4.30) for other symbols.

Backlash. In general, backlash is the lost motion between mating gear teeth. It may be measured along the line of action or on the pitch cylinders of the gears (transverse backlash) and, in the case of helical gears, normal to the teeth.

In a set of meshing gears, the backlash that exists is the result of the actual center distance at which the gears operate, and the thickness of the teeth. Changes in temperature, which may cause differential expansion of the gears and mountings, can produce appreciable changes in backlash.

When establishing the backlash that a set of gears will require, the following should be considered:

1. The minimum and maximum center distance. These values are the result of tolerance buildups in the distance between the bores supporting the bearings on which the gears are mounted as well as the basic or design values. Antifriction bearings, for example, have runout eccentricity between the outside diameter and the outer ball path and runout between the bore and the inner ball path. As the shaft and the inner race of the bearing rotate, and the outer race creeps in the housing, the center distance will vary by the amount of the eccentricities of these bearing elements.

2. The thickness of the teeth, as measured at a fixed distance from the center on which the gear rotates, will vary because of gear runout. Also the workman cutting the gear is given a tooth thickness tolerance to work to, which introduces tooth thickness variations from one gear to the next.

The minimum backlash will occur when all the tolerances react all at the same time to give the shortest center distance and the thickest teeth with the high points of gear runout. The maximum backlash will occur when all the tolerances move in the opposite direction.

Design backlash is incorporated into the mesh to ensure that contact will not occur on the nondriving sides of the gear teeth. Although backlash may be introduced by increasing the center distance, it is usually introduced by thinning the teeth. The minimum value should be at least sufficient to accommodate a lubricating film on the teeth.

Sometimes a statistical approach is used, since there are a sufficient number of tolerances involved; thus the design backlash introduced may not be numerically as large as the possible adverse buildup of tolerances. This approach is particularly handy in instrument gearing, since the maximum backlash allowable usually has to be held to a minimum.

By definition, backlash cannot exist in a single gear. Backlash is a function of the actual center distance on which the gears are operated, and the actual thicknesses of the teeth of each gear.

It is customary to use generally recognized values of center-distance tolerance and gear tooth tolerances for power gearing. If this is done, and the calculated tooth thicknesses are reduced by the amounts of design backlash as shown in Table 4.11, satisfactory gears should result. In the case of instrument gearing, either the values in Table 4.11 may be used or the tooth thickness actually required and the resulting maximum backlash may be calculated.

The backlash that is measured in gears under actual operation will in all probability be considerably larger than the values given in Table 4.11, since these values are for design and do not include a correction for normal machining tolerances.

Limit and Form Diameters. These dimensions, shown in Fig. 4.2, are the distances from the axis of spur, helical, and internal gears at which the active profiles of the teeth begin. Form diameter is the lowest point (spur and helical gearing) at which the mating tooth can contact the active profile (see Fig. 4.5).

4.2 STANDARD SYSTEMS OF GEAR TOOTH PROPORTIONS

A standard or system of gear tooth proportions provides a means of achieving engineering interchangeability for gears of all numbers of teeth of a given pitch and pressure angle. Because of the large variety of tooth proportions that are possible, it has been found desirable to standardize on a limited number of tooth systems. These systems specify the various relationships among tooth thickness, addendum, working depth, and pressure angle.

The American Gear Manufacturers Association and the American Standards Association have provided standard systems for various types of gear teeth. When undertaking the design of new equipment, designers should be sure that they are working with the latest standards. The earlier systems are shown in the following sections to aid designers confronted with replacement gear problems. Designers should also be aware of the limitations of each of the systems. These are clearly stated in the discussion of each system. When a design falls outside these limitations, designers will have to employ tooth modifications required by the design. Since these standards are not intended to be absolutely rigid, designers are at liberty to modify the tooth proportions within limits to achieve an optimum design. Previously, information was provided to indicate the more common tooth modifications and the ends achieved by these modifications. As much of the standard system as is applicable should be used. When departing from the proportions given in the standard, consideration should be given to the reasons giving rise to these departures and also the consequences. In some cases,

GEAR TOOTH DESIGN   4.29

departures will require special tooling, whereas others, if properly handled, will not. The following sections provide the basic data covered by most of the systems in present-day use. For additional information about these systems, reference should be made to the original standards listed. In each of the following systems, the tooth proportions are shown in terms of the basic rack of that system. In each case, all gears designed to the basic tooth proportions will have engineering interchangeability.

4.2.1 Standard System for Spur Gears

The following data are based on the information contained in AGMA standard 207.05, "20-Degree Involute Fine-Pitch System for Spur and Helical Gears"[4] and AGMA standards 201.02 and 201.02A, "Tooth Proportions for Coarse-Pitch Involute Spur Gears."[5]

The standards just mentioned are typical of standards used in the time period 1960–1990. As this text is being written some revisions are being made to the AGMA standards and new standards can be expected in the 1990s. Also, AGMA has changed its numbering system for standards. In the opinion of the authors of this book, the older data should be maintained for reference and for instruction in the many things that need to be considered when designing the shape of gear teeth. Then, as a designer works on a job, appropriate decisions can be made.

For instance, there may be a new standard that is mandatory for an application—or designers may be free to develop their own standard for a product application. Of course, those making their own standard should be aware of the things covered in this part of Chap. 4—to avoid making mistakes in tooth design.

Limitations in Use of Standard Tables. Caution should be exercised in using the data contained in Table 4.1. The items shown apply only to gears that meet the following requirements:

1. Standard-addendum gears must exceed the minimum numbers of teeth shown in Table 4.3.
2. Long- and short-addendum designs, as derived from Tables 4.1, 4.5, and 4.8, are to be used for *speed-reducing drives only.*

The tooth proportions that result from the data shown in this section or from the application of the original standards will be suitable for most speed-reducing (torque-increasing) applications. All gears, including pinions with small numbers of teeth, designed in accordance with the procedure shown will be free from undercut. In order to avoid the problems of undercut in pinions having fewer than the minimum standard number of teeth, each system shows proportions for long- and short-addendum teeth. Gears designed with long- and short-addendum teeth cannot be operated interchangeable on standard centers. In general, such gears should be designed to operate only as pairs. These gears can be cut with the same generating-type cutters and checked with the same equipment as standard-addendum gears. The proportions of long and short addendum shown in Table 4.9 are based on avoiding undercut. Such teeth will not have the optimum balance of strength and wear. Slightly different proportions can be used to achieve equal sliding balanced strength or a reduction in the tendency to score. Table 4.10 shows tooth proportions that have a good balance of strength and a minimum tendency to score.

Standard Tooth Forms That Have Become Obsolete. Because industry design standards are continuously reviewed by the sponsoring organizations to ensure that the standards embody the most modern technology, there now exists a group of obsolete tooth form standards. On occasion, a designer is confronted with a situation in which a replacement gear must be made to mesh in a gear train conforming to one of these earlier standards.

Table 4.15 shows basic data for some of the obsolete standards. The use of this data for new designs is not recommended.

Brown and Sharpe System. This system, see Table 4.15, was developed by the Brown and Sharpe Company to replace the cycloidal tooth system. It was therefore given similar tooth proportions. It was intended to be cut by form-milling cutters. The departure from the true involute curve in this system is made to avoid the problems of undercut in pinions having small numbers of teeth. Backlash is achieved by feeding the cutter deeper than standard.

The American Gear Manufacturers Association adopted the principal features of this system in 1932 as the "composite system." Present-day form-milling cutters are normally made to this system.

AGMA $14\frac{1}{2}°$ Composite System. This system, based on the Brown and Sharpe system, was standardized in 1932 as an AGMA standard. It is interchangeable with the B and S system. It has been made obsolete by AGMA standard 201.02.[5] Replacement parts may be designed from the data in Table 4.15, which are extracted from AGMA 201.02A.[5]

Fellows 20° Stub Tooth System. In order to achieve a stronger tooth form for special drives, the Fellows Gear Shaper Company developed a stub tooth system in 1898. This system avoided the problem of tooth interference by the combined means of higher pressure angle and smaller values of addendum and dedendum. The pitch was specified as a combination of two standard diametral pitches; thus 10/12 (read ten-twelve). The circular pitch, pitch diameter, and tooth thickness are based on the first number, that is, 10, and are the same as for a standard 10-diametral-pitch gear. The addendum, dedendum, and clearance, however, are based on the second number (12) and are the same as for a 12-diametral-pitch gear.

AGMA $14\frac{1}{2}$ Full-Depth System. In this system, the tooth proportions are identical with those of the 14½ composite system. The sides of the rack teeth are straight lines and therefore produce involute gear tooth profiles in the generating process. The minimum number of teeth to obtain full tooth action is 31 unless the teeth are modified.

Cycloidal Tooth Profiles. The cycloidal tooth profile is no longer used for any types of gears except clockwork and certain types of timer gears. A combination of involute and cycloidal tooth profile is found in the now-obsolete "composite system." *Clockwork gearing* is based on the cycloid but has been greatly modified for practical reasons.

The cycloidal tooth is derived from the trace of a point on a circle (called the describing circle) rolling without slippage on the pitch circle of the proposed gear. The addendum portions of the tooth are the trace of a point on the describing circle rolling on the outside of the pitch circle. This trace is an epicycloid. The dedendum portion of the gear tooth is formed by a trace of a point on the describing circle rolling on the inside of the pitch circle. This trace is a hypocycloid.

Interchangeable systems of gears must have describing circles that are identical in diameter, and teeth that have the same circular pitch. The faces and flanks of the teeth must be generated by describing circles of the same size. The tooth proportions for

GEAR TOOTH DESIGN					4.31

cycloidal teeth were made to the same proportions as were adopted by the B and S system for gears.

The cycloidal system does not have a standard pressure angle. The operating pressure angle varies from zero at the pitch line to a maximum at the tips of the teeth.

In order to achieve correct meshing, the gears must be operated on centers that will maintain the theoretical pitch circles in exact contact. This was one of the major disadvantages of the cycloidal tooth form.

If the diameter of the describing circles were made equal to the radius of the pitch circle of the smallest pinion to be used in the system, the flanks of the teeth would be radial. This member, in effect, establishes the describing circle diameters of each system. In general, the systems for industrial gears were based on pinions of 12 or 15 teeth.

Clockwork and Timer Tooth Profiles. The tooth form of most clockwork and timer gearing involving some type of escapement differs from other types of gearing because of the peculiar requirements imposed by the operating conditions. Two requirements of this type of gearing are outstanding.

1. The gearing must be of the highest possible efficiency.
2. The gearing is speed increasing, i.e., large gears driving small pinions, with high ratios (between 6:1 and 12:1) due to the need for minimizing the number of gear wheels and economizing on space.

As is discussed in Sec. 4.1.3 under "Low-Friction Gearing," the most efficient mode of tooth engagement is found in the arc of recess. In the case of speed-increasing involute profile drives, this means short-addendum pinions. This requirement may be in direct conflict with the need to make the pinion long addendum to avoid undercut. Clockwork gearing also does not have the requirement that it transmit smooth angular motion. A typical watch train will come to a complete stop five times a second. Furthermore, the motion and force are in one direction, minimizing the need for accurate control of backlash.

The modified cycloidal tooth form is the most commonly used form for the going train of clocks, timers, and watches. Experience has shown that the various modifications to this basic tooth have little effect on the performance because of the scale effect of tolerances required.

There are no American standard tooth forms for clockwork gears.

Specific Spur Gear Calculation Procedure. The following directions give a step-by-step procedure for determining spur gear proportions:

1. The application requirements should give the requirements of ratio, input speed, and kind of duty to be performed. The duty may be one of power transmission or of motion transmission.
2. Based on application requirements, decide what pressure angle to use and plan to use a standard system (if possible) (see Table 4.1).
3. Pick approximate number of pinion teeth.
   a. Consult Table 4.2 for general information.
   b. Check Tables 4.3 and 4.4 for undercut limitations.
   c. If power gearing, check Table 4.5 for data on balancing strength and wear capacity.
4. Having approximate number of pinion teeth, determine appropriate center distance and face width. If power gearing, estimate size from data in Secs. 11.3 and 11.4.

TABLE 4.1 Basic Tooth Proportions of Spur Gears, AGMA and ASA Standard Systems

(Tooth proportions for fully interchangeable gears operated on standard center distances)

Symbol	Item	Coarse pitch (coarser than 20P), full depth		Fine pitch (20 and finer), full depth	Explanation No.
ϕ	Pressure angle	20°	25°	20°	1
a	Addendum (basic*)	$\dfrac{1.000}{P}$	$\dfrac{1.000}{P}$	$\dfrac{1.000}{P}$	2
b	Dedendum (min.) (basic†)	$\dfrac{1.250}{P}$	$\dfrac{1.250}{P}$	$\dfrac{1.200}{P} + 0.002''$	3
h_k	Working depth	$\dfrac{2.000}{P}$	$\dfrac{2.000}{P}$	$\dfrac{2.000}{P}$	4
h_t	Whole depth (min.) (basic†)	$\dfrac{2.25}{P}$	$\dfrac{2.25}{P}$	$\dfrac{2.200}{P} + 0.002''$	5
t	Circ. tooth thk. (basic*)	$\dfrac{\pi}{2P}$	$\dfrac{\pi}{2P}$	$\dfrac{1.5708}{P}$	6
r_f	Fillet radius (in basic rack)	$\dfrac{0.300}{P}$	$\dfrac{0.300}{P}$	Not standardized	7
c	Clearance (min.) (basic†)	$\dfrac{0.250}{P}$	$\dfrac{0.250}{P}$	$\dfrac{0.200}{P} + 0.002''$	8
c	Clear., shaved or grd. teeth	$\dfrac{0.350}{P}$	$\dfrac{0.350}{P}$	$\dfrac{0.3500}{P} + 0.002''$	9
N'_P	Min. numbers of teeth:* Pinion	18	12	18	10
	Pair	36	24	....	11
t_o	Min. width of top land	$\dfrac{0.25}{P}$	$\dfrac{0.25}{P}$	Not standardized	12
	Reference standards: AGMA	201.02	201.02	207.04	13
	ASA			B6.19	14

* These values are basic, for equal-addendum gearing. When the gearing is made long- and short-addendum these values will be altered.

† These values are minimum. Shaved or ground teeth should be given proportions suitable to these processes.

‡These explanations start on page 4.36.

5. Based on number of pinion teeth and center distance, determine approximate pitch. Check Table 4.6 for standard pitches, and, if possible, use a standard pitch. Readjust pinion tooth numbers, center distance, and ratio to agree with pitch chosen. See Chap. 2 for basic relations of pitch, center distance, and ratio. Also use Eqs. (4.8), (4.9), (4.10), and (4.11) if special "operating" center distance is used.
6. Determine whole depth of pinion and gear. Use Tables 4.7 and 4.1. Divide tabular value for 1 diametral pitch by actual diametral pitch to get design whole depth.
7. Determine addendum of pinion and gear. Consult Tables 4.8, 4.9, 4.10, and 4.1. Follow these rules:
   a. Use enough long addendum to avoid undercut (Table 4.9).
   b. If a critical power job—speed decreasing—balance addendum for strength (Table 4.10).
   c. If no undercut or power problems, use standard addendum (Table 4.1).
8. Determine operating circular pitch. If standard pitch is used, consult Table 4.12. If

TABLE 4.2 Numbers of Pinion Teeth, Spur Gearing

No. of teeth	Remarks
7-9	Smallest number of teeth recognized by AGMA 207.04 a. Requires long addendum to avoid undercut on all pressure angles b. If 20°, outside diameter should be reduced in proportion to tooth thickness to avoid pointed teeth c. Should be made with 25° pressure angle if feasible d. May result in poor contact ratio in very fine diametral pitches because of accumulation of tolerances e. See Table 4.4 for minimum number of teeth in mating gear f. Subject to high specific sliding and usually have poor wear characteristics
10	Smallest practical number with 20° pressure angle. a. Requires long addendum to avoid undercut if 20° pressure angle or less b. Contact ratio may be critical in very fine pitches c. See Table 4.4 for minimum number of teeth in mating gear
12	Smallest practical number for power gearing of pitches coarser than 16 diametral pitch. a. Requires long addendum to avoid undercut if 20° pressure angle or less b. Smallest number of teeth that can be made "standard" if 25° pressure angle c. About the minimum number of teeth for any good fractional-horsepower gear design where long life is important d. See Table 4.4 for minimum number of teeth in mating gear
15	Used where strength is more important than wear a. Requires long addendum to avoid undercut if 20° pressure angle or less b. See Table 4.4 for minimum number of teeth in mating gear
19	Can be made standard addendum if 20° pressure angle or greater
25	Allows good balance between strength and wear for hard steels. Contact (contact diameter) is well away from critical base-circle region
35	If made of hard steels, strength may be more critical than wear. If made of medium-hard (Rockwell C 30) steels strength and wear about equal
50·	Excellent wear resistance. Favored for high-speed gearing because of quietness. Critical on strength on all but low-hardness pinions

TABLE 4.3 Minimum Number of Pinion Teeth vs. Pressure Angle and Helix Angle Having No Undercut

Helix angle, deg	Min. No. of teeth to avoid undercut			
	Normal pressure angle, ϕ_n			
	$14\tfrac{1}{2}$	20	$22\tfrac{1}{2}$	25
0 (spur gears)	32	17	14	12
5	32	17	14	12
10	31	17	14	12
15	29	16	13	11
20	27	15	12	10
23	25	14	11	10
25	24	13	11	9
30	21	12	10	8
35	18	10	8	7
40	15	8	7	6
45	12	7	5	5

NOTE: Addendum $1/P_d$; whole depth $2.25/P_d$.

TABLE 4.4 Numbers of Teeth in Pinion and Gear vs. Pressure Angle and Center Distance

No. of teeth in pinion	No. of teeth in gear and pressure angle			
	14½ coarse pitch*	20 coarse pitch†	20 fine pitch†	25 coarse pitch†
7			42‡	
8			39‡	
9			36‡	
10		25	33	15
11		24	30	14
12	52	23	27	12
13	51	22	25	
14	50	21	23	
15	49	20	21	
16	48	19	19	
17	47	18	18	
18	46			
19	45			
20	44			
21	43			
22	42			
23	41			
24	40			
25	39			
26	38			
27	37			
28	36			
29	35			
30	34			
31	33			

* Gears of this pressure angle not recommended for new designs.
† Gears with these numbers of teeth can be made standard addendum and operated on standard center distances.
‡ Not recommended; but if essential, use these values.
NOTES: Pinions having fewer than 10 teeth are not recommended in any AGMA standard. The fine-pitch standard AGMA 207.04 gives data for pinions from seven through nine teeth but strongly discourages their use. Gears having fewer teeth than shown for any given pinion-gear combination will require an enlarged center distance for proper operation.

enlarged center distance is used, determine *operating* circular pitch from *operating* pitch diameters [Eqs. (4.8), (4.9), (4.10), and (4.11)].

9. Determine design tooth thickness. Decide first on how much to thin teeth for backlash. Table 4.11 gives recommended amounts for *power gearing* having in mind normal accuracy of center distance and normal operating temperature variations between gear wheels and casings. If special designs with essentially "no backlash" or unusual materials and accuracy are involved, consult the last part of this chapter, titled "Center Distance," for special calculations. If the pinion and gear have equal addendum, make the theoretical tooth thicknesses equal, and do this by dividing the circular pitch by 2. If a *long*-addendum pinion is used with a *short*-addendum gear, adjust tooth thicknesses by the following:

$$\Delta t = \Delta a\, 2 \tan \Phi_n \quad (4.33)$$

If the pinion only is enlarged and the center distance is enlarged to accommodate a standard gear, enlarge the pinion tooth thickness only (see Table 4.13).

GEAR TOOTH DESIGN 4.35

TABLE 4.5 General Recommendations on Numbers of Pinion Teeth for Spur and Helical Power Gearing

The maximum number of teeth in the range is based on providing a suitable balance between strength and wear capacity. The minimum number of teeth is intended to require either no enlargement to avoid undercutting or no more enlargement than necessary.

Range of No. of pinion teeth	Ratio m_G	Diametral pitch P_d	Hardness
19–60	1–1.9	1–19.9	200–240 BHN
19–50	2–3.9		
19–45	4–8		
19–45	1–1.9	1–19.9	Rockwell C 33–38
19–38	2–3.9		
19–35	4–8		
19–30	1–1.9	1–19.9	Rockwell C 58–63
17–26	2–3.9		
15–24	4–8		

NOTE: Small numbers of teeth in the pinion require special tooth proportions to avoid undercut.

Gears containing prime numbers of teeth of 101 and over may be difficult to cut on the gear-manufacturing equipment available. In general, prime numbers over 101 and all numbers over 200 should be checked with the shop to be sure the teeth can be made.

TABLE 4.6 Recommended Diametral Pitches

Coarse pitch			Fine pitch		
2	4	12	20	48	120
2.25	6	16	24	64	150
2.5	8		32	80	200
3	10		40	96	

These diametral pitches are suggested as a means of reducing the great amount of gear-cutting tooling that would have to be inventoried if all possible diametral pitches were to be specified.

After the theoretical tooth thicknesses are obtained, subtract one-half the amount the teeth are to be thinned for backlash from the pinion and gear theoretical tooth thicknesses. This is the *maximum design* tooth thickness. Obtain the *minimum design* tooth thickness by subtracting a reasonable tolerance for machining from the maximum design tooth thickness.

10. Recheck load capacity using design proportions just obtained. Use Secs. 16.6, 11.7, and 11.9. If not within allowable limits, change design.
11. If the gear design is for critical power gears, additional items will need calculation:
    a. Root fillet radius [Eq. (4.28)]
    b. Form diameter

TABLE 4.7 Equations and References for Whole-Depth Calculations

Pitch range	Equation	Ref.	Application	Remarks
Coarse pitch up to 19.99	$\dfrac{2.25}{P_d}$	Table 5-1	General application	AGMA standard
	$\dfrac{2.35}{P_d}$	....	Gears to be given finishing operations such as shave or grind	AGMA standard
	$\dfrac{2.40}{P_d}$	....	Max. strength for full-radius (tip) hobs	
Fine* pitch $20\,P_d$ and finer	$\dfrac{2.20}{P_d} + 0.002$	Table 5-1	Standard whole depth for general applications	AGMA standard
	$\dfrac{2.35}{P_d} + 0.002$	....	Whole depth for preshave cutters for shaving or grinding	AGMA standard

* Does not apply to pinions containing nine teeth or less.

    c. Modification of profile [Eqs. (4.26), (4.27)]
    d. Diameter over pins (Sec. 5.8)
12. Certain general dimensions must be calculated and toleranced.
    a. Outside diameter, $D + 2(a \pm \Delta a)$
    b. Root diameter, $D_o - 2h_t$
    c. Face width
    d. Chordal addendum and chordal thickness (Sec. 5.5)
13. It may be necessary to specify
    a. Tip round (at outside diameter) (Table 4.14)
    b. Edge round (Table 4.14)
    c. Roll angle (Sec. 5.3)
    d. Base radius [Eq. (5.4)]
14. In some unusual designs, it may be necessary to check the following to assure a sound design:
Diameter at which teeth become pointed [Eq. (4.32)]
Width of top land [Eq. (4.29)]
Effective contact ratio [Eqs. (4.3) and (4.4), Sec. 7.4]
Undercut diameter (Sec. 5.9)

Explanation and Discussion of Items in Table 4.1. Table 4.1 shows the "basic" or standard tooth proportions as given in the AGMA or ASA standards listed at the end of the table.

    1. *Pressure angle.* The pressure angle and numbers of teeth in a given pair of gears are related in that low pressure angles (14½°) should be avoided for low numbers of teeth (see Table 4.3 and Table 4.4).
    2. *Addendum.* For general applications, use equal-addendum teeth when the numbers of teeth in the pinion and in the pair exceed the values shown in items 10 and 11. Use long addendums for pinions having numbers of teeth, pressure angles, and diametral pitches as shown in Table 4.9. The amount of decrease of addendum in the mating gear is to be limited to values that will not produce undercut.

TABLE 4.8 Equations and References for Addendum Calculations, Spur Gears

Type of Tooth Design	Operating Condition	Pinion	Gear
Standard or equal addendum	Numbers of teeth in gear and pinion greater than minimum numbers shown in Tables 4.4 and 4.3	(a) $\dfrac{1.00}{P_d}$; see Table 4.1 (b) Value from Table 4.12	(a) $\dfrac{1.00}{P_d}$; see Table 4.1 (b) Value from Table 4.12
Long and short addendum	Numbers of teeth in pinion *less* than minimum numbers shown in Table 4.3 and numbers of teeth in gear *more* than minimums shown in Table 4.4	(c) $\dfrac{\text{Table 4.9}}{P_d}$ or (d) $\dfrac{\text{Table 4.10}}{P_d}$ if not less than (c)	(c) $\dfrac{\text{Table 4.9}}{P_d}$ or (d) $\dfrac{\text{Table 4.10}}{P_d}$
	Numbers of teeth in pinion *less* than minimum numbers shown in Table 4.4 and numbers of teeth in gear *less* than minimum numbers of teeth shown in Table 4.4	*Increase center distance* sufficiently to avoid undercut, in pinion *and* gear (e) $\dfrac{\text{Table 4.9}}{P_d}$	(e) See Sec. 4.3
	Gearing designed for balanced strength	(f) $\dfrac{\text{Table 4.10}}{P_d}$ NOTE: If value is *less* than given by Table 4.9 check for undercut	(f) $\dfrac{\text{Table 4.10}}{P_d}$ NOTE: If value is *less* than given by Table 4.9 check for undercut
	Designed for nonstandard center distance	(g) Calculate tooth thickness to meet required center distance from Eq. (4.56).	

NOTE: The values in this table are for gears of 1 diametral pitch. For other sizes divide by the required diametral pitch.
The values of addendum shown are the minimum increase necessary to avoid undercut. Additional addendum can be provided for special applications to balance strength. See Table 4.10.
*These values are less than the proportional amount that the tooth thickness is increased (see Table 4.13) in order to provide a reasonable top land.

TABLE 4.9 Values of Addendum

No. of teeth in pinion	Coarse-pitch teeth (1 through 19 P_D) per AGMA 201.02 20°			Coarse-pitch teeth (1 through 19 P_D) per AGMA 201.02 25°			Fine-pitch teeth (20 P_D and finer) per AGMA 207.04		
	a_P pinion	a_G gear	Recommended min. No. of teeth	a_P pinion	a_G gear	Recommended min. No. of teeth	a_P pinion	a_G gear	Recommended min. No. of teeth
7	.....	.....	..	.....	.....	..	1.4143*	0.4094	42
8	.....	.....	..	.....	.....	..	1.4369*	0.4679	39
9	.....	0.532	..	.....	.....	..	1.4190*	0.5264	36
10	1.468	0.532	25	.....	.....	15	1.4151	0.5849	33
11	1.409	0.591	24	1.184	0.816	14	1.3566	0.6434	30
12	1.351	0.649	23	1.095	0.905	12	1.2982	0.7019	27
13	1.292	0.708	22	1.000	1.000	12	1.2397	0.7604	25
14	1.234	0.766	21	1.000	1.000	12	1.1812	0.8189	23
15	1.175	0.825	20	1.000	1.000	12	1.1227	0.8774	21
16	1.117	0.883	19	1.000	1.000	12	1.0642	0.9358	21
17	1.058	0.942	18	1.000	1.000	12	1.0057	0.9943	19
18	1.000	1.000	..	1.000	1.000	12	1.000	1.0000	18

NOTE: The values in this table are for gears of 1 diametral pitch. For other sizes divide by the required diametral pitch. The values of addendum shown are the minimum increase necessary to avoid undercut. Additional addendum can be provided for special applications to balance strength. See Table 4.10.

*These values are less than the proportional amount that the tooth thickness is increased (see Table 4.13) in order to provide a reasonable top land.

TABLE 4.10 Values of Addendum for Balanced Strength

M_G (gear ratio) N_G/N_P		a (addendum)		M_G (gear ratio) N_G/N_P		a (addendum)	
From	To	Pinion a_P	Gear a_G	From	To	Pinion a_P	Gear a_G
1.000	1.000	1.000	1.000	1.421	1.450	1.240	0.760
1.001	1.020	1.010	0.990	1.451	1.480	1.250	0.750
1.021	1.300	1.020	0.980	1.481	1.520	1.260	0.740
1.031	1.040	1.030	0.970	1.521	1.560	1.270	0.730
1.041	1.050	1.040	0.960	1.561	1.600	1.280	0.720
1.051	1.060	1.050	0.950	1.601	1.650	1.290	0.710
1.061	1.080	1.060	0.940	1.651	1.700	1.300	0.700
1.081	1.090	1.070	0.930	1.701	1.760	1.310	0.690
1.091	1.110	1.080	0.920	1.761	1.820	1.320	0.680
1.111	1.120	1.090	0.910	1.821	1.890	1.330	0.670
1.121	1.140	1.100	0.900	1.891	1.970	1.340	0.660
1.141	1.150	1.110	0.890	1.971	2.060	1.350	0.650
1.150	1.170	1.120	0.880	2.061	2.160	1.360	0.640
1.170	1.190	1.130	0.870	2.161	2.270	1.370	0.630
1.190	1.210	1.140	0.860	2.271	2.410	1.380	0.620
1.210	1.230	1.150	0.850	2.411	2.580	1.390	0.610
1.231	1.250	1.160	0.840	2.581	2.780	1.400	0.600
1.251	1.270	1.170	0.830	2.781	3.050	1.410	0.590
1.271	1.290	1.180	0.820	3.051	3.410	1.420	0.580
1.291	1.310	1.190	0.810	3.411	3.940	1.430	0.570
1.311	1.330	1.200	0.800	3.941	4.820	1.440	0.560
1.331	1.360	1.210	0.790	4.821	6.810	1.450	0.550
1.361	1.390	1.220	0.780	6.811	∞	1.460	0.540
1.391	1.420	1.230	0.770				

NOTE: Do not select values from this table for the pinion member that are smaller than those given in Table 4.9

TABLE 4.11 Recommended Backlash Allowance for Power Gearing

Normal diam. pitch P_{nd}	Center distance, in.						
	0–5″	5–10″	10–20″	20–30″	30–50″	50–80″	80–120″
½	.....	.....	.....	.....	0.045	0.060	0.080
1	.....	.....	.....	0.035	0.040	0.050	0.060
2	.....	.....	0.025	0.030	0.035	0.045	0.055
3	.....	0.018	0.022	0.027	0.037	0.042	
4	.....	0.016	0.020	0.025	0.030	0.040	
6	0.008	0.010	0.015	0.020	0.025		
8	0.006	0.008	0.012	0.017			
10	0.005	0.007	0.010				
12	0.004	0.006					
16	0.004	0.005					
20	0.004						
32	0.003						
64	0.002						

TABLE 4.12 Standard Tooth Parts

Diam. pitch P_d (1/in.)	Circ. pitch p, in.	Module $1/P_d$ addendum a, in.	Whole depth* Shave or grind h_t, in.	Whole depth* Cut teeth h_t, in.*	Tooth thickness $(p/2)$ t, in.	Double depth Cut teeth $2\,h_t$, in.	Double depth Ground $2\,h_t$, in.
1	3.1415927	1.0000000	2.35000	2.25000	1.5707963	4.5000	4.7000
1¼	2.5132741	0.8000000	1.88000	1.80000	1.2566371	3.6000	3.7600
1½	2.0943951	0.6666667	1.56667	1.50000	1.0471980	3.0000	3.1333
1¾	1.7951958	0.57142857	1.34286	1.28571	0.8975979	2.5714	2.6857
2	1.5707963	0.5000000	1.17500	1.12500	0.7853982	2.2500	2.3500
2½	1.2566371	0.4000000	0.94000	0.90000	0.6283185	1.8000	1.8800
3	1.0471976	0.3333333	0.78333	0.75000	0.5235988	1.5000	1.5667
3½	0.8975979	0.2857143	0.67143	0.64286	0.4487990	1.2857	1.3429
4	0.7853982	0.2500000	0.58750	0.56250	0.3926991	1.1250	1.1750
5	0.6283185	0.2000000	0.47000	0.45000	0.3141593	0.9000	0.9400
6	0.5235988	0.1666667	0.39167	0.37500	0.2617994	0.7500	0.7833
8	0.3926991	0.1250000	0.29375	0.28125	0.1963495	0.5625	0.5875
10	0.3141593	0.1000000	0.23500	0.22500	0.1570796	0.4500	0.4700
12	0.2617994	0.0833333	0.19583	0.18750	0.1308997	0.3750	0.3917
16	0.1963495	0.0625000	0.14688	0.14063	0.0981748	0.2813	0.2938
20	0.1570796	0.0500000	0.11950	0.11200	0.0785398	0.2240	0.2390
24	0.1308997	0.0416667	0.09992	0.09367	0.0654499	0.1873	0.1998
32	0.0981748	0.0312500	0.07544	0.07075	0.0490873	0.1415	0.1509
40	0.0785398	0.0250000	0.06075	0.05700	0.0392699	0.1140	0.1215
48	0.0654498	0.0208333	0.05096	0.04780	0.0327249	0.0956	0.1019
64	0.0490874	0.0156250	0.03872	0.03638	0.0245437	0.0728	0.0774
80	0.0392699	0.0125000	0.03138	0.02950	0.0196350	0.0590	0.0628
96	0.0327249	0.0104167	0.02648	0.02492	0.0163625	0.0498	0.0530
120	0.0261799	0.0083333	0.02158	0.02033	0.0130900	0.0407	0.0432
150	0.0209440	0.0066667	0.01767	0.01667	0.0104720	0.0333	0.0353

* h_t for fine pitch (20 P_d and finer) is $2.2/P_d + 0.002''$.

The values of long addendum for pinions are limited to speed-decreasing drives. Data on speed-increasing drives are given in Sec. 4.1.3 but are beyond the scope of AGMA standards.

3. *Dedendum.* The values shown in the table are the minimum standard values. Shaved or ground teeth should be given a greater dedendum (whole depth and clearance values); see items 5 and 9. A constant value, 0.002 in., is added to the dedendum of fine-pitch gears, which allows space for the accumulation of foreign matter at the bottoms of the spaces. This provision is particularly important in the case of very fine diametral pitches.

4. *Working depth.* The working depth customarily determines the type of tooth; that is, a tooth with a 1.60/P working depth is a *stub tooth* whereas a tooth with a 2.0/P working depth is called a *full-depth tooth*.

5. *Whole depth.* The value shown is the standard minimum whole depth. It will increase in proportion to the increase in backlash cut into the teeth (unless the outside diameter is also correspondingly adjusted). It will also increase slightly if long- and

TABLE 4.13 Values of Tooth Thickness for Pinions t_P and Gears t_G with Small Numbers of Teeth (Long and Short Addendums)

No. of teeth in pinion	Coarse-pitch system				Fine-pitch system	
	20° pressure angle		25° pressure angle		20° pressure angle	
	Pinion	Gear	Pinion	Gear	Pinion	Gear
7					2.0007	1.1409
8					1.9581	1.1835
9					1.9155	1.2261
10	1.912	1.230	1.742	1.399	1.8730	1.2686
11	1.868	1.273	1.659	1.482	1.8304	1.3112
12	1.826	1.315	1.5708	1.5708	1.7878	1.3538
13	1.783	1.358	1.5708	1.5708	1.7452	1.3964
14	1.741	1.400	1.5708	1.5708	1.7027	1.4389
15	1.698	1.443	1.5708	1.5708	1.6601	1.4815
16	1.656	1.486	1.5708	1.5708	1.6175	1.5241
17	1.613	1.529	1.5708	1.5708	1.5749	1.5667
18	1.5708	1.5708	1.5708	1.5708	1.5708	1.5708
19	1.5708	1.5708	1.5708	1.5708	1.5708	1.5708

NOTE: These tooth thicknesses go with Table 4.9 addendums.
The above values are for 1 diametral pitch.
These basic tooth thicknesses do not include an allowance for backlash.

TABLE 4.14 Values for Tip Round and Edge Round and End Round

Diam. pitch	Edge round			Tip round and end round
	General applications	Medium strength	High strength	
20 and finer	Burr-brush edges	0.001–0.005	0.005–0.010	0.001–0.005
16	Burr-brush edges	0.003–0.015	0.010–0.025	0.003–0.010
12	Burr-brush edges	0.005–0.020	0.012–0.030	0.005–0.015
10	Burr-brush edges	0.010–0.025	0.015–0.035	0.005–0.015
8	Burr-brush edges	0.010–0.025	0.020–0.045	0.010–0.030
5	Burr-brush edges	0.010–0.025	0.025–0.060	0.010–0.030
3	Burr-brush edges	0.015–0.035	0.040–0.090	0.010–0.050
2	Burr-brush edges	0.015–0.035	0.060–0.125	0.010–0.050

TABLE 4.15 Tooth Proportions of Spur Gears, Obsolete Systems‡
(Tooth proportions for fully interchangeable gears operated on standard center distances)

Symbol	Item	Brown and Sharpe system	AGMA $14\frac{1}{2}°$ composite system	Fellows 20° stub system	AGMA full-depth composite system
ϕ	Pressure angle, degrees	$14\frac{1}{2}$	$14\frac{1}{2}$	20	$14\frac{1}{2}$
a	Addendum (basic*)	$\dfrac{1.00}{P}$	$\dfrac{1.00}{P}$	$\dfrac{0.800}{P}$	$\dfrac{1.00}{P}$
b	Dedendum (min.) (basic†)	$\dfrac{1.157}{P}$	$\dfrac{1.157}{P}$	$\dfrac{1.000}{P}$	$\dfrac{1.157}{P}$
h_k	Working depth	$\dfrac{2.00}{P}$	$\dfrac{2.000}{P}$	$\dfrac{1.60}{P}$	$\dfrac{2.00}{P}$
h_t	Whole depth (min.) (basic†)	$\dfrac{2.157}{P}$	$\dfrac{2.157}{P}$	$\dfrac{1.80}{P}$	$\dfrac{2.157}{P}$
t	Circular tooth thickness (basic*)	$\dfrac{\pi}{2P}$	$\dfrac{\pi}{2P}$	$\dfrac{\pi}{2P}$	$\dfrac{\pi}{2P}$
r_f	Fillet radius (in basic rack)	$\dfrac{0.157}{P}$	$\dfrac{0.157}{P}$	Not standardized	1.33 × clearance
c	Clearance (min.) (basic†)	$\dfrac{0.157}{P}$	$\dfrac{0.157}{P}$	$\dfrac{0.20}{P}$	$\dfrac{0.157}{P}$
c	Clearance, shaved or ground teeth				
N_P'	Min. numbers of teeth*				
	Pinion	32		14	32
	Pair	64			64
	Reference standards: AGMA		201.02A		201.02A

*These values are basic for equal-addendum gearing. When the gearing is made long- and short-addendum these values will be altered.
†These values are minimum. Shaved or ground teeth should be given proportions suitable to these processes.
‡See Table 4.1.

short-addendum teeth are generated with pinion-shaped cutters. See also items 3, 8, and 9 for increases due to manufacturing processing requirements.

6. *Circular tooth thickness, basic.* This is the basic circular tooth thickness on the standard pitch circle. These values will be slightly altered if backlash is introduced into the gears to allow them to mesh at a standard center distance. These values will be drastically altered in the case of long- and short-addendum designs. Table 4.13 gives values of tooth thickness corresponding to standard values of long and short addendum for small numbers of teeth.

7. *Fillet radius, basic rack.* The fillet radius shown is that in the basic rack. The tooth form standard (AGMA 201.02)[5] directs that the edge radius on hobs and rack-type shaper cutters should be equal to the fillet radius in the basic rack. It also directs that pinion-shaped cutters should be designed using the basic rack as a guide so that the gear teeth generated by these cutters will have a fillet radius approximating those produced by hobs and rack-type shaper cutters. It can be calculated approximately by the data shown in Eq. (4.33).

In the case of 25°-pressure-angle teeth, the fillet radius shown must be reduced for teeth having a clearance of $0.250/P$. This is discussed in Sec. 4.1.4.

In the case of fine-pitch teeth, the fillet radius usually will be larger than the clearance customarily given as $c = 0.157/P$ because of edge breakdown in the cutter.

GEAR TOOTH DESIGN 4.43

The effects of tool wear on fine-pitch gears are proportionally larger than the effects produced by coarse-pitch tools.

8. *Clearance.* The value shown is the minimum standard. Greater clearance is usually required for teeth that are finished by grinding or shaving. In general the value shown in item 9 will be suitable for these processes. See also items 3 and 5.

9. *Clearance for shaved or ground teeth.* This is the recommended clearance for teeth to be finished by shaving or grinding. In the case of 25°-pressure-angle teeth, the fillet radius shown must be reduced for teeth having a clearance of $0.250/P$.

10. *Minimum number of teeth in pinion.* These are the lowest numbers of teeth that can be generated in pinions having standard addendums and tooth thicknesses that will not be undercut. Pinions having fewer teeth should be made long addendum in accordance with Table 4.9.

11. *Minimum numbers of teeth in pair.* This is the smallest number of teeth in pinion and gear that can be meshed on a standard center distance without one member's being undercut. For pairs with fewer teeth the members will have to be meshed on a nonstandard (enlarged) center distance.

12. *Minimum width of top land.* This is the approximate minimum width of top land allowable in standard long-addendum pinions. Increases in addendum that cause the tops of the teeth to have less than this value should be generally avoided.

13. *Reference standard.* The source of the data in each column. AGMA standards are published by the American Gear Manufacturers Association.

14. *Reference standard.* The source of the data in each column. ASA standards are published by the American Society of Mechanical Engineers.

4.2.2 System for Helical Gears

The following data are based on the information contained in AGMA standard 207.05, "20-Degree Involute Fine-Pitch System, for Spur and Helical Gears."[5] In addition, tooth proportions are shown that are used by several of the larger gear manufacturers for the design of helical gears. These tooth proportions are shown since there are no AGMA standards for coarse-pitch helical gears.

In general, helical tooth proportions are based on either getting the most out of a helical gear design or on using existing tooling. The tooth proportions referred to above have been found to yield very good gears. In some cases, for less critical applications, tools on hand may be used. Very often hobs for spur gears are available, and gears for general service can be made with these tools. Since such hobs have no taper, they are not well suited to cut helix angles much over 30°.

In helical gear calculations, care should be taken to avoid confusion as to which plane the various tooth proportions are measured in. In some cases, the transverse plane is used. This plane is perpendicular to the axis of the helical gear blank. In some cases, it is desirable to work in the normal plane.

If a spur gear hob is used to cut helical teeth, the relationship between the transverse and normal pressure angles and the transverse and normal pitches as well as the base pitches should be established.

$$P_D = \frac{\cos \psi \times \pi}{p_n} \tag{4.34}$$

$$\tan \Phi = \frac{\tan \Phi_n}{\cos \psi} \tag{4.35}$$

$$p_N = p_n \cos \Phi_n \qquad (4.36)$$

where P_D = diametral pitch in transverse plane
 ψ = helix angle
 p_n = circular pitch in normal plane
 Φ_n = pressure angle in normal plane
 p_N = normal base pitch (normal to surface)

When gears mesh together or with hobs, racks, shaper, or shaving cutters, their normal base pitches must be equal.

When a gear shaper is to be used to cut helical teeth, special guides are installed in the machine. These impart the twist into the cutter spindle, which produces the proper helix angle and lead. The cutter used has the same lead ground into the teeth as is produced by the guides. The cutter and guides will in turn produce a given lead angle on the gear being cut. Practical considerations limit the lead angle of the guides to roughly 35°. In order to control costs, helical gears that are to be shaped should be designed either to a standard series of values of leads or to values of lead available in existing guides.

Equation (4.37) shows the relationship of lead, number of teeth, diametral pitch, and helix angle in a given cutter, guide, or helical gear.

$$\tan \psi = \frac{N\pi}{LP_d} \qquad (4.37)$$

where ψ = helix angle
 N = number of teeth, cutter, or gear
 P_d = diametral pitch
 L = lead of cutter

Selection of Tooth Form. The tooth forms shown in Table 4.16 are used for ranges of applications as follows:

Type of Gear	Suggested Tooth Form
Single helix, low allowable thrust reaction	Type 1
Single helix, moderate allowable thrust reaction	Types 2 and 3
Double helix, general purpose	Types 4 and 5
Double helix, minimum noise	Type 6
Double helix, high load	Types 20 and 21

The foregoing types each require a separate set of tooling. They have the advantage of getting the most out of good helical gear design.

When an existing spur gear hob is to be used to produce the gear, use types 18 and 19. The data in Table 4.17 will be found helpful in making calculations for gears having types 18 and 19 tooth forms. Type 19 is similar to type 18 except that its proportions are based on the use of hobs designed to cut fine-pitch gears (20 diametral pitch or finer).

Selection of Helix Angle. Single helical gears usually are given lower helix angles than double helical gears in order to limit thrust loads. Typical helix angles are 15 and 23°.

TABLE 4.16 Basic Tooth Proportions for Helical Gears

Tooth form No.	Diametral Normal P_{nd}	Diametral Transverse P_d	Circular Normal p_n	Circular Transverse p	Axial p_z	Pressure angle Normal ϕ_n	Pressure angle Transverse ϕ	Working h_k	Whole h_t	Helix angle ψ	Addendum a	Tooth thickness (arc) t	Edge radius of generating rack r_T
1	1.03528	1	3.03454	3.1416	11.72456	19°22′12″	20°	2.0000	2.3500	15°	1.0000	1.5710	0.350
2	1.0836	1	2.8919	3.1416	7.40113	18°31′22″	20°	2.0000	2.3500	23°	1.0000	1.4460	0.350
3	1.0836	1	2.89185	3.1416	7.40113	18°31′22″	20°	1.8400	2.2000	23°	0.9200	1.4460	0.350
4	1.15470	1	2.72070	3.1416	5.44140	17°29′43″	20°	1.6400	2.0500	30°	0.8700	1.3604	0.300
5	1.22077	1	2.57340	3.1416	4.48666	16°36′ 6″	20°	1.6400	1.9500	35°	0.8200	1.2867	0.300
6	1.41421	1	2.22144	3.1416	3.14159	14°25′58″	20°	1.4200	1.7000	45°	0.7100	1.1107	0.2500
7	1		3.1416			23°14′ 7″		2.0000	2.2500		1.0000		0.240
8	1		3.1416			23°14′ 7″		2.0000	2.3300		1.0000		0.270
9	1		3.1416			21°58′50″		2.0000	2.2700		1.0000		0.310
10	1		3.1416			21°58′50″		2.0000	2.3500		1.0000		0.300
11	1		3.1416			21°58′50″		2.0000	2.4000		1.0000		0.290
12	1		3.1416			21°58′50″		2.0000	2.4500		1.0000		0.275
13	1		3.1416			18°31′35″		1.6300	2.0180		0.8150		0.204
14	1		3.1416			24°39′57″		1.8280	2.1170		0.9140		0.158
15	1		3.1416			24°51′32″			2.1570		0.9260		0.220
16	1		3.1416			24°54′59″			2.1550		0.9230		0.264
17	1		3.1416			15° 0′ 0″			2.6150		1.1580		0.410
18	1		3.1416					2.000	2.3500		1.0000		0.350
19	1		3.1416				20°	2.000	$2.200 + 0.002 P_d$				
20	1.15470	1	2.72070	3.1416	5.44140	22°30′	25°33′41″	1.7400	2.0500	30°	0.8700		0.250
21	1.10338	1	2.84725	3.1416	6.73717	25°	27°13′35″	1.5000	1.7500	25°	0.7500		0.200

4.45

Low helix angles do not provide so many axial crossovers as can be achieved on high-helix-angle gears of a given face width. Double-helix gears have helix angles that typically range from 30 through 45°. Although higher helix angles provide smoother operation, the tooth strength is lower.

In order to get the most quiet gears and at the same time achieve good tooth strength, special cutting tools for each helix angle should be provided. Table 4.16 shows typical tooth proportions. Tooth types 1 through 5 require special cutters, and the helix angle shown should be used. These teeth are stubbed more and more as the helix angle is higher. Tooth types 6 through 12 can be made with any helix angle desired. Types 13 and 14 can also be made with almost any helix angle. These tooth proportions are based on the use of standard spur gear hobs. Since such hobs are not usually tapered, they do not do as good a job in cutting as do the specially designed helical hobs, and, as a result, they should be limited in use to helix angles less than 25°.

Face Width. In the design of helical gears, the face width is usually based on that needed to achieve the required load-carrying capacity. In addition, the face width and lead are interrelated in that it is necessary to obtain at least two axial pitches of face width ($F = 2p_x$) to get reasonable benefit from the helical action, and four or more if high speeds, noise, or critical designs are confronted.

Long- and short-addendum designs are not as common among helical gears as among spur gears. This is because a much lower number of teeth can be cut in helical pinions without undercut. In low-hardness helical gearing, pitting, which is little affected by changes in addendum, is usually the limiting feature. In high-hard-

TABLE 4.17 Tooth Proportions for Helical Gears

Helix angle, deg	Diam. pitch P_d	Circ. pitch P_t	Axial pitch p_x	Pressure angle*	Working depth h_k	Whole depth h_t†
0	1.000000	3.14159	∞	20°00'00"	2.000	2.250
5	0.996195	3.15359	36.04560	20° 4'13.1"	2.000	2.250
8	0.990268	3.17247	22.57327	20°10'50.6"	2.000	2.250
10	0.984808	3.19006	18.09171	20°17' 0.7"	2.000	2.250
12	0.978148	3.21178	15.11019	20°24'37.1"	2.000	2.250
15	0.965926	3.25242	12.13817	20°38'48.8"	2.000	2.250
18	0.951057	3.30326	10.16640	20°56'30.7"	2.000	2.250
20	0.939693	3.34321	9.18540	21°10'22.0"	2.000	2.250
21	0.933580	3.36510	8.76638	21°17'56.4"	2.000	2.250
22	0.927184	3.38832	8.38636	21°25'57.7"	2.000	2.250
23	0.920505	3.41290	8.04029	21°34'26.3"	2.000	2.250
24	0.913545	3.43890	7.72389	21°43'22.9"	2.000	2.250
25	0.906308	3.46636	7.43364	21°52'58.7"	2.000	2.250
26	0.898794	3.49534	7.16651	22° 2'44.2"	2.000	2.250
27	0.891007	3.52589	6.91994	22°13'10.6"	2.000	2.250
28	0.882948	3.55807	6.69175	22°24' 9.0"	2.000	2.250
29	0.874620	3.59195	6.48004	22°35'40.0"	2.000	2.250
30	0.866025	3.62760	6.28318	22°47'45.1"	2.000	2.250

* Pressure angle based on 20° normal pressure angle.
† The values shown for whole depth are for coarse-pitch gears. If the gears are to be shaved or ground, use $h_t = 2.35$.
For fine-pitch gears, use $2.2/P_{nd} + 0.002$ for general-purpose gearing and $2.35/P_{nd} + 0.002$ for gear to be shaved or ground.

ness designs, the same proportions as those used for spur gear addendums should be employed.

Specific Calculation Procedure for Helical Gears. The procedure for helical gears is very similar to that given in Sec. 4.2.1 for spur gears.

1. Determine application requirements: ratio, power, speed, etc.
2. Decide on basic tooth proportions and helix angle (see Table 4.16).
3. Pick appropriate number of pinion teeth (see Tables 4.2, 4.3, and 4.5).
4. Determine approximate center distance and face width. If power gearing, estimate from Sec. 11.4.
5. Determine pitch of teeth.
6. Determine whole depth.
7. Determine addendum of pinion and gear. Use equal addendum for pinion and gear except in special cases where addendum must be increased to avoid undercut or special consideration must be given to increasing the pinion strength.
8. Determine operating circular pitch, operating helix angle, and operating normal pitch if special center distance was used.
9. Determine design tooth thickness. If long and short addendum, adjust theoretical tooth thicknesses accordingly. Thin teeth for backlash. Consult Table 4.11 for power gearing backlash allowance. See Chap. 4 if special "no-backlash" design.
10. Recheck load capacity. Use Sec. 11.6 and 11.7. Check number of axial crossovers [Eq. (4.5)]. Check ability of thrust bearings to handle thrust reactions (Chap. 10). If results of these checks are unsatisfactory, change proportions.
11. If gear design is for critical power gears, additional items may need calculation.
    a. Root fillet radius [Eq. (4.28)]
    b. Form diameter
    c. Modification of profile [Eqs. (4.26), (4.27)]
    d. Diameter over balls (Sec. 5.8)
12. General dimensions to be determined and toleranced.
    a. Outside diameter
    b. Root diameter
    c. Face width
    d. Chordal addendum and chordal tooth thickness (Sec. 5.5)

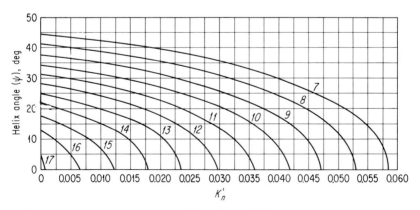

FIGURE 4.10 Addendum increment K'_n for 1 diametral pitch

TABLE 4.18 Equations and References for Addendum Calculations, Helical Gears

Type of Tooth Design	Operating Condition	Pinion	Gear
Standard or equal addendum	Numbers of teeth in gear and pinion greater than minimum numbers shown in Fig. 4.10	a = value from Table 4.16 $$a = \frac{\text{types 1-6}}{P_d}, \frac{\text{types 7-19}}{P_{nd}},$$ $$\frac{\text{types 20-21}}{P_d}$$	If standard center distance and values of aP as shown at (a) left, $a_G = h_k - a_P$ If nonstandard center distance, see item below
Long and short addendum		Value from special tools to be used	
	Numbers of teeth in pinion *less* than minimum numbers shown in Table 4.3 and numbers of teeth in gear more than minimums shown in Fig. 4.3	Increase addendum sufficiently to avoid undercut $a_P = a + K_n'/P_n$ See Fig. 4.10 for K_n'	Decrease addendum by amount pinion addendum is increased. If undercut, increase center distance. See item below
	Numbers of teeth in pinion *less* than minimum numbers shown in Table 4.3 and numbers of teeth in gear *less* than minimum numbers of teeth shown in Table 4.3	Increase addendum sufficiently to avoid undercut $a_P = a + K_n'/p_n$	Increase addendum sufficiently to avoid undercut $a_G = a + K_n'/P_n$
	Designed for nonstandard center distance	Increase center distance by amount sufficient to accommodate increased addendums of both pinion and gear Calculate tooth thickness to meet required center distance from Eq. (4.56), then calculate required addendum from Eq. (4.18).	

13. It may be necessary to specify
    a. Tip round or chamber (Table 4.14)
    b. Edge round or end bevel (Table 4.14)
    c. Roll angle (Sec. 5.3)
    d. Base radius [Eq. (5.4)]
14. Generally helical gears are not designed for such critical applications as to be in danger of pointed teeth or undercut.

4.2.3 System for Internal Gears

At present, there is no AGMA standard covering internal gears. The following material is based on information covered in Dudley.[3]

In general, both spur internal meshes and helical internal meshes may be calculated by the same methods as external meshes. However, in internal gear meshes, there are several problems unique to this type of gearing.

The first of these is tip interference. In this type of interference, the pinion member cannot be assembled radially with the gear. Only axial assembly is possible, and it should be provided for in the design. If a shaper cutter having a number of teeth equal to or greater than the pinion is used to cut the internal gear, it will cut its way into mesh but in so doing will remove some material from the flanks of a few of the teeth that should have been left in place in order to have good tooth operation. This cutting action is also known as trimming. Such teeth will have poor contact and will tend to be noisy. If the proper shaper cutter or a broach is used, this problem will not occur.

The second problem is sometimes known as *fouling*. In this case, the internal gear teeth interfere with the flanks of the external toothed pinion if there is too small a difference in numbers of teeth between the pinion and gear members.

Both these problems can be avoided in most gear designs by reducing the addendum of the internal gear (increasing its inside diameter). Tables 4.19 and 4.20 show a group of tooth proportions that will avoid these problems. A rather complicated graphical layout or involved calculations may be required to determine the exact proportions to avoid these problems. Since such calculations are beyond the scope of this chapter, Tables 4.19 and 4.20 are offered as a guide.

Special Considerations.

In general, the tip interference problem can be eliminated by providing more than a 17-tooth difference between gear and pinion.

The internal gear teeth have an addendum that extends toward the inside from the pitch circle. The critical dimension of addendum height is maintained by holding inside diameter.

In general, the addenda of internal gears are made considerably shorter than those for equivalent external gears to avoid interference. The effect of the gear wrapping around the pinion tends to increase contact ratio, and also the chances for pinion-fillet interference.

Internal gearsets may be designed so that the pinion can be introduced into mesh at assembly by a radial movement. For any given number of teeth in the pinion, there

TABLE 4.19 Addendum Proportions and Limiting Numbers of Teeth for Internal Spur Gears of 20° Pressure Angle

No. of Pinion Teeth	Pinion Addendum	Min. No. of Gear Teeth		Minimum Ratio*	Gear Addendum		
		Axial Assembly	Radial Assembly		Ratio of 2	Ratio of 4	Ratio of 8
12	1.350	19	26	0.472	0.510	0.582	0.616
	1.510	19	26	0.390	0.412	0.451	0.471
13	1.290	20	27	0.507	0.556	0.635	0.673
	1.470	20	27	0.419	0.445	0.488	0.509
14	1.230	21	28	0.543	0.601	0.688	0.729
	1.430	21	28	0.447	0.479	0.525	0.548
15	1.180	22	30	0.574	0.642	0.733	0.777
	1.400	22	30	0.470	0.506	0.554	0.577
16	1.120	23	32	0.608	0.688	0.786	0.834
	1.380	23	32	0.487	0.526	0.574	0.597
17	1.060	24	33	0.642	0.734	0.839	0.890
	1.360	24	33	0.505	0.546	0.594	0.617
18	1.000	25	34	0.676	0.779	0.892	0.947
	1.350	25	34	0.516	0.558	0.605	0.628
19	1.000	27	35	0.702	0.792	0.808	0.950
	1.330	27	35	0.539	0.578	0.625	0.648
20	1.000	28	36	0.713	0.802	0.903	0.952
	1.320	28	36	0.550	0.590	0.636	0.658
22	1.000	30	39	0.733	0.821	0.912	0.957
	1.290	30	39	0.577	0.621	0.666	0.688
24	1.000	32	41	0.750	0.836	0.920	0.960
	1.270	32	41	0.599	0.644	0.687	0.709
26	1.000	34	43	0.766	0.849	0.926	0.963
	1.250	34	43	0.620	0.666	0.709	0.729
30	1.000	38	47	0.792	0.870	0.936	0.968
	1.220	38	47	0.654	0.702	0.741	0.761
40	1.000	48	57	0.836	0.903	0.952	0.976
	1.170	48	57	0.718	0.764	0.797	0.814

*Minimum ratio of number of gear teeth divided by number of pinion teeth for axial assembly.

must be a number of teeth in the gear that is somewhat greater than required if axial assembly were to be allowed.

If a minimum difference in numbers of pinion and gear teeth is desired, the designer must design the gear casings and bearings so that the pinion can be introduced into mesh with the gear in an axial direction.

GEAR TOOTH DESIGN

TABLE 4.20 Addendum Proportions and Limiting Numbers of Teeth for Internal Spur Gears of 25° Pressure Angle

No. of Pinion Teeth	Pinion Addendum	Min. No. of Gear Teeth		Minimum Ratio*	Gear Addendum		
		Axial Assembly	Radial Assembly		Ratio of 2	Ratio of 4	Ratio of 8
12	1.000	17	20	0.699	0.793	0.900	0.934
	1.220	17	21	0.601	0.656	0.720	0.740
13	1.000	18	22	0.718	0.810	0.908	0.940
	1.200	18	22	0.622	0.680	0.742	0.761
14	1.000	19	24	0.734	0.824	0.915	0.944
	1.180	19	24	0.644	0.703	0.763	0.782
15	1.000	20	25	0.748	0.837	0.921	0.948
	1.170	20	25	0.659	0.719	0.776	0.794
16	1.000	21	26	0.761	0.847	0.926	0.951
	1.150	21	26	0.679	0.741	0.797	0.815
17	1.000	22	27	0.773	0.857	0.930	0.954
	1.140	22	27	0.694	0.755	0.809	0.826
18	1.000	23	28	0.783	0.865	0.934	0.957
	1.130	23	28	0.708	0.769	0.820	0.837
19	1.000	24	29	0.793	0.873	0.938	0.959
	1.120	24	29	0.721	0.782	0.832	0.848
20	1.000	25	30	0.802	0.879	0.941	0.961
	1.110	26	29	0.741	0.795	0.843	0.859
22	1.000	28	32	0.824	0.891	0.947	0.965
	1.090	28	32	0.765	0.820	0.866	0.881
24	1.000	30	34	0.837	0.900	0.951	0.968
	1.080	30	34	0.782	0.836	0.879	0.893
26	1.000	32	37	0.847	0.908	0.955	0.970
	1.060	32	37	0.806	0.859	0.900	0.914
30	1.000	36	40	0.865	0.921	0.961	0.974
	1.050	36	40	0.829	0.879	0.915	0.927
40	1.000	46	51	0.896	0.941	0.971	0.981
	1.020	46	51	0.880	0.923	0.952	0.961

*Minimum ratio of number of gear teeth divided by number of pinion teeth for axial assembly.

Tables 4.19 and 4.20 give the minimum numbers of teeth that a gear may have for any given pinion to achieve either radial or axial assembly.

The "hand" of helical internal gears is determined by the direction in which the teeth move, right or left, as the teeth recede from an observer looking along the gear axis. Whereas two external helical gears must be of opposite hand to mesh or parallel axes, an internal helical gear must be of the same hand as its mating pinion.

Specific Calculation Procedure for Internal Gears. Generally speaking, internal gears may be designed by the same procedures as outlined in Sec. 4.2.1 for spur gears or Sec. 4.2.2 for helical gears. However, there are some special considerations:

1. *Number of teeth.* The number of teeth in the pinion is based on the ratio required and also on tooth strength and wear considerations. In addition there must be sufficiently large difference between the numbers of teeth in the pinion and gear to avoid problems of tip interference. Tables 4.19 and 4.20 give the minimum numbers of teeth in gear and pinion that can be specified and still be able to achieve either axial or radial assembly.

2. *Number of teeth, gear.* The minimum number of teeth in the gear member that can be used without getting into problems of assembly techniques or tip interference problems is shown in Tables 4.19 and 4.20. The manufacturing organization that will produce the gears should be checked to determine the availability of suitable equipment when the number of teeth in the gear exceeds 200 or when prime numbers over 100 are used. If the internal gear is to be broached, consideration should be given to minimizing the number of teeth in the gear so as to keep broach costs down.

3. *Helix angle.* Since shaper guides are usually required for the cutting of the internal gear, first choice in helix angles should be based on existing guides. Because of the kinematics of the gear shaper, helix angles above about 30° should be avoided. The helix angle of the pinion member is of the same hand as that of the gear.

4. *Diametral pitch.* Since special tools are usually required to produce the internal gear, thought should be given to standardizing on ranges of numbers of teeth and diametral pitches for internal gearing. Table 4.6 shows suggested diametral pitches. Internal gearing can be made long and short addendum so that the need of nonstandard pitches to meet special center distances is virtually nonexistent.

5. *Normal circular pitch.* In helical internal gears the normal circular pitch may be specified on the internal gear cutter in the case of existing cutters.

4.2.4 Standard Systems for Bevel Gears

Over the years from 1950 to 1990, the Gleason Works in Rochester, New York, USA, has published recommended standard systems for bevel gears. AGMA has also published standard data for bevel gears. At the present time, the older standards are being revised and updated. The widespread use of the computer has tended toward more complex calculation procedures.

The reader is advised to obtain the latest standards from both AGMA and the Gleason Works. If manufacturing equipment built by German, Swiss, or Japanese companies is in use, appropriate standard data should be obtained from the manufacturer.

For the general guidance of the reader, the principal characteristics of systems now in use are discussed. The reader is also referred to Chapter 3 of Dudley.[3] Data on both bevel gear design and load rating are given.

Discussion of 20° Straight Bevel Gear System. The tooth form of the gears in this system is based on a symmetrical rack. In order to avoid undercut and to achieve approximately equal strength, a different value of addendum is employed for each ratio. If these gears are cut on modern bevel gear generators, they will have a localized tooth bearing. This resulting Coniflex tooth form is discussed in Sec. 4.1.4. The selection of

GEAR TOOTH DESIGN

addendum ratios and outside diameter is limited to a 1:1.5 ratio in width of top lands of pinion and gear. The face cone of the gear and pinion blanks is made parallel to the root cone elements to provide parallel clearance. This permits the use of larger edge radii on the generating tools with the attendant greater fatigue strength. Figure 4.11 shows data on the relation of dedendum angle to undercut for straight bevel gears.

Discussion of Spiral Bevel Gear System. Tooth thicknesses are proportioned so that the stresses in the gear and pinion will be approximately equal with the left-hand pinion driving clockwise or a right-hand pinion driving counterclockwise. The values shown apply for gears operated below their endurance limits. For gears operated above their endurance limits, special proportions will be required. Special proportions will also be required for reversible drives on which optimum load capacity is desired. The method of establishing these balances is beyond the scope of this chapter but may be found in the Gleason Works publication "Strength of Bevel and Hypoid Gears."[6]

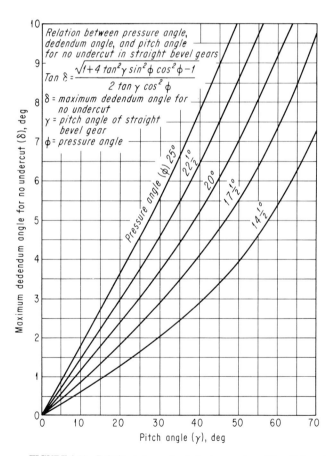

FIGURE 4.11 Relation between the dedendum angle and the pitch angle at which undercut begins to occur in generating straight bevel gears using sharp-cornered tools

The tooth proportions are based on the 35° spiral angle. A smaller angle may result in undercut and a reduction in contact ratio. See Fig. 4.12.

Discussion of Zerol Bevel Gear System. The considerations of tooth proportions to avoid undercut and loss of contact ratio as well as to achieve optimum balance of strength are similar in this system to those of the straight and spiral systems.

This system is based on the duplex cutting method in which the root cone elements do not pass through the pitch cone apex. The face cone of the mating member is made parallel to the root cone to produce a uniform clearance.

Special Tooth Forms. The teeth of bevel gears can be given special forms to reduce manufacturing costs or for engineering reasons.

Because of the flexibility of the machining system used to cut most bevel gears, many forms are possible. In general, it is customary to produce a desired profile on the teeth of the gear member and then to generate a conjugate profile on the teeth of the pinion.

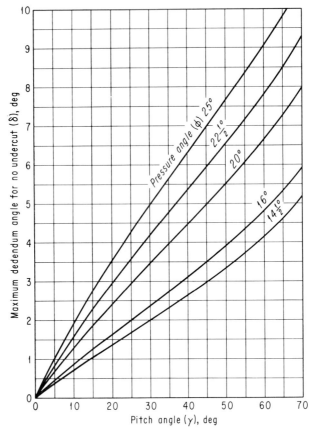

FIGURE 4.12 Relation between the dedendum angle and the pitch angle at which undercut begins to occur in generating spiral bevel gears at 35° spiral angle using sharp-cornered tools

GEAR TOOTH DESIGN 4.55

TABLE 4.21 Tooth Proportions of Standard Bevel Gears

Item	Type of tooth		
	Straight teeth	Spiral teeth	Zerol teeth
Pressure angle, deg.	20 standard[a]	20 standard[d]	20 basic, 22½, or 25, where needed
Working depth	$2.000/P_d$	$1.700/P_d$	$2.00/P_d$
Clearance	$0.188/P_d + 0.002''$	$0.188/P_d$	$0.188/P_d + 0.002''$
Face width	$F \leq A_o/3$ or $F \leq 10/P_d$ do not exceed smaller value[b]	$F \leq 0.3A_o$ or $F \leq 10/P_d$ whichever is smaller[b]	$F \leq 0.25A_o$ or $F \leq 10/P_d$ whichever is smaller. $F < 1.00$ on duplex Zerols[b,g]
Spiral angle	....	35°[e]	0°
Min. No. of teeth in system	13;[c] see Table 4.22	12	13
Whole depth	$\dfrac{2.188}{P_d} + 0.002''$	$\dfrac{1.888}{P_d}$[f]	$\dfrac{2.188}{P_d} + 0.002''$
Diametral-pitch range	....	12 and coarser	3 and finer
AGMA ref.	208.02	209.02	202.02

[a]20° is the standard pressure angle for straight-tooth bevel gears. Table 4.22 shows ratios that may be cut with a 14½° pressure angle.

[b]If the face width exceeds one-third the outer cone distance, the tooth is in danger of breakage in the event that tooth contact shifts to the small end of the tooth.

[c]This is the minimum number of teeth in the basic system (see Table 4.22 for equivalent minimum number of teeth in the gear member).

[d]20° is the standard pressure angle for spiral bevel gears. Table 4.22 shows ratios that may be cut with 14½ and 16° pressure angle teeth.

[e]35° is the standard spiral angle. If smaller spiral angles are used, undercut may occur and the contact ratio may be less.

[f]For gears of 10 diametral pitch and coarser, the teeth are often rough-cut 0.005 deeper to avoid having the finishing blades cut on their ends.

[g]On duplex Zerol bevel gears 1 in. is the maximum face width in all cases.

Although it is theoretically possible to use bevel gear teeth with an involute profile, this is rarely done. The most common curve for bevel gearing is the octoid, so called because of the shape of its line of action. This curve is much easier to generate than the spherical involute. It closely approximates the involute.

The Formate* tooth form is often used for gearing of high ratios because of its manufacturing economies. The teeth of the gear member are cut without generation, thus saving time, and the extra generation required to produce a conjugate pair is taken on the pinion. Since there are fewer pinion teeth, less time is spent in generation than if both pinion and gear teeth were generated.

Another tooth form suitable for high-speed manufacture is the Revacycle† straight bevel gear. This special tooth form is generated by a large disk cutter, one space being completed with each cutter revolution.

Both these forms are beyond the scope of this chapter.

---

*Registered trademark of the Gleason Works, Rochester, N.Y.
†Registered trademark of the Gleason Works, Rochester, N.Y.

Limitations in 20° Straight Bevel Gear System. The data contained in Table 4.21 apply only to gears that meet the following requirements:

1. The standard pressure angle is 20°; in certain cases, depending on numbers of teeth, other pressure angles may be used (see Table 4.22).
2. In all cases, full-depth teeth are used. Stub teeth are avoided because of the reduction in contact ratio, which may increase noise, and the reduction in wear resistance.

TABLE 4.22 Pressure Angle and Ratio, Minimum Number of Teeth in Gear and Pinion That Can Be Used with Any Given Pressure Angle

Pressure angle, deg	Type of bevel gear					
	Straight tooth		Spiral tooth		Zerol tooth	
	Pinion	Gear	Pinion	Gear	Pinion	Gear
20 (standard)	16	16	17	17	17	17
	15	17	16	18	16	20
	14	20	15	19	15	25
	13	30	14	20		
			13	22		
			12	26		
14½	29	29	28	28		
	28	29	27	29		
	27	31	26	30		
	26	35	25	32		
	25	40	24	33	Not used	
	24	57	23	36		
			22	40		
			21	42		
			20	50		
			19	70		
16			24	24		
			23	25		
			22	26		
			21	27		
	Not used		20	29	Not used	
			19	31		
			18	36		
			17	45		
			16	59		
22½	13	13	16	16	14	14
			to		13	15
			16	19		
			15	15		
			to			
			15	24		
			14	14		
			13	15		
25	12	12	13	13	13	13
			to			
			13	14		
			12	12		

GEAR TOOTH DESIGN   **4.57**

3. Long- and short-addendum teeth are used throughout the system (except for 1:1 ratios) to avoid undercut and to increase the strength of the pinion.
4. The face width is limited to between one-fourth and one-third cone distance. The use of greater face widths results in an excessively small tooth size at the inner ends of the teeth.

Limitations in Spiral Bevel Gear System. This system is more limited in its application than the straight tooth system. The data in this system do *not* apply to the following:

1. Automotive rear-axle drives
2. Formate pairs
3. Gears and pinions of 12 diametral pitch and finer, which are usually cut with one of the duplex spread-blade methods
4. Gear cut spread-blade and pinion cut single-side, with a spiral angle of less than 20°
5. Ratios with fewer teeth than those listed in Table 4.22
6. Large spiral bevel gears cut on the planning-type generators where the spiral angles should not exceed 30°

Designs that fall in these restrictions should be referred to the Gleason Works for assistance.

Limitations in Zerol Bevel Gear System. The data contained in Table 4.21 apply only to Zerol bevel gears that meet the following requirements:

1. The standard pressure angle (basic) is 20°. Where needed to avoid undercut, 22½ and 25° pressure angles are standard (see Table 4.22).
2. The face width is limited to 25 percent of the cone distance since the small-end tooth depth decreases even more rapidly as the face width increases because of the duplex taper. In duplex Zerols, 1 in. face is the maximum value in any case.

General Comments. Table 4.5 will be found helpful in selecting the proper numbers of teeth for most power gearing applications. Table 4.22 gives the minimum numbers of teeth in the gear for each number of pinion teeth and pressure angle. In general, the more teeth that are in the pinion, the more quietly it will run and the greater will be its resistance to wear. The equipment used to produce bevel gearing imposes upper limits on numbers of teeth. In general, if the gear is to contain over 120 teeth if an even number, or above 97 if a prime number, the manufacturing organization that is to produce the gears should be checked for capability.

Table 4.22 gives the minimum numbers of teeth in the pinion for each number of teeth in the gear and for each pressure angle. The minimum number of pinion teeth is based on considerations of undercut.

Table 4.23 shows addendum values that have been used in the past. These values are at the large end of the tooth (outer cone distance). These values show what was usually used in older designs that are in production. New standards that are now being developed may not agree with these values. (If a new standard is used, all values should be taken from the new standard—including addendum and backlash.)

Table 4.24 shows typical backlash values for bevel gears. In the field of bevel gears, mounting distance settings can change backlash and change the tooth contact pattern. Usually the desired contact is achieved when a bevel gear set is mounted at the specified mounting distances. Figure 4.13 shows a typical mounting surface and body dimensions for a bevel gear.

TABLE 4.23 Bevel Gear Addendum (for 1 Diametral Pitch)*

Ratio (m_g)		Addendum		Ratio (m_g)		Addendum	
From	To	Straight and Zerol	Spiral	From	To	Straight and Zerol	Spiral
1.00	1.00	1.000	0.850	1.52	1.56	0.730	0.620
1.00	1.02	0.990	0.840	1.56	1.57	0.720	0.620
1.02	1.03	0.980	0.830	1.57	1.60	0.720	0.610
1.03	1.04	0.970	0.820	1.60	1.63	0.710	0.610
1.04	1.05	0.960	0.820	1.63	1.65	0.710	0.600
1.05	1.06	0.950	0.810	1.65	1.68	0.700	0.600
1.06	1.08	0.940	0.800	1.68	1.70	0.700	0.590
1.08	1.09	0.930	0.790	1.70	1.75	0.690	0.590
1.09	1.11	0.920	0.780	1.75	1.76	0.690	0.580
1.11	1.12	0.910	0.770	1.76	1.82	0.680	0.580
1.12	1.13	0.900	0.770	1.82	1.89	0.670	0.570
1.13	1.14	0.900	0.760	1.89	1.90	0.660	0.570
1.14	1.15	0.890	0.760	1.90	1.97	0.660	0.560
1.15	1.17	0.880	0.750	1.97	1.99	0.650	0.560
1.17	1.19	0.870	0.740	1.99	2.06	0.650	0.550
1.19	1.21	0.860	0.730	2.06	2.10	0.640	0.550
1.21	1.23	0.850	0.720	2.10	2.16	0.640	0.540
1.23	1.25	0.840	0.710	2.16	2.23	0.630	0.540
1.25	1.26	0.830	0.710	2.23	2.27	0.630	0.530
1.26	1.27	0.830	0.700	2.27	2.38	0.620	0.530
1.27	1.28	0.820	0.700	2.38	2.41	0.620	0.520
1.28	1.29	0.820	0.690	2.41	2.58	0.610	0.520
1.29	1.31	0.810	0.690	2.58	2.78	0.600	0.510
1.31	1.33	0.800	0.680	2.78	2.82	0.590	0.510
1.33	1.34	0.790	0.680	2.82	3.05	0.590	0.500
1.34	1.36	0.790	0.670	3.05	3.17	0.580	0.500
1.36	1.37	0.780	0.670	3.17	3.41	0.580	0.490
1.37	1.39	0.780	0.660	3.41	3.67	0.570	0.490
1.39	1.41	0.770	0.660	3.67	3.94	0.570	0.480
1.41	1.42	0.770	0.650	3.94	4.56	0.560	0.480
1.42	1.44	0.760	0.650	4.56	4.82	0.560	0.470
1.44	1.45	0.760	0.640	4.82	6.81	0.550	0.470
1.45	1.48	0.750	0.640	6.81	7.00	0.540	0.470
1.48	1.52	0.740	0.630	7.00	00	0.540	0.460

*NOTE: In the earlier AGMA Standards and Gleason handbooks, the equations for addendum for all types of bevel gears were a_{oG} = Table 4.23/P. The values used in this table should be used only when checking calculations based on earlier standards.

4.2.5 Standard System for Worm Gears

There have been AGMA standards for worm gear tooth proportions in the past. Like other types of gear teeth, the old standards are being modified. For general reference, Table 4.25 presents the tooth proportions for single-enveloping worm gears and for double-enveloping worm gears. These proportions are typical of past design practice.

GEAR TOOTH DESIGN

TABLE 4.24 Design Backlash for Bevel Gearing, in.

Diametral pitch	Range of design backlash	Diametral pitch	Range of design backlash
1.00–1.25	0.020–0.030	3.50–4.00	0.007–0.009
1.25–1.50	0.018–0.026	4.0–5.0	0.006–0.008
1.50–1.75	0.016–0.022	5.0–6.0	0.005–0.007
1.75–2.00	0.014–0.018	6.0–8.0	0.004–0.006
2.00–2.50	0.012–0.016	8.0–10.0	0.003–0.005
2.50–3.00	0.010–0.013	10.0–12.0	0.002–0.004
3.00–3.50	0.008–0.011	12 and finer	0.001–0.003

* *Note:* In the earlier AGMA Standards and Gleason handbooks, the equations for addendum for all types of bevel gears were $a_oG = \dfrac{\text{Table 4.23}}{P}$. The values used in this table should be used only when checking calculations based on earlier standards.

General Rules. The following rules apply to conventional single-enveloping worm gears.

1. The worm axis is at a right angle to the worm gear axis (90°).
2. The worm gear is hobbed. Except for a small amount of oversize, the worm gear hob has the same number of threads, the same tooth profile, and the same lead as that of the mating worm. (A slight change in lead may be made to compensate for oversize effects.)
3. Fine-pitch worms are usually milled or ground with a double-conical cutter or grinding wheel with an included angle of 40° (tool pressure angle is 20°).
4. Coarse-pitch worms are often made with a straight-sided milling or grinding tool—in American practice. Some European countries favor making the worm member an involute helicoid. Functionally, either type will work well in practice providing that the accuracy and fit of mating parts is of equal quality.

For double-enveloping worm gears, the defined profile is that of the worm gear. The worm is made with a special tool that has a shape similar to that of the worm gear. For the double-enveloping worm gear, the usual shape is straight-sided in the axial section and tangent to defined basic circle.

Basic Tooth Forms for Worm Gearing. The tooth forms of worm gearing have not been standardized to the same degree as tooth forms for spur gearing, for example. Since a special hob has to be made to cut the gears that will mesh with each design of worm and since the elements that constitute good worm gearing design are less well understood, there has been little incentive to standardize.

The most general practice in worm gear design and manufacture is to establish the shape of the worm thread and then to design a hob that will generate teeth on the gear

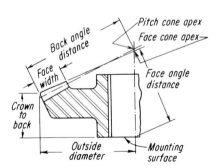

FIGURE 4.13 Bevel gear dimensions

TABLE 4.25 Typical Tooth Proportions for Worms and Worm Gears

Tooth Proportions for Single-Enveloping Worms and Worm Gears

	No. Worm Threads	Cutter Pressure Angle	Addendum	Working Depth	Whole Depth
Index or holding mechanism	1 or 2	14.50°	1.000	2.000	2.250
Power gearing	1 or 2	20°	1.000	2.000	2.250
	3 or more	25°	0.900	1.800	2.050
Fine-pitch (instrument)	1–10	20°	1.000	2.000	2.200 + 0.002 in. English
	1–10	20°	1.000	2.000	2.200 + 0.05 mm Metric

Tooth Proportions for Double-Enveloping Worms and Worm Gears

	No. Worm Threads	Pressure Angle	Addendum	Working Depth	Whole Depth
Power gearing	1–10	20°	0.700	1.400	1.600

NOTES: The addendum, working-depth, and whole-depth values are for 1 normal diametral pitch. (The normal diametral pitch of the worm is equal to 3.141593 divided by the axial pitch of the worm and then the result divided by the cosine of the worm lead angle.)

In the English system, for normal diametral pitches other than 1, divide by the normal diametral pitch. In the metric system the values are for 1 normal module. For modules other than 1, multiply by the normal module.

that are conjugate to those on the worm. This practice is not followed in the case of double-enveloping gearing since special tooling based on the shape of the gear member is required to generate the worm.

The shape of the teeth of the worm member is dependent on the size of the tool and the method used to cut the threads. These threads may be cut with a straight-sided V-shaped tool in a lathe or milled with double conical milling cutters of up to 6-in. diameter in a thread mill or hobbed with a special hob, or ground in a thread-grinding machine with a grinding wheel having a diameter up to 20 in. In each case the worm thread profile will be noticeably different, especially in the case of the higher lead angles. Worms can also be rolled, in which case the thread profile will be still different. It is essential, therefore, for the designer of worm gearing either to specify the manufacturing method to be used to make the worm or to specify the coordinates of the worm profile.

Specific Calculations for Worm Gears. In designing worm gear units and making calculations several considerations should be kept in mind.

1. The ratio is the number of worm gear teeth divided by the number of worm threads. (The pitch diameter of the gear divided by the pitch diameter of the worm is almost never equal to the ratio!)
2. Due to the tendency of the worm gear to wear-in to best fit the worm, common factors between the number of worm threads and the number of worm gear teeth should be avoided.
3. The normal circular pitch of the worm and the worm gear must be the same. Likewise, the normal pressure angle of worm and the normal pressure angle of the worm gear must be the same.
4. The axial pitch of the worm and transverse circular pitch of the worm gear must be the same. In a like manner, the axial pressure angle of the worm must be the same as the transverse pressure angle of the worm gear.
5. The lead of the worm equals the worm axial pitch multiplied by the number of threads. The lead can be thought of as the axial advance of a worm thread in one turn (360°) of the worm.
6. It has been customary to thin the worm thread to provide backlash but not thin the worm gear teeth.
7. There is a small difference between the pressure angle of a straight-sided tool used to mill or grind worm threads and the normal pressure angle of the worm thread.

Table 4.25 shows typical tooth proportions for single-enveloping worm gear sets and for double-enveloping worm gear sets. This table shows the addendum is one-half the working depth. In some cases it may be desirable to make the worm addendum larger than the addendum of the worm gear.

Table 4.26 shows recommended minimum numbers of teeth for single-enveloping worm gears. Since the pressure angle changes going across the face width of the worm gear, larger numbers of teeth are needed to avoid undercut problems in worm gears than in spur gears.

Table 4.27 shows that lead angle needs to increase with the numbers of threads. If,

TABLE 4.26 Minimum Number of Worm Gear Teeth for Standard Addendum

Pressure Angle, °	Min. No. of Teeth
14.50	40
20	25
25	20
30	15

TABLE 4.27 Suggested Limits on Lead Angle

No. Worm Threads	Lead angle,°
1	Less than 6
2	3 to 12
3	6 to 18
4	12 to 24
5	15 to 30
6	18 to 36
7 and higher	(not over 6° per thread)

TABLE 4.28 Some Recommended Normal Circular Pitches

Fine Pitch		Coarse Pitch	
English, in.	Metric, mm	English, in.	Metric, mm
0.030	1.000	0.200	5.000
0.040	1.250	0.250	6.500
0.050	1.500	0.300	8.000
0.065	2.000	0.400	10.000
0.080	2.500	0.500	12.000
0.100	3.000	0.625	15.000
0.130	3.500	0.750	20.000
0.160	4.000	1.000	25.000
		1.250	30.000
		1.500	40.000

NOTE: The normal circular pitch of the wormgear is equal to the axial pitch of the worm multiplied by the cosine of the lead angle.

for instance, a lead angle of 20° is desired to get good efficiency, the designer should plan to use 4 or 5 threads on the worm.

Worm gears are normally sized by picking the numbers of threads and teeth and then choosing a normal circular pitch that will give enough center distance to have the necessary load-carrying ability. Table 4.28 shows suggested normal circular pitches.

Table 4.29 shows nominal backlash values for worm gear sets. These values represent thinning of worm threads and tolerances on worm thread thickness and gear tooth thickness.

Table 4.30 shows recommended cutter or grinding wheel outside diameters for the making of single-enveloping worms.

Further details in worm gear design and rating are given in Dudley.[3]

GEAR TOOTH DESIGN

TABLE 4.29 Recommended Values of Backlash for Single-Enveloping Worm Gearing and Double-Enveloping Worm Gearing

Center distance	Backlash (amount worm should be reduced)	
2	0.003	0.008
6	0.006	0.012
12	0.012	0.020
24	0.018	0.030

TABLE 4.30 Recommended Values of Cutter Diameter (Single-Enveloping Gear Sets)

Type of worm	Suggested cutter diameter D_c, in.	Process
Low-speed power............	4	Milled
High-speed power...........	20	Ground
Fine-pitch:		
Commercial quality.......	3	Milled
Precision quality.........	20	Ground

4.2.6 Standard System for Face Gears

Fine-pitch (20 diametral pitch and finer) face gears have been standardized by the American Gear Manufacturers Association. Coarse-pitch face gears have not been standardized. The following data are arranged to provide a logical method of calculating the tooth proportions of face gears. Such gears will be suitable for most applications. For a more complete treatment of the subject the reader is referred to current AGMA standards. Table 4.31 shows basic tooth proportions for face gear sets.

Caution should be exercised in using Table 4.31. The items shown apply only to gears that meet the following requirements:

1. The axes of the gear and the pinion must intersect at an angle of 90°.
2. The gear must be generated by means of a reciprocating pinion-shaped cutter having the same diametral pitch and pressure angle as the mating pinion and must be of substantially the same size.
3. The pinion should be sized to meet Chap. 11 requirements on load-carrying capacity.
4. The minimum number of teeth in the pinion in this system is 12, and the minimum cutter pitch diameter is 0.250 in.

TABLE 4.31 Tooth Proportions for Pinions Meshing with Face Gears

Item	Coarse Pitch $P_d = 20$		Fine Pitch $P_d = 20$	
No. of teeth in pinion N_p	$N_p < 18$	$N_p \geq 18$	$N_p < 18$	$N_p \geq 18$
Pressure angle $\phi,°$	20	20	20	20
Addendum a	See Table 4.33	$\dfrac{1}{P_d}$	See Table 4.33	$\dfrac{1}{P_d}$
Standard pitch diameter D_p	$\dfrac{N}{P_d}$	$\dfrac{N}{P_d}$	$\dfrac{N}{P_d}$	$\dfrac{N}{P_d}$
Working depth h_k		$\dfrac{2}{P_d}$	See Table 4.34	$\dfrac{2.0}{P_d}$
Whole depth h_t		$\dfrac{2.25}{2P_d}$	See Table 4.34	$\dfrac{2.2}{P_d} + 0.002$
Circular tooth thickness* t		$\dfrac{\pi}{2P_d}$	See Table 4.34	$\dfrac{\pi}{2P_d}$
Clearance c		$\dfrac{0.25}{P_d}$	$\dfrac{0.2}{P_d} + 0.002$	$\dfrac{0.2}{P_d} + 0.002$
Reference standard	None		AGMA 203.01	

*Thin teeth for backlash. See Sec. 4.5 for general information on spur gear backlash.

TABLE 4.32 Minimum Numbers of Teeth in Pinion and Face Gear

Diametral pitch range	Min.* numbers of teeth	
	Pinion	Gear
20– 48	12	18
49– 52	13	20
53– 56	14	21
57– 60	15	23
61– 64	16	24
65– 68	17	26
69– 72	18	27
73– 76	19	29
77– 80	20	30
81– 84	21	32
85– 88	22	33
89– 92	23	35
93– 96	24	36
97–100	25	38

*The minimum numbers of teeth in the pinion are limited by the design requirements of the gear cutter. These requirements, in addition to the minimum-gear-ratio limitation, limit the numbers of teeth in the gear.

GEAR TOOTH DESIGN

TABLE 4.33 Addendum of Face Gears and Pinions

No. of teeth in pinion	Coarse pitch Pinion addendum	Coarse pitch Gear addendum	Fine pitch pinion addendum
12	1.120	0.700	$\dfrac{1.1215}{P_d} - 0.002$
13	1.100	0.760	$\dfrac{1.1050}{P_d} - 0.002$
14	1.080	0.820	$\dfrac{1.0865}{P_d} - 0.002$
15	1.060	0.880	$\dfrac{1.0650}{P_d} - 0.002$
16	1.040	0.940	$\dfrac{1.0420}{P_d} - 0.002$
17	1.020	0.980	$\dfrac{1.0175}{P_d} - 0.002$
18 and 19	1.000 1.250	1.000 0.750	
20 and 21	1.000 1.250	1.000 0.750	
22 and 29	1.000 1.200	1.000 0.800	$\dfrac{1.0000}{P_d} - 0.002$
30 through 40	1.000 1.150	1.000 0.850	
40 and higher	1.000 1.100	1.000 0.900	

5. The minimum gear ratio is 1.5:1 and the maximum ratio is 12.5:1.
6. The long- and short-addendum designs for the pinion with less than 18 teeth should not generally be used on speed-increasing drives.

Pinion Design. The pinion member in cases of numbers of teeth less than 18 has an enlarged tooth thickness to avoid undercut when cut with a standard cutter. Compared with the pinions designed in accordance with data in Sec. 4.2.1 the outside diameter is somewhat less. This is to avoid the necessity of cutting the gear with shaper cutters that have excessively pointed teeth. The limit set is top land width $\geq 0.40/P_d$. Table 4.34 gives the tooth proportions for face gear pinions.

Face Gear Designs. The face-gear member is generated by a cutter having proportions based on the pinion with which the face gear will operate. The most important specification for the shape of the teeth of a face gear is a complete specification of the cutter to be used to cut it or, next best, a detailed specification of the mating pinion.

The face gear has two dimensions unique to face gears which control the face width of the teeth: the outer and inner diameter of the face gear.

The maximum usable face width may be estimated from Table 4.35. The outside

TABLE 4.34 Tooth Proportions for Fine-Pitch Face Pinions

No. of teeth	Tooth thickness at standard pitch dia.	Whole depth	Working depth*
12	1.7878	2.0234	1.8234
13	1.7452	2.0654	1.8604
14	1.7027	2.1054	1.9054
15	1.6601	2.1424	1.9424
16	1.6175	2.1778	1.9778
17	1.5749	2.2118	2.0118
18	1.5708	2.2000	2.0000
19 and up	1.5708	2.2000	2.0000

* Divide by diametral pitch and subtract 0.002".

TABLE 4.35 Tooth Proportions and Diameter Constants for 1 Diametral Pitch Face Gears, 20° Pressure Angle

No. of pinion teeth N_P	Gear dia. constants							
	$m_g = 1.5$		$m_g = 2$		$m_g = 4$		$m_g = 8$	
	m_o	m_i	m_o	m_i	m_o	m_i	m_o	m_i
12	.....	.....	1.221	1.020	1.221	0.960	1.221	0.945
13	1.202	1.064	1.202	1.015	1.202	0.959	1.202	0.945
14	1.187	1.062	1.187	1.011	1.187	0.958	1.187	0.944
15	1.174	1.052	1.174	1.007	1.174	0.957	1.174	0.944
16	1.161	1.051	1.161	1.004	1.161	0.956	1.161	0.944
17	1.156	1.041	1.156	1.000	1.156	0.955	1.156	0.944
18	1.150	1.039	1.150	0.997	1.150	0.954	1.150	0.943
	1.176	1.042	1.176	0.999	1.176	0.955	1.176	0.943
20	1.144	1.030	1.144	0.991	1.144	0.953	1.144	0.943
	1.166	1.032	1.166	0.993	1.166	0.953	1.166	0.943
22	1.140	1.022	1.140	0.987	1.140	0.952	1.140	0.943
	1.156	1.024	1.156	0.988	1.156	0.952	1.156	0.943
24	1.133	1.015	1.133	0.983	1.133	0.951	1.133	0.942
	1.150	1.017	1.150	0.984	1.150	0.951	1.150	0.943
30	1.121	1.001	1.121	0.975	1.121	0.949	1.121	0.942
	1.131	1.001	1.131	0.975	1.131	0.949	1.131	0.942
40	1.109	0.986	1.109	0.966	1.109	0.946	1.109	0.941
	1.113	0.986	1.113	0.966	1.113	0.946	1.113	0.941

GEAR TOOTH DESIGN 4.67

diameter of the gear should not exceed $m_o D$. The inside diameter should not be less than $m_i D$. Thus

$$\text{Outer diameter} = m_o D = D_o$$
$$\text{Inner diameter} = m_i D = D_i$$
$$\text{Pitch diameter} = D$$
$$\text{Face width} = \frac{D_o - D_i}{2}$$

4.2.7 System for Spiroid and Helicon Gears

At the present time there are no AGMA standards covering Spiroid, Helicon, or Planoid* gears. The following material is based on information covered in "Spiroid Gearing," paper 57-A-162 of the ASME,[7] and on additional information supplied by its author, W. D. Nelson.

Although specialized designs can be developed that will best suit a given application, the standardized procedure presented here will yield designs that will meet the needs of most Spiroid and Helicon gear applications.

Planoid gears are often associated with Spiroid and Helicon gearing but are an entirely different form of gearing, used for ratios generally under 10:1 where maximum strength and efficiency are required. Their design is beyond the scope of this section.

The basic approach to the establishment of the Spiroid or Helicon tooth form is to establish the pinion tooth form and then develop a gear tooth form that is conjugate to it.

The tooth form of the Spiroid gear is comprised of teeth having different pressure angles on each side of the teeth; a "low"-pressure-angle side and a "high"-pressure-angle side. The choice of pressure angle will control the extent of the fields of conjugate action of the teeth. The values of pinion taper angle have been standardized by the Illinois Tool Works in order to achieve the best general-purpose gearing. These standard values are shown in Table 4.36.

The design of Helicon gearing is closely related to Spiroid gearing. Table 4.37 shows the general equations used in the design of both kinds of gearing. The general Spiroid formulas become Helicon formulas when the pinion taper angle τ is set equal to zero. Figure 4.14 shows a Spiroid gearset. A Helicon pinion and gear are similar except that the pinion taper angle τ and the gear face angle γ are both *zero* in Helicon gearing.

Spiroid Gearing. Spiroid gearing is suitable for gear ratios of 10:1 or higher. The numbers of teeth in the gear can range from 30 to 300.

The design procedure for Spiroid gearing differs somewhat from other types of gearing discussed in this chapter. The following are the basic steps:

---
*Spiroid, Helicon, and Planoid are trademarks registered by the Illinois Tool Works, Chicago, Ill.

TABLE 4.36 Standard Tooth and Gear Blank Relationships, Spiroid and Helicon Gearing

Spiroid gearing:
1. "Sigma" angle, $\sigma_P = 40°$ (standard)
2. Pinion taper angle τ vs. gear face angle γ

Pinion taper angle,°	Gear face angle,*
τ	γ
5	8 preferred
7	11
10	14

3. Gear ratio $m_G = 10:1$ higher
4. No. of teeth in gear:
   Varies with center distance (see Fig. 4.17)
   Hunting ratios are desirable with all multiple-thread pinions
5. Tentative pressure angle selection

Ratio	Pressure angle selection	
	Low side	High side
$M_G \leq 16:1$	15°	35°
$M_G > 16:1$	10°	30°

*Approximate values, for $\sigma_P = 40°$.

1. Center distance is the basic starting point for the calculation of Spiroid tooth proportions. Center distance and gear ratio are first established based on information given in Chap. 11.
2. Lead and the pinion "zero" plane radius are then calculated.
3. The tooth proportions and the blank proportions are based on the pinion zero plane radius and are then calculated.

From this procedure it can be seen that the pitch of the teeth is an end result of the calculating procedure for Spiroid gearing, rather than the beginning as on other types of gearing.

Center distance governs the horsepower capacity of the gear set. Chapter 11 gives information on the center distances required for the various load ratings of Spiroid gears.

Table 4.36 shows the basic tooth proportions for Spiroid gearing. Table 4.37 gives the general equations used to determine Spiroid tooth proportions.

If optimum efficiency is desired, the size of the pinion relative to the gear should be kept small by using values of R_G/R_P which are larger than those shown in Fig. 4.15. This has the effect of producing an increased lead angle for a given center distance and gear ratio. Figures 4.16 and 4.17 show recommended data for Helicon and Spiroid calculations.

If control gear applications are being considered, in which runout is critical, the driving side should be the low-pressure-angle side of the teeth. This minimizes the

TABLE 4.37 General Equations for Spiroid and Helicon Gearing

Spiroid gearing:

$$\sin \phi_P = \frac{\tan \tau}{\tan \sigma_P} \qquad (4.38)$$

$$R_P = \frac{C}{\sin \phi_P + (R_G/R_P) \cos \sigma_P} \qquad (4.39)$$

$$R_G = \frac{R_G}{R_P} R_P \qquad (4.40)$$

$$L = \frac{2\pi R_G \cos \sigma_P}{m_G - (R_G/R_P) \sin \sigma_P \cos \phi_P} \qquad (4.41)$$

$$x_P = R_G \sin \sigma_P \qquad (4.42)$$

$$r_o = R_P - x_P \tan \tau \qquad (4.43)$$

$$D_W = \frac{0.6 \, (L/N_P) \sec \tau}{[\sin \psi_1 / \cos (\psi_1 + \tau)] + [\sin \psi_2 / \cos (\psi_2 - \tau)]} \qquad (4.44)$$

$$D_N = D_W \cos \tau \qquad (4.45)$$

$$CLR = 0.07 \frac{L}{N_P} + 0.002 \qquad (4.46)$$

$$\psi_{2L} = \tan^{-1} \left[\frac{R_P}{x_P} + \frac{CZ_P}{R_P(ky_P - x_P)} \right] \qquad (4.47)$$

$$\tan \lambda_m = \frac{L \sec \tau}{2\pi r_m} \qquad (4.48)$$

$$\psi_2 = \psi_{2L} + 5° \qquad (4.49)$$

Helicon gearing:

$$R_P = \frac{C}{(R_G/R_P) \cos \sigma_P} \qquad (4.50)$$

$$L = \frac{2\pi R_G \cos \sigma_P}{m_G - (R_G/R_P) \sin \sigma_P} \qquad (4.51)$$

$$D_w = \frac{0.6(L/n)}{\tan \psi_1 + \tan \psi_2} \qquad (4.52)$$

$$\psi_{2L} = \tan^{-1} \frac{R_P}{x_P} \qquad (4.53)$$

$$\psi_{2L} = \tan^{-1} \frac{1}{(R_G/R_P) \sin \sigma_P} \qquad (4.54)$$

$$\psi_2 = \psi_{2L} + 7° \qquad (4.55)$$

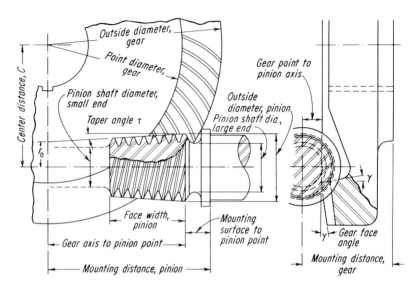

FIGURE 4.14 Spiroid gear set—mounting and gear blank nomenclature

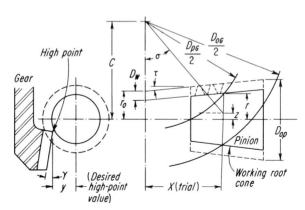

FIGURE 4.15 Gear tooth high point—spiroid gearing

GEAR TOOTH DESIGN 4.71

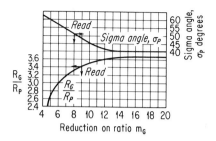

FIGURE 4.16 Sigma angle for different ratios—Helicon gears

effects of runout in both members. A 10° taper angle τ may be used, since it requires only one-half the axial movement of the pinion to produce a given change in backlash.

For gear sets of the highest strength, the number of teeth in the gear should be kept low (30 to 40 for lower ratios) and a stub tooth form employed. Such designs are beyond the scope of this section.

Straddle-mounted pinions may require a shaft of larger size (greater stiffness) than would result from the data shown in Fig. 4.14. In such cases, the values of R_G/R_P are decreased until a shaft of the desired size can be achieved. There is a definite limit to the maximum size of pinion for a given pressure angle. The size of the pinion may be increased until a "limiting pressure" angle condition is reached. Limiting pressure angle can be calculated by means of Eq. (4.49).

Helicon Gearing. Helicon gearing is suitable for gear ratios from 4:1 to 400:1 in light- to medium-loaded applications.

The design procedure for Helicon gears is somewhat similar to that for Spiroid gears but starts with the gear outside diameter as the basis of load capacity instead of the center distance as is used in Spiroid gearing.

The following are the basic steps:

1. The gear ratio m_G, pinion r/min, and horsepower are used to establish the necessary gear outside diameter as shown in Chap. 11.
2. Pinion pitch radius, center distance, lead, and working depth are next established.
3. From these values, the gear and pinion blank dimensions are established.

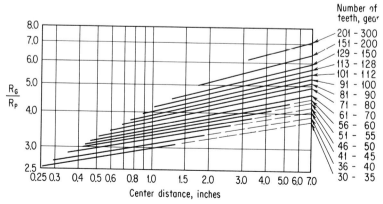

FIGURE 4.17 R_G/R_P vs. center distance and numbers of teeth—Spiroid gears

Detailed Calculations of Spiroid and Helicon Tooth Data. In the first edition of Dudley,[1] Tables 5.41 and 5.42 were presented to show detail calculations of 59 items of specific tooth data relating to the specific geometry of the teeth. Several pages of instruction were also given on how to use or interpret the various items.

Computer programs for gear tooth data were not in use in the 1960s. Now in the 1990s, computer programs are in general use. The "software" of these programs is established and kept up to date by the patent holders and developers of special gear types.

A lengthy detail calculation procedure is not presented in this book. For Spiroid and Helicon gears, the reader is advised to obtain the latest technical publications of the Spiroid Division of the Illinois Tool Works. When a gear application gets down to final design calculations, arrangements can be made to have a computer run made with the latest and best computer software.

4.3 GENERAL EQUATIONS RELATING TO CENTER DISTANCE

This section of Chap. 4 deals with the distance between the shafts of meshing gears. This distance is called *center distance* in the case of gearing operating on nonintersecting shafts. Spur and helical gearing, both external and internal, worm gearing, hypoid gearing,* spiral gearing, Spiroid† gearing, and Helicon‡ gearing must all operate at specific center distances. Certain types of gears which operate on intersecting shafts, such as bevel gearing, do not have a center-distance dimension. However, their pitch surfaces must be maintained in the correct relationship; hence, the axial position of these gears along their shafts is critical. Such gears, therefore, have a *mounting distance* that defines the axial position of the gears.

Certain types of gears, such as hypoid gears, Spiroid gears, throated worms and worm gears, face gears, Helicon gears, and Planoid** gears, have both a center distance and a mounting distance.

When establishing the center distance for a set of gears, it is customary first to determine the theoretical center distance for the gears. The actual operating center distance is next determined. This center distance includes considerations of tolerance and the interrelationship of the various parts that may be included in the final assembly such as bearing, mounting brackets, and housings.

In spur and helical gears, it may be necessary or desirable to operate the gears on an appreciably larger center distance than theoretical to get improved load-carrying capacity. For instance, gears can be designed so that 20° hobs cut the tooth; but the "spread" center is calculated so as to make the gears operate at a 25° pressure angle. See Secs. 4.3.4, 4.3.5, and 4.3.6.

The next article covers the calculation of the theoretical center distances at which various types of gearing will operate. The most generally used equations are summa-

---

*In the case of hypoid gearing, *offset* is the correct term for the distance between gear and pinion shafts.
†Spiroid, Helicon, and Planoid are trademarks of the Illinois Tool Works, Chicago, Ill.
‡Spiroid, Helicon, and Planoid are trademarks of the Illinois Tool Works, Chicago, Ill.
**Spiroid, Helicon, and Planoid are trademarks of the Illinois Tool Works, Chicago, Ill.

TABLE 4.38 Center-Distance Equations

Standard center distance C

Spur, helical, and worm gears

$$C = \frac{d + D}{2} \tag{4.56}$$

$$C = \frac{n + N^*}{2P} \tag{4.57}$$

Internal gears

$$C = \frac{D - d}{2} \tag{4.58}$$

$$C = \frac{N - n}{2P} \tag{4.59}$$

Operating center distance C'

Spur, helical, and internal gears

$$C' = \frac{C \cos \phi}{\cos \phi'} \tag{4.60}$$

TABLE 4.38 Center-Distance Equations *(Continued)*

Spur, helical

$$C' = \frac{d' + D'}{2} \quad (4.61)$$

Internal

$$C' = \frac{D' - d'}{2} \quad (4.62)$$

General equations relating tooth thickness and center distance for parallel axis gearing

External spur and helical gears

$$\text{inv } \phi' = \frac{n(t_P + t_G + B) - \pi d}{(n + N)d} + \text{inv } \phi \quad (4.63)$$

Internal gears

$$\text{inv } \phi' = \frac{\pi d - n(t_P + t_G + B)}{(N - n)d} + \text{inv } \phi \quad (4.64)$$

Note: To obtain C' use ϕ' from Eq. (4.63) or (4.64) in Eq. (4.60).

General equations relating tooth thickness and center distance for nonparallel nonintersecting axis gearing

$$K_a = \frac{n}{N} = \frac{1}{m_G} \quad (4.65)$$

$$K_b = \text{inv } \phi_G + K_a \text{ inv } \phi_P - \frac{\pi(p_n - t_{nP} - t_{nG})}{Np_n} \quad (4.66)$$

$$K_d = \frac{\text{inv } \phi_G}{\text{inv } \phi_P} \quad (4.67)$$

$$K_c = K_a + K_d \quad (4.68)$$

$$\text{inv } \Phi_P = \frac{K_b}{K_c} \quad (4.69)$$

$$\text{inv } \Phi_G = K_b - K_a \text{ inv } \Phi_P \quad (4.70)$$

Check equations: To check equation series (4.65) through (4.70) above.

$$K_d' = \frac{\text{inv } \Phi_G}{\text{inv } \Phi_P} \quad (4.71)$$

$$K_d' = K_d \text{ (check equation)} \quad (4.72)$$

$$D_P' = \frac{D_{bP}}{\cos \Phi_P} \quad (4.73)$$

$$D_G' = \frac{D_{bG}}{\cos \Phi_G} \quad (4.74)$$

TABLE 4.38 Center-Distance Equations *(Continued)*

Use values from Eqs. (4.71) and (4.72) in Eq. (4.61) where

$$K_c = \frac{\sin \phi_G}{\sin \phi_P} = \frac{\sin \Phi_G}{\sin \Phi_P} \tag{4.75}$$

P = transverse diametral pitch
C = "standard" center distance [defined by Eqs. (4.56) and (4.57)]. See also Sec. 4.13.
C' = operating center distance
d = "standard" (n/P) pitch diameter of pinion. See also Sec. 4.14.
d' = operating pitch diameter of pinion. See also Sec. 4.15.
D = "standard" (N/P) pitch diameter of gear. See also Sec. 4.14.
D' = operating pitch diameter of gear. See also Sec. 4.15.
n = number of teeth in pinion
N = number of teeth in gear
ϕ = cutting pressure angle
B = design backlash, the total amount the teeth in both members are thinned for backlash considerations
ϕ' = operating pressure angle. See also Sec. 4.16.
t_P = circular (transverse) thickness of pinion tooth
t_G = circular (transverse) thickness of gear tooth
t_{nP} = circular (normal) thickness of pinion tooth
t_{nG} = circular (normal) thickness of gear tooth
ψ_P = helix angle of pinion
ψ_G = helix angle of gear
Φ_P = rolling pressure angle of pinion
Φ_G = rolling pressure angle of gear

*Does not apply to worm gearing.

rized in Table 4.38. In articles that follow, the special considerations on which the equations are based are covered in detail.

4.3.1 Center-Distance Equations

Table 4.38 is a summary of equations convenient to use to obtain the values of center distance for the various types of gearing shown in Fig. 4.18.

Center distance and tooth thickness are inseparable. In most cases, however, *standard* tooth proportions are used, and the simplified equations of center distance [Eqs. (4.56) through (4.59)] are adequate. *Standard* tooth proportions are defined in Sec. 4.3.2. When the sum of the tooth thickness of pinion and gear does not equal the circular pitch, nonstandard center distances, as obtained from equations such as Eqs. (4.60) and (4.63), will be required. Most problems involving center distance and tooth thickness fall into one of the following categories:

1. The thickness of the teeth of the gear and of the pinion is fixed. The center distance at which the gears will mesh properly is to be established.
    a. The sum of the tooth thicknesses of both members plus the design backlash is equal to the circular pitch. In this case, standard center distance is correct. Use Eqs. (4.56) through (4.59).
    b. The sum of the tooth thickness of both members plus the design backlash is *not* equal to the circular pitch. In this case, a nonstandard center distance is required. Use Eqs. (4.60) and (4.63). This problem is covered in greater detail in Sec. 4.3.9.
2. The center distance is fixed. The tooth thicknesses for gears that will operate on the given center distance are to be established. If the center distance for the given diametral pitch and number of teeth is different from that obtained from Eq. (4.57), Eq. (4.60) must be solved and the operating pressure angle thus found used in Eq. (4.63), which is then solved for the term $(t_P + t_G + B)$. In Sec. 4.1.4, the ways in which this total tooth thickness can be divided between gear and pinion are discussed.
3. Neither the tooth thickness nor the center distance is fixed; the best values for both are to be established. In this happy case, the tooth thickness is usually established first on the basis of strength or some other appropriate consideration and the center distance required is then based on Eqs. (4.63) and (4.60).
4. Both the center distance and the tooth thickness are fixed. The amount of backlash or the degree of tooth interference is to be established. This is a frequent "check problem" of an existing gear design. Equations (4.60) and (4.63) are used and solved for B. Minus values of B indicate tooth interference.

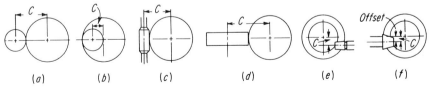

FIGURE 4.18 Center distance, shown for various kinds of gears, is the distance between the axes of the shafts measured on a common normal. *(a)* Spur and helical gearing—external; *(b)* spur and helical gearing—internal; *(c)* worm gearing—single- and double-enveloping; *(d)* spiral gearing; *(e)* Spiroid, Planoid, and Helicon gearing; *(f)* Hypoid gearing

4.3.2 Standard Center Distance

Most gear designs are based on standard tooth proportions. Such gears are intended to mesh on standard center distances. The equations that establish standard center distances are based on the following assumptions:

1. The sum of the circular tooth thicknesses (effective) equals the circular pitch minus the backlash.

$$t_P + t_G = p_t - B \qquad (4.76)$$

This rule covers the case of long and short addendums in which the gear teeth are corrected for such things as undercut or balanced strength. In such cases, the tooth thicknesses of both members are altered sufficiently to permit the pair to mesh on standard center distances.

2. The tooth thickness of an individual gear is one-half the circular pitch (transverse) minus one-half the backlash.

$$t = \frac{p - B}{2} \qquad (4.77)$$

This covers the more common case. It represents the approach used by most makers of "catalogue"-type gears. Such gears are all expected to operate on standard center distances regardless of the number of teeth in the pinion. The center distances on which such gears will operate may be calculated by means of Eqs. (4.56) and (4.57).

The tolerance on standard center distance can be bilateral (thus 5.000 in. $\pm$ 0.002 in.) if the value of B was chosen sufficiently large. This is the most convenient way from the standpoint of manufacturer. If B is not large enough, a unilateral tolerance (thus 5.000 in. $+$ 0.004 in. $-$ 000 in.) may be used.

4.3.3 Standard Pitch Diameters

The standard pitch diameter of a gear is a dimension of a theoretical circle. It is given for each type of gear by the following relations:

Spur gearing:

$$\frac{N}{P_d}$$

where P_d is the diametral pitch.

Helical gearing:

$$\frac{N}{P_t}$$

where P_t is the transverse diametral pitch.

Bevel gearing:

$$\frac{N}{P_d}$$

Worm gear:

$$\frac{N_G p}{\pi}$$

where N_G equals the number of worm gear teeth and p equals the circular pitch of the worm gear.

Since the diametral pitch of a gear is fixed by the tool that is used to cut it (hob, shaper cutter, shaving cutter, etc.), and since the number of teeth in the gear is a whole number, the *standard* pitch diameter is an imaginary circle. It can have no tolerance, regardless of the variations in tooth thickness or in the center distance on which it is to operate.

4.3.4 Operating Pitch Diameters

As shown elsewhere, it is entirely practical to operate involute gears of specific diametral pitch and numbers of teeth at various center distances. It is convenient in such cases to calculate *operating* pitch diameters for such gears. See Eqs. (4.78), (4.79), (4.80), and (4.81).

External spur and helical pinions:

$$d' = \frac{2C'}{m_G + 1} \qquad (4.78)$$

Internal spur and helical pinions:

$$d' = \frac{2C'}{m_G - 1} \qquad (4.79)$$

External spur and helical gears:

$$D' = \frac{2C'}{m_G + 1} \qquad (4.80)$$

Internal spur and helical gears:

$$D' = \frac{2C'}{m_G - 1} \qquad (4.81)$$

where C' = operating center distance
m_G = gear ratio $\left(\dfrac{N}{n}\right)$
d' = operating pitch diameter of pinion
D' = operating pitch diameter of gear

These equations define the operating pitch diameters as being proportional to the transmitted ratio and the instantaneous center distance.

Since gears with thicker-than-standard teeth must operate on enlarged center distances, they will run on operating pitch diameters that are larger than their standard pitch diameters. The operating pitch diameter should not be specified as a drawing dimension on the detail drawing of the gear since it will vary for every different gear and center distance. The best place to show the operating center distance is on an assembly drawing that shows both the pinion and gear that mesh together.

4.3.5 Operating Pressure Angle

The pressure angle of an individual gear is based on the diameter of the base circle of the gear and on the identification of the specific radius at which the pressure angle is to be considered. In standard gears, this radius is customarily the standard pitch radius. It is convenient to consider the pressure angle of gears operating on nonstandard center distances at the point of intersection of the line of action and line of centers. This is the definition of the operating pressure angle which may be calculated by means of Eq. (4.82).

$$\cos \Phi' = \frac{C \cos \Phi}{C'} \qquad (4.82)$$

where Φ' = operating pressure angle
C' = operating center distance
C = standard center distance [see Eq. (4.57)]
Φ = standard pressure angle

It can be seen from the foregoing that a gear can be cut with a cutter of one pressure angle and operated at a different pressure angle. This flexibility causes much of the confusion in gear design. It is necessary to define accurately each of the standard elements of a nonstandard gear—pitch and pressure angle. A specification of base-circle diameter is highly desirable.

4.3.6 Operating Center Distance

The actual center distance at which a gear will operate will have a large influence on the way in which the gear will perform in service. The actual operating center distance is made up of the combined effects of manufacturing tolerances, the basic center distance, differential expansion between the gears and their mountings, and deflections in the mountings due to service loads.

The items that should be considered when determining the minimum and maximum operating center distance for any given gear design are discussed in detail in Secs. 4.3.9 and 4.4.1.

In any critical evaluation of a gear design, particularly in the field of control gearing, the minimum and the maximum operating center distance should be used in equations covering backlash, contact ratio, tooth tip clearance, etc.

In this chapter the concept *operating center distance* is used in two ways. In one case, it is considered to be the center distance that results from the buildup of all tolerances that influence center distance in any one case. This concept is illustrated in case II in Sec. 4.4.1. Thus it is the largest or smallest actual center distance that could be encountered in a given design. In the other case, the one covered by Eqs. (4.60) and (4.63). For example, it is the center distance at which gears of a specified tooth thickness and backlash will operate. Depending on the problem encountered, the correct concept to use will have to be selected by the designer.

GEAR TOOTH DESIGN 4.81

4.3.7 Center Distance for Gears Operating on Nonparallel Nonintersecting Shafts

The most frequently encountered example of this type of gear operation is in the shaving or lapping of gears. Here, a cutter, usually a helical-toothed member, is run in tight mesh with a spur or helical gear. The operating center distance is dependent on the tooth thickness of each member. Equations (4.65) through (4.74) should be solved in sequence and the results used in Eq. (4.60) to obtain the operating center distance. Equation (4.75) gives the ratio constant for the transverse pressure angles for crossed axis gearing.

4.3.8 Center Distance for Worm Gearing

The center distance for worm gearing is based on the sum of the standard pitch diameters of the worm and worm gear Eq. (4.83). The standard pitch diameter of the worm is obtained from Eq. (4.84). This circle has no kinematic significance but is the basis for worm tooth proportions.

$$C = \frac{d_w + D}{2} \tag{4.83}$$

$$d_w = \frac{L}{\pi \tan \psi} \tag{4.84}$$

where d_w = pitch diameter of worm
C = center distance
D = pitch diameter of gear
L = lead
ψ = lead angle

Reasons for Nonstandard Center Distances. On some occasions a set of gears must operate on a center distance that is not one-half the sum of the standard pitch diameter of the meshing gears. The designer is confronted with nonstandard center distance in several situations, the more important of which are

1. Gear trains in which the teeth are made to standard tooth thickness and backlash is introduced by increasing the standard center distance slightly.
2. Gear trains in which the number of teeth and pressure angle relationship requires a long addendum (enlarged tooth thickness design) to avoid undercut, and yet the number of teeth in the gear is so small that to make the gear sufficiently short addendum to compensate for the pinion enlargement would cause undercut. In such cases, an enlarged center distance is usually indicated.
3. Gear trains in which the sum of the tooth thicknesses of pinion and gear is not equal to the circular pitch for reasons of tooth strength, wear, or scoring.
4. Gear trains in which a minor change in ratio (total number of teeth in mesh) has been made without a change in center distance.

In each of these cases, the calculation of operating center distance is performed using Eqs. (4.60) and (4.63) or (4.64).

4.3.9 Nonstandard Center Distance

By proper adjustments of the thickness of the teeth on each gear of a meshing pair, it is possible to achieve gear designs that will meet most nonstandard center distances. In Sec. 4.1.3, the limitations governing tooth thickness are outlined. In cases where maximum tooth strength is not critical, it will be found that gears not exceeding these limitations will satisfy most nonstandard center distance problems encountered. This is discussed more fully in method 3 below.

Sometimes designers who do not realize the possibilities of gear design will select such things as fractional diametral pitches (10.9652, etc.) in order to make gears fit a nonstandard center distance. This is poor practice in that it necessitates special tooling, which may be costly. It is better to attempt to utilize standard-pitch tooling and, by proper adjustments of tooth thickness, to establish gear designs that will meet the nonstandard distances.

In cases wherein the center distance is *nonstandard,* the designer has four possible methods of designing gears to meet the given center distance.

1. A number of teeth in gear and pinion different from the numbers originally selected may be found which will more nearly meet the given center distance. A Brocot table is very useful in finding the numbers of teeth that will give different gear ratios $(1/m_G)$. In this case, it is assumed that any of several ratios may be "good enough." If not, the following alternatives, which all assume that a specific ratio must be obtained, may be considered.

2. Equation (4.56), (4.57), (4.58), or (4.59) may be rearranged to solve for P_d. Thus a diametral pitch that will meet the gear ratio and center distance is found. This method will yield nonstandard diametral pitches for spur gears, which will almost always lead to large manufacturing costs because of the need for nonstandard tooling, cutters, and master gears. This method should be used only if methods 1, 3, and 4 do not meet the needs of the application.

3. The gear and pinion can be made with teeth that are thicker or thinner than standard. The easiest approach when using this method is to follow method 2 above, and then use the results as the basis for a selection of the standard diametral pitch nearest to the one found. Equation (4.60) is then used, which will yield the operating pressure angle for the gears based on the new selection of diametral pitch and the desired center distance C'. The value of C in this equation is determined from Eq. (4.56) or (4.57). Equation (4.63) is then used and solved for the term $(t_p + t_G)$. The selection of values for each term, t_P and t_G, should be based somewhat on the ratios of their number of teeth. In general, pinions cannot be changed from standard as much as gears with large numbers of teeth. The selected value of tooth thickness should be checked for undercut in the event that t_P is less than $p/2$. If the thickness is appreciably greater than $p/2$ the pinion should be checked for pointed teeth. The same should be done for the gear, especially if the changed tooth thickness was not based on the gear ratio. Once the values of C', P, n, N, t_P, and t_G have been established, the remainder of the tooth proportions can be calculated from these values. If the addendum and whole depth are properly adjusted from the tooth thickness, standard tools can be used to cut and to inspect these gears.

4. If helical gears can be used, the helix angle can be adjusted to obtain the required tooth thickness–center distance relationship. In effect, the designer is following method 3 above but is adjusting the transverse tooth thickness by the selection of the required helix angle. The procedure in this case is to use Eq. (4.57) or (4.59) to establish the diametral pitch required in the transverse plane P to suit the given center distance. The nearest finer (smaller) diametral pitch for which tools are available is selected (P_n), and these values are used in Eq. (4.85) to establish the required helix angle.

$$\cos \psi = \frac{P}{P_n} \qquad (4.85)$$

GEAR TOOTH DESIGN   **4.83**

Although not an essential part of this problem, somewhat smoother-running gears will result if the designer can manage to pick diametral pitches that will result in a helical overlap of at least 2. After the helix angle has been found, the designer can determine the remaining tooth proportions from data found in this chapter. This method can be used economically only if the hobbing process of cutting gears is to be used. Special guides and shaper cutters are required for every different lead of helical gear for gears cut by the shaping process. This then fixes limits on the values of helix angle that any given shop could cut.

4.4 ELEMENTS OF CENTER DISTANCE

Up to this point, this chapter has considered only the means of determining the theoretical center distance that is required by a set of gears. The theoretical center distance is either the *basic* center distance, as established by Eq. (4.56), (4.57), (4.58), or (4.59), or the theoretical *tight-mesh* center distance, as established by Eqs. (4.60) through (4.75). In an actual gear set, however, the gears will be forced to operate on a real center distance that may be larger or smaller than the theoretical center distance by amounts that are based on the way that the part tolerances (bearings, casing, bores, etc.) will add up.

4.4.1 Effects of Tolerances on Center Distance

In all but the very simplest gear casing designs, there are many tolerances that will govern the operating position of the gear and pinion shafts. Usually power gearing is not designed to small values of backlash. It is customary, therefore, to apply a generous amount of backlash to the gear teeth, a value that has been found by experience sufficient to prevent binding of the teeth. In control gearing, however, particularly in those applications that are operated in both directions and are intended to have a minimum of lost motion, a critical study of the effects of part tolerances is usually required. The tooth thickness can then be made as large as possible, ensuring a minimum of backlash with little risk of binding. The following two examples indicate a very simple case and a more complex case of center-distance calculations.

Case I. In Fig. 4.19 is shown a spur gear pair mounted on a cast-iron gear case. The loads and speeds are such that the bearings are simply smooth bores in the cast iron with shafts running in them. Figure 4.20 is a vector diagram showing the forces produced on the bearings as a result of the gear tooth loads. Figure 4.21 shows the displacements of the shafts in the bearings resulting from these loads.

In this example, it is desired to establish the largest and smallest center distance that the gears will ever experience in order to determine the range of backlash existing in the gear set. Table 4.39 outlines the calculations made to establish the various center distances at which the gears may be expected to operate in service. Part A of Table 4.39 shows the calculation of the maximum distance that the axis of either shaft can move away from the axis of its bore. This is given as the eccentricity of the shaft within the bearing. It is assumed that the bearing is dry. For a very rough approximation of maximum and minimum possible center distances, these values of eccentricity can be added to and subtracted from the boring center distance plus its tolerance. This is shown in part B of Table 4.39. This assumes the force acting on the shafts are tooth-load reactions that act along the line of centers. In the case of journal bearings that are properly lubricated, the calculated value of eccentricity of operation can be used. This calculation is discussed in texts on journal bearing design. In most cases if the designer

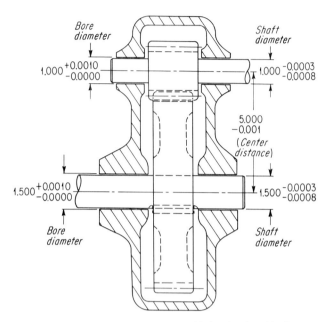

FIGURE 4.19 Spur or helical gears operated in bearings bored in the gear casing exhibit the simplest case of tolerance buildup on center distance

can select the tolerances for the gear tooth thickness, and also the tolerances on the various parts of the casing and bearings that control the center distance, such that the backlash is not excessive when the calculations shown thus far are carried out, no further calculations need be made. However, when the design is to achieve an absolute minimum of lost motion, the more accurate evaluation of center distance should be carried out.

Part C shows a calculation assuming that the gear tooth reactions will control the position of the journal in the bearing. The minimum center distance will occur when the minimum boring center distance is achieved and the minimum bearing clearance occurs. The tooth loads tend to increase the center distance in this example so that distances less than the minimum boring center distance cannot occur.

In actual service, the maximum and minimum operating center distances will rarely reach these limits. In the case of the maximum center distance, a properly lubricated shaft will develop an oil wedge that will keep the journal away from the walls of the bore. Second, if the only separating loads on the shafts are those produced by gear tooth reactions, the radial movement of the shaft along the line of centers will be e (sin Φ). The proper evaluation of operating center distance will depend on whether the application is intermittent in rotation and on the direction of the rotation.

Case II. A more complex case occurs when the pinion and gear shafts are mounted in separate units. The following example illustrates a case in which a motor having a pinion cut integral with the shaft is mounted by means of a rabbet in a gear case. This type of construction is often found in power hand tools and in aircraft actuators. This example also illustrates a critical evaluation of antifriction bearings. It is desired to determine the largest and smallest center distance that can occur in order to establish

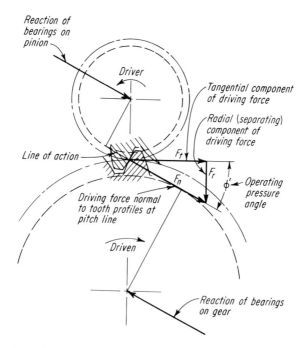

FIGURE 4.20 The reaction forces developed by the bearings to overcome the gear teeth driving forces are parallel to the line of action

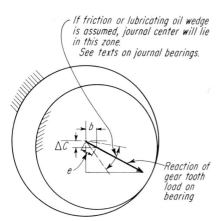

ΔC is increase in center distance due to movement of journal in bearing.
e is eccentricity of bearing
$\left(\dfrac{\text{bore diameter} - \text{journal diameter}}{2} \right)$

FIGURE 4.21 The position occupied by the journal in a bearing is a function of the gear tooth loads along the shaft and the effects of friction or of the oil wedge in the bearing

TABLE 4.39 Summary of Calculations for Case I
A. *Calculation of Bearing Clearance and Shaft Eccentricity. See Fig. 4.19 for Drawing Dimensions*

Bore for pinion shaft	Max. 1.0010	Min. 1.0000
Pinion-shaft diameter	Min. 0.9992	Max. 0.9997
Clearance in bearing (c)	Max. 0.0018	Min. 0.0003
Eccentricity of shaft position in bearing ($c/2$)	Max. 0.0009	Min. 0.00015
Bore for gear shaft	Max. 1.5010	Min. 1.5000
Gear shaft diameter	Min. 1.4992	Max. 1.4997
Clearance in bearing (c)	Max. 0.0018	Min. 0.0003
Eccentricity of shaft position in bearing ($c/2$)	Max. 0.0009	Min. 0.00015

B. *Calculation of Theoretical Maximum and Minimum Center Distance (assuming reaction forces acting along line of centers; values from above)*

Center distance of bores (machining tolerance)	Max. 5.0010	Min. 5.000
Eccentricity of pinion bearing	0.0009	−0.0009
Eccentricity of gear-bearing center distance	0.0009	−0.0009
	Max. 5.0028	Min. 4.9982

C. *Calculation of Theoretical Maximum and Minimum Center Distance (assuming reaction forces acting along line of action; values from above, part B)*

Center distance of bores (machining tolerance)	Max 5.0010	Min 5.0000
Eccentricity, radial component of pinion bearing		
(0.0009) sin 20°	0.0003	
(0.00015) sin 20°	—	0.00005
Eccentricity, radial component of gear bearing		
(0.0009) sin 20°	0.0003	
(0.00015) sin 20°	—	0.00005
	Max. 5.0016	Min. 5.0001

the maximum and minimum tooth thickness and the amounts of backlash that result (see Fig. 4.22).

Table 4.40, which summarizes the calculations, shows a total of 16 different elements that combine to control the maximum and minimum actual center distance. Since there are so many tolerances, the use of the root mean square of these values is suggested as a reasonable approximation to operating conditions, and it is used here to establish the maximum and minimum probable center distance. This method gives over 90-percent assurance that the values shown will not be exceeded.

GEAR TOOTH DESIGN

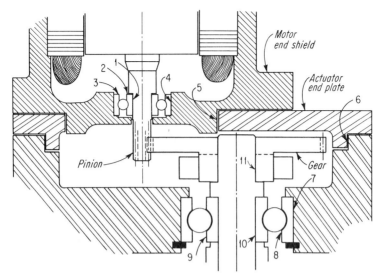

FIGURE 4.22 Cross section of the first stage of gearing in an aircraft-type actuator. The numbers refer to the surfaces discussed in Table 4.40

4.4.2 Machine Elements That Require Consideration in Critical Center-Distance Applications

As illustrated in cases I and II, the operating center distance for a pair of gears is made up of several elements, each of which contributes to the overall center distance. The accumulation of tolerances on each of these elements must be considered in applications in which a minimum of backlash is to be established. The following is a consideration of the elements that have the largest contribution to variations in center distance:

1. *Rolling-element bearings.* Ball-and-roller bearings consist of three major elements that contribute to backlash and to changes in center distance. Some of these are illustrated in case II. The outer race has machining tolerances that cause eccentricity of the axes of the inner raceway and the outer bore. Depending on how this member is installed, the center distance established by the bores in which this element is fitted will be increased or decreased. It is customary to let this element creep; thus the eccentricity will go through all positions. There is also a tolerance on the outer diameter of this member. This tolerance, plus the tolerance on the bore in which this element is fitted, can cause varying degrees of looseness and therefore changes in operating center distance.

The inner race has machining tolerances that cause eccentricity in the same manner as in the outer race member. In cases where the shaft rotates, it is customary to use a light interference fit between the inner race and the shaft. In such cases, the center of the gear will move in a path about the center of the ball or roller path of the bearing. The eccentricity of the gear will be that of the inner race of the bearing. This will cause a once-per-revolution change in center distance.

The clearance in the bearing is controlled by the selection of the diameter of the balls. This clearance allows greater or lesser changes in center distance of the gears supported by the bearings.

TABLE 4.40 Summary of Calculations for Case II

Item Discussed in Footnotes	Surface Designation (see Fig. 4.22) (1)	Tolerance (clearance), in. (2)	Equivalent Change in Center Distance (3)	Min. Center Distance Col. (3)[c] $\times 10^{-6}$ (4a)	(4b)	Max. Center Distance Col. (3)[2] $\times 10^{-6}$ (5a)	(5b)
1. Fit of motor shaft in inner race of ball bearing[a]	(1)	0.0000	0.0000	−0.00	−0.00	+0.00	+0.00
2. Eccentricity of inner ring of bearing[b]	(1–2)	0.0002	0.0001	−0.01	−0.01	+0.01	+0.01
3. Radial clearance in bearing[c]	(2–3)	0.0004–0.0000	0.0002	−0.04	+0.04	−0.04	+0.04
4. Eccentricity of outer ring of bearing[d]	(3–4)	0.0004	0.0002	−0.04	−0.04	+0.04	+0.04
5. Fit of outer race of bearing into motor end shield[e]	(4)	0.0008–0.0000	0.0004–0.0000	−0.16	−0.16	+0.16	+0.16
6. Concentricity of axis of bore with axis of rabbet on end shield[f]	(4–5)	0.0020	0.0010	−1.00	−1.00	+1.00	−1.00
7. Clearance between rabbet of motor end shield and bore of actuator end plate[g]	(5)	0.0025	0.00125	−1.56	−1.56	+1.56	+1.56
8. Concentricity of motor bore in actuator end plate and rabbet[h]	(5–6)	0.0010	0.0005	−0.25	−0.25	+0.25	+0.25
9. Distance between axis of gear shaft bore and rabbet in actuator housing[i]	(6–7)	0.0020	0.0010	−1.00	−1.00	+1.00	+1.00
10. Clearance between bore for bearing and bearing outer race[e]	(7)	0.0008	0.0004	−0.16	−0.16	−0.16	+0.16
11. Eccentricity, outer race of bearing[d]	(7–8)	0.0004	0.0002	−0.04	−0.04	+0.04	+0.04
12. Radial clearance in bearing[c]	(8–9)	0.0004–0.0000	0.0002–0.0000	−0.04	+0.04	+0.04	+0.04
13. Eccentricity, inner race of bearing[b]	(9–10)	0.0002	0.0001	−0.01	−0.01	+0.01	+0.01
14. Fit of gear shaft in inner race of bearing[a]	(10)	0.0000	0.0000	−0.00	−0.00	+0.00	+0.00

15. Eccentricity, bearing journal and surface locating gear[j]	(10–11)	0.0002	0.0001	−0.01	−0.01	+0.01
16. Fit gear on shaft	(11)	0.0002	0.0001	−0.01	−0.01	+0.01
			Totals	−4.33	−4.17	+4.33

Probable values $\sqrt{4.17} = -0.0020''$ $\sqrt{4.33} = +0.0021''$

NOTE: The maximum values of center distance (Columns 4b and 5b) are based on the assumption that the separating forces between pinion and gear act to hold the shafts at their maximum separation within the limits of bearing clearance.

The minimum values of center distance (Columns 4a and 5a) are based on the assumption that the external forces on gear and pinion act to hold the shafts at their minimum separation within the limits of bearing clearance.

The minimum center distance (Column 4) is based on the assumption that all parts, made to their maximum clearances, are assembled to achieve the minimum possible center distance. The maximum center distance (Column 5) is based on the assumption that all parts, made to their maximum clearances, are assembled to achieve the maximum possible center distance.

Column (1) shows the surface designated on Fig. 4.22 by a number in a circle; (2) shows the tolerance (usually on a diameter); (3) shows the amount this tolerance can contribute to center distance (±); (4) shows the square of the tolerance contributing to the minimum center distance; and (5) shows the square of the tolerance contributing to the maximum center distance. The values in columns (4) and (5) are added algebraically and their square roots are shown at the foot of the column.

[a]Most bearing catalogues indicate bearing and shaft tolerances which will produce an interference fit. This fit cannot, therefore, contribute to the maximum or minimum center distance at which the gears will operate.

[b]Most bearing catalogues indicate a tolerance on runout between the bore and the raceway of the inner ring. This eccentricity contributes to the instantaneous center distance when the inner face is the moving portion of the bearing.

[c]Most bearings include a certain radial clearance between the inner and the outer race of a bearing. This clearance usually acts to allow an increase in the center distance due to the reaction between the two meshing gears.

[d]Most bearing catalogues indicate a tolerance on runout between the raceway and the outer diameter of the outer ring. This eccentricity contributes to the center distance at which the gears will operate. This value will change during running since the outer race is often permitted to creep.

[e]Most bearing catalogues indicate a bore tolerance on the hole that the bearing is fitted in. Usually the tolerance will provide clearance between the bearing and housing. When clearance exists, the gear reaction acts to increase the center distance to the limit allowed by the clearance.

[f]Many manufacturers of pilot-mounted motors specify a value of runout between the motor shaft and the pilot surface. It is customary to measure this by holding the motor shaft fixed (as between centers) and to rotate the motor about its shaft. A dial indicator running on this surface will show the runout (twice eccentricity).

[g]A clearance fit is usually specified on the hole into which the pilot of the motor is to fit. This clearance allows the motor to be shifted somewhat and, as a result, has a direct effect on center distance. The value shown assumes a tolerance of $+0.0005$ in. on the bore and a tolerance of $+0.000$ in. on the diameter of the pilot on the motor.

[h]In an assembly, such as shown in this example, a tolerance must be placed on the location of the bore to receive the motor relative to the rabbet on the end plate.

[i]This, in effect, is the same type of tolerance as discussed under g above.

[j]Usual drafting practice is to specify a tolerance on the concentricity of the several cylindrical surfaces of a shaft. If these are all machined at one "setup," they will have little or no eccentricity relative to each other. In some cases, they must be machined in separate "setups."

2. *Journal bearings.* Journal bearings can change the center distance at which the gear was intended to operate by an amount that is a function of the clearance designed into the bearing, its direction or rotation, the speed of rotation, and the lubricant used. This change from the distance actually established by the distance between the centers of the bores is due to the oil wedge developed. Handbooks give convenient methods of evaluating the eccentricity that can be expected of a given bearing design. When necessary, these refinements can be introduced into the calculations illustrated by cases I and II.

3. *Sleeve bearings.* The term *sleeve bearing* is used here to identify the journal-type bearing in which a sleeve of some bearing material is pushed into the gear casing or frames. The tolerance on the eccentricity of the outside diameter and the inside bearing surface is the element to consider in this type of bearing. In some cases, the clearance between the sleeve and the shaft is affected because of the interference fit between the sleeve and the bore in the casing.

4. *Casing bores.* The distance between bearing bores in a gear casing can be made up of several elements. In the simplest case, a single frame or casing has two holes drilled at a specified center distance. The tolerance is selected to give the machine operator the necessary working allowance (see Fig. 4.19). This tolerance has a direct effect on the operating center distance of the gears.

In some designs, the casing is made up of several parts bolted together. One part may carry the bore for the gear shaft, and another may carry the bore for the pinion shaft. The center distance is then made up of a series of parts each having dimensions and tolerances that can add up in various ways to give maximum and minimum center distance. Case II illustrates this type of gearing.

4.4.3 Control of Backlash

In cases where it is necessary to control the amount of backlash introduced by the mountings, two courses of action may be employed. In the first, the center distance may be made adjustable. That is, provisions may be made so that, at assembly, the centers may be moved until the desired mesh is obtained. This entails a method of moving the parts through very small distances, and then being able to fix them securely when the desired distance has been reached. Provision must also be made for assemblers to see what they are doing when adjusting the mesh. In this approach, the difference between the smallest and the largest backlash in the adjusted mesh is only the effect of total composite error (runout) in both meshing parts. Size tolerance on the teeth can be generous since this is one of the tolerances adjusted out of the mesh by assembler.

After the center-distance adjustment is completed, the parts may be drilled and doweled, although this causes difficulties if replacement gears are ever to be used on these centers.

In the second approach, very close tolerances are held both in the mountings (boring center distance, bearings, etc.) and on the tooth thickness of the gears. This approach is used whenever interchangeability is specified and in mass production in which assembly costs are to be minimized.

4.4.4 Effects of Temperature on Center Distance

Many gear designs consist of gears made of one material operating in mountings made of another material having a markedly different coefficient of expansion. Since it is customary to manufacture and assemble gears at room temperature of about 68°F, an analytical check of the effective center distance present in the gears at their extremes of operating temperature is desirable.

Power gears may heat up because of the frictional losses in the mesh and may achieve an operating temperature above that of their casing. If made of the same materials as their mountings, such gears would have a larger apparent pitch diameter* relative to the center distance and would therefore have a smaller effective center distance at their running temperature. Some gears, particularly those in military applications, may be subjected to extreme cold for periods of time. Often these are steel gears operating in aluminum or magnesium casings. In this case, the center distance will shrink to a greater degree than the apparent pitch diameter of the gears, and the resulting backlash will be less than at room temperature.

In evaluating the effective center distance of a mesh at an operating condition other than room temperature, the temperature of the gear blanks as well as that of the mountings, which may be considerably different, must be considered, as well as the coefficients of expansion of the gear and mounting materials. There are six operating conditions in which temperature can have an important effect on gear performances:

1. Gears operating at temperatures *higher* than their mountings:
   a. Operating temperatures *lower* than assembly-room temperature
   b. Operating temperatures *higher* than assembly-room temperature
2. Gears operating at temperatures *lower* than their mountings:
   a. Operating temperatures *lower* than assembly-room temperature
   b. Operating temperatures *higher* than assembly-room temperature
3. Gears operating at temperatures essentially the *same* as their mountings:
   a. Operating temperatures *lower* than assembly-room temperature
   b. Operating temperatures *higher* than assembly-room temperature

Equations (4.86) and (4.87) evaluate each of these possibilities and establish the most critical extremes at which minimum or maximum effective center distance will occur.

The significance of these possibilities can be appreciated by considering the following service conditions. The first case is typical of power gearing under steady-state conditions. Condition 1a is a possibility at start-up under a cold operating environment. The second case is usually a transitory condition where the gears may have to perform adjacent to an external source of heat. The third condition is most typical of control gears that do not transmit enough energy to be at a temperature greatly different from their mountings.

The procedure recommended here is to calculate a minimum and a maximum effective center distance. These are based on the extremes of operating temperatures. From these values, the designer can calculate either a value of tooth thickness that will give the necessary operating backlash under the tightest mesh conditions, or the amounts of backlash that will be found in a given gearset.

Equations (4.86) and (4.87) give minimum and maximum effective center distances.

$$C_e \min = C_d + C_t' = \Delta C_{T_T} \quad (4.86)$$

$$C_e \max = C_d + C_t' = \Delta C_{T_L} \quad (4.87)$$

where $C_e \min$ = minimum effective center distance that occurs under temperature extreme

$C_e \max$ = maximum effective center distance that occurs under temperature extreme

---

Apparent pitch diameter is the diameter at which a given value of tooth thickness is found.

C_d = basic or nominal center distance
C'_t = minimum tolerance on center distance
C''_t = maximum tolerance on center distance
ΔC_{T_L} = change in center distance; see note following Eqs. (4.88) and (4.89)
ΔC_{T_T} = change in center distance; see note following Eqs. (4.88) and (4.89)

Equations (4.88) and (4.89) indicate the amount that the center distance (effective) will change because the gears and their mountings shrink or expand at different rates.

$$\Delta C'_T = \frac{D}{2}(\Delta T'_M K_M - \Delta T'_G K_G) + \frac{d}{2}(\Delta T'_M K_M - \Delta T'_G K_P) \quad (4.88)$$

$$\Delta C''_T = \frac{D}{2}(\Delta T''_M K_M - \Delta T''_G K_G) + \frac{d}{2}(\Delta T''_M K_M - \Delta T''_G K_P) \quad (4.89)$$

Note: Compare $\Delta C'_T$ and $\Delta C''_T$. Assign the smallest positive number or the largest negative number to ΔC_{T_T}. Assign the largest positive number or the smallest negative number to ΔC_{T_L}. If signs are opposite, use this negative value for C_{T_T} and use the positive value for ΔC_{T_L}. These values are used in Eqs. (4.86) and (4.87) to obtain effective center distance.

$$\Delta T'_M = T_M - T_R$$
$$\Delta T''_M = T'_M - T_R$$
$$\Delta T'_G = T_G - T_R$$
$$\Delta T''_G = T'_G - T_R$$

where T_R = assembly room temperature (68°F)
T_M = minimum mounting temperature (operating)
T'_M = maximum mounting temperature (operating)
T_G = minimum gearing temperature (operating)
T'_G = maximum gearing temperature (operating)
K_P = coefficient of expansion of pinion material, in. per in. per °F
K_G = coefficient of expansion of gear material
K_M = coefficient of expansion of mounting material

4.4.5 Mounting Distance.*

Bevel gearing, worm gearing, face gearing, and Spiroid gearing must be given close control on axial positioning (mounting distance) if good performance is to be achieved (see Fig. 4.23). Most of the elements that must be given control to achieve proper center distance must also be given close control to achieve proper mounting distance. If any

---

*See Chap. 10 for general information gear mounting tolerances and practices.

GEAR TOOTH DESIGN 4.93

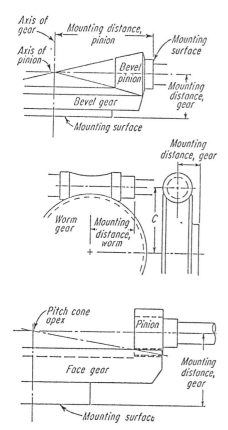

FIGURE 4.23 Mounting distance, shown for various kinds of gears, is the distance from the axis of a gear to the mounting surface of its mating gear. This distance is measured parallel to the shaft of the subject gear

of the above types of gearing can move along the axis of its shaft relative to the mating gear, the teeth will either bind or, in the opposite case, have excessive backlash.

Straight-toothed bevel gears are critical in respect to axial position, since backlash and tooth bearing pattern are affected by changes in axial position of the pinion and gear. Spiral bevel gears, hypoid gears, and Spiroid gears are also critical in that they tend to "screw" into mesh in one direction of rotation, which can cause binding, and will "unscrew" in the other direction of rotation, causing backlash, unless the mounting system is stiff enough to prevent axial shift. All properly designed mountings for gears of this type have bearings capable of withstanding all axial thrust loads imposed both by the gears and by possible external loads. Occasionally a design is attempted in which the designer relies on the separating forces produced in the mesh to keep the gears away from a binding condition. Only a few such designs are truly successful.

Worm gearing must also be accurately positioned if of the single- or double-enveloping design. If a throated member is permitted to move axially, the frictional force developed in the mesh will tend to move the member, and the geometry of the throated design will cause a reduction in backlash, sometimes to the point of binding.

Mounting distance is usually specified for each gear member as the distance from a specific mounting surface on the gear blank, a face of a hub, for example, to the axis of the mating gear. Bevel gears are usually stamped with the correct mounting distance for the pair. This distance was established in a bevel gear test fixture at the time of manufacture.

The design of mountings for gearing requiring a mounting distance should include provisions for shimming or otherwise adjusting the position of the members at assembly. The calculation of mounting distance is accomplished in two steps. The design of the gear teeth (earlier in this chapter) includes a distance from the axes or pitch cone apex to a given surface of the gear. To this distance is added the distance to the closest bearing face that is used to provide the axial location of the gear and its shaft. Provision should be made in the design to secure the bearing seats that provide axial location at the specified mounting distances.

In the previous articles, radial clearance in rolling-element bearings was considered. In gear designs requiring a mounting distance, the axial clearance in one bearing on each shaft must be controlled. Reference to a bearing catalogue or handbook is recommended.

REFERENCES

1. Dudley, D. W., *Gear Handbook*, McGraw-Hill, New York, 1962.
2. Buckingham, E., *Analytical Mechanics of Gears*, McGraw-Hill, New York, 1949.
3. Dudley, D. W., *Handbook of Practical Gear Design*, 2nd ed., McGraw-Hill, New York, 1984.
4. AGMA 207.05, "20-Degree Involute Fine-Pitch System for Spur and Helical Gears," Alexandria, Va., 1967. See also 207.06–1977.
5. AGMA 201.02 and 201.02A, "Tooth Proportions for Coarse-Pitch Involute Spur Gears," American Gear Manufacturers Association, Alexandria Va., August 1968 (R 1974).
6. Gleason Works, "Strength of Bevel and Hypoid Gears," Rochester, New York.
7. Nelson, W. D., "Spiroid Gearing," American Society of Mechanical Engineers paper 57-A-162.
8. Dudley, D. W., *Practical Gear Design*, McGraw-Hill, New York, 1954.

CHAPTER 5
SPECIAL CALCULATIONS OF SPUR AND HELICAL GEARS

Dr. Charles Chao
Superior Gear
Gardena, California

5.1 NOTATION

The notation used in this chapter is as follows:

Symbols

- B = backlash
- C = center distance
- F = tooth width
- L = helical lead
- m_F = face contact ratio
- m_p = transverse contact ratio
- m_t = total contact ratio
- N = number of teeth
- P = diametral pitch
- P_n = normal diametral pitch
- p_b = base pitch
- p = circular pitch
- p_n = normal circular pitch
- p_x = axial base pitch
- R = pitch radius
- R_b = base radius
- R_I = inside radius of internal gear
- R_o = outside radius
- R_R = root radius

TABLE 5.1 Fundamental Gear Equations

$$p = \frac{2\pi R}{N} \tag{5.1}$$

$$P = \frac{N}{2R} = \frac{\pi}{p_t} \tag{5.2}$$

$$p_n = \frac{2\pi R \cos \psi}{N} \tag{5.3}$$

$$p_b = \frac{2\pi R \cos \phi}{N} = p \cos \phi \tag{5.4}$$

$$P_n = \frac{N}{2R \cos \psi} = \frac{P}{\cos \psi} \tag{5.5}$$

$$p_n P_n = pP \tag{5.6}$$

$$L = 2\pi R \cot \psi = 2\pi R_b \cot \psi_b \tag{5.7}$$

$$p_x = \frac{2\pi R \cos \phi}{N} \cdot \cot \psi_b \tag{5.8}$$
$$= p_b \cdot \cot \psi_b = p \cos \psi$$

$$t_n = t \cos \psi \tag{5.9}$$

$$\tan \phi_n = \tan \phi \cos \psi \tag{5.10}$$

$$\sin \phi_n = \sin \phi \cos \psi_b \tag{5.11}$$

$$\cos \phi_n = \sin \psi_b \csc \psi \tag{5.12}$$

r = radius to involute profile
s = space width
t = circular tooth thickness
t_n = normal circular tooth thickness
Z = contact length
ϵ = roll angle
θ = involute polar angle
ρ = radius of curvature
ϕ = pressure angle
ϕ_n = normal pressure angle
ϕ_o = operating pressure angle
ψ = helix angle
ψ_b = base helix angle

Subscript

1 = pinion
2 = gear

All angles are expressed in radians.

5.2 FUNDAMENTAL GEAR EQUATION

Fundamental equations relating the common gear terms are given in Table 5.1. Since a spur gear is a special case of helical gear, all equations are applicable to both spur and helical gears.

5.3 INVOLUTE FUNCTION AND ROLL ANGLE

The mechanical way to construct an involute is using a circular disk with a string wrapped around it, unwinding the taut string to produce a series of lines tangent to this *base circle,* and tracing the ends of these tangent lines to form an involute curve (Fig. 5.1). The equation of involute is

$$\theta = \frac{\sqrt{r^2 - R_b^2}}{R_b} - \tan^{-1}\left(\frac{\sqrt{r^2 - R_b^2}}{R_b}\right) \qquad (5.13)$$

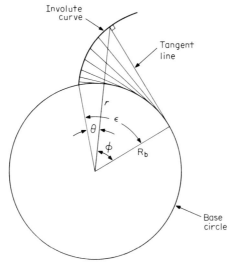

FIGURE 5.1 Construction of involute curve.

Define ϕ as pressure angle at radius r

$$\tan \phi = \frac{\sqrt{r^2 - R_b^2}}{R_b} \tag{5.14}$$

then $\theta = \tan \phi - \phi = \text{inv } \phi$ (5.15)

Also, from Fig. 5.1

$$r = \frac{R_b}{\cos \phi} \tag{5.16}$$

Equations (5.15) and (5.16) are the basic relationships of an involute by means of which many characteristics of involute gear teeth are readily calculated.

In practice, roll angle is often used to locate the point in an involute. Roll angle is defined by rewriting Eq. (5.15) as

$$\epsilon = \theta + \phi = \tan \phi \tag{5.17}$$

where ϵ is the roll angle.

5.4 LINE OF ACTION AND CONTACT RATIO

When the gears are put into mesh with center distance C (Figs. 5.2 and 5.3), the line tangent to both base circles is defined as the *line of action*. The contact starts at point A, where the outside diameter circle of the gear intersects the line of action, passes through pitch point P, and ends at point B where the outside diameter of the pinion intersects the line of action. The total length Z of action is

$$Z \text{ (external)} = \sqrt{R_{o_1}^2 - R_{b_1}^2} + \sqrt{R_{o_2}^2 - R_{b_2}^2} - C \cdot \sin \phi \tag{5.18}$$

$$Z \text{ (internal)} = \sqrt{R_{o_1}^2 - R_{b_1}^2} - \sqrt{R_{I_2}^2 - R_{b_2}^2} + C \cdot \sin \phi \tag{5.19}$$

A contact zone can be visualized as a rectangle with length of action Z being one side and tooth width F being another side (Fig. 5.4). As contacts move from point A to point B, the average number of teeth moving across Z is defined as the transverse contact ratio

$$m_p = \frac{Z}{p_b} \tag{5.20}$$

and that moving across F is the face contact ratio

$$m_F = \frac{Z}{p_x} \tag{5.21}$$

SPUR AND HELICAL CALCULATIONS

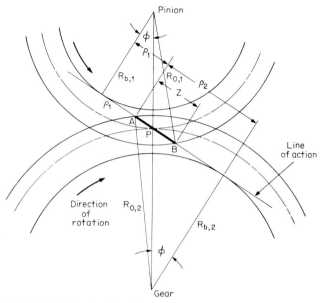

FIGURE 5.2 Line of action of external gear.

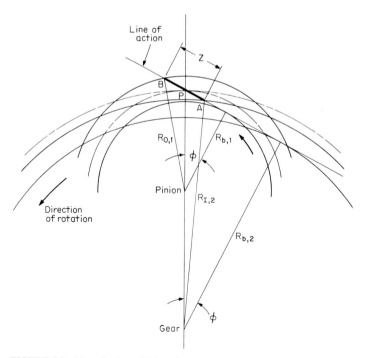

FIGURE 5.3 Line of action of internal gear.

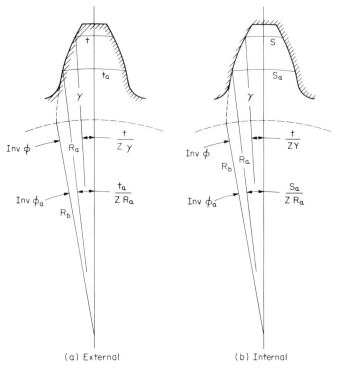

(a) External   (b) Internal

FIGURE 5.4 Tooth thickness and space.

where p_b is base pitch and p_x is axial base pitch. The total contact ratio m_t is

$$m_t = m_p + m_F \qquad (5.22)$$

5.5 TOOTH THICKNESS CALCULATION

Once the tooth thickness t_a at the given radius γ_a is known, the tooth thickness at any other radius can be determined from Fig. 5.4:

$$\frac{t}{2r} + \text{inv } \phi = \frac{t_a}{2r_a} + \text{inv } \phi_a \qquad (5.23)$$

For internal gears, when the space width s_a at radius r_a is known, then the space width at any radius is given by

$$\frac{s}{2r} + \text{inv } \phi = \frac{s_a}{2r_a} + \text{inv } \phi_a \qquad (5.24)$$

5.6 HERTZIAN STRESS ALONG CONTACT PATH

The radius of the pinion at the initial contact point A is ρ_1 (as shown in Fig. 5.2) and that of the gear is ρ_2. The Hertzian contact stress at point A is

$$\sigma_{max} = \sqrt{\frac{W_n \left(\frac{1}{\rho_1} + \frac{1}{\rho_2}\right)}{\pi F \left(\frac{1 - \theta_1^2}{E_1} + \frac{1 - \theta_2^2}{E_2}\right)}} \qquad (5.25)$$

for external spur gears, and

$$\sigma_{max} = \sqrt{\frac{W_n \left|\frac{1}{\rho_1} - \frac{1}{\rho_2}\right|}{\pi F \left(\frac{1 - r_1^2}{E_1} + \frac{1 - r_2^2}{E_2}\right)}} \qquad (5.26)$$

for internal spur gears. Where W_n is normal force, E is Young's modulus, and r is the poisson ratio.

For a gear design with contact ratio m between 1 and 2, the maximum stress occurs either at the *lowest single tooth contact point* (**LSTCP**) or the *highest single tooth contact point* (**HSTCP**) where

$$\rho_2 = \sqrt{R_{b_2}^2 + R_{o_2}^2} - p_b (m - 1) \qquad \text{(HSTCP)} \qquad (5.27)$$

$$\text{or } \rho_2 = \sqrt{R_{b_2}^2 + R_{o_2}^2} - p_b (m + 1) \qquad \text{(LSTCP)} \qquad (5.28)$$

$$\text{and } \rho_1 = C \cdot \sin \phi - \rho_2 \qquad (5.29)$$

for external gears.

$$\rho_2 = \sqrt{R_{b_2}^2 + R_{I_2}^2} + p_b (m - 1) \qquad \text{(LSTCP)} \qquad (5.30)$$

$$\text{or } \rho_2 = \sqrt{R_{b_2}^2 + R_{I_2}^2} + p_b (m + 1) \qquad \text{(HSTCP)} \qquad (5.31)$$

$$\text{and } \rho_1 = \rho_2 - C \cdot \sin \phi \qquad (5.32)$$

for internal gears.

Since helical gears have inclined contact lines (see Fig. 5.5), the radii along the contact lines are not constant and the load may not be uniformly distributed along the contact line. The radius at any given contact location x is

$$\rho_2 = \frac{\sqrt{R_{o_2}^2 - R_{b_2}^2} - x}{\cos \psi} \qquad (5.33)$$

$$\rho_1 = \frac{C \cdot \sin \phi}{\cos \psi} - \rho_2 \qquad (5.34)$$

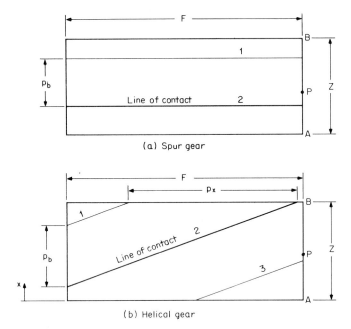

FIGURE 5.5 Contact zone.

for external gears, and

$$\rho_2 = \frac{\sqrt{R_{I_2}^2 - R_{b_2}^2} + x}{\cos \psi} \quad (5.35)$$

$$\rho_1 = \rho_2 - \frac{C \cdot \sin \phi}{\cos \psi} \quad (5.36)$$

for internal gears.

5.7 BACKLASH AND CENTER DISTANCE

Backlash for a gear set is required to prevent gear tooth binding under running conditions. For high-speed gears, backlash also functions as a passage to prevent any lubricant being trapped between meshes.

When a set of gears is cut with transverse tooth thickness t_a and t_b at manufactured pitch diameter, the manufactured backlash for the gear set is

$$B = p - t_a - t_b \quad (5.37)$$

SPUR AND HELICAL CALCULATIONS

This gear set can be operated at different center distance C_o so the operating pitch diameter is different from manufactured pitch diameter, and the change of backlash under such situation is

$$\Delta B = 2 \cdot \Delta C \cdot \tan \phi \qquad (5.38)$$

where $\Delta C = C_o - C$ and C is manufactured center distance.

In practice, the dial indicator is used to measure backlash, where the reading is taken in a normal direction. The comparison between dial reading B_n and backlash at transverse pitch circle B is

$$B = \frac{B_n}{\cos \phi_n \, \cos \psi} \qquad (5.39)$$

5.8 TOOTH THICKNESS FROM PIN OR BALL MEASUREMENT

Actual tooth thickness and actual space width are usually measured with two pins or two balls. The diameters of pins or balls are such that the pins contact the tooth approximately at the midpoint of the profile while diamter over pin/ball (external gear) or diameter between pin/ball (internal gear) is cleared by tooth outside or inside diameter.

Figure 5.6 shows measurements on external gears. From the geometry, the following equation is obtained.

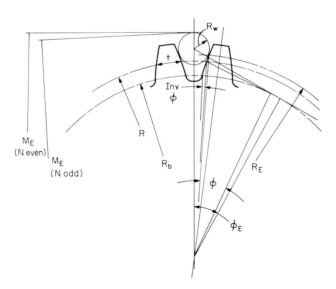

FIGURE 5.6 Measurement over pins/balls on external spur gear.

$$\tan \phi_E - \phi_E = \text{inv } \phi_E = \frac{R_W}{R_b} - \left(\frac{\pi}{N} - \frac{t}{2R} - \text{inv } \phi\right) \quad (5.40)$$

After solving ϕ_E, the measurements over pins are calculated as follows:

$$M_E \ (N \text{ even}) = 2 R_W + \frac{2R_b}{\cos \phi_E} \quad (5.41)$$

$$M_E \ (N \text{ odd}) = 2R_W + \frac{2R_b}{\cos \phi_E} \cos \left(\frac{\pi}{2N}\right) \quad (5.42)$$

and the contact radius is

$$R_E = \sqrt{(R_b \tan \phi_E - R_w)^2 + R_b^2} \quad (5.43)$$

for internal gears, the measurements between pins are (Fig. 5.7):

$$\tan \phi_I - \phi_I = \text{inv } \phi_I = \text{inv } \phi + \frac{s}{2R} - \frac{R_w}{R_b} \quad (5.44)$$

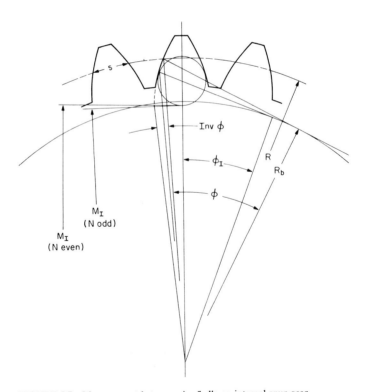

FIGURE 5.7 Measurement between pins/balls on internal spur gear.

SPUR AND HELICAL CALCULATIONS

Again, after solving ϕ_I, the measurements between pins are calculated as follows:

$$M_I \ (N \text{ even}) = \frac{2R_b}{\cos \phi_I} - 2R_w \tag{5.45}$$

$$M_I \ (N \text{ odd}) = \frac{2R_b}{\cos \phi_I} \cdot \cos\left(\frac{\pi}{2N}\right) - 2R_w \tag{5.46}$$

and contact radius is

$$R_I = \sqrt{(R_b \cdot \tan \phi_I + R_w)^2 + R_b^2} \tag{5.47}$$

The procedures are reversed to calculate tooth thickness for given pin measurements. That is, solve ϕ_E or ϕ_I first by Eq. (5.41), (5.42), (5.45), or (5.46) with known measurements M_E or M_I, and then get t or s by Eq. (5.40) or (5.44).

The thickness factor K_m is the ratio

$$K_m = \frac{\Delta M}{\Delta t} = \frac{\text{change of pin measurement}}{\text{change of tooth thickness (space width)}}$$

For external gears,

$$K_m = \frac{R_b}{R \sin \phi_E} \quad \text{or} \quad \frac{R_b \cdot \cos\left(\dfrac{\pi}{2N}\right)}{R \cdot \sin \phi_E} \tag{5.48}$$

For internal gears,

$$K_m = \frac{R_b}{R \sin \phi_I} \quad \text{or} \quad \frac{R_b \cdot \cos\left(\dfrac{\pi}{2N}\right)}{R \cdot \sin \phi_I} \tag{5.49}$$

The value of K_m depends on the pressure angle, the number of teeth, the pin diameter, and the pitch.

For helical gears, the measurements over balls are shown in Fig. 5.8 for external gears. ϕ_E is solved by

$$\tan \phi_E - \phi_E = \text{inv } \phi_E = \frac{R_w}{R_b \cos \psi_b} - \left(\frac{\pi}{N} - \frac{t}{N} - \text{inv } \phi\right) \tag{5.50}$$

and then

$$M_E \ (N \text{ even}) = 2R_w + \frac{2R_b}{\cos \phi_E} \tag{5.51}$$

$$M_E \ (N \text{ odd}) = 2R_w + \frac{2R_b}{\cos \phi_E} \cdot \cos\left(\frac{\pi}{2N}\right) \tag{5.52}$$

$$R_E = \sqrt{(R_b \cdot \tan \phi_E - R_w \cos \psi_b)^2 + R_b^2} \tag{5.53}$$

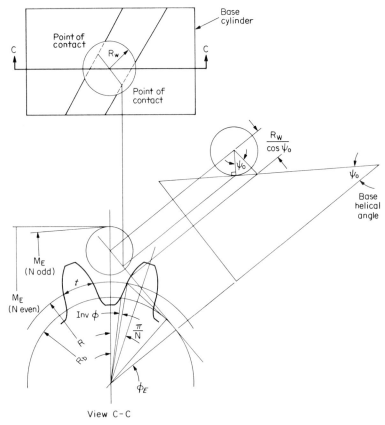

FIGURE 5.8 Measurement over balls on external helical gear.

Similarly, the measurements between balls for internal helical gears can be derived,

$$\tan \phi_I - \phi_I = \operatorname{inv} \phi_I = \operatorname{inv} \phi + \frac{s}{2R} - \frac{R_w}{R_b \cos \psi_b} \quad (5.54)$$

$$M_I \ (N \text{ even}) = \frac{2R_b}{\cos \phi_I} - 2 R_w \quad (5.55)$$

$$M_I \ (N \text{ odd}) = \frac{2R_b}{\cos \phi_I} \cdot \cos \left(\frac{\pi}{2N}\right) - 2R_w \quad (5.56)$$

$$R_I = \sqrt{(R_b \cdot \tan \phi_I + R_w \cos \psi_b)^2 + R_b^2} \quad (5.57)$$

Accurate measurements with pins for helical gears are much more complicated. Because of the skew motion of helical gears, the pin does not contact with teeth along a continuous line. For external helical gears, the pins usually drop to the highest points and make a single point contact with each tooth, just like a ball making contact with a tooth. The measurements over pins in such a situation are identical to the measurements over balls. For internal helical gears, the pins are picked up at the end points and contact with only one tooth at midpoint due to the helical inward skew motion. Therefore, the measurements are not true indications of tooth thickness. Because of the convenience of pin measurement, however, it is normal practice to establish the difference of measurements between pins and measurements between balls with both pins and balls of the same diameter for the same gear. Once the difference is known, the internal helical gears can be checked with pins with the same pin length.

5.9 INTERFERENCE LIMIT

Interference occurs when gears contact at some point other than along the line of action and cause nonuniform motion or gear teeth lock. To check for possible interference, the following equations are used for external gears (Figs. 5.9 and 5.10).

$$R_{o_2}^2 \leq R_{b_2}^2 + (R_2 + R_1)^2 \sin^2 \phi_o \qquad (5.58)$$

$$R_{o_1}^2 \leq R_{b_1}^2 + (R_2 + R_1)^2 \sin^2 \phi_o \qquad (5.59)$$

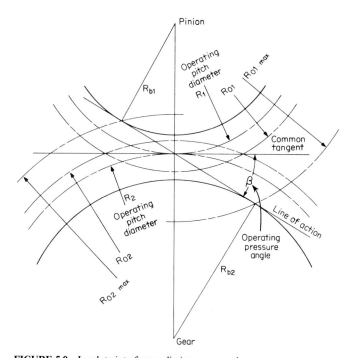

FIGURE 5.9 Involute interference limits on external gear.

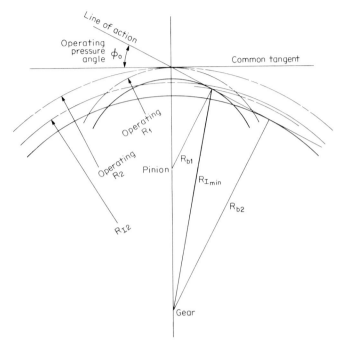

FIGURE 5.10 Involute interference limits on internal gear.

$R_2 + R_1$ is the operating center distance; ϕ_o is the operating pressure angle. For the internal gear

$$R_{I_2}^2 \geq R_{b_2}^2 + (R_2 - R_1)^2 \sin^2 \phi_o \quad (5.60)$$

$$R_{o_1} \leq R_{R_2} - (R_2 - R_1) \quad (5.61)$$

R_{R_2} is the root radius of the gear, and $R_2 - R_1$ is the operating center distance.
When the equal sign holds in the foregoing equations, R_{o_1} and R_{o_2} are the maximum outside diameters without interference, and R_{I_2} is the minimum inside diameter without interference. It is noted in Fig. 5.9 that the pinion tooth length can be increased appreciably more than gear tooth length without interference. Hence, the gear tooth is the one causing the trouble.

For a standard involute gear where addendum = $1/P$,

$$R_{o_2} = R_2 + \frac{1}{P}$$

rewrite Eq. 5.58

$$R_{o_2 \max}^2 = R_2^2 + 2 R_1 R_2 \sin^2 \phi_o + R_1^2 \sin^2 \phi_o$$

$$P \max = \frac{1}{R_2 \sqrt{1 + (2m_G + m_G^2) \cdot \sin^2 \phi_o} - 1)} \quad (5.62)$$

where m_G is R_1/R_2.

This equation gives the maximum diametral pitch with given gear ratio, pressure angle, and center distance.

There are several different ways to avoid gear interference. One method is to undercut the flank of the pinion teeth to remove the interfering contact. This often results in reduced bending strength and decreased contact ratio. Another method is to use long-addendum pinion teeth and short-addendum gear teeth. Short addendum on gear teeth is used to avoid interference and long addendum is used on the pinion teeth to maintain or increase contact ratio. The third method is to increase the pressure angle. As shown in Figs. 5.9 and 5.10, the larger the pressure angle is, the longer the addendum can be without interference. The increased pressure angle will shorten the contact ratio, however. It is desirable to combine any of the three methods to achieve required contact ratio, bending strength, and contact without interference.

For internal gears, Eqs. (5.60) and (5.61) are used to check interference. The minimum inside diameter without interference occurs when it extends to the intersection point between the line of action and the base circle of the pinion (see Fig. 5.10), and the maximum outside diameter of the pinion is smaller than the root diameter of the internal gear less operating center distance.

5.10 HIGH CONTACT RATIO GEAR

High contact ratio gears are defined as those that have at any given moment at least two pairs of teeth in contact as compared with only one pair of teeth in contact for regular gears. For a gear with minimum m (m is an integer) pairs of teeth in contact, the bending and contact stresses shall be calculated at the highest m teeth contact point,

$$r_1 = \sqrt{(pm + C \sin \phi_o - \sqrt{R_{o_2}^2 - R_{b_2}^2})^2 + R_{b_1}^2} \quad (5.63)$$

$$r_2 = \sqrt{(pm + C \sin \phi_o - \sqrt{R_{o_1}^2 - R_{b_1}^2})^2 + R_{b_2}^2} \quad (5.64)$$

Since high contact ratio gear designs are often achieved by enlarged addendum, the tip thickness often becomes the limit to how long an addendum can be used. Therefore, a minimum tip thickness is required to avoid brittleness at the tip when the gear is case hardened and to prevent tooth breakage due to thin section. Once the minimum tip thickness t_a is determined, the maximum radius from gear center to tip can be calculated by solving ϕ_a from

$$\text{inv } \phi_a + \frac{t_a \cos \phi_a}{2R_b} = \text{inv } \phi + \frac{t}{2R} \quad (5.65)$$

and then the maximum radius to tip is

$$r_a = \frac{R_b}{\cos \phi_a} \quad (5.66)$$

pointed tooth is the extreme case; when $t_a = 0$, Eq. (5.65) becomes

$$\text{inv } \phi_a = \text{inv } \phi + \frac{t}{2R} \quad (5.67)$$

and the radius to the tip of the pointed tooth is

$$r_{a_{\text{pointed}}} = \frac{R_b}{\cos \phi_a}$$

5.11 SPEED RATIO OF EPICYCLIC GEARS

There are many arrangements for epicyclic gears. The speed ratios of commonly used ones are listed in Table 5.2. Figure 5.11 shows these arrangements.

5.12 COMPOUND EPICYCLIC GEAR INDEX

Since a compound epicyclic gear has two planet gear members attached on a common shaft, the relative position of the teeth on these two gears varies around the shaft. An index is therefore required to hold certain relative positions between these two gears so that proper gear assembly can be insured. The condition for proper timing (Fig. 5.12) is

$$\alpha = \beta$$

$$\frac{N_R N_{p_1} + N_s N_{p_2}}{k} = (I_R - M)N_{p_1} + (I_s + L)N_{p_2} \qquad (5.68)$$

where I_R = an integer $I_R = N_R/k$ or next whole number larger than N_R/k
I_s = an integer $I_s = N_s/k$ or nearest whole number smaller than N_s/k
M = number of teeth from base line on planet 2
L = number of teeth from tooth/space on planet 1 in mesh with sun gear in position 1 that is to mesh with tooth on sun gear
k = number of equally spaced planet gears
N_R = number of teeth of ring gear
N_{p_1} = number of teeth of planet gear 1
N_{p_2} = number of teeth of planet gear 2
N_s = number of teeth of sun gear

L and M must be integer values. If such a set of integers is not found, the gear set will not assemble properly.

5.13 NONSTANDARD-ADDENDUM GEAR

The AGMA and ASA systems use $1.00/P$ as standard addendum. The nonstandard-addendum gear systems usually have a long-addendum (greater than $1.00/P$) pinion to mate with a short-addendum (less than $1.00/P$) gear. In order to maintain a proper tooth contact, it is necessary to adjust the tooth thickness for nonstandard-addendum

TABLE 5.2 Epicyclic Gear Ratio

	First Stage			Second Stage			
Arrangement	Sun N_1	Planet/Carrier N_2	Ring N_3	Sun N_4	Planet/Carrier N_5	Ring N_6	Total Ratio
a	Input	Fixed	Output				N_3/N_1
b	Input	Output	Fixed				$1 + N_3/N_1$
c	Fixed	Output	Input				$1 + N_1/N_3$
d	Input	Output	Coupled	Input	Coupled/output	Fixed	$1 + \dfrac{N_6}{N_4}\left(1 + \dfrac{N_3}{N_1}\right)$
e	Input	Coupled	Output	Input	Coupled/output	Fixed	$1 - \dfrac{N_3 N_6}{N_1 N_4}$
f	Input	Output	Coupled	Input	Fixed	Coupled/output	$\left(1 + \dfrac{N_3}{N_1}\right)\left(1 - \dfrac{N_6}{N_4}\right)$
g	Input	Coupled	Output	Input	Fixed	Coupled/output	$1 + \dfrac{N_3}{N_1}\left(1 + \dfrac{N_6}{N_4}\right)$
h	Input	Compound	—	—	Compound/output	Fixed	$1 + \dfrac{N_6 N_2}{N_1 N_5}$
i	Input	Compound	—	—	Compound/fixed	Output	$\dfrac{N_2 N_6}{N_1 N_5}$
j	—	Compound	Input	Fixed	Compound/output	—	$1 + \dfrac{N_4 N_2}{N_3 N_5}$
k	Fixed	Input/compound	—	Output	Compound	—	$\dfrac{1}{1 - \dfrac{N_1 N_5}{N_2 N_4}}$
l		Input/Compound	—	Fixed	Compound/output	—	$1 - \dfrac{N_2 N_4}{N_1 N_5} - \dfrac{N_3}{N_1}$

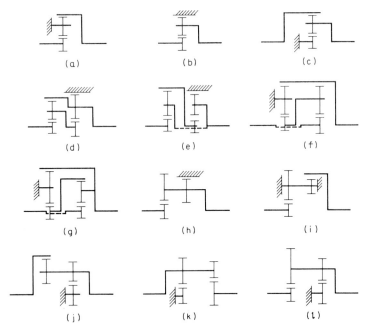

FIGURE 5.11 Common epicyclic gear arrangements.

gears. This adjustment is usually done by allowing the standard hob to cut the modified gears. The following formula is used to calculate this tooth thickness adjustment:

$$t = t_{\text{standard}} + \Delta_a \cdot (2 \cdot \tan \phi) \qquad (5.69)$$

where t_{standard} = tooth thickness for standard-addendum gears.
$\Delta a = a - a_{\text{standard}}$
a_{standard} = standard addendum

The reasons for using nonstandard addendum are:

1. *To avoid undercut:* The undercut of a pinion with a low number of teeth is usually big enough to substantially reduce the bending strength and sometimes introduces the interference with the mating gear. The use of a long-addendum pinion can eliminate undercut by moving the lowest contact point farther away from the pinion base circle.
2. *To balance strength between pinion and gear:* Gear and pinion are usually designed in equal strength. However, pinion teeth tend to be weaker in operating life for being subjected to more load cycles than gear teeth. A long-addendum pinion design with proper tooth thickness adjustment, making the pinion stronger and the gear weaker, makes a balanced strength in terms of operating life between gear and pinion.
3. *To minimize scoring potential:* Scoring potential is based on sliding velocity and compressive stress. The maximum compressive stress and the highest sliding velocity along the contact path occur between the lowest single tooth contact point and

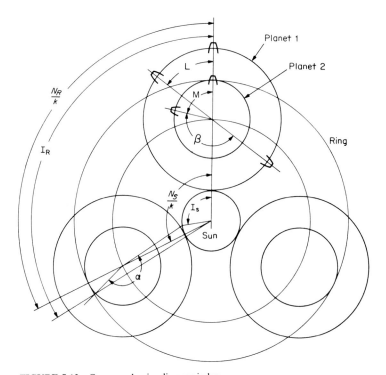

FIGURE 5.12 Compound epicyclic gear index.

the lowest contact point in the pinion. With long-addendum design, these contact points are held farther away from the pinion base circle. Therefore, the compressive stress is reduced by having a longer effective contact radius and the sliding velocity is also decreased by moving the contact point closer to the pitch point.

If pinion addendum is overmodified, however, the pinion may become stronger than the gear in operating life, and scoring potential may also increase due to the highest sliding velocity and the maximum compressive stress being changed to the tip of the pinion.

REFERENCES

1. AGMA 112.06, "Gear Nomenclature," American Gear Manufacturers Association, Washington, D.C.
2. Buckingham, E., *Analytical Mechanics of Gears,* McGraw-Hill, New York, 1949.
3. Chironis, N. P., "Gear Design and Applications," McGraw-Hill, New York, 1967.
4. Dudley, D., "Handbook of Practical Gear Design," McGraw-Hill, New York, 1984.
5. Fellows Gear Shaper Co., "The Involute Curve" and "Involute Gearing," Fellows Gear Shaper Co., Springfield, Vermont, 1950.
6. Myers, W., "Compound Planetaries," *Machine Design,* September 2, 1965.

CHAPTER 6
GEAR TOOTH CALCULATIONS

Robert Errichello
Geartech
Albany, California

Dennis P. Townsend
NASA Lewis Research Center
Cleveland, Ohio

Gear tooth profiles and other gear tooth calculations may be performed on a personal computer by the use of a small computer program or by a purchased program developed by a gear consultant or others.

This chapter shows some of the basic equations used for calculating various gear tooth problems. These include involute profile coordinates, gear tooth fillet coordinates, inverse of an involute, addendum modification, Hertzian stresses, bending stresses, and interference of internal gears. Computer program flow charts are given for some of the calculations.

Gear computer programs may be user friendly and thus very easy to use to perform gear calculations. However, a designer using such programs without a good understanding of gear design may arrive at a design that does not meet the expected requirements.

6.1 CALCULATING INVOLUTE TOOTH PROFILE COORDINATES

The following algorithm calculates the coordinates of an involute profile based on a known (datum) tooth thickness and a known base radius and base helix angle.

Figure 6.1 is a transverse-plane view of an involute gear tooth showing a rectangular coordinate system $\xi - \eta$ with its origin located at the center of the gear and its η axis directed along the centerline of the gear tooth.

Selecting the radius r_i to an arbitrary point i on the involute profile as the independent parameter, the transverse pressure angle at radius r_i is

$$\phi_i = \cos^{-1}\left(\frac{R_b}{r_i}\right) \quad (6.1)$$

where R_b is the base radius (a constant for a given gear).

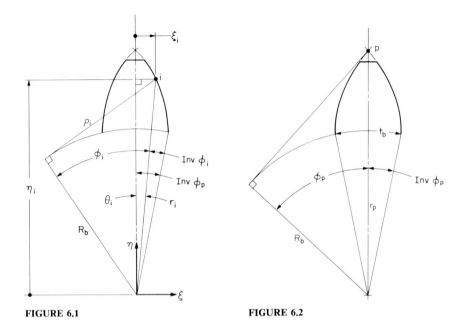

FIGURE 6.1   **FIGURE 6.2**

The involute function is

$$\text{inv } \phi_i = \tan \phi_i - \phi_i \tag{6.2}$$

The polar angle is

$$\theta_i = \text{inv } \phi_p - \text{inv } \phi_i \tag{6.3}$$

where inv ϕ_p is the involute function at radius r_p where the gear tooth is pointed. The involute function is shown in Fig. 6.2 and is given by

$$\text{inv } \phi_p = \frac{t_b}{2R_b} \tag{6.4}$$

where t_b is the transverse circular base tooth thickness. It is selected as a convenient datum thickness that can be determined experimentally by span measurement. Alternatively, it can be calculated from any known thickness t_i from

$$t_b = 2R_b \left(\frac{t_i}{2r_i} + \text{inv } \phi_i \right) \tag{6.5}$$

where t_i is the known transverse circular tooth thickness at known radius r_i, and ϕ_i is the transverse pressure angle at radius r_i [see Eq. (6.1)].

The transverse circular tooth thickness at radius r_i is found from the polar angle θ_i [Eq. (6.3)] from

$$t_i = 2r_i \theta_i \tag{6.6}$$

GEAR TOOTH CALCULATIONS

The coordinates of point i are

$$\xi_i = r_i \sin \theta_i \qquad (6.7)$$

$$\eta_i = r_i \cos \theta_i \qquad (6.8)$$

The radii of curvature of the profile at point i in the transverse and normal planes are

$$\rho_i = r_i \sin \phi_i \qquad (6.9)$$

$$\rho_{n_i} = \frac{\rho_i}{\cos \psi_b} \qquad (6.10)$$

where ψ_b is the base helix angle (a constant for a given gear).
The helix angle at radius r_i is

$$\psi_i = \tan^{-1}\left(\frac{\tan \psi_b}{\cos \phi_i}\right) \qquad (6.11)$$

The normal circular tooth thickness is

$$t_{n_i} = t_i \cos \psi_i \qquad (6.12)$$

The normal pressure angle at radius r_i is

$$\phi_{n_i} = \tan^{-1}(\tan \phi_i \cos \psi_i) \qquad (6.13)$$

The coordinates of an internal gear tooth can be calculated from the coordinates of the tooth space of an external gear. This is equivalent to rotating the coordinate system so that the η axis is directed along the centerline of the tooth space rather than the centerline of the tooth. This can be incorporated in the algorithm by redefining θ_i for an internal gear as

$$\theta_i = \frac{\pi}{N} - \theta_i = \frac{\pi}{N} - \text{inv } \phi_p + \text{inv } \phi_i \qquad (6.14)$$

where N is the number of teeth in the internal gear, and inv ϕ_p is based on the thickness t_b of the space.

6.2 CALCULATING GEAR TOOTH FILLET COORDINATES

The following algorithm calculates the coordinates of gear tooth fillets that are generated by rack-type tools (hobs, rack cutters, or generating grinding wheels) with or without protuberance.

Figure 6.3 is a normal-plane view of a rack-type tool with protuberance δ_o. A tool without protuberance is a particular case for which $\delta_o = 0$. The $\xi_{n_o} - \eta_{n_o}$ coordinate system has its origin at the intersection of the centerline of the tool space and the generating pitch line. The coordinates of the center of the tip radius, point S are

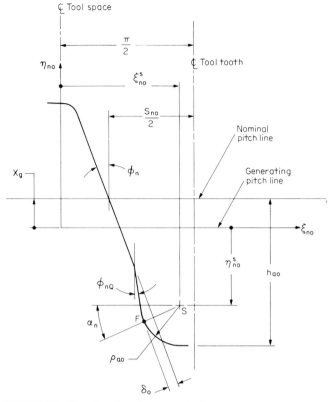

FIGURE 6.3 Normal section of rack-type tool.

$$\xi_{n_o}^s = \frac{\pi - S_{n_o}}{2} + (h_{a_o} - \rho_{a_o}) \tan \phi_n + \frac{\rho_{a_o} - \delta_o}{\cos \phi_n}$$
$$\eta_{n_o}^s = X_g - (h_{a_o} - \rho_{a_o})$$
(6.15)

An arbritary point F on the tool tip radius is defined by angle α_n. Its coordinates in the coordinate system $\xi_{n_o} - \eta_{n_o}$ are

$$\xi_{n_o} = \xi_{n_o}^s - \rho_{a_o} \cos \alpha_n$$
$$\eta_{n_o} = \eta_{n_o}^s - \rho_{a_o} \sin \alpha_n$$
(6.16)

To obtain the coordinates of the gear tooth fillet, we consider the relative motion between the rack-type tool and the generated gear tooth. Figure 6.4 is a transverse-plane view showing the coordinate system $\xi - \eta$ connected with the gear, and the coordinate system $\xi_o - \eta_o$ connected with the tool. In the "zero-position," the origin of the system $\xi_o - \eta_o$ is located at the pitch point P, and the axis η is coincident with the axis η_o. The transformation of the normal-plane tool coordinates ξ_{n_o}, η_{n_o} to the transverse plane is obtained by

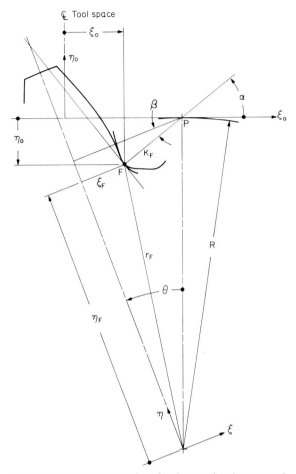

FIGURE 6.4 Transverse section of tool generating the gear tooth fillet.

$$\xi_o = \frac{\xi_{n_o}}{\cos \psi}$$
$$\eta_o = \eta_{n_o}$$
(6.17)

In the position shown in Fig. 6.4, the tool is generating an arbritrary point F on the gear tooth fillet. From the law of conjugate gear tooth action, point F lies on the contact normal that extends through the generating pitch point P. Salamoun and Suchy[1] show that the angle α of the contact normal in the transverse plane is related to the angle α_n in the normal plane by

$$\tan \alpha = \frac{\tan \alpha_n}{\cos \psi} \qquad (6.18)$$

The fillet coordinates ξ, η are obtained by selecting the angle α_n as the independent parameter. Then for $\alpha_n = \pi/2$, generation starts at the lowest point on the fillet and proceeds up the fillet as α_n is diminished corresponding to movement of the tool to the left and counterclockwise rotation of the gear. The distance from the pitch point to point F is

$$\kappa_F = \frac{\eta_o}{\sin \alpha} \qquad (6.19)$$

The relation between α and the gear rotation is

$$\theta = \frac{1}{R} (\xi_o - \kappa_F \cos \alpha) \qquad (6.20)$$

The slope of the line tangent to the fillet at point F is

$$\beta = \alpha - \theta \qquad (6.21)$$

The coordinates of point F are

$$\xi_F = R \sin \theta + \kappa_F \cos \beta$$
$$\eta_F = R \cos \theta + \kappa_F \sin \beta \qquad (6.22)$$
$$r_F = (\xi_F^2 + \eta_F^2)^{1/2}$$

Figure 6.5 is a flow chart for calculating the fillet coordinates of p number of points on the fillet over the range of $\alpha_n = \pi/2$ to $\alpha_n = \phi_{nQ}$.

6.3 CALCULATING THE INVERSE OF AN INVOLUTE

The involute function is defined as

$$\text{inv } \phi = \tan \phi - \phi \qquad (6.23)$$

where ϕ is the pressure angle in radians.

Calculating the involute function inv ϕ from Eq. (6.23) is a simple matter if ϕ is known. But geometry calculations for involute gears often require that the inverse of the involute function be calculated; that is, given inv ϕ, find ϕ. It is difficult to obtain ϕ from Eq. (6.23) because the equation involves a transcendental term, and ϕ is usually obtained by interpolating from a table of involute functions.

The following numerical technique can be used for computer programs. It is based on Newton's method of iteration and is very efficient, usually requiring only three to five iterations for convergence. Newton's method solves for ϕ with an iterative process in which an initial estimate for ϕ is successively improved on using

$$\phi_{i+1} = \phi_i - \frac{f(\phi_i)}{f'(\phi_i)} \qquad (6.24)$$

GEAR TOOTH CALCULATIONS

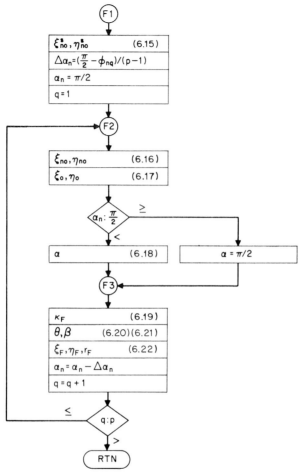

FIGURE 6.5 Flow chart for fillet coordinates.

To apply Newton's method, rearrange Eq. (6.23) and define

$$f(\phi_i) = \tan \phi_i - \phi_i - \text{inv } \phi = 0 \qquad (6.25)$$

The derivative of Eq. (6.25) is

$$f'(\phi_i) = \sec^2 \phi_i - 1 - 0 = \tan^2 \phi_i \qquad (6.26)$$

Substituting Eqs. (6.25) and (6.26) in Eq. (6.24) gives

$$\phi_{i+1} = \phi_i + \frac{\text{inv } \phi + \phi_i - \tan \phi_i}{\tan^2 \phi_i} \qquad (6.27)$$

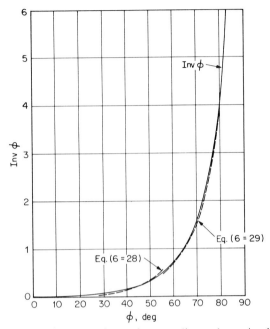

FIGURE 6.6 Curves of approximate equations and exact inv ϕ.

To minimize the number of iterations, the initial estimate for ϕ_i should be calculated as follows. If inv $\phi < 0.5$, let

$$\phi_i = 1.441 \, (\text{inv } \phi)^{1/3} - 0.366 \, (\text{inv } \phi) \qquad (6.28)$$

If inv $\phi \geq 0.5$, let

$$\phi_i = 0.243 \, \pi + 0.471 \, \tan^{-1}(\text{inv } \phi) \qquad (6.29)$$

Figure 6.6 compares curves described by the approximate Eqs. (6.28) and (6.29) with the exact curve for inv ϕ. For involute gear teeth, a practical range for the involute function is inv $\phi = 0$ (for a point on the base circle) to inv $\phi = \pi$ (for a point at the pointed tip of a single-tooth pinion). The corresponding range for ϕ is $\phi = 0$ to $\phi = 77.4534°$.

Figure 6.7 is a flow chart that summarizes the subroutine for calculating the inverse of an involute.

6.4 CALCULATING ADDENDUM MODIFICATION

6.4.1 Basic Gear Geometry

$$m_G = \frac{n_2}{n_1} \qquad (6.30)$$

GEAR TOOTH CALCULATIONS

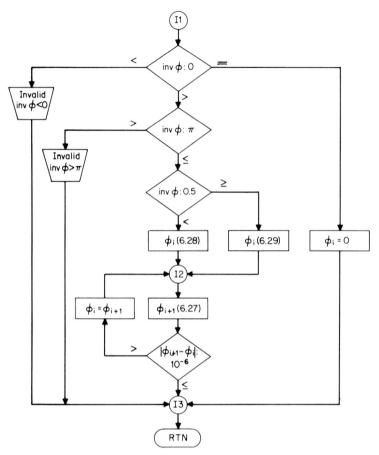

FIGURE 6.7 Flow chart for calculating the inverse of an involute.

$$R_1 = \frac{n_1}{2 P_n \cos \psi} \quad (6.31)$$

$$R_2 = R_1 m_G \quad (6.32)$$

$$C = R_2 \pm R_1 \quad (6.33)$$

$$\phi = \tan^{-1}\left(\frac{\tan \phi_n}{\cos \psi}\right) \quad (6.34)$$

$$\phi_r = \cos^{-1}\left(\frac{C}{C_r} \cos \phi\right) \quad (6.35)$$

$$\text{inv } \phi = \tan \phi - \phi \quad (6.36)$$

$$\text{inv } \phi_r = \tan \phi_r - \phi_r \quad (6.37)$$

6.4.2 List of Symbols and Subscripts

Symbols

B_n = normal operating circular backlash
C_1 = distance to SAP
C_5 = distance to EAP
C_6 = distance between interference points
C = standard center distance
C_r = operating center distance
h_{a_1}, h_{a_2} = addenda
J_1, J_2 = bending strength geometry factors
k_s = tip-shortening coefficient
m_G = gear ratio
n_1, n_2 = number of teeth in pinion, gear
P_n = normal diametral pitch
R_1, R_2 = standard pitch radii
R_{o_1}, R_{o_2} = outside radii
s_{n_1}, s_{n_2} = reference normal circular tooth thicknesses
s_{n_t} = bending strength
x_1, x_2 = addendum modification coefficients
x_{g_1}, x_{g_2} = generating rack shift coefficients
$(x_1)_{min}$ = minimum addendum mod. coeff. to avoid undercut
ΔC = center distance modification
$\Delta s_{n_1}, \Delta s_{n_2}$ = tooth thinning for backlash
Σx = sum of addendum modification coefficients
Σx_g = sum of generating rack shift coefficients
ϕ = standard transverse pressure angle
ϕ_n = standard normal pressure angle
ϕ_r = operating transverse pressure angle
ψ = standard helix angle

Subscripts

1 = pinion
2 = gear
n = normal
r = operating or running
(*no subscript indicates transverse*)

6.4.3 Sum of Addendum Modification Coefficients for Zero Backlash

$$\Sigma x = \frac{C P_n (\text{inv } \phi_r - \text{inv } \phi)}{\tan \phi} \quad (6.38)$$

$$\Sigma x = x_2 \pm x_1 \quad (6.39)$$

GEAR TOOTH CALCULATIONS   6.11

Selecting Values for x_1 and x_2.   Values of x_1 and x_2 are selected such that their sum satisfies Eq. (6.39) while considering several criteria:

- Avoiding undercut
- Balanced specific sliding
- Balanced bending fatigue life
- Balanced flash temperature
- Avoiding narrow toplands

Avoiding Undercut.   The minimum addendum modification to avoid undercut for the pinion is given by:

$$(x_1)_{\min} = 1.1 - n_1 \left(\frac{\sin^2 \phi}{2 \times \cos \psi} \right) \tag{6.40}$$

Balanced Specific Sliding.   Maximum pitting and wear resistance is obtained by balancing the specific sliding ratio at the ends of the path of contact. This is done by iteratively varying the addendum modification coefficients of the pinion and gear until the following equation is satisfied:

$$\left(\frac{C_6}{C_1} \mp 1 \right) \left(\frac{C_6}{C_5} \mp 1 \right) = m_G^2 \tag{6.41}$$

where C_1 = distance to SAP (see Fig. 6.8)
C_5 = distance to EAP (see Fig. 6.8)
C_6 = distance between interference points (see Fig. 6.8)

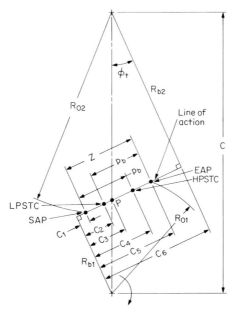

FIGURE 6.8

Balanced Bending Fatigue Life. Maximum bending fatigue resistance is obtained by iteratively varying the addendum modification coefficients of the pinion and gear until the ratio of the bending strength geometry factors equals the ratio of bending strengths, that is,

$$\frac{J_1}{J_2} = \frac{s_{n_{t2}}}{s_{n_{t1}}} \tag{6.42}$$

Balanced Flash Temperature. Maximum scuffing resistance is obtained by minimizing the contact temperature. This is done by iteratively varying the addendum modification coefficients of the pinion and gear, while calculating the flash temperature by Blok's equation, until the flash temperature peaks in the approach, and recess portions of the line of action are equal. The flash temperature should be calculated at the points SAP, LPSTC, HPSTC, EAP, and at several points in the two pair zones (between points SAP and LPSTC and between points HPSTC and EAP, see Fig. 6.8).

Avoiding Narrow Toplands. The maximum permissible addendum modification coefficients are obtained by iteratively varying the addendum modification coefficients of the pinion and gear until their topland thicknesses are equal to the minimum allowable (usually $\frac{0.3}{P_n}$).

6.4.4 Tooth Thinning for Backlash

The small adjustments of the position of the cutting tool to thin the gear teeth for backlash are considered independently of the addendum modification coefficients (x_1 and x_2) by specifying the amount the pinion and gear teeth are thinned for backlash, Δs_{n_1} and Δs_{n_2}. This way, the outside diameters are independent of tooth thinning for backlash, and are based solely on the addendum modification coefficients x_1 and x_2. The tooth thinning coefficients are selected such that

$$\Delta s_{n_1} + \Delta s_{n_2} = B_n P_n \left(\frac{C}{C_r}\right) \tag{6.43}$$

6.4.5 Tip-Shortening Coefficient for External Gear Sets

For gears operating on extended centers ($C_r > C$), the outside radii of the gears are shortened to maintain adequate tip-to-root clearance. The amount of adjustment of outside radii is proportional to the tip-shortening coefficient

$$k_s = \Sigma x - \Delta C\, P_n \tag{6.44}$$

where

$$\Delta C = C_r - C \tag{6.45}$$

6.4.6 Tip-Shortening Options

1. *Full-length teeth*

$$h_{a_1} = \frac{1 + x_1}{P_n} \tag{6.46}$$

GEAR TOOTH CALCULATIONS

$$h_{a2} = \frac{1 + x_2}{P_n} \quad (6.47)$$

Caution: Option 1 (full-length teeth) may give insufficient tip-to-root clearance if $C_r >> C$. Check clearances or use option 3 to be safe.

2. *Standard working depth*

$$h_{a1} = \frac{1 + x_1 - \frac{k_s}{2}}{P_n} \quad (6.48)$$

$$h_{a2} = \frac{1 + x_2 - \frac{k_s}{2}}{P_n} \quad (6.49)$$

Caution: Option 2 (standard working depth) may give insufficient tip-to-root clearance if $C_r >> C$. Check clearances or use option 3 to be safe.

3. *Standard tip-to-root clearance*

$$h_{a1} = \frac{1 + x_1 - k_s}{P_n} \quad (6.50)$$

$$h_{a2} = \frac{1 + x_2 - k_s}{P_n} \quad (6.51)$$

6.4.7 Tip-Shortening Coefficient for Internal Gear Sets

For internal gear sets, several requirements must be met in addition to those that apply to external gear sets. There must be no tip interference between the pinion and gear or between the cutter and gear. Also, there must be no rubbing between the cutter and gear during the return stroke of the cutter. A likely place for interference is between the tooth root fillets of the pinion and the tips of the gear teeth, and it is common practice to shorten the teeth of the internal gear to prevent this. Likewise, the tip radius of the pinion must be selected to ensure that the pinion tips do not interfere with the root fillets of the gear. As with external gear sets, undercut should be avoided and adequate tip-to-root clearances must be maintained. Colbourne[2] describes a design procedure for a generalized form of addendum modification which includes all of the foregoing considerations.

Outside Radii

$$R_{o1} = R_1 + h_{a1} \quad (6.52)$$

$$R_{o2} = R_2 \pm h_{a2} \quad (6.53)$$

Generating Rack Shift Coefficients

$$x_{g1} = x_1 - \frac{\Delta s_{n1}}{2 \tan \phi_n} \quad (6.54)$$

$$x_{g2} = x_2 \mp \frac{\Delta s_{n2}}{2 \tan \phi_n} \quad (6.55)$$

CHAPTER SIX

Normal Circular Tooth Thicknesses

$$s_{n_1} = \frac{\frac{\pi}{2} + 2 x_{g_1} \tan \phi_n}{P_n} \quad (6.56)$$

$$s_{n_2} = \frac{\frac{\pi}{2} \pm 2 x_{g_2} \tan \phi_n}{P_n} \quad (6.57)$$

6.4.8 Determining Addendum Modification Coefficients of Existing Gears

If the normal circular tooth thicknesses are known, the generating rack shift coefficients are found from Eqs. (6.58) and (6.59):

$$x_{g_1} = \frac{s_{n_1} P_n - \frac{\pi}{2}}{2 \tan \phi_n} \quad (6.58)$$

$$x_{g_2} = \pm \left(\frac{s_{n_2} P_n - \frac{\pi}{2}}{2 \tan \phi_n} \right) \quad (6.59)$$

Sum of Generating Rack Shift Coefficients

$$\Sigma x_g = x_{g_2} \pm x_{g_1} \quad (6.60)$$

Normal Operating Circular Backlash

$$B_n = \pm \frac{2 C_r \tan \phi_n}{CP_n} (\Sigma x - \Sigma x_g) \quad (6.61)$$

6.4.9 Tooth Thinning for Backlash

The tooth thinning coefficients must satisfy Eq. (6.43). However, it is usually impossible to determine the ratio $\Delta s_{n_1}/\Delta s_{n_2}$ that was used for existing gears. The following analysis is based on common practice where $\Delta s_{n_1} = \Delta s_{n_2}$, in which case

$$\Delta s_{n_1} = \Delta s_{n_2} = \frac{B_n P_n}{2} \left(\frac{C}{C_r} \right) \quad (6.62)$$

GEAR TOOTH CALCULATIONS

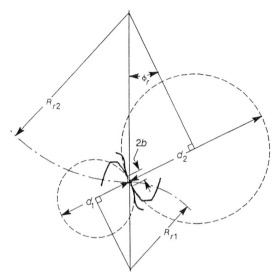

FIGURE 6.9 Tooth contact stress area.

6.4.10 Addendum Modification Coefficients

From Eqs. (6.54) and (6.55)

$$x_1 = x_{g1} + \frac{\Delta s_{n_1}}{2 \tan \phi_n} \tag{6.63}$$

$$x_2 = x_{g2} \pm \frac{\Delta s_{n_2}}{2 \tan \phi_n} \tag{6.64}$$

6.5 DERIVATION OF AGMA HERTZIAN STRESS EQUATION*

6.5.1 Derivation

The AGMA pitting resistance formula is based on the Hertz contact stress equation for cylinders with parallel axes (see Fig. 6.9). The load applied to the cylinder is the load normal to the tooth flank, and the length of contact is the minimum total length of lines of contact in the contact zone of the gear set. The radii of the cylinders are the radii of curvature of the teeth at the point of contact for the mating pair of gears. Depending on the face contact ratio of the gear set, this point can be the mean diameter of the pinion or the lowest point of single-tooth contact on the pinion. Additional rating factors are also added to the basic equation to adjust the stress due to factors peculiar to gearing. Starting with the general Hertz equation

---

*Extracted from AGMA Standard 908-B89[3] with the permission of the publisher.

$$S_c = \frac{2W}{\pi b L} \quad (6.65)$$

$$b = \sqrt{\left(\frac{2W}{\pi L}\right)\frac{\left(\dfrac{1-\mu_1^{\,2}}{E_1} + \dfrac{1-\mu_2^{\,2}}{E_2}\right)}{\left(\dfrac{1}{d_1} + \dfrac{1}{d_2}\right)}} \quad (6.66)$$

where S_c = maximum contact stress of parallel axis cylinders, lb/in.^{2}
W = contact load normal to the cylinders, lb
b = semi-width of contact between cylinders, in.
L = length of contact between cylinders, in.
μ_1, μ_2 = Poisson's ratio of material in cylinders 1 and 2
E_1, E_2 = modulus of elasticity of material in cylinders 1 and 2
d_1, d_2 = contact diameter of cylinders 1 and 2

Figure 6.10 shows the oblique contact line between helical gear tooth profiles. It can be represented as the line of two contacting cones. The radii of curvature, ρ_{n_1}, ρ_{n_2}, of the pinion and gear teeth at any point along the contact line are perpendicular to the line of contact and the involute profiles at the point of contact. They are contained in the plane of action and are inclined at the base helix angle ψ_b, relative to the transverse plane.

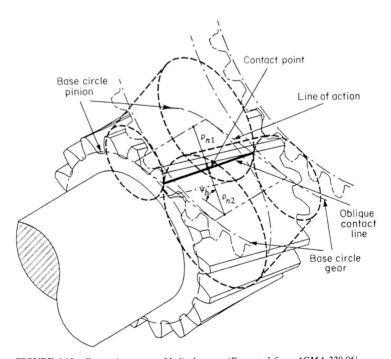

FIGURE 6.10 Contacting cones of helical gears. *(Extracted from AGMA 229.06)*

GEAR TOOTH CALCULATIONS 6.17

Converting the general nomenclature in Eqs. (6.65) and (6.66) to those used in AGMA 908-B89:

$$W_N = W \qquad L_{\min} = L$$
$$2\rho_{n_1} = d_1 \qquad 2\rho_{n_2} = d_2$$
$$\mu_P = \mu_1 \qquad \mu_G = \mu_2$$
$$E_P = E_1 \qquad E_G = E_2$$

where W_N = transmitted load in the plane of action (normal to the tooth flank)
$L_{\min}$ = minimum total length of lines of contact in contact zone
ρ_{n_1}, ρ_{n_2} = radius of curvature normal to the profile at point of stress calculation for pinion and gear, respectively
μ_P, μ_G = Poisson's ratio for pinion and gear material, respectively
E_P, E_G = modulus of elasticity for pinion and gear material, respectively

Rearranging Eqs. (6.65) and (6.66), then using AGMA 908–B89 terms

$$s_c = \sqrt{\left(\frac{W_N}{L_{\min}}\right) \frac{\left(\dfrac{1}{\rho_{n_1}} \pm \dfrac{1}{\rho_{n_2}}\right)}{\pi\left(\dfrac{1-\mu^2_P}{E_P} + \dfrac{1-\mu^2_G}{E_G}\right)}} \qquad (6.67)$$

Note: Double signs are used in Eq. (6.67), that is, $\pm$, to generalize the equation for both external and internal gears. The upper sign applies to external gear sets and the lower sign applies to internal gear sets.

By defining C_p as

$$C_p = \sqrt{\frac{1}{\pi\left(\dfrac{1-\mu^2_P}{E_P} + \dfrac{1-\mu^2_G}{E_G}\right)}} \qquad (6.68)$$

also transferring the load and radii of curvature from the normal to the transverse plane where the geometry is more readily defined:

$$W_N = \frac{W_t}{\cos \phi_t \, (\cos \psi_b)} \qquad (6.69)$$

$$\rho_{n_1} = \frac{\rho_1}{\cos \psi_b} \qquad (6.70)$$

$$\rho_{n_2} = \frac{\rho_2}{\cos \psi_b} \qquad (6.71)$$

where W_t = transmitted tangential load at the operating pitch diameter of the pinion
ϕ_t = operating transverse pressure angle

6.18 CHAPTER SIX

ψ_b = base helix angle
ρ_1, ρ_2 = radii of curvature of profile in transverse plane at the point of stress calculation

Eq. (6.67) can now be shown as

$$s_c = C_p \sqrt{\frac{W_t}{(\cos \phi_t) L_{\min}} \left(\frac{1}{\rho_1} \pm \frac{1}{\rho_2} \right)} \qquad (6.72)$$

Equation (6.76) was originally developed before calculators and computers, and it has some terms that made calculation simpler. These terms are not necessarily needed today, but data and values have been developed over the years, and changing terms could cause a great deal of confusion. The load sharing ratio m_N and the pitting resistance geometry factors I are these terms.

They are developed by multiplying Eq. (6.72) by

$$\sqrt{\frac{F}{F}} \quad \text{and} \quad \sqrt{\frac{d}{d}}$$

and then recombining terms so the m_N and I can be defined as follows:

$$m_N = \frac{F}{L_{\min}} \qquad (6.73)$$

$$I = \frac{(\cos \phi_t) C_\psi^2}{\left(\dfrac{1}{\rho_1} \pm \dfrac{1}{\rho_2} \right) d m_N} \qquad (6.74)$$

where F = effective face width of gear set
 d = operating pitch diameter of pinion
 C_ψ = helical overlap factor

The helical overlap factor C_ψ is added to the equation for I to give a smooth transition between the I factor of spur and *low axial contact ratio* (LACR) helicals. Now Eq. (6.72) can be rewritten as

$$s_c = C_p \sqrt{\frac{W_t}{dFI}} \qquad (6.75)$$

To adapt Eq. (6.75) to actual gears, AGMA has added additional factors to the equation. These factors are:

C_a = application factor for pitting resistance
C_v = dynamic factor for pitting resistance
C_s = size factor for pitting resistance
C_m = load distribution factor for pitting resistance
C_f = surface condition factor for pitting resistance

Putting these factors in Eq. (6.75) gives

$$s_c = C_p \sqrt{\frac{W_t\, C_a}{C_v}\left(\frac{C_s}{dF}\right)\frac{C_m\, C_f}{I}} \qquad (6.76)$$

Equation (6.76) is the fundamental formula for pitting resistance and is identical to Eq. (5.1) in ANSI/AGMA 2001–B88 and Eq. (5.1) in AGMA 218.01.

6.6 GEAR BENDING STRESS ANALYSIS

Consider a cantilever beam of arbitrary height h and uniform width b, Fig. 6.11

$$\sigma = \frac{Mc}{I} = \frac{6Wx}{bh^2} \qquad (6.77)$$

For constant stress,

$\dfrac{x}{h^2}$ = constant → parabolic beam

σ = constant (ignoring stress concentration)

Lewis (1892) found the critical section of a gear tooth by inscribing a parabola tangent to the root fillets (Fig. 6.12).

Load normal to tooth profile (Fig. 6.13):

$$W_N = \frac{W_t}{\cos \phi_{nr} \cos \psi_r} \qquad (6.78)$$

Load/inch of minimum contact length (Fig. 6.13):

$$w_N = \frac{W_N}{L_{\min}} \qquad (6.79)$$

Radial load:

$$w_r = w_N \sin \phi_{nL} \qquad (6.80)$$

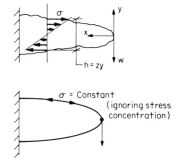

FIGURE 6.11 Cantilever beam.

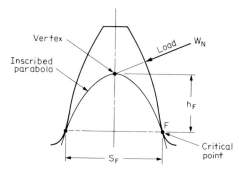

FIGURE 6.12 Inscribed parabola.

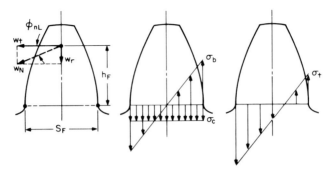

FIGURE 6.13 Tooth bending stress.

Tangential load:

$$w_t = w_N \cos \phi_{nL} \tag{6.81}$$

Direct compressive stress:

$$\sigma_c = \frac{-w_r}{S_F} \tag{6.82}$$

Bending stress:

$$\sigma_b = \frac{6 w_t h_F}{S_F^2} \tag{6.83}$$

Net tensile stress:

$$\sigma_t = \sigma_b + \sigma_c \tag{6.84}$$

$$= \frac{6 w_t h_F}{S_F^2} - \frac{w_r}{S_F} \tag{6.85}$$

Subtstitute for w_t, w_r

$$\sigma_t = w_N \cos \phi_{nL} \left(\frac{6 h_F}{S_F^2} - \frac{\tan \phi_{nL}}{S_F} \right)$$

$$= \frac{W_t \cos \phi_{nL}}{\cos \phi_{nr} \cos \psi_r L_{\min}} \left(\frac{6 h_F}{S_F^2} - \frac{\tan \phi_{nL}}{S_F} \right) \tag{6.86}$$

By convention, h_F and S_F are based on $P_n = 1.0$ therefore multiply by

$$P_n = \frac{P}{\cos \psi} \tag{6.87}$$

GEAR TOOTH CALCULATIONS

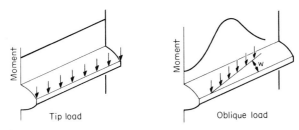

FIGURE 6.14 Helical gear tip and oblique load.

Also multiply by the stress correction factor K_f

$$\sigma_t = \frac{W_t \cos \phi_{nL}}{\cos \phi_{n_r} \cos \psi_r \cos \psi L_{\min}} \left(\frac{6 h_F}{S_F^2} - \frac{\tan \phi_{nL}}{S_F} \right) P \cdot K_f \quad (6.88)$$

Wellauer and Seireg in 1960 empirically found the effects of an oblique line of contact in helical gears (Fig. 6.14). They introduced the helical factor

$$C_h = \frac{\text{bending moment due to tip load}}{\text{bending moment due to oblique load}} \quad (6.89)$$

$$C_h = \frac{1}{1 - \left[\frac{w}{100} \left(1 - \frac{w}{100} \right) \right]^{1/2}} \quad (6.90)$$

$$w = \tan^{-1} (\tan \psi \sin \phi_n)° \quad (6.91)$$

Helical gears are analyzed by assuming tip loading (Fig. 6.15). The bending moment is reduced by the helical factor

$$\sigma_b = \frac{6 w_t h_F}{S_F^2 C_h} \quad (6.92)$$

FIGURE 6.15 Tip loading.

Define

Load-sharing ratio:

$$m_N = \frac{F}{L_{\min}} \quad (6.93)$$

Form factor

$$Y = \frac{K_\psi}{\dfrac{\cos \phi_{nL}}{\cos \phi_{n_r}} \left[\dfrac{6 h_F}{S_F^2 C_h} - \dfrac{\tan \phi_{nL}}{S_F} \right]} \quad (6.94)$$

Helix angle factor: $\quad K_\psi = \cos \psi_r \cos \psi \quad (6.95)$

Bending strength geometry factor:

$$J = \frac{Y\,C_\psi}{K_f\,m_N} \tag{6.96}$$

where C_ψ = helical overlap factor
K_f = stress correction factor from photoelastic experiments of Dolan and Broghamer in 1942

$$K_f = H + \left(\frac{S_F}{P_F}\right)^L \left(\frac{S_F}{h_F}\right)^M \tag{6.97}$$

$$\begin{aligned} H &= 0.331 - 0.436\,\phi_n \\ L &= 0.324 - 0.492\,\phi_n \\ M &= 0.261 + 0.545\,\phi_n \end{aligned} \tag{6.98}$$

ϕ_n = radians
ρ_F = minimum radius of curvature of root fillet

Substituting Eqs. (6.93) through (6.96) in Eq. (6.88) gives

$$\sigma_t = \frac{W_t P}{FJ} \tag{6.99}$$

Adding derating factors gives the AGMA bending stress equation

$$S_t = \frac{W_t\,K_a}{K_v} \cdot \frac{P_d}{F} \cdot \frac{K_s K_m}{J} \tag{6.100}$$

6.7 INTERFERENCE OF INTERNAL GEARS

By Dr. ir. J.W. POLDER
Neunen 5671.ED.5, the Netherlands

6.7.1 Involute Interference

Involute interferences concern imperfect generation of involutes, directly described by the geometry of the line of action. Two cases may be distinguished:

Pinion tooth tip interferes with root fillet of internal gears
Internal gear tooth tip interferes with root fillet of pinion

The first case means that the active profile of an internal gear tooth, determined by the pinion, exceeds its usable profile, determined by the cutter.
The active profile is determined by the path of contact, bounded between the tip radius of curvature of the internal gear and the radius.

6.7.2 Nomenclature

a = center distance
a_{dam} = center distance at which maximum overcut originates
f_a = deviation of center distance
j = backlash
j_1 = backlash component for pinion
j_2 = backlash component for internal gear
k_1 = tip-shortening coefficient of pinion
m_n = normal module
Q_1 = auxiliary quantity
Q_2 = auxiliary quantity
r_{a_1} = tip radius of pinion
r_{a_2} = tip radius of internal gear
r_{b_1} = base radius of pinion
r_{b_2} = base radius of internal gear
s_{k_1} = tip-to-tip chord of pinion
s_{k_2} = tip-to-tip chord of internal gear
x_1 = addendum modification coefficient of pinion
x_2 = addendum modification coefficient of internal gear
z_1 = number of teeth of pinion
z_2 = number of teeth of internal gear
α_n = normal pressure angle
α_t = transverse pressure angle
α_t' = transverse working pressure angle
α^* = auxiliary pressure angle
β = helix angle
ϵ_α = contact ratio
ρ_{a_1} = tip radius of curvature of pinion
ρ_{a_2} = tip radius of curvature of internal gear
ρ_{u_1} = radius of curvature of pinion at lowest point of contact
ρ_{u_2} = radius of curvature of internal gear at highest point of contact
ψ = tip-to-tip chord half angle

$$\rho_{u_2,\text{ active}} = -\sqrt{r_{a_1}^2 - r_{b_1}^2} - \sqrt{a^2 - (r_{b_2} - r_{b_1})^2} \quad (6.101)$$

The usable profile is determined by the same Eq. (6.101), substituting the quantities of the mating pinion and its center distance by those of the pinion-type cutter that produces the internal gear. The condition (for the absolute values of the radii)

$$|\rho_{u_2,\text{ active}}| < |\rho_{u_2,\text{ usable}}| \quad (6.102)$$

may be considered a condition for the cutting process of the internal gear.

The second case means that the active profile of a pinion tooth, determined by the internal gear, exceeds its usable profile, determined by the production process of the pinion.

The active profile is determined by the path of contact, bounded between the tip radius of curvature of the pinion and the radius

$$\rho_{u_1, \text{active}} = +\sqrt{r_{a_2}^2 - r_{b_2}^2} - \sqrt{a^2 - (r_{b_2} - r_{b_1})^2} \quad (6.103)$$

The condition

$$\rho_{u_1, \text{active}} > \rho_{u_1, \text{usable}} \quad (6.104)$$

may be considered a condition for the cutting process of the pinion. If nothing is known about the production process of the pinion, it may be stated that the usable profile of a pinion tooth should at least include the profile defined by the standard basic rack tooth profile.

$$m_n \left(\frac{z_1 \cdot \sin \alpha_t}{2\cos \beta} - \frac{1-x_1}{\sin \alpha_t} \right) > \rho_{u_1, \text{usable}} \quad (6.105)$$

and the condition

$$\rho_{u_1, \text{active}} > m_n \left(\frac{z_1 \cdot \sin \alpha_t}{2\cos \beta} - \frac{1-x_1}{\sin \alpha_t} \right) \quad (6.106)$$

may be considered a severe condition for the internal gear, often incuding tip shortening of the internal gear teeth. However, tip shortening may be avoided if the pinion is produced by adequately adapted cutters.

6.7.3 Cycloidal Overcut

Cycloidal overcut concerns the interference of tooth tips far outside the line of action. It is explained by considering the center distance as variable along the central line, while the pinion penetrates into the gear blank. Initially, at low center distances ($a < r_{b_2} - r_{b_1}$), only cycloids are cut by the tip of a pinion tooth. Envelopes of these cycloids create cycloidal penetrations that are known as overcut. The maximum overcut penetration always occurs at an early stage when the center distance, denoted by a_{dam}, is still too low for involute generation.

Later on, involutes of the internal gear are generated by the complete involute pinion flanks at increasing center distances ($a < r_{b_2} - r_{b_1}$). The center distance has to increase until the previous overcut penetration is recovered by newly generated involutes. The relationship between the new involutes and the previous cycloids will be expressed by a conjugate function of the center distances as a helpful mathematical tool.

The cycloid cut by a pinion tooth tip at a center distance A, is recovered by an involute cut by a pinion tooth flank at a center distance conjugate (A).

$$\text{inv } \alpha^* = \frac{z_2}{z_2 - z_1} (\text{inv } \alpha_{a_2} + 2 \arctan \sqrt{Q_1 \cdot Q_2})$$

$$- \frac{z_1}{z_2 - z_1} \left(\text{inv } \alpha_{a_1} + \pi - 2 \arctan \sqrt{\frac{Q_1}{Q_2}} \right) \quad (6.107)$$

GEAR TOOTH CALCULATIONS

$$Q_1 = \frac{-A + r_{a_1} + r_{a_2}}{+A + r_{a_1} + r_{a_2}} \qquad (6.108)$$

$$Q_2 = \frac{+A + r_{a_1} - r_{a_2}}{+A - r_{a_1} + r_{a_2}} \qquad (6.109)$$

If inv $\alpha^* < 0$, then inv α^* is taken inv $\alpha^* = 0$ \qquad (6.110)

$$\text{conjugate } (A) = (r_{b_2} - r_{b_1}) \frac{1}{\cos \alpha^*} \qquad (6.111)$$

The condition for avoidance of cycloidal overcut depends on the path to be traversed by the center distance.

If the center distance need not traverse all the lower values before reaching its true value, except for a certain path f_a, that means axial assembly between $a - f_a$ and a, then the condition will be

$$a > \text{conjugate } (a - f_a) \qquad (6.112)$$

If the center distance has to traverse all the lower values before reaching its true value, that means radial assembly, then a safe condition will be

$$a > \text{conjugate } (a_{\text{dam}}) \qquad (6.113)$$

in which a_{dam} is the "damage center distance" at which the penetration of cycloids reaches its maximum

$$a_{\text{dam}} = \sqrt{\frac{z_2 - z_1}{z_2 + z_1} (r_{a_2}^2 - r_{a_1}^2)} \qquad (6.114)$$

With no other sense than to prevent computation problems in the calculation of conjugate (a_{dam}), the gear data may satisfy

$$x_2 \neq 1 + \frac{z_2}{z_1}(1 + x_1 - k_1) \qquad (6.115)$$

In a computing procedure it is not necessary to calculate Eq. (6.111). The comparison of center distances can be replaced by a comparison of the involute of the transverse working pressure angle α'_t and the involute of the auxiliary pressure angle α^* [see Eqs. (6.107) to (6.110).].

For radial assembly between $a - f_a$ and a, the safe condition (6.112) may be replaced by Eq. (6.107) to Eqs. (6.110) and (6.116).

$$\text{inv} Q\alpha'_t > \text{inv } \alpha^* \qquad \text{for } A = a - f_a \qquad (6.116)$$

For axial assembly the condition (6.113) may be replaced by Eq. (6.107) to Eqs. (6.110), (6.114), and (6.117).

$$\text{inv } \alpha'_t > \text{inv } \alpha^* \qquad \text{for } A = a_{\text{dam}} \qquad (6.117)$$

6.7.4 Radial Assembly

Radial assembly means a continuous increase of the center distance together with a certain freedom in the angular positions of the pinion and the internal gear. Condition (6.110), closely approximating the true limits, is on the safe side since for radial assembly and angular positions are not restricted to those of gear generation and the influence of backlash is not considered.

If condition (6.110) is satisfied and condition (6.111) is not, then radial assembly is restricted to exceptional cases, depending on tooth thicknesses and angular positions. To escape from extremely complicated mathematical solutions, the position of the displacing pinion will be supposed symmetrical with respect to the internal gear. Parallel tip-to-tip chords of the pinion have to pass the corresponding tip-to-tip chords of the internal gear. The decisive tip-to-tip chords approach an imaginary tip-to-tip chord with a tip chord half angle.

$$\psi = \arccos \sqrt{\frac{1 - \left(\dfrac{r_{a_1}}{r_{a_2}}\right)^2}{1 - \left(\dfrac{z_1}{z_2}\right)^2}} \qquad (6.118)$$

The numbers of teeth in the span of tip-to-tip chords to be checked are either

$$z_s = \text{trunc}\left(\frac{\psi}{\pi} z_2\right) \quad \text{and} \quad z_s = \text{trunc}\left(\frac{\psi}{\pi} z_2\right) + 2 \qquad (6.119)$$

or

$$z_s = \text{trunc}\left(\frac{\psi}{\pi} z_2\right) - 1 \quad \text{and} \quad z_s = \text{trunc}\left(\frac{\psi}{\pi} z_2\right) + 1 \qquad (6.120)$$

For at least one of the combinations (6.112) or (6.113) the two tip-to-tip chords of the pinion have to be smaller than that of the internal gear.

$$s_{k_1} < s_{k_2} \qquad (6.121)$$

$$s_{k_1} = d_{a_1} \sin\left[\frac{\pi}{z_1}(z_s - 0.5) + \frac{1}{z_1}\left(2x_1 \tan \alpha_n - \frac{j_1}{m_n \cos \alpha_t}\right)\right.$$
$$\left. + \text{inv } \alpha_t - \text{inv } \alpha_{a_1}\right] \qquad (6.122)$$

$$s_{k_2} = d_{a_2} \sin\left[\frac{\pi}{z_2}(z_s - 0.5) + \frac{1}{z_2}\left(2x_2 \tan \alpha_n + \frac{j_2}{m_n \cos \alpha_t}\right)\right.$$
$$\left. + \text{inv } \alpha_t - \text{inv } \alpha_{a_2}\right] \qquad (6.123)$$

If the distribution of the backlash j in parts j_1 and j_2 is not known, then the condition on the safe side takes $j_1 = 0$ and $j_2 = j$.

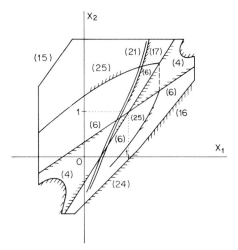

FIGURE 6.16 Fictitious representation of interference.

6.7.5 Selection of Addendum Modification Coefficients

The above-mentioned conditions for avoidance of involute interference, cycloidal overcut, and exceptionally radial assembly may be answered by a proper selection of the addendum modification coefficients of pinion and internal gear.

To ensure that the transverse working pressure angle is large enough, the gear data may satisfy

$$\frac{x_2 - x_1}{z_2 - z_1} > -\frac{\text{inv } \alpha_t}{2 \tan \alpha_n} > -0.02 \text{ (approx.)} \qquad (6.124)$$

The contact ratio may have an agreed limit value

$$\epsilon_\alpha = \frac{1}{2\pi} [z_1(\tan \alpha_{a1} - \tan \alpha'_t) - z_2(\tan \alpha_{a2} - \tan \alpha'_t)] \qquad (6.125)$$

The various conditions will seldom appear all together in the same diagram, as fictitiously shown in Fig. 6.16 for fixed values of the helix angle and numbers of teeth.

REFERENCES

1. Salamoun, C., and Suchy, M., "Computation of Helical or Spur Gear Fillets," *Mechanism and machine Theory,* Vol. 8, 1973, pp. 305–323.
2. Colbourne, J. R., "The Geometric Design of Internal Gear Pairs," AGMA Paper No. 87 FTM 2.
3. AGMA Standard 908–389, "Geometry Factors for Determining the Pitting Resistance and Bending Strength of Spur, Helical, and Herringbone Teeth," *Information Sheet,* 1500 King Street, Suite 201, Alexandria, Virginia 22314.

4. Lewis, W., "Investigation of the Strength of Gear Teeth," Proc. of the Engineers Club, Philadelphia, PA, 1893, pp. 16–23.
5. Wellauer, E. J., and Seireg, A., "Binding Strength of Gear Teeth by Cantilever Plate Theory," *Journal of Engineering for Industry*, Trans. ASME, Series B, Vol. 82, 1960, pp. 213–222.
6. Dolan, T. J., and Broghamer, E. L., "A Photoelastic Study of the Stresses in Gear Tooth Fillets," University of Illinois, Engineering Experiment Station, Bulletin No. 335, 1942.

CHAPTER 7
GEAR TOLERANCES

Robert E. Smith
Consultant, R.E. Smith & Co., Inc.
Rochester, NY

The contents of this chapter, some of which were excerpted from Michalec,[1] are devoted to the subject of gear tolerances, which are related to the information contained in all the previous and remaining chapters. Tolerances constitute a complex area of gear specification which directly affects gear performance, materials and finishes, fabrication and inspection techniques, and cost. Gear tolerances must be considered in the design and specification of gears and will also influence the design of associated elements such as bearings, shafts, and housing. Because of this broad effect, the engineer must comprehend and appreciate tolerances with a competent knowledge of proper specification. For this reason, the subject of gear tolerances is given paramount significance with detailed coverage. In this chapter are a definition and complete review of tolerances for all types of gears and applications. Tolerance effect of function, fabrication, and cost is presented. As a design guide, numerous standardized tolerance values are presented.

7.1 TERMINOLOGY AND DEFINITIONS PERTAINING TO GEAR TOLERANCES

7.1.1 Terms and Definitions[2,3,4]

Accumulated pitch variation, total, V_{ap}: Total accumulated pitch variation is equal to the algebraic difference between the maximum and minimum values obtained from the summation of successive values of pitch variation and is the same as total index variation.

Allowance: Allowance is the minimum specified difference between two dimensions.

Bellmouth: Bellmouth is an extreme taper, with the largest diameter at the ends of the gear bore.

Blank, gear: A gear blank is the material shape used for the manufacture of a gear, prior to machining the gear teeth.

Body dimension: A body dimension is a dimension of the gear pertaining to a portion of the gear blank that carries the teeth.

Concavity: Concavity pertains to the degree of departure from flat to concave of a surface in the plane of rotation.

Convexity: Convexity pertains to the degree of departure from flat to convex of a surface in the plane of rotation.

Deviation: (1) (General) Deviation is a departure from a standard or norm. (2) (MIL Std. 109B) A deviation is a written authorization, granted prior to the manufacture of an item, to depart from a particular performance or design requirement of a contract, specification, or referenced document, for a specific number of units or specific period of time.

Form diameter: See Profile control diameter.

Fullness: Fullness is an excessive amount of material beyond what is desired. Referring to a gear tooth profile, it is extra material on the space side of the desired profile.

Functional element: A functional element is a gear dimension or other criterion that directly bears on gear performance, for example, tooth thickness, which directly relates to backlash.

Functional profile: Functional profile is that portion of the tooth flank between the profile control diameter and the addendum circle or start of tip round or chamfer.

Gear element dimension: Gear element dimension refers to a dimension that pertains to the gear features such as teeth or profiles, as contrasted to a dimension pertaining to the gear blank such as the hub or bore.

Gear function: Gear function pertains to an operational feature of the gear, such as noise, accuracy, or strength.

Gear size: Gear size is the precise tooth thickness or testing radius. In this use, it is not meant as a general term for gear pitch or pitch diameter.

Generating mounting surface: The generating mounting surface is the diameter and/or plane of rotation surface or shaft center that is used for locating the gear blank during fabrication of the gear teeth.

Hollow: Hollow is a deficiency of material. Referring to a gear tooth profile, it is the space or emptiness between the actual and desired profiles.

Index variation, V_x: Index variation is the displacement of any tooth from its theoretical position relative to a datum tooth.

Index variation, total: Total index variation is the maximum algebraic difference between the extreme values of index variation for a given gear. Total index variation is also equivalent to total accumulated pitch variation, as measured by a two-probe spacing system.

Journal: A journal is the portion of a shaft-type gear blank that engages a bearing within which the gear rotates.

Lay: Lay is the direction of the predominant pattern of a finished surface.

Lead: Lead is the axial advance of a helix for one revolution (360°) as in a screw thread and the teeth of a helical gear. (*See* Tooth alignment.)

Microinch: A microinch is one millionth of an inch (0.000001 in.).

Micrometer: A micrometer is one millionth of a meter (μm).

Mounting surface: A mounting surface is the diameter and/or plane of rotation surface that is used for locating the gear in the application assembly.

Nonconformance: Nonconformance is the failure of a unit of product to meet specified requirements for any quality characteristic.

Nonfunctional element: A nonfunctional element is a gear dimension or other criterion that does not affect gear performance, such as keyways or bolt holes.

Non-gear-element dimension: A non-gear-element dimension refers to a dimension that does not pertain to the gear features of teeth or profiles. As examples, the hub or bore dimensions.

Over-pins measurement: Over-pins measurement pertains to measurement, radial or diametrical, made across cylindrical pins or balls inserted between the teeth.

Parallelism: Parallelism is the degree of departure from parallel of two surfaces in the plane of rotation.

Pitch variation, V_p: Pitch variation is the algebraic plus or minus (+ or −) difference in the transverse plane, between the true position pitch P_m and an actual pitch measurement.

Profile: Profile is the shape of the tooth flank from its root to its tip. The profile of standard spur, helical, and herringbone gears is normally an involute and cannot extend below the base circle.

Profile control diameter: The profile control diameter is the diameter of the circle outside of which the tooth profile must conform to the specified involute curve. This diameter is sometimes referred to as the form diameter.

Runout: Runout is the maximum variation of the distance between a surface of revolution and a datum surface, measured perpendicular to that datum surface.

Runout, axial: Axial runout (wobble) is the runout measured in a direction parallel to the datum axis of rotation.

Runout, radial, V_r: Radial runout is the runout measured in a direction perpendicular to the datum axis of rotation.

Surface finish: Surface finish is the degree of surface roughness and waviness pertaining to relatively finely spaced irregularities produced by the cutting or abrading action of machine tools, or the result of a casting or forming process.

Surface roughness: Surface roughness pertains to the relatively finely spaced surface irregularities produced by machining and abrasive operations. Roughness may be considered as superimposed on a "wavy" surface.

Surface waviness: Surface waviness pertains to surface irregularities that have a longer wavelength than the roughness.

Test radius: The test radius is the radial distance from the gear center to the theoretical pitch radius or pitch plane of the mating master gear or rack.

Tolerance: Tolerance is the amount by which a specific dimension is permitted to vary. The tolerance is the difference between the maximum and minimum limits and is an absolute value without sign.

Tolerance, bilateral: A tolerance in which variation is permitted in both directions from the specified dimension.

Tolerance, close: A tolerance magnitude that is small for a particular dimension and therefore relatively difficult to achieve.

Tolerance, open: A tolerance magnitude that is large for a particular dimension and therefore relatively easy to achieve.

Tolerance, unilateral: A tolerance in which variation is permitted in one direction from the specified dimension.

Tooth alignment (formerly lead): Tooth alignment is the alignment, normal to the helix, between the line of intersection of actual and theoretical tooth surfaces on the pitch cylinder.

Top generation: Top generation is the gear fabrication process that finishes the outside diameter of the gear teeth simultaneously with cutting of the tooth profiles.

Undercut: Undercut is a condition in generated gear teeth when any part of the fillet curve lies inside of a line drawn tangent to the working profile at its point of juncture with the fillet. Undercut may be deliberately introduced to facilitate finishing operations.

Variation, V: Variation is the measured plus or minus change from the specified value.

Variation, allowable, V_A (subscript such as V_{pA}): Allowable variation is the permissible plus or minus departure from the specified value.

Waiver: A waiver is a written authorization to accept a configuration item or other designated items that, during production or after having been submitted for inspection, are found to depart from specified requirements, but nevertheless are considered suitable for use "as is" or after rework by an approved method.

7.1.2 Symbols

The following symbols are used in this chapter:[5, 6]

a = addendum
B = backlash
b = dedendum
C = center distance
d = pin or ball diameter
DM = over-pin measurement
D_m = diameter of circle through center of measuring pin
V_p = pitch variation
V_{ap} = accumulated pitch variation
V_r = runout
m = module
N_G = number teeth, gear
N_P = number teeth, pinion
P_d = diametral pitch, transverse
p = circular pitch
R_G = pitch radius, gear
R_P = pitch radius, pinion
R_{bG} = base radius, gear
R_{bP} = base radius, pinion

GEAR TOLERANCES   7.5

t = tooth thickness, circular
V_ϕ = profile variation
V_ψ = tooth alignment variation
ϕ = pressure angle

7.2 INTRODUCTION OF GEAR TOLERANCES

7.2.1 Definition of Gear Tolerance

As a general description, a tolerance can be thought of as the magnitude of permissible variation of a dimension or control criterion from the specified value.

Tolerances constitute an engineering legality for variation from the ideal values, and, as in legal rulings, much good thought and planning must go into tolerance formulation. Tolerances must be specified because of inevitable human failings and machine limitations that prevent perfection during fabrication. No part dimension can be made to an absolute value, nor can absolute reproducibility be achieved. The tolerance establishes a necessary permissible range of acceptance of variation.

7.2.2 Tolerance, Deviation, Variation, and Error

Dimensional variations, within the band of tolerance, constitute *allowable variations*. If the variation is beyond tolerance, the gear has not been made correctly and can be described quite rightfully as "in error." This would also be known as *nonconformance*. There has been a tendency, in the gear trade, to describe any departure from theoretical perfection of profile, pitch (spacing), or tooth alignment (lead) as an *error*. The American Gear Manufacturers Association (AGMA) considers an *error* to be an amount of variation that is beyond the agreed-on tolerance for the gear element.

The word *deviation* might be used interchangeably with *variation,* in the foregoing discussion. It is commonly used in quality control, however, with a different meaning, as described in MIL Std. 109B and the definitions at the start of this chapter.

Therefore, the word *variation* is preferred for the description of the departure from ideal or specified dimension.

It should be pointed out that some variations in gear dimensions have no effect, or almost none, on gear performance, and some most directly affect performance. The gear designer has the problem of constantly weighing the effect of any proposed tolerance on gear performance and cost, and of specifying proper values.

7.2.3 Allowances and Relation to Tolerance

Allowance has been defined in Sec. 7.1 and is the intentional clearance between mating parts. In the case of gears this would be

1. Clearance between gear bore and mating shaft
2. Clearance between integral gear shaft and mating bearing
3. Clearance between mating gear teeth, or "backlash allowance"

Allowance and tolerance differ in that *allowance* is an intentional and fixed minimum difference between two independent but related dimensions, whereas *tolerance* is the permissible variation of a dimension. They are related in that tolerance value adds to the allowance effect. This means that there must be consistency between the two. For instance, if the allowance on backlash is intended to be very small, then a small tolerance (or high degree of accuracy) is needed in the concentricity of the gears and in the center distance. (See Chap. 4.)

7.2.4 Examples of Tolerance

Tolerances are applicable to measurements regardless of their nature. Thus, they apply to measurements of length, weight, force, velocity, etc. The toleranced quantities of gears are predominantly measures of length or angle, and some measures of force such as associated with material hardness specification. Linear and angular dimensions are directly located on the drawing and toleranced in an equally direct manner. Some control measurements and parameters, however, are shown in the form of notes. Examples are runout, squareness, and hardness specifications. In the past, traditional drawings were done in this manner (see Fig. 7.1). Today, many drawings are done in the manner shown in Fig. 7.2. This is in accordance with ANSI Y14.5M Dimensioning and Tolerancing Standard.[7] Geometric tolerance symbols are used in this system, rather than notes, for the control of characteristics such as runout and squareness. Feature control frames, which include a characteristic symbol, tolerance, and datum reference letters, are used to express a geometric tolerance.

7.2.5 Gear Tolerance Magnitudes

No typical tolerance values can be stated as applying to the various gear dimensions. Each dimension has a unique tolerance that varies with many gear considerations such as size, precision, material, and application. As an indication of the general tolerance

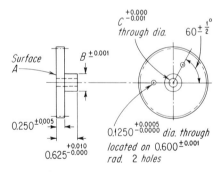

FIGURE 7.1 Old style gear drawing.

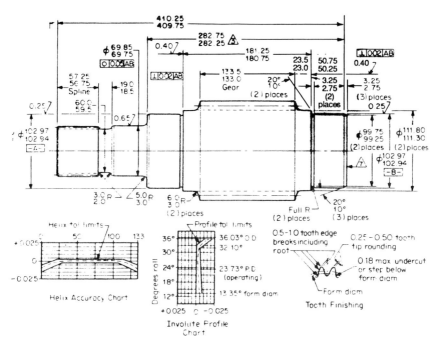

EXTERNAL HELICAL GEAR MANUFACTURING DATA

Module, normal	5.0
Module, transverse	5.1764
Pressure angle, normal	20°
Pressure angle, transv	20.647°
Helix angle and hand	15° R H
Number of teeth	24
Pitch diameter	124.233
Base circle diameter	116.254
Base helix angle	14.076°
Helix lead	1456.58
Major diameter	137.33 137.07
Root diameter	112.60 111.88
Form diameter max	119.37
Roll angle at form	13.356°
Roll angle at O.D	36.033°
Root fillet radius min	1.75
Arc tooth thickness, norm	8.928 8.801
Pin diameter	8.6400
Dimension over pins	138.67 138.40

OPERATING CONDITIONS (REF.)

No. teeth in mating gear	80
Center distance	272.637
Pitch diameter	125.832
Module, transverse	5.2430
Pressure angle, transv	22.500°
Addendum	5.7500
Backlash	0.265 0.530

GEAR QUALITY

Material grade	1
Quality level, tooth geometry	A
Tooth to tooth spacing limits	0.0061
Cumulative tooth to tooth	0.0127
Total composite variation	0.0256
Profile tolerance	see chart
Helix accuracy	see chart
Profile surface finish, microns	0.5
Root fillet surface finish, micr.	2.0

EXTERNAL INVOLUTE SPLINE DATA
FLAT ROOT SIDE FIT

Number of teeth	26
Spline module	2.54 1.27
Pressure angle	30°
Pitch diameter	66.040
Base diameter	63.924
Major diameter	68.58 67.82
Form diameter, max	63.348
Minor diameter, min	62.230
Circular tooth thickness	
Max effective	3.990
Min. actual	3.909
Max. Actual	3.945
Pin diameter	4.8768
Min. measurement over pins	73.482

NOTES

1. Dias -A- and -B- form axis of gear
2. Material: AISI 4320, make from forging no. 1002.2
3. Heat treatment
   Carburize and HDN. noted area
   Case hardness HRC 59 minimum
   Effective case depth after grinding 1.0 - 1.5
   Core hardness HRC 35 42
4. Grinding prohibited in roots of teeth
5. Magnetic particle inspect. Indications prohibited in gear or spline teeth
6. Nital etch inspect. grind burns or retempered areas prohibited
7. Brand part no. and serial no
8. (1) required per assembly

FIGURE 7.2 New style gear drawing.

range, so as to fix the art of gearing in the full spectrum of measurements, pertinent gear tolerances range from several hundreds, 250 μm (0.010 in.), down to 2.5 μm (0.0001 in.) or less. This covers only the majority of gear tolerances, as there are exceptions. The several hundredfold variation affords a wide selection ranging from possibilities of wholly inadequate design to overdesign and is the reason tolerances constitute an important subject.

7.3 TOLERANCE EFFECTS

7.3.1 Tolerance Effects on Fabrication

The specific magnitude of the gear tolerances has a determining effect on the method of manufacture, tooling, jigs, and fixtures, and inspection techniques. The basic fabrication method—hobbing, grinding, molding, etc.—will be established by the general gear quality.

7.3.2 Tolerance Effects on Cost

The cost of fabricating the whole gear set is a more or less direct function of the tolerances specified. The magnitude of the tolerances must be set so as to permit as low manufacturing costs as possible and still achieve the necessary gear functional performance. This requires a thorough understanding of critical and noncritical dimensions and controls and the interrelationship of tolerances.

The closer the tolerance, regardless of the particular dimension, the greater fabrication effort required, which results in increased manufacturing cost. The relationship between tolerance and cost is not linear, for as the tolerance approaches zero, the required effort becomes enormous. Similarly, at the other extreme, fabrication cost decreases with relaxation of tolerance, but a value is reached beyond which little is gained. Tolerance versus cost is a nonlinear function (Fig. 7.3). Numerical values cannot be stated because of the dependence on specific dimension. For example, shaft diameter can be held to a closer tolerance than shaft length for approximately the same effort and cost. Fabrication costs may also vary among shops, making most comparative dollar values meaningless without additional qualifying information. However, in general, the curve of Fig. 7.3 becomes asymptotic to 2.5 μm (0.0001 in.) for most critical gear tolerances.

7.3.3 Tolerance Effects on Function

Many toleranced gear dimensions bear directly on functional operation of the gear. Gear function is defined as the intended action of the gear, that is, the transmission of power at a constant velocity, at specified speed, with other additional qualifications such as backlash magnitude, life, and noise. Since the function of a gear is its primary purpose, tolerances affecting function are the most critical. Typical functional tolerance examples are tooth thickness, profile deviation, bore or journal diameter, hardness, surface finish, etc. The foregoing tolerances have a very direct influence on gear function. In addition there are tolerances that have a secondary or negligible functional effect. An example is the tolerance on face width, which remotely affects the strength of the gear.

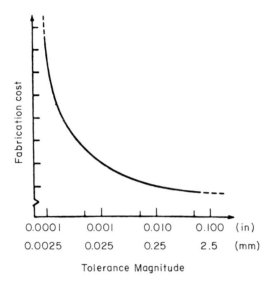

FIGURE 7.3 General relationship of tolerance and fabrication cost.

7.3.4 Nonfunctional Tolerance Effects

The functional elements of a gear are concentrated at the periphery with a relation to the center of rotation. To be practical, these elements must be carried by a structure, or gear body. The dimensions of this gear body, that is, web and hub design, mounting arrangements, etc., do not affect gear function; and associated tolerances are therefore noncritical in regard to gear function. Typical nonfunctional tolerances are those on hub diameter and length, keyway or equivalent mount design, etc.

7.4 TYPES OF TOLERANCES

7.4.1 Introduction

Gear tolerances can be divided into three general classifications; functional tooth tolerances, functional blank tolerances, and nonfunctional blank tolerances on body dimensions and other non-gear-element controls. In a complete gear design all tolerances are interwoven to specify the gear. Descriptions of the three tolerance groupings follow.

7.4.2 Functional Tooth Tolerances (Gear Element)

Functional tooth dimensions fall into two categories; basic and/or reference dimensions, usually not toleranced, and control dimensions that are toleranced and measured during manufacture.

Basic and/or Reference Dimensions

Number of teeth
Module (diametral pitch)
Pressure angle
Standard pitch diameter (ref)
Addendum (ref)
Whole depth (ref)
Circular tooth thickness on standard pitch diameter (ref)*
Base circle diameter (ref)
Helix angle
Base helix angle

Control Dimensions (Toleranced)

Pitch variation
Runout
Profile
Tooth alignment (lead)
Outside diameter
Face width
Surface finish
Material hardness

7.4.3 Functional Blank Tolerances

Functional blank tolerances pertain to dimensions and control criteria not associated with the gear teeth but nevertheless having direct effect on gear performance. They are:

Mounting bore or journal diameter (size and runout)
Mounting shoulder (squareness and runout)
Parallelism
Convexity
Concavity

7.4.4 Nonfunctional Dimension Tolerances

These tolerances pertain to dimensions relating to the tooth carrier or body and do not directly affect gear function. They are:

---

*This is a toleranced reference dimension, although it is not measured directly. It is controlled by over-pins, span, or other measurements.

Hub design dimensions
Web design dimensions
Mounting/fastening design—keyway, setscrew, boltholes, etc.
Special body dimensions

7.5 FACTORS AFFECTING CHOICE OF TOLERANCE

Since the setting of tolerances is not an arbitrary matter, consideration must be given to the influences bearing on choice. The most significant are the functional requirements. Often these primary considerations require compromises with standardization, methodizing, and manufacturing needs.

7.5.1 Functional Requirements

The primary tolerance purpose is to permit variation in dimensions and to control parameters without degradation of gear performance beyond the limits established by the specification of the design. In high-performance designs the functional requirements will be the dominating factor in setting tolerances. When the functional performance is relatively open, the tolerance choice may be influenced and determined by the other factors such as standards, methods of tooling, and available fabrication equipment.

7.5.2 Standardization

Although functional requirements can be used to determine every tolerance, using them in all cases may not be practical because it would result in:

1. Too many different tolerances—essentially a continuous spectrum
2. Unnecessarily excessive amounts of calculations and studies to establish tolerance values from the functional requirements
3. Excessive amount of special tooling
4. Too loose tolerances for broad functional requirements
5. Needless complication and prolonging of inspection

Thus, very often it is better to establish a limited number of standard tolerances and obtain the benefits of fewer varieties of tooling, fewer calculations, and increased unit quantities because of repeated uses of the same designs.

When standardized tolerances are used, the choice of specific tolerances must be such that the functional gear requirements are satisfied. Generally, the tolerances chosen will be slightly closer than function would dictate. The degree of excessiveness can be kept reasonable by establishing a rational number of standard tolerances.

7.5.3 Methodizing

Tolerance choices reflect on machines, tools, fixtures, and operations. Achievement of tolerances depends on the methodizing, which is centered about the tools and machines.

The methodizing procedures will establish normal fabrication procedures and tolerance patterns.

7.5.4 Manufacturing Needs

Often a tolerance is not set by functional requirements but by manufacturing needs. One machine operation may affect the next. Jigs and fixtures may require tolerances for proper applications, but otherwise the tolerances may not be needed. For example, lateral runout may be irrelevant with regard to gear performance. However, in order to fabricate the gear and obtain a close functional tolerance on total composite variation, it may be necessary to control the lateral runout closely. In a similar manner many other tolerances may be dictated by fabrication needs.

7.5.5 Ranges of Numerical Tolerances

The numerical values of gear tolerances range across the entire spectrum of measurements. The broad open tolerances are in the order of 1 mm ($\frac{1}{32}$ in.), and the finest are down to fractions of a millimeter (millionths of an inch). In this broad area of choice, all the influencing factors mentioned in the preceding section combine and interplay in setting tolerance values.

The gear elements and other dimensions affecting function utilize the more precise tolerances, generally in the order of micrometers (ten-thousandths of an inch) to 30 to 50 μm (one or two thousandths). Noncritical dimensions not affecting function are generally larger-valued tolerances, often on the order of several tenths of a millimeter (hundredths of an inch) or greater.

7.6 DRAWINGS AND TOLERANCES FOR VARIOUS GEAR TYPES

After the details and size of a design have been worked out, the gear designer is left with the task of determining dimensional details and tolerances of the various gears in the train. The problem is to commit all the complex details of a gear design to paper in such a manner that the manufacturer can produce the gear to perform its required function, at a reasonable cost.

7.6.1 Dimensional Data and Tolerances

The dimensional data and tolerances that may be required to make a gear drawing can be broken down into two categories; blank dimensions and tooth data.[8] The blank dimensions are usually shown in cross-sectional views, while the tooth data are either tabulated or shown directly on an enlarged view of one or more teeth. Some of the common blank dimensions are:

Outside diameter	Root diameter
Face width	Bore diameter
Outside cone angle (bevel gears)	Mounting distance (bevel gears)

GEAR TOLERANCES 7.13

Back cone angle (bevel gears)
Throat diameter (worm gears)
Throat radius (worm gears)

Inside rim diameter
Web thickness
Journal diameter

The tabulated gear tooth data will cover such items as:

Number of teeth
Module (or diametral pitch)
Pitch diameter
Circular pitch
Linear pitch (worm gears)
Pressure angle
Normal pressure angle (helical gears)
Normal circular pitch (helical gears)
Normal module or pitch (helical gears)
Addendum
Whole depth

Helix angle
Hand of helix
Lead angle (worm gears)
Pitch cone angle (bevel gears)
Root cone angle (bevel gears)
Tooth thickness
Lead (worm gears)
Pitch tolerance
Runout tolerance
Tooth alignment tolerance (lead)
Profile tolerance

Some of the foregoing are reference dimensions and some are toleranced. Special views of gear teeth may be used to show things like root fillet radius, form diameter, tip radius, end radius, and surface finish. Notes, sketches, or symbols may also be used to specify the heat treatment and to define other aspects of accuracy and form of the gear teeth. Notes and symbols are also used to define the reference axes and shoulders used for inspection of the gears.

7.6.2 General Drawing Practices

General drawing practices are covered in ANSI Y14.1 through ANSI Y14.5M[8] of the American National Standards for Engineering Drawing and Related Documentation Practices. See Chap. 9 for additional gear drawing information.

7.6.3 Gear Drawing Practices

Gear drawings have a complexity that requires special formats. A major effort, over several years by interested parties, culminated in the development of the gear and spline standards known as ANSI Y14.7.[9] The following drawing examples are taken from ANSI Y14.7.1 through Y14.7.3.

Spur and Helical Gears. Figure 7.4 shows examples of spur and helical gear drawing formats. See ANSI Y14.7.1 for other special versions of these such as internals and racks.

Bevel Gears. Figures 9.11 and 9.12 show straight and spiral bevel gear drawing formats. See ANSI Y14.7.2 for special versions such as hypoid gears.

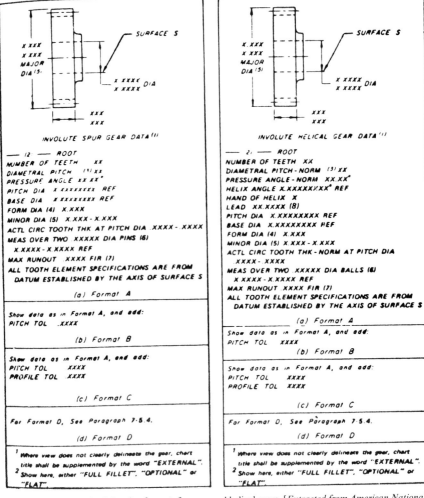

FIGURE 7.4 Standard drawing formats for spur and helical gears. [Extracted from American National Standard Drafting Practices, Gear Drawing Standards (ANS Y14.7.1) with the permission of the publisher, The American Society of Mechanical Engineers, United Engineering Center, 345 East 47th Street, New York, NY 10017]

Worm Gears. Figures 9.14 and 9.15 show examples of single- and double-enveloping worm gears. See ANSI Y14.7.3.

Special Gears. Figure 9.16 shows an example of a Spiroid* pinion and gear. See ANSI Y14.7.3 for further details as well as Helicon† type gears.

---

*Registered trademark of Spiroid Div., Illinois Tool Works, Chicago, Illinois.
†Registered trademark of Vinco Corp., Detroit, Michigan.

GEAR TOLERANCES 7.15

7.7 TOLERANCE STANDARDS

Some companies, especially large automotive firms, establish their own in-house tolerance standards or criteria. However, national and international standards have been developed to provide a "bridge" between gear designers, manufacturers, and users. These standards provide different quality levels to cover gears of all types from cast or molded, to cut, shaved, or ground. Tolerances of all elements, within a certain quality level, are chosen to be compatible with the manufacturing process or level of craftsmanship. It should be noted that, as the teeth get coarser or the diameter gets larger, the tolerance increases.

In general, the designer should look at the functional requirements of a gear, and then select the quality grade that will meet these requirements. Conditions may also require that one or more of the individual elements or composite tolerances be of a lower or higher quality number than the other tolerances.

7.7.1 American Gear Standards

American gear standards are developed by the American Gear Manufacturers Association (AGMA) and approved by the American National Standards Institute (ANSI). The current standards are ANSI/AGMA 2000-A88 for Unassembled Spur and Helical Gears and AGMA 390.03a (1980) for Unassembled Bevel, Fine Pitch Wormgears, and Racks. These standards cover tolerances and measuring methods. AGMA 2000-A88 uses conventional inch units as well as SI (metric) units. They address tooth tolerances only, not blank tolerances.

Anyone involved in the design, manufacture, or purchase and use of gears should have a copy of this standard as a reference. Some details are explained below.

The quality of a gear is usually specified by a number that describes various characteristics, for example, Q8 A-HA 14. The Q is a designator that indicates the tolerance is from AGMA 390.03 or ANSI/AGMA 2000-A88. Without the Q designator, the quality level 8 would refer to earlier AGMA 390.01 or 390.02 standards. The quality number 8 identifies the accuracy level of the elemental or composite tolerances. (This number ranges from 3 through 15.) The letter A refers to a tooth thickness code. (This ranges from A through D.) The two-letter group HA refers to a material type. The last two digits (14) refer to a heat treatment and hardness range.

The standard includes two different groups of tolerances; elemental and composite. *Elemental* tolerance includes runout tolerance, allowable pitch variation, profile tolerance, and tooth alignment tolerance (formerly lead tolerance). Composite includes total composite tolerance and tooth-to-tooth composite tolerance.

The evaluation of AGMA quality can be made by either of the two foregoing methods but not both. When composite action tolerance is specified, it shall be in lieu of any elemental inspection.

Other criteria of quality can be used, in addition to the foregoing, such as index variation or surface finish, but these must be specified separately on drawings or purchase specifications.

7.7.2 ANSI/AGMA 2000-A88 Unassembled Spur and Helical Gears

Examples of tolerance tables are shown in Tables 7.1 through 7.4. For a complete list of tables, or formulas used to generate them, refer to the actual standard, AGMA 2000-A88.

TABLE 7.1 Element Tolerances

Runout Tolerance V_{rT}; Allowable Pitch Variation V_{pA}, Profile Tolerance $V_{\phi T}$; in Micrometers (μm)

Number of Teeth (z); for Helical Gears, Use Adjusted Number of Teeth ($z_i = z/\cos\beta$)

Q No.	Module m_n	Tolerance Type	6	8	10	12	16	20	24	28	32	36	40	44	48	52	56	60	80	100	200
Q10	1.25	Runout	13.	14.	15.	16.	17.	18.	18.	19.	20.	20.	21.	21.	22.	22.	22.	23.	24.	26.	30.
		Pitch+/−	5.4	5.7	6.0	6.2	6.5	6.7	7.0	7.2	7.3	7.5	7.6	7.7	7.9	8.0	8.1	8.2	8.6	9.0	10.
		Profile	6.2	6.5	6.8	7.0	7.3	7.5	7.7	7.9	8.1	8.2	8.4	8.5	8.6	8.7	8.8	8.9	9.3	9.6	11.
	1.50	Runout	15.	16.	17.	18.	19.	20.	21.	22.	22.	23.	24.	24.	25.	25.	26.	26.	28.	29.	35.
		Pitch+/−	5.9	6.2	6.4	6.6	7.0	7.2	7.5	7.7	7.9	8.0	8.2	8.3	8.5	8.6	8.7	8.8	9.3	9.6	11.
		Profile	7.0	7.3	7.5	7.7	8.1	8.4	8.6	8.8	9.0	9.2	9.3	9.4	9.6	9.7	9.8	9.9	11.	11.	12.
	2.0	Runout	19.	20.	21.	22.	23.	25.	26.	27.	28.	28.	29.	30.	30.	31.	31.	32.	34.	36.	43.
		Pitch+/−	6.6	6.9	7.2	7.4	7.8	8.1	8.4	8.6	8.8	9.0	9.2	9.3	9.5	9.6	9.8	9.9	11.	11.	12.
		Profile	8.2	8.6	8.9	9.2	9.6	9.9	10.	11.	11.	11.	11.	11.	12.	12.	12.	12.	12.	13.	14.
	3.0	Runout	25.	27.	28.	29.	31.	33.	34.	36.	37.	38.	39.	40.	41.	41.	42.	43.	46.	48.	57.
		Pitch+/−	7.7	8.1	8.5	8.7	9.2	9.6	9.9	10.	11.	11.	11.	11.	11.	12.	12.	12.	12.	13.	15.
		Profile	11.	11.	11.	12.	12.	13.	13.	13.	14.	14.	14.	14.	15.	15.	15.	15.	16.	16.	18.
	4.0	Runout	31.	33.	34.	36.	38.	41.	42.	44.	45.	47.	48.	49.	50.	51.	52.	53.	56.	59.	70.
		Pitch+/−	8.7	9.1	9.5	9.8	10.	11.	11.	12.	12.	12.	12.	13.	13.	13.	13.	13.	14.	14.	16.
		Profile	13.	13.	14.	14.	15.	15.	16.	16.	16.	16.	17.	17.	17.	17.	18.	18.	19.	19.	21.
	6.0	Runout	41.	44.	46.	48.	51.	54.	57.	59.	61.	62.	64.	65.	67.	68.	69.	70.	75.	80.	94.
		Pitch+/−	10.	11.	11.	12.	12.	13.	13.	14.	14.	14.	14.	15.	15.	15.	15.	16.	16.	17.	19.
		Profile	16.	17.	17.	18.	18.	19.	20.	20.	21.	21.	21.	22.	22.	22.	22.	23.	24.	24.	27.
	8.0	Runout	50.	54.	57.	59.	63.	67.	70.	72.	75.	77.	79.	81.	82.	84.	85.	87.	93.	98.	120.
		Pitch+/−	12.	12.	13.	13.	14.	14.	15.	15.	16.	16.	16.	16.	17.	17.	17.	17.	18.	19.	21.
		Profile	19.	20.	20.	21.	22.	23.	23.	24.	24.	25.	25.	25.	26.	26.	26.	27.	28.	29.	32.

Q10																								
12.	Runout	67.	72.	76.	79.	85.	89.	93.	97.	100.	110.	110.	110.	110.	110.	110.	120.	120.	120.	130.	130.	130.	160.
	Pitch +/−	14.	14.	15.	15.	16.	17.	17.	18.	18.	19.	19.	19.	20.	20.	20.	20.	20.	20.	21.	22.	22.	25.
	Profile	24.	25.	26.	26.	28.	29.	29.	30.	31.	31.	32.	32.	33.	33.	33.	33.	34.	35.	35.	37.	37.	41.
16.	Runout	83.	89.	93.	97.	110.	110.	120.	120.	130.	130.	130.	130.	140.	140.	140.	140.	140.	150.	150.	160.	160.	190.
	Pitch +/−	15.	16.	17.	17.	18.	19.	20.	20.	21.	21.	22.	22.	22.	23.	23.	23.	24.	24.	25.	25.	28.
	Profile	28.	29.	30.	31.	33.	34.	35.	36.	36.	37.	38.	38.	39.	39.	40.	40.	42.	43.	43.	48.
20.	Runout	97.	110.	110.	120.	120.	130.	140.	140.	150.	150.	150.	160.	160.	160.	160.	170.	170.	180.	190.	230.
	Pitch +/−	17.	18.	18.	19.	20.	21.	22.	22.	23.	23.	23.	24.	24.	24.	25.	25.	26.	27.	31.
	Profile	32.	34.	35.	36.	37.	39.	40.	41.	41.	42.	43.	43.	44.	44.	45.	45.	46.	48.	49.	55.
25.	Runout	120.	120.	130.	130.	140.	150.	160.	170.	170.	180.	180.	180.	190.	190.	190.	200.	200.	210.	220.	260.
	Pitch +/−	18.	19.	20.	21.	22.	23.	23.	24.	24.	25.	25.	26.	26.	27.	27.	27.	29.	30.	34.
	Profile	37.	38.	40.	41.	43.	44.	45.	46.	47.	48.	49.	50.	50.	51.	52.	52.	54.	56.	63.
50.	Runout	190.	200.	210.	220.	240.	250.	260.	270.	280.	290.	300.	300.	310.	320.	320.	330.	350.	370.	430.
	Pitch +/−	24.	25.	26.	27.	29.	30.	31.	31.	32.	33.	34.	34.	35.	35.	36.	36.	38.	39.	45.
	Profile	55.	57.	59.	61.	64.	66.	68.	70.	71.	72.	73.	75.	76.	77.	77.	78.	82.	85.	94.

Tooth Alignment Tolerance (Lead) $V_{\psi T}$

Q10	Facewidth	1	25	30	40	50	60	80	100	125	150	200	250
	Tolerance	8.2	8.2	9.3	12.	14.	15.	19.	22.	26.	30.	37.	43.

SOURCE: ANSI/AGMA 2000-A88.[3]

TABLE 7.2 Composite Tolerances

Total Composite Tolerance V_{cqT}, Tooth-To-Tooth Composite Tolerance V_{qT}, Micrometers (μm)

Number of Teeth (z); for Helical Gears, Use Adjusted Number of Teeth ($z_i = z/\cos\beta$)

Q No.	Module m_n	Tolerance Type	6	8	10	12	16	20	24	28	32	36	40	44	48	52	56	60	70	80	90	100	150	200
Q10	0.20	Total	18.	18.	17.	17.	16.	16.	16.	16.	16.	16.	16.	17.	17.	17.	17.	17.	18.	18.	18.	18.	19.	20.
		T-to-T	15.	14.	13.	12.	12.	11.	11.	11.	11.	10.	10.	11.	10.	10.	10.	10.	10.	10.	10.	10.	10.	10.
	0.25	Total	19.	19.	19.	18.	18.	17.	17.	17.	18.	18.	18.	18.	18.	19.	19.	19.	19.	20.	20.	20.	21.	22.
		T-to-T	15.	14.	14.	13.	12.	12.	12.	11.	11.	11.	11.	11.	11.	11.	11.	11.	11.	11.	11.	11.	11.	11.
	0.30	Total	21.	20.	20.	20.	19.	19.	18.	19.	19.	19.	20.	20.	20.	20.	20.	21.	21.	21.	22.	22.	23.	24.
		T-to-T	16.	15.	14.	14.	13.	12.	12.	12.	12.	11.	11.	11.	11.	11.	11.	11.	11.	11.	11.	11.	11.	11.
	0.40	Total	23.	23.	22.	22.	21.	21.	21.	21.	22.	22.	22.	22.	23.	23.	23.	23.	24.	24.	25.	25.	26.	27.
		T-to-T	17.	16.	15.	15.	14.	13.	13.	13.	12.	12.	12.	12.	12.	12.	12.	12.	12.	12.	12.	12.	12.	12.
	0.50	Total	25.	24.	24.	24.	24.	23.	23.	23.	24.	24.	24.	25.	25.	25.	26.	26.	26.	27.	27.	28.	29.	30.
		T-to-T	18.	17.	16.	15.	14.	14.	13.	13.	13.	13.	13.	13.	13.	13.	13.	13.	13.	13.	13.	13.	13.	13.
	0.60	Total	26.	26.	26.	26.	25.	25.	25.	26.	26.	26.	27.	27.	27.	28.	28.	28.	29.	29.	30.	30.	32.	33.
		T-to-T	19.	18.	17.	16.	15.	14.	14.	14.	14.	13.	13.	13.	13.	13.	13.	13.	13.	13.	13.	13.	13.	13.
	0.80	Total	29.	29.	29.	29.	29.	28.	28.	29.	29.	30.	30.	31.	31.	31.	32.	32.	33.	33.	34.	35.	37.	38.
		T-to-T	20.	19.	18.	17.	16.	15.	15.	15.	14.	14.	14.	14.	14.	14.	14.	14.	14.	14.	14.	14.	14.	14.
	1.0	Total	32.	32.	32.	32.	32.	32.	31.	32.	33.	33.	34.	34.	35.	35.	35.	36.	37.	37.	38.	39.	41.	43.
		T-to-T	21.	20.	19.	18.	17.	16.	16.	15.	15.	15.	15.	15.	15.	15.	15.	15.	15.	15.	15.	15.	15.	15.
	1.25	Total	35.	35.	35.	35.	35.	35.	35.	36.	36.	37.	38.	38.	39.	39.	40.	40.	41.	42.	42.	43.	46.	48.
		T-to-T	23.	21.	20.	19.	18.	17.	17.	16.	16.	16.	16.	16.	16.	16.	16.	16.	16.	16.	16.	16.	16.	16.

Q10																									
2.0	Total	44.	44.	44.	44.	44.	44.	44.	44.	44.	45.	46.	47.	48.	48.	49.	50.	50.	51.	52.	53.	54.	55.	59.	62.
	T-to-T	25.	24.	22.	21.	20.	19.	18.	18.	18.	18.	18.	18.	18.	18.	18.	18.	18.	18.	18.	18.	18.	18.	18.	18.
3.0	Total	54.	54.	54.	54.	54.	54.	54.	54.	54.	55.	57.	58.	59.	60.	61.	62.	62.	63.	65.	66.	68.	69.	74.	78.
	T-to-T	28.	26.	25.	24.	22.	21.	20.	20.	20.	20.	20.	20.	20.	20.	20.	20.	20.	20.	20.	20.	20.	20.	20.	20.
6.0	Total	79.	79.	79.	79.	79.	79.	79.	79.	79.	81.	83.	85.	87.	88.	90.	91.	92.	94.	97.	99.	100	110	110	120
	T-to-T	33.	31.	29.	28.	26.	25.	24.	24.	23.	23.	23.	23.	23.	23.	23.	23.	23.	23.	23.	23.	23.	23.	23.	23.
12.	Total	120	120	120	120	120	120	120	120	120	120	130	130	130	140	140	140	140	140	150	150	160	160	180	190
	T-to-T	39.	36.	34.	33.	31.	29.	28.	27.	27.	27.	27.	27.	27.	27.	27.	27.	27.	27.	27.	27.	27.	27.	27.	27.
20.	Total	160	160	160	160	160	160	160	170	170	170	180	180	190	190	190	200	200	200	210	220	220	230	250	270
	T-to-T	44.	41.	39.	37.	35.	33.	31.	31.	31.	31.	30.	31.	31.	31.	31.	31.	31.	31.	31.	31.	31.	31.	31.	31.
25.	Total	190	190	190	190	190	190	190	190	200	200	210	210	220	220	230	230	230	240	250	260	260	270	300	320
	T-to-T	46.	43.	41.	39.	36.	35.	33.	32.	32.	32.	32.	32.	32.	32.	32.	32.	32.	32.	32.	32.	32.	32.	32.	32.
50.	Total	310	310	310	310	310	310	310	310	310	330	340	350	360	370	380	380	390	400	410	430	440	460	510	550
	T-to-T	54.	51.	48.	46.	43.	41.	38.	38.	38.	38.	37.	38.	38.	38.	38.	38.	38.	38.	38.	38.	38.	38.	38.	38.

Source: ANSI/AGMA 2000-A88.[3]

TABLE 7.3 Element Tolerances

Runout Tolerance V_{rT}, Allowable Pitch Variation $\pm V_{pA}$, Profile Tolerance V_ϕ, Ten-Thousandths of an Inch (0.0001 in.)

Number of Teeth (N); for Helical Gears, Use Adjusted Number of Teeth ($N_i = N/\cos\psi$)

Q No.	Pitch P_{nd}	Tolerance Type	6	8	10	12	16	20	24	28	32	36	40	44	48	52	56	60	80	100	200
Q10	100.5	Runout	75.	80.	85.	88.	95.	100.	110.	110.	110.	120.	120.	120.	120.	130.	130.	130.	140.	150.	170.
		Pitch+/−	9.5	10.	11.	11.	11.	12.	12.	13.	13.	13.	13.	14.	14.	14.	14.	14.	15.	16.	18.
		Profile	22.	23.	24.	24.	25.	26.	27.	28.	28.	29.	29.	30.	30.	31.	31.	31.	33.	34.	38.
	1.0	Runout	46.	49.	51.	54.	57.	61.	63.	66.	68.	70.	71.	73.	75.	76.	77.	79.	84.	89.	110
		Pitch+/−	7.2	7.5	7.8	8.1	8.5	8.9	9.2	9.4	9.6	9.8	10.	10.	11.	11.	11.	11.	12.	12.	14.
		Profile	15.	15.	16.	16.	17.	18.	18.	19.	19.	19.	20.	20.	20.	20.	21.	21.	22.	22.	25.
	1.5	Runout	34.	36.	38.	40.	43.	45.	47.	49.	51.	52.	53.	55.	56.	57.	58.	59.	63.	66.	78.
		Pitch+/−	6.1	6.4	6.7	6.9	7.2	7.5	7.8	8.0	8.2	8.4	8.5	8.7	8.8	8.9	9.0	9.2	9.6	10.	12.
		Profile	12.	12.	13.	13.	13.	14.	14.	15.	15.	15.	15.	16.	16.	16.	16.	16.	17.	18.	20.
	2.0	Runout	28.	30.	31.	33.	35.	37.	38.	40.	41.	42.	43.	44.	45.	46.	47.	48.	51.	54.	64.
		Pitch+/−	5.4	5.7	5.9	6.1	6.5	6.7	6.9	7.1	7.3	7.5	7.6	7.7	7.8	8.0	8.1	8.2	8.6	8.9	10.
		Profile	9.6	10.	11.	11.	11.	12.	12.	12.	13.	13.	13.	13.	13.	14.	14.	14.	15.	15.	17.
	3.0	Runout	21.	22.	23.	24.	26.	28.	29.	30.	31.	32.	32.	33.	34.	34.	35.	36.	38.	40.	47.
		Pitch+/−	4.6	4.9	5.1	5.2	5.5	5.7	5.9	6.1	6.2	6.3	6.5	6.6	6.7	6.8	6.9	6.9	7.3	7.6	8.6
		Profile	7.6	7.9	8.2	8.4	8.8	9.1	9.4	9.6	9.8	10.	10.	10.	11.	11.	11.	11.	11.	12.	13.
	4.0	Runout	17.	18.	19.	20.	21.	22.	23.	24.	25.	26.	26.	27.	28.	28.	29.	29.	31.	33.	39.
		Pitch+/−	4.1	4.3	4.5	4.7	4.9	5.1	5.3	5.4	5.5	5.7	5.8	5.9	5.9	6.0	6.1	6.2	6.5	6.8	7.6
		Profile	6.4	6.7	6.9	7.1	7.5	7.7	7.9	8.1	8.3	8.4	8.6	8.7	8.8	8.9	9.0	9.1	9.5	9.9	11.
	6.0	Runout	13.	14.	14.	15.	16.	17.	17.	18.	19.	19.	20.	20.	21.	21.	21.	22.	23.	24.	29.
		Pitch+/−	3.5	3.7	3.8	4.0	4.2	4.3	4.5	4.6	4.7	4.8	4.9	5.0	5.1	5.1	5.2	5.3	5.5	5.8	6.5
		Profile	5.1	5.3	5.5	5.6	5.9	6.1	6.2	6.4	6.5	6.7	6.8	6.9	7.0	7.0	7.1	7.2	7.5	7.8	8.7

		1	2	3	4	5	6	7	8	9	10	11	12	13	14	15	16	17	18	19
Q10	8.0	Runout	10.	11.	12.	12.	13.	14.	14.	15.	15.	16.	16.	16.	17.	17.	17.	18.	19.	20.	23.
		Pitch+/−	3.1	3.3	3.4	3.5	3.7	3.9	4.0	4.1	4.2	4.3	4.4	4.4	4.5	4.6	4.6	4.7	4.9	5.1	5.8
		Profile	4.3	4.5	4.6	4.7	5.0	5.1	5.3	5.4	5.5	5.6	5.7	5.8	5.9	5.9	6.0	6.1	6.3	6.6	7.3
	10.	Runout	8.6	9.2	9.7	10.	11.	12.	12.	13.	13.	13.	14.	14.	14.	15.	15.	15.	16.	17.	20.
		Pitch+/−	2.9	3.0	3.1	3.2	3.4	3.5	3.7	3.8	3.8	3.9	4.0	4.0	4.1	4.1	4.2	4.2	4.3	4.5	4.7	5.3
		Profile	3.7	3.9	4.0	4.2	4.4	4.5	4.6	4.7	4.8	4.9	5.0	5.1	5.1	5.2	5.3	5.3	5.6	5.8	6.4
	12.	Runout	7.6	8.1	8.5	8.9	9.5	10.	11.	11.	11.	12.	12.	12.	13.	13.	13.	13.	14.	15.	18.
		Pitch+/−	2.7	2.8	2.9	3.0	3.2	3.3	3.4	3.5	3.6	3.6	3.7	3.8	3.8	3.9	3.9	4.0	4.2	4.4	4.9
		Profile	3.4	3.5	3.6	3.7	3.9	4.0	4.2	4.3	4.3	4.4	4.5	4.6	4.6	4.7	4.7	4.8	5.0	5.2	5.8
	16.	Runout	6.1	6.6	6.9	7.2	7.8	8.2	8.5	8.9	9.1	9.4	9.6	9.9	10.	10.	11.	11.	12.	12.	14.
		Pitch+/−	2.4	2.5	2.6	2.7	2.8	2.9	3.0	3.1	3.2	3.3	3.3	3.4	3.4	3.5	3.5	3.6	3.7	3.9	4.4
		Profile	2.8	3.0	3.1	3.2	3.3	3.4	3.5	3.6	3.7	3.7	3.8	3.9	3.9	4.0	4.0	4.0	4.2	4.4	4.9
	20.	Runout	5.2	5.6	5.9	6.2	6.6	7.0	7.3	7.5	7.8	8.0	8.2	8.4	8.6	8.7	8.9	9.0	9.7	10.	12.
		Pitch+/−	2.2	2.3	2.4	2.5	2.6	2.7	2.8	2.8	2.9	3.0	3.0	3.1	3.1	3.2	3.2	3.3	3.4	3.6	4.0
		Profile	2.5	2.6	2.7	2.8	2.9	3.0	3.1	3.2	3.2	3.3	3.3	3.4	3.4	3.5	3.5	3.5	3.7	3.8	4.3

Tooth Alignment Tolerance (Lead) $V_{\phi T}$

Q10	Facewidth	0.1	1.0	1.2	1.5	2.0	2.5	3.0	4.0	5.0	6.0	8.0	10.0
	Tolerance	3.3	3.3	3.7	4.4	5.3	6.3	7.2	8.8	11.0	12.0	15.0	17.0

SOURCE: ANSI/AGMA 2000-A88.[3]

TABLE 7.4 Composite Tolerances

Total Composite Tolerance V_{cqT}, Tooth-to-Tooth Composite Tolerance V_{qT}
Ten-Thousandths of an Inch (0.0001 in.)
Number of Teeth (N); for Helical Gears, Use Adjusted Number of Teeth ($N_i = N/\cos\phi$)

Q No.	Pitch P_{nd}	Tolerance Type	6	8	10	12	16	20	24	28	32	36	40	44	48	52	56	60	70	80	90	100	150	180	200	220
Q10	0.5	Total	120	120	120	120	120	120	130	130	140	140	140	150	150	150	160	160	170	170	180	180	200	200	220
		T-to-T	22.	20.	19.	18.	17.	16.	15.	15.	15.	15.	15.	15.	15.	15.	15.	15.	15.	15.	15.	15.	15.	15.	15.
	1.0	Total	75.	75.	75.	75.	75.	75.	76.	79.	81.	83.	86.	88.	89.	91.	93.	94.	98.	100	110	110	120	120	130
		T-to-T	18.	17.	16.	16.	15.	14.	13.	13.	13.	13.	13.	13.	13.	13.	13.	13.	13.	13.	13.	13.	13.	13.	13.
	2.0	Total	48.	48.	48.	48.	48.	48.	48.	50.	51.	53.	54.	55.	56.	57.	58.	59.	61.	63.	64.	66.	72.	76.	
		T-to-T	16.	15.	14.	13.	12.	12.	11.	11.	11.	11.	11.	11.	11.	11.	11.	11.	11.	11.	11.	11.	11.	11.	
	4.0	Total	32.	32.	32.	32.	32.	32.	32.	33.	34.	35.	35.	36.	37.	37.	38.	38.	39.	40.	41.	42.	46.	48.	
		T-to-T	13.	12.	12.	11.	10.	9.8	9.4	9.2	9.0	9.2	9.2	9.2	9.2	9.2	9.2	9.2	9.2	9.2	9.2	9.2	9.2	9.2	
	8.0	Total	22.	22.	22.	22.	22.	22.	22.	23.	23.	24.	24.	24.	25.	25.	25.	26.	26.	27.	28.	28.	30.	32.	
		T-to-T	11.	10.	9.8	9.4	8.7	8.3	8.0	7.9	7.7	7.8	7.8	7.8	7.8	7.8	7.8	7.8	7.8	7.8	7.8	7.8	7.8	7.8	
	12.	Total	18.	18.	18.	18.	18.	18.	18.	18.	19.	19.	19.	20.	20.	20.	20.	21.	21.	22.	22.	23.	24.	25.	
		T-to-T	10.	9.4	8.9	8.5	7.9	7.5	7.3	7.2	7.0	7.0	7.0	7.0	7.0	7.0	7.0	7.0	7.0	7.0	7.0	7.0	7.0	7.0	
	16.	Total	16.	16.	16.	16.	16.	16.	15.	16.	16.	16.	17.	17.	17.	17.	18.	18.	18.	19.	19.	19.	21.	22.	
		T-to-T	9.4	8.7	8.3	7.9	7.4	7.0	6.8	6.7	6.6	6.6	6.6	6.6	6.6	6.6	6.6	6.6	6.6	6.6	6.6	6.6	6.6	6.6	
	20.	Total	14.	14.	14.	14.	14.	14.	14.	14.	15.	15.	15.	15.	15.	16.	16.	16.	16.	17.	17.	17.	18.	19.	
		T-to-T	8.9	8.3	7.8	7.5	7.0	6.6	6.5	6.4	6.3	6.2	6.2	6.2	6.2	6.2	6.2	6.2	6.2	6.2	6.2	6.2	6.2	6.2	
	24.	Total	13.	13.	13.	13.	13.	13.	13.	13.	13.	14.	14.	14.	14.	14.	14.	15.	15.	15.	15.	16.	17.	17.	
		T-to-T	8.5	7.9	7.5	7.2	6.7	6.4	6.2	6.1	6.0	6.0	6.0	6.0	6.0	6.0	6.0	6.0	6.0	6.0	6.0	6.0	6.0	6.0	

Q10																											
32.	Total	12.	12.	12.	12.	11.	11.	11.	11.	12.	12.	12.	12.	12.	12.	12.	12.	13.	13.	13.	13.	13.	13.	13.	14.	14.	15.
	T-to-T	7.9	7.4	7.0	6.7	6.3	5.9	5.8	5.7	5.6	5.6	5.6	5.6	5.6	5.6	5.6	5.6	5.6	5.6	5.6	5.6	5.6	5.6	5.6	5.6	5.6	5.6
40.	Total	11.	11.	11.	11.	10.	10.	9.9	10.	11.	11.	11.	11.	11.	11.	11.	11.	11.	11.	12.	12.	12.	12.	12.	12.	13.	13.
	T-to-T	7.5	7.0	6.6	6.4	5.9	5.6	5.5	5.4	5.3	5.3	5.3	5.3	5.3	5.3	5.3	5.3	5.3	5.3	5.3	5.3	5.3	5.3	5.3	5.3	5.3	5.3
48	Total	9.9	9.8	9.7	9.6	9.4	9.3	9.2	9.4	9.5	9.7	9.8	9.9	9.9	10.	10.	10.	11.	11.	11.	11.	11.	11.	11.	12.	12.	12.
	T-to-T	7.2	6.7	6.4	6.1	5.7	5.4	5.3	5.2	5.1	5.1	5.1	5.1	5.1	5.1	5.1	5.1	5.1	5.1	5.1	5.1	5.1	5.1	5.1	5.1	5.1	5.1
64.	Total	8.9	8.8	8.7	8.6	8.4	8.2	8.1	8.2	8.4	8.5	8.6	8.7	8.8	8.9	9.0	9.1	9.3	9.4	9.6	9.7	10.	11.
	T-to-T	6.7	6.3	5.9	5.7	5.3	5.0	5.0	4.9	4.8	4.7	4.7	4.7	4.7	4.7	4.7	4.7	4.7	4.7	4.7	4.7	4.7	4.7
80.	Total	8.2	8.1	8.0	7.9	7.7	7.4	7.3	7.5	7.6	7.7	7.8	7.9	8.0	8.1	8.2	8.2	8.4	8.5	8.6	8.8	9.2	9.5
	T-to-T	6.4	5.9	5.6	5.4	5.0	4.8	4.7	4.6	4.6	4.5	4.5	4.5	4.5	4.5	4.5	4.5	4.5	4.5	4.5	4.5	4.5	4.5
96.	Total	7.7	7.6	7.5	7.4	7.1	6.9	6.8	6.9	7.0	7.1	7.2	7.3	7.4	7.5	7.5	7.6	7.7	7.9	8.0	8.1	8.5	8.8
	T-to-T	6.1	5.7	5.4	5.2	4.8	4.6	4.5	4.4	4.4	4.3	4.3	4.3	4.3	4.3	4.3	4.3	4.3	4.3	4.3	4.3	4.3	4.3
120.	Total	7.2	7.0	6.9	6.8	6.5	6.3	6.2	6.3	6.4	6.5	6.6	6.6	6.7	6.8	6.8	6.9	7.0	7.1	7.2	7.3	7.7	7.9
	T-to-T	5.8	5.4	5.1	4.9	4.6	4.3	4.3	4.2	4.1	4.1	4.1	4.1	4.1	4.1	4.1	4.1	4.1	4.1	4.1	4.1	4.1	4.1

Source: ANSI/AGMA 2000-A88.[3]

7.7.3 AGMA 390.03a Unassembled Bevel, Fine Pitch Worms and Worm Gears, and Racks

AGMA 390.03a is an interim standard for the aforementioned types of gears. It was the result of removing spur and helical gears when the new inch/metric version (ANSI/AGMA 2000-A88) was published. Refer to this standard for applicable tolerance tables.

7.7.4 Other Gear Types

Coarse-Pitch Worms. The worm deviates from other gears in that it resembles a screw thread and its teeth have cylindrical form rather than plane. In addition to the fundamental tooth proportion tolerances, the worm specification must control the axial pitch and any special form factors. No current standard exists, in the United States, for coarse-pitch cylindrical worms. AGMA 234.01 was the previous standard but is now obsolete. Table 7.5 is from the old AGMA 234.01 standard for coarse-pitch worms and worm gears, and can still be used as a guide today. It should be noted that with today's manufacturing techniques and equipment, it is possible to produce cylindrical worms to tolerances better than the class 3 precision limits. Any use of this table should be by prior agreement between maker and buyer.

Coarse-Pitch Worm Gears. Worm gears would be very similar to helical gears, except for throat design. For this reason, tolerances from the AGMA 2000-A88 standard could be used as a guide for pitch variation and runout. Motion-transmission requirements are the same, regardless of gear type. Table 7.6 is from the old AGMA 234.01 standard for coarse-pitch worms and worm gears, and can also be used as a guide. As noted in the foregoing, it is possible to produce worm gears to tolerances better than the class 3 precision limits. Any use of this table should be by prior agreement between maker and buyer.

Special Gear Types. Tolerances for the more special gear types such as face gears, Spiroid,* and Beveloid† are established in a fashion similar to those for spur, helical, worm, and bevel types. The differences between these types is mainly in the geometry of the tooth form. Allowable pitch variation and runouts would be similar for all types of gears. Refer to AGMA 2000-A88 and AGMA 390.03a for guidance to tolerances. Any use of these tolerances should be by prior agreement between maker and buyer.

7.7.5 Other National and International Standards

There are other national and international standards that anyone involved in international trade should be familiar with. West Germany uses the DIN 3962 & 3963[10] standards for spur and helical gears and DIN 3965 for bevel gears. Japan uses JIS B 1702[11] for spur and helical gears and JIS 1704 for bevel gears. An international standard for spur and helical gears, ISO 1328,[12] has also been published and is currently under revision.

---

*Registered trademark of Spiroid Div., Illinois Tool Works, Chicago, Illinois.
†Registered trademark of Vinco Corp., Detroit, Michigan.

TABLE 7.5 Tolerances for Coarse-Pitch Cylindrical Worms*

Class	Axial Pitch, In.	Runout Measurement on Pitch Circle, Total Indicator Reading; Pitch Diam., In.					Pitch Error in Axial Plane, Multiple Threads Only; Pitch Diam., In.					Profile Error Total Variation, Not Plus and Minus: Lead Angle, °			Lead Error Total Variation per Convolution: Lead Angle, °		Lead Error Total Variation per Active Length of Worm: Lead Angle, °		
		¾	1½	3	6	12	¾	1½	3	6	12	0–15	15–30	30–45	0–15	15–30	0–15	15–30	30–45
Class 1 light duty	4	—	—	—	—	80	—	—	—	—	50	30	40	45	40	80	80	100	120
	2	—	—	50	50	60	—	—	—	35	40	13	15	17	25	35	35	40	45
	1	—	—	40	50	60	—	—	35	35	40	11	13	15	15	22	28	32	35
	½	—	30	40	50	60	—	25	35	35	40	10	12	14	15	20	25	28	32
	¼	30	30	40	—	—	20	25	25	—	—	10	11	12	12	18	22	25	28
Class 2 power transmission	2	—	—	—	30	30	—	—	20	20	25	10	12	14	17	22	23	25	30
	1	—	—	23	23	30	—	—	18	20	25	8	10	12	9	15	19	21	23
	½	—	18	23	23	30	—	12	18	20	25	7	9	11	9	14	17	19	21
	¼	15	18	23	—	—	12	12	15	—	—	7	8	9	8	12	15	17	19
Class 3 precision	2	—	—	—	15	—	—	—	15	15	—	8	10	12	15	20	18	20	24
	1	—	—	12	12	—	—	—	12	12	—	6	8	10	7	12	14	16	18
	½	—	8	12	12	—	—	10	12	12	—	5	7	9	7	11	12	14	16
	¼	8	8	12	—	—	10	10	12	—	—	5	6	7	6	9	10	12	14

*For worms ¾ to 12 in. diameter; ¼ to 4 in. axial pitch; in ten-thousandths of an inch.
SOURCE: Taken from AGMA standard 234.01.

TABLE 7.6 Tolerances for Coarse-pitch Worm Gears*

Class Pitch Diam., In.	Cir. Pitch, In.	Runout Measurement on Pitch Circle, Total Indicator Reading							Pitch Variation Measured on Pitch Circle in Plane of Rotation							Max. Index Error between Any Two Teeth Exclusive of Runout Effect							Required Min. Initial Area of Contact, %
		¾	1½	3	6	12	25	50	100	¾	1½	3	6	12	25	50	100	¾	1/12	3	6	12	25	50	100	
Class 1 light duty	4	—	—	—	—	—	—	90	100	—	—	—	—	—	—	60	80	—	—	—	—	—	—	—	—	20
	2	—	—	—	—	60	80	90	—	—	—	—	—	35	40	70	60	—	—	—	—	—	—	—	—	20
	1	—	—	—	60	60	80	80	90	—	—	30	35	35	30	50	50	—	—	—	—	—	—	—	—	20
	½	—	—	50	60	60	80	80	—	—	25	25	30	30	30	40	—	—	—	—	—	—	—	—	—	20
	¼	—	30	50	60	60	80	—	—	20	20	20	25	25	25	35	—	—	—	—	—	—	—	—	—	20
Class 2 power transmission	2	—	—	—	—	25	30	35	40	—	—	—	15	15	20	25	30	—	—	—	—	—	—	—	—	30
	1	—	—	—	20	25	30	35	40	—	—	10	10	15	15	15	20	—	—	—	—	—	—	—	—	30
	½	—	—	20	20	25	30	35	—	—	10	10	10	10	12	15	—	—	—	—	—	—	—	—	—	30
	¼	—	15	20	20	25	30	—	—	9	9	9	10	10	10	—	—	—	—	—	—	—	—	—	—	30
Class 3 precision	2	—	—	—	—	20	25	—	30	—	—	—	7	9	9	—	8	—	—	—	—	20	27	—	—	40
	1	—	—	—	20	20	25	25	30	—	—	7	6	7	7	8	7	—	—	15	15	15	15	35	55	40
	½	—	10	15	20	25	25	—	—	—	6	6	6	6	6	—	—	—	15	15	15	15	15	30	45	40
	¼	10	10	15	20	25	—	—	—	3½	4	4	4	4	4	—	—	15	15	15	15	15	—	—	—	40

*For worm gears ¾ to 100 in. diameter; ¼ to 4 in. axial pitch; in ten-thousandths of an inch.
SOURCE: Taken from AGMA standard 234.01.

GEAR TOLERANCES 7.27

7.7.6 Comparison of Standards (Unassembled Spur and Helical Gears)

It is impossible to come up with a good correlation among the various standards. To begin with, elemental tolerances, for such specifications as profile, are not analyzed in the same way; for example. AGMA profile tolerance relates to a "K" chart evaluation while DIN looks at three separate parts of the profile variation. They evaluate average slope or shape, waviness from this average line, and the total variation.

If one chooses a given gear size or specification, the tolerance for runout of an AGMA Q9 gear might be close to the tolerance of a DIN class 9, but tolerances for other elements in the same class can differ considerably. Also, when comparing gears of other diameters or module (DP) within the same class, differences in tolerance levels will be noticed. Classes in the AGMA system run from 3 (loosest tolerance) to 15 (tightest tolerance). Classes in the DIN system run just the opposite; 12 (loosest tolerance) to 1 (tightest tolerance).

7.7.7 Comparison Charts

Charts shown in Figs. 7.5 through 7.10 show the comparison of elemental and composite tolerances for a gear of 3 module (8DP), 150-mm diameter (6 in.), and 32-mm face width (1.25 in.). They compare AGMA, DIN, ISO, and JIS standards. The charts were established by aligning DIN and ISO class 9 with AGMA class 9 and JIS class 4 for runout. Except for ISO, the comparison isn't too bad for this element. The comparison is not as good when looking at other parameters with the same class alignment.

It is obvious from the foregoing discussion that gear quality cannot be compared from one system to another on a simple class basis. One must look at each parameter of gear quality separately. Caution should also be observed when looking at English translations of foreign standards. Such terms as *pitch error, single pitch error, spacing variation,* and *pitch variation* can be confusing. Comparing symbols and definitions can be helpful when trying to find the proper tables. Refer to AGMA 2000-A88, App. B for a table of symbol comparisons. DIN and ISO symbols are essentially the same.

7.8 SPECIAL CONSIDERATIONS FOR TOLERANCE VALUES

7.8.1 Cost

As noted in Sec. 7.3.2, tolerances cost money. It is, therefore, wise to choose them carefully. Emphasis should be put on the parameters that have the greatest influence on function, relative to the application.

7.8.2 Tolerances Relative to Application

In regard to application, gears divide into three general groupings based on the fundamental functions of power, motion transmission (accuracy), and noise. Power transmissions are the many cases of transmitting torque for the purpose of performing useful work. This grouping is divided into two types of applications: general power transmission, such as household appliances, machine tools, etc.; and high-power and high-speed

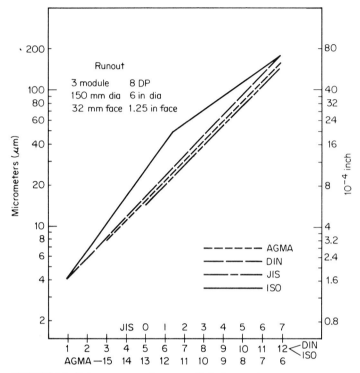

FIGURE 7.5 Tolerance comparison—spur/helical: runout.

applications as exemplified by turbine gearboxes. Motion transmission as a primary gear function usually is with negligible or light power transmission, although there are significant exceptions. Instruments, controls, index mechanisms, and robots are examples of motion transmission. Gearboxes for which noise is of primary concern are generally running at higher speeds, and can be either lightly or heavily loaded. Applications are many, such as automotive, household appliances, power tools, etc.

Tolerance Recommendations. The AGMA 390.03 Handbook has an extensive gear classification table with recommended quality levels for various industries and applications. These are only suggestions, but do provide a starting point. Certain designs or operating conditions may require or tolerate gears of a higher or lower quality number. Table 7.7 also suggests some tolerance ranges, per application, in the three general categories. In the power category, tooth alignment (lead) is probably one of the most critical parameters. In the motion-transmission category, accumulated pitch variation is the most critical. For the noise category, profile (involute) or conjugacy, is the most critical.

7.8.3 Tolerance per Speed of Operation

Another criterion for determining gear tolerances is speed of operation. This is a condition that is somewhat independent of application, general quality class, transmit-

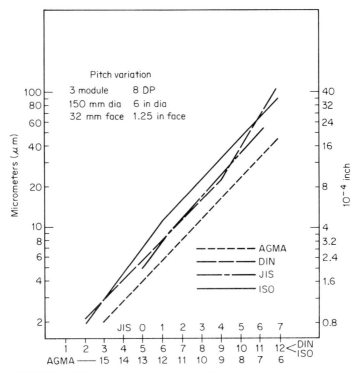

FIGURE 7.6 Tolerance comparison—spur/helical: pitch variation.

ting power, and size. With increasing speed the dynamics of operation become paramount, and regardless of size, power, etc., critical qualities of the gear must be maintained. Comparative tolerances for basic gear criteria are given in Table 7.8. If noise is a criteria, profile slope variation may have to be held to a smaller tolerance.

7.8.4 Tolerances for Various Gear Materials

Gear dimension tolerances are related to material in regard to practicability and limitations of achievement. Stable, hard materials such as metallic alloys can be finished to a much higher degree of accuracy and surface finish compared with soft, nonmetallic materials. The nonmetallic materials are subject to greater warpage, deflections, temperature variations, moisture absorption, and deterioration with aging. The nature of the material also limits the possible techniques of fabrication. For example, nonmetallics cannot be conveniently ground; some ferrous alloys machine better than others; etc. Table 7.9 offers general tolerance ranges for various materials.

7.8.5 Tolerances for Various Fabrication Methods

The method of fabrication directly affects the tolerance limits that can be placed on gear dimensions and controls. A comprehensive listing of fabrication methods is presented in the following.

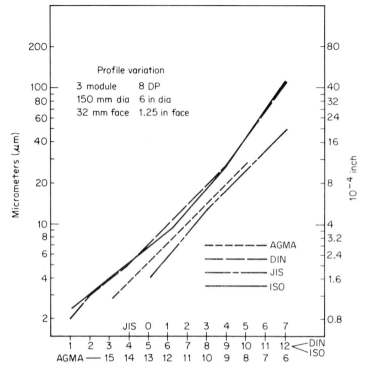

FIGURE 7.7 Tolerance comparison—spur/helical: profile variation.

Gear Fabrication Methods

1. General machining:
   Machine-generated
   Machine-form-milled or planed
2. Refinement methods:
   Grinding
   Shaving
   Burnishing
   Lapping
3. Forming—mechanical:
   Extruded
   Rolled
   Stamped
4. Forming—hot:
   Powdered-sintered
   Molded—plastic
5. Casting:
   Sand
   Plaster
   Precision investment
   Die

GEAR TOLERANCES 7.31

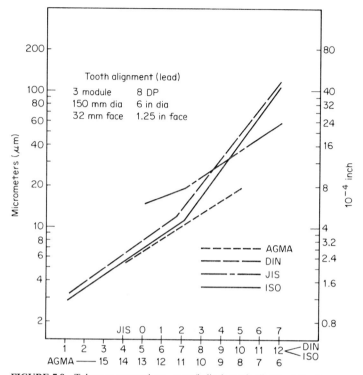

FIGURE 7.8 Tolerance comparison—spur/helical: tooth alignment (lead).

For comparative purposes the foregoing fabrication methods are relisted in Table 7.10 in descending order of gear quality results. That is, the listing indicates the relative quality in terms of producible tolerance ranges that can be obtained by the various fabrication methods.

7.8.6 Tooth Thickness Tolerance (Size Control)

A tight tooth thickness tolerance should not be used unless it is absolutely necessary. In order to control a tight tolerance, it is necessary to make a much higher-class gear, with the resulting drastic increase in cost.

Suggested Tolerances. ANSI/AGMA 2000-A88 "Gear Classification and Inspection Handbook"[3] has a table of tooth thickness tolerance. The tolerance ranges are defined by a letter code (A through D) for various diametral pitches and AGMA Q number classes. In most applications, allowing a larger range of tooth thickness tolerance or operating backlash will not affect the noise, performance, or load-carrying capacity of gears and does not make them less accurate.

Tooth Thickness Calculations. Refer to ANSI/AGMA 2002-B88[13] for a complete treatment of tooth thickness specification and measurement.

Design tooth thickness is usually established from engineering considerations of

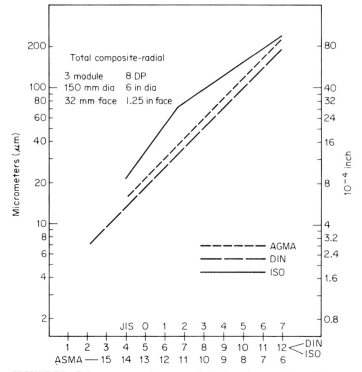

FIGURE 7.9 Tolerance comparison—spur/helical: total composite—radial.

geometry, tooth strength, mounting, and backlash. Calculations are usually made in the transverse plane and measurements are made in the normal plane.

Tooth Thickness Measurement. Tooth thickness varies around a gear, and is influenced by runout, profile variation, spacing, and tooth alignment. It is customary to assume that the entire gear is characterized by the measured data from as few as one or two measurements.

Tooth thickness is commonly measured by one of four different means: chordal, over pins or wires, span, or composite action test. Each of these methods is influenced differently by the above mentioned effects of runout, profile variation, spacing, and tooth alignment. If a given gear is measured by each of these methods, different results will be observed.

Functional tooth thickness is the value that is obtained on a composite action test (while rolling with a master gear), and encompasses all the aforementioned effects (similar to the concept of maximum material condition). It is probably the most reliable method.

Effective tooth thickness, similar to functional tooth thickness, can be determined by adding the combined effects of tooth element variation and mounting to the measured tooth thickness, as determined by the other measuring methods. The maximum effective tooth thickness is the thickness of the thickest tooth, with reference to the mounting surfaces, at the operating pitch diameter with its mating gear. In most

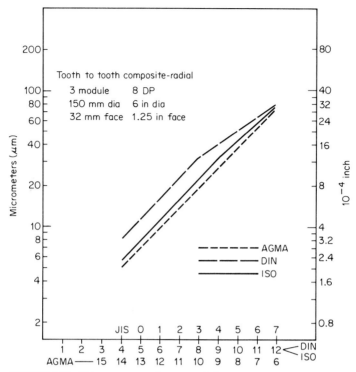

FIGURE 7.10 Tolerance comparison—spur/helical: tooth-to-tooth composite—radial.

designs, it is desirable to establish the maximum effective thickness equal to the maximum design thickness. Figure 7.11 illustrates the foregoing concepts.

7.8.7 Position Tolerance (Tangential Variation)

Gears require a consideration that until recently has been almost neglected. Much emphasis has been placed on the total composite tolerance control for achievement of linearity in angular motion. This has been a subtle fallacy, particularly with high-accuracy gearing. The double flank composite action technique is an adequate control of center distance variation when rolling with a master, but is incapable of detecting all angular variation. For example, as an extreme case, the double flank composite method cannot detect the huge angular transmission error which occurs when a gear has one more or less tooth than is proper.

Transmission Error. Transmission error is defined as the deviation of the position of the driven gear, for a given angular position of the driving gear, from the position that the driven gear would occupy if the gears were geometrically perfect. Today the most common method of measurement makes use of optical encoders, as shown in Fig. 7.12. The encoders measure the angular position of the input and output shafts. The data

TABLE 7.7 Tolerance Ranges per Application

Application	Tolerance Level (Q number)	Type of Tolerance (emphasis)
Power		
Farm machinery	5–7	Tooth thickness and runout
Household appliance	6–10	Double flank composite
Automotive	10–11	Double flank composite
Machine tools	8–12	Composite/elemental
Steam turbine	10–12	All gear elements
Aircraft engine	11–13	All gear elements
Rocket engine	13–14	All gear elements
Motion Transmission		
Clock gears	6–8	Sample elemental
Industrial timers	7–8	Sample composite
Governor and valve gears in engines	10–11	DF composite (all gears)
Radar data gears	11–13	DF composite (all gears)
Robots	12–14	Single flank composite Elemental/acc. pitch var.
Index (machine tools)	12–14	Single flank composite Elemental/acc. pitch var.
Critical missile guidance	13–14	SF composite/all elements
Noise		
Household appliance	8–10	DF/SF composite elemental/profile
Power tools	8–10	DF/SF composite elemental/profile
Automotive	10–11	DF/SF composite elemental/profile
Machine tools	10–13	SF composite elemental/profile

is processed by the electronics, to compensate for gear ratio, and then converted to analog form for recording on a strip chart. The once-per-revolution-type wave forms relate to runout or accumulated pitch variation, while the once-per-tooth mesh waveforms relate to profile shape; spacing; and, in the case of long face helicals, to tooth alignment errors. For spur and short face helicals (less than one axial pitch) tooth alignment has little effect on tooth to tooth transmission errors.

The equipment used for this measurement is often referred to as a single flank composite tester. The gears are mounted at their actual operating center distance, are running with the normal backlash, and have only one set of flanks in contact.

Transmission error is useful for the evaluation of gears used in applications requiring positional accuracy. In this case, the once-per-revolution-type data are most useful. It is also the best way to evaluate the causes of gear noise. In this application, the

TABLE 7.8 Comparative Tolerances for Various Pitch Line Velocities

Tolerance Item	Pitch Line Velocity, ft/min			
	500	5000	20,000	40,000
Maximum allowable runout	0.1 (0.004)	0.05 (0.002)	0.03 (0.0012)	0.0175 (0.0007)
Maximum allowable pitch tolerance	0.025 (0.001)	0.01 (0.0004)	0.0075 (0.0003)	0.005 (0.0002)
Maximum profile slope	+/−0.0125 +/−(0.0005)	+/−0.01 +/−(0.0004)	+/−0.0075 +/−(0.0003)	+/−0.005 +/−(0.0002)
Surface finish of active profile Ra mm or (μin)	0.0015 (63)	0.001 (32)	0.0005 (20)	0.0005 (20)
Double flank composition tolerance:				
Maximum tooth to tooth	0.05 (0.002)	0.0175 (0.0007)	0.01 (0.0004)	0.0075 (0.0003)
Maximum total composite	0.125 (0.005)	0.06 (0.0025)	0.04 (0.0015)	0.025 (0.001)

NOTE: Comparison based on steel gears 125- to 500-mm diameter (5 to 20 in.) and 6 to 0.4 module (4 to 10 DP). Dimensions are given in millimeters or (inches).

TABLE 7.9 General Tolerance Ranges for Various Gear Materials

Material	Limiting Tolerance Range				
	Functional Linear Dimensions	Gear Body Linear Dimensions	Bore Diameter	Journal Diameter	Surface Finish mm Ra (μin. Ra)
Metallic:					
Ferrous soft	0.005–0.025 (0.0002–0.001)	0.025–0.125 (0.001–0.005)	0.005–0.0125 (0.0002–0.0005)	0.0025–0.0075 (0.0001–0.0003)	0.0002–0.0015 (8–63)
Ferrous hard	0.0025–0.025 (0.0001–0.001)	0.0125–0.05 (0.0005–0.002)	0.005–0.0125 (0.0002–0.0005)	0.0025–0.0075 (0.0001–0.0003)	0.0001–0.0004 (4–16)
Nonferrous	0.005–0.05 (0.0002–0.002)	0.025–0.125 (0.001–0.005)	0.005–0.0125 (0.0002–0.0005)	0.005–0.0125 (0.0002–0.0005)	0.0004–0.003 (16–125)
Nonmetallic	0.0125–0.075 (0.0005–0.003)	0.05–0.4 (0.002–0.015)	0.025–0.125 (0.001–0.005)	0.025–0.075 (0.001–0.003)	0.0008–0.003 (32–125)

NOTE: Tolerance limits given as a range, since exact limit is a function of the material and size of gear elements. Dimensions are given in millimeters or (inches).

TABLE 7.10 Quality Listing of Fabrication Methods in Descending Order

Fabrication Methods	Comments	Achievable Tolerance Range, General Degree of Precision of Final Product	Order of Use*
Grinding	Excellent finish and closest tolerances AGMA Q11 to Q14	0.0025–0.0125 (0.0001–0.0005)	SP
Shaving	Excellent finish, refinement on machine cut accuracy AGMA Q10 to Q11	0.0025–0.0125 (0.0001–0.0005)	SP
Lapping	Fair finish. Some correction of dimensional variation AGMA Q9 to Q10 (hypoid gears)	0.005–0.0175 (0.0002–0.0007)	SP
Burnishing	Good finish. Gives uniformity of surface condition	0.0025–0.005 (0.0001–0.0002) dimensional variation on previous operation	R
Machine cut generated	Good general precision AGMA Q7 to Q10	0.0125–0.05 (0.0005–0.002)	N
Sintered	Good commercial quality AGMA Q6 to Q7	0.025–0.075 (0.001–0.003)	HPL
Molded plastic	Good commercial quality AGMA Q6 to Q7	0.025–0.075 (0.001–0.003)	HPL
Rolled	Good commercial quality AGMA Q7 to Q9	0.025–0.075 (0.001–0.003)	SP
Extruded	Commercial quality AGMA Q5 to Q7	0.05–0.125 (0.002–0.005)	SP
Stamped	Commercial quality AGMA Q5 to Q7	0.05–0.125 (0.002–0.005)	HPL
Die cast	Commercial quality AGMA Q5 to Q7	0.05–0.125 (0.002–0.005)	HPL
Form machine cut	Good to poor quality	0.025–0.075 (0.001–0.003)	R
Precision investment	Fair quality, high cost	0.025–0.25 (0.001–0.010)	R
Plaster cast	Poor quality AGMA Q3 to Q5	0.375–0.75 (0.015–0.03)	R
Sand cast	Very poor quality AGMA Q3 to Q4	0.75–1.5 (0.03–0.06)	R

*N = normally used; SP = special applications; HPL = high-production, low-cost applications; R = rarely used.
Dimensions are given in millimeters or (inches).

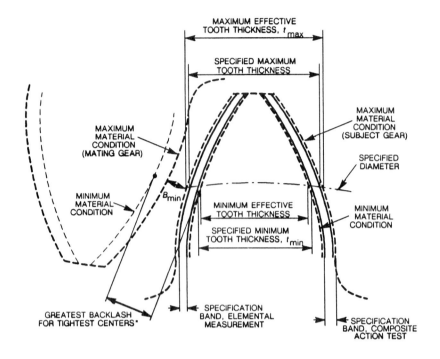

* THIS FIGURE IS DRAWN AT THE POSITION OF TIGHTEST CENTER DISTANCE; IF CENTER DISTANCE IS INCREASED BACKLASH WILL INCREASE.

FIGURE 7.11 Tooth thickness transverse plane.

amplitude and shape characteristics of the tooth-to-tooth transmission error, after filtering out the once-per-revolution error, are most useful.

Tolerance (Transmission Error). Standards do not exist in this country for the tolerancing of transmission error. However, in the case of positional accuracy, the permissible amount of allowable arcseconds of error in the application will dictate the tolerance of the gear pair. For noise control applications, the subject is much more complex, because of the amplification effects of the housing dynamics. A good place to start, however, is to limit filtered tooth-to-tooth transmission error to 5 μm (0.0002 in.), at the pitch radius of the driven gear. The allowable tolerance may be slightly more or less than this value, depending on housing dynamics and application.

Tolerance (Accumulated Pitch or Index Variation). The AGMA 2000-A88 *Gear Classification and Inspection Handbook* does not provide tolerances for accumulated pitch or index variation. However, the amplitude of the once-per-revolution sine wave of the total transmission error curve, could be related to runout tolerance by the following approximation; runout equals accumulated pitch variation times the cosine of the pressure angle at the standard pitch circle (approximate point of measurement).

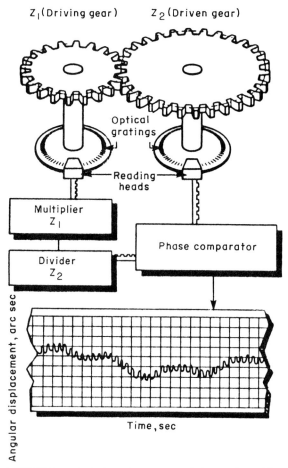

FIGURE 7.12 Optical encoders measure relative angular position of gears. Encoder data is processed to yield analog waveform of transmission error.

Accumulated pitch variation or index variation can also be measured by any of the precision index devices common today, such as CNC involute and lead checkers.

7.8.8 Relationship of Blank Quality to Tooth Quality

A gear blank is a material shape around which some teeth are machined. The teeth are complex shapes or geometry, requiring great precision, that are spaced around a circle on the gear blank. Unfortunately, many designers get overly concerned with the accuracy specifications of the tooth geometry, such as profile shape and pitch variation, but completely ignore the necessity of providing an accurate gear blank around which the teeth are placed. Blank tolerances are the important foundation on which the teeth are built; for example, bore or journal tolerances for size, roundness, and runout should be at least three to five times smaller than the specified tolerance for runout or accumulated pitch variation of the gear teeth.

It is not uncommon for a designer to specify involutes and "spacing" to be held to 5 μm (0.0002 in.) to 10 μm (0.0004 in.), while allowing 25 μm (0.001 in.) on bore size and squareness of shoulders. These tolerances are not compatible. Certainly it is more difficult for the machinist to produce around a gear blank 60 teeth that have complex involute form, and are accurately spaced apart to within a few micrometers (ten-thousandths of an inch), than it is for another machinist to make a simple round hole to a tolerance of a few micrometers (ten-thousandths of an inch).

Functional Surfaces. For inspection purposes, always try to use the same bores and journals, as well as shoulders, that will be used in the final application. After all, what is important from a functional standpoint is whether the teeth are accurate relative to the surfaces from which they will be used.

Bores. In addition to size, bores should be controlled for shape and form errors such as out of roundness, barreled, waisted, tapered, as well as the squareness of axis to shoulder.

Shafts. With shaft-type gears, the worst problem, relative to inspection and resulting quality, is centers. A lot of inspection, in the parallel axis field, is done from centers. In the bevel field, it is more common to inspect from functional journals. Never assume that journals run true with the centers of a shaft within the accuracy necessary to properly inspect gear teeth. In critical applications, it may be necessary to regrind centers true with journals. It is also important to be sure that shoulders run true, as they will cause runout in assembly.

Proofspots. Axial and radial bands or reference surfaces, machined concentric to bores or journals, can be of great help in the manufacture, inspection, and assembly of very accurate gears. Figure 7.13 shows an example of such proofspots or reference bands. If checked at all stages of manufacturing, inspection, and assembly, they guarantee proper location of the gear, relative to the axis of rotation.

7.8.9 Surface Texture

The practical achievement of close tolerances is very much influenced by surface texture. Critical surfaces in this regard are tooth surfaces and mounting and registering surfaces.

Close tolerances require finer finishes because surface roughness and waviness would become a significant portion of the tolerance and create difficulty of control and measurement. Also, the tolerance and finish combination influence the choice of fabrication method as well as feed and speeds.

Surface texture of gear teeth can be evaluated according to two basic parameters: surface finish (microfinish) and waviness.

Surface finish is the roughness of the surface texture related to random scratches and tears that occur at small wavelengths. Surface finish, or microfinish, is generally of concern when considering the life or wear characteristics of gear teeth. It is not usually of concern relative to noise. Today, it is measured in terms of microinches Ra (rms is an obsolete term).

Surface waviness is a more regular or systematic undulation in the surface, caused

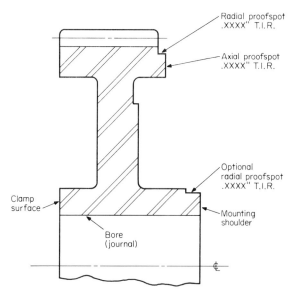

FIGURE 7.13 Axial and radial proofspots or reference bands.

by machine or cutter kinematics. It occurs at longer wavelengths than surface roughness. If waviness is oriented roughly parallel to the instantaneous line of contact, it results in transmission error, and often noise.

Standards. ANSI B46 is a general standard relating to surface roughness and waviness. AGMA 118.01 is an information sheet relating to surface texture on gear teeth.
ANSI B46 talks about standard cutoff wavelengths of 0.003, 0.010, 0.030, 0.100, and 0.300 in. All commercial surface measuring instruments are designed around these filter cutoff frequencies.
AGMA 118.01 information sheet on surface texture talks of wavelengths of less than 1/20th of face length for roughness and longer than 1/20th that of face length for waviness. General practice is to consider a wavelength of 0.030 inches as the dividing line between roughness and waviness.
Table 7.11 relates surface finish, gear tolerances, machining method, and relative cost.

7.9 ASSEMBLY AND APPLICATION CONSIDERATIONS

7.9.1 Introduction

The previous sections were concerned with tolerances related to individual gears. In addition, there are significant tolerances that apply to the assembly of gears into a housing and the meshing of gear pairs.

TABLE 7.11 Surface Finish and Applicable Dimension Tolerance

Quality Class	Description	Maximum Ra Value mm or μ in.	Suitable Range of Total Tolerance	Typical Fabrication methods	Approximate Relative Cost to Produce
Extremely rough	Extremely crude surface produced by rapid removal of stock to nominal dimension	0.025 (1,000)	1.5–3.0 (0.06–0.125)	Rough sand casting, flame cutting	1
Very rough	Very rough surface unsuitable for mating surfaces	0.0125 (500)	0.4–1.5 (0.015–0.06)	Sand casting, contour sawing	2
Rough	Heavy toolmarks	0.006 (250)	0.25–0.4 (0.010–0.015)	Very good sand casting, saw cutting, very rough machining	3
Fine	Machined appearance with consistent toolmarks	0.006 (125)	0.125–0.25 (0.005–0.010)	Average machining-turning, milling, drilling; rough hobbing and shaping; die casting, stamping, extruding	4
Fine	Semismooth without objectionable toolmarks	0.0015 (64)	0.05–0.125 (0.002–0.005)	Quality machining-turning, milling, reaming; hobbing, shaping; sintering, stamping, extruding, rolling	6
Smooth	Smooth, where toolmarks are barely discernible	0.0008 (32)	0.0125–0.05 (0.005–0.002)	Careful machining; quality hobbing, and shaping; shaving; grinding; sintering	10
Ground	Highly smooth finish	0.0004 (16)	0.005–0.0125 (0.0002–0.0005)	Very best hobbing and shaping; shaving, grinding, burnishing	15
Polish	Semi-mirror-like finish without any discernible scratches or marks	0.0002 (8)	0.0025–0.005 (0.0001–0.0002)	Grinding, shaving, burnishing, lapping	20
Superfinish	Mirrorlike surface without tool, grinding, or scratch marks of any kind	0.0001 (4)	0.001–0.0025 (0.00004–0.0001)	Grinding, lapping, and polishing	25

Note: Dimensions are given in millimeters or (inches).

7.9.2 Functional Requirements or Concerns

The tolerances in this section should be evaluated in relation to their effect on the functional requirements of maintaining adequate backlash and contact ratio of the gear pair as well as avoidance of interferences.

Backlash. The center distance directly affects backlash by the approximate equation

$$B = \Delta C \times 2 \tan \phi$$

where B = backlash
 ΔC = change in center distance
 ϕ = pressure angle

The center distance tolerance must be consistent with the backlash requirement so as not to contribute backlash in excess of the stipulated maximum value or a disproportionate share.

Contact Ratio and Interference. Variation in center distance can also affect loss of contact ratio or the occurrence of interference. If the center distance is too large (external gears), loss of active profile and contact ratio will result. Conversely, if the center distance is too small (external gears), fillet interferences may be picked up.

7.9.3 Influential Factors

Many factors involving the choice of components, tolerances, materials, and environment can have an influence on the resulting operating center distance of a gear pair.

Center Distance Tolerance. Gear center distance is of primary concern because it directly affects gear function in terms of backlash, contact ratio, and interference. Care must be exercised so as to dimension and tolerance center distance properly. For external gears, it is good practice to tolerance center distances only in the plus direction from the nominal specified center distance. If the center distance is toleranced in the minus direction, the gear teeth would have to be thinned an additional amount to avoid possible interferences and loss of backlash.

Thermal Growth. The difference in thermal coefficient of expansion of materials for the gears, as well as the housing, can result in a change of center distance. This would be especially true of plastic gears in a metallic housing. A change in environmental or operating temperature can result in a relatively large center distance change. If all parts are ferrous, this would not be a problem.

Shaft and Bore Fits. The tolerances on shaft journals and bores, as well as those designed in clearances are important to the aforementioned center distance considerations. If the shafts or bores have clearance relative to their mounting, they will be eccentric, and this has the same effect as losing center distance, as far as backlash and contact ratio are concerned. Precision gears must have a close fit to utilize the inherent gear quality. Lower-quality gears have relaxed tolerances on gear fit. The fit of bore and shaft is a combination of allowance plus the tolerances on the bore and shaft. It is preferable to use standard sizes for the bore and vary the shaft diameter so as to require minimum bore tooling and gages. Generally, shaft diameters can be fabricated to closer tolerances than bore diameters.

TABLE 7.12 Center Distance Tolerance for Various Magnitudes and Quality

	Center Distance				
Gear Train Quality	Under 25 mm (1.00)	25–150 mm (1.0–6.0)	150–300 mm (6.0–12.0)	300–600 mm (12.0–24.0)	Over 600 mm (24.0)
AGMA Q7, Q8	0.1 (0.004)	0.15 (0.006)	0.25 (0.010)	0.50 (0.020)	0.50 per 300 mm (0.020 per ft)
AGMA Q9 to Q11	0.025 (0.001)	0.05 (0.002)	0.10 (0.004)	0.10 (0.004)	0.10 per 300 mm (0.004 per ft)
AGMA Q12 to Q14	0.005 (0.0002)	0.01 (0.0004)	0.01 (0.0004)	0.015 (0.0006)	0.025 per 300 mm (0.001 per ft)

Note: Dimensions are given in millimeters or (inches).

Bearing Runout. Low-quality rolling-element bearings can have runout that is sufficient to affect instantaneous center distance, along with its other detrimental effects. Also, solid sleeve-type bearings have a built in clearance that can affect center distance.

Gear Quality. Gear tolerance levels, such as the AGMA Q number, allow varying degrees of tolerances that are related to center distance tolerance in that there must be consistency between gear tooth size, runout, and center distance tolerances. For example, it would be illogical to place a stringent requirement on center distance and simultaneously to allow a liberal tolerance on gear size. Presumably good-quality gears are specified for a good reason and will also require good centers of equivalent tolerance and vice versa.

Module (Diametral Pitch) and Center Distance Magnitude. Module (diametral pitch) bears on center distance tolerance in that with fineness it demands closer control of center distance variation so as to prevent loss of contact ratio. Also, the center distance magnitude influences the tolerances in that problems, negligible over short distances, become significant for large dimensions. This is due to temperature effects, work distortions and deflection, and magnified machine and tooling errors. See Table 7.12 for guidelines to center distance tolerances relative to magnitude and AGMA Q level. In applying these tolerances it is necessary to give due consideration to the items listed in the earlier portions of this section as influential in determining tolerances of gear centers. This is only an approximate guide and one should refer to AGMA 2002-B88 for a more complete and accurate treatment of the subject.

BIBLIOGRAPHY

1. Michalec, G. W., *Gear Handbook,* McGraw-Hill, New York, Chap. 9, 1962.
2. AGMA 390.03a, "Gear Classification, Materials and Measuring Methods for Unassembled Gears," American Gear Manufacturers Association, March 1980.
3. ANSI/AGMA 2000–A88, "Gear Classification and Inspection Handbook," American Gear Manufacturers Association, March 1988.
4. Mil Std. 109B, "Military Standard Quality Assurance Terms and Definitions," Department of Defense, Washington, D.C. 20301, April 1969.

GEAR TOLERANCES **7.45**

5. AGMA 112.05, "Gear Nomenclature, Terms, Definitions, Symbols, and Abbreviations," American Gear Manufacturers Association, June 1976.
6. AGMA 1012–F90, "Gear Nomenclature—Definition of Terms with Symbols," American Gear Manufacturers Association, 1990.
7. ANSI Y14.1–Y14.5M, "Dimensioning and Tolerancing," 1982.
8. Dudley, D. W., *Handbook of Practical Gear Design,* McGraw-Hill, New York, 1984.
9. ANSI Y14.7.1–Y14.7.3, "Gear Drawing Standards for Spur, Helical, Double Helical, Racks, Bevel and Hypoid, and Wormgears," 1971, 1978, 1979.
10. DIN 3962, 3963, & 3965, "Tolerances for Cylindrical Gear Teeth and Bevel Gears," 1978, 1978, 1979.
11. JIS B1702 & B1704, "Accuracy for Spur, Helical and Bevel Gears," 1976, 1978.
12. ISO 1328, ISO, "System of Accuracy for Parallel Involute Gears," 1975.
13. ANSI/AGMA 2002–B88, "Tooth Thickness Specification and Measurement," American Gear Manufacturers Association, October 1988.

CHAPTER 8
GEAR MATERIALS

Roy J. Cunningham
*Chief Metallurgist, Boeing Helicopters,
Philadelphia, Pennsylvania*

The gear trade uses a wide variety of steels, cast irons, bronzes, aluminum, phenolic laminates, plastics, and other materials for gears [1]. This chapter covers the basic information on the more commonly used gear materials and their processing treatment. Some of the newer, more advanced materials are also discussed.

The use of a specific gear material should be based on several factors, chief of which is the service application for which the component is designed. The properties of the material should be tailored, that is, heat treated, finish processed, etc., only in a manner that is relevant to this application. Other considerations for the use of particular material would be material availability, raw stock cost, load-carrying capacity, environmental considerations such as corrosion–corrosion protection and manufacturing requirements.

8.1 FERROUS GEAR MATERIALS

The "ferrous" gear materials are those that contain mainly the element iron. Iron with larger percentages of carbon such as 2 to 4 percent is customarily called *cast iron*. The whole family of ferrous materials used in gear work comprises steels, gray cast iron, Meehanite,* ductile iron, malleable iron, and sintered iron.

The ferrous materials are of great importance in the gear trade. Generally speaking, they represent the lowest-cost raw material in terms of cents per pound. The aircraft-type hardened steels can carry the greatest load in terms of horsepower per unit volume. The ferrous materials represent the greatest tonnage of any kind of material used to make either toothed gears or gear casings.

---
*Registered trademark of the Meehanite Metals Corp., New Rochelle, New York.

8.2 GEAR MATERIALS

8.2.1 General Characteristics of Steels and Cast Irons

The properties of steels and cast irons vary rather widely depending on composition and heat treatment. Table 8.1 shows in outline from the general range of properties obtainable.

Some interesting general principles can be noted from a study of Table 8.1. The steels shown vary from 0.2 to 0.6 percent carbon. NOTE: "XX" after numbers, for steel, indicates carbon percentage; that is, 1040 indicates 0.40 percent carbon. All except AISI 1020 can be treated to 350 BHN. The hardness of a steel is generally a function, not of its composition, but rather of its carbon level and heat treatment. It is true that the maximum obtainable hardness for a given steel is determined by

TABLE 8.1 General Properties of Steels and Cast Irons

Material	Approx. % Carbon	Hardness, BHN	Modulus of Elasticity,		Yield Strength,		Tensile Strength,	
			GPa	(Mlb/in.^{2})	MPa	(klb/in.^{2})	MPa	(klb/in.^{2})
Steels:								
AISI 1020	0.2	180	197	(28.5)	373	(54)	614	(89)
AISI 1040	0.4	200	197	(28.5)	441	(63.8)	683	(99)
		350	197	(28.5)	641	(93)	848	(123)
AISI 4140	0.4	200	197	(28.5)	417	(60.5)	655	(95)
		350	197	(28.5)	1069	(155)	1172	(170)
AISI 4340	0.4	220	197	(28.5)	472	(68.5)	745	(108)
		350	197	(28.5)	1145	(166)	1207	(175)
AISI 1060	0.6	350	197	(28.5)	1160	(168)	1207	(175)
		550	197	(28.5)	1828	(265)	1897	(275)
Cast iron:								
AGMA 20	3.5	150	83	(12)	—		138	(20)
AGMA 40	3.5	175	111	(16)	—		276	(40)
AGMA 60	3.5	200	138	(20)	—		414	(60)
Ductile iron:								
ASTM 80-60-3	3.5	200	166	(24)	414	(60)	552	(80)

carbon content level, that is, the higher the carbon content, the higher the attainable hardness.

Most of the materials used for gears, other than the steels, are not subject to hardness control by heat treatment. Composition rather than heat treatment determines the hardness of most bronzes, zincs, plastics, and laminates.

The ultimate tensile strength of steel can be correlated with the hardness. However, the tensile strength values in conversion tables are averages and as such should never be used in place of tensile testing.

The yield strength of steels tends to range from about 60 percent of the tensile strength in a low-alloy steel to about 90 percent of the tensile strength for a high-alloy steel.

Gray cast iron is a brittle material that really does not have a yield point like steel. A very slight yielding may occur at around 25 percent or more of the tensile strength. Ductile iron does have a yield point somewhat like steel.

Significance of the Hardness of Steel or Iron. Since the steels and irons vary so widely in hardness, it is desirable to form some mental impressions of the significance of different hardnesses. Table 8.2 shows some general levels of hardness and their meaning in the gear trade. A relative rating of machinability is also given in this table. Machinability can be optimized by annealing, normalizing and tempering, or quenching and tempering the material prior to cutting teeth. There is a wide variety of hardness-testing instruments that are used to measure gear hardness. There are also approximate numerical conversions/tables used to convert from one scale to another. Table 8.3 is an example. In general discussions of hardness it is common practice to speak of Brinell hardness number (BHN) or Rockwell C hardness. These two scales are the best known in the trade.

Table 8.2 skips the region of 400 to 500 BHN, which is seldom used. The steel is too hard to cut; and if one is going to the trouble of grinding, the part might just as well be made still harder so that more surface load and wear capacity is obtained.

8.2.2 Heat Treatment of Steels and Cast Irons

The basic process for hardening a steel material consists of heating to its austenitizing temperature, quenching in oil or water, and then reheating to temper the steel. The austenitizing temperature is generally in the range of 790 to 870°C (1450 to 1600°F). The quench must be fast enough to insure a maximum amount of "martensite." This makes the gear very hard and somewhat brittle. However, the reheat for tempering at a temperature below the critical, that is, austenitizing, temperature results in a much tougher and more ductile material than the original quenched material.

Figure 8.1 shows the approximate maximum hardness that can be obtained for different carbon contents. A very important practical fact about the hardness of martensite in steels is that in all of the so-called low-alloy steels (less than about 5 percent total alloying elements) the hardness of martensite can be assumed to depend only on the carbon concentration of the metal. Consequently, one who knows the carbon content of any low-alloy steel is able to determine (with the aid of one of the curves such as in Fig. 8.1) the approximate hardness of the steel when it has a martensitic structure. Fig. 8.1 shows the approximate maximum hardness that can be obtained for different carbon contents.

Figure 8.1 indicates that it takes about 0.40–0.50 percent carbon to obtain a full hardness of R/C 60. Greater amounts of carbon only result in more carbides in the gear surface and make the part more wear resistant but they do not appreciably increase the maximum hardness. The "eutectoid" of carbon and iron is at 0.85 percent carbon. This

TABLE 8.2 Relative Hardness Levels for Steel or Iron Gears

Hardness		Machinability	Comments
Brinell	Rockwell C		
150–200	—	Very easy	Very low hardness; minimum load-carrying capacity
200	—	Easy	Low hardness, moderate load capacity; widely used in industrial gear work
250	24		
250	24	Moderately hard to cut	Medium hardness; good load capacity; widely used in industrial work
300	32		
300	32	Hard to cut, often considered limit of machinability	High hardness; excellent load capacity; used in lightweight, high-performance jobs
350	38		
350	38	Very hard to cut. Many shops cannot handle	High hardness; load capacity excellent provided heat treatment develops proper structure
400	43		
500	51	Requires grinding to finish	Very high hardness; wear capacity good; may lack tooth bending strength
550	55		
610	58	Requires grinding	Full hardness; usually obtained as a surface hardness by case carburizing; very high load capacity for aircraft gears, automobile gears, trucks, tanks, etc.
710	63		
	65	May be surface hardened after final machining	Superhardness; generally obtained by nitriding; very high load capacity
	70		

amount of carbon is considered an optimum for the carbon content of a case-carburized part.

Carbon contents in the range of 0.10 to 0.20 percent considerably reduce the maximum hardness that can be obtained. A gear made from low-carbon-content steel, that is, 0.15 percent, might only be R/C 40 on the surface and would exhibit poor wear resistance because of not having enough carbides in the surface. Carburizing this component would provide the necessary increased amount of carbon to the surface of the gear and thus significantly improve its hardness and wear resistance. Low carbon content, such as 0.10- or 0.20-percent carbon, is useful, though, as the core material for a gear that is case carburized after the teeth are cut. The case is enriched with carbon up to around 0.85 percent. After quenching and tempering, the carbon-rich case should be R/C 60 and the core material—because of the lean amount of carbon—will develop a hardness of only around R/C 30–40. The core is much tougher and more ductile than the case. The cased tooth has better bending fatigue strength and less sensitivity to notches than a gear tooth would have if it were made from a homogeneous material having 0.85 percent carbon all the way through the tooth.

TABLE 8.3 Approximate Tensile Strength for Equivalent Hardness Numbers of Steel

Brinell Indentation Diameter, mm	Brinell Hardness Number, 3006-Kg 10 mm Tungsten Carbide BaN	Rockwell Hardness Number* B-Scale 100-Kg Lead 1/16 in. Ball	Rockwell Hardness Number* C-Scale 150-Kg Lead Brale Penetrater	Vickers Diamond Pyramid Hardness Number	Shore Sclorescope Hardness Number	Approx. Tensile Strength 1000 lb/in.^{2}
2.25	745	—	65.3	840	91	—
—	710	—	63.3	780	87	—
2.35	682	—	61.7	737	84	—
2.40	653	—	60.0	697	81	—
2.45	627	—	58.7	667	79	—
2.50	601	—	57.3	640	77	—
2.55	578	—	56.0	615	75	—
2.60	555	—	54.7	591	73	298
2.65	534	—	53.5	569	71	288
2.70	514	—	52.1	547	70	274
2.75	495	—	51.0	528	68	264
2.80	477	—	49.6	508	66	252
2.85	461	—	48.5	491	65	242
2.90	444	—	47.1	472	63	230
2.95	429	—	45.7	455	61	219
3.00	415	—	44.5	440	59	212
3.05	401	—	43.1	425	58	202
3.10	388	—	41.8	410	56	193
3.15	375	—	40.4	396	54	184
3.20	363	—	39.1	383	52	177
3.25	352	(110.0)	37.9	372	51	170
3.30	341	(109.0)	36.6	360	50	163
3.35	331	(108.5)	35.5	350	48	158
3.40	321	(108.0)	34.3	339	47	152
3.45	311	(107.5)	33.1	328	46	147
3.50	302	(107.0)	32.1	319	45	143
3.55	293	(106.0)	30.9	309	43	139
3.60	285	(105.5)	29.9	301	—	136
3.65	277	(104.5)	28.8	292	41	131
3.70	269	(104.0)	27.6	284	40	128
3.75	262	(103.0)	26.6	276	39	125
3.80	255	(102.0)	25.4	269	38	121
3.85	248	(101.0)	24.2	261	37	118
3.90	241	100.0	22.8	253	36	114

TABLE 8.3 Approximate Tensile Strength *(Continued)*

Brinell Indentation Diameter, mm	Brinell Hardness Number, 3006-Kg 10 mm Tungsten Carbide BaN	Rockwell Hardness Number*		Vickers Diamond Pyramid Hardness Number	Shore Sclorescope Hardness Number	Approx. Tensile Strength 1000 lb/in.^{2}
		B-Scale 100-Kg Lead 1/16 in. Ball	C-Scale 150-Kg Lead Brale Penetrater			
3.95	235	99.0	21.7	247	35	111
4.00	229	98.2	20.5	241	34	109
4.05	223	97.3	(18.8)	234	—	104
4.10	217	96.4	(17.5)	228	33	103
4.15	212	95.5	(16.0)	222	—	100
4.20	207	94.6	(15.2)	218	32	99
4.25	201	93.8	(13.8)	212	31	97
4.30	197	92.8	(12.7)	207	30	94
4.35	192	91.9	(11.5)	202	29	92
4.40	187	90.7	(10.0)	196	—	90
4.45	183	90.0	(9.0)	192	28	89
4.50	179	89.0	(8.0)	188	27	88
4.55	174	87.8	(6.4)	182	—	86
4.60	170	86.8	(5.4)	178	26	84
4.65	167	86.0	(4.4)	175	—	83
4.70	163	85.0	(3.3)	171	25	82
4.80	156	82.9	(0.9)	163	—	80
4.90	149	80.8	—	156	23	—
5.00	143	78.7	—	150	22	—
5.10	137	76.4	—	143	21	—
5.20	131	74.0	—	137	—	—
5.30	126	72.0	—	132	20	—
5.40	121	69.8	—	127	19	—
5.50	116	67.6	—	122	18	—
5.60	111	65.7	—	117	15	—

Source: Courtesy Republic Steel Corp. The indentation and hardness values in the foregoing table are taken from Table 2. Approximate Equivalent Hardness Numbers for Brinell Hardness Numbers for Steel, pages 122 and 123 of 1952 SAE Handbook, Society of Automotive Engineers, Incorporated.
*The values shown in parentheses are beyond the normal range of the test scale and are given only for comparison with other values.

An important aspect of choosing a gear material for a particular application is its hardenability. Hardenability is basically related to the ability of a material to fully harden to martensite through to the center of the part. Hardenability varies for different steels and is primarily a function of its alloying element content.

GEAR MATERIALS 8.7

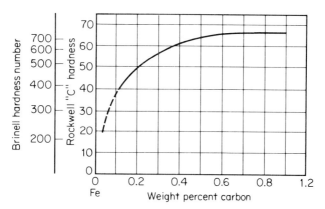

FIGURE 8.1 Maximum hardness of steel variation of the hardness of martensite as a function of carbon content.

Insight on the effect of alloy content on hardenability can be gained by studying Jominy test results. The standard *Jominy test* is made by quenching one end of a 1-in. rod in water under closely controlled conditions. After the part has cooled, hardness readings are taken every $\frac{1}{16}$ in. from the spot where the quench occurred. Figure 8.2 shows Jominy data for AISI 1045 and 4340. Figure 8.3 shows Jominy data for AISI 1020 AND 9310. In each case the highly alloyed steels, that is, 4340 and 9310, show good hardness for a considerable distance from the end of the quench. The plain carbon steels 1045 and 1020 develop about the same hardness right at the quench line, but they drop off in hardness rapidly behind the quench line. This drop-off is related to their hardenability. The curves indicate that the better the hardenability of the alloy, the less the hardness will drop off from the quenching line. Hardenability becomes important when through-the-thickness strength is important. The better the hardenability, the greater will be the thickness of the gear that can be fully hardened.

Figure 8.4 indicates the effect of various alloying elements on the hardenability of steel. The multiplying factor principle is of importance not only as a means of predicting the approximate hardenability of a steel from its composition, but also because it shows that, in general, the addition of relatively small amounts of several alloying elements

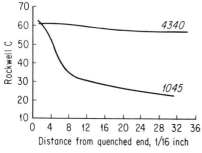

FIGURE 8.2 Jominy end-quench hardenability for typical through-hardening steels. *(Courtesy of International Nickel Co., New York)*

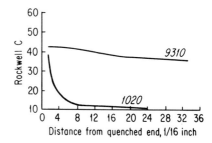

FIGURE 8.3 Joining end-quench hardenability for typical carburizing steels. *(Courtesy of International Nickel Co., New York)*

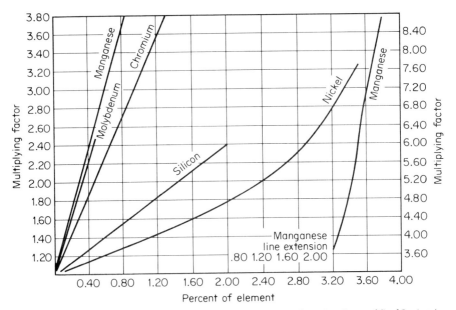

FIGURE 8.4 Multiplying factors for a variety of alloying elements. *(American Iron and Steel Institute)*

is more effective in increasing hardenability than a relatively large amount of a single element. From the graph it can be seen that small amounts of manganese have a much greater effect on hardenability than do large amounts of nickel. Hardenability is an important consideration, particularly in the use of large gears.

Figure 8.5 gives an example of the properties of 1045 steel that can be obtained in various section sizes by water quenching and tempering.

The data presented in this article so far show that a wide range of properties can be obtained by proper heat treatment of a steel. It is also clear that the alloy content and the carbon content of a steel profoundly affect the response of the steel to a given heat treatment. The different compositions of gear steels shown later in this chapter have become popular mainly because each has the composition to respond properly to a reasonable heat-treating cycle and then fill a given gear application well.

Gear metallurgists use a variety of special terms in their work. Some of these terms have already been used in this chapter. Others will be needed in the remaining part. In view of this, a glossary of terms is given.

Glossary of Metallurgical and Heat-Treating Terms

Aging: A change in the properties of an alloy that generally occurs slowly at room temperature or may be accelerated by an increase in temperature.

Annealing: A broad term used to describe the heating and cooling of steel usually applied to induce softening. The term annealing usually implies relatively slow cooling. Annealing is also used to remove stresses, to alter physical and mechanical properties, to remove gases, to change the crystalline structure, and to produce a desired microstructure.

GEAR MATERIALS

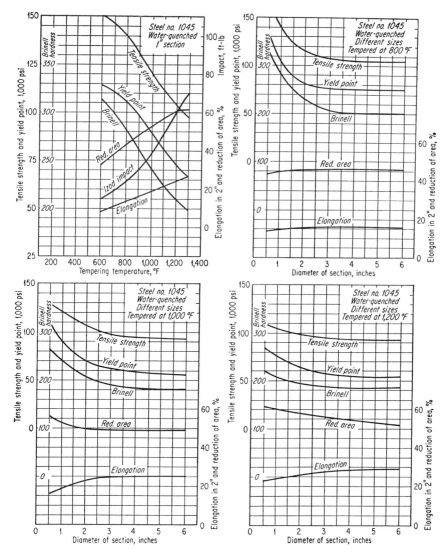

FIGURE 8.5 Properties of water-quenched and tempered carbon steel 1045 in different sizes. (In sizes over 1 in. the properties represent those at the section midway between the surface and the axis on samples cut longitudinally.) In sections ½ to 2 in., inclusive, quenched from 1475 to 1525°F; over 2 to 4 in., inclusive, from 1550 to 1550°F; over 4 in., from 1525 to 1575°F. *(Courtesy of International Nickel Co., New York)*

Austempering: A process for the transformation of austenite by quenching in a special bath, as a molten salt or molten lead bath, to allow transformation at a constant temperature to produce a structure other than martensite.

Austenite: In steels, the gamma form of iron with carbon in solid solution. Austenite is the structure from which all quenching heat treatments must start.

Austenitic steels: Steels that are austenitic at room temperature, typically the 300 series stainless alloys.

Blue brittleness: Brittleness occurring in steel when in the temperature range of 200 to 370°C (400 to 700°F), or when cold after being worked within this temperature range.

Bright annealing: A process of annealing usually carried out in a controlled atmosphere to protect the surface of the material from oxidation or discoloration.

Brinell hardness: A hardness number determined by applying a known load to the surface of the material to be tested through a hardened steel ball of known diameter. The diameter of the resulting permanent indentation is measured. This method is unsuitable for measuring the hardness of sheet or strip metal.

Carburizing: Diffusing carbon into the surface of iron-base alloys by heating such alloys in the presence of carbonaceous materials at high temperature. Such treatments, followed by appropriate quenching and tempering, harden the surface of the metal to a depth proportional to the time of carburizing.

Case depth: That portion of a carburized or nitrided, iron-base-alloy part in which the carbon or nitrogen content has been noticeably increased.

Case hardening: A process of surface hardening involving a change in the composition of the outer layer of an iron-base alloy by inward diffusion by carbon, etc., from a gas or liquid followed by appropriate thermal treatment. Included are carburizing, nitriding, carbonitriding, and cyaniding.

Cementite: A microconstituent of steel consisting of Fe3C (iron carbide, containing 6.69 percent C). It is hard, brittle, and weakly magnetic.

Cold working: Permanent deformation of a metal below its recrystallization temperature. Some metals can be hot worked at room temperature, while others can be cold worked at temperatures in excess of 540° (1000°F).

Controlled cooling: A process by which a steel object is cooled from an elevated temperature in a predetermined manner to avoid hardening, cracking, or internal damage, or to produce a desired metallurgical structure.

Creep: The slow deformation of steel under stress at elevated temperatures.

Critical range: The range of temperature required for completion of phase transformations in alloys.

Crystal: A physically homogeneous solid in which the atoms are arranged in a three-dimensional repetitive pattern.

Decarburization: The loss of carbon from the surface of solid steel during heating.

Deoxidizing: Removal of oxygen from molten metal or reduction of scale on metal.

Ductility: The property of a metal that allows it to be permanently deformed before final rupture. It is commonly evaluated by tensile testing in which the amount of elongation or reduction of area of the broken test specimen, as compared with the original, is measured.

Elastic limit: Maximum stress to which a metal can be subjected without permanent deformation.

Endurance limit: The maximum stress to which material may be subjected an infinite number of times without failure.

Ferrite: Iron in the alpha form in which alloying constituents may be dissolved. Ferrite is magnetic, soft, and acts as solvent for manganese, nickel, and silicon.

Flakes: Internal fissures in steel with a bright scaly appearance. Flakes can be minimized in forgings and rolled products by thermal treatment.

Flame hardening: A process consisting of heating a desired area, usually localized, with an oxyacetylene torch or other type of high-temperature flame and then quenching to produce a desired hardness.

Full annealing: A softening process in which metal is heated to a temperature above the transformation range and, after being held for a proper time, is cooled slowly to a temperature below the transformation range. The rates of cooling are very slow.

Grain size: The grain size number is determined by a count of a definite microscopic area, usually at 100 magnifications. The larger the grain size number, the smaller the grains.

Grains: Individual crystals in metals.

Hardenability: The property that determines the depth and distribution of hardness induced by quenching. The higher the hardenability value, the greater is the depth to which the material can be hardened and the slower is the quench that can be used.

Heat treatment: One or more operations involving the heating and cooling of a metal in the solid state for the purpose of obtaining certain desired conditions or properties. Heating and cooling for the prime purpose of mechanical working are not included in the meaning of the definition.

Impact test: A test in which one or more blows are suddenly applied to a specimen. The results are usually expressed in terms of energy absorbed or number of blows of a given intensity required to break the specimen.

Inclusions: Particles of nonmetallic materials, usually silicates, oxides, or sulfides, which are mechanically entrapped or are formed during solidification or by subsequent reaction within the solid metal. Impurities in metals.

Isothermal annealing: A method of annealing that consists of heating to the proper temperature above the critical range for austenitizing, followed by rapid cooling to a suitable temperature and holding for sufficient time at that temperature to complete the transformation.

Isothermal transformation: A process of transforming austenite in any given steel to ferrite or a ferrite-carbide aggregate at any constant temperature within the transformation range.

Jominy test: The Jominy test is used for determining end-quench hardenability of steel. It consists of water quenching one end of a 1-in. diameter bar under closely controlled conditions and measuring the degree of hardness at regular intervals along the side of the bar from the quenched end up.

Killed steel: Steel deoxidized with a strong deoxidizer like silicon or aluminum to reduce the oxygen content of the steel to a minimum to prevent any gaseous reaction between carbon and oxygen during solidification.

Martempering or marquenching: A hardening process in which the steel part is quenched from the austenitizing temperature in a salt or molten bath held at a temperature above or slightly below the temperature at which martensite begins to form. The part is held at this temperature long enough to equalize in temperature throughout, but not long enough to permit any transformation to occur. The steel is then cooled in air through the martensite range to accomplish full transformation.

Martensite: A microconstituent in quenched steel characterized by an acicular, or needle-type, structure. It has the maximum hardnesses of any structure obtained from the decomposition of austenite.

Modulus of elasticity: The ratio, within the elastic limit, of the stress to the corresponding strain. The stress in pounds per square inch is divided by the elongation in inches per inch of the original gauge length of the specimen. Also known as Young's modulus.

Nitriding: Adding nitrogen to solid iron-base alloys by heating the steel in contact with ammonia gas or other suitable nitrogenous material. Nitriding develops a very hard case by exposure to the atmosphere after a long time at comparatively low temperature, without quenching.

Normalizing: A process in which a steel is heated to a temperature above the transformation range and is subsequently cooled in still air to room temperature. The purpose is to refine grain structure and help reduce segregation in castings or forgings.

Pearlite: The lamellar aggregate of ferrite and cementite resulting from the direct transformation of austenite at the lower critical point.

Proof stress: The greatest static stress at which a material exhibits an arbitrarily permanent deformation. No standard has been adopted and published values may not be comparable.

Proportional limit: The greatest load per square inch of original cross section for which the elongation is proportional to the load applied.

Residual stress: Stresses remaining in a part after the completion of working, heat treating, welding, etc., due to phase changes, expansion, contraction, and other phenomena.

Rockwell hardness: A hardness number as determined by a Rockwell hardness tester, a direct-reading machine which may use a steel ball or a diamond penetrator.

Secondary hardening: Increase in hardness developed by tempering high-alloy steels after quenching, usually associated with precipitation reactions.

Stress relief: A process of reducing internal residual stresses in a metal part by heating to a suitable temperature and holding the part at that temperature for a proper time. Stress relief may be applied to parts that have been welded, machined, cast, heat treated, or worked.

Temper brittleness: Brittleness resulting from the slow cooling of certain steels through certain temperature ranges (below transformation). Notched-bar impact tests are used to reveal brittleness.

Tempering: A process of reheating quench-hardened or normalized steels to a temperature below the transformation range and then cooling at any rate desired.

Tensile strength: The maximum load per unit of original cross-sectional area carried by a material during a tensile test.

Yield point: The load per unit of original cross section at which a marked increase in deformation occurs without any increase in load.

8.2.3 Carbon and Alloy Steels

The plain carbon steels have only small amounts of alloy. In the AISI series these steels have 10 for the first two digits, for example, AISI 1020 or AISI 1045. The last two digits signify the carbon content. AISI 1020 has 0.20 percent carbon.

Table 8.4 shows the compositions of the more popular gear steels. Those having numbers starting with 2 or higher have relatively high alloy content and are considered alloy steels.

GEAR MATERIALS

TABLE 8.4 Composition of Typical Gear Steels

AISI No.	Chemical-composition limits, %				
	C	Mn	Ni	Cr	Mo
1015	0.13/0.18	0.30/0.60			
1025	0.22/0.28	0.30/0.60			
1045	0.43/0.50	0.60/0.90			
1060	0.55/0.65	0.60/0.90			
1118	0.14/0.20	1.30/1.60	(0.08/0.13 sulfur)		
1137	0.32/0.39	1.35/1.65	(0.08/0.13 sulfur)		
1320	0.18/0.23	1.60/1.90			
1335	0.33/0.38	1.60/1.90			
2317	0.15/0.20	0.40/0.60	3.25/3.75		
2340	0.38/0.43	0.70/0.90	3.25/3.75		
3140	0.38/0.43	0.70/0.90	1.10/1.40	0.55/0.75	
3250	0.48/0.53	0.40/0.60	1.65/2.00	0.90/1.20	0.20/0.30
4047	0.45/0.50	0.70/0.90	....	....	0.20/0.30
4023	0.20/0.25	0.70/0.90	....	....	0.20/0.30
4130	0.28/0.33	0.40/0.60	....	0.80/1.10	0.15/0.25
4140	0.38/0.43	0.75/1.00	....	0.80/1.10	0.15/0.25
4320	0.17/0.22	0.45/0.65	1.65/2.00	0.40/0.60	0.20/0.30
4340	0.38/0.43	0.60/0.80	1.65/2.00	0.70/0.90	0.20/0.30
4620	0.17/0.22	0.45/0.65	1.65/2.00	....	0.20/0.30
4640	0.38/0.43	0.60/0.80	1.65/2.00	....	0.20/0.30
4820	0.18/0.23	0.50/0.70	3.25/3.75	....	0.20/0.30
5132	0.30/0.35	0.60/0.80	....	0.75/1.00	
5145	0.43/0.48	0.70/0.90	....	0.70/0.90	
E52100	0.95/1.10	0.25/0.45	....	1.30/1.60	
6120	0.17/0.22	0.70/0.90	....	0.70/0.90	(Va 0.10 min.)
6150	0.48/0.53	0.70/0.90	....	0.80/1.10	(Va 0.15 min.)
8620	0.18/0.23	0.70/0.90	0.40/0.70	0.40/0.60	0.15/0.25
8640	0.38/0.43	0.75/1.00	0.40/0.70	0.40/0.60	0.15/0.25
E9310	0.08/0.13	0.45/0.65	3.00/3.50	1.00/1.40	0.08/0.15
9840	0.38/0.43	0.70/0.90	0.85/1.15	0.70/0.90	0.20/0.30

NOTES: The prefix E means a grade of steel made by the basic electric-furnace process. The phosphorus and sulfur are limited to 0.040 maximum for all the steels except AISI 1118 and 1137—which are resulfurized for free machining—and E9310 and E52100—which are held to a maximum of 0.025 phosphorus and sulfur.

Many steel companies sell nonstandard steels under their own brand names or composition numbers. Frequently these nonstandard steels make excellent gears for a given job and can be obtained at lower prices than the standard AISI steels. Usually the nonstandard steel compares quite closely with the specification of some standard steel. The behavior of the nonstandard steel can usually be predicted quite well by comparison with the behavior of the standard steels having the nearest composition.

8.3 CARBON AND ALLOY STEELS

For most gear work it is usually advisable to work to recognized AISI steel specifications. The limits on composition are set as closely as is practical. The standard steel can be given a precise heat-treating cycle that has been established to give the desired

TABLE 8.5 Chemical Composition (H^3)—
Comparison of High Hot Hardness Gear Steels

Alloy	Nominal Composition (%)						
	C	Mn	Si	Cr	Ni	Mo	V	Other
Vasco X2m	0.15	0.30	0.90	5.00	—	1.40	0.45	1.35 W
Pyrowear 53*	0.10	0.35	1.00	1.00	2.00	3.25	0.10	2.00 Cu
CBS 600†	0.19	0.60	1.10	1.45	—	1.00	—	—
M50 NiL	0.13	0.28	0.18	4.21	3.44	4.30	1.19	0.05 Cu 0.01 Co

High hot hardness refers to material exhibiting and retaining their room temperature hardness and properties at high (>150°C) temperatures.
*Tradename of Carpenter Steel Co., Reading, Pennsylvania.
†Tradename of Timken Steel Co., Timken, Ohio.

results. The user can be confident that each mill heat of standard steel will behave almost exactly like the previous mill heat.

For critical applications, certain of the steels listed in Table 8.4 are produced by *single vacuum melting* (SVM) rather than air melt quality as is shown in the table. Vacuum melting of the steel results in phosphorous and sulphur contents held to a maximum of 0.015%. Both cleanliness and fatigue properties of the gear steel are significantly improved by using SVM material. In many aerospace applications, double vacuum melting of the steel is required. Double vacuum melting further enhances fatigue properties as well as fracture toughness properties of the gear. It is noted, however, that single or double vacuum melting increases the cost of the new material and for large parts may not be justified.

Table 8.5 lists the chemical composition limits of various new *high hot hardness* (HHH) steels. These steels are typically double vacuum melted and are used in aerospace applications. These gear steels are known for their ability to operate at high transmission operating temperatures, for example, up to 316°C (600°F). A hot hardness curve for each of these alloys is given in Fig. 8.6. For comparative purposes, 9310 steel hot hardness characteristics is also given.

It is noted that hot hardness properties not only relate to operation at higher temperatures but also result in excellent scuffing and scoring resistance, that is, surface contact fatigue properties. As an example of this, one of the alloys detailed in Table 8.5, Vasco X2M, has been utilized in production aerospace components for a number of years without experiencing a single spalling-pitting fatigue failure. Also tests conducted with M50NiL gears have shown that M50NiL has excellent resistance to pitting fatigue failures.[2]

Table 8.6 shows nominal heat-treating temperatures for the typical gear steels. The "reheat temperature" column applies only to those steels that are ordinarily used for case carburizing. The "Ms temperature" is significant when a marquench is used.

Table 8.7 shows in a general fashion which steels might be considered for different size and hardness of gear parts. It is assumed that the teeth are to be through-hardened to a relatively uniform* hardness. Table 8.7 is limited in showing only a sampling of steels that might be used for the gear sizes listed. Many other compositions not shown have been used with good results in certain applications.

---

*See Sec. 8.3.1 for a discussion on localized hardening.

GEAR MATERIALS   8.15

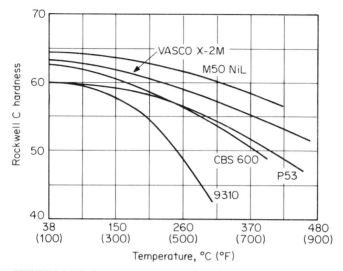

FIGURE 8.6 Hot hardness at temperature for various high-hot-hardness and AIAI 9310 steels.

How to Choose a Gear Steel. With so many gear steels available, designers are often quite uncertain as to which steel they should be using. Here are some general guides:

1. Use a steel with no more alloy content than is necessary for the size of part to harden properly.
2. If wear is a problem, a higher carbon content may be helpful. (AISI 1060 will outwear AISI 1025.)
3. If machinability is a problem, a lower carbon content may be helpful. (AISI 1025 will machine better than AISI 1060 at the same hardness.)
4. Consider the price of the raw steel and the price of processing it into gears and select the material that will give the lowest-priced gear that is adequate to do the job at hand.

How to Process the Steel for a Gear. Steel obtained from the mill may be in an as-forged condition. It may be "hot rolled" or it may be "cold rolled." The material should always be normalized or normalized and annealed. The following list gives an example of heat treatment operations that might be needed on a tough job.

1. Normalize the forging.
2. Rough-machine the part.
3. Heat the part to hardening temperature, quench, and draw to R/C 30 hardness range. Finish journals and cut teeth.
4. Stress relief temper at a temperature just slightly below that used for the draw in step 3. Refinish journals or centers (if necessary).

TABLE 8.6 Heat-Treating Data for Typical Gear Steels

AISI No.	Normalizing temp., °F	Annealing temp., °F	Hardening temp., °F	Carburizing temp., °F	Reheat temp., °F	M_s temp., °F
1015	1700	1600	....	1650–1700	1400–1450	
1025	1650–1750	1600	1575–1650	1500–1650		
1040	1650–1750	1450	1525–1575			
1045	1600–1700	1450	1450–1550			
1060	1550–1650	1400–1500	1450–1550	....	....	555
1118	1700	1450	....	1650–1700	1650–1700	
1320	1600–1650	1500–1700	....	1650–1700	1450–1500	740
1335	1600–1700	1500–1600	1500–1550	....	....	640
2317	1560–1750	1575	....	1650–1700	1450–1500	725
2340	1600–1700	1400–1500	1425–1475	....	....	555
3140	1600–1700	1450–1550	1500–1550	....	....	590
4047	1550–1750	1525–1575	1475–1550			
4130	1600–1700	1450–1550	1550–1650	....	....	685
4140	1600–1700	1450–1550	1525–1625	....	....	595
4320	1600–1800	1575	....	1650–1700	1425–1475	720
4340	1600–1700	1100–1225	1475–1525	....	....	545
4620	1700–1800	1575	....	1650–1700	1475–1525	555
4640	1600–1700	1450–1550	1450–1550	....	....	605
4820	1650–1750	1575	....	1650–1700	1450–1500	685
5145	1600–1700	1450–1550	1475–1525			
E52100	....	1350–1450	1425–1600	....	....	485
6120	1700–1800	1600	....	1700	1475–1550	760
6150	1650–1750	1550–1650	1550–1650	....	....	545
8620	1600–1800	1575	....	1700	1425–1550	745
E9310	1650–1750	1575	....	1650–1700	1425–1550	650
9840	1575–1725	1475–1700	1475–1525	....	....	575

5. Harden (including surface harden, where required) and draw the part to hardness requirements.
6. Shave, lap, or grind gear teeth as necessary.

Sometimes the stress relief temper may be used after rough-cutting the teeth and before finish-cutting. Quite often the pitch of the teeth is small enough not to require this operation. The normalize treatment's main value lies in cutting down the distortion after a quench-and-draw treatment. Gears that are to be carburized after cutting the teeth are critical from the standpoint of distortion after quenching since the warpage may be enough to cause most of the case to be ground away in certain spots.

When gears are made by welding, it is common practice to weld a quenched and tempered ring to low-hardness web plates and shaft. After welding, the complete weldment is given a stress relief temper. If the weldment is quite large and complicated, a second stress relief treatment may be needed after the weldment is machined and the teeth are rough-cut. The stress relief treatment takes out localized machining and welding stresses and help to prevent cracking of the component.

GEAR MATERIALS

TABLE 8.7 General Recommendations on Choice of Steel for Different Gear Sizes

Pitch of Teeth	Wall Thickness,* cm (in.)	Hardness†	Gear Steels
		Through-Hardening	
10-30	1.3 (½)	200 BHN	1045, 1137, 1335, 4047
		300 BHN	1045, 1060, 3140, 4047
5-15	2.5 (1)	200 BHN	1045, 1060, 3140, 4047
		300 BHN	2340, 3140, 3250, 4140, 4340, 4640
2½-8	5.1 (2)	200 BHN	1060, 2340, 3250, 4340, 5145, E52100
		300 BHN	2340, 3250, 4340, 4640, 8640, 9840
1¼-4	10.3 (4)	200 BHN	2340, 3250, 4140, 4340, 4640, 9840
		300 BHN	3250, 4340
		Carburizing	
10-30	1.3 (½)	58 Rockwell C	1015, 1025, 1118, 1320, 4023, E9310
5-15	2.5 (1)	58 Rockwell C	2137, 4620, 6120, 8260, E9310
2½-8	5.1 (2)	58 Rockwell C	4620, E9310

*"Wall thickness" is based on thickest section of rim, web, or shaft that must develop the minimum hardness throughout.
†Hardnesses are minimum; 300 BHN would normally be a range of 300 to 350 BHN.

8.3.1 Localized Hardening of Steel Gears

The surface durability of gear teeth is roughly proportional to the square of the surface hardness. This means that a gear tooth with 600 BHN hardness on the surface may be able to carry as much as 9 times the power of a gear with only a surface hardness of 200 BHN! Generally speaking, the limit of machinability is at about 350 BHN. In addition to this bending strength of through-hardened gear teeth is a maximum in the 350- to 400-BHN range and slightly decreases at higher hardnesses. This situation poses the problem of how to take advantage of the high surface durability of gears in hardness ranges above 350 BHN and at the same time have good bending strength and a means of making the gears with accurate teeth. The answer to this problem lies in localized hardening of gear teeth.

Table 8.8 shows five commonly used methods for the localized hardening of gear teeth. In each case the surface of the tooth is made very hard and the center of the tooth, the gear rim, and the gear web are at a medium or low hardness. Generally speaking, teeth that are given a localized hardening treatment are brought up to a full hardness in the neighborhood of 650 BHN (or 60 Rockwell C).

The details of the different methods of localized hardening are given in Secs. 8.5 to 8.8

TABLE 8.8 Methods for Localized Hardening of Gear Teeth

Method	Description of Method
Carburizing	The teeth are usually finish-cut, carburized, and ground. Sometimes gears less than 12 in. in diameter are finished before carburizing and distortion is low enough to permit using the gears without grinding.
Nitriding	The teeth are usually finished before nitriding. If properly done, nitrided gears do not distort much. Grinding of the teeth is not necessary.
Induction hardening	The teeth are usually finished before hardening. If properly done and the pitch is not too coarse, distortion is very low. These gears can be ground.
Flame hardening	The teeth may or may not be ground.
Carbonitriding	The teeth are usually finished before carbonitriding.

Some important considerations that affect the performance of surface-hardened teeth, and should be kept in mind from a processing standpoint are listed below:

Lack of case depth in root fillet

High residual tensile stress from an improper quenching or grinding technique

Lack of hardness in root fillet

Grinding burns in root fillet

Decarburization of root fillet

Notches, tears, or scratches in root fillet

Intergranular oxidation in root fillet

8.3.2 Carburizing Gears

Carburizing is one of the most widely used methods for hardening gear teeth. The gear to be carburized is hobbed, shaped, or cut by some other method. A low-carbon steel having a carbon content in the range of 0.10 to 0.25 percent is used for the gear blank. The cut gear is placed in a carburizing medium and heated above its critical range. Carbon is absorbed into the surface layer of the gear tooth, and after one or more hours at temperature, the carbon has penetrated to give the required case depth. Rim areas, web areas, and hub areas that are not to be carburized are masked off during the carburizing cycle. (In some cases an area may be carburized and then the carbon case removed by machining or grinding after carburizing.)

The carburized gear may be quenched immediately after removal from the carburizing chamber; or it may be slow-cooled so as to not develop its hardness and/or undesirable residual stresses, and then, after some further processing, reheated and quenched.

The carburizing medium may be a gas atmosphere or a liquid like a salt containing free carbon. In very simple carburizing setups the gear is placed with charcoal* and heated in a chamber to carburizing temperature.

---

*Certain chemicals are added to the charcoal.

Carburizing has also been accomplished in vacuum, by ion deposition and by plasma. These methods all produce free carbon to the surface of the gear and result in very uniform carburized layers on the surface. In certain cases, they have been shown to reduce the carburizing time by as much as half when compared with conventional gas techniques. Better carburized case uniformity has also been demonstrated using these new processes.

1. Vacuum carburizing has been shown to give precision to the dynamics of carbon absorption and diffusion such that controllable rates for carburizing times and diffusion have been developed. In addition, use of vacuum eliminates nonequilibrium gas states, temperature differentials, and surface abnormalities.[3]
2. Ion carburizing has been shown to virtually eliminate detrimental grain boundary oxides[4] in the surface of the material. Improved distortion control has also been documented. However, at present, more processing skill is required to achieve good uniformity and case deth control than with vacuum carburizing since, for example, loading the parts into the furnace must be accomplished in a manner that does not result in uneven concentrations of energy.[5]
3. Plasma carburizing[6] has been used successfully to carburize blind holes with length-to-depth ratios of 12:1 and to eliminate sooting problems.

The concentration of carbon in the case may go as high as 1.40 percent. For the best hardness and fatigue strength, the carbon content should be in the 0.80- to 0.95-percent range. The control of carbon content is achieved by regulating the richness of the carburizing medium. Special equipment has been developed by builders of heat-treating furnaces to control carbon potential, for example, dew point analyzers, oxygen sensors, and infared analyzers. In some cases shim analysis is used.

Figure 8.7 shows an example of carburizing time to obtain different case depths. The curves are based on data obtained from gas carburizing tests.

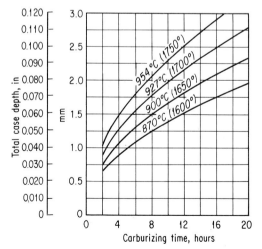

FIGURE 8.7 Nominal carburzing time and temperatures for different total case depths, gas carburized. *(Courtesy of INCO, New York)*

Carburizing rates vary rather appreciably. It is common practice for the heat treater to put several test coupons in with the gears to be carburized. Coupons are pulled from the furnace at regular intervals and checked immediately for case depth. The carburizing is stopped as soon as the coupons indicate the case depth is within the required limits.

Figure 8.8 shows several heat treatment techniques that may be used on carburized gears. Some treatments favor the case and some the core. The choice of treatment depends on grain size and carbon content as well as the end use of the gears. Tables 8.4 and 8.5 list the compositions of carburizing steels. Table 8.6 shows normal carburizing temperatures for different steels.

There is fairly general agreement among gear experts that a case-hardened gear should have a core hardness of about 35 Rockwell C. If the core is very hard, such as 50 Rockwell C, the teeth tend to be too brittle, that is, the bending fatigue strength of the gear is severely reduced! If the core is of low hardness, for example, 20 Rockwell

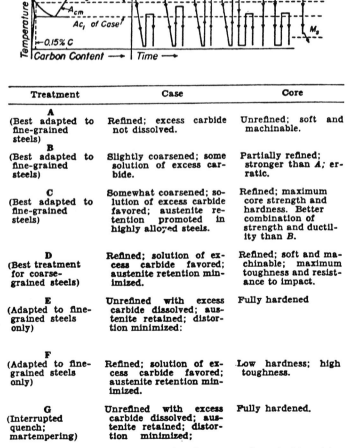

Treatment	Case	Core
A (Best adapted to fine-grained steels)	Refined; excess carbide not dissolved.	Unrefined; soft and machinable.
B (Best adapted to fine-grained steels)	Slightly coarsened; some solution of excess carbide.	Partially refined; stronger than A; erratic.
C (Best adapted to fine-grained steels)	Somewhat coarsened; solution of excess carbide favored; austenite retention promoted in highly alloyed steels.	Refined; maximum core strength and hardness. Better combination of strength and ductility than B.
D (Best treatment for coarse-grained steels)	Refined; solution of excess carbide favored; austenite retention minimized.	Refined; soft and machinable; maximum toughness and resistance to impact.
E (Adapted to fine-grained steels only)	Unrefined with excess carbide dissolved; austenite retained; distortion minimized:	Fully hardened
F (Adapted to fine-grained steels only)	Refined; solution of excess carbide favored; austenite retention minimized.	Low hardness; high toughness.
G (Interrupted quench; martempering)	Unrefined with excess carbide dissolved; austenite retained; distortion minimized;	Fully hardened.

FIGURE 8.8 Diagrammatic representation of treatments after carburizing, giving resultant case/core microstructure constituents.

GEAR MATERIALS 8.21

C, the teeth lack bending strength and the case may tear loose from the core. For most high-capacity gears, the core hardness can be held to a range of 30 to 40 Rockwell C with good results. A close control on core hardness is difficult because the core is heat treated purely as a secondary operation to the heat treatment for the case. Also, the core is very sensitive to small differences in carbon content. (Figure 8.1 showed the sensitive relation between maximum hardness and carbon content.) In actuality, the heat treatment of the case-hardened material should be developed so that case and core properties are optimized.

An optimum case hardness is 60 Rockwell C. For different gear applications, it is advisable to consider several specifications on case hardness. Table 8.9 shows a recommended practice. This agrees quite well with the practices of a majority of gear manufacturers.

The case depth required for a carburized tooth is primarily a function of pitch. The larger the tooth, the more case is needed to carry the loads that will be imposed on the tooth. For each size of tooth there is an optimum case depth. Too much case makes the tooth brittle with a tendency for the top of the tooth to shatter off. This is called *case-core separation*. Too thin a case reduces tooth strength and resistance to pitting; case crushing can result. Table 8.10 shows the general practice on case depth for different diametral pitches.

For extremely critical gears, a smaller range of case depth may be used than that shown in Table 8.10. The change is made by raising the minimum case depth. For instance, a critical 10-pitch job might be held to 0.64 to 0.89 mm (0.025 to 0.035 in.) case depth. The case depth specified is the *effective case*, that is, the portion of the carburized case where the hardness exceeds 50 Rockwell C. The effective case is typically about 75–90 percent of the total case.

TABLE 8.9 Recommended Case Hardness

Application	Drawing Specification
General-purpose industrial gearing	55 Rockwell C min.
High-capacity industrial gearing	58 Rockwell C min.
Aircraft gearing	59–64 Rockwell C
Max.-capacity aircraft or rocket gearing	60–63 Rockwell C

TABLE 8.10 Recommended Case Depths

Normal Module	(diametral pitch)	Case-depth Specification, mm	(in.)
1.3	(20)	0.25–0.46	(0.010–0.018)
1.6	(16)	0.30–0.58	(0.012–0.023)
2.5	(10)	0.51–1.0	(0.020–0.035)
3.2	(8)	0.64–1.0	(0.025–0.040)
4.2	(6)	0.77–1.3	(0.030–0.050)
6.4	(4)	1.0–1.5	(0.040–0.060)
12.7	(2)	1.8–2.6	(0.070–0.100)

8.3.3 Nitriding Gears

Nitriding[7] is a process for case hardening alloy-steel gears, whereby very little distortion of the finished gear occurs. The gear to be nitrided is given a quench-and-draw treatment first to establish a medium hardness such as R/C 30–40. The gear blank is finish-machined to size and the teeth are cut and finished. The parts of the gear that are not to be nitrided are masked off with copper plate or some other suitable medium. The gear is put in a nitriding furnace and heated to about 540°C. It is important to note that the nitriding temperature must always be below the draw temperature of the alloy so that reduction in core hardness (strength) does not result from the nitriding operation. The core hardness also has an effect on the surface hardness, in that the higher the core hardness, the higher will be the surface hardness. Nitriding is accomplished using ammonia gas, which is broken down into atomic nitrogen and hydrogen at the surface of the steel. Various techniques for nitriding exist, the most popular being gas nitriding. Ion nitriding utilizing the glow discharge process, has been shown to significantly reduce nitriding time, provide efficient use of gas and electrical energy, and virtually eliminate pollution (of the processing gas). However, this form of nitriding is more expensive and requires precision fixturing.

The atomic nitrogen slowly penetrates the steel surface and combines with elements like aluminum, chromium, molybdenum, tungsten, and vanadium to form very hard nitrides. The useful case on a gear is formed by the nitrogen and alloying elements present in the gear.

The nitriding is carried out at a temperature that is well below the critical temperature of the steel. This means that the gear blank is not red hot and no phase changes are being made in the body of the gear—provided that, prior to nitriding, the gear blank was drawn at a temperature at least as hot as the nitriding temperature. A gear properly nitrided will exhibit little or no distortion. This makes it possible to finish nitrided gears prior to nitriding. Generally the only work done after nitriding is to strip the copper plate and polish journals or bores to exact size.

Nitrided parts have a surface layer of super-rich nitrides. This layer is called a *white layer* because it etches out white in a micrograph. This surface layer is very brittle and, if not subsequently removed, can have a negative effect on gear tooth bending fatigue life. With a special two-cycle nitriding process, often called *diffusion cycle,* it is possible to hold the white layer to 0.0005 in. or less. If just a simple nitriding cycle is used with a rich ammonia gas, the white layer may build up to 0.002 or 0.003 in. Quite often gears with white layer are lapped, ground, blast cleaned, or chemically etched to remove the white layer. Table 8.11 shows the compositions of some typical nitriding steels; Table 8.11 also shows typical nitriding temperatures and hardnesses for these steels.

Nitride cases are formed at a much slower rate than carburized cases. Figure 8.9 shows the nominal time required for different nitride case depths.

Generally speaking, nitrided gears do not require as much case depth as case-hardened gears. Nitrided gears with rather thin cases have performed excellently in many critical aircraft applications. It appears that the lack of case depth can be compensated for by the extra-high hardness of the outer case and a harder core underneath the case. Table 8.12 shows some nominal case depth specifications for nitrided parts.

8.3.4 Induction Hardening Gears

Gear teeth may be surface hardened by high-frequency alternating currents. The process involves wrapping an induction coil around the piece. Generally the part is rotated

GEAR MATERIALS 8.23

TABLE 8.11 Nominal Compositions and Temperatures

Steel	Nominal Compositions of Nitriding Steels						
	Carbon	Manganese	Silicon	Chromium	Aluminum	Molybdenum	Nickel
Nitralloy®* 135	0.35	0.50	0.30	1.20	1.15	0.20	
Nitralloy 135 modified	0.41	0.55	0.30	1.60	1.0	0.35	
Nitralloy N	0.23	0.55	0.30	1.25	1.15	0.25	3.5
Nitralloy N 5-2	0.23	0.35	0.25	0.5	2.0	0.25	5(Va 0.10)
AISI 4340	0.40	0.70	0.30	0.80	—	0.25	1
AISI 4140	0.40	0.90	0.30	0.95	—	0.20	

	Nominal Temperatures Used in Nitriding and Hardness Obtained			
Steel	Temp. before Nitriding, °C (°F)	Nitriding, °C (°F)	Hardness, Rockwell C	
			Case	Core
Nitralloy 135	675 (1250)	525 (975)	65–70	30–34
Nitralloy 135 modified	675 (1250)	525 (975)	65–70	32–36
Nitralloy N	650 (1200)	525 (975)	65–70	40–44
Nitralloy N 5-2	690 (1275)	525 (975)	65–70	45–50
AISI 4340	590 (1100)	525 (975)	48–53	27–33
AISI 4140	590 (1100)	525 (975)	50–55	23–33

*Nitralloy is a trade name of the Nitralloy Corp., NY.

inside the coil. In a few seconds the teeth are brought up to a red heat (above the critical temperature). The gear is withdrawn from the coil and given a controlled spray quench by a special quench ring or agitated bath.

Before induction hardening, the gear blank is heat treated to a medium hardness such as 33 to 38 Rockwell C. The teeth are finish-cut and finish-shaved or lapped for final accuracy. If the induction-hardening cycle has been properly developed for the

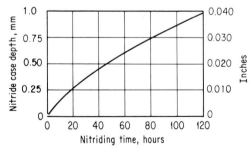

FIGURE 8.9 Nominal time required for different nitride case depths.

TABLE 8.12 Recommended Nitride Case Depths

Normal Module	(Diametral Pitch)	Case Depth Specification,	
		mm	(in.)
1.3	(20)	0.13–0.16	(0.005–0.010)
1.6	(16)	0.20–0.33	(0.008–0.013)
2.5	(10)	0.31–0.46	(0.012–0.018)
3.2	(8)	0.36–0.51	(0.014–0.020)
4.2	(6)	0.41–0.56	(0.016–0.022)
6.4	(4)	0.51–0.72	(0.020–0.028)

part, almost no distortion occurs. Since only the surface layers of the teeth are heated, the bulk of the gear blank serves as a cold, strong fixture to maintain the accuracy of the portion of the part that is heated to red heat. The depth to which the heated zone extends depends on the frequency of the current and on the duration of the heating cycle.

After induction hardening the gear is given a low-temperature draw (in the 150 to 200°C range). The teeth may be brushed lightly to remove the slight amount of scale that develops. Since a slight dimensional change may occur in bores or journals, it may be desirable to finish the bores or journals to size after induction hardening.

If there is a very high intensity of electric power per square inch of piece, it is possible to get a fairly uniform case somewhat analogous to that obtained by carburizing or nitriding. Normally, though, an induction-hardened piece is hardened almost all the way through the tooth at the pitch line. At the bottom of the tooth, there may be an unhardened area with the coarser-pitch teeth.

Figure 8.10 shows a comparison of typical hardness patterns obtained by different surface-hardening treatments. Note views *b* and *c* showing typical induction-hardening patterns.

In an induction-hardened tooth that requires high bending strength, it is necessary to get a reasonable depth of hardened stock in the center of the root fillet. Table 8.13 shows how much is needed.

Even though the right amount of hardened material is obtained in the root region, it is difficult to obtain high bending strength with an induction-hardened tooth. As the

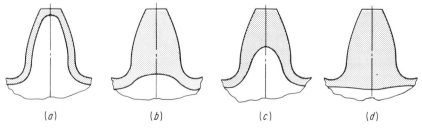

FIGURE 8.10 Typical hardness patterns obtained through various surface-hardening techniques: *(a)* carburized or nitrided; *(b)* induction hardened or shallow hardened; *(c)* induction hardened or shallow hardened; *(d)* flame hardened.

TABLE 8.13 Recommended Depth* of Hardness in Root Region for Induction-Hardened Teeth

Normal Module	(Diametral Pitch)	Recommended Hardness Depth, mm	(in.)
1.3	(20)	0.26–1.0	(0.010–0.040)
1.6	(16)	0.38–1.5	(0.015–0.060)
2.5	(10)	0.51–2.6	(0.020–0.100)
3.2	(8)	0.77–3.2	(0.030–0.125)
2.2	(6)	1.2–3.9	(0.045–0.150)
6.4	(4)	1.5–4.5	(0.060–0.175)

*Depth reading taken in center of root fillet. Depth should be greater at sides of root fillet.

outer parts of the gear tooth are induction heated they expand rapidly and at high temperature can plastically deform. Upon cooling, the layers contract, and cracking can occur due to the presence of undesirable residual tensile stresses. One undesirable effect of increasing depth of hardening is an increase in the magnitude of subsurface tensile stresses. Tempering reduces the stresses near this surface.

Induction hardening is a good process for low-cost, uniform-quality, high-production gears when a sound development program can be used before the gear is put into production. It is a risky process for job-shop work where a few parts are made without benefit of several development samples and fatigue tests on some of these samples.

The frequency of electric power has a relation to the pitch of the teeth. Table 8.14 shows the general relation.

A general processing procedure involves a preheat cycle, a final heat cycle, and a temper. The actual process can require either seconds or minutes. Preheating insures a reasonable heated depth at the roots of the gear, enabling the attainment of the desired metallurgical results and decreasing distortion in some materials.

The final heat position[8] may be the same coil used to preheat the part if the same frequencies are used for preheat and final. Once positioned correctly, the part is heated (while rotating) to hardening temperatures, which may be as high as 900°C (1650°F). Final heat times can range from less than a second to several seconds. Parts are rotated

TABLE 8.14 Frequencies Generally Recommended for Induction Hardening

Normal Module	(Diametral Pitch)	Frequency, kHz
1.3	(20)	500–1,000
2.5	(10)	300–500
3.2	(8)	300–500
2.2	(6)	10–500
6.4	(4)	6–10
12.7	(2)	6–10

during heating to ensure an even distribution of energy across them. Rotation rates are chosen to suit process requirements. After the heat is off, quenchant is applied to the hot part in place; that is, no repositioning is required. This practically instantaneous quench provides a consistent metallurgical response. During quenching there is minimal or no rotation to ensure that the quench fluid penetrates all areas of the gear evenly.

Induction tempering reduces locked-in stress and lowers the Rockwell hardness a few points to achieve metallurgical specification. Tempering is accomplished by heating to between 204 to 316°C (400 to 600°F). A medium frequency is used, with heat times depending on the size and shape of the part.

The hardened part is finished, machined, inspected, and moved to assembly.

Medium-carbon steels are used for induction hardening. Some of the popular steels are AISI 1040, 1050, 4340, and 4350.

8.3.5 Flame Hardening Gears

Gears of larger sizes are typically flame hardened by oxyacetylene flames using special burner heads or high-velocity combustion product gases. To achieve uniform hardening the gear is usually rotated in the flame. Like induction hardening, the gear blank is made medium hard and the teeth are cut and finished prior to hardening. With a good flame-hardening setup the distortion is low enough to permit the gear to be used without grinding after hardening.

In a flame-hardened gear, the heat travels inward by conduction. By comparison, the induction-hardening process develops heat by electronic action inside the surface of the tooth. The hardness pattern cannot be controlled in flame hardening the same as it is controlled in induction heating. See Fig. 8.10 for a typical pattern. Case depths of 1.5–6.5 mm. (0.050–0.250 in.) are possible. An advantage of this process is that it can follow any contour and, since it leaves the core untouched, distortion is minimized.

Medium-carbon steels are used for flame hardening. The depth requirements for flame hardening in the root fillet are similar to those for induction hardening. To obtain good bending fatigue strength it is necessary to develop and establish burner design and intensity, heating time, and quenching procedure on each part.

In large sizes, gear teeth may be shallow hardened by using a steel with limited hardenability and a special heat-quench-and-draw cycle. For instance, an AISI steel of 1040 composition will not harden all the way through a three-pitch tooth. By using a controlled spray quench it is possible to flame harden coarse-pitch teeth and get a hardness pattern similar to that obtained in induction hardening (see Fig. 8.10, view c).

Large gear teeth that have been properly flame hardened have good wear resistance and fatigue strength. It is a low-cost production method, but it is not recommended for use in job-lot production of a few gears. For very critical applications, flame-hardened gears require grinding. Generally speaking, their distortion on hardening is less than that of a carburized gear but more than that of a nitrided or induction-hardened gear.

Cyaniding. Still another method of hardening gears is called *cyaniding*. One way of using this process is to harden gears, whose teeth have already been finished, by putting them in a bath of molten sodium cyanide. A shallow case is formed, which contains both nitrides and carbides [9].

A variety of treatments may be used to produce a combined carburized and nitrided type of case. More properly these treatments are called *carbonitriding* treatments.

The carbonitriding treatments have the advantages that distortion is low and cycle time is much shorter than in nitriding. The case produced is not so hard as a nitride case. The depth is shallow, being in the order of 0.25 to 0.65 mm (0.010 to 0.025 in.) If this case is backed up with a very hard core and the pitch is not too coarse (6 pitch or finer), the teeth tend to have good wear resistance and good fatigue strength.

Gears to be cyanided are made with more carbon in the core than case carburizing gears. They do not require aluminum or chromium like a nitriding gear. AISI 4640 is typical of one of the popular steels for cyanided parts. AISI 5132 is popular in automotive work.

The case on cyanided gears runs on the order of 57 to 63 Rockwell C. Generally, it is possible to get good cyanided gears without extensive development work. This process has been used widely in both job-lot and production-lot applications.

8.3.6 Gray-Cast-Iron Gears

Gray cast iron has long been used as a gear material. Cast iron is low in cost and it can be cast easily into any desired shape for the rim, web, and hub of a gear. Cast iron machines easily.

Cast iron gears generally show good resistance to wear and are often less sensitive to lubrication inadequacies than are steel gears. Cast iron has good dampening qualities.

Cast-iron gear teeth have about three-quarters of the surface load-carrying capacity of steel gears of the same pitch diameter and face width. Their bending strength capacity is about one-third that of steel gears of the same normal diametral pitch. By making cast-iron gears somewhat coarser in pitch for the same pitch diameter, it is possible to design cast-iron gear sets about the same size as steel gear sets for a given application.

Gray cast iron has low impact strength and should not be used where severe shock loads occur.

Gray cast iron has free carbon in the form of flakes. The matrix of the material is a combined carbon and iron that is very similar to steel. Gray cast iron may be heat treated like steel. Even the cooling of the casting in the mold can be considered a heat treatment, and the gears are used with only the "as-cast" heat treatment.

AGMA standard 242.02 lists six classes of gray irons according to tensile strength. The amount of alloy in the composition is not specified, but it is assumed that enough alloy will be used to meet the requirements specified for standard test bars and also to meet the minimum hardness specified for the teeth. Table 8.15 shows the standard AGMA classes for gray cast iron. Those planning to use gray cast iron should study AGMA 242.02 for full detail requirements of the standard.

TABLE 8.15 AGMA Standard Classes of Gray Cast Iron

Class No.	Minimum Tensile Strength, MPa (klb/in.^{2})	Minimum Brinell Hardness on Toothed Portion
20	140 (20)	155
30	205 (30)	180
35	240 (35)	205
40	275 (40)	220
50	345 (50)	250
60	415 (60)	285

Meehanite. This is a special kind of gray iron made in licensed[10] foundries. The proprietary process controls the graphite flakes so that they are smaller and more uniformly dispersed. The process tends to produce a matrix in the pearlitic condition. Several grades of Meehanite are available, which cover a range of tensile strengths similar to that of regular gray cast iron. Although low in impact strength, Meehanite can be treated to have rather good impact strength for a cast iron. A slight amount of ductility can also be obtained with Meehanite.

8.3.7 Ductile Iron

Ductile iron is a cast iron having the free carbon in the form of spheroidal nodules instead of flakes. Ductile iron is often called *nodular iron.*

The fact that the carbon is in little round balls instead of flakes explains why this cast iron behaves a lot like steel. The flakes in gray cast iron act as stress risers and make the impact strength and fatigue strength low. The nodules in ductile iron exhibit less of a stress-riser effect. Ductile iron has good impact strength and enough ductility to have elongations in the range of 2 to 15 percent, depending on the class. The fatigue strength of ductile iron can approach that of steel of equal hardness.

Ductile iron may be used as cast or it may be quenched and tempered to a wide range of hardness levels. Gears may be cut from ductile-iron blanks with the minimum hardness specified all the way from 165 to 350 BHN. Ductile-iron gears may be surface hardened by flame- or induction-hardening methods up to a surface hardness of 450 BHN minimum.

Ductile iron for gears is often specified by specifications developed by the American Society of Testing Materials. Table 8.16 shows some of the more popular classes for gear use.

A ductile-iron gear made with a blank material specified at 830 MPa (120 klb/in.^{2}) minimum tensile strength should have a minimum hardness on the gear teeth of 255 BHN. At a minimum tensile strength of 415 MPa (60 klb/in.^{2}), the minimum tooth hardness should be 155 BHN.

8.4 NONFERROUS GEAR MATERIALS

A wide variety of bronzes, aluminum alloys, zinc alloys, and nonmetallic plastics and laminates are used to make gears. In many cases the nonferrous gear is driven by a steel

TABLE 8.16 Ductile-Iron Specifications

Specifying Body	Class	Minimum Tensile Strength		Yield Strength,		Elongation, %
		MPa	(klb/in.^{2})	MPa	(klb/in.^{2})	
ASTM*	60-40-18	415	(60)	275	(40)	18.0
	80-55-06	550	(80)	380	(55)	6.0
ASTM	100-70-03	690	(100)	480	(70)	3.0
	120-90-02	830	(120)	620	(90)	2.0

*See ASTM A536 or SAE J434 for further information.

TABLE 8.17 General Characteristics* of Nonferrous Gear Materials

Material	Modulus of Elasticity, GPa (klb/in.^{2})		Density, lb/in.^{3}	Ultimate Strength,		Yield Strength,	
				M Pa	(klb/in.^{2})	M Pa	(klb/in.^{2})
Phosphor bronze	103	(15)	0.32	275	(40)	124	(18)
Manganese bronze	97–124	(14–18)	0.31	450	(65)	207	(30)
Aluminum bronze, heat treated	134	(19.4)	0.27	690	(100)	345	(50)
Silicon bronze	103	(15)	0.31	310	(45)	138	(20)
Aluminum alloy	73	(10.6)	0.10	470	(68)	331	(48)
Zinc alloy	97	(14)	0.24	275	(40)	179	(26)
Magnesium alloy	45	(6.5)	0.065	300	(44)	221	(32)
Brass alloy	103	(15)	0.306	525	(76)	310	(45)
Nylon (73°F)	3	(0.4)	0.036	77	(11.2)	58.6	(8.5)
Phenolic laminates:							
Paper 10	(1.5)(length)		0.042	93	(13.5)	69	(10)
7	(1.0)(crosswise)						
Fabric 7	(1.0)(length)		0.042	93	(13.5)	69	(10)
5.5	(0.8)(crosswise)						

*The values shown in this table are typical. Individual compositions may vary appreciably.

pinion. When loads are light and the parts are small, it is quite often possible to make both members of the pair from nonferrous material.

The nonferrous materials are used for a variety of reasons. Certain bronzes will stand high sliding velocity with a steel worm better than any other material. The nonferrous materials tend to be more corrosion resistant than the ferrous materials. Some of the nonferrous materials have less mass per unit of volume than ferrous material. This is important in certain instrument gear work where inertia of rotating parts must be kept extremely low. Some nonferrous materials lend themselves to very low cost mass production techniques. For instance, stamped bronze gears and injection-molded plastic gears are about as low in cost as is possible for a toothed gear.

Table 8.17 shows some general characteristics of the nonferrous gear materials. Detail data are given in the articles that follow.

8.4.1 Gear Bronzes

A family of four bronzes accounts for much of the nonferrous gear materials chiefly because of their wear resistance characteristics. These characteristics result from the alloying of small amounts of tin, zinc, lead, and traces of other elements and the manner in which they mix or position themselves with the base element copper. The purpose of this section is to emphasize the important advantages contributed by the different alloys and to identify the resulting materials by recognized specifications. The four gear bronzes are:

Phosphor bronze
Manganese bronze
Aluminum bronze
Silicon bronze

The metallurgy of bronze[11] is based on the ability of copper to form solid solutions with the various alloy elements. The extent of solubility varies widely, usually forming more than one type of solid solution with copper. This accounts for the effect on the physical properties produced by varying the amount of alloy element.

When a bronze gear is worn to a good contact, the hard constituent of the microstructure has become aligned to the rubbing surface. The softer constituent, it is then believed, allows this aligning process.[1] The major alloying elements are described with brief explanations of their contributions:

Zinc (Zn) is added to the basic copper-tin alloys to increase strength properties, but to some extent at the sacrifice of bearing properties.

Tin (S) is the principal alloying agent in phosphor bronze. It is a major contributor used to affect strength, but, if improper melting is made, the tin may oxidize, resulting in very hard crystals of tin that could be very abrasive.

Lead (Pb) is used to improve machinability, and, as with each contributing element, the compromise must be recognized. Although the limited amount of lead has little effect on tensile strength, it does cause a lowering of shear strength and ductility. Lead does not combine with the base alloy, but it does act as a "lubricant."

Iron (Fe) is another major contributor for strength, even in the materials called "nonferrous."

Phosphor Bronze. These bronzes are tough and hard and capable of high fatigue resistance. This group covers a wide range of copper-tin and copper-tin-lead bronzes, all of which are deoxidized with phosphorus. A long-time gear material of this group is 89-11 gear bronze, used particularly for worn wheels that operated with hardened-steel worms for moderate-to-high sliding velocities and moderate tooth loads. Adding a trace of nickel provides fine grain structure. Another of this group is 88-10-2 G bronze, or gun metal, having good physical qualities and good resistance to wear. Both these phosphor bronzes have good resistance to the corrosive action of sea water, have good machinability, and make excellent castings.

Manganese Bronze. This is tough, having about the tensile strength and ductility of ordinary cast steels, one of the toughest without benefit of heat treatment. Its high physical characteristics are due to the hardening and deoxidizing agents iron, manganese, aluminum, and tin. It makes excellent close-grain castings that result in a smooth surface quality.

Aluminum Bronze. Another tough strong casting material like manganese bronze, it is lighter in weight and capable of higher physical characteristics because of added elements of iron, manganese, and nickel, plus the fact that it can be heat treated. The attainment of highest strength will be at the sacrifice of the most important property of the gear tooth, ductility. It would be best to produce gear blanks as centrifugally cast parts.

Silicon Bronze. This casting material produces smooth golden-yellow gears capable of accurate thin sections but requiring special foundry technique. For gears of fine pitch, it is advisable to cast blanks oversize to assure the removal of surface defects. Centrifugal castings will help in this respect also. This material has importance in many electrical applications because of its nonmagnetic quality.

Table 8.18 shows the compositions of several kinds of bronzes and general data on the properties. Table 8.19 shows a cross-referencing of different bronzes with Table 8.18 and uses of the various bronzes.

8.4.2 Lightweight Metal Alloys for Gears

Lightweight gears contribute not only to weight reduction in gearboxes but to the low-inertia effect of rotating parts. The smoothest running gear trains consist of well-balanced, low-inertia gears.

Alloyed aluminum is the most common light metal for these gears. Three prominent grades are 6061-T6, 2024-T4, and 7075. Occasionally blanks are made from 356-T6 castings. Centrifugal casting reduces the problem of porosity.

Operating tests have shown that gears made of these materials having light loads and suitable lubrication can be used in the as-machined condition, giving a few million cycles of operating life.

Greater wear life can be obtained by having the gears anodized. When this is done with the proper control, the dimensional buildup may be as small as 0.0001 in. per surface. The result is a hard aluminum-oxided surface resistant to salt atmosphere corrosion and wear. It is noted, however, that anodizing can have a detrimental affect on bending fatigue life. This process, then, would be applicable to gear sets where only light tooth loading is applied.

Die-Cast Materials. [12] Die-cast gears provide great savings when large quantities are involved since the parts can be finished integral with keys, splines, ratchets, pawls, etc. In many cases, the only machining necessary is to remove the flash on one edge. In most alloys, thin walls and high tolerances are easily obtainable. For example, both zinc and aluminum work very well in a thin-wall application (1.0 mm and under) with the as-cast surface intact. All die castings accept many surface treatments such as hard surfacing, plating, and impregnation.

From a design standpoint CAD/CAM concepts work very well. It is noted that the best casting is obtained when the die caster participates in the design stage. Some of the more important considerations that influence the use of die cast components are:

Quantities warrant the making of the die.

Nonferrous metals meet requirements.

Cast teeth have requisite strength and can be used as cast.

Various operating devices can be included in the die to reduce subsequent machining costs.

If the gear teeth forms involve undercuts, parts of the teeth have to be finished machine-cut. Helical gears can be cast and must be formed by part of the die that can be turned to clear at the time of ejection, or the gear must be turned a few degrees to make ejection possible.

Trade Standards in Die-Casting Materials. The American Die Casting Association has standardized the commonly used zinc-alloy, aluminum-alloy, magnesium-alloy, and copper-base-alloy materials. Tables 8.20 to 8.23 show the data for these materials. Chemical compositions and physical property data are both given.

TABLE 8.18 Gear Bronzes

Common Name and Trade Term	Principal Elements in Composition, %					Ultimate Strength		Yield Strength	
	Cu	Sn	Zn	Fe	Others for Total of 100%	M Pa	(klb/in.²)	M Pa	(klb/in.²)
A. Gun metal, cast G bronze 88-10-2	88	10	2	0.15	Pb 0.3, Ni 1.0	275	(40)	124	(18)
B. Phosphor bronze, wrought	88	10	0.75	—	Pb 1–2.5, Ni 1.0	241	(35)		
C. G bronze	88	8	4	0.15	Pb 0.3, Ni 1.0	275	(40)		
D. Phosphor bronze, cast 89-11 gear bronze	89	11	—	—		241	(35)	124	(18)
E. Nickel-tin bronze, cast	87	10	—	—	Ni 2	345	(50)		
F. Manganese bronze, cast	60	1	Remain.	2	Mn 1.5, Al 1.0, Pb 0.4	448	(65)	172	(25)
G. Manganese bronze, cast	68	0.2	Remain.	4.5	Mn 5, Al 7.5	620	(90)	310	(45)
H. Silicon bronze, cast	88	2	5	2.5	Mn 1.5, Si 4	310	(45)	138	(20)
J. Aluminum bronze, cast	86	—	—	1	Al 10	552*	(80)	275	(40)
K. Aluminum bronze, cast	83	—	—	4	Al 11, Ni 2.5	620*	(90)	310	(45)
L. Aluminum bronze, cast	78	—	—	4	Al 11, Ni 5, Mn 3.5	760*	(110)	414	(60)

Notes: This table in conjunction with Table 8.19 lists gear bronzes as diminishing in order of preference for machinability and resistance to wear, but increasing in order of toughness. It is important to note that the physical properties can be met only by careful proportioning of elements and the practice of good foundry technique.

TABLE 8.19 Typical Uses for the Various Bronzes Detailed in Table 8.18

Bronze type*	Use
A, B, C	For gears of low to moderate service, free machining, resistant to corrosive action of sea water
D	Predominantly for moderate strength and toughness of worm wheels used with hard steel worms and moderate to high pitch line velocity
E, F, G	Sound castings, good machinability, moderate hardness, good ductility
H	Good strength, toughness, corrosion resistance, nonmagnetic
J, K, L	High strength, good fatigue resistance for gears under repeated stresses, good strength at higher-than-usual temperatures

*See Table 8.18.

Zinc Die Castings. The majority of die castings are made from zinc alloys because of the advantages of mechanical properties and casting characteristics: high production speeds, greater variation in section thicknesses, and closer dimensional tolerances than are obtainable with any other commonly used die-casting alloys. Because of lower pressures and temperatures, zinc die castings afford less die maintenance. Only brass die castings exceed the impact strength of zinc die castings.

In many applications zinc alloy die castings are used without any applied surface finish or treatment. However a variety of chemical or organic finishes may be applied when required. Also, zinc-alloy die castings are readily electroplated for a bright finish and/or protection from corrosion.

Typical zinc-alloy die cast materials commonly used are given in Table 8.20. Also given are typical properties and constants. The values indicated are for separately die-cast test bars and do not represent values for specimens cut from die castings.

Aluminum-Alloy Die Castings. The aluminum alloys most generally used are shown in Table 8.21. The chemical compositions of standard aluminum-base-alloy die castings and related typical values for physical properties and constants are shown in the table. The typical values indicated are for separately die-cast test bars and do not represent values for specimens cut from die castings. Die-casting and other processing characteristics data involved in alloy selection are covered in ADCI M4. Special alloys are shown in ADCI M3.

For special alloys, the following is given:

Alloy 43 is sufficiently ductile for special applications requiring forming or upsetting during assembly operations but is difficult to die cast.

Alloy 218 provides the best combination of strength, ductility, corrosion resistance, and finishing qualities but is more difficult to die cast.

Alloys 360 and A360 provide somewhat better ductility and corrosion resistance than either of the standard alloys.

Alloy 383 provides a superior combination of properties for complex thin-wall castings where pressure tightness is a requirement.

Alloy 384 has excellent fluidity or die-filling capacity which is of advantage in producing large, thin-walled, intricate die castings.

TABLE 8.20 Composition and Physical Properties of Zinc Pressure Die Casting Alloys Casting Limits, % by Weight

Element	2	3	5	7	ZA-8	ZA-12	ZA-27
Al	3.5–4.3	3.5–4.3	3.5–4.3	3.5–4.3	8.0–8.8	10.5–11.5	25.0–28.0
Mg	0.02–0.05	0.02–0.05	0.03–0.08	0.005–0.020	0.015–0.030	0.015–0.030	0.010–0.020
Cu	2.5–3.0	0.25 max	0.75–1.25	0.25 max	0.8–1.3	0.5–1.2	2–2.5
Fe (max)	0.10	0.10	0.10	0.075	0.075	0.075	0.075
Pb (max)	0.005	0.005	0.005	0.003	0.006	0.006	0.006
Cd (max)	0.004	0.004	0.004	0.002	0.006	0.006	0.006
Sn (max)	0.003	0.003	0.003	0.001	0.003	0.003	0.003
Ni	—	—	—	.005–.020	—	—	—
Zn*	Balance	Balance	Balance	Balance	Balance	Balance	Balance
Source	ASTM B86	ASTM B86	ASTM B86	ASTM B86	ASTM B791	ASTM B791	ASTM B791

Property							
Density							
kg/m^3 AT 21°C	6600	6600	6700	6600	6300	6000	5000
lb/in^3 AT 70°F	0.24	0.24	0.24	0.24	0.227	0.218	0.181
Solidification shrinkage, %†	1.25	1.17	1.17	1.17	1.1	1.25	1.25
Solidification temperature (melting)							
Range, °C‡	390-379	387-381	386-380	387-381	404-375	432-377	484-376
Range, °F	734-715	728-719	727-717	728-718	759-707	810-710	903-708
Thermal expansion							
μm/mm/°C AT 20–100°C	27.8	27.4	27.4	27.4	23.3	24.2	26.0
in/in °F AT 68–212°F	15.4	15.2	15.2	15.2	12.9	13.4	14.4
Specific heat capacity							
J/kg/°C AT 200-100°C	418.7	418.7	418.7	418.7	435.4	448.0	534.4
BTU/lb/°F AT 68–212°F	0.10	0.10	0.10	0.10	0.104	0.107	0.125
Thermal conductivity							
W/m/hr/m^2°C AT 70–140°C	104.7	113.0	108.9	113.0	114.7	116.1	125.5
BTU/ft/hr/ft^2/°F AT 158–252°F	60.5	65.3	62.9	65.3	66.3	67.1	72.5
Electrical conductivity, % IACS	25	27	26	27	27.7	28.3	29.7
Electrical resistivity							
μ ohm-cm at 20°C		6.3694	6.5359		6.2,	6.1	5.8
μ ohm-in at 68°F		2.5	2.6		2.4	2.4	2.3

Source: Data courtesy International Lead Zinc Research Organization, Inc. 2525 Meridian Parkway, Research Triangle Park, North Carolina 27709-2036.
*Zinc-alloy die castings may contain nickel, chromium, silicon, and manganese for amounts of 0.2, 0.2, 0.035; harmful effects have never been noted because of these elements in these concentrations.
†Solidification Shrinkage: The decrease in volume of a metal during solidification.
‡Temperature: unless otherwise noted all values shown are at 68°F (English) or 20°C (S.I.) units.

TABLE 8.21 Aluminum-Base-Alloy Die Castings
Chemical Composition and Properties: Aluminum-alloy Die Castings

Designation	Standard Alloys*	
Commercial	13	380
ASTM	S12B	SC84B
Composition, %:†		
Copper	0.6	3.0–4.0
Iron	2.0	2.0
Silicon	11.0–13.0	7.5–9.5
Manganese	0.35	0.50
Magnesium	0.10	0.10
Zinc	0.50	1.0
Nickel	0.50	0.50
Tin	0.15	0.35
Total others	0.25	0.50
Aluminum	Remainder	Remainder
Properties and constants:‡		
Tensile strength, MPa (klb/in.^{2})	248 (36)	331 (43)
Yield strength (0.2% offset), psi MPa (klb/in.^{2})	145 (21)	172 (25)
Elongation, % in 2″	2.0	2.0
Shear strength, MPa (klb/in.^{2})	172 (25)	193 (28)
Fatigue strength, MPa (klb/in.^{2})¶	131 (19)	138 (20)
Specific gravity	2.65	2.71
Weight g/cm^3 (lb/in.^{3})	2.66 (0.096)	2.7 (0.098)
Melting point (liquid), °C (°F)	582 (1080)	593 (1100)
Thermal conductivity, cgs	0.29	0.23
Thermal expansion, in. per in. per °F§	11.9	12.1
Electrical conductivity, % of copper standard	31	23

Source: Reprinted by permission of the American Die Casting Institute, Inc., Des Plaines, Illinois.
*These standard alloys are also available with 1.3-percent maximum iron content (designated 13X and A380) having 3-percent elongation.
†Composition in percent, maximum unless shown as range. Same as ASTM B85.
‡Data furnished by Aluminum Company of America and ASTM B85. Revisions are in progress and will be made from time to time.
¶R. R. Moore rotating-beam test data at 500,000,000 cycles.
§To be multiplied by 10^{-6}. Temperature range: 20 to 200°C.

Note: Reference to ADCI-M2 and M4 is advised before specifying "Special Aluminum-Alloy Die Castings."

Magnesium-Alloy Die Castings. Magnesium-alloy die castings are the lightest in weight of all die castings. Mechanical properties approximate those of aluminum-alloy die castings and the strength to weight ratio is higher. They have excellent machinability.

Because of magnesium's affinity for oxygen, special precautions must be taken when machining or grinding castings.

Magnesium-alloy die castings are usually treated by the die caster with a chrome pickle, a chrome manganate treatment, or an oil dip coating. The treatment or coating protects against the tarnishing or slight surface corrosion that can occur on unprotected die castings during storage in moist atmospheres. Proper selection of treatment depends on subsequent finishing methods, a variety of which are available where required.

Alloy AZ91B is the most commonly used magnesium die-casting alloy. This standard alloy is available in high-purity composition with 0.10 maximum copper content (designated AZ91A), having improved corrosion resistance.

Alloy AM60 is used in applications requiring good elongation and toughness combined with reasonably good yield strength and tensile strength.

Alloy AS41 is used for its superior elevated-temperature creep resistance.

The chemical compositions of the alloys noted in the foregoing are given, along with certain physical properties and constants in Table 8.22. The values indicated in the table are relevant to separately die-cast bars and do not represent values for specimens cut from die castings.

Brass Die Castings. Brass die castings have the highest mechanical properties and corrosion resistance of all die castings. While the standard copper-base alloys generally used (ADCI-M5) are readily die cast in intricate shapes, the high temperatures and pressures at which they are die cast result in comparatively short die life. Thus, brass die castings are relatively high in cost as compared with die castings of other metals.

However, where added strength, corrosion resistance, wear resistance, and higher hardness are required, the possible economies of brass die castings should be carefully considered.

The brass alloys most generally used are shown in Table 8.23. Die-casting and machining characteristics involved in alloy selection and the basis for specifying brass die castings are covered in ADCI-M6.

The chemical compositions and related typical values for physical properties and constants are shown in the table. The typical values indicated are for separately die-cast test bars and do not represent values for specimens cut from die castings.

Alloy 858 is a general-purpose, low-cost, yellow brass alloy, with good machinability and soldering characteristics. It has the lowest mechanical strength of the three standard alloys.

Alloy 878 has the highest mechanical strength, hardness, and wear resistance of the three alloys but is most difficult to machine. It is generally used only when the application requires high strength or resistance to wear.

Alloy 879 is a general-purpose alloy having higher strength than 858. It is somewhat easier to die cast but slightly more difficult to machine.

8.4.3 Nonmetallic Gears

Nonmetallic or "plastic" gears are being used more frequently in many applications (small power tools, cameras, printers, etc.). The designer must consider both pros and cons of plastic material before choosing it as a gear material. Some of the pros are: noise reduction, resistance to corrosion, relatively low cost, approximate 15-percent reduction in weight when compared with steel, and self-lubrication capacity. Some of the disadvantages are: lower strength, size change with moisture absorption, and greater thermal expansion/contraction characteristics.

In the past decade, many new plastic materials have been introduced and utilized

TABLE 8.22 Composition and Properties of Magnesium Alloy Die Castings (ADCI-M7)

Designation	Standard Alloys		
Commercial and ASTM	AZ91B	AM60A	AS41A
Composition-per cent*			
Aluminum	8.3–9.7	5.0–6.5	3.5–5.0
Zinc	0.4–1.0	0.22 max.	0.12 max.
Manganese	0.13 min.	0.13 min.	0.20–0.50
Silicon	0.50 max.	0.50 max.	0.50–1.5
Copper, maximum	0.35	0.35	0.06
Nickel, maximum	0.03	0.03	0.03
Others, total, maximum	0.3	0.3	0.3
Magnesium	Remainder	Remainder	Remainder
Properties and Constants†		§	§
Tensile strength, MPa (klb/in.²)	235 (34)		
Tensile yield strength (0.3% offset) MPa (klb/in.²)	160 (23)		
Elongation, % in 2 in. (51 mm)	3		
Shear strength, MPa (klb/in.²)	138 (20)		
Compressive yield strength, MPa (klb/in.²)	152 (22)		
Ultimate compressive strength, MPa (klb/in.²)	400 (58)		
Fatigue strength, MPa (klb/in.²)‡	97 (14)		
Specific gravity	1.80		
Weight per cu. in., lb.	0.066		
Thermal conductivity, CGS	0.17		
Thermal expansion, in./in./°F¶	15.2		
Electrical conductivity, % of copper standard	10		

Source: Reprinted by permission of the American Die Casting Institute, Inc., Des Plaines, Illinois.
*Composition in percent by weight. Same as ASTM B94.
†This data is furnished for information only and does not constitute part of this specification.
‡R. R. Moore rotating-beam test at 500,000,000 cycles.
¶To be multiplied by 10^{-4}. Temperature range: 20 to 200°C.
§Data being established by ASTM Committee B6.
See ADCI-M8 for characteristics of magnesium-alloy die castings.

for gearing applications. However, the common materials utilized are still the acetals and nylons. Table 8.24, shown in the following, lists some of the pertinent physical properties of these materials. Also shown in the table is polycarbonate material which, along with thermoplastic polyurethanes and thermoplastic polyesters, has been used in certain applications.

Engineering characteristics of the common plastics materials follow:[13] (See Chap. 17 for additional information on plastic gears.)

Acetal: Strong, stiff plastic, has exceptional dimensional stability because of low moisture absorption, resistance to creep and vibration fatigue; has low coefficient of friction; high resistance to abrasion and chemicals; retains most properties even when immersed in hot water; low tendency to stress-crack.

TABLE 8.23 Composition and Properties of Brass Alloy Die Castings, ADCI-M5

Designation Commercial and ASTM	Standard Brass Alloys		
	858	878	879
Composition—percent*			
Copper	57.0 min.	80.0 to 83.0	63.0 to 67.0
Silicon	0.25 max.	3.75 to 4.25	0.75 to 1.25
Lead, maximum	1.50	0.15	0.25
Tin, maximum	1.50	0.25	0.25
Manganese, maximum	0.25	0.15	0.15
Aluminum, maximum	0.25	0.15	0.15
Iron, maximum	0.50	0.15	0.15
Magnesium, maximum	—	0.01	—
Other elements, maximum	0.50	0.25	0.50
Zinc	30.0 min.	Remainder	Remainder
Properties and constants†			
Tensile strength, MPa (klb/in.^{2})	380 (55)	586 (85)	483 (70)
Yield strength (0.2% offset) MPa (klb/in.^{2})	207 (30)	345 (50)	241 (35)
Elongation, % in 2 in. (51 mm)	15	25	25
Impact strength, charpy, k6-M (ft.lb)	5.5 (40)	9.7 (70)	6.9 (50)
Hardness, Rockwell B scale	55–60	85–90	68–72
Modulus of elasticity, psi‡	15	20	15

Source: Reprinted by permission of The American Die Casting Institute, Des Plaines, Illinois.
*Composition in percent as indicated. Same as ASTM B176.
†This data is furnished for information only and does not constitute part of this specification.
‡To be multiplied by 10^6.

Nylon: Family of resins that have outstanding toughness and wear resistance, low coefficient of friction, excellent electrical properties and chemical resistance. Resins are hygroscopic; however, dimensional stability is poorer than that of most other plastics.

Polycarbonate: Highest impact resistance of all rigid plastics; has excellent stability and resistance to creep; fair chemical resistance; stress cracks in hydrocarbons; typically used with addition of glass fiber reinforcement and PTFE lubricant.

Thermoplastic polyester: Excellent dimensional stability and electrical properties; has excellent toughness and chemical resistance, except to strong acids or bases; notch sensitive, is not suitable for outdoor use and/or for service in hot water. Material is relatively soft and has the potential for tooth damage.

Thermoplastic polyurethane: Tough, extremely abrasive and impact resistant material; good electrical properties and chemical resistance. Difficult to injection mold small parts due to the material's elastic properties.

Polyester elastomer: Sound dampening; resistance to flex fatigue and impact.

All of the aforementioned base materials can be formulated with fillers, such as glass fibers, for added strength, and PTFE, silicone, and molybdenum disulphide for added lubricity.

TABLE 8.24 Physical Properties of Nonmetallic Gears

Property	Polycarbonate	Polyamide	Acetal	Phenolic Fabric, LE Crosswise	Lengthwise
MPa	62–72.5	59–76	69	65.5	93
Tensile strength, (klb/in.^{2})	9–10.5	8.5–11	10	9.5	13.5
MPa	76–90	100	97	93	103
Flexural strength, (klb/in.^{2})	11–13	14.6	14	13.5	15
Elongation, %	60–100	60–300	15–75		
Impact strength, ft.lb per in.	12–16	0.9–2.0	1.4–2.3	1	1.25
				Thickness $\frac{1}{8}''\ \frac{1}{4}''\ \frac{1}{2}''\ 1''$ up	
Water absorption, % per 24 hr	0.3	1.5	0.4	1.3 0.95 0.70 0.55	
Coefficient of thermal linear expansion, μ-in. per in. per °F	3.9	5.5	4.5	1.1	
Heat resistance, °F (continuous)	250–275	250	—	250	
°C	121–135	121		121	
Representative trade names	Lexan	Nylon, Zytel	Delrin	Phenolite, Textolite	

Source: This table is courtesy of General Electric Company, Plastics Division, Pittsfield, Massachusetts. The tabulated data are average values and should not be used for specifications.

Phenolic: Stability and strength when reinforced with long-fiber fillers.

Polyimide: Outstanding resistance to heat (500°F, continuous; 900°F, intermittent) and to heat aging. High-impact strength and wear resistance; low coefficient of thermal expansion; excellent electrical properties; difficult to process by conventional methods; high cost.

Polyphenylene sulfide: Outstanding chemical and heat resistance (450°F, continuous); excellent low-temperature strength; inert to most chemicals over a wide temperature range; inherently flame retardant; requires high processing temperature.

Since plastic gears are more sensitive to moisture absorption and temperature changes, each design application should have prototype models made.

For critical applications or where sections of a gear may provide stress raisers it may be advisable to stress relieve or anneal the part. This applies to gears machined from nylon stock and molded gears. Residual stresses in the latter can be severe.

Delrin, an acetal resin, is a material used for gears showing some improvements over nylon. The same kinds of considerations for dimensional stability must be given as when designing with nylon.

Laminates as gear materials provide several advantages: quietness, little lubrication, and manufacturing ease. Sheet stock made from phenolic resins that are combined with fine-weave linen provide the best condition.

Blanks may be punched out or circles may be obtained to fit into quantity manufacturing. Whether one blank or several are used, a metal backing-up disk is required to obtain a "clean" edge on the end piece.

Gears from laminate stock are usually drive members of a gear mesh.

If pitch line velocities are high, a steel pinion driving a phenolic gear should be hardened to prevent scoring and premature wear on the pinion.

A precaution frequently omitted is consideration for the high elasticity of nonmetallic gears. Too much deflection because of this allows the entering driven tooth to scrape over the flank of the driver and scuff the surface. To avoid this where the pinion is metal and the driven tooth nonmetallic, it is always adviseable to use the long- and short-addendum system for teeth proportions, except if undercutting would result.

REFERENCES

1. Dudley, D. W., *Practical Gear Design,* McGraw-Hill, New York, 1954, pp. 152–189.
2. Townsend, D. P., and Bamberger, E. N., "Surface Fatigue Life of Carburized and Hardened M50NiL and AISI 9310 Sour Gears and Rolling-Contact Test Bars," *Journal of Propulsion and Power,* vol. 7, no. 1, 1991.
3. Herring, D. H., "Why Vacuum Carburizing Is Effective for Today and Tomorrow: II," *Industrial Heating,* September 1987, pp. 22–26.
4. Goodman, D., and Simons, T., "Constant Velocity Joints Being Produced with World's First Production Ion Carbonizing System," *Industrial Heating,* November 1987, pp. 14–16.
5. Titus, J. W., "A Comparison-Traditional Batch Atmosphere Carburizing Versus Vacuum and Ion Carburizing," *Industrial Heating,* September 1987, pp. 23–26.
6. Legge, G., "Plasma Carburizing—Facility Design and Operating Data," *Industrial Heating,* March 1988, pp. 26–30.
7. The Nitralloy Corporation, "Nitralloy and the Nitriding Process," The Nitralloy Corp., New York, 1954.
8. Inductoheat, "Contour Gear Hardening Using Induction Heating with RF and Thermographic Control," *Industrial Heating,* July 1988, pp. 15–17.
9. National Broach and Machine Co., *Modern Methods of Gear Manufacture,* 3rd ed., National Broach and Machine Co., Detroit, Michigan, 1950.
10. Meehanite Metals Corp., *Handbook of Meehanite Metals,* Meehanite Metals Corp., New Rochelle, New York, 1948.
11. American Manganese Bronze Co., *Reference Book Bronze Casting Alloys,* American Manganese Bronze Co., Philadelphia, 1953.
12. The New Jersey Zinc Co., *Practical Considerations in Die Casting Design,* The New Jersey Zinc Co., New York, 1955.
13. Paquet, R. M., "Systematic Approach to Designing Plastic Spur and Helical Gears," *Gear Technology,* November/December 1989, pp. 12–28.

CHAPTER 9
GEAR DRAWINGS

Dezi J. Folenta,
President, Transmission Technology Co., Inc.
Fairfield, New Jersey

9.1 GENERAL COMMENTS CONCERNING PREPARATION OF GEAR DRAWINGS

This chapter presents information on how to prepare various types of detail gear manufacturing drawings. These examples show typical dimensioning practices and tolerancing. The chapter briefly addresses the need to include or to delete from the drawing various gear manufacturing instructions and specifications.

The required data for a given gear drawing varies from a bare minimum to a comprehensive treatment of many design parameters. For example, high performance aircraft and aerospace gear drawings require detail definition of gear tooth profile and include many specifications. These specifications contain detail instructions covering such areas as material processing, tooling design, quality control, and manufacturing sequences. On the other hand, gear manufacturing is possible using a few major dimensions and some basic gear data. The minimum gear data required are the number of teeth, module (diametral pitch), and pressure angle.

Gear drawings form the basis for purchase orders and contractual agreements. Accordingly, the information presented on the drawings must be clear, and, if possible, of singular interpretation.

The illustrations presented in this chapter cover a wide variety of gear drawing types and applications. Selection of a given drawing format is a function of the design specification, manufacturing considerations, service requirements, and cost.

While the drawing examples illustrate typical dimensioning for gear teeth and related mounting features, the detail drawing must define the entire gear. The drawing needs to address the dimensional compatibility of the gear with mating components and the integrity of the supporting structure. The structure must have adequate strength and stiffness to maintain good gear tooth contact throughout the gear's operating range.

All drawings presented in this chapter use metric dimensions with English conversions included in notes or in brackets [ ].

9.1.1 Gear Drawing Principles

Although there is a wide variety of gear types, there are some fundamental principles that govern preparation of detail gear drawings.

The primary requirement is that the information and data presented on a drawing are complete and clear. Ideally, the drawing defines in a clear, succinct style the required manufacturing, inspection, and acceptance or rejection criteria. From a practical point of view, the gear trade has established some rules to ease preparation of gear drawings. These rules cover a typical gear drawing format (see Sec. 9.2.1), gear data tabulations, material specifications, and gear inspection.

9.1.2 Manufacturing Considerations versus Design Considerations

A wide variety of gear manufacturing processes and production techniques are in use today. These processes include stamping, forging, hobbing, shaping, cutting, grinding, broaching, and honing. Each of these manufacturing techniques has specific limitations and implications. The final detail gear drawing must address such parameters as quality, cost, lot size, application, strength, and service life. Before preparing gear drawings, the designer must give serious consideration to the functional requirements of the gear and to the manufacturing processes.

As a general rule, drawings that depict standard commercial quality steel gears should not specify the gear manufacturing process. To do so may limit manufacturing sources and could adversely affect the cost and delivery of the gears. On the other hand, special and high-performance gears require detailed definition of manufacturing processes.

9.2 GEAR DRAWING PRACTICE

The designer may use a single or several views to describe the desired gear features. A typical gear drawing shows a cross-sectional view and an end view. Figures 9.1 and

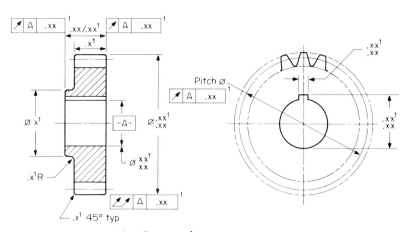

FIGURE 9.1 Commercial quality external spur gear.

GEAR DRAWINGS   9.3

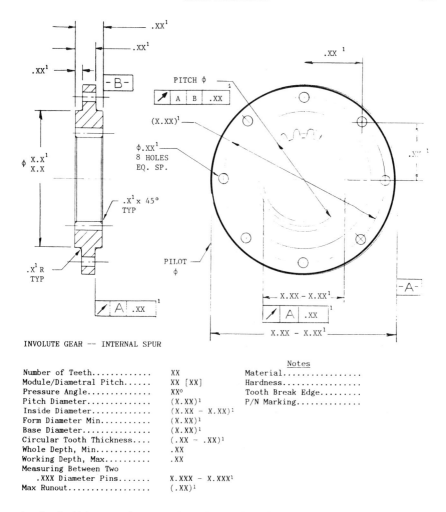

INVOLUTE GEAR -- INTERNAL SPUR

Number of Teeth.............	XX	Notes	
Module/Diametral Pitch......	XX [XX]	Material.................	
Pressure Angle..............	XX°	Hardness.................	
Pitch Diameter..............	(X.XX)[1]	Tooth Break Edge.........	
Inside Diameter.............	(X.XX - X.XX)[1]	P/N Marking..............	
Form Diameter Min...........	(X.XX)[1]		
Base Diameter...............	(X.XX)[1]		
Circular Tooth Thickness....	(.XX - .XX)[1]		
Whole Depth, Min............	.XX		
Working Depth, Max..........	.XX		
Measuring Between Two			
.XXX Diameter Pins.......	X.XXX - X.XXX[1]		
Max Runout..................	(.XX)[1]		

[1] For English system drawings, the number of decimal places to the right of decimal point should be increased by one.

FIGURE 9.2 Internal spur gear.

9.2 illustrate this drawing practice. These illustrations represent only a part of the drawing. Additional information, such as part name, title block, and various notes, are needed to complete the drawing. For more complicated shapes and for gears requiring special gear tooth geometry, the designer can use additional views and illustrations.

Many gear designers and manufacturers include a detailed shape of the gear tooth cutter on the gear drawing. The specific tool shape assists designers, manufacturers, and inspectors to meet the requirements of the given specifications. When specifying standard gear tooth profiles, there is no need to draw the involute curve accurately. Figure 9.3 illustrates the approximate method of drawing the involute curve. The designer can also use drafting templates to draw the involute. Both of these methods are sufficient for most drawings.

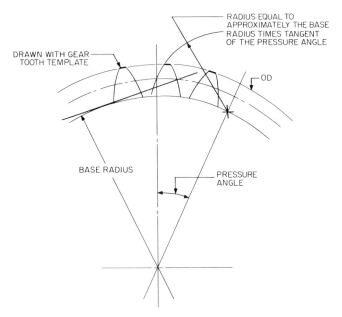

FIGURE 9.3 Method of drawing approximate tooth profile.

9.2.1 Arrangements

In typical detail gear drawing arrangements, the tabulated gear data are put on the lower left hand side of the drawing. The lower center of the drawing contains diagrams showing special data relative to the involute profile and lead. Data relative to gear material selection and material processing are usually found in the lower part of the drawing near the title block. The upper center of the drawing typically contains the plan view, elevation view, and section views. Normally, the right side of the drawing contains the title block, revisions block, and notes. It is customary to place the title block in the lower right hand corner of the drawing. The revision block is typically put in the upper right hand corner. Figure 9.20 depicts a detail gear drawing using this arrangement.

9.2.2 Items to Be Specified

This section presents the typical information required to prepare a detail gear drawing. For convenience, the data are divided into three groups. The first group addresses the gear blank dimensions and associated notes concerning gear material, heat treating, etc. The second group contains information covering gear design and quality control considerations. The third group includes special notes and the title block information. The cross-sectional views usually define the gear blank shape and its dimensions. The gear tooth data are normally tabulated.

Some common gear blank dimensions are:

Outside diameter	Outside cone angle (bevel gears)
Face width	Back cone angle (bevel gears)
Journal diameters	Mounting distance (bevel gears)
Web thickness	Throat diameter (worm gears)
Hub diameters	Throat radius (worm gears)
Bearing bores	Mounting surfaces or holes
Inside rim diameter	Break edges and various radii

For further information and references covering gear blank dimensions, see applicable drawings in this chapter.

All gear blank dimensions and features form the basis for manufacturing and inspection. Typical drawing features include runouts, perpendicularity, true position, concentricity, and flatness.

Further, the gear drawing must identify and define the datum axis or datum surfaces. The datum axis or surfaces selected can be shaft centers, pilot diameters, mounting holes, and bearing diameters. The designer must take into consideration gear blank manufacturing and gear tooth cutting when selecting the datum axis or surfaces. These datum surfaces form the basis for gear manufacturing and inspection.

Datum selection can affect cost, quality, and performance of the gear. Accordingly, designers need to coordinate their efforts not only within engineering, but also with marketing, manufacturing, inspection, and service.

The accuracy of the gear blank has a direct impact on the accuracy and the quality level of the finished gear. Depending on a gear's end use, designers need to select the accuracy level and number of decimal places for each dimension. See Chap. 7, "Gear Tolerances." Some drawing title blocks define tolerance limits applicable to dimensions with two and three decimal places. When using these tolerances, the designer does not need to impose further limits. Selection of tolerances different from those given in the title block requires defining both the upper and lower limits.

Three types of gear dimensions are common: basic, reference, and manufacturing. A basic dimension is a numerical value used to describe the theoretically exact size, profile, orientation, or location of a feature. It is the basis from which permissible variations are made. These allowable variations include tolerances or other dimensions in notes or in feature control frames.

Dimensions without tolerances, used as the basis for other toleranced dimensions, are reference (ref) dimensions. The designer needs to identify all reference dimensions. In this chapter, dimensions designated as "ref" or shown in parentheses are reference dimensions.

Toleranced dimensions are working dimensions for manufacturing and inspection. Sometimes a toleranced dimension may become a reference dimension. For example, circular tooth thickness is used as the basis for determining chordal tooth thickness or measurements with pins.

Figures 9.1 and 9.2 illustrate a gear drawing format suitable for commercial quality spur gears. For illustrations of other gear drawings, see applicable figures in this chapter. In addition to presenting various gear drawing formats, sketches illustrating gear tooth surface roughness and involute profile control are also shown. For example, Figure 9.4 presents several methods used in defining gear tooth surface roughness.

Drawing data used in this chapter are mostly spelled out. Certain abbreviations are permissible provided that their interpretation is clear. When in doubt, spell it out.

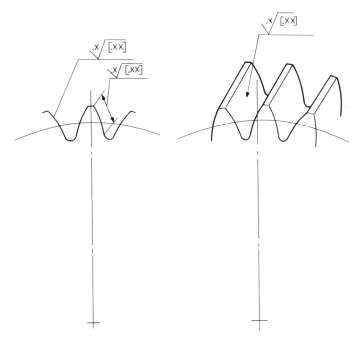

FIGURE 9.4 Typical methods of showing surface roughness.

9.2.3 Degree of Control

It is imperative that the selected datum surfaces have a numerical relationship to the surfaces on which the gear ultimately runs. For example, when using shaft centers as datums for gear cutting, the drawing needs to show a dimensional relationship to the gear's pitch diameter. Using this dimensioning arrangement, the pitch diameter now needs to show a dimensional relationship to the bearing diameters. The use of geometry symbols such as runout, total runout, concentricity, etc. achieves this dimensional control. Sometimes the designer can use an appropriate note instead of a geometric control symbol. For example, the reference axis note might say: "For inspection of teeth, establish reference axis concentric with diameters A and B within .XXX." The cost of gear production is a function of gear tooth accuracy and gear quality. Accordingly, the selection of datum surfaces and the accuracy levels specified requires great care.

9.3 GEAR DRAWING SPECIFICS

All gear tooth dimensions and features specified on a drawing are subject to inspection. The gear designer must define the gear quality and give specific information concerning manufacturing tolerances. Various gear types require different manufacturing and

GEAR DRAWINGS   **9.7**

inspection procedures. The following is a brief outline of the more common gear manufacturing and inspection processes presently used in the gear industry. This outline also presents information on the preparation of detail gear drawings. For more information on gear inspection, see Chap. 23, "Gear Inspection Devices."

9.3.1 Spur Gears

Some of the common methods used to manufacture steel spur gears are milling, shaping, hobbing, and grinding. Figure 9.1 illustrates gear dimensioning when circular tooth thickness is used as the basis for chordal measurements. Here, the designer may make a correction in the chordal addendum to account for variations of the outside diameter.

A more accurate method of spur gear tooth inspection is a measurement with pins. Since pins contact the surfaces of the gear teeth, this inspection technique is usable for either intermediate manufacturing inspection or final inspection. When this method of gear inspection is contemplated, the data on the drawing should include:

MEASURING PIN DIAMETER XX   [.XXX]
MEASUREMENT OVER PINS. XX   [.XXX]
                                            .XX   [.XXX]

If desired, the designer may use measurements over pins as a final inspection dimension. See Fig. 9.5. The drawing would specify measurements over pins for external gears and measurements between pins for internal gears. The pins are put 180° apart for gears with even numbers of teeth. For gears with odd numbers of teeth, the second pin is placed half the circular pitch away from the 180° location. When the drawing calls for measurements with pins, the toleranced tooth thickness dimension is marked "ref."

Another common spur gear inspection technique is via master gears or master racks. Here, a spring loaded master gear rolls in tight mesh with the gear being inspected. The inspector measures and records variations in center distances. This process is known as *composite error inspection*. It measures the combined effect of runout, pitch error, tooth thickness variations, profile error, and lateral runout. The drawing specification for this inspection method should read:

MAX TOOTH-TO-TOOTH COMPOSITE TOL XXX   [.XXXX]
MAX TOTAL COMPOSITE TOL XXX   [.XXXX]

In volume production of precision spur gears when the use of master gears is economically justified, it might be desirable to set limits on center distance variations. This is an economical measurement of gear tooth thickness variation and a quick determination of the meshing accuracy of the gear. If the designer intends to use a special manufacturing or inspection tool, the drawing must define it.

Inspection data for gear sets where part interchangeability is not a design require-

9.8 CHAPTER NINE

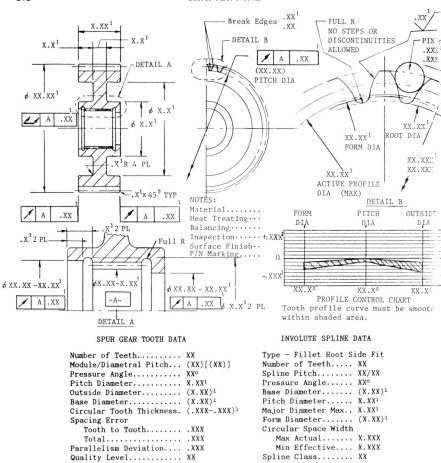

FIGURE 9.5 Precision spur gear.

ment should be so defined on the drawing. This is true when the designer selects a particular gear set for quiet operation. Typical data might be as follows:

PART NUMBER OF MATING GEAR 00000
NUMBER OF TEETH IN MATING GEAR XX
BACKLASH (AT CENTER DISTANCE XX.XX [X.XXX]) . . .XX [.XXX]
. . . . .XX [.XXX]

Chapter 7, "Gear Tolerances," and Chap. 23, "Gear Inspection Devices," present further information on tolerancing and inspection. The designer must keep in mind that drawing tolerances signify permissible deviations from a given nominal dimension. This total deviation is a sum of several independent factors. The factors include inherent inaccuracies of the manufacturing equipment and variations caused by setup, tools, temperature fluctuations, and the operator.

Aside from runout and profile error checks, detail gear inspection is usually specified only for high-precision gears. The costs of quality control and inspection can be a significant portion of the overall gear manufacturing cost. Where required, permissible geometric variations may be called out as follows:

MAX RUNOUT	.XX	[.XXX]
MAX PROFILE TOLERANCE	+.XX to −.XX	[+.XXX −.XXX]
MAX PITCH TOLERANCE	.XXX	[.XXXX]
MAX INDEX TOLERANCE	.XXX	[.XXXX]
MAX TOLERANCE IN PARALLELISM	.XXX	[.XXXX]

In designs requiring involute profile control, the drawing needs to include the involute chart. This chart defines the permissible deviation from the theoretical shape. Figure 9.5 shows an illustration of one way of specifying the profile control chart. When this chart is used, the following dimensions are added:

FORM DIA MAX (EXTERNAL) MIN (INTERNAL). . . .	X.XX	[X.XXX]
MINIMUM FILLET RADIUS	X.XX	[X.XXX]
ROOT DIA.	X.XX	[X.XXX]
. . . .	X.XX	[X.XXX]

Figure 9.6 illustrates additional profile control charts. For highly loaded gears and for high-performance gears, design requirements may call for a specific modification to the involute profile. The desired involute profile modification is closely defined on the profile control chart. Since it is desirable to have a smooth curve within the shaded area, add the following note to emphasize this requirement:

TOOTH PROFILE MUST BE A SMOOTH CURVE WITHIN THE SHADED AREA.

Where operating conditions or a contractual specification calls for it, add a note covering the gear tooth contact area. A typical note might be:

MINIMUM TOOTH CONTACT AREA WITH MASTER XX%.

To relieve end loading, gear drawings for certain spur and helical gears specify crowning of teeth. Crowning, by definition, is the gear face end relief. Crowning is measured as a lead variation. (See Fig. 9.7.) Where the crown is offset from the center of the tooth, a dimension locating the center of the crown is given.

For precision and high-performance gears, the design engineer needs to include many notes and specifications on the drawing. Care must be exercised to avoid duplication and the use of specifications that are contradictory.

For a typical illustration of a spur gear rack, see Fig. 9.8. Racks are normally dimensioned from the end of the blank and from the datum surface. The datum surface is used for set up and inspection.

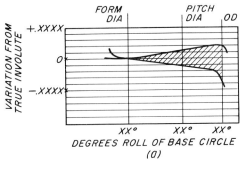

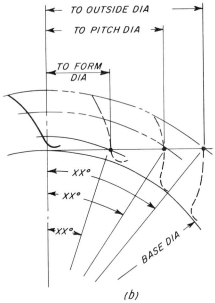

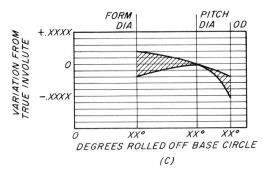

[1] For English system drawings, the number of decimal places to the right of decimal point should be increased by one.

FIGURE 9.6 Profile control chart.

GEAR DRAWINGS

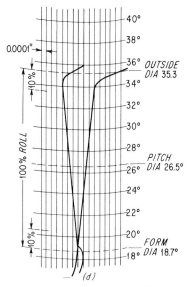

Note: Undercut is permitted in first 10% of roll and tip radius fall off is permitted in last 10%. Read profile slope from highest point in first 10% of roll to highest point in last 10% of roll.

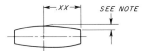

NOTE: Crown to relieve teeth on each side and each end .XXX[1]
.XXX

For English system drawings, the number of decimal places to the right of decimal point should be increased by one.

FIGURE 9.7 Dimensioning of crown teeth.

9.3.2 Helical Gears

Most of the comments and drawing practices used for spur gears are also applicable to helical gears. Helical gears can be single-helical, double-helical, or herringbone gears. However, certain differences between spur and helical gears should be noted.

Normal and Transverse Planes. Helical gear teeth may be dimensioned in a transverse plane (plane perpendicular to the axis of rotation), the same as in spur gears, or in a normal plane (plane perpendicular to the gear teeth). In a spur gear, the normal plane and transverse plane are the same. With helical gears, however, the drawing must specify whether the various dimensions are in the transverse plane or the normal plane. (See Fig. 9.9.)

9.12 CHAPTER NINE

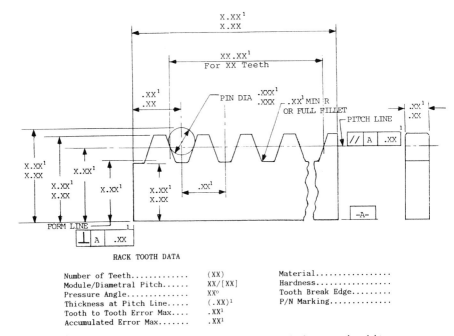

RACK TOOTH DATA

Number of Teeth............	(XX)	Material................
Module/Diametral Pitch......	XX/[XX]	Hardness................
Pressure Angle.............	XX°	Tooth Break Edge.........
Thickness at Pitch Line.....	(.XX)[1]	P/N Marking.............
Tooth to Tooth Error Max....	.XX[1]	
Accumulated Error Max.......	.XX[1]	

[1]For English system drawings, the number of decimal places to the right of decimal point should be increased by one.

FIGURE 9.8 Dimensioning of rack teeth.

Gear Cutting. It is a common practice to use spur-type gear cutters to cut helical gears. These cutters, when generating a helical gear, produce spur-type gear geometry in the normal plane. In the transverse plane, however, such gear design parameters as module (diametral pitch), pressure angle, circular pitch, etc., which are a function of the helix angle, must be clearly defined. See Fig. 9.9 for a typical single-helical gear data sheet.

Inspection. Helical gear tooth manufacturing, inspection, and tooling are similar to those used with spur gears, with several exceptions. Gear tooth thickness measurements use balls rather than pins. Figures 9.9 and 9.10 show drawing format and data sheet for a commercial quality and a precision quality single-helical gear, respectively.

Profile control techniques used for spur gears are also applicable to helical gears. The drawing may show profile tolerances, runout limits, pitch diameter, and ball measurements or may present them in tabular form. For helical gear tooth data in the normal plane, the principal items are:

NORMAL MODULE . . . . .XX/NORMAL DIAMETRAL PITCH [XX.XXX]
NORMAL PRESSURE ANGLE . . . . X.XX
NORMAL CIRCULAR PITCH. . . . . .XXX [.XXXX]
NORMAL CIRCULAR THICKNESS . . . . .XXX [.XXXX]
NORMAL CHORDAL ADDENDUM . . . . .XX [.XXX]

GEAR DRAWINGS   9.13

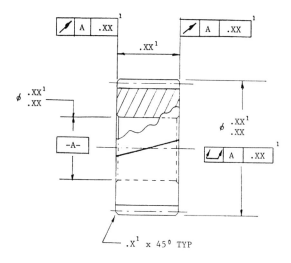

INVOLUTE HELICAL GEAR TOOTH DATA

Number of Teeth.............	XX	**Notes**
Normal Module/Normal		Material...........
Diametral Pitch..........	XX [XX]	Hardness...........
Transverse Module/Transverse		Break Edge.........
Diametral Pitch..........	(XX)[(XX)]	P/N Marking........
Normal Pressure Angle.......	XX°	
Transverse Pressure Angle...	(XX.XX°)	
Helix Angle.................	XX.XX°	
Hand of Helix...............	RH or LH	
Lead........................	XX.XXX[1]	
Pitch Diameter..............	(X.XX)[1]	
Base Diameter...............	(X.XX)[1]	
Form Diameter...............	(X.XX)[1]	
Root Diameter...............	X.XX − X.XX[1]	
Actual Circ Tooth Thickness		
at Normal Pitch Diameter.	.XXX − .XXX[1]	
Ball Diameter...............	.XXX[1]	
Measurement Over Two Balls..	X.XXX − X.XXX[1]	
Max Runout..................	.XXX[1]	

[1]For English system drawings, the number of decimal places to the right of decimal point should be increased by one.

FIGURE 9.9 Helical gear—commercial quality.

NORMAL CHORDAL THICKNESS. XX   [.XXX
. XX   .XXX]

Helical gear drawings use *lead deviation* or *lead error* instead of *parallelism deviation*.

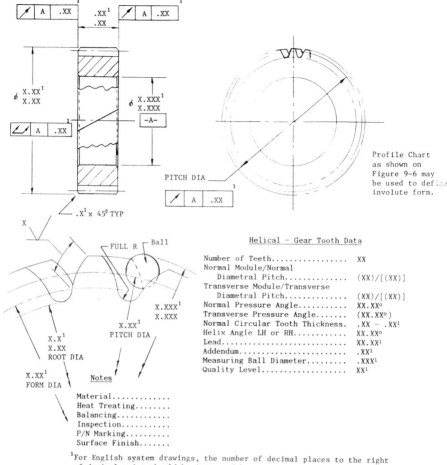

FIGURE 9.10 Precision helical gear.

9.3.3 Bevel and Hypoid Gears

In bevel gears, the axes usually intersect. The shaft angle, although usually set at 90°, can be set at other desired angles. Common bevel gears include straight bevels, spiral bevels, and Zerol* gears. In hypoid gears, the axes are offset.

For general dimensions and gear data for straight bevel gears, see Fig. 9.11. All data are in the plane that contains the pinion and gear axis.

Bevel Gear Tooth Proportions. In a cross-sectional view of bevel gear teeth, the gear proportions are similar to those of an equivalent spur gear that has the same curvature of pitch surface and the same circular pitch. It is customary when designing long-addendum pinion and short-addendum gear teeth to adjust the tooth thickness to

---

*Registered trademark of Gleason Works, Rochester, New York.

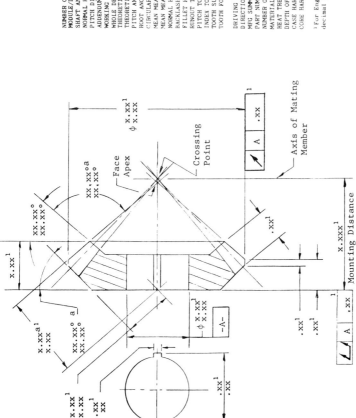

NUMBER OF TEETH	XX
MODULE/DIAMETRAL PITCH	XX.XX/1XX.XX[1]
SHAFT ANGLE	(XX.XX°)
NORMAL PRESSURE ANGLE	XX.XX°,
PITCH DIAMETER	(X.XXX)[1]
ADDENDUM	.XX[1]
WORKING DEPTH	.XX[1]
WHOLE DEPTH	.XX − .XX[1]
THEORETICAL OUTSIDE DIAMETER	(X.XX)[1]
THEORETICAL CROWN TO BACK	(X.XX)[1]
PITCH ANGLE	(XX.XX°)[1]
ROOT ANGLE	(XX.XX°)[1]
CIRCULAR THICKNESS	(.XXX)
MEAN MEASURING ADDENDUM	.XX[1]
MEAN MEASURING THICKNESS	.XX − .XX[1]
NORMAL BACKLASH WITH MATE	.XX − .XX[1]
BACKLASH VARIATION TOLERANCE	XXX[1]
FILLET RADIUS	.XX − .XX[1]
RUNOUT TOLERANCE	.XXX[1]
PITCH TOLERANCE	.XXX[1]
INDEX TOLERANCE	.XXX[1]
TOOTH SURFACE TEXTURE	XXX[1]
TOOTH FORM	X AA or R
DRIVING MEMBER	CONIFLEX(R) or REVACYCLE(R)
DIRECTION OF ROTATION	PINION or GEAR
MFG SUMMARY NUMBER	CW and/or CCW
PART NUMBER OF MATE	00000
NUMBER OF TEETH IN MATE	00000
MATERIAL SPECIFICATION	XX
HEAT TREAT SPECIFICATION	
DEPTH OF CASE	
CASE HARDNESS	
CORE HARDNESS	

[1] For English drawings, the number of decimal places to the right of the decimal point should be increased by one.

a. When face angle distance and back angle distance are used for dimensioning the gear blank, the face angle and the back angle should be given as reference dimensions on the drawing without a tolerance.

FIGURE 9.11 Straight bevel pinion.

balance the gear tooth strength. The module [diametral pitch] and other tooth dimensions apply to the outer ends of the bevel gear teeth unless otherwise noted.

9.3.4 Spiral Bevel Gears

Spiral bevel gears are similar to straight bevels with the exception that spiral bevel gear teeth have a spiral angle. The pitch diameter and module [diametral pitch] are normally given at the outer ends of the gear teeth. The spiral angle specified is at the mean cone distance and it varies along the tooth. The pressure angle is in the normal plane. For a typical illustration of a spiral bevel gear drawing and associated gear data, see Fig. 9.12.

Detail gear tooth specifications vary according to the cutting equipment used. Accordingly, designers must coordinate their designs with manufacturing to insure drawing compatibility with the manufacturing equipment used. This coordination allows the designer to define the geometry of the cutters and to determine machine settings. These settings are customarily recorded on summary sheets. For additional information, see Chap. 20, "Bevel and Hypoid Gear Manufacturing."

Hypoid Gears. Figure 9.13 illustrates detail drawings of a hypoid gear and a hypoid pinion. These gears are similar to spiral bevel gears except for the shaft offset.

Bevel and Hypoid Gear Manufacturing. In the production of bevel gears and hypoid gears, important differences can be noted in such areas as:

- Tooth form of bevel gears as produced by various manufacturers
- Cutting tools used
- Cutting or generating methods
- Type of cutting or generating machines used
- Setup methods and operation

Bevel and Hypoid Gear Tolerances. Gear blank and gear tooth tolerances are applied to the bevel gear dimensions in a manner similar to those for spur or helical gears. There are, however, no simple direct methods of checking bevel tooth surfaces, as there are for spur and helical gears. Figures 9.11 and 9.12 illustrate typical tolerances for bevel gears.

Gear designers and manufacturers have developed many alternate and acceptable methods of dimensioning and tolerancing bevel gears. One method is zone tolerancing. Here, the untoleranced dimensions define the basic contour. Phantom lines and tolerances depict the permissible deviations.

Bevel and Hypoid Gear Inspection. In general, the final inspection of a bevel gear set is made by running it in a gear checker using specific mounting distances. The selected gear mounting distance is the one that shows the best contact, lowest noise, and proper backlash. Following the gear checker spin test, the gear set is installed in the gear housing using the mounting distance established in the gear checker. Gear tooth contact, sound, and backlash form the acceptance or rejection criteria.

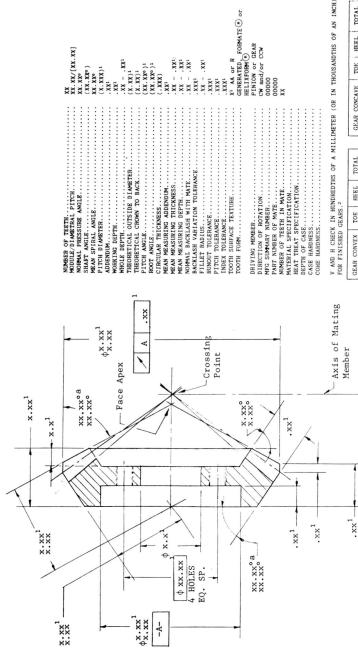

FIGURE 9.12 Spiral bevel gear.

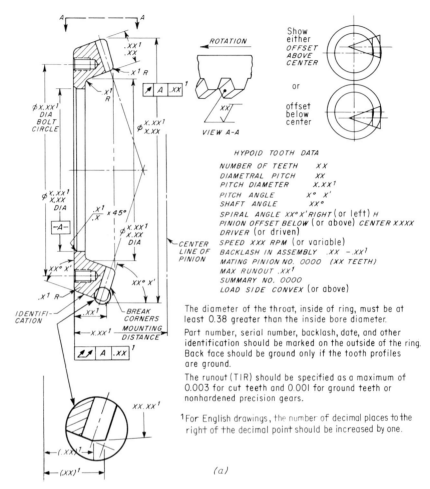

FIGURE 9.13 Hypoid gear and hypoid pinion.

9.3.5 Worm Gears

Worm gears are characterized by one member of a gear set having a screw-type thread. The thread angle or lead can be as low as a few degrees. With low lead angles, the worm has an appearance of a thread on a bolt. At the other extreme, the worm can have a lead angle approaching 45°. Worm gears mount on nonintersecting shafts, usually at a 90° shaft angle. Worm gears are made in sets. It is recommended that the same manufacturer make both the worm and the worm gear.

Axial pitch designates the size of the worm threads and is the distance from thread to thread measured in the axial direction. Some popular axial pitches have evolved over the years in both the metric and the English systems. In the metric system, the pitches are based on even millimeter designations. In the English system, the popular axial

GEAR DRAWINGS 9.19

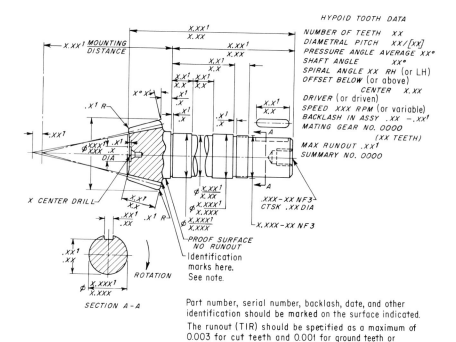

(b)

FIGURE 9.13 (Continued)

pitches include various inch fractions. Figure 9.14 illustrates a typical manufacturing drawing of a worm gear set and associated data.

9.3.6 Special Gear Types

Cone Drive*

These double-enveloping, double-throated worm gears have straight-sided teeth. The tooth profiles are tangent to a common circle and contact the full depth of the teeth. The final geometry of the gear teeth is a function of the cutter design. The designer must coordinate the cutter configuration with the manufacturer before finalizing the detail drawing. Figure 9.15 illustrates the gear drawing and gear data for this type of gear set.

---

*Registered trademark, Michigan Tool Co., Detroit, Michigan.

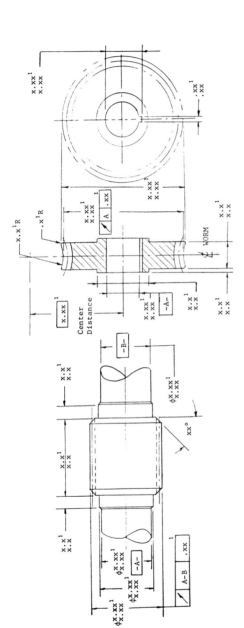

FIGURE 9.14 Single-enveloping worm gears.

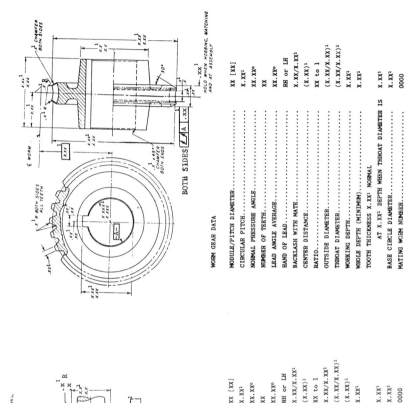

FIGURE 9.15 Double-enveloping worm gears.

Spiroid Gears*

The pinion of a Spiroid gear set resembles a screw thread on a tapered blank. The designer should coordinate the final drawing with the manufacturer. The manufacturer furnishes much of the detail gear data including tool design, tool setting, machine identification, gear material, and heat treating. Figure 9.16 represents a typical drawing format for Spiroid gears.

Face Gears. A face gear set consists of a spur gear pinion driving a face gear. Figure 9.17 presents an illustration of a face gear detail drawing. Face gears can mesh either on center or off center. In the on-center design, the axis of the pinion lies in the same plane as the axis of the face gear. Conversely, in the off-center design, the pinion axis is offset in much the same way as it is in hypoid gears.

9.4 PREPARATION OF DETAIL MANUFACTURING DRAWINGS USING COMPUTER-AIDED DESIGN AND COMPUTER-AIDED DRAFTING EQUIPMENT (CAD)

Many engineering companies, small and large, are finding that the use of CAD is essential to their survival. The practicality of using computer processing to prepare detail gear drawings is clearly established. Although many companies manufacture and sell CAD systems, the basic concept of preparing detail gear drawings is similar. Figures 9.18 and 9.19 show typical installations of a CAD system. The system shown contains a computer module, a graphics terminal, a magnetic tape system for data storage, and a hard-copy plotter. Figure 9.20 illustrates a typical gear drawing prepared on a CAD system.

The single, most compelling reason why many companies are using computer graphics to prepare detail gear drawings is to increase engineering productivity. Once the basic format of a given gear type is input into a computer, the task of preparing another similar drawing is reduced significantly. Further, since the gear format can be standardized, the number of drawing errors or omissions is greatly reduced. The quality and uniformity of drawings produced on CAD systems are excellent. It is not uncommon for the accuracy of the line work to be within a fraction of a millimeter. To produce a drawing in a metric or an English system, operators simply need to define the system in which they want to work. The CAD system automatically plots the dimensions in the selected system.

A very natural extension of CAD is Computer-Aided Manufacturing (CAM). Many companies and industries are presently using CAD/CAM systems to manufacture their products. Thus, designers can go directly from the computer to manufacturing without producing paper drawings.

---

*Registered trademark, Illinois Tool Works, Inc., Chicago, Illinois.

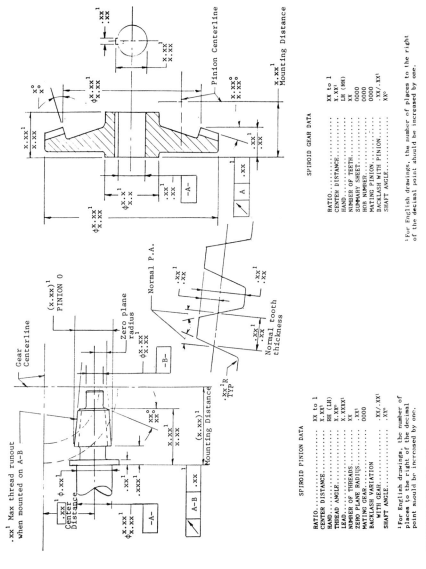

FIGURE 9.16 Spiroid gears.

9.24   CHAPTER NINE

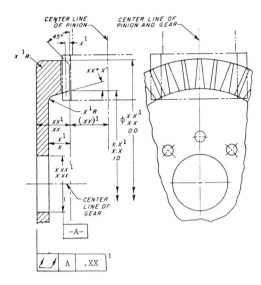

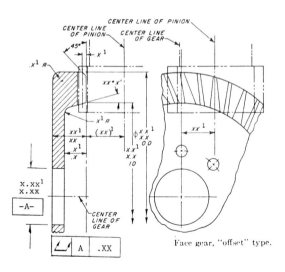

Face gear, "offset" type.

[1] For English drawings, the number of decimal places to the right of the decimal point should be increased by one.

FIGURE 9.17 Face gear.

GEAR DRAWINGS 9.25

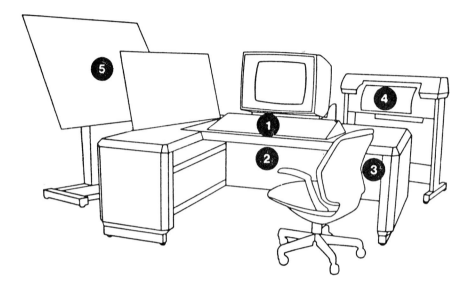

① GRAPHICS WORKSTATION
GRAPHICS TERMINAL — 21" vector stroke refresh with 4096 x 4096 resolution.
KEYBOARD — standard 95-key ASCII for text entry.
GRAPHICS PROCESSOR — high performance minicomputer with bipolar bit slice technology.
DATA TABLET — 11" x 11" active surface for all drafting commands.
WORKSTATION TABLE
② WINCHESTER DISK STORAGE
Provides on-line storage of designs and drawings constructed.
③ MAGNETIC TAPE CARTRIDGE UNIT
Provides a duplicate set of drawing files — for backup and to install software.
④ PLOTTERS (OPTIONAL)
Produces up to D or E size, multi-color ink drawings on paper, vellum, or mylar.
⑤ DIGITIZER (OPTIONAL)
Drawings already in existence can be readily digitized and incorporated into CADMAX-II storage.

FIGURE 9.18 Typical CAD system installation.

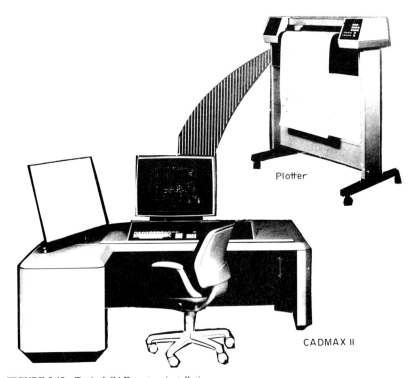

FIGURE 9.19 Typical CAD system installation.

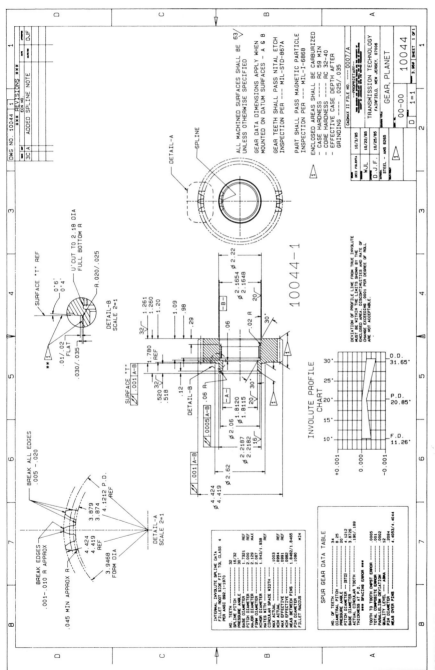

FIGURE 9.20 Typical gear drawing prepared on a CAD system.

REFERENCES

1. ANSI Y14.5M, "Dimensioning and Tolerancing," American Society of Mechanical Engineers, New York, 1982.
2. Dudley, D., *Handbook of Practical Gear Design,* McGraw-Hill, New York, 1984.
3. ANSI Y14.7.1, "Gear Drawing Standards—Part 1 for Spur, Helical, Double Helical and Rack," American Society of Mechanical Engineers, New York, 1971.
4. ANSI Y14.7.2, "Gear and Spline Drawing Standards—Part 2—Bevel and Hypoid Gears," American Society of Mechanical Engineers, New York, 1978.
5. McCain, G. L., Chap. 11 in D. Dudley, ed., *Handbook of Practical Gear Design,* McGraw-Hill, New York, 1984.

CHAPTER 10
GEAR REACTIONS AND MOUNTINGS

Martin A. Hartman
Consultant
Thousand Oaks, California

Darle W. Dudley
President, Dudley Engineering Company, Inc.
San Diego, California

It is often assumed that gear engineers are primarily concerned with toothed wheels that mesh with each other. The gear engineer, of course, has to understand and deal with the design application of gear teeth. In a broad sense, though, the gear engineer has to be concerned with the whole gearbox. For instance, the bearings that support the toothed gear wheels and the mounting of gears on shafts is a very important part of the technology of gear units. This chapter covers how to determine the reaction forces that come from typical loaded gears and it also covers typical gear mounting arrangements.

10.1 MECHANICS OF GEAR REACTIONS

The main function of a gear is to transmit motion and/or power. The main function of the supporting body is to neutralize or create a state of equilibrium. Since a gear is a rotating or moving body, a state of *dynamic equilibrium* must be obtained.

To be in dynamic equilibrium, all the reactions from the rotating gear must be neutralized by equal and opposite forces supporting the gear shaft. All couples and moments from the gear or power source must also be contained. In essence, the total work forces and moments in must equal the total forces and work out. Only the most basic form of dynamics is considered. It is realized that shaft unbalance, accelerating and decelerating changes, and various other dynamic forces exist. Normally these are not an important factor. Gears generally rotate about a near-uniform center axis which

---

In the first edition, this chapter was prepared by Martin A. Hartman, then with the Rocketdyne Division of North American Aviation, Canoga Park, California.

TABLE 10.1 Units Used for Gear Reactions

Item	Metric	English
Force	Newtons	pounds
Distance	meters, or millimeters	inches or feet
Time	minutes or seconds	minutes or seconds
Work	Newton·meters	inch·pounds
Power	kilowatts	horsepower
Velocity	meters per second	feet per minute

always tends to pass through its center of gravity and percussion; gears and the connecting bodies reach a state of relative constant velocity. Only short time periods of extreme acceleration or deceleration exist. However, these conditions must be considered and taken into account when they form the major loading conditions. They will not be considered further here.

In this article, a brief review is made of the basic mechanics used in gear reaction calculations. The units to be used may be either English units or metric units. Table 10.1 shows the units normally used.

To have dynamic equilibrium, all the laws of dynamics must be obeyed and all forces must have direction, magnitude, and a point of application. This is shown by a graphical or pictorial representation by means of a line with an arrowhead. The arrow denotes direction; the length of the line, magnitude; and the tail, the point of origin.

10.1.1 Summation of Forces and Moments

Any number of lines of force in three planes can be resolved into one resultant. Figure 10.1 shows two forces in an XY plane and their resultant.

$$L = \sqrt{X^2 + Y^2} \tag{10.1}$$

$$\tan \theta_x = \frac{Y}{X} \tag{10.2}$$

FIGURE 10.1
Vector diagram.

Conversely,

$$X = L \cos \theta_x \tag{10.3}$$

$$Y = L \sin \theta_x \tag{10.4}$$

A moment of a force is equivalent to the sum of the individual moments of its components. This is true in either coplanar or three planes.

In Fig. 10.2 a moment is shown in the XY plane

$$M = L \cdot d \tag{10.5}$$

$$M = X(dy) + Y(dx) \tag{10.5a}$$

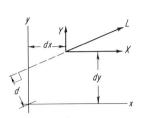
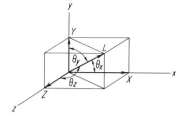

FIGURE 10.2 Moment diagram. $M = L \cdot d = X\,dy + Y\,dx$.

FIGURE 10.3 Three-dimensional vector diagram.

These same laws hold for three planes, and for dynamic equilibrium their total sum must equal zero except for the power transmitted or lost.

In Fig. 10.3 for forces

$$L = \sqrt{X^2 + Y^2 + Z^2} \qquad (10.6)$$

where

$$\begin{aligned} X &= L\cos\theta_x \\ Y &= L\cos\theta_y \\ Z &= L\cos\theta_z \end{aligned} \qquad (10.7)$$

Summation of moments is as shown in Fig. 10.4

$$M = M_x \leftrightarrow M_y \leftrightarrow M_z \qquad (10.8)$$

or

$$M = \sqrt{M_x^2 + M_y^2 + M_z^2} \qquad (10.8a)$$

$$\begin{aligned} \cos\theta_x &= \frac{M_x}{M} \\ \cos\theta_y &= \frac{M_y}{M} \\ \cos\theta_z &= \frac{M_z}{M} \end{aligned} \qquad (10.9)$$

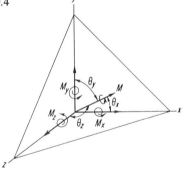

FIGURE 10.4 Three-dimensional moment diagram.

It is also important to know two other facts pertaining to couples and moments:

1. Figure 10.5 shows a couple that has no resultant; the two forces are equal but act in opposite directions. This produces rotation, and the moment is the product of one force multiplied by the entire distance between them. Only an equal and opposite couple can balance another couple. It cannot be balanced or held stationary by a single force or reaction (see Fig. 10.5b).

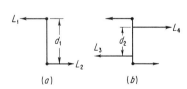

FIGURE 10.5 A couple: (a) couple, (b) couple in equilibrium.

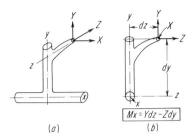
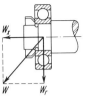

(a)　(b)　　$M_x = Y\,dz - Z\,dy$

FIGURE 10.6 Force vectors at a point in space.

FIGURE 10.7 Load reactions at a bearing. W = combined load. W_r = radial load at 90° to axis of rotation. W_x = thrust load directly along axis of rotation.

2. All gears rotate about an axis, and it is here that the application of moments in space can be simplified. No force has a moment about a parallel axis.[2*]

Figure 10.6a and b shows how the moments in space are made coplanar and the X axis is a point. Therefore, the moment of force about the X axis is due only to the sum of the Y and Z forces and arms.

To have equilibrium,

$$\Sigma X = 0 \quad \Sigma Y = 0 \quad \Sigma Z = 0 \quad (10.10)$$

$$\Sigma M_x = 0 \quad \Sigma M_y = 0 \quad \Sigma M_z = 0 \quad (10.11)$$

Since this shows only three independent variables, there can exist a maximum of three unknown values. To have dynamic equilibrium, there cannot be more than six unknown quantities to be solved, and both sets of Eqs. (10.10) and (10.11) must be completely satisfied. These rules always hold true for gears rotating about any axis.

To reduce gear reactions and their resultant bearing load calculations to the simplest form, no matter how many moments or forces there are on a gear, always resolve them into two basic bearing loads, one of which is axial and the other radial, as shown in Fig. 10.7.

The one basic rule that must be applied to all gear mounting and mechanics analysis is, *the summation of all forces must equal zero and the summation of all moments must equal zero.*

10.2 BASIC GEAR REACTIONS, BEARING LOADS, AND MOUNTING TYPES

There are various sources of gear reactions and bearing loads. These are further complicated by the manner in which the gear and its shaft are mounted.

The main sources of loads are torque, reactions, weight, centrifugal forces, and vibrations. In most cases the gear torque is the main applied load and is usually caused by power input and work being done at the output.

To determine torque or twisting moment and loads from horsepower, the amount

---

*Superscript numbers refer to references at the end of the chapter.

of horsepower, the revolutions per minute, the pitch diameters of gear and pinion, and the reduction ratio must be known.

$$T_1 = \frac{P \times 9549.3}{n_1} \quad (10.12a)$$

$$T_P = \frac{P_h \times 63{,}025}{n_P} \quad (10.12b)$$

where P (P_h) = power [metric (English)], kW (hp)
T_1 (T_P) = pinion torque, N · m (in · lb)
n_1 (n_P) = speed, r/min
d_1 (d) = pitch diameter, mm (in.)

The gear torque is

$$T_2 = \frac{T_1 \times \text{number of gear teeth}}{\text{number of pinion teeth}} \quad (10.13a)$$

$$T_G = \frac{T_p \times \text{number of gear teeth}}{\text{number of pinion teeth}} \quad (10.13b)$$

The tangential load at the pinion pitch diameter is designated for both metric and English units.

$$W_t = \frac{2 \times \text{pinion torque}}{\text{pitch diameter of pinion}} \quad (10.14)$$

10.2.1 Gear Reactions to Bearings

With W_t known it is possible to calculate the separating force W_r' and the axial force W_x. Once these are determined, the values may be added vectorially so that W_t and W_r' determine the total radial load W_r on the gear and bearing and the axial force W_x.

As stated in Sec. 10.1 and shown in Fig. 10.8, the total radial (W_r) and axial load (W_x) determined in space are all that is necessary to determine total loads W on the gear teeth and reactions on bearings.

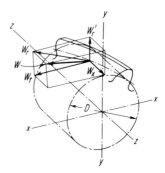

$$W_r = W_r' \leftrightarrow W_t \quad (10.15)$$
$$W = W_r \leftrightarrow W_x \quad (10.16)$$

FIGURE 10.8 Tooth reactions on a gear loaded in three planes.

10.2.2 Directions of Loads

The method of finding the calculating W_r', W_r, W_x is found later on and is shown for spur and helical gears. There are also general rules that hold true for directions of loads on driving and driven members.

The radial load on the bearing supporting a gear tooth that is driving is always opposite to its direction of rotation.

The radial load on the bearing supporting a gear tooth that is being driven is always the same as its direction of rotation.

The rules for determining axial loads tend to be more complicated:

When viewed from the axis of rotation and when the hand of spiral is left on a counterclockwise-rotating driving gear, its axial bearing thrust is away from the viewer.

When viewed from the axis of rotation and when the hand of spiral is right on a counterclockwise-rotating driving gear, the axial bearing thrust is toward the viewer.

When viewed from the axis of rotation and when hand of spiral is left on a clockwise-rotating driving gear, the axial bearing thrust is toward the viewer.

When viewed from the axis of rotation and when the hand of spiral is right on a clockwise-rotating driving gear, the direction of axial bearing thrust is away from the viewer.

The direction of the driven gears' axial thrust is always directly opposite the axial thrust of the driving gear.

This can all best be shown in tabular form in Table 10.2.

It should always be kept in mind that the tangential, radial, axial, and total vectorial sums of any of these same components on a driving gear are always directly opposite to those on the driven gear.

10.2.3 Additional Considerations

Consideration must be given, not only for the axial and radial bearing loads generated from the geometry of the gears, but for the other external forces that must be supported by the shaft and bearings. For example, spur gears should not normally generate any axial thrusts, but, on a common shaft, a pump, fan, or propeller could be attached, generating substantial axial or centrifugal forces; and these would have to be restrained as well as the gear forces.

Transfer of power in any mechanism always results in losses due to inefficiency. In the calculations presented here, 100-percent efficiency is assumed. In types of gearing with high rubbing rates, such as worms, the output power should be reduced because of the higher losses incurred.

Weights of actual gears are usually so small that they are of no consequence. Direction of gravity and weight must be considered when large gears or when large

TABLE 10.2 Direction of Axial Thrusts on Driving and Driven Gears*

Hand of Spiral	Direction of Rotation	Driving	Driven
Left	Clockwise	Toward viewer	Away from viewer
	Counterclockwise	Away from viewer	Toward viewer
Right	Clockwise	Away from viewer	Toward viewer
	Counterclockwise	Toward viewer	Away from viewer

*For thrust directions on bevel and hypoid gears, see Sec. 10.7.

GEAR REACTIONS AND MOUNTINGS   10.7

masses are attached to the gear shaft. Care must also be taken in allowing for the effect of g, or gravity loadings. In some cases values as high as 15 g's are encountered.

Forces caused by the rotation of a mass around a center different from its own center of gravity are called centrifugal forces. Parts that are out of balance, have large external masses, or rotate in a planetary field have centrifugal forces that must be considered. The most common formula for calculation is

$$W_{cf} = \frac{\bar{r}wn^2}{36,128} \qquad (10.17)$$

where W_{cf} = centrifugal force, lb
$\bar{r}$ = radius from rotational center to center of gravity, in.
n = rotations per minute
w = weight, lb

or, where in · oz of unbalance are known,

$$W_{cf} = 1.73 \left(\frac{n}{1000}\right)^2 \times \text{in · oz of unbalance} \qquad (10.18)$$

For those needing an answer in kilograms of force, the simple English formula shown above can be used, and the answer in pounds can be converted to kgf by dividing by 2.20462.

All gear reactions can be resolved into tangential force, separating force, and axial thrust. The relative values of these varies in intensity depending on the type of gears; and the loads applied to the bearings are extremely dependent on the type of mounting used.

10.2.4 Types of Mountings

There are really only two basic types of mountings—straddle and overhung.

In straddle mounting, as shown in Fig. 10.9, the radial load is divided in inverse proportion to the distance from points of application to the total distance between shaft supports. The bearing reactions act opposite to the direction of the load from the gearing.

In an overhung mounting as in Fig. 10.10, the load is applied outside the two supports. This produces a greater reaction or load on the bearing nearest the applied load. The load on the bearing farther away is smaller. The bearing loads are not in the same directions. The bearing reaction nearest the applied load is opposite to the direction of the load. The reaction on the other bearing, farthest away, is in the same direction as the applied load.

Since, in an overhung-mounted gear, the largest bearing load is nearest the mesh, this assumes greater importance in heavily loaded optimum-designed bearings mountings. To compensate for this, generally the bearing nearest the mesh is of a type that has the greatest radial load-carrying ability. The bearing farthest away is often a ball or axial resistance bearing and constrains both the axial and the smaller radial loads. This method is used when there are no great temperature or deflection problems. In some cases two different-sized bearings are used, with the greater-capacity bearing nearest the overhung member. It is wise to locate the axial-thrust retaining bearing next to the gear when large temperature variations or deflections exist. Change of rotation and potential thrust-direction changes must be taken into account.

While it is necessary to balance out reactions loads, from now on whenever a load

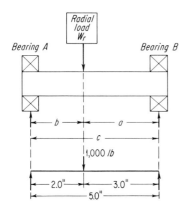

Radial load on $A = \dfrac{W_r a}{c}$

$= \dfrac{1{,}000 \times 3.0}{5.0} = 600$ lb

Radial load on $B = \dfrac{W_r b}{c}$

$= \dfrac{1{,}000 \times 2.0}{5.0} = 400$ lb

(600 + 400 = 1,000. Check.)

FIGURE 10.9 Example of straddle mounting.

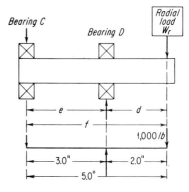

Radial load on $C = \dfrac{W_r d}{e}$

$= \dfrac{1{,}000 \times 2.0}{3.0} = 667$ lb

Radial load on $D = \dfrac{W_r f}{e}$

$= \dfrac{1{,}000 \times 5.0}{3.0} = 1{,}6670$ lb

(1,667 − 667 = 1,000. Check.)

FIGURE 10.10 Example of overhung mounting.

is calculated it will be the actual load applied to the bearing or support in the correct direction. It will not be the bearing reaction, which opposes the load and is in the opposite direction of the bearing load.

There are relative merits between the overhung- and straddle-mounted gears and shafts. The most commonly used and preferred method is the straddle mounting. It balances loads, more efficiently utilizes the bearings, and in most cases reduces misalignment and deflection problems.

When there are extreme or unusual axial-thrust loads or special mounting problems, the overhung mounting has its merits.

10.2.5 Efficiencies

Gear efficiencies and resultant power losses can cause great variation in the actual power delivered and the load transmitted. These loss values vary from ½ percent per mesh to values as high as 80 percent per mesh, depending on the type of gear, lubrication, bearings, and degree of accuracy of manufacturing. It is a poor gear design or application that has an efficiency less than 50 percent. It is generally a nonplanar application that has great losses. Some designers allow a set value for each mesh. While 100-percent efficiency is assumed in all these calculations, it is always necessary at least to consider the effect of efficiency. If the power loss is low, generally a few percentage

GEAR REACTIONS AND MOUNTINGS   **10.9**

TABLE 10.3 Approximate Gear Losses for Various Types

Different Types of Gear	Total Range of Losses per Mesh, %
Spur and helical external	½–3
Spur and helical internal	½–3
Bevel gears	½–3
Worms	2–50
Spiroids*	2–50
Hypoids	2–50
Crossed helicals	5–50

*Registered trade mark of Spiroid Division, Illinois Tool Works, Chicago, Illinois.

NOTE: All nonplanar gears have great variation due to ratios, sizes, materials, lubricants, and relative sliding.

points or less, it is best to ignore this effect on bearings and gears. Coplanar gears can be from 97 to 80 percent efficient. When critical, experience or actual test data should be used to determine losses, and reference should be made to Chap. 12.

As a quick guide for efficiencies or if useful data are not available, the values in Table 10.3 are suggested for good conditions.

10.3 BASIC MOUNTING ARRANGEMENTS AND RECOMMENDATIONS

Once the bearing loads are calculated for any gear, experience, art, and common sense give some pointed recommendations and rules. These rules apply to all simply mounted gears no matter what type or shape they take. Various other rules for each specific type of gear are shown later under their appropriate heading.

Keep bearing mountings for gears as close as possible to their faces, allowing reasonable space for lubrication and arrangements. This eliminates large moments and reduces vibration problems.

Whenever possible, use only two bearings for each gear shaft. When making a center-distance layout for gear proportions, also consider shaft, spline, and bearing sizes and load-carrying abilities. These often are limiting conditions.

Whenever possible, straddle mount both members of a mesh. For straddle-mounted gears a spread between bearings of approximately 70 percent of the pitch diameter is a minimum. When only one gear can be straddle-mounted, it should be the gear with the highest radial load.

Overhung gears should have a spread between bearings of 70 percent of the pitch diameter, and the spread between bearings should be at least twice the overhang. The shaft supporting the overhung gear should be greater in diameter than the overhang (see Fig. 10.14).

Idlers can often be positioned to offset and minimize bearing loads. Consideration must be given to the potential for the two meshes on the gear train setting up undesirable torsional vibration and overloads.

In general, for all types of gears, the tangential forces applied to the bearing are opposite to the direction of rotation of the driving member. The separating forces are away from the tooth surfaces. The driven gear has its tangential forces and separating forces always opposite to the driving pinion or driving force. Reverse direction of rotation, and there is a reversal of tangential driving loads, but the separating forces tend to remain in the same direction away from the teeth (toward center).

Axial thrusts vary for all types of gears, but they also reverse direction with reversal of rotation except in the case of Zerol* and straight bevel gears.

If the total force on any pinion or any gear of a single mesh is found, the force on the mating gear of that mesh is directly opposite to it. When calculations are laborious, this fact can be used to establish the direction and magnitude of the forces on the mating gear for that individual meshing point in space.

All shafting should be checked for torsional and moment load-carrying ability. The distance between supports and shaft diameter should preferably be designed so that operating speed is below the first critical speed. The shaft should not operate near any critical vibration frequency.

10.3.1 Bearing and Shaft Alignment

It is extremely important to minimize gear tooth mounting displacements. The bearing sizes and shaft dimensions are determined only in part by strength and life considerations. The above-mentioned rules are good examples that should be followed in addition to utilizing strength and stress knowledge. The bearing spacing and shaft stiffness, as well as the housing and methods of attachments between gear, shaft, bearing, and housing, must be considered. In most applications the deflections should not exceed 0.001 at the working surfaces between the gear and pinion. Total amount of allowable misalignment between gear surfaces should be further investigated and calculated. The method of calculating increased stresses in spur and helical gears due to misalignment should be investigated by the formulas shown in Chap. 11. The total contributing factors to misalignment should not exceed the calculated or known allowable value.

10.3.2 Bearings

From the tentative gear layout, calculate the bearing loads and establish the preferred types and sizes of bearings. Actual calculations should be based on data that follow in detail each type of gearing and mounting that is involved. It is an accepted practice to preload the bearings carrying the thrust loads in order to obtain mountings with minimum total axial displacements under operating loads. The amount of preload depends on the mounting, load, and operating speed and should be established in collaboration with the bearing manufacturer. At this time the design of the gearbox with complete speed, load, and operating data should be submitted to the bearing manufacturer for approval of bearing sizes, types, bearing mountings, lubrication, and bearing life. The calculation of the bearing load capacities and life is not included here, but such data and information are readily available from any first-rate bearing manufacturer. Many bearing manufacturers have engineering design books that are given to users of their products. Selection of dimensional fits can also be learned. Generally, the most important rule is for the tight or interference fit to be between the inner race and

---
*Registered trademark of Gleason Works, Rochester, New York.

GEAR REACTIONS AND MOUNTINGS	10.11

shaft, and a loose or controlled slippage to be between the outer race and the housing or bearing liner.

10.3.3 Mounting Gears to Shaft

Whenever possible the gear should be made integral with the shaft. However, because of complex shapes and bearing and mounting arrangements, and with several gears and components mounted on a single shaft, this is not always possible. When gears must be assembled onto the shaft, involute splines are often used. In mounting gears with splined bores, a cylindrical centering fit is recommended to avoid excessive eccentricity in the final mounting. The most satisfactory method is to provide a suitable length of cylindrical fit at one or both ends of the gear hub for centering purposes. The splines are then used for driving only. This method is particularly applicable in the case of involute side-fit splines that cannot be adequately centered from the major or minor diameters (see Fig. 10.35). When space or length allowable is critical, the major-diameter-fit involute spline is often used. The internal spline is often broached and the major diameter of the external member is ground to fit the broached part. When used this way, the spline is stressed from both tangential driving torques and possible tip loading from interference fits caused by assembly or temperature contractional effects.

Present practice is to avoid the use of minor-diameter-fit involute splines whenever possible. In many applications straight-sided splines and serrations are used. Serrations and straight-sided splines are generally treated in the same manner and their rules of usage are interchangeable. Hardened gears with straight-sided splines in the bore should be centered in assembly by the bore or minor diameter of the splines, which, after hardening, should be ground concentric with the teeth. Unhardened gears with straight-sided splines should be centered in cutting and in assembly by the major diameter of the splines. Since heat treatment may introduce distortion in all splines, it is important that the splines be of no greater length than is actually required for load transmission.* In blanks with long hubs, the splines should be located as nearly as possible to the gear teeth and, if possible, directly under the gear teeth.

Keys, pins, and press-fit diameters are used in many gear applications. In data gearing, loads normally are light and the mounting method is picked to give the least amount of running-index error. The integral gear and shaft is preferred, with the cylinder fit of gear to shaft next. When heavy loads and extreme accuracy requirements are needed, ingenious combinations of various basic mounting methods must be employed.

10.3.4 Housing

The function of the housing is to form a strong base in which to mount the bearings and support the gears and shafts and to create an environment and space where a satisfactory lubricant may be introduced to lubricate and cool the gears. The housing is also used to mount and support various other components such as accessories and parts near or common to the gearing. In many installations the lubrication system is contained entirely with the housing; in some oil systems, lines into and out of the housing must be supplied.

The actual design and calculation of most gearboxes can be so complicated that a

---

*Generally the length of a spline is equal to or less than the spline pitch diameter.

general practice is to employ experience or test data or actually to submit the gear housing to loads and measure the results by dial or strain gauges. In simple-type shapes, basic stress analysis is used. With the bearings selected and the general arrangement established, it is possible to proceed with the design of the supporting structure. The design should ensure, whenever possible, that loads from external sources such as belts, propellers, and pumps are carried entirely independently of the gear mountings; otherwise the design must be made sufficiently rigid to carry these external loads without affecting the operating position of the gears. Flexible-type couplings are preferred to connect the assembly to the input and output loads in order to protect the gears from outside forces resulting from shaft deflections or misalignments.

The housing and supports should be of adequate section and of suitable shape, with ribbing to support the bearings rigidly against the forces, as indicated by the analysis of their magnitude and direction. The choice of material for the housing depends to a great extent on the application. If weight is not a factor, ferrous castings or a combination of ferrous castings with weldments may be used. Where weight is very important, aluminum or magnesium castings may be used for the housing. When light metal housings are used, special design considerations must be given to minimize the effect of temperature changes on the mountings and to provide adequate rigidity under all operating conditions. Best practice calls for mounting the bearings in steel liners, either fitted and pinned in the housing bores or cast in and machined in place. Also, the thrust bearings that control the position of the gears on their axes should be located close to the gear teeth. Two opposed bearings should be arranged so that temperature changes will not seriously affect preloading. It should be noted that in order to obtain equal rigidity in a light metal casting it may be necessary to use sections of two to four times the thickness of those used in steel or iron.

10.3.5 Inspection Hole

As an aid to assembly and for periodic inspection in service, an opening should be provided in the housing to permit the inspection of the tooth surfaces of at least one member of the pair without disassembly. This is especially desirous in nonplanar axis gears or in bevel and hypoid gears.

10.3.6 Break-in

While gears and bearings are designed, mounted, and manufactured to carry their rated loads without trouble, the initial period of operation is most critical, and preferably a "break-in" run at lighter loads or slower speed should be made. The load should gradually be increased until full operating load and speed are reached to check the complete gearbox.

10.4 BEARING LOAD CALCULATIONS FOR SPUR GEARS

The most common gear used is the spur-type. This is true not only because of the simplicity of stress calculations, design, and manufacturing but largely because of the ease of mounting, calculations, and retention of bearing loads. Spur gears are almost

GEAR REACTIONS AND MOUNTINGS   **10.13**

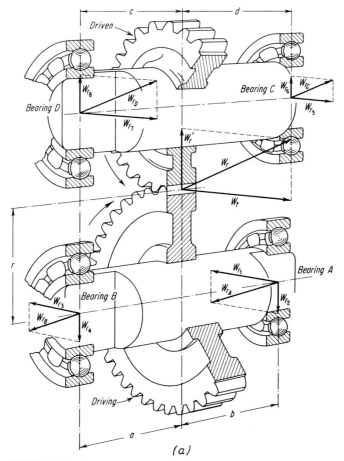

FIGURE 10.11 Spur gear bearing loads.

always devoid of any large self-generated axial thrusts. Their only components are the tangential driving load and its separating components, which results in a radial load only, which is the vectorial sum of these two forces [see Eq. (10.15)].

In a simple spur gear, only a radial load is present and W_r can be used interchangeably with the total load W. Because this fact assumes greater importance in other types of gears such as helical, W_r is used here.

The tangential driving load can be calculated by Eq. (10.14). The separating load is designated W'_r. For universal usage, the separating load is a function of the transverse pressure angle or that pressure angle formed by a plane slicing through the gear tooth at 90° to the axis of rotation. In a spur gear, the normal pressure angle ϕ and the transverse pressure angle ϕ_t are one and the same. To simplify matters later on, the transverse pressure angle ϕ_t will always be used in separating and radial-load calculations. (For metric calculations use $\tan \alpha_t$ and use newtons for W_t instead of pounds.)

Forces		Bearing A	Bearing B
Tan force W_t		$W_{r_1} = \dfrac{W_t a}{a+b}$	$W_{r_3} = \dfrac{W_t b}{a+b}$
Separating force W_r'		$W_{r_2} = \dfrac{W_t' a}{a+b}$	$W_{r_4} = \dfrac{W_r' b}{a+b}$
Total load		$W_{r_A} = \sqrt{(W_{r_1})^2 + (W_{r_2})^2}$	$W_{r_B} = \sqrt{(W_{r_3})^2 + (W_{r_4})^2}$
		Bearing C	Bearing D
Tan force W_t		$W_{r_5} = \dfrac{W_t c}{c+d}$	$W_{r_7} = \dfrac{W_t d}{c+d}$
Separating force W_r'		$W_{r_6} = \dfrac{W_r' c}{c+d}$	$W_{r_8} = \dfrac{W_r' d}{c+d}$
Total load		$W_{r_C} = \sqrt{(W_{r_5})^2 + (W_{r_6})^2}$	$W_{r_D} = \sqrt{(W_{r_7})^2 + (W_{r_8})^2}$

(b)

FIGURE 10.11 (*Continued*)

$$W_r' = W_t \tan \phi_t \tag{10.19}$$

Let W_r designate the total radial load

$$W_r = \sqrt{W_t^2 + W_r'^2} \tag{10.20}$$

If W_t is designated in the X plane and W_r in the Y plane, Eq. (10.20) agrees exactly with Eq. (10.1), which is for forces in one plane.

The values of W_r, W_t, and W_r' will all be considered as originating from the pitch diameter of the pinion or gear and in the center of its contact face, as will W and W_x later on.

Figure 10.11 shows a perspective of two spur gears and their respective driving or driven load.

The pressure angle of the gear controls the effect of the separating component. The smaller pressure angle gives lesser separating and total forces, as shown in Fig. 10.12. Equation (10.20) is plotted against the tangent of the pressure angle to calculate W_r

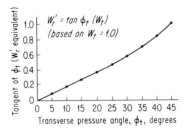

FIGURE 10.12 Chart of relation W_r for $W_t = 1.0$.

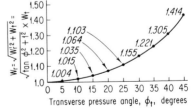

FIGURE 10.13 Chart of total radial load versus ϕ_t.

GEAR REACTIONS AND MOUNTINGS   10.15

proportions shown in Fig. 10.13. A pressure angle of 0 to 45° represents one unit of tangential driving load in pounds.

Figures 10.9 and 10.10 and sample problems show how to calculate overhung- and straddle-mounted gear resultant bearing reactions.

Spur gears have no calculable axial thrust; however, they tend to walk or to be displaced by torque and other components. It is wise to allow for small amounts of axial movements or to restrain the gear. In heavily loaded spurs, or when spurs are common to shafts carrying other axial loads, it is a necessity to restrain or position them.

10.4.1 Internal Gears

All data shown so far have applied to external gears. However, internal gears are used in many applications. In external gears, the direction of rotation is changed in the mesh.

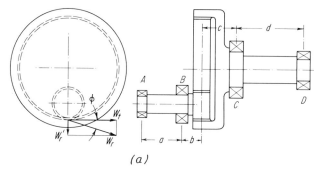

(a)

Forces	Pinion loads	
	Bearing A	Bearing B
Tan force W_t	$W_{r1} = \dfrac{W_t b}{a}$	$W_{r3} = \dfrac{W_t(a+b)}{a}$
Separating force W_r'	$W_{r2} = \dfrac{W_r' b}{a}$	$W_{r4} = \dfrac{W_r'(a+b)}{a}$
Total load	$W_{rA} = \sqrt{(W_{r1})^2 + (W_{r2})^2}$	$W_{rB} = \sqrt{(W_{r3})^2 + (W_{r4})^2}$
	Gear loads	
	Bearing C	Bearing D
Tan force W_t	$W_{r5} = \dfrac{W_t(c+d)}{d}$	$W_{r7} = \dfrac{W_t c}{d}$
Separating force W_r'	$W_{r6} = \dfrac{W_r'(c+d)}{d}$	$W_{r8} = \dfrac{W_r' c}{d}$
Total load	$W_{rC} = \sqrt{(W_{r5})^2 + (W_{r6})^2}$	$W_{rD} = \sqrt{(W_{r7})^2 + (W_{r8})^2}$

(b)

FIGURE 10.14   Internal spur gear bearing loads.

In internal gears, the rotation between the two gears is in the same direction. In most cases one or both of the gears are overhung. While it is possible to arrange straddle mounting, normally space and length do not permit it. Internal gears offer the advantage of larger reduction or step-up changes in smaller center distances, but at the expense of allowable space for bearing mounting arrangements.

Procedures for calculating bearing loads are shown in Fig. 10.14; W'_r, W_t, and W_r are calculated as before.

10.4.2 Gears in Trains

Many times gears are in trains, either in one continuous plane with idlers or in offset planes on intermediate shafts and in planetary drives. Planetary gears are considered as a separate type of gear train.

10.4.3 Idlers

The idlers may be used to change direction of rotation, to gain additional distance between centers with smaller-sized and greater quantity of parts, or to provide additional mounting pads at various r/min for driven components, shafts, or accessories.

If only the input and output shafts are used and a constant horsepower is driven through the train, the load on all the gear teeth in the train is the same. When the idlers are all arranged in one straight line, the two separating components W'_r cancel out, but the tangential driving loads are added directly. Under the aforementioned conditions, the resultant bearing load is twice W_t.

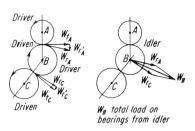

FIGURE 10.15 Vector analysis of idler loads.

When the gear train is offset at an angle between centers, the total forces tend to combine or cancel out. For accuracy of results, a graphical or vector analysis should be made. Figure 10.15 shows a vector analysis of idler loads. If in Fig. 10.15 the angle on BC in relation to AB had been acute instead of obtuse, the reaction W_B would have been much smaller in magnitude. If it is possible, proper selection of rotation and angles of intersection can be adjusted to reduce the total bearing loads generated in an idler gear train.[3]

10.4.4 Intermediate Gears

The other most common gear train in spur gears is the intermediate shaft arrangement. Its main advantage is the combination of greater gear ratio reductions and a gain in center distances for mounting various combinations.

There are basically two gear meshes but only three shafts instead of four. Also, the planes of mesh are parallel but separated (see Fig. 10.16).

The total loads that are applied to and must be reacted against by bearing D are found by a summation of moments about bearing C. The same is true as to finding the forces on bearing C. Summation of forces, however, is taken about bearing D. When contact is at an angle or several meshes are concerned, both vertical and horizontal planes of moments must be taken. The bearings must resist a load to make the total

GEAR REACTIONS AND MOUNTINGS 10.17

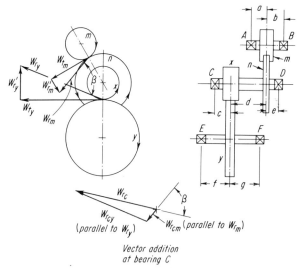

FIGURE 10.16 Intermediate gear trains.

moment summation equal zero. The proportions of spacing between bearings are very critical. The distances or moment arms should be selected to favor the F bearing with the lesser load-carrying capacity. For example, bearing F in Fig. 10.16 would, in most gearboxes, be the smaller because of lack of space. Therefore, the distance g on the y gear was intentionally made longer to reduce the bearing load on F and put the greater load on E, as it should have more space available for a larger bearing with greater load-carrying capacity. Also, whenever possible, the distance between gears and bearings is as small as is feasible to eliminate large moment loads in the shafts. The shaft on $x - n$ gears will have a coupling tendency to cock or walk, and consideration should be given to axial-thrust bearings to resist the coupling effect. To aid the positioning of planes of intersection on the gear teeth, avoid large axial moments and overhangs.

10.4.5 Planetary Gears

Planetary gears are used most often for maximum horsepower transmitted in the smallest space or when large changes in speed ratios are needed. The increased carrying capacity is generally possible because the gear reactions tend to equalize each other, reducing bearing loads, size required, and total weight. Helical, bevel, and spur gears are used in planetary drives, but spurs are often preferred because of ease of manufacturing and elimination or reduction of axial-thrust loads.

For simplicity's sake, spur gears have been assumed in these discussions, but helical and other types of gears can just as easily be used.

In a simple planetary arrangement there is generally a sun gear, a planet gear, an internal ring gear, and a cage or arm (see Fig. 10.17).

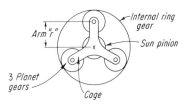

FIGURE 10.17 Simple planetary arrangement.

10.4.6 One Planet

If the planetary drive has one planet gear, load reactions are different from those in the case where there are two or more planet gears. The one-planet-gear calculations are basically the same as the combination of an external idler and an internal gear and external pinion. For example, the sun-and-one-mesh planetary drive are the same as two external gears meshing. The reaction on the internal gear is the same as in a simple one meshing with an external pinion. However, on the planet gear, consideration must be given for the reactions on its center of rotation from the cage or arm.

10.4.7 Several Planets

If two or more planets are used, preferably three or more, the reactions are different because three planets mate with one sun gear and one ring gear; and the bearing reactions tend to balance out. There are basically zero resultant separating bearing forces on the sun pinion and the internal ring gear. The planet gear bearing has the loads from the two meshes and the cage driving arm, plus the weight and centrifugal force from the planet and its bearings. When there are high-speed planets, the effect of centrifugal force must be considered. If speeds and weights are low, centrifugal force may be ignored. If they are high, compute them by means of Eq. (10.17) or (10.18).

Centrifugal force may become the prime factor in the final decision of bearing design or type of planetary gear drive used.

When the total gear reaction W and W_{cf} (centrifugal force) are found, they are added vectorially or computed as follows:

$$L = \sqrt{(W_{cf})^2 + (W)^2} \qquad (10.21)$$

where L = total load of all forces
  W = total gear reactions
  W_{cf} = centrifugal force

However, helix forces, unbalance, and driven propeller or fan reactions must be considered when they are tied into the planetary drive system. When two or three planets are used, the loads should be considered shared equally by all planets. When there are more than three planets, it is not always possible to share the load equally, and some allowance should be made for unequal loading of all the individual planets.

The following special recommendations apply to idlers. Preferably use two or more planets to carry the load. When possible use three planets. If more than three are needed, many designers feel it is wise to go to five or more because of a lack of load-sharing tendencies.

Always mount the entire system as flexibly as possible. For example, always mount the planets on spherical floating bearings or mounts. Make all mountings or actual holding surfaces for bearings as rugged as is allowable. Keep planets and outer bearing cages or surfaces as light in weight as can be utilized to reduce centrifugal forces.

10.5 BEARING-LOAD CALCULATIONS FOR HELICALS

So far all data have been based on spur gears and all radial values calculated are directly applicable to helical gears if the transverse pressure angle is considered.

Helical gears have thrust and reactions in all three planes. Besides the tangential driving and separating forces, there is also an axial-thrust load. This thrust is due to the helix angle and is normally considered emanating from the tooth contact at the pitch diameter.

Since the thrust is applied at the distance of the pitch radius from the axis of rotation a radial moment is produced. In spur gears no bearing moments had to be considered until the gear was mounted. However, in helicals the moment is produced within the gear.

This information on helicals can be transcribed into the laws of mechanics. Three forces are shown in Fig. 10.8.

W_t = tangential driving load
W'_r = separating force
W_x = axial thrust
W = total load
M_a = moment produced by axial thrust

W_t and W'_r are found as before. It is important, though, to consider a new effect on the pressure angle.

$$\tan \alpha_t = \frac{\tan \alpha_n}{\cos \beta} \quad \text{metric} \quad (10.22a)$$

$$\tan \phi_t = \frac{\tan \phi_n}{\cos \psi} \quad \text{English} \quad (10.22b)$$

where α_n (ϕ_n) = normal pressure angle [metric (English)]
α_t (ϕ_t) = transverse pressure angle
β (ψ) = helix angle

The separating load is a function of transverse pressure angle, which in a helical gear is always greater than ϕ_n, the normal pressure angle. Therefore, W'_r will always be a slight amount larger than in a spur gear but is calculated as before in Eq. (10.19).

The helical angle ψ assumes great importance in calculating the axial and separating load, and in the equations presented, the helix angle will be used for all calculations.

$$\tan \psi = \frac{\pi \times \text{pitch diameter}}{\text{lead}} \quad (10.23)$$

Lead is the distance a point would advance axially across the face of the gear for one revolution measured at the pitch diameter. A small helix angle has a high lead and a large helix angle has a low lead.

W_x increases with a higher helix angle and, whenever possible, should be kept to the smallest angle. However, for a helix angle to be of any value, it should have a face overlap of a little over 1. This means the contacting distance of face advance along the pitch-diameter periphery of a helical tooth is equal to or greater than the transverse circular pitch (see Fig. 10.18).

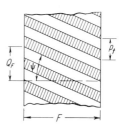

FIGURE 10.18 Helical gearface advance. Face advance = $Q_F + F \tan \psi$. Contact ratio of faces Q_F/P_t ≥ 1.

Normally, for high-speed gears, a face advance of over 2 is preferred.

$$\text{Face contact ratio} = \frac{\text{face advance}}{\text{circular pitch}} \quad (10.24)$$

As gears become more heavily loaded with higher speeds, and quietness is a requirement, a greater overlap ratio is needed. Therefore, the selection of the helix angle becomes a direct function of face advance.

$$\tan \text{ helix angle} = \frac{\text{face advance}}{\text{face width}} \quad (10.25)$$

With the selection of the helix angle, the helical gear forces can be determined as shown in Fig. 10.8. In its simplest form,

$$W_x = W_t \times \tan \text{ helix angle} \quad (10.26)$$

The total load W is the vectorial summation of W_t, W_x and W'_r.

$$W = \sqrt{W_t^2 + W_r'^2 + W_x^2} \quad \text{for } X, Y, \text{ and } Z \text{ axes} \quad (10.27)$$

More consideration must now be given to summation of forces in various planes. In some cases, values are additive and in other cases they are subtractive.

The moment produced from the helix is now calculated.

$$M_a = W_x \times \text{pitch radius of gear} \quad (10.28)$$

where M_a is the axial moment and W_x is the axial thrust. M_a produces an additional moment and, in turn, an additional radial load, which must be contained by the bearings and supports. Also notice that, if the vectors representing W_t, W'_r, and W_x are assumed to be, respectively, the X, Y, and Z axes; Eq. (10.27) conforms to Eq. (10.6).

From the standpoint of load produced axially, it is always desired to keep it to the absolute minimum value. Since a helical gear always requires a bearing to resist an axial force, this can be a disadvantage in some installations. The forces can become so large as to be detrimental in large helix angles. See Fig. 10.19 and Table 10.4 for details of bearing load calculations.

10.5.1 Double Helical

To overcome axial thrust, the forces are sometimes counteracted or nullified by double-helical gears. The total face is made into two equal halves, both with the same helix angle but opposite hands. The axial thrusts oppose one another and the forces are contained in the gear and not transmitted to the bearing. The double-helical gears are often made of two separate halves of helical gears bolted or joined together, or else a space between the gears is allowed in the middle for tool clearances. There are some gear machine tools that permit a continuous herringbone when face width is limited.

Only one gear in a double-helical train is positioned or restrained axially by a bearing, and the rest of the gears then get their location from this one part. For speed-increasing gears it is often wise to use or consider either helical or double-helical

TABLE 10.4 Equations for Helical Gear Bearing Loads with Straddle Mounting

Force	Gear bearing loads		Pinion bearing loads	
	Bearing A	Bearing B	Bearing C	Bearing D
Tan force W_t	$W_{r_1} = \dfrac{W_t a}{a+b}$	$W_{r_4} = \dfrac{W_t b}{a+b}$	$W_{r_7} = \dfrac{W_t c}{c+d}$	$W_{r_{10}} = \dfrac{W_t d}{c+d}$
Separating force W_r'	$W_{r_2} = \dfrac{W_r' a}{a+b}$	$W_{r_5} = \dfrac{W_r' b}{a+b}$	$W_{r_8} = \dfrac{W_r' c}{c+d}$	$W_{r_{11}} = \dfrac{W_r' d}{c+d}$
Thrust W_z	$W_{r_3} = \dfrac{W_z r_1}{a+b}$	$W_{r_6} = \dfrac{W_z r_1}{a+b} = W_{r_3}$	$W_{r_9} = \dfrac{W_z r_2}{c+d}$	$W_{r_{12}} = \dfrac{W_z r_2}{c+d} = W_{r_9}$
Total radial load	$W_{r_A} = \sqrt{(W_{r_1})^2 + (W_{r_2} - W_{r_3})^2}$	$W_{r_B} = \sqrt{(W_{r_4})^2 + (W_{r_5} + W_{r_6})^2}$	$W_{r_C} = \sqrt{(W_{r_7})^2 + (W_{r_8} + W_{r_9})^2}$	$W_{r_D} = \sqrt{(W_{r_{10}})^2 + (W_{r_{11}} - W_{r_{12}})^2}$
Total thrust	W_z (may be applied to either bearing A or B)		W_z (may be applied to either bearing C or D)	

Note: See Fig. 10.19 for definition of terms and bearing locations.

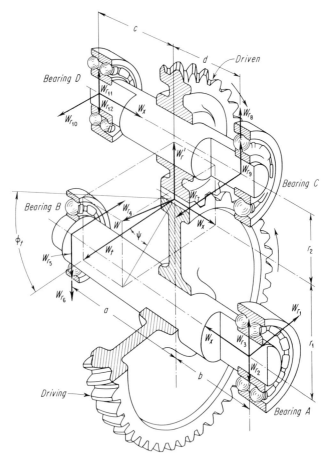

FIGURE 10.19 Helical gear bearing loads.

gears. As axial leads and helix angles increase, so also do the merits of double-helical gears increase.

10.5.2 Skewed or Crossed Helicals

Besides helical gears on parallel axes, it is possible to have the axes of rotation skewed or at an angle with each other. The latter gears are called *crossed helical* or *skewed helical gears*. When the individual gear reactions are found, the same methods are used. In helical gears, the total reaction for the pinion is equal and opposite to the mating gear. In a pair of crossed helicals, each gear and pinion must be considered and calculated separately.

The direction of W_x is generally obvious.

In crossed helical gears one of the mating parts could be a helical or spur gear and it would mate with a helical gear. All spiral helicals tend to have only point contact, are used to transmit light loads, and have small bearing reactions. Allowance must be made for the helix angle in relation to the axis of rotation of each gear. In external helical gears the hands of the mating parts are equal and opposite, but in crossed helical gears all types of hands and angles of helix can and must be considered. Calculations of the bearing loads from the gear reactions are the same as for helical gears. More care and consideration must be given to the angles of helix in relation to the shaft angles. The hand of the helices and shaft angles can be varied so that direction of rotation is reversed. Once the true helix angle and the actual direction of rotation between the driver and driven gears are determined, the reactions are easily found. Figure 10.20 shows a pair of crossed-helical gears. The light lines on the driving gear show the teeth as they would appear to the viewer. The meshing point considered in the diagram is *underneath* the driver. After allowing for the direction of rotation, shaft angle, and hand of spirals, the direction of the forces can be found using Table 10.1. The angles ψ_P and ψ_G can be found, and, knowing W_t, the axial thrusts for the gear and pinion are determined as in any helical gear. See Fig. 10.19 for details of bearing load calculations. In the metric system, the symbol for pinion helix angle is β_1 instead of ψ_P. For the gear the symbol is β_2 instead of ψ_G.

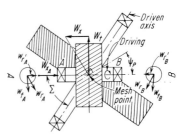

FIGURE 10.20 Bearing loads on crossed helical gears. Regardless of shaft angle between crossed helical pinion and crossed helical gear, the directions of the pinion tangential force W_t and the pinion thrust force W_x are as shown above for the mesh point.

10.6 MOUNTING PRACTICE FOR BEVEL AND HYPOID GEARS*

All gears operate best when their axes are maintained in correct alignment, and, although bevel and hypoid gears have the ability to absorb reasonable displacements without detriment to the tooth action, excessive misalignments reduce the load capacity and complicate the manufacture and assembly of the gears. Accurate alignment of the gears under all operating conditions may be accomplished only by good design, accompanied by accurate manufacture and assembly of both mountings and gears. The following recommendations are given to aid in the design of bevel and hypoid gearbox assemblies in order to obtain the best performance of the gears.

10.6.1 Analysis of Forces

In the design of the mountings for bevel and hypoid gears, the first step is to establish the magnitude and directions of the axial and separating forces, from Figs. 10.21 and

---

*The original material for Secs. 10.6 and 10.7 was prepared by M. L. Fallon, Senior Application Design Analyst and Warner T. Cowell, Chief Engineer, Gear Engineering Test Center, Gleason Works, Rochester, New York.

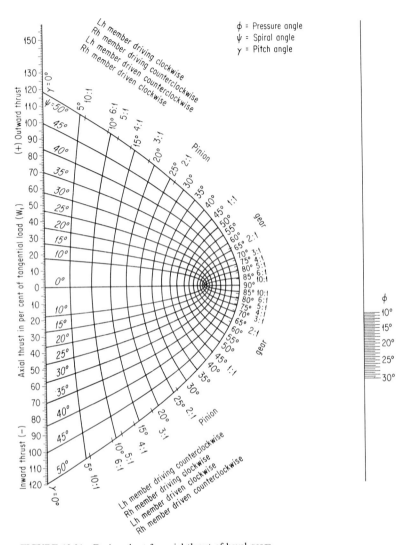

FIGURE 10.21 Design chart for axial thrust of bevel gears.

10.22, and draw a diagram of the resultant force on the gears as shown in Figs. 10.23 to 10.26. These provide the basis for the design of the gear blank and the mountings and for the selection and arrangement of the bearings. In case the gear must operate in both directions of rotation or if there is reversal of torque, the force diagram should include both conditions of operation. Clockwise and counterclockwise rotation are as viewed from the back of the gear looking toward cone center. Values shown are for shaft angles at right angles, but values and determination of directions also apply to shaft angles other than 90°.

GEAR REACTIONS AND MOUNTINGS

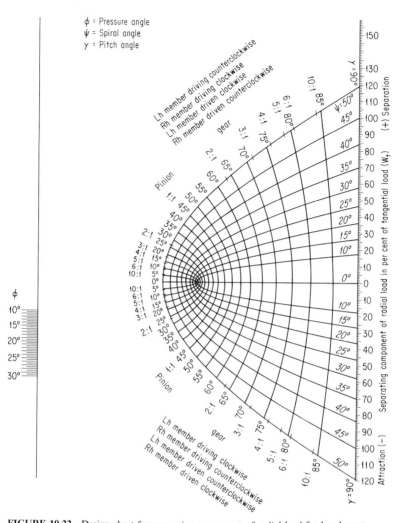

FIGURE 10.22 Design chart for separating component of radial load for bevel gears.

10.6.2 Rigid Mountings

Gear mountings should be designed for maximum rigidity since the problems involved in producing satisfactory gears multiply rapidly when deflections in the mountings cause excessive gear displacements. It is necessary to modify the standard cutting in order to narrow and shorten the tooth contacts to suit the flexible mounting. The decrease in bearing area raises the unit tooth pressure and reduces the number of teeth in contact, which results in problems of noise and increases the danger of surface failure and tooth breakage.

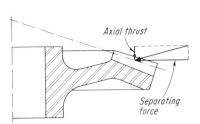
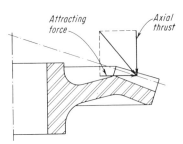

FIGURE 10.23 Resultant force in axial plane due to right-hand gear being driven counterclockwise or driving clockwise; also left-hand gear being driven clockwise or driving counterclockwise.

FIGURE 10.24 Resultant force in axial plane due to right-hand gear being driven clockwise or driving counterclockwise; also left-hand gear being driven counterclockwise or driving clockwise.

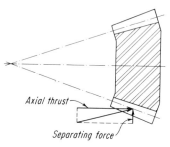
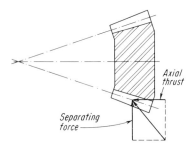

FIGURE 10.25 Resultant force in axial plane due to left-hand pinion driving clockwise or being driven counterclockwise; also right-hand pinion driving counterclockwise or being driven clockwise.

FIGURE 10.26 Resultant force in axial plane due to left-hand pinion driving counterclockwise or being driven clockwise; also right-hand pinion driving clockwise or being driven counterclockwise.

10.6.3 Maximum Displacements

For gears from 6- to 15-in. diameter (150 to 375 mm), the desirable limits of gear and pinion displacements at maximum load are listed as follows:

1. The pinion offset should not exceed ± 0.003 in. from correct position (± 0.076 mm).
2. The pinion should not yield axially more than 0.003 in. in either direction (± 0.076 mm).
3. The gear offset should not exceed ± 0.003 in. from correct position (± 0.076 mm).
4. The gear should not yield axially more than 0.003 in. (0.076mm) in either direction on miters or near miters, or more than 0.010 in. (0.25mm) away from the pinion on higher ratios.

The aforementioned limits are for average applications in which the gears operate over a range of loads and at maximum load approximately 10 percent of the time. When

GEAR REACTIONS AND MOUNTINGS   **10.27**

maximum load occurs for longer periods, reduce the limits accordingly. Somewhat wider tolerances are allowable for larger gears.

10.6.4 Rolling-Element Bearings

While plain bearings may be used in bevel or hypoid gear mountings, it is usually much easier to maintain the gears in satisfactory alignment with rolling-element bearings. Either ball or roller bearings are satisfactory. However, it is important that the type and size of the bearings be carefully selected to suit the particular application, considering especially the operating speeds, loads, the desired life of the bearings, and the rigidity required for the best operation of the gears.

10.6.5 Straddle Mounting

The preferred design of a bevel or hypoid gearbox provides straddle mountings for both gear and pinion, and this arrangement is generally used for industrial and other heavily loaded applications. Usually, the desired bearing life and mounting rigidity may be obtained more economically with a straddle mounting. When it is not feasible to use this arrangement for both members of a pair, the gear member that usually has the higher radial load should be straddle-mounted.

In straddle mounting the gear or larger member of a pair, usually two tapered roller bearings or angular-contact ball bearings are mounted opposed with sufficient spacing between them to provide adequate control of the gear (Fig. 10.27).

A second straddle-mounting arrangement (Fig. 10.28), used mostly for pinions, provides an inboard or pilot bearing for pure radial support of one end of the pinion shaft; and, on the opposite end, bearings suitable for carrying both radial and thrust

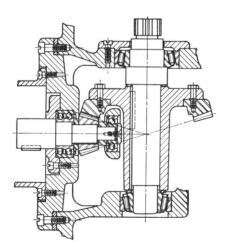

FIGURE 10.27 Webless-type gear member, straddle-mounted using two opposed tapered roller bearings widely spaced to provide good control of gear.

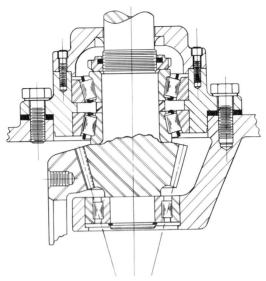

FIGURE 10.28 Straddle pinion mounting for short shafts, showing use of combined thrust and radial bearings.

loads. In this location several different types of bearings are used: a single double-row, deep-groove angular-contact ball bearing; or two single-row angular-contact bearings mounted DB or DF; or two tapered roller bearings, either direct- or indirect-mounted.

A similar arrangement is used also for straddle mounting the gear or larger member of a pair in large applications, and in an aluminum or magnesium housing, where temperature changes would seriously affect the preloading or thrust bearings if spaced one on either end of the gear shaft (Fig. 10.29).

10.6.6 Overhung Mounting

When the pinion is mounted overhung on tapered roller bearings, the indirect mounting should be used (Fig. 10.30). This arrangement provides greater stability to the mounting for a given spacing of the bearings, and the thrust load for normal operation is thus carried by the bearing adjacent to the pinion, adding further to its stability.

The bearing adjacent to the pinion should be located as closely as possible to the pinion teeth to reduce the overhang. See Fig. 10.31.

All spiral bevel and hypoid gears should be locked against thrust in both directions. Provision against inward thrust on straight and Zerol bevel gears is often omitted, provided the conditions of operation are such that inward thrust cannot occur. One of the advantages of Zerol and straight bevel gears is that change of rotation does not change direction of axial thrust.

Since the contact pattern of spiral bevel and hypoid gears is controlled by their axial position, special attention must be given to the selection of the type and to the preloading of the bearings.

The lubrication of bevel and hypoid gears is most important; for specific data refer to Chap. 15 or to instructions of the Gleason Works.

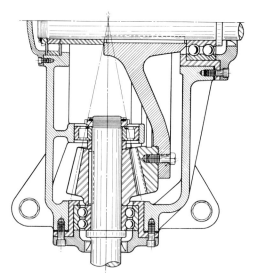

FIGURE 10.29 Straddle mounting for both members of a spiral bevel pair, using a straight roller radial bearing on one end of the shaft and combined thrust and radial bearing on other end. Shim-type adjustments for positioning both gear and pinion.

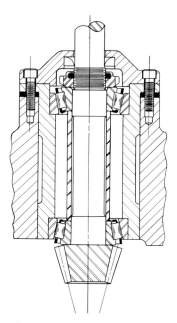

FIGURE 10.30 Typical overhung pinion mounting with shim adjustment for positioning pinion, and selected spacer for preloading bearings.

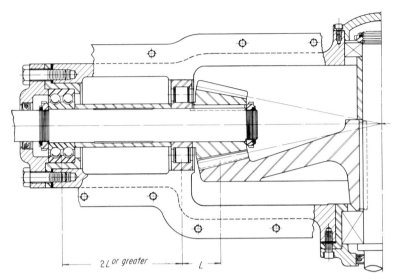

FIGURE 10.31 Typical overhung pinion mounting and straddle gear mounting with shim adjustment for positioning both gear and pinion.

10.6.7 Gear Blank Design

The gear blank should be designed to avoid excessive localized stresses and serious deflections within itself. The direction of the resultant force should be considered in the design of the blank and in the method employed in mounting it on its shaft or centering hub (see Figs. 10.32 to 10.37).

Ring gears are made in three general types: the webless ring, web-type ring, and counterbored type. Of these, the bolted-on webless ring design shown in Fig. 10.27 is best for hardened gears larger than 7 in. (180 mm) in diameter. The fit of the gear on its centering hub should be either size-to-size or a slight interference. These gears should be clamped to the centering hub with fine-thread cap screws as shown in Figs. 10.27, 10.33, and 10.37, or with bolts. Several methods of locking screws and nuts in place are indicated in the illustrations. Self-locking-type nuts and cap screws are widely used for this purpose. For severe operating conditions, cap screws should be locked more positively than is possible by wiring.

The method of centering the gear on the counterbore, shown in Fig. 10.37, is recommended for large gears, especially near miters. Figure 10.33 illustrates mounting for a gear that operates with an inward thrust. Designs in which screws or bolts are subjected to added tension from gear forces should be avoided. On severe reversing or vibrating installations, separate dowel drives may be used with the designs or splines as in Figs. 10.35 and 10.39.

The use of dowels or dowel-fit bolts has been found unnecessary in most automotive and industrial drives because, with bolts or cap screws drawn tightly, the friction of the ring gear on its hub prevents shearing of the screws.

Hardened gears smaller than 7 in. in diameter (180 mm) may best be designed with integral hubs. A front hub on a Zerol bevel, spiral bevel, or hypoid gear should in no case intersect the extended root line so as to interfere with the path of the cutter toward the root cone apex as shown in Fig. 10.38. On hub-type gears the length of the hub

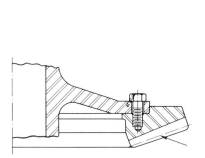 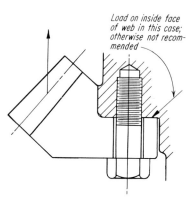

FIGURE 10.32 Method of mounting gear to absorb the thrust with minimum amount of deflection.

FIGURE 10.33 Method of mounting gear when thrust is inward.

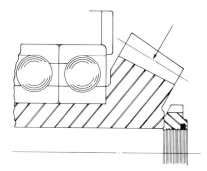

FIGURE 10.34 Blank sections beneath the teeth should be sufficiently rigid in the direction of thrust so that deflections will be minimized.

should be at least equal to the bore diameter, and the end of the hub should be securely clamped against a shoulder on the shaft.

Whether the gear is mounted on a flanged hub or is made integral with the hub, the supporting flange should be made sufficiently rigid to prevent serious deflections in the direction of the gear axis at the mesh point. For larger gears the web preferably should be made conical and without ribbing to permit machining for balance, to eliminate oil churning when dip lubrication is used and to lessen the danger of stress concentrations being set up with the castings. The gear hub and bearing arrangement should be designed with sufficient rigidity that the use of a thrust button behind the gear mesh is unnecessary. The area of a thrust button is too small to provide durable support.

For hardened gears, the blanks should be designed with sections and shape suitable for hardening with minimum distortion.

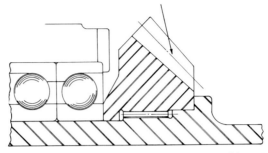

FIGURE 10.35 Gear blank designed to take care of inward thrust. Also a preferred mounting with spline drive.

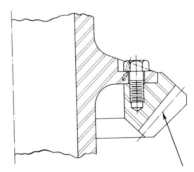

FIGURE 10.36 Method of mounting gear to hub to absorb high outward thrust.

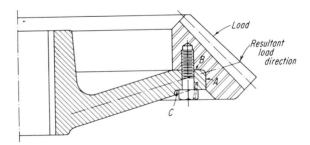

FIGURE 10.37 Webless-type miter gear—counterbored type.

GEAR REACTIONS AND MOUNTINGS 10.33

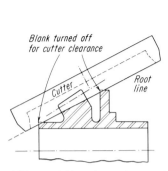

FIGURE 10.38 When using Zerol bevel, spiral bevel, and hypoid gears, clearance must be provided for the cutter as shown.

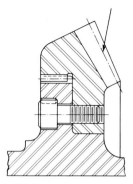

FIGURE 10.39 Positive drive for severe reversing or shock loads.

10.6.8 Gear and Pinion Adjustments

Provision should be made for adjusting both gear and pinion axially at assembly. Shim-type adjustments as shown in Figs. 10.28 and 10.29 or threads and nuts with locks as shown in Fig. 10.40 may be used. Shims less than 0.015 in. (0.38 mm) thick may "pound out" in service. When used adjacent to a bearing, the shims should be placed next to the stationary member. If an assortment of shims varying in thickness in steps of 0.003 in. (0.076 mm) is used [such as 0.020 in. (0.50 mm), 0.023 in. (0.60 mm), 0.026 in. (0.65 mm)], the member to be adjusted may be positioned within ± 0.0015 in. (± 0.04 mm).

10.6.9 Assembly Procedure

The following is a recommended procedure for the assembly of bevel and hypoid gearboxes:

Thoroughly clean housing and parts, following the bearing manufacturer's instructions in handling the rolling-element bearings.

Keep lapped or matched gears in their original sets.

Assemble even-ratio pairs with the teeth mating as lapped. Tooth and mating spaces are usually designated with X markings.

Set pinion on its correct axial position by measurement.

Adjust gear along its axis until the specified backlash is obtained.

Securely lock gears and thrust bearings in this position. Note, for spiral bevel and hypoid gears, both members should be locked against thrust in both directions.

As a final check, paint the teeth with marking compound and observe the tooth contact pattern under light load. This may indicate the need for further adjustment.

Check to ensure that there is an adequate supply of the specified lubricant to the mesh and to the bearings.

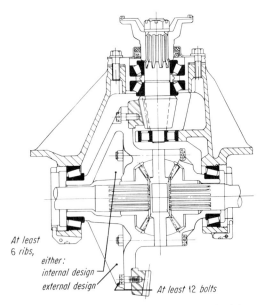

At least
6 ribs,
  either:
  internal design
  external design
At least 12 bolts

FIGURE 10.40 Truck axle—straddle-mounted pinion, and with threaded-nut adjustments for the gear bearings.

10.7 CALCULATION OF BEVEL AND HYPOID BEARING LOADS

The normal load on the tooth surfaces of bevel or hypoid gears may be resolved into two components: one in the direction along the axis of the gear, and the other perpendicular to the axis. The axial force produces an axial thrust on the bearings, while the force perpendicular to the axis produces a radial load on the bearings. The direction and magnitude of the normal load depends on the ratio, pressure angle, spiral angle, hand of spiral, direction of rotation, and whether the gear is the driving or driven member.

10.7.1 Hand of Spiral

The hand of spiral on spiral bevel and hypoid gears is denoted by the direction in which the teeth curve; that is, left-hand teeth incline away from the axis in the counterclockwise direction when an observer looks at the face of the gear, and right-hand teeth incline away from the axis in the clockwise direction. The hand of spiral of one member of a pair is always opposite to that of its mate. It is customary to use the hand of spiral of the pinion to identify the combination; that is, a left-hand combination is one with the left-hand spiral on the pinion and a right-hand spiral on the gear. The hand of spiral has no effect on the smoothness and quietness of operation or on the efficiency. Attention, however, is called to the difference in the effect of the thrust loads as stated in the following paragraph.

GEAR REACTIONS AND MOUNTINGS   10.35

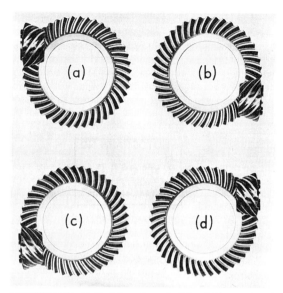

FIGURE 10.41 Both pinions shown in *(a)* and *(b)* are referred to as having an offset "below center," while those in *(c)* and *(d)* have an offset "above center." In determining the direction of offset, it is customary to look at the face of the gear with the pinion at the right.

A left-hand spiral pinion driving clockwise (viewed from the back) tends to move axially away from the cone center, while a right-hand pinion tends to move toward the center because of the oblique direction of the curved teeth. If there is excessive end play in the pinion shaft because of faulty assembly or bearing failure, the movement of a right-hand pinion driving clockwise will take up the backlash under load, and the teeth of the gear and pinion may wedge together, while a left-hand spiral pinion under the same conditions would back away and introduce additional backlash between the teeth, a condition that would not prevent the gears from functioning. When the ratio, pressure angle, and spiral angle are such that doing so is possible, the hand of spiral should be selected to give an axial thrust that tends to move both the gear and the pinion out of mesh. Otherwise, the hand of spiral should be selected to give an axial thrust that tends to move the pinion out of mesh. Often the mounting conditions dictate the hand of spiral to be selected.

In a reversible drive there is, of course, no choice unless the pair performs a heavier duty in one direction a greater part of the time.

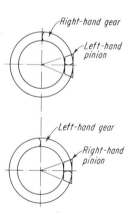

FIGURE 10.42 Diagram illustrating hand of Zerol. The upper drawing shows a left-hand, and the lower, a right-hand Zerol bevel gear combination.

In the calculation of reactions, Zerol and straight bevel gears may be treated as a special case of spiral bevel gears in which the spiral angle is zero degrees. In these, the direction of the axial thrust is always outward on both members regardless of the direction of rotation and whether the pinion is the driving or driven member.

On hypoids when the pinion is below center and to the right when facing the front of the gear, the pinion hand of spiral should always be left. With the pinion above center and to the right, the pinion hand should always be right (see Fig. 10.41).

10.7.2 Spiral Angle

The spiral angle, if possible, should be selected to give a face contact ratio of at least 1.25. For maximum smoothness, the face contact ratio should be between 1.50 and 2.0. On straight or Zerol bevel gears the spiral angle will, of course, be zero.

10.7.3 Tangential Load*

The tangential load on a bevel or hypoid gear is given by

$$W_t = \frac{19,098,600 \, P}{d_{2m} \times n_2} \quad (10.29a)$$

$$W_t = \frac{126,050 \, P_h}{d_{mG} \times n_G} \quad (10.29b)$$

where W_t (W_t) = tangential load [metric (English)], N (lb)
P (P_h) = power, kW (hp)
d_{2m} (d_{mG}) = mean pitch diameter of gear, mm (in.)
n_2 (n_G) = speed, r/min

The mean pitch diameter of a bevel gear may be determined by subtracting the term (bevel gear face width times the sine of the bevel gear pitch cone angle) from the bevel gear pitch diameter at the large end of the bevel gear.

The tangential load on the mating bevel pinion will be equal to the tangential load on the gear. The tangential load on the mating hypoid pinion, however, must be determined as follows:

$$W_{tP} = W_{tG} \frac{\cos \text{ spiral angle of pinion}}{\cos \text{ spiral angle of pinion}} \quad (10.30)$$

10.7.4 Axial Thrust

The value of the axial thrust is dependent on the tangential tooth load, spiral angle, pressure angle, and pitch angle. This may be determined with the aid of Fig. 10.21. Figure 10.21 is symmetrical about a horizontal center line; therefore, there are two points of intersection of the pitch-angle and the spiral-angle curves. Selection of the proper point is dependent on the hand of spiral, the direction of rotation, and whether the member is the driving or driven one. This is given in Fig. 10.21. Note that the

---

*In all calculations herein, the tangential load is calculated at the mean pitch radius.

intersection points for the pinion and its mating gear are always on opposite sides of the horizontal centerline. The pressure angle ϕ is given on the scale at the right. A straight line connecting the pressure angle with the intersection point of the spiral-angle and pitch-angle curves can now be extended to the scale on the left. This scale gives the axial thrust W_x in percent of the tangential tooth load W_t. The spiral angle on straight and Zerol gears is zero.

For determining the axial thrust in hypoid gears, the pressure angle of the driving side should be used. Instead of using the pitch angles, the face angle of the pinion and the root angle of the gear should be used.

Note: The spiral angle, pitch angle, and pressure angle for the corresponding member must be selected in each case.

The actual axial-thrust load may be determined by multiplying the tangential load in pounds for the corresponding member by the percent given in Fig. 10.21.

$$W_x = \text{(percent from Fig. 10.21)} \times W_t \qquad (10.31)$$

Figure 10.43 illustrates the direction of rotation and direction of thrust.

10.7.5 Radial Load

The radial load caused by the separating force between the two members may be determined in a similar manner to that used for the axial thrust with the aid of Fig. 10.22. The same procedure is followed here as outlined in the foregoing section, "Axial Thrust." The scale on the right gives the component part of the radial load W_r in percent of the tangential tooth load W_t.

This component of the actual radial load may be determined by multiplying the

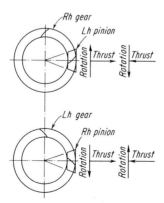

FIGURE 10.43 Diagram illustrating hand of spiral and its effect on the direction of axial thrust in spiral bevel gears. The upper drawing shows a left-hand; and the lower, a right-hand spiral bevel combination.

tangential load in pounds for the corresponding member by the percent given in Fig. 10.22

$$W'_r = (\text{percent from Fig. 10.22}) \times W_t \qquad (10.32)$$

In addition to the foregoing element, there is a radial component caused by the tangential load itself, and a third component caused by the couple which the axial thrust on the given member produces.

The total radial load produced on the bearings will therefore depend on the resultant of these three components.

For the overhung mounting the radial components on bearing A will be (see Fig. 10.44)

$$W_{r_1} = W'_r \frac{L + M}{M} \qquad (10.33)$$

$$W_{r_2} = W'_t \frac{L + M}{M} \qquad (10.34)$$

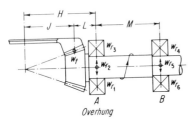

where W'_r is the separating component from Fig. 10.22, and W_t is the tangential tooth load.

FIGURE 10.44 Diagram showing load components on a pinion and bearings—overhung mounting.

$$W_{r_3} = W_x \frac{d_m}{2M} \qquad (10.35)$$

where W_x = axial-thrust load
L = distance along the axis from center of the face width to center of bearing A*
M = spacing between centers of bearings A and B
$d_m = d - F \sin \gamma$ = mean pitch diameter†
d = pitch diameter
F = face width
γ = pitch angle

The total radial component on bearing A will be

$$W_{r_A} = \sqrt{W_{r_2}^2 + (W_{r_1} - W_{r_3})^2} \qquad (10.36)$$

In like manner, the radial components on bearing B will be

---

*For a hypoid pair use Eq. (10.52).
†For a hypoid pair use Eq. (10.49) for d_m and N by directions given for Eq. (10.52).

GEAR REACTIONS AND MOUNTINGS

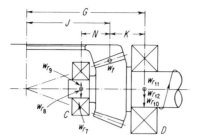

FIGURE 10.45 Diagram showing load components on a pinion and bearings—straddle mounting.

$$W_{r_4} = W'_r \frac{L}{M} \tag{10.37}$$

$$W_{r_5} = W_t \frac{L}{M} \tag{10.38}$$

$$W_{r_6} = W_x \frac{d_m}{2M} \tag{10.39}$$

The total radial component on bearing B will be

$$W_{r_B} = \sqrt{W_{r_5}^2 + (W_{r_4} - W_{r_6})^2} \tag{10.40}$$

For the straddle mounting shown in Fig. 10.45 the radial components on bearing C will be

$$W_{r_7} = W'_r \frac{K}{N + K} \tag{10.41}$$

$$W_{r_8} = W_t \frac{K}{N + K} \tag{10.42}$$

$$W_{r_9} = W_x \frac{d_m}{2(N + K)} \tag{10.43}$$

where N is the distance along the axis from center of face width to center of bearing C,* and K is the distance along the axis from center† of face width to center of bearing D.

The total radial component on bearing C will be

$$W_{r_c} = \sqrt{W_{r_8}^2 + (W_{r_7} - W_{r_8})^2} \tag{10.44}$$

In like manner, the radial components on bearing D will be

$$W_{r_{10}} = W'_r \frac{K}{N + K} \tag{10.45}$$

---

*For a hypoid pair use Eq. (10.49) for d_m and find N by directions given for Eq. (10.52).
†For a hypoid pair find K by Eq. (10.51). See Fig. 10.45.

$$W_{r_{11}} = W_t \frac{K}{N+K} \qquad (10.46)$$

$$W_{r_{12}} = W_x \frac{d_m}{2(N+K)} \qquad (10.47)$$

The total radial component on bearing D will be

$$W_{r_D} = \sqrt{W_{r_{11}}{}^2 + (W_{r_{10}} + W_{r_{12}})^2} \qquad (10.48)$$

For a hypoid pair, change the preceding equations as follows:

d_{m_G} = gear pitch diameter − face width × sin gear root angle (10.49)

$$d_{m_P} = d_{m_G} \times \frac{\text{number of pinion teeth}}{\text{number of gear teeth}} \times \frac{\cos \text{ gear spiral angle}}{\cos \text{ pinion spiral angle}} \qquad (10.50)$$

$$K = G - J \qquad (10.51)$$

$$L = H - J \qquad (10.52)$$

$$J_P = \frac{d_{m_G}}{2} \cos \epsilon \quad \text{for hypoid pinion}$$

$$J_G = \frac{d_{m_P}}{2} \quad \text{for hypoid gear}$$

where G = distance along axis from center line of mate to center of bearing D
H = distance along axis from center line of mate to center of bearing A
ϵ = arctan [tan (spiral angle pinion − spiral angle gear) × sin gear root angle]
M = spacing between centers of bearings A and B
N = spacing between centers of bearings C and D minus dimension K.

See Figs. 10.44 and 10.45.

Example of Spiral-Bevel-Gear Bearing Load Calculation. A sample problem for a 16-tooth pinion driving a 49-tooth gear demonstrates how the equations given in this section are used.

TABLE 10.5 Spiral Bevel Gear Sample Data and Bearing Arrangement

	Pinion		Gear	
	Symbol	English Units	Symbol	English Units
Pitch diameter	d	3.200″	D	9.800″
Pitch angle	γ	18°5′	Γ	71°55′
	$\sin \gamma$	0.3104	$\sin \Gamma$	0.9506
Mean pitch diameter	$d_{m_p} = d - F \sin \gamma$	2.734″	$d_{m_G} = D - F \sin \Gamma$	8.374″
Overhung mounting:				
L		1.500″		
M		5.000″		
Straddle mounting:				
K				2.500″
N				3.500″

GEAR REACTIONS AND MOUNTINGS

Given: a $\frac{16}{49}$ combination, 5 diametral pitch, 1.500-in. face width, 20° pressure angle, 35° spiral angle. (See Table 10.5 for further data relative to this problem.)

Load data: 71 hp at 1800 r/min of pinion; left-hand pinion driving clockwise; gear is straddle mounted; pinion, overhung.

The tangential load on this spiral bevel gear will be

$$W_{tG} = \frac{126{,}050 P_h}{d_{m_G} n} = \frac{126{,}050 \times 71}{8.374 \times 588} = 1818 \text{ lb}$$

where $P_h = 71$ hp
$d_{m_G} = 8.374$ in.
$n = \frac{16}{49} \times 1800 = 588$ r/min of gear

The tangential load on the aforementioned spiral bevel pinion will be equal to the tangential load on the gear:

$$W_{tP} = W_{tG} = 1818 \text{ lb}$$

The axial-thrust loads for both gear and pinion are obtained from Fig. 10.21.

$$W_x = (\text{percent from Fig. 10.21}) \times W_t$$
$$W_{x_P} = +0.80 \times 1818 = 1454 \text{ lb (outward)}$$
$$W_{x_G} = +0.20 \times 1818 = 364 \text{ lb (outward)}$$

The separating force for both gear and pinion is obtained from Fig. 10.22.

$$W'_r = (\text{percent from Fig. 10.22}) \times W_t$$
$$W'_{rP} = +0.20 \times 1818 = 364 \text{ lb (separating)}$$
$$W'_{rG} = +0.80 \times 1818 = 1454 \text{ lb (separating)}$$

In Table 10.6 are tabulated the formulas and sample calculations for the resultant bearing loads on the aforementioned pair of gears.

Note: Since an overhung mounting is used on the pinion, the formulas for bearings *A* and *B* are used; and, since a straddle mounting is used on the gear, the formulas for bearings *C* and *D* are used. If both members had been overhung-mounted, the formulas for bearings *A* and *B* would have been used for both gear and pinion; or if both members had been straddle-mounted, the formulas for *C* and *D* would have been used for both gear and pinion. In any case, the dimensions and loads for the corresponding member must be used.

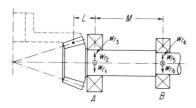

FIGURE 10.46 Overhung mounting.

TABLE 10.6 Solution to Sample Problem on Spiral Bevel Bearing Loads (English units)

Forces	Load calculations	Bearing loads			
		Overhung mounting		Straddle mounting	
		Bearing A	Bearing B	Bearing C	Bearing D
Tangential W_t	$W_t = \dfrac{126{,}050 P}{d_m{}_g n}$	$W_{r_2} = W_t \dfrac{L+M}{M}$ $W_{r_2} = 1{,}818 \dfrac{1.5+5.0}{5.0}$ $= 2{,}363$	$W_{r_5} = W_t \dfrac{L}{M}$ $W_{r_5} = 1{,}818 \dfrac{1.5}{5.0}$ $= 545$	$W_{r_8} = W_t \dfrac{K}{N+K}$ $W_{r_8} = 1{,}818 \dfrac{2.5}{3.5+2.5}$ $= 758$	$W_{r_{11}} = W_t \dfrac{N}{N+K}$ $W_{r_{11}} = 1{,}818 \dfrac{3.5}{3.5+2.5}$ $= 1{,}060$
Thrust W_z	$W_z = W_t \times$ thrust factor See Fig. 10.21		$W_z =$ thrust factor $\times W_t$ $W_z = +0.80 \times 1{,}818$ $= 1{,}454$		$W_z =$ thrust factor $\times W_t$ $W_z = +0.20 \times 1{,}818$ $= 364$
Separating $W_r{}'$	$W_r{}' = W_t \times$ separating factor See Fig. 10.22	$W_{r_1} = W_r{}' \dfrac{L+M}{M}$ $W_{r_1} = 364 \dfrac{1.5+5.0}{5.0}$ $= 473$	$W_{r_4} = W_r{}' \dfrac{L}{M}$ $W_{r_4} = 364 \dfrac{1.5}{5.0}$ $= 109$	$W_{r_7} = W_r{}' \dfrac{K}{N+K}$ $W_{r_7} = 1{,}454 \dfrac{2.5}{3.5+2.5}$ $= 606$	$W_{r_{10}} = W_r{}' \dfrac{N}{N+K}$ $W_{r_{10}} = 1{,}454 \dfrac{3.5}{3.5+2.5}$ $= 848$
Thrust couple	$W_z \dfrac{d_m}{2}$	$W_{r_3} = W_z \dfrac{d_m}{2M}$ $W_{r_3} = 1{,}454 \dfrac{2.734}{2 \times 5.0}$ $= 398$	$W_{r_6} = W_z \dfrac{d_m}{2M}$ $W_{r_6} = 1{,}454 \dfrac{2.734}{2 \times 5.0}$ $= 398$	$W_{r_9} = W_z \dfrac{d_m}{2(N+K)}$ $W_{r_9} = 364 \dfrac{8.374}{2(3.5+2.5)}$ $= 254$	$W_{r_{12}} = W_z \dfrac{d_m}{2(N+K)}$ $W_{r_{12}} = 364 \dfrac{8.374}{2(3.5+2.5)}$ $= 254$
Total radial load	$W_r = \sqrt{\Sigma(\text{loads})^2}$	$W_{r_A} = \sqrt{(W_{r_2})^2 + (W_{r_1} - W_{r_3})^2}$ $W_{r_A} = \sqrt{(2{,}363)^2 + (473 - 398)^2}$ $= 2{,}364$	$W_{r_B} = \sqrt{(W_{r_5})^2 + (W_{r_4} - W_{r_6})^2}$ $W_{r_B} = \sqrt{(545)^2 + (109 - 398)^2}$ $= 617$	$W_{r_C} = \sqrt{(W_{r_8})^2 + (W_{r_7} - W_{r_9})^2}$ $W_{r_C} = \sqrt{(758)^2 + (606 - 254)^2}$ $= 836$	$W_{r_D} = \sqrt{(W_{r_{11}})^2 + (W_{r_{10}} + W_{r_{12}})^2}$ $W_{r_D} = \sqrt{(1{,}060)^2 + (848 + 254)^2}$ $= 1{,}536$

$W_t =$ tangential tooth load, lb
$P =$ power transmitted, hp
$d_{m_g} = D - F \sin \Gamma =$ mean pitch diameter of gear, in.
$D =$ gear pitch diameter, in.
$F =$ gear face width, in.
$\Gamma =$ gear pitch angle
$n =$ speed of gear, rpm
$W_z =$ axial-thrust component
$W_r{}' =$ separating component
$L =$ axial distance from center of face width to center line of bearing A
$M =$ spacing between center line of bearings A and B
$N =$ axial distance from center of face width to center line of bearing C
$K =$ axial distance from center of face width to center line of bearing D
$d_m = d - F \sin \gamma$
$d =$ pitch diameter, in.
$\gamma =$ pitch angle

bearing as shown, or by a set of duplex bearings mounted in tandem.

Example of Hypoid Gear Bearing Load Calculation. A sample problem of a 14-tooth hypoid pinion driving a 47-tooth hypoid gear is given in the following:

Given: a $\frac{14}{47}$ combination, 5.013 gear diametral pitch, 1.438-in. face width, 1.750-in. pinion offset below center. Pressure angles 15°20′ on forward drive; 27°10′ on reverse drive. Spiral angles 21°28′ on the gear; 45°3′ on the pinion. Pinion face angle = 21°47′. Gear root angle = 66°27′.

Load data: 60 hp at 1500 rpm of pinion; left-hand pinion driving clockwise; gear is straddle-mounted; pinion, overhung.

The tangential load on the aforementioned hypoid gear will be

$$W_{t_G} = \frac{126{,}050 P_H}{d_{m_G} n} = \frac{126{,}050 \times 60}{8.057 \times 447} = 2100 \text{ lb}$$

where $P_H = 60$ hp
$d_{m_G} = 8.057$
$n = \frac{14}{47} \times 1500 = 447$ r/min of gear

The tangential load on the hypoid pinion will be

$$W_{t_P} = W_{t_G} \frac{\cos \psi_P}{\cos \psi_G} = 2100 \times \frac{0.70649}{0.93063} = 1594 \text{ lb}$$

$\psi_P = 45°3′ \quad \cos \psi_P = 0.70649$

$\psi_G = 21°28′ \quad \cos \psi_G = 0.93060$

The axial-thrust loads for both gear and pinion for forward drive are obtained from Fig. 10.21.

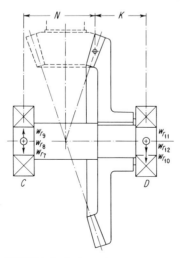

FIGURE 10.47 Straddle mounting.

TABLE 10.7 Hypoid Gear Sample Data and Bearing Arrangement (English units)

	Pinion		Gear	
Pitch dia............			$D = \dfrac{N}{\text{diam. pitch}}$	9.375"
Pitch angle: Pinion face angle..	γ_o	21°47'		
Gear root angle....			Γ_R	66°27'
sin γ_o		0.37110	sin Γ_R	0.91671
Mean pitch dia	$d_{m_P} = d_{m_G} \dfrac{n \cos \psi_G}{N \cos \psi_P}$	3.161"	$d_{m_G} = D - F \sin \Gamma_R$	8.057"
Distance from center line of mate to center of bearing.....	H (from layout)	5.615"	G (from layout)	4.581"
	$J_P = \dfrac{d_{m_G}}{2} \cos \varepsilon$	3.740"	$J_G = \dfrac{d_{m_P}}{2}$	1.581"
Overhung mounting (see Figs. 10.44 and 10.46):				
$L = H - J_P$.....		1.875"		
$M =$		4.750"		
Straddle mounting (see Figs. 10.45 and 10.47):				
$K = G - J_G$.....				3.000"
$N =$				6.250"

$$W_x = (\text{percent from Fig. 10.21}) \times W_t$$
$$W_{x_G} = +0.11 \times 2100 = 231 \text{ lb (outward)}$$
$$W_{x_P} = +1.08 \times 1594 = 1722 \text{ lb (outward)}$$

The separating forces for both gear and pinion are obtained from Fig. 10.22.

$$W'_r = (\text{percent from Fig. 10.22}) \times W_t$$
$$W'_{r_G} = +0.48 \times 2100 = 1008 \text{ lb (separating)}$$
$$W'_{r_P} = -0.01 \times 1594 = -16 \text{ lb (attracting)}$$

In Table 10.8 are tabulated the formulas and sample calculations for the resultant bearing loads on the hypoid pair.

10.7.6 Required Data for Bearing Load Calculations

To calculate the bearing loads, the following data must be available:

Bevel Gears

Gear ratio (number of teeth in gear and pinion)
Gear pitch diameter or diametral pitch

TABLE 10.8 Solution to Sample Problem on Hypoid Gear Bearing Loads (English units)

		Bearing loads			
		Overhung mounting		Straddle mounting	
Forces	Load calculations	Bearing A	Bearing B	Bearing C	Bearing D
Tangential W_t	$W_t = \dfrac{126{,}050 P}{d_m n}$	$W_{r_2} = W_t \dfrac{L+M}{M}$ $W_{r_2} = 1{,}594 \dfrac{1.875 + 4.750}{4.750}$ $= 2{,}223$	$W_{r_5} = W_t \dfrac{L}{M}$ $W_{r_5} = 1{,}594 \dfrac{1.875}{4.750}$ $= 629$	$W_{r_8} = W_t \dfrac{K}{N+K}$ $W_{r_8} = 2{,}100 \dfrac{3.0}{6.25 + 3.00}$ $= 681$	$W_{r_{11}} = W_t \dfrac{N}{N+K}$ $W_{r_{11}} = 2{,}100 \dfrac{6.25}{6.25 + 3.00}$ $= 1{,}419$
Thrust W_x	$W_x = W_t \times$ thrust factor See Fig. 10.21	$W_x = $ thrust factor $\times W_t$ $W_x = +1.08 \times 1{,}594$ $= 1{,}722$	....	....	$W_x = $ thrust factor $\times W_t$ $W_x = +0.11 \times 2{,}100$ $= 231$
Separating W_r'	$W_r' \times W_t \times$ separating factor See Fig. 10.22	$W_{r_1} = W_{r'} \dfrac{L+M}{M}$ $W_{r_1} = -16 \dfrac{1.875 + 4.750}{4.750}$ $= -22$	$W_{r_4} = W_{r'} \dfrac{L}{M}$ $W_{r_4} = -16 \dfrac{1.875}{4.750}$ $= -6$	$W_{r_7} = W_{r'} \dfrac{K}{N+K}$ $W_{r_7} = 1{,}008 \dfrac{3.0}{6.25 + 3.00}$ $= 327$	$W_{r_{10}} = W_{r'} \dfrac{N}{N+K}$ $W_{r_{10}} = 1{,}008 \dfrac{6.25}{6.25 + 3.00}$ $= 681$
Thrust couple	$W_x \dfrac{d_m}{2}$	$W_{r_3} = W_x \dfrac{d_m}{2M}$ $W_{r_3} = 1{,}722 \dfrac{3.161}{2 \times 4.750}$ $= 573$	$W_{r_6} = W_x \dfrac{d_m}{2M}$ $W_{r_6} = 1{,}722 \dfrac{3.161}{2 \times 4.750}$ $= 573$	$W_{r_9} = W_x \dfrac{d_m}{2(N+K)}$ $W_{r_9} = 231 \dfrac{8.057}{2(6.25 + 8.00)}$ $= 101$	$W_{r_{12}} = W_x \dfrac{d_m}{2(N+K)}$ $W_{r_{12}} = 231 \dfrac{8.057}{2(6.25 + 3.00)}$ $= 101$
Total radial load	$W_r = \sqrt{\Sigma(\text{loads})^2}$	$W_{r_A} = \sqrt{W_{r_2}^2 + (W_{r_1} - W_{r_3})^2}$ $W_{r_A} = \sqrt{(2{,}223)^2 + (-22 - 573)^2}$ $= 2{,}301$	$W_{r_B} = \sqrt{W_{r_5}^2 + (W_4 - W_6)^2}$ $W_{r_B} = \sqrt{(629)^2 + (-6 - 573)^2}$ $= 855$	$W_{r_C} = \sqrt{W_{r_8}^2 + (W_{r_7} - W_{r_9})^2}$ $W_{r_C} = \sqrt{(681)^2 + (327 - 101)^2}$ $= 718$	$W_{r_D} = \sqrt{W_{r_{11}}^2 + (W_{r_{10}} + W_{r_{12}})^2}$ $W_{r_D} = \sqrt{(1{,}419)^2 + (681 + 101)^2}$ $= 1{,}620$

W_t = tangential tooth load, lb
P = power transmitted, hp
n = speed, rpm (values for corresponding member)
W_x' = separating component
W_x = axial-thrust component
L = axial distance from center of gear face width at point of contact to center line of bearing A
M = spacing between center line of bearings A and B
N = axial distance from center of gear face width at point of contact to center line of bearing C
K = axial distance from center of gear face width at point of contact to center line of bearing D
$d_m = D - F \sin \Gamma_R = d_{m_G}$ = mean pitch diameter of hypoid gear
$= d_{m_G} \dfrac{n}{N} \dfrac{\cos \psi_G}{\cos \psi_P}$ = mean pitch diameter of hypoid pinion
D = gear pitch diameter, in.
F = gear face width, in.
Γ_R = gear root angle
n = number of teeth in hypoid pinion
N = number of teeth in hypoid gear
ψ_G = gear spiral angle
ψ_P = pinion spiral angle

Pitch angles (both gear and pinion)
Face width
Pressure angle
Spiral angle
Horsepower or gear torque to be transmitted
Gear r/min
Hand of spiral
Direction of rotation (viewed from the back)
Sketch showing method of mounting and bearing spacing
Shaft angle

Hypoid Gears

Gear ratio (number of teeth in gear and pinion)
Gear pitch diameter or diametral pitch
Gear root angle
Pinion face angle
Gear face width
Pressure angles (both sides of tooth)
Spiral angles (both gear and pinion)
Horsepower or gear torque to be transmitted
Gear r/min
Hand of spiral
Direction of rotation (viewed from the back)
Sketch showing method of mounting and bearing spacing
Shaft angle

10.8 BEARING LOAD CALCULATIONS FOR WORMS

Basically, the worm-and-gear bearing load calculations are similar to helical gear reactions except for three main differences. All the members of the worm gear family run on *crossed axes*. Generally, the axes are crossed at 90°.

Besides the shaft arrangement, worm gears have "overlap." There may be no overlap, overlap in one plane, or overlap in two planes. *Overlap* is that characteristic whereby one tries to envelop the mating part by being so curved as to tend to wrap around the mating part. The overlap gives a greater area of tooth engagement. It also makes the mounting more sensitive. See Fig. 10.48 for worm types.

The third unusual feature of worm gears is that of ratio. The ratio is defined as the number of gear teeth divided by the number of worm threads (or worm "starts"). A single-start worm may have several turns of thread; but, if it meshes with a gear with 100 teeth, the *ratio* is 100:1. If the worm had two starts and the gear had 100 teeth, the ratio would be 50:1.

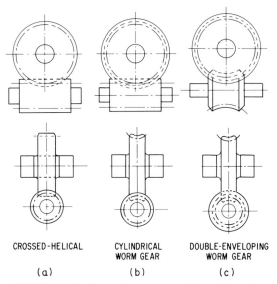

CROSSED-HELICAL	CYLINDRICAL WORM GEAR	DOUBLE-ENVELOPING WORM GEAR
(a)	(b)	(c)

FIGURE 10.48 Types of worm gears: *(a)* nonthroated crossed helical, *(b)* single-throated cylindrical worm gear, *(c)* double-throated, double-enveloping worm gear. *(a)* Point contact. *(b)* Line contact. *(c)* Line contact or area contact.

10.8.1 Calculation of Forces in Worm Gears

Figure 10.3 showed the basics of a force that acted in three planes. This concept is helpful in understanding how the force at the worm gear mesh acts.

The efficiency of worm gears is quite variable. A single-thread worm may have an efficiency as low as 50 percent. A multiple-thread worm with at least 15° lead angle and a reasonable rubbing speed will usually have an efficiency in the range of 90 to 95 percent. Since worm gear efficiency is so variable, it cannot be neglected in calculating load reactions. Section 12.4 may be used to estimate a relatively accurate efficiency. In particular see Eq. (12.27).

The worm gear overlap gives a relatively wide zone of contact. The pressure angle and the thread angle vary quite appreciably throughout this area. It is customary to calculate the worm gear load reactions if all the contact were at the theoretical pitch point. This permits use of the theoretical pressure angle, the theoretical lead angle, etc., as variables in the equations. Although this method is not really exact, it is good enough to get load reactions to size bearings and design casing structures.

Figure 10.49 shows the case of a worm and gear mounted on 90° axes and each straddle-mounted. Load reactions at the mesh and at each of the bearings are shown.

The tangential force on the worm may be calculated by

$$W_t = \frac{19{,}098{,}600 P}{n_1 d_{p_1}} \quad \text{metric} \quad (10.53a)$$

$$W_t = \frac{126{,}050 P_h}{n_w d} \quad \text{English} \quad (10.53b)$$

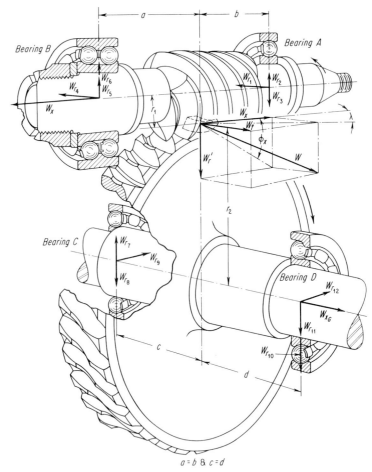

FIGURE 10.49 Worm gear bearing loads.

where W_t (W_t) = tan driving force [metric (English)] N (lb)
P (P_h) = input power, kW (hp)
n_1 (n_w) = rotations per minute of worm
d_{p_1} (d) = pitch diameter of worm

The separating force is

$$W'_r = \frac{W_t \tan \alpha_n}{\tan \gamma} \quad \text{metric} \quad (10.54a)$$

$$W'_r = \frac{W_t \tan \phi_n}{\tan \lambda} \quad \text{English} \quad (10.54b)$$

where W'_r (W'_r) = Separating force [metric (English)] N (lb)
γ (λ) = worm lead angle
α_n (ϕ_n) = normal pressure angle

The worm thrust force is

$$W_x = \frac{W_t}{\tan \text{lead angle}} \qquad (10.55)$$

The details for worm gear bearing load calculations are shown in Table 10.9. Figure 10.49 illustrates the bearing loads in a worm gear set.

The method of worm gear bearing load calculations just described may be used for crossed helical gears on right-angle axes, for cylindrical worm gears, and for double-enveloping worm gears.

10.8.2 Mounting Tolerances

The worm gear types vary all the way from the nonenveloping crossed helical to the double-enveloping type. In the crossed helical, alignment is not critical in any direction. Axial movement merely shifts the contact area along the length of the part. Center-distance error changes backlash and the amount of tooth contact, but neither of these things is highly critical in most applications.

The cylindrical worm gear set is critical on center distance, shaft angle, and axial position of the gear. Slight errors in any of these three things will tend to change a full-face-width contact pattern to only a small amount of contact at the end of the tooth. The contact area in a worm gear set may drop from 100 to 15 percent with an error in alignment of as little as 0.010 in. in a foot.

The double-enveloping worm gear set is critical in the three directions of the cylindrical set plus a fourth direction of the axial position of the worm.

Table 10.10 shows some typical mounting tolerances for high-capacity worm gear sets. The most precise worm gears used in timing devices, radar work, etc., are mounted with about half the error shown in Table 10.10. On the other hand, worm gear sets for commercial service and relatively light load service are mounted with about twice the error shown in Table 10.10.

10.8.3 Worm Gear Blank Considerations

The worm member is usually made of steel. Generally this member offers no problem in design. In a few cases the worm may be long and slender with a wide span between its bearings. In this case there may be danger of serious "bowing" of the worm between supports. Calculations should be made in such a case to make sure that the worm shaft is not too highly stressed. Also, the deflection should be calculated. This deflection acts as a change in center distance and should be judged on this basis.

The worm gear is generally made of bronze in the toothed area. Quite often the hub is made of steel or cast iron. The hub should have a large enough hole in it to assure

TABLE 10.9 Equations for Worm Gear Bearing Loads with Straddle Mounting

Forces	Worm bearing loads		Worm-gear bearing loads	
	Bearing A	Bearing B	Bearing C	Bearing D
Worm tan force W_t	$W_{r_1} = \dfrac{W_t a}{a+b}$	$W_{r_4} = \dfrac{W_t b}{a+b}$	$W_{r_7} = \dfrac{W_t r_2}{c+d}$	$W_{r_{10}} = \dfrac{W_t r_2}{c+d} = W_{r_7}$
Separating force W'_r	$W_{r_2} = \dfrac{W'_r a}{a+b}$	$W_{r_5} = \dfrac{W'_r b}{a+b}$	$W_{r_8} = \dfrac{W'_r d}{c+d}$	$W_{r_{11}} = \dfrac{W'_r c}{c+d}$
Thrust W_x	$W_{r_3} = \dfrac{W_x r_1}{a+b}$	$W_{r_6} = \dfrac{W_x r_1}{a+b} = W_{r_3}$	$W_{r_9} = \dfrac{W_x d}{c+d}$	$W_{r_{12}} = \dfrac{W_x c}{c+d}$
Total radial load	$W_{r_A} = \sqrt{(W_{r_1})^2 + (W_{r_2} - W_{r_3})^2}$	$W_{r_B} = \sqrt{(W_{r_4})^2 + (W_{r_5} + W_{r_6})^2}$	$W_{r_C} = \sqrt{(W_{r_9})^2 + (W_{r_7} - W_{r_8})^2}$	$W_{r_D} = \sqrt{(W_{r_{12}})^2 + (W_{r_{10}} + W_{r_{11}})^2}$
Total thrust	W_x (may be applied to either bearing A or B)		$W_{x_G} = W_t$ (may be applied to either bearing C or D)	

NOTE: See Fig. 10.49 for definition of terms and bearing locations.

TABLE 10.10 Typical Mounting Tolerances for High-Capacity Worm Gear Sets

Center Distance	Tolerance on Center Distance	Tolerance on Axial Position	Tolerance on Alignment
Metric values, mm			
0–75	±0.013	±0.025	±0.650 per m
75–150	±0.025	±0.050	±0.400 per m
150–375	±0.050	±0.075	±0.225 per m
375 up	±0.075	±0.100	±0.150 per m
English values, in.			
0–3	±0.0005	±0.001	±0.008 per ft
3–6	±0.001	±0.002	±0.005 per ft
6–15	±0.002	±0.003	±0.003 per ft
15 up	±0.003	±0.004	±0.002 per ft

a strong shaft. The length of the hub should be about one-third the gear pitch diameter to assure a stable mount.

Figure 10.50 shows a typical worm gear with a bronze ring bolted to a steel hub.

10.8.4 Run-in of Worm Gears

All types of worm gears benefit by a "run-in." This is particularly true when the gear member is made of bronze.

The as-cut worm gear generally has generating flats on its surface, and the surface itself is somewhat imperfect because of "oversize" effects of the hob. As the gear is broken in, its surface wears a slight amount to conform to that of the worm. In addition the surface of the gear becomes smooth and polished. Tests on bronze gears have shown that the amount of load that will cause seizure of the surface may go up as much as tenfold after a careful break-in!

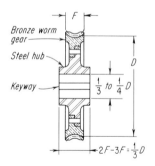

FIGURE 10.50 Worm gear blank proportions.

In breaking in worm gears, the load is gradually increased and the speed is kept low. The surfaces are watched carefully. If small amounts of bronze start adhering to the worm surface, it is necessary to stop the break-in and dress down the worm threads with fine abrasive paper to remove the bronze. The break-in may then continue.

After the break-in is complete and the tooth surfaces polished, and the contact pattern has been obtained to a satisfactory level of quality, the gears are stamped as a *matched* set. Exact axial-distance settings used in the break-in are stamped on the set to aid in field installation.

10.9 BEARING LOAD CALCULATIONS FOR SPIROIDS

The Spiroid gear is a crossed-axes-type gear somewhat intermediate between a worm gear and a hypoid gear Its originators, the Illinois Tool Works of Chicago, Illinois, consider it to be a screw-type gear. The pinion of the Spiroid set is like a tapered worm. The pinion meshes on the side of the gear rather than the outside diameter of the gear. The gear does not wrap around the worm as a conventional worm gear does. The Spiroid gear is coned.

Another member of the Spiroid family is the Helicon gear. A Helicon pinion is cylindrical rather than coned.

The Spiroid pinion has two effective pressure angles, which are called the "high" pressure angle and the "low" pressure angle. Figure 10.51 shows the high side and the low side of a typical Spiroid pinion. The low side of the Spiroid pinion is preferred for driving because it has the largest amount of tooth contact area and it exerts the least separating force on the gear.

Spiroid and Helicon gears may be designed for a right-hand or left-hand system. The choice of hand on the pinion determines the direction of rotation of the gear. The position of the pinion (normally in one of four quadrants) and the offset between the axes of the pinion and the gear have much to do with the direction of bearing reactions. Figure 10.52 illustrates the configurations just described for Helicon gear sets.

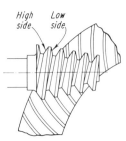

FIGURE 10.51 High and low side of spiroid gear.

Before bearing load calculations can be made for a Spiroid set, it is necessary to establish design details like ratio, cone angle, offset, and pressure angle. The pinion has a constant lead; but, because of the taper, the lead angle is variable. Figure 10.53 shows how the lead angle varies along the length of the cone.

Bearing load calculations are based on the lead angle, the pressure angles, the pitch diameters, and the center distance at the mean length of the pinion.

The exact calculation of Spiroid bearing loads is a complicated process. Generally, it is satisfactory to calculate approximate bearing loads using special constants. The error is usually less than 10 percent. If exact load data are needed, the Illinois Tool Works is willing to furnish assistance on a specific request.

The approximate tangential driving force may be obtained from the output horsepower or output torque as follows:

$$W_t = \frac{19{,}098{,}600 P}{n_2 \, d_{p_2}} \qquad \text{metric} \qquad (10.56a)$$

$$W_t = \frac{126{,}050 P_h}{n_G \, D} \qquad \text{English} \qquad (10.56b)$$

where W_t (W_t) = Tan driving force [metric (English)] N (lb)
n_2 (n_g) = Speed of gear, r/min
d_{p_2} (D) = Gear pitch diameter, mm (in.)
P (P_h) = Output power, kw (hp)

The approximate radial force on the gear is

GEAR REACTIONS AND MOUNTINGS 10.53

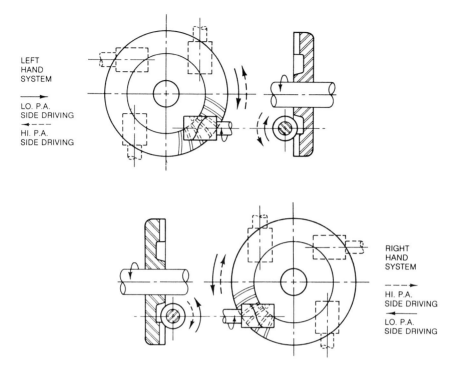

FIGURE 10.52 Examples of mounting positions for left-hand and right-hand Helicon gear sets.

$$W_{r_G} = 0.91 W_t \tag{10.57}$$

The radial force on the pinion is

$$W_{r_P} = f W_t \tag{10.58}$$

where $f = 0.60$ for 35° pressure angle on high side
$= 0.52$ for 30° pressure angle on high side
$= 0.35$ for 15° pressure angle on low side
$= 0.21$ for 10° pressure angle on low side

The thrust force on the pinion is equal to the radial force of the gear. Likewise the thrust force on the gear is equal to the radial force on the pinion. Thus

$$W_{x_P} = W_{r_G} \tag{10.59}$$
$$W_{x_G} = W_{r_P} \tag{10.60}$$

The point of application of these forces should be taken at the midface of the gear and the middepth of the tooth.

For a straddle-mounted Spiroid set, the bearing loads may be determined by a procedure similar to that shown for worm gears in Table 10.8. Substitute W_{r_P} for the

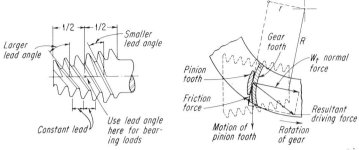

FIGURE 10.53 Spiroid pinion lead angle.

FIGURE 10.54 Spiroid gear bearing loads.

worm tan force W_t. Assume zero for the separating force. Substitute W_{x_P} for the worm thrust W_x. in the Spiroid gear calculation, use zero for the separating force and substitute W_{r_G} for the W_x term in the worm gear bearing load part of the table. The radii r_1 and r_2 should be taken to the midpoint of the Spiroid mesh.

Figure 10.54 shows the Spiroid gear meshing point and the direction of the tangential driving force.

The Spiroid Division of the Illinois Tool Works in Chicago, Illinois has established computer programs to determine bearing reactions for the commonly used Spiroid and Helicon arrangements. "Spiroid Gearing"[9] is an example of a design booklet available at this time (1990). Those designing gear sets in the Spiroid family of gears should obtain data of this type.

10.10 BEARING LOAD CALCULATIONS FOR OTHER GEAR TYPES

Space does not permit detail bearing load equations for several additional types of gears. The Planoid,* Helicon,† face gear, Beveloid,† and others are not covered. However, the principles used in this chapter make it relatively easy for designers to set up equations for any special types. The first step in solving the problem is to find the radial, separating, and thrust forces at some center point of the contact area. Then calculations are made to determine the final radial and axial reaction at each bearing location. If there is more than one load point on the same shaft, proper vector addition must be made to get the overall resultant.

In solving complex bearing load problems, the following rules should be kept in mind:

1. When all the gearing forces have been transferred to a given bearing mount, the final result is one resultant radial load and (depending on the bearing type) one resultant thrust (or axial) load.
2. The summation of all forces must equal zero.
3. The summation of all moments must equal zero.

---

*Registered trademarks of the Spiroid Division of Illinois Tool Works, Chicago, Illinois.
†Registered trademark of Vinco Corp., Detroit, Michigan.

GEAR REACTIONS AND MOUNTINGS   10.55

In the case of new gear types, the manufacturer of the type or the inventor of the type is usually glad to furnish potential customers with detail calculations for the forces at the meshing point.

10.11 DESIGN OF THE BODY OF THE GEAR

To conclude this chapter, some comments should be made relative to the structure of the gear body that supports the gear teeth. It sometimes happens that a gear designer does an excellent job of designing gear teeth and then making bearing selections and calculations to adequately support the gear, but does almost no engineering work relative to the gear body. In a gear set the pinion is often small and essentially amounts to teeth cut on a section of shafting. The gear member of the pair, though, tends to be much larger in diameter and therefore is apt to involve a hub, a web, and a rim. Each of the three things just mentioned need to be so designed as to have adequate strength, adequate rigidity, and be manageable from the standpoint of resonance and vibration.

The prime usage of a gear set may be the transmission of power, or it may be the transmission of motion. The following kinds of gear applications need to be recognized before design work is done to settle details of a rim, a web, or a hub:

Data gears: These gears are generally sized to fit an arrangement scheme and a need to bridge the distance from one shaft to another shaft. Some examples are tachometer drives, governor drives, or instrument drives. The amount of power transmitted through the gearing is usually relatively insignificant for the size of gears chosen to meet arrangement requirements.

Power gears: These gears are primarily used to transmit power and change speed. They may be either speed-reducing or speed-increasing. The size of the gears is usually determined by making the gear large enough to carry the required power. For weight reasons and cost reasons, the tendency is to design them as small as possible but still have adequate power capacity and reliability.

Accessory gears: In a power-producing engine or motor there may be a train of power gearing that takes the main power from the prime mover to the driven device. This might be a turbine driving gear connected to a pump, a compressor, or a propellor. In addition to the main power drive there is often a need to have a package of gears driving accessories. These accessories can involve oil pumps, fuel pumps, pumps for hydraulic power, and so forth. The arrangement of the accessory package frequently requires gears larger in diameter and wider in face width than what is needed for the transmitted power. This means that the body structure of accessory gears can often be less rugged than that needed for main drive power gears.

A further consideration in gear body structures has to do with the kind of application. For instance, rocket engines and aircraft engines have to be very lightweight. This means the use of very-high-strength materials and extra cost in manufacture to achieve the ultimate capacity in power per kilogram of metal or per pound of metal. At the other end of the spectrum, industrial gearing used in mills and in processing plants can be much heavier in weight, but there is much concern to keep the cost relatively low. Relatively heavy structures are used. The allowable stresses are held to much lower values than in aerospace work.

Table 10.11 has been put together to give some general guidance relative to rim thickness, web thickness, holes in the web, and hub thickness. Those using this table

TABLE 10.11 Approximate Ratio of Certain Gear Body Dimensions to the Whole Depth of the Gear Tooth

Gear Drive Application	Rim Thickness	Web Thickness	Lightening Holes Recommended?	Hub Thickness	Reference*
Data gears					
Instrument or data accessory	0.6 to 0.7 (0.07 in. min)	0.5 to 0.7 (0.07 in. min)	Normally if space available	0.6 to 1.0 (0.07 in. min)	Experience
Power gears					
Rocket engine main power	1.0 (0.125 in. min)	1.0 (0.125 in. min)	Not recommended	1.5–2.0	9 (3.2.6)
Accessory rocket	0.7	0.5	Recommended odd number of holes	1.5–2.5	9
Accessory aircraft	1.0 to 1.24 (0.125 in. min)	0.7 to 1.0 (0.125 in. min)		1.5–2.5	2, 5
Aircraft and helicopter main drives	1.25–1.50	1.0–1.35	Normally gears only	—	2, 9
Planets	1.5 avg.	1.35	Gear only	—	5, 6 (Fig. 1)
Accessory	1.25	1.00	Gear only	—	2, 5
Commercial hypoid	2.0	2.0	Yes	8	
Commercial spiroid	1.0–2.5	2.0	Yes	11	7 (Figs. 61, 62)
Marine, submarine	3 to 4	3 to 4	Holes required for welding, cleaning etc.	—	Experience
Stationary power	4 to 6	depends on design			

Notes: Antibacklash: Use values appropriate for data gears or for power gears.
Ratio of 1.0 means value equals whole depth; 2.0 is twice whole depth.
*Superscript numbers refer to references at end of chapter.

should consider the table as primarily historical with regard to what is commonly done rather than as positive design values that must be used. For instance, a well-designed aircraft gear may be acceptable with a rim thickness underneath the gear teeth that is only 1.25 times the gear tooth height. If the face width happened to be rather wide and the rim underneath the gear lacked stiffness due to a thin web or the tendency to have "spokes" connecting the rim to the hub, it may be necessary to increase the rim thickness to 2 or 3 times the tooth height to get adequate stiffness in the gear body.

The tendency to have a spoked condition is quite common in marine applications and gearing for stationary power. Gears of this type often have a rim thickness that is as much as 4 times the tooth height.

At times, rather thin webs are used that tend to go beyond Table 10.11 recommendations. This helps save weight, but there may be failures of the web due to resonant vibrations. This risk can be handled by using damper rings under the rim of the gear to stop vibration. If the stresses induced by vibrations are not too high, the shot-peening of a thin web in critical areas may give the added capacity that is needed to handle stresses from vibrations.

Another general area of concern is resonance in vibration. A webbed gear tends to have up to 3 or more critical speeds. If the gear is not running too fast and the web is fairly heavy, the resonant frequencies tend to be very high compared with the frequency that the teeth are meshing. In aerospace applications, though, it often happens that a critical frequency can be rather close to the operating frequency. This is a bad situation and should be avoided. Sometimes the first critical comes at about two-thirds of the operating frequency and the second critical is well above the operating frequency. This situation may be tolerable if the gear unit tends to run at a constant speed and the gear unit comes up to full speed rather quickly so that there is almost no time that the gear is operating at the first critical.

To sum things up, the design of the body structure of the gear should be handled just as thoroughly as the design of the gear teeth. It is beyond the scope of this book to present many pages of engineering data on the design of structures such as gear bodies, turbine wheel bodies, and so forth. The main point is that structural design is just as important in gearing as it is in the design of other types of machinery.

REFERENCES

1. New Departure Division of General Motors Corporation, *New Departure Handbook*, Vol. 2, pp. 27–73, 145–158, Bristol, CT
2. ANSI/AGMA standard 2001-B88, "Fundamental Rating Factors and Calculation Methods for Involute Spur and Helical Gear Teeth," American Gear Manufacturers Association, September 1988.
3. Barish, T., "Idler Gears," *Machine Design*, Cleveland, Ohio, November 28, 1957.
4. Drago, R. J., AGMA P229.24, "An Improvement in the Conventional Analysis of Gear Tooth Bending Fatigue Strength," October 1982.
5. Drago, R. J., AGMA 86FTM9, "On the Strength of a Planet Gear with an Integral Bearing," October 1986.
6. Gleason Works, "Bevel and Hypoid Gear Design," Gleason Works, Rochester, New York, 1971, pp. 1–104.
7. Gleason Works, "Gleason Computer Programs," Series D90/3 (2023), Gleason Works, Rochester, New York, 1990, pp. 9–8 to 9–14, page 13, III, A1-2-3 and A-4.
8. Hartman, M. A., and Butner, M. F., "Liquid Rocket Engine Turbo Pump Gears," NASA SP

8100, NASA Space Vehicle Design Criteria, NASA/Lewis Research Center, Cleveland, Ohio, March 1974.
9. Illinois Tool Works, "Spiroid Gearing," *Design Handbook No. 6*, Spiroid Division of Illinois Tool Works, Chicago, Illinois 1986.
10. AGMA 112.05, "Gear Nomenclature (Geometry), Terms, Definitions, Symbols and Abbreviations," American Gear Manufacturers Association, Washington, D.C., 1976.
11. Marlin Rockwell Corporation, *Ball Bearing Load Computational Manual*, Jamestown, New York.
12. Merritt, H. E., *Gears*, Pitman Publishing Corporation, New York, 1955, Chaps. 20, 23.
13. Michigan Tool Company, "Cone Drive Gears," Catalog 60, Cone Drive Gear Division, Michigan Tool Co., Detroit, 1953, pp. 1–16, 38–52.
14. Shigley, J. E., and Mischke, C. R., *Standard Handbook of Machine Design*, McGraw-Hill, New York, 1986, Chaps. 33–36, 43.

CHAPTER 11
LOAD RATING OF GEARS

Giovanni Castellani
Studio Progettazione Macchine e Trasmissioni
Modena, Italy

11.1 CONSIDERATIONS

In the present state of the art, gear capacity ratings must be considered only as an estimation. This is true because the gear metallurgy and the application conditions, such as load histograms, gear dynamics, and tooth alignment can differ from estimations and influence both the tooth strength and the surface resistance of gears. Roughness and irregularities of the tooth fillet may reduce the tooth strength, whereas tooth surface finish and lubrication conditions influence the surface resistance in various ways.

The present rating procedures cover most influences by means of proper "factors" so that they can serve as a guideline for designers and enable them to achieve reliable designs. The rating systems are only a means for applying the designer's competence, not a substitute for it.

For instance, detailed calculations of the geometry factors for pitting resistance are given. They may be complicated, but with a computer they can be calculated in a few seconds at most. They should not be considered more important because they are detailed. A proper choice of the application factor is often much more important for the final result and for the success of the gear pair than the method used.

The choices for design application factors must be based on experience or searched for in specific books for gear design, such as Dudley.[1]

Here it is emphasized that the choices must be consistent. For example, common spur gears are usually inconsistent with high speed and helical gears are inconsistent with low tooth accuracy. These types of inconsistencies can lower the reliability of gear ratings.

In this chapter:

    RH = resistance to hertzian pressure, that is, in general, pitting resistance. But the pressure limitation usually averts or limits other surface failures too, such as progressive wear or cold scoring. An approximate estimation for hot scoring for industrial gears is given.

    RF = resistance of the fillet of the tooth root, or so-called tooth strength.

RHA and RFA = ratings based fundamentally on AGMA's method reported in the Standards[2,3] that substantially coincide except for a pair of items that are clarified.

RHI and RFI mean that the original ISO method is taken as a basis.[4] The DIN method[5] has the same general conception as ISOs; some peculiarities are reported when important. The ISO method is described in Niemann and Winter,[6] Henriot,[7] and Castellani and Zanotti.[8]
All cited methods relate to involute gear teeth.
Simplifications or adaptations of any syntheses are specified.

11.2 MAIN NOMENCLATURE

AGMA symbols[3] are used as far as possible for ISO factors also, when the meaning is the same. Symbols that occur in a single item and ISO symbols are specified when necessary. The units are indicated in the formulas.

A, B = generic coefficients or exponents
A_H, A_F = adaptation factors for RH, RF
C = center distance
C_H, C_f, C_s, C_L, C_R = RH factors for hardness, finish, size, life, reliability
C_p = pressure ratings: elastic coefficient
C_{SF} = service factor for RH
d_b = base diameter
d_o = tooth outside diameter, that is, gear outside diameter for external, gear internal diameter for internal gears
d_P, d_G = operating pitch diameters of pinion and gear, respectively
E = Young's elasticity modulus
F = net face width, that is, overlap face width
f = surface finish, that is, roughness, arithmetic—average
f_d = ratio of the face width over the pinion pitch diameter, $f_d = F/d_P$
G_H = unified geometry factor for RH
I, J = AGMA's geometry factors for RH and RF
J_n = unified geometry factor for RF, related to normal module or diametral pitch
K = synthetic surface loading factor for RH
K_{sh}, K_a, K_v, K_m = overload derating factors, each one ≥ 1, for power sharing, application, gear dynamics (velocity), load distribution (misalignment)
K_f, K_s, K_L, K_R = RF factors for fillet notch, size, life, reliability
K_{SF} = service factor for RF
K_y = coefficient of the resistance to yielding for RF ($K_y < 1$)
m, m_n = transverse and normal module, nominal

LOAD RATING OF GEARS

m_G = gear ratio, $m_G = N_G/N_p$
m_p, m_F = transverse (profile) and face contact ratios
m_{pE}, m_{pA} = operating addendum contact ratios of pinion and gear [Eqs. (11.30) and (11.31)]
m_e, m_N = contact line coefficient, load-sharing ratio
$\max(x, y), \min(x, y)$ = maximum or minimum of two numbers x, y
N = number of tooth loading cycles or tooth number
N_G, N_P = tooth numbers of gear and pinion ($N_G \geq N_P$)
n = revolutions per minute
P = power
p_{b_t} = transverse base pitch
P_d, P_{nd} = transverse and normal diametral pitch, nominal
Q_H, Q_F = loading factors for RH and RF
S_2 = $S_2 = 1$ for external; $S_2 = -1$ for internal gear or gear pair
s_c = contact pressure, that is, hertzian pressure
s_t = stress at tooth fillet (notch effect included)
T = torque
U_L = unit load for RF
V = overall overload derating factor ($V \geq 1$)
v_t = tangential velocity
W_t = operating tangential load
X_B = curvature coefficient at the lower point of single contact of pinion
x_P, x_G = addendum modification coefficients of pinion and gear with reference to m_n or P_{n_d}
ν_{40} = kinematic viscosity in cSt at 40°C
ϕ_{ns} = normal standard pressure angle
ϕ_s = transverse standard pressure angle
ϕ_t = transverse operating pressure angle
ψ_b = base helix angle
ψ_s = standard helix angle
ψ = operating helix angle

Subscripts:

G, P = gear or pinion
H, F = RH or RF
lim = relating to a conventional fatigue limit
m = relating to the middle of the net face width for bevel gears
v = for bevel gears, relating to virtual gears with parallel axes
V = relating to the vertex of the life curve for the conventional fatigue limit

W = relating to the maximum load or pressure or stress of a life curve
y = relating to yielding

Note: RH, RHA, RHI, RF, RFA, RFI: see Sec. 11.1.

11.3 COPLANAR GEARS (INVOLUTE PARALLEL GEARS AND BEVEL GEARS)

The ratings are basically presented for involute gears with parallel axes. Adaptations are given for bevel gears.

11.3.1 Power, Torque, and Tangential Load

The pinion torque T_P can be given directly, otherwise it is deduced from the power:

$$T_P \text{ [N·m]} = \frac{9549.3 \ P \text{ [kW]}}{n_P \text{ [r/min]}} \tag{11.1}$$

$$T_P \text{ [lb·in.]} = \frac{63,025 \ P \text{ [hp]}}{n_P \text{ [r/min]}}$$

The ratings are based on the operating tangential load W_t. For gears with parallel axes, the operating pitch diameters are

$$d_P = \frac{2C}{m_G + S_2} \qquad d_G = m_G \ d_P \tag{11.2}$$

Hence:

$$W_t \text{ [N]} = 2000 \ \frac{T_P \text{ [N·m]}}{d_P \text{ [mm]}} \tag{11.3}$$

$$W_t \text{ [lb]} = 2 \ \frac{T_P \text{ [lb·in.]}}{d_P \text{ [in.]}}$$

Loading Levels. It is easy to calculate the tangential load, but the definition of the power or torque to be introduced in the formulas is sometimes the most difficult task for gear designers. The loading conditions are often variable. If load histograms of similar applications are available, or if they can be foreseen even summarily, it is useful to consider various loading levels where not only the load, but also the rotation velocity can change. In this case a so-called equivalent load sometimes is calculated, but this is risky. In fact, if a nondamaging load is applied for a higher number of cycles, this leads to underestimating the equivalent load. Therefore it is better to assess separate damages and cumulate them by an approximate approach according to Miner's rule.

If, on the contrary, a single main loading condition is identified, then the simple calculation of the "service factor" is often preferred. This includes the consideration of possible short higher loads and rules out further estimations of life and reliability.

LOAD RATING OF GEARS   11.5

11.3.2 RH and RF, Resistance to Yielding

Both flank and root of a gear tooth can yield or break if an excessive load induces plastic deformations.

11.3.3 RH and RF, Life Curves

Both surface and root resistances are fatigue resistances and can be approached in similar ways provided that no yielding occurs.

In Fig. 11.1 a generic life curve is schematized.[9] The initial maximum load W_{tW} (i.e., W_{tW_H} or W_{tW_F}) should mean yielding, although separate estimates are preferred for practical reasons.

A two-slope curve WV, VS is used for life predictions. The second sloped stretch has been introduced by AGMA 218.01 and is not defined by ISO, but the concept can be used for RHI and RFI life ratings. AGMA does not define the vertex S but there are some who consider that no failure occurs below a certain load W_{tS}. This is not confirmed experimentally, but it can be adopted as a criterion in the life range specified for a gear pair.

In the case of slow gears in boundary lubrication regime, the progressive wear can become a more important item than pitting for tooth flank damage, but the load limitation implied by the reference to life curves also helps in reducing wear.

11.3.4 RH and RF, Conventional Fatigue Limits

Conventional fatigue limits of both Hertz pressure and root bending stress are obtained either from field experience or from laboratory tests. These lead to a conventional

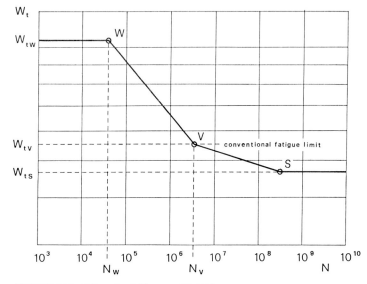

FIGURE 11.1 Schematized life curve (N, W_t).

fatigue limits $W_{t_{\lim H}}$ or $W_{t_{\lim F}}$ of the load by excluding reliability and life factors from the ratings. They usually correspond to vertex V except for RFA where a small difference follows the data of the life factor according to AGMA 218.01 or ANSI/AGMA 2001-B88.

The gear capacity rating stops here if a single loading level is considered and then service factors that exclude life calculations can be assessed. Otherwise the numerical definition of the life curves provides damage, life, and reliability ratings.

11.3.5 RH—Synthetic Surface Loading Factor K

A factor K, whose square root is proportional to the hertzian pressure, is defined as follows:

$$K \ [\text{N/mm}^2] = \frac{W_t \ [\text{N}] \left(1 + S_2 \dfrac{N_P}{N_G}\right)}{d_P \ [\text{mm}] \ F \ [\text{mm}]} \tag{11.4}$$

$$K \ [\text{lb/in}^2] = \frac{W_t \ [\text{lb}] \left(1 + S_2 \dfrac{N_P}{N_G}\right)}{d_P \ [\text{in.}] \ F \ [\text{in.}]}$$

11.3.6 RF—Unit Load U_L

A unit load U_L can be used for tooth bending strength

$$U_L \ [\text{N/mm}^2] = \frac{W_t \ [\text{N}]}{F \ [\text{mm}] \ m_n \ [\text{mm}]} \tag{11.5}$$

$$U_L \ [\text{lb/in}^2] = \frac{W_t [\text{lb}] \ P_{nd} \ [\text{in.}^{-1}]}{F \ [\text{in.}]}$$

ISO allows the F value to be increased for a gear (pinion or gear) that has a face width greater than the net one. The maximum permitted increase is equal to 1 module at each gear side.

11.3.7 Adaptations for Bevel Gears (W_t, K, U_L)

The ratings are related to the middle of the net face width. This contributes a small supplementary margin for reliability, as the resultant of the distributed load applies to a point somewhat more external (for perfectly aligned teeth). The corresponding mean pitch diameters are shown in Fig. 11.2:

$$d_{P_m} = d_P - F \sin \gamma_P \qquad d_{G_m} = d_G - F \sin \gamma_G \tag{11.6}$$

where γ_P and γ_G are the pitch angles of pinion and gear. The mean pitch diameter of pinion, d_{P_m}, must be introduced in Eq. (11.3) instead of d_P for calculating W_t.

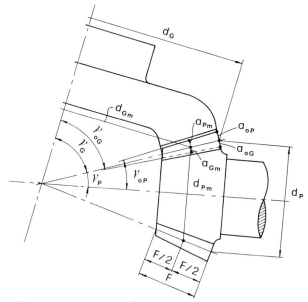

FIGURE 11.2 Geometrical data of a bevel gear pair for load capacity ratings.

A virtual gear pair with parallel axes must be considered for capacity ratings (pressure angle and helix angle are maintained). Virtual tooth numbers N_{P_v}, N_{G_v} and virtual pitch diameter $d_{P_{mv}}$ are introduced in the K Eq. (11.4) for RH instead of N_P, N_G, d_P:

$$N_{P_v} = \frac{N_P}{\cos \gamma_P} \qquad N_{G_v} = \frac{N_G}{\cos \gamma_G} \qquad (11.7)$$

$$d_{P_{mv}} = \frac{d_{P_m}}{\cos \gamma_P} \qquad (11.8)$$

Note that $N_{P_v} / N_{G_v} = \left(\dfrac{N_P}{N_G}\right)^2$ if $\gamma_P + \gamma_G = 90°$.

A mean normal module m_{n_m} or diametral pitch P_{nd_m} are introduced in the U_L Eq. (11.5) for RF instead of m_n or P_{nd}

$$m_{n_m} = \frac{m \cos \psi_s \, d_{P_m}}{d_P} \qquad (11.9)$$

$$P_{nd_m} = \frac{P_d \, d_P}{d_{P_m} \cos \psi_s}$$

where $m = d_P / N_P$ and $P_d = N_P / d_P$ are the usual nominal module or diametral pitch with reference to Fig. 11.2 for the nominal pinion pitch diameter d_P, as operating and standard diameters coincide for bevel gears.

Example—Bevel Gear Pair with

$$N_P = 13, N_G = 38, m = 5.08$$

or $P_d = 5$, $\psi_s = 35°$, $F = 30.48$ mm $= 1.2$ in., $T_P = 320$ N·m $= 2832$ lb·in

$\gamma_P = 18.8861°$, $\gamma_G = 71.1139°$ $- d_P = 66.04$ mm $= 2.6$ in. $- d_{P_m} = 56.174$ mm $= 2.2116$ in. $- W_t = 11,393$ N $= 2561$ lb $- d_{P_{mv}} = 59.37$ mm $= 2.337$ in. $- K = 7.03$ N/mm^2 $= 1020$ lb/in^2 $- m_{n_m} = 3.5396$

or $P_{nd_m} = 7.1759$ $- U_L = 105.6$ N/mm^2 $= 15,320$ lb/in^2

A simpler approach to the K calculation for 90° bevel gear pairs is given in Dudley,[1] which calculates W_t at the middle point of the face width as in the foregoing, but introduces outer d_P diameter and real tooth numbers in the K formula. It is most useful especially for immediate estimations in connection with special tables for bevel gears and gives results that are not too different, especially for higher gear ratios.

The K rating based on the virtual tooth numbers is useful especially for programmed computations. The real d_{P_m} diameter must be used for calculation of W_t! The same program can be used for bevel as well as for cylindrical gears for unification reasons, and the same K tables can be consulted. The special tooth alignment conditions of bevel gears must be taken into account when assessing the load distribution factor K_m.

Specific methods for bevel gears are given by ANSI[10] and in the standard DIN 3991 and are being elaborated by ISO.

11.3.8 Use of the Synthetic Factors K, U_L

The K and U_L factors serve for the first step in a detailed capacity analysis, as well as for simplified estimates or for the first tentative assumptions of the size of gear pairs in new designs.

11.4 COPLANAR GEARS: SIMPLIFIED ESTIMATES AND DESIGN CRITERIA

11.4.1 Direct Assumption of the Synthetic Factors

Table 11.1 gives values of K as used in industrial gears with case or induction surface hardening. The table can serve as an estimation. The K values as used in industry often fail to satisfy detailed capacity analyses. Wider tables are given in specific books for gear design such as Dudley.[1]

Gears without surface hardening have far lower K values and are somewhat more uncertain as they are more sensitive to both tooth flank roughness and load histogram. A summary estimation of acceptable K values can be made in advance if specific field experience is available. Otherwise, detailed ratings of similar cases can give useful indications.

The unit load U_L for RF varies in ampler ranges. For instance, U_L values in the range from 40 to 170 N/mm^2 (6000 to 25,000 lb/in^2) can be found for case-hardened gears. Note that the highest U_L values of this range often do not satisfy the AGMA estimations. It is advisable not to exceed U_L values of 120 N/mm^2 (17,500 lb/in^2) even

LOAD RATING OF GEARS   11.9

for the best nitrided teeth and for induction-hardened teeth with the tooth root correctly hardened. Lesser values are opportune for larger nitrided gears and far lesser values for induction hardening not extended to the tooth root. Industrial gears without surface hardening usually have lower U_L because their general size is greater in order to ensure RH. Thus they do not create problems for RF.

The K and U_L values work in two ways as far as simplified estimations are concerned:

1. They enable a rough assessment of the gear capacity for given gear data. (In this case, they can also be the first step for a detailed analysis.)
2. They are the basis for new gear designs.

11.4.2 Design Procedure

The usual procedure consists in adopting a provisional K value, establishing the general gear size, and choosing module or diametral pitch such that an acceptable U_L is obtained. Pinion torque T_P and desired gear ratio m_G are given. Thus, two cases must be distinguished:

1. The center distance C is given.
2. All geometric parameters are free.

In the first case the pitch diameter of the pinion is calculated from Eq. (11.2) and the necessary face width is determined from Eq. (11.4).

In the second case a ratio of the face width divided by the pinion pitch diameter can be adopted according to the kind of gear and application:

$$f_d = \frac{F}{d_P} \tag{11.10}$$

Thus the necessary pitch diameter is determined

$$d_P \text{ [mm]} = \sqrt[3]{\frac{2000 \, T_P \, [\text{N} \cdot \text{m}] \left(1 + S_2 \dfrac{N_P}{N_G}\right)}{f_d \, K \left[\dfrac{\text{N}}{\text{mm}^2}\right]}} \tag{11.11}$$

$$d_P \text{ [in.]} = \sqrt[3]{\frac{2T_P \, [\text{lb} \cdot \text{in}] \left(1 + \dfrac{S_2 N_P}{N_G}\right)}{f_d K \left[\dfrac{\text{lb}}{\text{in}^2}\right]}}$$

Both cases are shown in Fig. 11.3.

For bevel gears obvious adaptations are necessary.

Finally a standard module or diametral pitch is chosen on the basis of Eq. (11.5) and the tooth numbers are established. This may necessitate some slight retouching of the previous calculations according to the actual m_G ratio.

An alternative way for choosing the tooth numbers is given by Fig. 11.4 for case-

TABLE 11.1 Guideline Surface Factors K as Used in Industry for Case- and Induction-Hardened Gears*

Application	Materials P	Materials G	Finish P	Finish G	K, N/mm² (lb/in²)	Speed	Service
Ordinary industrial speed reducers	C	C	sh. or gr.	sh. or gr.	5.6–10 (810–1450)	Low–mean	Nom.
	C	I	gr.	cut.	3.6–6.3 (520–910)		
	I	I	cut.	cut.	2.8–5 (400–720)		
Planetary speed reducers sun/planets	C	C	sh. or gr.	sh. or gr.	4.5–8 (650–1150)	Low–mean	Nom.
planets/ring	C	I	sh. or gr.	cut.	2.8–5 (400–720)		
Parallel gear pairs for high power	C	C	gr.	gr.	2–6.3 (290–910)	High	Nom.
Machine tools for metal cut. or gr.	C	C	gr.	gr.	4–7.1 (580–1030)	Low–mean	Mean
Various types of machines	C	C	sh. or gr.	sh. or gr.	4.5–8 (650–1150)	Low–mean	Mean
					3.2–5.6 (460–810)		Heavy

Various types of machines						
C	C	cut.	4.5–7.1 (650–1030)	Low-mean		Mean
			2.8–5 (400–720)			Heavy
C	I	cut.	2.8–5 (400–720)			Mean
		gr.	2–3.6 (290–520)			Heavy

*Abbreviations and notes:
P = pinion
G = gear
C = case-hardened alloy steel, HRC = 58–62
I = induction-hardened alloy steel, HRC ≈ 55
Tooth finishing:
cut. = hobbing or cutting
sh. = shaving
gr. = grinding
Service:
nom. = nominal: $C_{SF} = 1$
mean: $C_{SF} = 1.25$–1.6
heavy: $C_{SF} = 1.8$–2.25

Maximum K values in the table are usually adopted for good precision helical gears, good tooth alignment, mean speed. Misalignment problems often imply lower values to be adopted for bevel gears.

Higher K values are sometimes allowed if a short life, i.e., a small overall number of cycles is required.

Design, manufacturing, or operating anomalies oblige adoption of lower K values—or they may be dictated by imposed general sizes.

Lower K values are often used for high-power gears for reasons not related to pitting resistance, e.g., bearing life—or gear efficiency, scoring resistance, and, indirectly, tooth breakage resistance. In fact lower modules (higher diametral pitches) may be advisable, which will signify an increase in general size.

Other material and manufacturing combinations may be adopted: For instance, specialist gear manufacturers may grind induction-hardened gears, which requires a great deal of specialized experience.

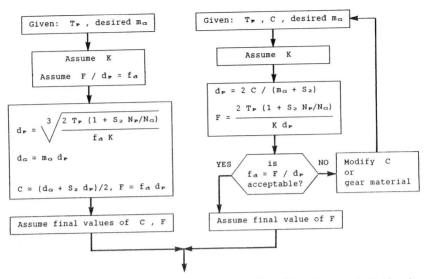

FIGURE 11.3 Design flow chart for the tentative choice of overall size of a gear pair. Metric units: T_P [N·mm], K [N/mm^2], F, d_P, d_G, C [mm]. English units: T_P [lb·in.], K [lb/in^2], F, d_P, d_G, C [in.].

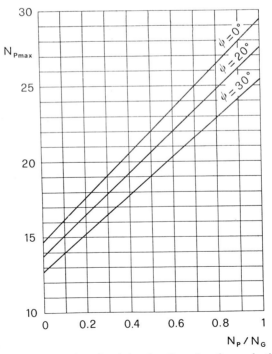

FIGURE 11.4 Tentative choice of tooth numbers for case-hardened gears: $N_{P_{max}}$, maximum pinion tooth number for RF if RH is fully exploited. External gears, monodirectional loading.

LOAD RATING OF GEARS   **11.13**

hardened gears subjected to unidirectional load. The diagram gives the maximum tooth number of the pinion that is compatible with RF if RH is fully exploited, that is, for higher K values. The $N_{P_{\max}}$ must be multiplied by 0.70 for an alternating load such as that of planet gears. Intermediate values can be chosen for bidirectional loading, for example, for translation of a bridge crane. The diagram can be used for bevel gears by considering virtual tooth numbers for both abscissa and ordinate. Induction-hardened gears with hardened tooth root are also covered by the diagram, but even more caution is necessary, especially if special tools are not used and if a final check of every tooth root is not performed.

The diagram indications, as well as the preliminary limitation of U_L, serve only as provisional guidance, and a subsequent detailed RF analysis is necessary, especially if AGMA's rating is taken as a basis.

11.5 COPLANAR GEARS: DETAILED ANALYSIS, CONVENTIONAL FATIGUE LIMITS, AND SERVICE FACTORS

The conventional fatigue limits are calculated for both RH and RF. The analysis can stop here with the determination of the equivalent service factors in the case of 1 loading level, or it can go on by rating RH and RF fatigue "damages" and gear life, as well as reliability factors.

11.5.1 RH and RF, Overload Derating Factors

The actual tooth load generally increases with regard to W_t and reaches a maximum value in a given instant and in a given tooth point for various reasons that are taken into account by means of specific overload factors. They all derate the load capacity of the gear pair and each one of them is assumed ≥ 1.

1. The power sharing may not be uniform in the case of multiple drivings like that of planet gears,[11, 12] or that of a wheel driven by two pinions.
   Power sharing factor: K_{sh}. (Note: This factor is not directly included in the methods of AGMA and ISO.)
2. The operating conditions of driving motor or of the driven machine can cause systematic overloads. Application factor: K_a.
3. The dynamic behaviour of the gear pair itself can cause overloads. Dynamic factor: K_v (velocity factor).
4. Tooth misalignment or profile and tooth spacing errors cause local overloads. Load distribution factor: K_m.

The application factor and the dynamic factor should be assumed $= 1$ and substituted by a complete dynamic analysis of the system (motor/gears/driven machine) for a rational approach, which would lead to a greater W_t value. But separate factors K_a and K_v are usually maintained for practical reasons.

An overall overload factor is defined as:

$$V = K_{sh}\, K_a\, K_v\, K_m \qquad (11.12)$$

11.5.2 Power Sharing Factor, K_{sh}

Assume that the tangential load W_t is calculated by considering a uniform power sharing. Thus the K_{sh} factor accounts for the actual differences by referring to the most loaded tooth meshing.

The best power sharing for common planetary speed reducers is obtained by the adoption of three planets that make the structure nearly isostatic, see Le Borzec.[12] Then a $K_{sh} = 1.1$ to 1.2 can be assumed. Greater values should be adopted for two, or for four or more planets. Increase K_{sh} by 10 to 40 percent for drive types with two planets on each axis if no special manufacturing equipment is used to enable perfect tooth phasing.

For ordinary gearing, $K_{sh} = 1$ to 1.2 for double driving by two electric motors according to the motor type. For example, $K_{sh} = 1.1$ to 1.2 for usual motors; $K_{sh} = 1.05$ to 1.1 for high-sliding motors (only used for lower powers); $K_{sh} = 1$ for motors with electronic control of torque balance as used for double driving of big mills. (Note that, in the latter case, possible different amount of failures of the two pinions usually depends on other causes.)

Double driving with single input shaft and single output shaft: $K_{sh} = 1$ is allowed if there is some phasing device or if the manufacturing equipment enables perfect tooth phasing. Otherwise, $K_{sh} = 1.1$ to 1.4 and more.

11.5.3 Application Factor, K_a

K_a is usually assumed in the range 1 to 2, sometimes more. Consult specific design books or load capacity standards.

11.5.4 Dynamic Factor, K_v

Tangential Velocity

$$v_t \ [\text{m/s}] = \frac{\pi \, d_P \ [\text{mm}] \, n_P \ [\text{r/min}]}{60{,}000} \quad (11.13)$$

$$v_t \ [\text{ft/min}] = \frac{\pi d_P \ [\text{in.}] \, n_P \ [\text{r/min}]}{12}$$

RHA and RFA. For $Q_v = 6$ to 11

$$B = (12 - Q_v)^{0.667} \quad (11.14)$$

$$K_v = \left(1 + \frac{\sqrt{v_t \left[\frac{\text{m}}{\text{s}}\right]}}{7.56 - B}\right)^{B/4} \quad (11.15)$$

$$K_v = \left(1 + \frac{\sqrt{v_t \left[\frac{\text{ft}}{\text{min}}\right]}}{106 - 14B}\right)^{B/4}$$

LOAD RATING OF GEARS   11.15

Q_v is generally meant as the AGMA gear quality number, but lesser values are suggested if any cause of vibration or tooth resonance is suspected.

11.5.5 RHI and RFI, K_v

A simplified method is given here, based on the so-called method C of the DIN standard[5] rather than on the original ISO method, as it gives a wider validity range. The dynamic factor can be made to depend on a factor A.

$$A = \frac{v_t \left[\dfrac{m}{s}\right] \dfrac{N_P}{100}}{\sqrt{1 + \left(\dfrac{N_P}{N_G}\right)^2}} \qquad (11.16)$$

$$A = \frac{v_t \left[\dfrac{ft}{min}\right] \dfrac{N_P}{19{,}700}}{\sqrt{1 + \left(\dfrac{N_P}{N_G}\right)^2}}$$

$$K_v = 1 + A \left[\frac{55\, B_1\, (0.65)^{Q_v - 7}}{\dfrac{K_{sh}\, K_a\, W_t[N]}{F\,[mm]}} + B_2\right] \qquad (11.17)$$

$$K_v = 1 + A \left[\frac{314\, B_1\, (0.65)^{Q_v - 7}}{\dfrac{K_{sh}\, K_a\, W_t\,[lb]}{F\,[in]}} + B_2\right]$$

where Q_v = the AGMA quality number
$B_1 = 1$, $B_2 = 0.02$ for spur gears
$B_1 = 0.89$, $B_2 = 0.009$ for helical gears with $m_F \geq 1$

For helical gears with $m_F < 1$, K_v has to be obtained by interpolation

$$K_v = K_{v_{hel}} + (1 - m_F)(K_{v_{spur}} - K_{v_{hel}}) \qquad (11.18)$$

The method refers properly to steel gear pairs with $N_p < 50$, and is applicable with a worse approximation for metal gears with a lower Young's module. General application is allowed for $A < 3$. The calculation can be accepted for higher A values insofar as the gear pair does not enter a resonance condition. Simplified analyses show resonance risk in the range $A = 10$ to 14 for steel gears (see the following section), but this can be altered by any peculiarity of the dynamic system. The AGMA rating with a lower Q_v number may be preferable in any doubtful case.

Note: The K_v equations are estimations. Even if the computer gives us 16 digits, the first decimal digit itself is uncertain and in fact its ISO or DIN value often differs from the AGMA one. Equation (11.17) often gives lower (more optimistic) results than AGMA, especially for heavy loaded helical gears. Past experience says it is acceptable, in its validity field, if the tooth profile errors are not the contrary kind of tip or root reliefs. Otherwise Eq. (11.15) is preferred with a cautious choice of Q_v, especially if the A parameters are rather high as a rough indication of vibration risk.

11.5.6 Dynamic Problems of Gears: Resonance and Vibration Conditions

The teeth act as a spring connecting pinion and gear masses: Therefore the gear pair itself has a natural vibration frequency and may incur a resonance condition according to tooth contact frequency, that is, according to speed and tooth number. A main load peak occurs at resonance speed and lower peaks at submultiples and multiples of it. In the schematic Fig. 11.5, n_P is pinion speed in $r/\min$; n_{r_P} the pinion resonance speed.[8]

The subject has been thoroughly investigated, both theoretically and experimentally, by Kubo,[13] who gives a synthetic description of dynamic problems in an appendix to Dudley's book.[1]

A procedure for the calculation of n_{r_P} is given by ISO[4] and DIN[5] but does not consider masses external to the net face width of the gear pair. The following equations indicate a very simple way of generalizing the calculation. The polar moments of inertia of pinion and gear are J_P and J_G, in kg · m² ≡ N m · s² (lb in. s²); but any mass rigidly connected with pinion or gear must be included in the calculation J_P or J_G. Then equivalent masses of pinion and gear are defined as though on tooth contact line, in kg ≡ N s²/m (lb s²/in.):

$$m_{c_P} = \frac{4 J_P}{d^2_{b_P}} \qquad m_{c_G} = \frac{4 J_P}{d^2_{b_G}} \qquad (11.19)$$

where the base diameters d_{b_P} and d_{b_G} are in m (in.), and an equivalent single mass is determined:

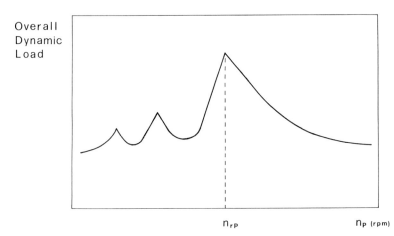

FIGURE 11.5 Pinion resonance speed.

LOAD RATING OF GEARS   **11.17**

$$m_c = \cfrac{1}{\cfrac{1}{m_{c_P}} + \cfrac{1}{m_{c_G}}} \qquad (11.20)$$

Finally,

$$n_{r_P} = \frac{30}{\pi N_P} \sqrt{\frac{10^6 G \left[\dfrac{N}{\mu m\ mm}\right] \cdot F\ [mm]}{m_c\ [kg]}} \qquad (11.21)$$

$$n_{r_P} = \frac{30}{\pi N_P} \sqrt{\frac{G\left[\dfrac{lb}{in^2}\right] \cdot F\ [in.]}{m_c \left[lb\ \dfrac{s^2}{in.}\right]}}$$

where G is the overall tooth stiffness. For steel gears, ISO and DIN give either a detailed method for its calculation, or a mean value $G = 20$ N/μm mm $\cong 3 \times 10^6$ lb/in^2. AGMA indicates $G = 1.5$ to 2×10^6 lb/in.$^2 \cong 10$ to 14 N/μm mm, but limits its validity to load distribution analysis, where deflections other than those of the gears are influential. German researchers show a certain stiffness diminution due to both gear and external causes.[14] The proper value for Eq. (11.21) may be situated between 14 and 20 N/μm mm, or between 2×10^6 and 3×10^6 lb/in^2, and varies, of course, with tooth proportions and the cooperation of various tooth pairs, depending on tooth spacing errors, etc.

Example. Consider a gear pair with $C = 16$ in. $= 406.4$ mm, $N_P/N_G = 27/43$, total face widths $F_P = 5.9$ in. $= 149.86$ mm, $F_G = 5.5$ in. $= 139.7$ mm, P_{nd} 2.5 in.$^{-1}$, that is, $m_n = 10.16$ mm, $\phi_{n_s} = 20°$, $\psi_s = 29°$; steel gears with a specific mass density of 0.0007345 lb $\cdot$ s^2/in.$^4 = 7850$ kg/m^3; $d_{b_P} = 11.400$ in. $= 289.57$ mm, $d_{b_G} = 18.156$ in. $= 461.17$ mm.

Assume the gears to be full cylinders with diameters equal to the pitch diameters 12.343 in. $= 313.51$ mm and 19.657 in. $= 499.29$ mm, and a tooth stiffness $G = 2.5 \times 10^6$ lb/in$^2 = 17$ N/μm mm.

Calculation in English units: Real masses $m_P = 0.5186$, $m_G = 1.226$; polar moments of inertia $J_P = 9.875$, $J_G = 59.22$; equivalent masses $m_{c_P} = 0.3040$, $m_{c_G} = 0.7186$; equivalent single mass $m_c = 0.2136$; $n_{r_P} = 2840$ r/min.

Metric units: $m_P = 90.81$, $m_{r_P} = 214.7$; $J_P = 1.116$, $J_G = 6.691$; $m_{c_P} = 53.23$, $m_{c_G} = 125.8$; $m_c = 37.4$; $n_{r_P} = 2820$ r/min.

Note that the resonant tangential velocity is $v_t = 9100$ ft/min $= 46$ m/s, and the factor A of Eq. (11.16) is equal to 10.6, that is, it is in the range of 10 to 14, as mentioned above.

Provided that the assessment of $n_{r_P} = 2820$ r/min is correct and that there is no cause of dynamic overloads external to the gear pair, the standard K_v factors may be considered as valid in a field of $n_P = 0$ to $0.25\ n_{r_P}$, that is, for pinion speeds less than about 700 r/min. For greater speeds it is advisable to adopt the AGMA method with a cautious choice of Q_v: For instance, diminish Q_v by 1 or 2 units in the range $0.25\ n_{r_P} < n_p < 0.7\ n_{r_P}$, and by 2 or 3 units for $n_P > 0.7\ n_{r_P}$. Of course, the actual situation may be far better, as the speed may be in the interval between two load peaks, but the assessment of resonant or vibration conditions presents a number of uncertainties.

The main problems are:

1. The real overall tooth stiffness.
2. How masses influence actual resonance speed when they are not adjacent and rigidly connected with pinion or gear. For instance, the polar moment of inertia of the gear of the previous stage, adjacent to a pinion, certainly must be added to that of the pinion, but the effect of an external coupling is uncertain.
3. External excitations can cause resonance. As mentioned above, a rational approach should substitute the dynamic analysis of the entire system for the product $K_a K_v$. Long and elastic external transmission shafts often exclude effects of masses that are far from gears, but in particular cases can themselves originate vibration conditions. Fig. 11.6 is determined from Castellani and Meneghetti[15] and refers to a starting process, that is, to a transient regime: The vibrations are much stronger in the case (b), and the only difference consists in a more elastic final transmission. The figure was computed, but the investigation was promoted by some anomalous failures of industrial gears.

If there is any doubt regarding the third problem—or any inexplicable failure has occurred—it is advisable to interpose elastic couplings or hydraulic couplings.

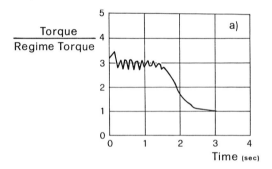

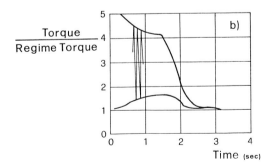

FIGURE 11.6 Gear vibration in transient condition depending on the stiffness of the external output shaft.

LOAD RATING OF GEARS   11.19

Special design features must be adopted when calculations show that the speed approaches resonance. Gears can work in a resonance condition (it should be avoided if possible); a number of marine gears do, as their speed is variable and can incur resonance. But they must be built with very good precision and must be helical gears with higher overlap ratios.

Note that when the calculation indicates a resonance risk it may be easier to avoid it by diminishing rather than increasing the resonance speed. This has also the advantage that uncertainties of load peaks corresponding to resonance submultiples are avoided. This can be accomplished by increasing either pinion mass or pinion tooth number or both, with the aim of operating above the resonance range. Higher tooth numbers, that is, lower modules, also reduce the risk of scoring. On the other hand, this often makes RF more restrictive than RH and requires careful verifications.

Gears that are to operate above the resonance speed should be precision gears. Unground tool-finished marine gears operated in the past above the resonance speed. In this case medium precision may be insufficient as the gear pair may incur a multiple of the resonance speed. Then it will be subjected to a load peak, even though it may be lower than the one corresponding to the main resonance condition.

For gear speeds either in the resonance range or above it, high-precision gears with AGMA number not less than 11, and AGMA K_v rating with a Q_v factor not greater than 10, are advisable as a summary general criterion. More optimistic estimates, as considered by AGMA for very accurate gearing, may be allowed if the system is simple enough and the calculation indicates that the speed is far enough from both the main resonance speed and its multiples.

Spur gears should not operate in the resonance range, although they do in particular cases. Then specific tests are advisable.

The procedure for the calculation of n_{r_p} cannot be considered valid for planetary gears, although the general concepts are the same.[16]

11.5.7 Load Distribution Factor, K_m

Consult AGMA and ISO documents for the original methods. For AGMA, $K_m = C_m = C_{m_f} C_{m_t}$, respectively, for face and transverse load distribution factors, but usually $C_{m_t} = 1$. ISO defines products $K_{H_\beta} K_{H_\alpha}$ for tooth surface and $K_{F_\beta} K_{F_\alpha}$ for tooth fillet where K_{F_β} is somewhat less than K_{H_β}.

Both standards give simplified and analytical methods. (Note that ANSI/AGMA 2001-B88 changes the "analytical" rating for helical gears with regard to AGMA 218.01, and obtains larger, more severe values of K_m—in fact, this is the only important difference between the ratings of the two standards.)

Warning: the "empirical" or simplified methods of ISO and AGMA standards largely differ from the analytical ones, especially when the shaft of the wheel deforms elastically, as often happens in multistage gear units.[17] However, the analytical methods require preliminary computation of the tooth misalignment, see Castellani and Moretti.[18]

Consult Dudley[1] for alternative choices.

A preliminary indication only is given in the following section. Assume that K_m varies from 1.1 to 1.8; greater values would imply bad general design or unacceptable manufacturing errors. Table 11.2 helps for the choice of K_m.

Hösep[19] gives detailed effects of helix corrections. Special attention must be given to bevel gears: see Winter and Paul[20] for the effects of incorrect axial location of spiral bevel gears.

Carter deformations often are an important item for tooth alignment.[21] They should

TABLE 11.2 Guideline for Assumption of the Load Distribution Factor K_m*

Item (a):† stiffness of the shafts of pinion and wheel				
Good:	$A = 0$			
Average:	$A = 1$			
Bad:	$A = 2$			
Item (b):‡ gear helix precision and shaft parallelism in the housing				
Good:	$B = 0$			
Average:	$B = 1$			
Bad:	$B = 2$			
Item (c):§ choice of K_m				
$A + B$	0	1	2	3	4
K_m	1.1	1.25	1.4	1.6	1.8

*NOTE: The table is valid for gears designed and manufactured following normal criteria and for normal housing stiffness. A low housing stiffness may produce opposite effects: (1) It may compensate tooth misalignment for one-stage gear pairs, or (2) for multistage gear trains, it may help some gear pairs, especially those under most load, and worsen the misalignment of others.
†Note to item (a):
 In normal multistage gear trains, deflections of the second pinion shaft (which carries the first wheel) cause misalignment of the first gear pair, and so on.
‡Notes to item (b):
 Helix precision: what is of most interest is the difference in pinion and wheel helices
 Shaft eccentricity at journal is due to elastic deflections of the bearing and to clearances: Internal bearing clearance and possible clearances between bearing and housing and between bearing and shaft. Remember that overhung gear mounting causes shaft eccentricity and gear misalignment even in the case of preloaded bearings, because of elastic deflections in the bearing, especially when an overhung pinion is the gear on a shaft under greatest load. This is typical of most bevel pinions.
§Notes to item (c):
 Lower K_m of 10–20% (minimum: $K_m = 1.1$) for appropriate helix correction or longitudinal tooth flank correction or crowning. (Warning: inappropriate corrections or crowning may cause worse misalignment!)
 Increase K_m of 10–30% for bad tooth spacing or profile precision, especially in helical gears.
 Increase K_m of 10–20% because of torsional pinion deflection for the first gear pair of multistage speed reducers if the gear ratio is high and the pinion is sited at the side of the power input.

be investigated by F.E.M. at least for large, important gear units or for mass production. This is not an easy task because of the complex form of the carters, usually obtained by the assembling of different parts. Experimental checks are useful when possible.

Note on overload factors: The choice of the single overload factors or the direct assumption of their synthetic product V is within the competence of the gear designer, even if the indications and the formulas of the standards are helpful. Their assessment is generally more important for the final results than the geometry calculations that follow, even if the latter are also necessary.

AGMA and ISO overload factors should be interchangeable, according to their

LOAD RATING OF GEARS   11.21

definitions, and in fact it may be necessary to alternate them our to make direct assumptions, for example, when the validity field of an equation is limited or when the standard indications are doubtful for a given application.

11.6 RH—CONVENTIONAL FATIGUE LIMIT OF FACTOR K

A unified formula of $K_{\lim}$ for both RHA and RHI is as follows:

$$K_{\lim} = G_H \left[\frac{s_{c_{\lim}}}{C_p}\right]^2 \frac{A_H}{V} \qquad (11.22)$$

where all parameter are in consistent units: $K_{\lim}$ and $s_{c_{\lim}}$ in N/mm^2 and C_p in (N/mm^2)$^{1/2}$, or $K_{\lim}$ and $s_{c_{\lim}}$ in lb/in^2 and c in (lb/in^2)$^{1/2}$. The overall overload factor V has been defined [Eq. (11.12)]. The other factors are given in the following.

11.6.1 RH—Preliminary Geometric Calculations

Pressure and helix angles:

$$\phi_s = \arctan\left(\frac{\tan \phi_{n_s}}{\cos \psi_s}\right) \qquad (11.23)$$

$$\psi_b = \arcsin\left(\frac{\sin \psi_s}{\cos \phi_{n_s}}\right) \qquad (11.24)$$

$$B = \frac{2\ C\ [\text{mm}] \cos \psi_s}{m_n\ [\text{mm}]\ (N_G + S_2 N_P)}$$

$$= \frac{2\ C\ [\text{in}]\ P_{nd}\ [\text{in}^{-1}] \cos \psi_s}{N_G + S_2\ N_P} \qquad (11.25)$$

$$\phi_t = \arccos\left(\frac{\cos \phi_s}{B}\right)$$

$$\psi = \arctan\ (B\ \tan \psi_s) \qquad (11.27)$$

Base diameters:

$$d_{b_P} = d_P \cos \phi_t \qquad d_{b_G} = d_G \cos \phi_t \qquad (11.28)$$

Profile (transverse) contact ratio:

$$m_p = m_{p_A} + m_{p_E} \qquad (11.29)$$

where m_{p_A} and m_{p_E} are the operating addendum contact ratios relating to the stretches AC, CE in Fig.s 11.7 and 11.8

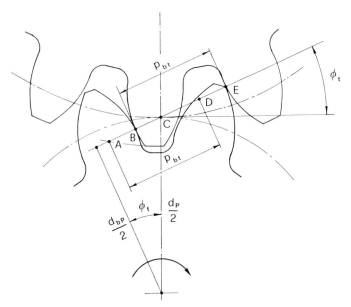

FIGURE 11.7 Involute meshing of an external gear pair in point B, Lower point of single contact of pinion or higher point of single contact of gear.

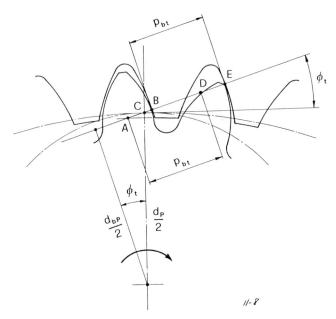

FIGURE 11.8 Involute meshing of an internal gear pair in point B, LPSC of pinion or HPSC of gear.

$$m_{PE} = \left(\sqrt{\left(\frac{d_{oP}}{d_{bP}}\right)^2 - 1} - \tan \phi_t\right) \frac{N_p}{2\pi} \qquad (11.30)$$

$$m_{PA} = \left(\sqrt{\left(\frac{d_{oG}}{d_{bG}}\right)^2 - 1} - \tan \phi_t\right) \frac{S_2 N_G}{2\pi} \qquad (11.31)$$

Outside and base diameters d_o and d_b can be given in any consistent units. If the contact does not extend as far as the tooth tip because of semitopping or other reasons, then the diameters at the contact limit must be introduced instead of d_{oP}, d_{oG}. Design criteria of gear teeth with semitopping can become an important item for the surface load capacity, as they can greatly affect the contact ratio.[22] This involves a variation of the overall contact length of helical teeth.

Curvature coefficient, X_B, relating to the lower point of *single contact* (LPSC) of the pinion involutes, B in Figs. 11.7 and 11.8:[8]

$$M_E = \frac{2\pi(1 - m_{PE})}{\tan \phi_t} \qquad (11.32)$$

$$X_B = \left(1 - \frac{M_E}{N_P}\right)\left(1 + S_2 \frac{M_E}{N_G}\right) \qquad (11.33)$$

The X_B formula is valid for usual gears with $1 < m_p < 2$.

The *LPSC B* depicts a real step in tooth meshing for spur gears, whereas it works as a reference point for helical gears and enables the trend of the profile curvature to be followed. In unusual design cases with a lower gear ratio and a longer path CE with regard to AC (Fig. 11.11), the relative curvature of the involutes may be worse in the *higher point of single contact* (HPSD) D. Then a coefficient X_d becomes lesser than X_B and should be substituted for X_B in every formula of the *Geometry factor G_H*:

$$M_A = \frac{2\pi(1 - m_{PA})}{\tan \phi_t} \qquad (11.32a)$$

$$X_D = \left(1 - S_2 \frac{M_A}{N_G}\right)\left(1 + \frac{M_A}{N_P}\right) \qquad (11.33a)$$

Face contact ratio (overlap ratio):

$$m_F = \frac{F[\text{mm}] \sin \psi_s}{\pi \, m_n[\text{mm}]} = \frac{F[\text{in.}] \, P_{nd}[\text{in}^{-1}] \sin \psi_s}{\pi} \qquad (11.34)$$

A single face width must be considered here for double-helical gear pairs.

The following contact line coefficient m_1 and load-sharing ratio m_N are for RHA only (as well as for RFA). For spur gears:

$$m_1 = 1 \qquad m_N = 1 \qquad (11.35)$$

For helical gears, m_{P_i}, m_{F_i} the integer, m_{P_d}, m_{F_d} the decimal parts of m_p, m_F:[8]

$$m_1 = \frac{m_p m_{F_i} + m_{P_i} m_{F_d} + \max(m_{P_d} + m_{F_d} - 1, 0)}{m_p m_F} \qquad (11.36)$$

$$m_N = \frac{\cos \psi_b}{m_1 \, m_p} \qquad (11.37)$$

11.6.2 Adaptation for Bevel Gears

Pressure and helix angles: Equations (11.23) and (11.24) are valid. Standard and operating angles coincide for bevel gears, then $\phi_t = \phi_s$, $\psi = \psi_s$. Addenda at the middle point of the face width, Fig. 11.2

$$a_{P_m} = a_{oP} - (F/2) \tan (\gamma_{oP} - \gamma_P),$$
$$a_{G_m} = a_{oG} - (F/2) \tan (\gamma_{oG} - \gamma_G) \qquad (11.38)$$

Virtual outside diameters:

$$d_{oP_v} = d_{Pmv} + 2 \, a_{P_m} \qquad d_{oG_v} = d_{G_{mv}} + 2 \, a_{G_m} \qquad (11.39)$$

Virtual base diameters:

$$d_{bP_v} = d_{Pm_v} \cos \phi_t \qquad d_{bG_v} = d_{G_{mv}} \cos \phi_t \qquad (11.40)$$

Then Eqs. (11.30) and (11.31) can be applied by introducing all virtual data, virtual tooth number included, while in Eq. (11.34) the mean normal module m_{n_m} or the mean normal diametral pitch P_{nd_m} must be introduced. No variation affects Eqs. (11.29), (11.32), (11.33), (11.35), (11.36), and (11.37).

11.6.3 RH—Unified Geometry Factor G_H

RHA

Spur gears:

$$G_H = X_B \, \frac{\sin (2\phi_t)}{4} \qquad (11.41)$$

Helical gears:

$$G_H = \frac{X_B}{0.9} \, \frac{\sin (2\phi_t)}{4 \, m_N} \qquad (11.42)$$

Note: AGMA's original geometry factor I is, as a concept

$$I = \frac{G_H}{1 + S_2 \, \dfrac{N_P}{N_G}} \qquad (11.43)$$

Thus Eqs. (11.41) and (11.43) give exactly the same I factor as the standards 218.01 or 2001-B88 for spur gears.

For helical gears Eq. (11.42) is simplified with regard to the standards but maintains the same trend depending on the profile curvature, whereas the results are somewhat more conservative in most common cases, for example, for small tooth numbers of the pinion, see comments below [17]. No modification is adopted for the so-called LCR gears, as the m_N ratio accounts for their load sharing. (*LCR* is an AGMA term that means *low contact ratio*—which is disputable, as it refers to overlap or face contact ratio and not to transverse contact ratio.)

RHI—Spur and Helical Gears

$$G_H = \frac{X_B}{0.9} \frac{\sin(2\phi_t)}{4 \cos \psi_b \cos \psi_s Z_\epsilon^2} \qquad (11.44)$$

where Z_ϵ is a factor of the contact ratios according to ISO

$$Z_\epsilon^2 = \frac{4 - m_p}{3} [1 - \min(m_F, 1)] + \frac{\min(m_F, 1)}{m_p} \qquad (11.45)$$

Note: Equation (11.44) leads to the same final results as the original ISO factors,[4] except for the effect of the correction coefficient $X_B/0.9$ that is introduced as a simple criterion for taking into account the trend of the variation of the relative profile curvature. The DIN method[5] of 1987 is equivalent to introducing the factor X_B only for spur gears and only when it is less than 1, without dividing it by 0.9. It is possible that the final draft of the ISO method coincides with DIN's. See the following comparisons and discussion.

In practice the X_B coefficient can be maintained as calculated from Eq. (11.33), also when $m_p > 2$ as an estimation, for both RHA and RHI. Specific tests must be performed if one wants to fully exploit the performances of such an unusual kind of gears, see Townsend, Baber, and Nagy.[23] Note that involute gear teeth with higher contact ratios can be obtained by adopting higher tooth numbers and longer teeth. Low tooth numbers are possible for the pinion if the tooth profiles differ from the involute, but this does not regard this chapter.

Tables 11.3 and 11.4 indicate G_H values for typical meshing cases of external and internal gear pairs with nominal addendum = m_n or $1/P_{nd}$. The addendum modification coefficients x_P and x_G are obtained by dividing the modifications of the nominal addendum by m_n or multiplying it by P_{nd} before any further correction of the tooth outside diameters, that the gear designer may adopt for any reason. The coefficients are positive if they involve an increase of the diametral size (note that the Germans adopt the opposite convention for internal gears, DIN 3960).

TABLE 11.3 Geometry Factors for RHA and RHI, External Gear Pairs

ψ_s	N_P/N_G	x_P	x_G	ϕ_t	m_{pA}	m_{pE}	m_p	G_{HA}	G_{HI}
0	13/82	0.4	0	21.240	0.592	0.893	1.485	0.149	0.198
	13/82	0.4	0.4	22.338	0.634	0.847	1.481	0.148	0.196
	13/41	0.4	0	22.085	0.587	0.858	1.444	0.152	0.199
	13/41	0.4	0.4	23.807	0.650	0.784	1.434	0.152	0.197
	13/21	0.4	0	23.132	0.573	0.813	1.386	0.161	0.205
	13/21	0.4	0.4	25.529	0.657	0.709	1.366	0.162	0.205
	26/82	0	0	20	0.914	0.810	1.725	0.146	0.214
	26/82	0.4	0	21.099	0.629	0.986	1.615	0.167	0.233

TABLE 11.3 Geometry Factors for RHA and RHI, External Gear Pairs *(Continued)*

ψ_s	N_P/N_G	x_P	x_G	ϕ_t	m_{pA}	m_{pE}	m_p	G_{HA}	G_{HI}
0	26/82	0.4	0.4	22.085	0.701	0.904	1.605	0.167	0.233
	26/41	0	0	20	0.859	0.810	1.670	0.152	0.217
	26/41	0.4	0	21.714	0.636	0.935	1.570	0.169	0.232
	26/41	0.4	0.4	23.173	0.736	0.811	1.547	0.173	0.234
	26/26	0	0	20	0.810	0.810	1.621	0.158	0.222
	26/26	0.4	0.4	23.929	0.746	0.746	1.493	0.182	0.242
15°	13/82	0.4	0	21.810	0.561	0.861	1.423	0.228	0.249
	13/82	0.4	0.4	22.851	0.602	0.817	1.419	0.226	0.246
	13/41	0.4	0	22.610	0.559	0.828	1.387	0.228	0.248
	13/41	0.4	0.4	24.254	0.620	0.757	1.377	0.225	0.245
	13/21	0.4	0	23.608	0.550	0.785	1.334	0.233	0.254
	13/21	0.4	0.4	25.912	0.631	0.684	1.315	0.230	0.251
	26/82	0	0	20.647	0.866	0.774	1.641	0.263	0.287
	26/82	0.4	0	21.677	0.597	0.946	1.542	0.281	0.307
	26/82	0.4	0.4	22.610	0.666	0.867	1.533	0.280	0.305
	26/41	0	0	20.647	0.818	0.774	1.592	0.267	0.291
	26/41	0.4	0	22.258	0.606	0.897	1.503	0.280	0.305
	26/41	0.4	0.4	23.646	0.703	0.779	1.482	0.280	0.305
	26/26	0	0	20.647	0.774	0.774	1.548	0.272	0.297
	26/26	0.4	0.4	24.370	0.716	0.716	1.432	0.286	0.312
30°	13/82	0.4	0	23.745	0.475	0.766	1.241	0.212	0.257
	13/82	0.4	0.4	24.616	0.510	0.728	1.238	0.209	0.254
	13/41	0.4	0	24.413	0.480	0.737	1.217	0.215	0.261
	13/41	0.4	0.4	25.818	0.534	0.675	1.209	0.211	0.257
	13/21	0.4	0	25.261	0.480	0.700	1.180	0.225	0.273
	13/21	0.4	0.4	27.274	0.554	0.610	1.163	0.220	0.268
	26/82	0	0	22.796	0.731	0.669	1.400	0.257	0.312
	26/82	0.4	0	23.635	0.505	0.827	1.332	0.272	0.331
	26/82	0.4	0.4	24.413	0.566	0.760	1.325	0.271	0.329
	26/41	0	0	22.796	0.699	0.669	1.368	0.265	0.322
	26/41	0.4	0	24.118	0.520	0.785	1.305	0.276	0.336
	26/41	0.4	0.4	25.294	0.607	0.682	1.290	0.275	0.334
	26/26	0	0	22.796	0.669	0.669	1.338	0.275	0.335
	26/26	0.4	0.4	25.919	0.627	0.627	1.254	0.284	0.346

General data:
$\phi_{ns} = 20°$
Nominal addendum $= m_n = 1/P_{nd}$
$x =$ addendum modification/$m_n =$ addendum modification $\cdot P_{nd}$
(x_P, x_G for pinion and gear, resp.),
i.e., $d_o = Nm_n/\cos\psi_s + 2 m_n (1 + x)$
Helical gear pairs:
$m_1 = 0.95$ for G_{HA}, $m_F \geq 1$ for G_{HI}.

LOAD RATING OF GEARS   11.27

TABLE 11.4 Geometry Factors for RHA and RHI, Internal Spur Gear Pairs

N_P/N_G	x_P	x_G	ϕ_t	m_{pA}	m_{pE}	m_p	G_{HA}	G_{HI}
13/82	0.4	0	17.959	0.343	1.027	1.369	0.153	0.194
13/82	0.4	0.4	20	0.412	0.944	1.356	0.147	0.185
13/41	0.4	0.4	20	0.431	0.944	1.376	0.145	0.185
26/82	0	0	20	0.863	0.810	1.673	0.135	0.193
26/82	0.4	0	17.406	0.204	1.285	1.490	0.186	0.247
26/82	0.4	0.4	20	0.412	1.077	1.489	0.172	0.228

General data:
$\phi_{ns} = 20°$
$\psi_s = 0$
Nominal addendum $= m_n = 1/P_{nd}$
$x =$ addendum modification$/m_n =$ addendum modification $\cdot P_{nd}$
(x_P, x_G for pinion and gear resp.).

NOTES: addendum modifications are assumed > 0, for both pinion and internal gear, when they involve an increase of the diametral size, according to ISO R1122. Therefore:
For the pinions: $d_o = Nm_n + 2m_n (1 + x_P)$
For the internal gears of the table the addenda are shortened by $0.2\ m_n = 0.2/P_{nd}$, thereby avoiding false contacts and widening the range of usable cutters, i.e., $d_o = Nm_n - 2\ m_n (1 - x_G - 0.2)$.

11.6.4 RH—Comments and Comparisons on the Unified Geometry Factor G_H

The trend of the ISO and DIN geometry factors, as depending on the helix angle according to Eq. (11.44), is based on German researches reported by Niemann in his original book.[24] On the other hand, more important differences between the various methods depend on the choice of the point along the contact line where the relative profile curvature is rated.

Figures 11.9 and 11.10 show the geometry factor as depending on the addendum modification coefficients (see above). When $x_G = -x_P$, the center distance is unmodified; otherwise it is modified. No addendum shortening has been considered as it is unnecessary in both cases.

AGMA-V and ISO-V refer to the "variants" that are adopted in this chapter, Eqs. (11.42) and (11.44). They have been presented in Castellani,[17] where other examples are investigated. Equation (11.41) for spur gears agrees with all the AGMA standards mentioned in the following.

For RHA, AGMA 210 and AGMA 211 refer to the previous standards 210.02/1965 and 211.02/1969, respectively. AGMA 218 refers to the standard 218.01 that coincides with ANSI/AGMA 2001-B88.[2,3] For RHI see Organisation Internationale de Normalisation[4] and Deutsches Institut für Normung.[5]

Both the variants for RHA and RHI show a very similar trend to that of AGMA 218 with regard to the influence of x_P and x_G coefficients, but G_H is far higher according to AGMA 218 in the case of helical pinions with low tooth numbers, Fig. 11.10, that are requested for surface-hardened gears if surface and root resistances have to be balanced. Of course, the success of gears such as these that have been designed according to AGMA 218.01 depends on other items, for example, on a conservative choice of the allowable hertzian pressure. But a more cautious choice of G_H, as given by Eq. (11.42), becomes opportune especially if higher values of the hertzian pressure are allowed according to ANSI/AGMA 2001-B88.

On the whole, a far greater gap between the G_H factors of helical and spur gears

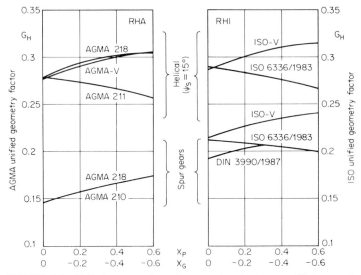

FIGURE 11.9 Unified geometry factor for external teeth with unmodified center distance. $\phi_{ns} = 20°$, $N_P/N_G = 26/82$, $m_F = 1$ for the helical gears.

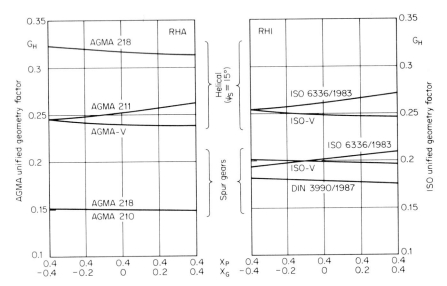

FIGURE 11.10 Unified geometry factor for external teeth with modified center distance and a low tooth number of the pinion. $\phi_{ns} = 20°$, $N_P/N_G = 13/82$, $m_F = 1$ for the helical gears.

can be observed for AGMA ratings with regard to ISOs. The DIN method reduces G_H as opposed to ISO for spur gears, when $X_B < 1$. The field experience of the author of this chapter disagrees with such a gap. Cases are known where helical gears, substituting for spur ones, did not avoid pitting. A lot of common industrial planetary units built with case-hardened spur gears, which have been working successfully for many years, would soon have failed if they behaved according to AGMA or even to DIN.

11.6.5 RH—Elastic Coefficient C_p and Conventional Fatigue Limit $s_{c_{lim}}$ of the Hertzian Pressure

Table 11.5 gives the elastic coefficient C_p (ISO symbol: Z_E) for gear pairs made from the same category of materials. Otherwise C_p must be calculated.

For metal materials with a Poisson's ratio $\mu = 0.3$

$$C_p \ [(N/mm^2)^{1/2}] = \sqrt{0.175 \ E_m} \ [N/mm^2]$$
$$C_p \ [(lb/in^2)^{1/2}] = \sqrt{0.175 \ E} \ [lb/in^2] \quad (11.46)$$

where E_m is the mean Young's modulus of elasticity of pinion (E_P) and gear (E_G)

$$E_m = \frac{2 \ E_P \ E_G}{(E_P + E_G)} \quad (11.47)$$

Table 11.6 gives values near to the maximum indicated by the standards for $s_{c_{lim}}$ (AGMA symbol: s_{ac} — ISO symbol: $\sigma_{H_{lim}}$). Warning: the notes to the table are essential!

TABLE 11.5 Elastic Coefficient C_p for RH

Pairing of Materials	E_m, N/mm² (lb/in²)	C_p, N½/mm (lb½/in.)
Steel/	207,000	190
steel	(30,000,000)	(2290)
Nodular or malleable iron/	172,000	173
nodular or malleable iron	(25,000,000)	(2090)
Cast iron/	138,000	155
cast iron	(20,000,000)	(1870)
Steel/	188,000	180
nodular or malleable iron	(27,000,000)	(2170)
Steel/	165,000	170
cast iron	(24,000,000)	(2050)
Nodular or malleable iron/	153,000	163
cast iron	(22,000,000)	(1960)

TABLE 11.6 RH and RF Conventional Fatigue Limits

Material	Flank and Root Hardness	$s_{c_{lim}}$, N/mm² lb/in.² RHA	$s_{c_{lim}}$, N/mm² lb/in.² RHI	$s_{f_{lim}}$, N/mm² lb/in.² RFA	$s_{f_{lim}}$, N/mm² lb/in.² RFI
Cast iron	175 HB	470 (68,000)	360 (52,000)	55 (8,000)	150 (22,000)
	200 HB	530 (77,000)	400 (58,000)	80 (12,000)	165 (24,000)
Nodular iron	180 HB	560 (81,000)	500 (73,000)	190 (27,000)	370 (54,000)
	240 HB	680 (98,000)	580 (84,000)	230 (33,000)	410 (59,000)
Malleable iron	180 HB	510 (74,000)	480 (69,000)	80 (12,000)	370 (54,000)
	240 HB	620 (89,000)	550 (80,000)	130 (19,000)	410 (59,000)
Through-hardened and tempered steel	220 HB	700 (102,000)	690 (100,000)	240 (35,000)	550 (80,000)
	260 HB	795 (115,000)	745 (108,000)	265 (38,000)	580 (84,000)
	300 HB	890 (129,000)	800 (115,000)	290 (42,000)	610 (88,000)
Induction-hardened steel	≈ 55 HRC	1,250 (180,000)	1,300 (190,000)	340 (50,000)	740 (105,000)
Carburized case-hardened steel	58–62 HRC	1,500 (215,000)	1,550 (225,000)	430 (63,000)	930 (135,000)
Gas-nitrided steel, 2–3% Cr	700–750 HV	1,300 (190,000)	1,380 (200,000)	400 (58,000)	840 (120,000)

NOTES:
For cast steels ISO suggests lowering $s_{c_{lim}}$ by 10% and $s_{f_{lim}}$ by 20%.
Data for surface-hardened steels require proper hardened depth: The tabulated data are not applicable to large-sized gears, induction or case hardened, or to medium- to large-sized gears, gas nitrided. Core hardness is a determining factor.
$s_{f_{lim}}$ values for induction hardening refer to root hardening.
$s_{f_{lim}}$ values must be reduced by 30% for alternating load. Intermediate values should be adopted for bidirectional loads according to the number of cycles in each direction: max reduction 30% for one load application.
Tabulated data take no account of shock sensitivity.
All mentioned steels are alloy steels.
All materials are considered tested and in correct metallurgical condition.
ANSI/AGMA 2001-B88 allows as much as $s_{c_{lim}} = 1910$ N/mm² (275,000 lb/in²) and $s_{f_{lim}} = 520$ N/mm² (75,000 lb/in²) for carburized case-hardened steels with good metallurgical quality and certified cleanliness (AGMA Grade 3).

11.6.6 RH—Adaptation Factor A_H

The $s_{c_{lim}}$ values are meant for gears similar to those of previous field experience or of laboratory tests. Adaptation factors aim to adjust the calculation to the manufacturing

LOAD RATING OF GEARS   11.31

and operating peculiarities of the examined gear pair. In this item AGMA and ISO present great differences.

RHA

$$A_H = \frac{C_H^2}{C_s\, C_f\, C_T{}^2} \tag{11.48}$$

The factors at the denominator, that is, for size, flank finish, and temperature, are usually assumed equal to unity. They should be > 1 if anomalous conditions arise. The hardness factor C_H equals 1 except in two cases for the gear:

1. For non-surface-hardened gears, if the Brinell hardnesses H_{BP} and H_{BG} of pinion and gear are in the range:

$$1.2 \le \frac{H_{BP}}{H_{BG}} \le 1.7$$

Then:

$$C_H = 1 + \frac{(1.08\,\dfrac{H_{BP}}{H_{BG}} - 1)\,(m_G - 1)}{120} \tag{11.49}$$

2. If the pinion teeth are surface hardened with minimum $HRC_P = 48$ Rockwell hardness, and the gear Brinell hardness is in the range $180 \le H_{BG} \le 400$

$$C_H = 1 + \frac{450 - H_{BG}}{1333\, e^{0.0125 f_P}} \tag{11.50}$$

where $e = 2.718$ (base of the natural logarithms), and f_P is the flank finish of the pinion (roughness, arithmetic average, μin.).

Note: The factor C_H implies work hardening of the gear, but field experience shows that in case (2) the pinion teeth have a file effect rather than a work-hardening effect on the gear teeth if f_P is higher, with disastrous wear. The pinion of such gear pairs should not be finished with $f_P > 40$ μin. or 1 μm.

RHI

$$A_H = (Z_X\, Z_W\, Z_L\, Z_v\, Z_R)^2 \tag{11.51}$$

The size factor Z_X is usually assumed equal to unity.
The hardness factor Z_W is > 1 only if the pinion has surface-hardened teeth with a flank finish $f_P < 1$ μm or 40 μin. (roughness, arithmetic average) and the gear is in the range $130 < H_{BG} \le 400$. Then

$$Z_W = 1.2 - \frac{H_{BG} - 130}{1700} \tag{11.52}$$

The factors Z_L, Z_v, and Z_R for lubrication, velocity, and roughness are the most typical ISO factors. Simplified equations are given for them, and some modifications for Z_R are suggested.

RHI Lubrication Factor

$$Z_L = \left(\frac{v_{40}}{167}\right)^A \qquad (11.53)$$

where v_{40} is the nominal kinematic viscosity of the lubricant at 40°C (104°F) in cSt (1 cSt = 10^{-6} m²/s). Assume $A = 0.05$ for surface-hardened, $A = 0.1$ for non-surface-hardened teeth (maximum: $Z_L = 1.25$).

RHI Velocity Factor

$$Z_v = \left(\frac{v_t \ [\text{m/s}]}{10}\right)^B \qquad (11.54)$$

$$Z_v = \left(\frac{v_t \ [\text{ft/min}]}{1970}\right)^B$$

where $B = 0.025$ for surface-hardened teeth (minimum: $Z_v = 0.93$), $B = 0.05$ for non-surface-hardened teeth (minimum: $Z_v = 0.85$).

RHI Roughness Factor. Tooth flank roughness is meant as the mean value of some teeth of pinion f_P, and of gear f_G. Then a reduced value f_r is calculated, which relates to the size of test gears

$$f_r \ [\mu\text{m}] = \frac{f_P + f_G [\mu\text{m}]}{2} \sqrt[3]{\frac{200}{d_P + d_G \ [\text{mm}]}} \qquad (11.55)$$

$$f_r \ [\mu\text{in.}] = \frac{f_P + f_G [\mu\text{in.}]}{2} \sqrt[3]{\frac{7.87}{d_P + d_G \ [\text{in.}]}}$$

Consider f_r as arithmetic average, as above. Then

$$Z_R = \left(\frac{0.5}{f_r \ [\mu\text{m}]}\right)^C \qquad (11.56)$$

$$Z_R = \left(\frac{20}{f_r \ [\mu\text{in.}]}\right)^C$$

where $C = 0.08$ for surface-hardened, $C = 0.15$ for non-surface-hardened teeth. This agrees with the original ISO data with good approximation, but greater C exponents are advisable when tool-finished gears are subjected to a continuous and constant load, except in the case of case-hardened gears where $C = 0.08$ seems sufficiently derating.

LOAD RATING OF GEARS   11.33

Note: The ISO surface factors have an experimental basis, but surface fatigue tests always give a lot of scattering and ISO explicitly states that such factors are only an estimation. However, they account for the roughness disturbance with Z_R as well as for the lubrication regime, even if only summarily, by means of Z_L and Z_v. Z_L is somehow disputable as it refers to a nominal viscosity. Z_v apparently indicates a right trend as speedy gears in EHD conditions are favored whereas slow gears in boundary lubrication regime are routinely derated, as they will meet progressive wear even if not always pitting. See Dudley[1] for the lubrication regimes.

There is actually a surface condition effect, which is sometimes much more important than it is thought to be, which risks misleading one's appreciation of the real nature of some field experience. For instance, a lot of big gears probably work in much more severe application conditions than those considered by K_a and do not have pitting just because of a more effective work hardening than that allowed by C_H. It would be risky both to rely on K_a not directly checked and to assume optimistic C_H especially if not proven for different kinds of materials.

Sometimes the gear designer should take into account other favourable or unfavourable circumstances not considered by the standards. This task can accomplish by varying directly the A_H factor. Shot-peening increases the pitting resistance, see Townsend and Zaretsky.[25] The type of lubrication apparatus as well as the lubrication and cooling conditions are also influential parameters, see Townsend,[26] because the lubricant action and viscosity influence EHD conditions. Synthetic oils are generally favourable to pitting resistance according to their type and the gear material. Oil additives, even if conceived for other purposes, have sometimes a favorable influence,[27] and any operating condition affecting tooth temperature can be indirectly important.[28]

11.6.7 RH—Hertzian Pressure

The value of the hertzian pressure s_c (ISO symbol: σ_H) does not directly enter the ratings based on the K factor and is given merely for information. AGMA's and ISO's equations differ for RHA and RHI.

RHA

$$s_c = C_p \sqrt{\frac{V K C_s C_f}{G_H}} \qquad (11.57)$$

RHI

$$s_c = C_p \sqrt{\frac{V K}{G_H}} \qquad (11.58)$$

Both formulas are fictitious for various reasons, but this does not affect the reliability of the general calculations.

11.6.8 RH—Service Factor, C_{SF} (only for 1 loading level)

Two cases are distinguished for the procedure.

1. Gear sizes and load are given, K and $K_{\lim}$ have been calculated. Then such data are equivalent to a service factor

$$C_{SF} = \frac{K_a K_{\lim}}{K} \qquad (11.59)$$

which substitutes and summarizes application as well as life and reliability problems.

2. $K_{\lim}$ has been rated for a new design with provisional data and a given C_{SF} is required. Then the gear design must be newly sized by assuming

$$K = \frac{K_a K_{\lim}}{C_{SF}} \qquad (11.60)$$

The service factor is often mistaken for the application factor. This does not supply any margin for reliability, unless a cautious $s_{c_{\lim}}$ value has been chosen, but an explicit margin C_{SF}/K_a is preferable for the sake of clearness. The margin must be increased for a long gear life and may be reduced for a shorter one, even to the point where C_{SF} itself may be < 1. In such cases the direct estimation of tooth damage and gear life is preferable.

The C_{SF} procedure normally excludes life analysis, and vice versa.

11.6.9 Power Capacity Tables

Tables 11.7 to 11.12 give a general survey of performances of industrial case-hardened gears.

As the present rating standards require detailed geometrical data, Tables 11.7 and 11.9 give them for gear pairs with parallel axes, which can be built by metric metric standardized tools. Of course, the performances are not too different for gears built with

TABLE 11.7 Spur Gears: Data and General Assumptions for Capacity Table 11.8.

Gear Ratio m_G	Tooth Numbers N_P/N_G	Add. Modif. Coefficients		Operating Press.Angle ϕ_t	Center Distance C, mm			
		x_P	x_G		50	100	200	400
					Normal Module m_n, mm			
1	25/25	0	0	20°	2	4	8	
1.6	22/35	0.1394	−0.0673	20.3899°	1.75	3.5	7	
2.5	19/47	0.2849	0.0607	21.5190°	1.5	3	6	12
4	16/64	0.2981	−0.2981	20°	1.25	2.5	5	10
6.3	14/86	0.3628	−0.3628	20°	1	2	4	8

NOTES:
External teeth. Normal standard pressure angle $\phi_{ns} = 20°$.
Outside diameters $d_o = N m_n + 2 m_n (1 + x)$.
Carburized case-hardened alloy steel, 58–62 HRC. Ground teeth.
Service factor $C_{SF} = 1.5$ including application, reliability, and life factors.
Power-sharing factor $K_{sh} = 1$.
Dynamic factors K_v rated according to RHA or RHI methods for AGMA quality number = 11.
Load distribution factor $K_{m_2} = 1.25$.
RHA: $s_{c_{\lim}} = 1500$ N/mm²; $A_H = 1$.
RHI: $s_{c_{\lim}} = 1550$ N/mm²; A_H rated for f = 0.8 μm (32 μin.), $v_{40} = 200$ cSt, and pitch velocity (other influencing factors = 1).

TABLE 11.8 Surface Capacity of Case-Hardened Spur Gears

Gear Ratio m_G	Pinion Speed n_P, r/min	Center Distance C, mm/Net Face Width F, mm							
		50/17.5		100/35		200/70		400/140	
		Power, kW (multiply by 1.341 for power in hp)							
		RHA	RHI	RHA	RHI	RHA	RHI	RHA	RHI
1	100	0.6	0.7	4.7	5.8	37	50		
	250	1.5	1.8	12	15	91	128		
	500	2.9	3.6	23	31	178	257		
	710	4.1	5.2	32	44	250	(364)		
	1000	5.7	7.4	45	62				
	1420	8.0	11	63	88				
	1700	9.5	13	74	104				
1.6	100	0.4	0.5	3.4	4.2	27	35		
	250	1.1	1.3	8.3	11	66	91		
	500	2.1	2.6	16	22	129	185		
	710	2.9	3.7	23	32	182	264		
	1000	4.1	5.3	32	45	253	370		
	1420	5.8	7.6	45	64	354	(520)		
	1700	6.9	9.1	54	76				
2.5	100	0.28	0.32	2.2	2.7	18	22	139	191
	250	0.7	0.8	5.5	6.8	43	58	342	(492)
	500	1.4	1.6	11	14	85	118		
	710	1.9	2.4	15	20	120	170		
	1000	2.7	3.4	21	29	167	240		
	1420	3.8	4.8	30	41	235	340		
	1700	4.6	5.8	36	49	279	405		
4	100	0.15	0.17	1.2	1.4	9.2	12	73	99
	250	0.36	0.42	2.9	3.5	23	30	179	256
	500	0.7	0.8	5.7	7.2	45	62	352	(523)
	710	1.0	1.2	8.0	10	63	89	(494)	(746)
	1000	1.4	1.7	11	15	88	126		
	1420	2.0	2.5	16	21	123	180		
	1700	2.4	3.0	19	26	147	215		
6.3	100	0.08	0.09	0.6	0.7	4.8	6.0	38	50
	250	0.19	0.22	1.5	1.8	12	15	95	131
	500	0.38	0.44	3.0	3.7	24	32	186	268
	710	0.5	0.6	4.2	5.3	33	45	262	385
	1000	0.7	0.9	5.9	7.6	47	65	365	(546)
	1420	1.1	1.3	8.3	11	66	93	(512)	(778)
	1700	1.3	1.5	9.9	13	78	111		

NOTES: Gear and rating data: See Table 11.7. See the text for due reservations about table validity.

TABLE 11.9 Helical Gears: Data and General Assumptions for Capacity Table 11.10

Gear Ratio m_G	Tooth Numbers N_P/N_G	Helix Angle ψ_s	Add. Modif. Coefficients		Transv. Oper. Press. Angle ϕ_t	Center Distance C, mm			
			x_P	x_G		50	100	200	400
						Normal Module m_n, mm			
1	23/23	21°	0.1903	0.1903	23.3437°	2	4	8	
1.6	20/32	22°	0.3036	0.2566	23.9929°	1.75	3.5	7	
2.5	18/45	18°	0.2413	−0.0245	21.8755°	1.5	3	6	12
4	15/60	19°	0.3218	0.0270	22.2820°	1.25	2.5	5	10
6.3	13/82	17°	0.3736	−0.0366	21.8078°	1	2	4	8

NOTES:
External teeth. Normal standard pressure angle $\phi_{ns} = 20°$.
Outside diameters $d_o = N\,m_n/\cos\psi_s + 2\,m_n\,(1 + x)$.
Carburized case-hardened alloy steel, 58–62 HRC. Ground teeth.
Service factor $C_{SF} = 1.5$ including application, reliability, and life factors.
Power-sharing factor $K_{sh} = 1$.
Dynamic factors K_v rated according to RHA or RHI methods for AGMA quality number = 11.
Load distribution factor $K_m = 1.25$.
RHA: contact line coefficient m_1 rated according to the actual tooth data; $s_{c_{lim}} = 1500$ N/mm^2; $A_H = 1$.
RHI: $s_{c_{lim}} = 1550$ N/mm^2; A_H rated for f = 0.8 μm (32 μin.), $v_{40} = 200$ cSt, and pitch velocity (other influencing factors = 1).

standardized diametral pitches and general sizes and tooth numbers near to those of the tables. The addendum modifications are so chosen that the so-called Almen factors are balanced, that is, the product of theoretical hertzian pressure multiplied by sliding velocity and access or recess length is the same at the extreme points of the contact path. This criterion is suitable for both speed-reducing and speed-increasing gear pairs.

The addenda suggested by Gleason are adopted for bevel gears, Tables 11.11 and 11.12.

Tables 11.8, 11.10, 11.11, and 11.12 refer to surface capacity, that is, to pitting and generic wear. The tooth numbers are chosen such that tooth strength generally is no problem in the case of RFI rating and unidirectional loading. RFA rating and/or bidirectional loading may give more restrictive results.

Higher powers may require oil cooling especially if continuously transmitted, otherwise, the real power capacity will be reduced, apart from other drawbacks. Anomalous dynamic conditions, for example, big masses rigidly connected to pinions or external causes of vibration excitation, may invalidate the data (see Sec. 11.5).

Values in brackets are doubtful for scoring resistance, see Sec. 11.9. Otherwise, scoring should not be a problem, provided that good EP oils are used. If the scoring capacity of the lubricant is lower, problems may arise for the highest gear sizes and speeds indicated in the tables.

Greater powers can be transmitted at higher speeds, but lower modules, that is, greater tooth numbers are advisable, and this can make RF more restrictive than RH. Each case must be solved by itself according to the risk of operating in resonance ranges (see Sec. 11.5) and according to the loading conditions, the prescribed rating method, and the desired reliability against tooth breakage; no general tabulated indications can be given.

TABLE 11.10 Surface Capacity of Case-Hardened Helical Gears

Gear Ratio m_G	Pinion Speed n_P, r/min	Center Distance C, mm/Net Face Width F, mm							
		50/17.5		100/35		200/70		400/140	
		Power, kW (multiply by 1.341 for power in hp)							
		RHA	RHI	RHA	RHI	RHA	RHI	RHA	RHI
1	100	1.1	1.0	8.5	8.1	67	69		
	250	2.6	2.5	21	21	165	180		
	500	5.2	5.1	41	43	(323)	(367)		
	710	7.4	7.3	58	62				
	1000	10	10	81	89				
	1420	14	15	113	126				
	1700	17	18	135	152				
1.6	100	0.7	0.7	5.6	5.7	45	48		
	250	1.7	1.7	14	15	110	124		
	500	3.5	3.5	27	30	216	255		
	710	4.9	5.1	39	43	303	366		
	1000	6.9	7.2	54	62	(422)	(519)		
	1420	9.7	10	76	88				
	1700	12	13	90	106				
2.5	100	0.44	0.41	3.5	3.4	28	28	219	242
	250	1.1	1.0	8.6	8.6	68	74	(536)	(629)
	500	2.2	2.1	17	18	134	151		
	710	3.0	3.0	24	26	188	218		
	1000	4.3	4.3	34	37	263	310		
	1420	6.0	6.2	47	52	368	443		
	1700	7.2	7.4	56	63	438	(531)		
4	100	0.21	0.21	1.7	1.8	14	15	108	124
	250	0.53	0.53	4.2	4.4	34	38	265	323
	500	1.1	1.1	8.4	9.1	66	78	(520)	(664)
	710	1.5	1.5	12	13	93	112		
	1000	2.1	2.2	17	19	130	160		
	1420	3.0	3.2	23	27	183	230		
	1700	3.5	3.8	28	33	217	276		
6.3	100	0.10	0.10	0.8	0.8	6.6	7.0	52	58
	250	0.26	0.26	2.1	2.1	16	18	129	153
	500	0.51	0.51	4.1	4.3	32	37	254	315
	710	0.7	0.7	5.8	6.2	46	53	358	453
	1000	1.0	1.0	8.1	8.9	64	76	(499)	(646)
	1420	1.4	1.5	11	13	89	109		
	1700	1.7	1.8	13	15	107	131		

NOTE: Gear and rating data: See Table 11.9. See the text for due reservations about table validity.

TABLE 11.11 Surface Capacity of Case-Hardened Straight Bevel Gears

Gear Ratio m_G	Tooth Numbers N_P/N_G	Pinion Speed n_P r/min	Gear Pitch Diameter d_G mm (in. approx.)							
			50 (2)		100 (4)		200 (8)		400 (16)	
			Power, kW (multiply by 1.341 for power in hp)							
			RHA	RHI	RHA	RHI	RHA	RHI	RHA	RHI
1	25/25	100	0.32	0.34	2.5	2.8	19	24		
		500	1.5	1.7	11	14	86	111		
		1000	2.9	3.4	22	27	160	197		
		1420	3.9	4.7	30	36				
		1700	4.6	5.6	35	42				
1.6	18/29	100	0.12	0.13	0.9	1.0	7.3	8.5		
		500	0.6	0.6	4.4	5.3	33	44		
		1000	1.1	1.3	8.3	11	63	85		
		1420	1.5	1.8	12	15	86	116		
		1700	1.8	2.2	14	18	101	135		
2.5	15/37	100	0.05	0.05	0.39	0.43	3.1	3.5	24	29
		500	0.25	0.25	1.9	2.2	14	18	108	151
		1000	0.5	0.5	3.6	4.4	27	36	203	(293)
		1420	0.6	0.7	5.0	6.2	37	51		
		1700	0.8	0.9	5.8	7.5	44	61		
4	14/56	100			0.15	0.16	1.2	1.3	9.2	11
		500	0.09	0.10	0.7	0.8	5.5	6.8	42	58
		1000	0.18	0.19	1.4	1.6	11	14	80	115
		1420	0.25	0.28	1.9	2.4	15	20	109	161
		1700	0.30	0.34	2.3	2.8	17	24	128	190
6.3	13/82	100			0.06	0.06	0.47	0.53	3.7	4.4
		500			0.28	0.32	2.2	2.7	17	23
		1000	0.07	0.08	0.5	0.6	4.3	5.5	32	46
		1420	0.10	0.11	0.8	0.9	5.9	7.8	45	65
		1700	0.12	0.13	0.9	1.1	7.0	9.4	53	78

Notes:
Shaft angle $\Sigma = 90°$. Pressure angle $\phi_{ns} = 20°$.
Net face width $= 0.3 \cdot$ cone distance.
Tooth proportions recommended by Gleason Works, Rochester, New York.
Carburized and case-hardened alloy steel, 58–62 HRC. Tool finishing.
Service factor $C_{SF} = 1.5$ including application, reliability, and life factors.
$K_{sh} = 1$, K_v for AGMA quality number $= 8$, $K_m = 1.4$.
RHA: $s_{c_{lim}} = 1500$ N/mm^2; $A_H = 1$.
RHI: $s_{c_{lim}} = 1550$ N/mm^2; A_H rated for $f = 2$ μm (80 μin.), $v_{40} = 200$ cSt, and actual pitch velocity.
See the text for due reservations about table validity.

TABLE 11.12. Surface Capacity of Case-Hardened Spiral Bevel Gears

Gear Ratio m_G	Tooth Numbers N_P/N_G	Pinion Speed n_P r/min	Gear Pitch Diameter d_G, mm (in. approx.)							
			50 (2)		100 (4)		200 (8)		400 (16)	
			\multicolumn{8}{c}{Power, kW (multiply by 1.341 for power in hp)}			
			RHA	RHI	RHA	RHI	RHA	RHI	RHA	RHI
1	21/21	100	0.58	0.60	4.6	5.0	36	42		
		500	2.8	3.1	21	26	164	217		
		1000	5.4	6.3	41	52	310	423		
		1420	7.4	8.9	57	74	(425)	(583)		
		1700	8.8	11	67	87				
1.6	16/25	100	0.21	0.21	1.7	1.8	13	15		
		500	1.0	1.1	7.9	9.2	61	78		
		1000	2.0	2.2	15	19	116	158		
		1420	2.7	3.2	21	27	160	224		
		1700	3.3	3.9	25	32	188	267		
2.5	14/35	100	0.07	0.07	0.58	0.64	4.6	5.3	36	45
		500	0.36	0.38	2.8	3.3	22	28	166	236
		1000	0.7	0.8	5.4	6.7	41	57	315	(477)
		1420	1.0	1.1	7.5	9.6	57	81		
		1700	1.2	1.4	8.9	12	68	98		
4	13/52	100			0.23	0.24	1.8	2.0	14	16
		500	0.14	0.14	1.1	1.2	8.5	10	66	87
		1000	0.27	0.29	2.1	2.5	16	21	126	178
		1420	0.38	0.42	3.0	3.6	23	30	174	255
		1700	0.45	0.50	3.5	4.3	27	36	205	305
6.3	12/75	100			0.09	0.09	0.7	0.8	5.5	6.5
		500			0.43	0.47	3.3	4.0	26	34
		1000	0.11	0.11	0.8	0.9	6.5	8.2	50	70
		1420	0.15	0.16	1.2	1.4	9.1	12	70	100
		1700	0.18	0.19	1.4	1.7	11	14	82	120

NOTES:
Shaft angle $\Sigma = 90°$. Normal pressure angle $\phi_{ns} = 20°$. Spiral angle $\psi_s = 35°$.
Net face width $= 0.3 \cdot$ cone distance.
Tooth proportions recommended by Gleason Works, Rochester, New York.
Carburized and case-hardened alloy steel, 58–62 HRC. Lapped teeth.
Service factor $C_{SF} = 1.5$ including application, reliability, and life factors.
$K_{sh} = 1$, K_v rated for AGMA quality number $= 9$, $K_m = 1.4$.
RHA: contact line coefficient m_1 rated according to the actual tooth data; $s_{c_{lim}} = 1500$ N/mm^2; $A_H = 1$.
RHI: $s_{c_{lim}} = 1550$ N/mm^2; A_H rated for f = 1.25 μm (50 μin.), $\nu_{40} = 200$ cSt, and actual pitch velocity.
See the text for due reservations about table validity.

Service and load distribution factors, as well as quality numbers, tooth roughness, and lubricant viscosity, as indicated in Tables 11.7, 11.9, 11.11, and 11.12, must be considered only as reference values.

11.7 RF—CONVENTIONAL FATIGUE LIMIT OF FACTOR U_L

A unified formula for both RFA and RFI can be given as follows:

$$U_{L_{\lim}} = \frac{J_n A_F s_{t_{\lim}}}{K_s V} \quad (11.61)$$

where $U_{L_{\lim}}$ and $s_{t_{\lim}}$ are in consistent units, either N/mm^2 or lb/in^2. The overall overload factor V has been defined in the foregoing. See the following discussion for the other factors.

11.7.1 RF—Geometry Factor J_n

The geometry factor accounts for tooth form and stress concentration at fillet. The higher point of single contact, D for the pinion, Figs. 11.11 and 11.12, and B for the gear, Figs. 11.7 and 11.8, is taken as the rating point for spur gears in RFA and for both spur and helical gears in RFI, so-called method B, according to the procedure of Castellani and Parenti-Castelli[29] that is followed entirely here. Tip meshing is considered in RFA normal method for helical gears instead. Both methods introduce adjusting factors.

The detailed computation of the geometry factor requires programming. The equations can be deduced from the original standards or from the cited reference that defines the actual fillet form as generated by pinion cutter as well as by rack cutter or by hob. Then, a formally unified geometry factor can be defined for both RFA and RFI (although different for the values) J_n that can be deduced from the original factors by means of the following equations.

RFA:

$$J_n = \frac{J}{\cos \psi_s} \quad (11.62)$$

where J is the geometry factor as defined in standards 218.01 or 2001-B88.

RFI:

$$J_n = \frac{\cos \phi_t}{\cos \phi_s \, Y_F \, Y_S \, Y_\beta} \quad (11.63)$$

The RFI factors for tooth form (Y_F) stress correction ($Y_S = K_f$) and helix (Y_β) are clarified in the cited reference. The calculation of the helix factor Y_β is reported here because it shows how a lower face contact ratio m_F affects J_n according to ISO for helical gears

$$Y_\beta = 1 - \frac{\min(m_F, 1) \min(\psi_s, 30)}{120} \quad (11.64)$$

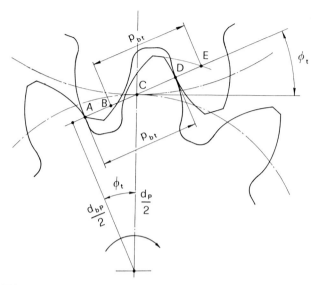

FIGURE 11.11 Involute meshing of an external gear pair in point D, pinion HPSC.

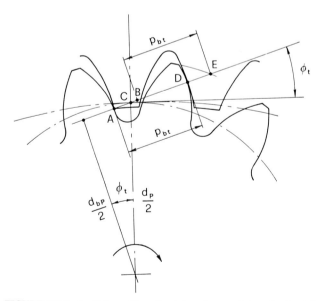

FIGURE 11.12 Involute meshing of an internal gear pair in point D, pinion HPSC.

This must be remembered when consulting the following J_n tables that refer to $m_F \geq 1$ for RFI.

The operating angular parameters, instead of the standard ones, have been introduced by AGMA 218.01 for both form and stress correction factors. However, the differences are small, especially for small center-distance modifications. The tables of the cited reference can be used for guidance in addition to the tables that are given here. The J_n factor is indicated by G_F/K_f there, for both RFA and RFI.

Warning: AGMA 218.01 or 2001-B88 give an alternative procedure for the so-called LCR helical gears with $m_F \leq 1$, but they often give more optimistic results. Instead, the normal procedure is followed in this chapter, as based on the load-sharing ratio m_N [see Eq. (11.37)] the same as for RH.

Tables 11.13 and 11.14 are supplied here. The stress correction factors K_f are added, not only for comparison purposes, but also because they must be considered again for excluding the stress concentration effect in the yielding calculations RFA (which follow).

TABLE 11.13 Geometry Factors for RFA and RFI, External Gear Pairs

ψ_s	N_P/N_G	x_P	x_G	J_{nAP}	K_{fAP}	J_{nAG}	K_{fAG}	J_{nIP}	K_{fIP}	J_{nIG}	K_{fIG}
0°	13/82	0.4	0	0.378	1.938	0.358	1.908	0.285	2.213	0.277	2.216
	13/82	0.4	0.4	0.380	1.906	0.388	1.935	0.282	2.208	0.261	2.547
	13/41	0.4	0	0.370	1.872	0.337	1.807	0.276	2.166	0.277	1.988
	13/41	0.4	0.4	0.372	1.817	0.388	1.857	0.271	2.155	0.266	2.380
	13/21	0.4	0	0.360	1.782	0.296	1.649	0.264	2.103	0.253	1.768
	13/21	0.4	0.4	0.362	1.698	0.377	1.733	0.256	2.082	0.261	2.187
	26/82	0	0	0.364	2.015	0.412	2.132	0.323	2.028	0.317	2.429
	26/82	0.4	0	0.434	2.120	0.386	2.008	0.311	2.527	0.297	2.322
	26/82	0.4	0.4	0.434	2.081	0.420	2.048	0.307	2.513	0.278	2.699
	26/41	0	0	0.353	1.972	0.379	2.035	0.313	1.991	0.319	2.144
	26/41	0.4	0	0.422	2.054	0.361	1.910	0.301	2.466	0.298	2.071
	26/41	0.4	0.4	0.419	1.988	0.417	1.980	0.294	2.435	0.285	2.512
	26/26	0	0	0.344	1.934	0.344	1.934	0.305	1.960	0.305	1.960
	26/26	0.4	0.4	0.406	1.909	0.406	1.909	0.283	2.368	0.283	2.368
15°	13/82	0.4	0	0.516	1.487	0.504	1.643	0.319	2.193	0.308	2.216
	13/82	0.4	0.4	0.525	1.458	0.542	1.642	0.316	2.188	0.290	2.522
	13/41	0.4	0	0.511	1.465	0.477	1.556	0.310	2.150	0.309	1.994
	13/41	0.4	0.4	0.524	1.419	0.541	1.563	0.304	2.139	0.296	2.361
	13/21	0.4	0	0.502	1.437	0.427	1.425	0.297	2.091	0.287	1.780
	13/21	0.4	0.4	0.518	1.374	0.530	1.451	0.288	2.070	0.291	2.172
	26/82	0	0	0.515	1.543	0.568	1.680	0.361	2.031	0.350	2.415
	26/82	0.4	0	0.573	1.602	0.545	1.647	0.343	2.488	0.329	2.318
	26/82	0.4	0.4	0.581	1.573	0.583	1.650	0.339	2.475	0.308	2.662
	26/41	0	0	0.499	1.543	0.527	1.613	0.351	1.996	0.354	2.143
	26/41	0.4	0	0.565	1.584	0.513	1.566	0.334	2.433	0.332	2.074

TABLE 11.13 Geometry Factors for RFA and RFI, External Gear Pairs *(Continued)*

ψ_s	N_P/N_G	x_P	x_G	J_{nAP}	K_{fAP}	J_{nAG}	K_{fAG}	J_{nIP}	K_{fIP}	J_{nIG}	K_{fIG}
15°	26/41	0.4	0.4	0.573	1.542	0.576	1.582	0.326	2.405	0.315	2.483
	26/26	0	0	0.486	1.543	0.486	1.543	0.341	1.966	0.341	1.966
	26/26	0.4	0.4	0.562	1.520	0.562	1.520	0.314	2.343	0.314	2.343
30°	13/82	0.4	0	0.522	1.558	0.504	1.672	0.339	2.106	0.326	2.197
	13/82	0.4	0.4	0.531	1.536	0.533	1.671	0.336	2.102	0.308	2.430
	13/41	0.4	0	0.519	1.541	0.485	1.606	0.331	2.077	0.331	1.998
	13/41	0.4	0.4	0.531	1.505	0.535	1.611	0.326	2.068	0.313	2.289
	13/21	0.4	0	0.512	1.519	0.450	1.504	0.320	2.035	0.317	1.800
	13/21	0.4	0.4	0.526	1.469	0.530	1.525	0.311	2.018	0.310	2.116
	26/82	0	0	0.520	1.596	0.557	1.699	0.381	2.011	0.360	2.350
	26/82	0.4	0	0.564	1.643	0.539	1.675	0.355	2.354	0.343	2.279
	26/82	0.4	0.4	0.571	1.621	0.568	1.677	0.351	2.346	0.323	2.535
	26/41	0	0	0.508	1.596	0.527	1.650	0.372	1.985	0.369	2.117
	26/41	0.4	0	0.559	1.630	0.517	1.614	0.347	2.320	0.351	2.064
	26/41	0.4	0.4	0.565	1.597	0.565	1.626	0.340	2.300	0.330	2.381
	26/26	0	0	0.497	1.596	0.497	1.596	0.364	1.962	0.364	1.962
	26/26	0.4	0.4	0.556	1.580	0.556	1.580	0.330	2.256	0.330	2.256

NOTES:
General data, ϕ_t and contact ratios: same as for Table 11.3.
Cutting data for $m_n = P_{nd} = 1$:
 Hob or rack cutter without protuberance, addendum = 1.25, radius of tip edge rounding = 0.2.
 Reduction of normal base thickness of pinion and gear = 0.02 for tooth backlash.

Note that the DIN method (see Deutsches Institut für Normung[5]) coincides with ISOs for spur gears, whereas it leads to more optimistic results for helical gears with regard to the RFI procedure considered here: With reference to Table 11.13, J_n increases by about 4 percent for $\psi_s = 15°$ and by as much as 17 to 21 percent for $\psi_s = 30°$.

In Table 11.13 a European standard hob or rack cutter has been considered (though tip edge rounding is considered rather small for cautionary reasons). On the other hand, the results are not too different for standard AGMA tools with smaller addendum and smaller tip rounding, respectively, or with greater addendum and full tip rounding. In fact the bending arm of the tooth load and the notch effect at the fillet are in some measure self-compensatory. Full rounding gives somewhat better J_n values; remember that the tooth root surface must be good to ensure a real improvement of tooth strength. The table can obviously serve only as a general guide. A programmed computation is necessary if specific manufacturing conditions must be taken into account.

As regards internal teeth, some comparisons with F.E.M. analyses in Castellani and Parenti-Castelli[29] show that the standard methods hardly agree with real fillet stresses. Thus Table 11.14 serves more as pure information than for practical purposes. A 45° angle of the tangent to the fillet (instead of 30°) has been considered for RFI according to provisional suggestions of the cited reference. (The original ISO and DIN methods simplify the procedure by considering racks instead of internal gears, but they fail to give adequate estimations of the fillet radius.)

TABLE 11.14 Geometry Factors for RFA and RFI, Internal Spur Gears

N_P/N_G	x_P	x_G	Z_C	ρ_{aC}	x_C	J_{nAG}	K_{fAG}	J_{nIG}	K_{fIG}
13/82	0.4	0	25	0	0.4	0.380	2.278	0.377	2.191
				0	0	0.339	2.533	0.276	2.894
				0	−0.4	0.280	3.060	0.195	4
				0.2	0.4	0.423	2.064	0.414	2.013
				0	0	0.406	2.115	0.334	2.411
				0	−0.4	0.394	2.177	0.276	2.854
			40	0	0	0.301	2.854	0.197	4
			40	0.2	0	0.398	2.155	0.298	2.667
13/82	0.4	0.4	25	0	0.4	0.420	2.305	0.362	2.461
				0	0	0.375	2.577	0.256	3.392
				0	−0.4	0.306	3.154	0.213	4
				0.2	0.4	0.469	2.076	0.407	2.204
				0	0	0.455	2.124	0.330	2.653
				0	−0.4	0.446	2.170	0.278	3.101
			40	0	0	0.319	3.035	0.214	4
			40	0.2	0	0.447	2.162	0.289	2.998
13/41	0.4	0.4	15	0	0	0.443	2.660	0.331	3.141
			15	0.2	0	0.520	2.266	0.389	2.666
26/82	0	0	25	0	0	0.445	2.662	0.309	3.330
			25	0.2	0	0.528	2.290	0.386	2.719
			40	0	0	0.397	2.934	0.253	4
			40	0.2	0	0.515	2.313	0.339	3.045
26/82	0.4	0	25	0	0	0.365	2.654	0.290	3.039
			25	0.2	0	0.440	2.219	0.355	2.513
			40	0	0	0.323	2.982	0.217	4
			40	0.2	0	0.431	2.251	0.315	2.792
26/82	0.4	0.4	25	0	0	0.417	2.688	0.269	3.644
			25	0.2	0	0.508	2.237	0.353	2.814
			40	0	0	0.354	3.144	0.242	4
			40	0.2	0	0.498	2.264	0.307	3.203

NOTES:
General data, notes, ϕ_t and contact ratios: the same as for Table 11.4.
Cutting data for $m_n = P_{nd} = 1$:
  Shaper cutter with Z_C tooth number, nominal addendum = 1.25, radius of tip edge rounding = ρ_{aC}, addendum modification = x_C relating to a given sharpening condition of the cutter.
Reduction of gear base thickness = 0.02 for tooth backlash.
Angle of fillet tangent = 45° for RFI, see Castellani and Parenti-Castelli.[29]
The occasional value $K_{fIG} = 4$ is arbitrary (ISO's rating is out of range).

In the case of Gleason bevel gears, if a computation is performed, the tooth thickness correction must be considered as independent of the addendum modification. When using tables, note that a slight increase in J_n for the pinion and decrease for the gear should be considered if the well-known Gleason geometrical coefficient K has been introduced for the thickness correction and $K > 0$.

11.7.2 RF—Adaptation Factor A_F

A factor A_F corrects the geometry factor by taking into account any peculiarity of geometry and material condition at fillet: $A_F > 1$ for favorable, $A_F < 1$ for unfavourable circumstances as a whole.

There is no similar overall factor in the standard methods, whereas there are some particular factors that may be thought of as a part of A_F.

Slim ring gears, both external and internal—the case is obviously more frequent for internal ones—have a stress concentration increase at the tooth root. The ANSI/ AGMA Standard 2001-B88[3] introduces a derating "rim thickness factor," K_B, and ("for informational purposes only") suggests $K_B > 1$ when the ratio of the rim thickness over the whole tooth depth is less than 1.2. This means that the adaptation factor A_F should be divided by K_B. On the other hand, the real root stress in such gears depends on many parameters, and direct analyses should be made if their tooth strength is central to the overall performance of a gear pair, see Faure[30] and Chong and Kubo.[31]

In the original AGMA method there is a factor $K_T \geq 1$ that derates strength for higher temperature. It may be included in A_F inversely, that is, it may lower A_F. There certainly is a derating temperature influence, but it is a problem to make a reliable assessment of it.

As for ISO, there are two factors that can be thought of as included in A_F: the notch sensitivity factor $Y_{\delta_{rel}T}$, and the roughness factor, $Y_{R_{rel}T}$. They are defined as *relative* factors, as they introduce modifications of the stress correction with regard to "test" gears.

The relative variation of the notch sensitivity has no great influence on steel gear strength according to ISO. However, a reduced notch sensitivity can be favourable for bad fillets, unfavourable for good ones (for example, for fillets better than those considered in the tables), that is, the computation of K_{fl} gives too optimistic values in such a case.

A higher fillet roughness is unfavorable and tool scores are worse. It may be necessary to diminish A_F by 5 to 10 percent.

A factor $A_F = 0.95$ to 0.9 can be suggested for teeth cut without tool protuberance and shaved, if no other peculiarity occurs.

More severe restrictions are necessary for grinding steps that can be originated by many grinding conditions, see Castellani and Zanotti.[32] $A_F = 0.9$ to 0.75 in common cases. But A_F should be as low as 0.3 in some extreme cases! See Winter and Wirth.[33]

It must be stressed that even teeth cut with tool protuberance may present fillet anomalies.[32]

The effect of shot-peening is doubtful for planet gears and in any case of bidirectional loading.[34] Otherwise it is generally favourable and suggests $A_F > 1$, especially if applied on low-roughness fillets free from any irregularities. Specific tests are advisable for important cases or for mass production gears.

Note: We may correct the previous indications for RFA ratings since the original AGMA values of fatigue limit stress are rather low (see the following). They have been deduced from field experience, contrary to ISO indications that originate from laboratory tests. Therefore, it may be supposed that AGMA's fatigue limits already take into

account the most common cases of bad fillet conditions and $A_F = 1$ may be adopted in such cases.
On the contrary, an accurate A_F assessment is necessary for RFI ratings.

11.7.3 RF—Size Factor, K_s

AGMA does not give any numerical indication for $K_s > 1$.
ISO prescribes $K_s = 1$ for any case if m_n [mm] ≤ 5, and $K_s > 1$ for $m_n > 5$. (Note that the original ISO factor is $Y_X = 1/K_s$.)
For RFI,

$$m_n \text{ [mm]} > 5 \quad K_s = \frac{1}{1 - A(B - 5)} \quad (11.65)$$

where

For steel or nodular iron gears, non-surface-hardened

$A = 0.006 \quad B = m_n$ [mm] for $m_n \leq 30 \quad B = 30$ for $m_n > 30$

For surface-hardened steel

$A = 0.01 \quad B = m_n$ [mm] for $m_n \leq 30 \quad B = 30$ for $m_n > 30$

For cast iron

$A = 0.015 \quad B = m_n$ [mm] for $m_n \leq 25 \quad B = 25$ for $m_n > 25$

Note: The meaning of the size factor K_s for ISO is as follows. It is known that equal materials with similar metallurgical structure and equal hardness have different fatigue resistance according to their sizes. Then K_s must affect the fatigue limit—as it does, see Eq. (11.61)—but not the stress value [Eq. (11.67)]. AGMA apparently considers it as a generic derating factor that takes into account any undetermined cause of stress increasing for big gear sizes: In fact, K_s is included in the calculation of s_t according to AGMA [see the following discussion, Eq. (11.66)].

11.7.4 RF—Conventional Fatigue Limit of the Fillet Stress, $s_{f\text{lim}}$

Table 11.6 reports values near the maximum indicated by the standards for $s_{t\text{lim}}$. (AGMA symbol: s_{at}. ISO and DIN symbol: σ_{FE}. Note that ISO intended a different parameter by the symbol $\sigma_{F\text{lim}}$, which is a nominal stress and amounts to one-half of σ_{FE}, as the ISO stress correction factor for test gears equals 2.)
The difference between AGMA and ISO indications is striking and is only in part compensated by the differences of K_f and J_n factors. For instance, ISO indications for case-hardened gears lead to greater $U_{L\text{lim}}$ so that Fig. 11.4 generally gains in significance. However the ISO RF assessments have been industrially tested, as they are a refinement of the original Niemann method[24] that has been applied for more than twenty years. What is necessary in RFI ratings is a careful choice of the adaptation factors (see the foregoing), as well as a proper assumption of all factors influencing the final results.

LOAD RATING OF GEARS   11.47

The indication given in Table 11.6 about reducing $s_{t_{\text{lim}}}$ for bidirectional load, as applied for a number of cycles in every direction, is determined from Japanese researches, see Aida and Oda.[34]

11.7.5 RF—Tooth Root Stress at Fillet, s_t

The tooth root stress does not directly enter the ratings based on the U_L factor, and is given only for information. Its value includes the stress correction factor K_f. Thus it would mean the actual root stress, provided that no plastic phenomena occur (which is generally not true, at least for steel gears).

$$\text{RFA: } s_t = \frac{V U_L K_s}{J_n} \tag{11.66}$$

$$\text{RFI: } s_t = \frac{V U_L}{J_n} \tag{11.67}$$

The difference between the two equations depends on the nature of the K_s factor (see the foregoing discussion).

11.7.6 RF—Service Factor, K_{SF} (only for 1 loading level)

Once the detailed gear data have been established, the loading data are equivalent to the following service factor for RF:

$$K_{SF} = \frac{K_a\, U_{L_{\text{lim}}}}{U_L} \tag{11.68}$$

If this way is adopted, the service factor summarizes application and reliability problems as well as those of gear life and excludes detailed ratings of the latter.

11.8 COPLANAR GEARS: DETAILED LIFE CURVES AND YIELDING

In the following paragraphs equations and data are given for defining the life curves in a unified way for RH and RF. The simple approach may be disputed, especially for RH for the calculation of the cumulative damage according to Miner's rule, but it is usually accepted for industrial gears as best estimated values and is preferable to no checking at all. The adopted procedure facilitates the assumption of life curves different from the indications of the standards, if the designer prefers to follow the indications of specific gear design books or those of an available field experience. In fact, many factors can affect the real slope of the life curve and the real life for a given load.[35]

More appropriate approaches are advisable for high-performance gears in special fields, see Coy, Townsend, and Zaretsky,[36] and require specific experimentation in conditions similar to the applications as regards sizes, materials, lubrication, loading, and velocity.

The procedure is based on the definition of a loading factor Q_H or Q_F that coincides with the life factor of the standards K_L for RF, or with the squared C_L factor for RH, but works in an opposite way. The value of the life factor depends on the cycle number N and leads to determining a reliability factor C_R or K_R. Such a procedure cannot deal with the cases of more than one loading level. Instead, the Q_H or Q_F factor is calculated as depending directly on the given load, so that cumulative damage and gear life are rated, and the reliability factor can be determined from the damage. Yielding ratings are facilitated.

Loading Factors Q_H, Q_F. For each loading level the following ratio is called *loading factor*:

$$\text{RH: } Q_H = \frac{K}{K_{\text{lim}}} \tag{11.69}$$

$$\text{RF: } Q_F = \frac{U_L}{U_{L_{\text{lim}}}} \tag{11.70}$$

11.8.1 Definition of the Life Curves and Gear Life Ratings for One Loading Level

As the loading factor is proportional to the tangential load W_t, the life curve of Fig. 11.1 can be redrawn with Q_H or Q_F as ordinates, see Fig. 11.13.

Q_W is the maximum initial value, Q_V corresponds to vertex V, and Q_S is a safety value below which some assume no failure will occur.

The choice of Q_S is up to the gear designer. It is advisable not to assume Q_{HS} greater than 0.8 or Q_F greater than 0.7, with a possible exception: Higher Q_{FS} values might

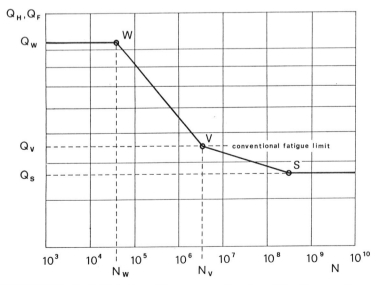

FIGURE 11.13 Unified definition of the characteristic points of the life curves for gear capacity ratings, RH curves (N, Q_H) or RF curves (N, Q_F).

LOAD RATING OF GEARS 11.49

be allowed for RFA, especially if a proper adaptation factor A_F is adopted, because a safety margin is implied in the fatigue limits given by the AGMA standards.
Lower values are necessary for an increased reliability, for example, when a gear failure may be dangerous for the operators (see Sec. 11.8.4 "Reliability"), for example, for hoists, marine applications, etc.
The vertex value Q_V equals unity with the exception of RFA, where $Q_{FV} = 1.04$.
The initial maximum value Q_W can be chosen by the designer, or it equals the maximum standard value of the life factor (squared for RH), C_{LW^2} for RH and K_{LW} for RF, reported in Table 11.15. In the same table the standard cycle numbers for points W and V are given, that is, N_W and N_V.
AGMA does not consider any correction of such maximum. Thus,

$$\text{RHA: } Q_{HW} = C_{LW^2} \tag{11.71}$$

$$\text{RFA: } Q_{FW} = K_{LW} \tag{11.72}$$

TABLE 11.15 Coordinates of Points Defining the Life Curve
The table refers to the following materials: through-hardened and tempered steel; carburized and case-hardened steel; gas-nitrided steel; induction-hardened steel, with root hardening for RF. The RHI and RFI indications include cast iron, malleable iron (pearlitic), and nodular (ductile) iron.

Initial point W, Fig. 11.13:
  number of cycles, N_W
  maximum loading factor, Q_{HW} for RH, Q_{FW} for RF

RHA: $N_W = 10^4$, all steels
      $Q_{HW} = C_{LW}{}^2 : C_{LW}{}^2 = 2.17$, all steels

RHI: $N_W = 10^5$, all materials

      $Q_{HW} = \dfrac{C_{LW}{}^2}{A_H} : C_{LW}{}^2 = 1.69$ for cast iron and for gas-nitrided steels

              $C_{LW}{}^2 = 2.56$ for all other materials

RFA: $N_W = 10^3$, all steels
      $Q_{FW} = K_{LW} : K_{LW} = 2.7$ for case-hardened steels
              $K_{LW} \approx 2.4 + (H_B - 250)/140$ for through-hardened steels in the range $210 \leq H_B \leq 320$

RFI: $N_W = 10^3$ for cast iron and surface-hardened steels
      $N_W = 10^4$ for all other materials

      $Q_{FW} = \dfrac{K_{LW} K_s}{A_F} : K_{LW} = 1.6$ for cast iron and for gas-nitrided steels

              $K_{LW} = 2.5$ for all other materials

Vertex V, Fig. 11.13:
  number of cycles, N_V
  conventional fatigue limit, Q_{HV} for RH, Q_{FV} for RF

RHA: $N_V = 10^7$ $Q_{HV} = 1$, all steels
RHI: $N_V = 2 \cdot 10^6$ $Q_{HV} = 1$ for cast iron and gas-nitrided steels
      $N_V = 5 \cdot 10^7$ $Q_{HV} = 1$ for all other materials
RFA: $N_V = 3 \cdot 10^6$ $Q_{FV} = 1.040$, all steels
RFI: $N_V = 3 \cdot 10^6$ $Q_{FV} = 1$, all materials

Note that the concept of Q_W is "yielding" (for direct yielding assessments see the following). ISO suggests some corrections of the life curves that can be extended by assuming approximately that the adaptation factors, as well as size factor for RF, influence fatigue, but not yielding limits. Thus, such factors must be excluded, as they have been introduced in the calculation of the fatigue limit

$$\text{RHI: } Q_{HW} = \frac{C_{LW^2}}{A_H} \tag{11.73}$$

$$\text{RFI: } Q_{FW} = \frac{K_{LW} K_s}{A_F} \tag{11.74}$$

Thus, the estimations are adjusted with regard to the gears that have been tested in the laboratory. Note that the corrections do not mean that the yielding assessment is modified: on the contrary, they mean that the yielding limit is considered as independent of the cited factors that influence the fatigue limit. The ratio Q_W is adjusted for this purpose. (A single further adjustment will be added for RFI for the direct assessment of the margin against yielding, in the case of reverse or bidirectional load, see the following.)

The sloped stretches WV and VS of the life curve follow the equations:

$$\text{RH: } N_{Hf} = \frac{N_{HV}}{(Q_H)^A} \tag{11.75}$$

$$\text{RF: } N_{Ff} = \frac{N_{FV}}{\left(\dfrac{Q_F}{Q_{FV}}\right)^A} \tag{11.76}$$

where $A = A_1$ for the first stretch
$A = A_2$ for the second one
N_f = gear life (cycle number) until failure (pitting or tooth breakage)

The curve stretch WV is fully defined by the previous choice of the coordinates of points W, V. For $N_W \leq N \leq N_V$

$$A = A_1 = \frac{\log\left(\dfrac{N_V}{N_W}\right)}{\log\left(\dfrac{Q_W}{Q_V}\right)} \tag{11.77}$$

The exponent A_2 for the stretch VS, that is, for $N > N_V$ and $Q_H > Q_{HS}$ or $Q_F > Q_{FS}$, must be chosen by the designer.

AGMA 218.01 and 2001-B88 only give ranges that correspond

For RHA, to $A_{2_{\min}} = A_1, A_{2_{\max}} = 22$
For RFA, to $A_2 = 31$ to 56

ISO does not give any indication. Of course A_2 cannot be less than A_1, whose value is rather different from AGMA's in its turn.

LOAD RATING OF GEARS 11.51

Lower A_2 leads to lower curve and more pessimistic results. There are a lot of influencing parameters, but for RH the velocity is probably the most important. A low velocity involves boundary lubrication and suggests a lower curve, that is, lower exponent A_2, possibly $A_2 = A_1$; whereas a high velocity allows higher A_2 values if it makes possible EHD lubrication of accurate gears with low surface roughness. See Dudley[37, 1] for the lubrication regimes.

For RF the choice is above all a matter of reliability from various points of view: A good and checked reliability of materials and manufacturing allows higher A_2, while a desired better reliability of gear resistance suggests lower A_2 values.

Different life curves can be established by adopting different coordinates of points W and V. Specific tests or indications of specific books for gear design can help. Any type of ordinates can be found there: pinion torque, tangential load, hertzian pressure, root stress, etc. It is easy to translate them into Q_H or Q_F factors: A point that serves as the conventional fatigue limit should be chosen and a loading factor $Q_V = 1$ attributed. Remember that the hertzian pressures must be squared when calculating the ratios Q_H. Finally, the exponent A_2 is chosen so that the desired curve is fully schematized.

Now, the cycle number per minute n_L may or may not equal the gear revolutions per minute n_f according to possible power sharing and uni- or bidirectional loading.

If N_f is the cycle number until failure for a given loading level, as rated by Eq. (11.75) or (11.76), the gear life until failure, in hours, is:

$$L_f = \frac{N_f}{60\, n_L} \qquad (11.78)$$

Note: n_L can be different for RH and RF. For example, consider a common planet pinion: For RF, it completes a load cycle for each revolution of the planet pinion, relative to the sun pinion and ring gear, and includes the load applications on both tooth sides. (Of course, its strength is derated because of the reverse bending load: This affects the preliminary assumption of $s_{t_{\lim}}$.) For RH, the surface K factor relating to the planet-ring mesh is usually nondamaging for the planet gear. One cycle per revolution must be considered if the load transmitted by the sun gear is unidirectional; but one half cycle per revolution, as an average, if the load is bidirectional and its application is borne equally by the two tooth sides.

Note: Uni- or bidirectional load is not the same as uni- or bidirectional motion. For instance, the speed reducer for a hoist has bidirectional motion but unidirectional loading, whereas the gears that control the traveling of a bridge crane have both bidirectional motion and load.

11.8.2 Yielding

As in the foregoing discussion, the Q_W value of the loading factor should mean *yielding*.

RH. AGMA does not say anything on this subject, whereas ISO explicitly attributes the meaning of surface yielding to the beginning of the life curve. No safety margin is defined in this regard. Field experience usually does not show any failures for occasional loads at this level, for well-rated gears. In all probability, possible initial plastic deformations of the tooth surface turn into work hardening. Thus a yielding loading factor Q_{Hy} is simply given by Eq. (11.79).

RHI

$$Q_{Hy} = Q_{HW} \qquad (11.79)$$

Loading factors Q_H higher than Q_{HW} should not be allowed. Plastic destruction of the tooth surface is not uncommon if excessive loads are applied even for a short time.

RF. AGMA gives an independent rating based on a yielding stress s_{ty} (original symbol: s_{ay}); whereas ISO attributes the meaning of yielding to the initial maximum of the life curve. However, a safety margin is necessary for tooth root yielding and it is obtained by means of a factor defined by AGMA: K_y that must be < 1. The definition of the loading factor facilitates the RF yielding ratings as follows. It must be $Q_F \leq Q_{Fy}$ where Q_{Fy} is a yielding loading factor that includes K_y.

RFA

$$Q_{Fy} = \frac{K_y \, s_{ty} \, K_f}{A_F \, s_{t_{\lim}}} \qquad (11.80)$$

The following equation corresponds approximately to AGMA's indications for through-hardened and tempered steels in the range $180 \leq H_B \leq 410$:

$$s_{ty} = A \, (H_B - 68) \qquad (11.81)$$

where A is 3.3 for metric units [N/mm^2], and A is 480 for English units [lb/in^2].

AGMA 218.01 and 2001-B88 do not give any indications for surface-hardened steels. Here the hardening depth is a determining factor. Besides, the yielding stress must not be thought of in general terms, as the AGMA geometry factor for bending strength refers to the tooth side where the load is applied. This has to do with normal kinds of fatigue failure, but the stress is higher at the opposite side because of compressive stress.

The following data can be assumed as a guideline, provided that the proper hardening depth be adopted and that a suitable K_y coefficient be introduced (see the following discussion):

Carburized case-hardened steels:

$$s_{ty} = 1200 \text{ N/mm}^2 \, (175{,}000 \text{ lb/in}^2)$$

Tooth root induction hardened:

$$s_{ty} = 950 \text{ N/mm}^2 \, (140{,}000 \text{ lb/in}^2)$$

Gas-nitrided steels:

$$s_{ty} = 700 \text{ N/mm}^2 \, (100{,}000 \text{ lb/in}^2)$$

(Same general specifications as in Table 11.6.)

RFI

$$Q_{Fy} = \frac{K_y \, Q_{FW} \, s_{t_{\lim} T}}{s_{t_{\lim}}} \qquad (11.82)$$

LOAD RATING OF GEARS   11.53

where $s_{t_{\lim}T}$ is the tabulated or test $s_{t_{\lim}}$, while the adopted $s_{t_{\lim}}$ can be lower because of reverse or bidirectional load or for any reason.

Note that the introduction of the K_y factor can make $Q_{Fy} < Q_{FW}$. In such cases Q_{FW} serves solely for determining the life curve, but the maximum load level must be kept lower than Q_{Fy}.

AGMA suggests $K_y = 0.75$ for general use and $K_y = 0.5$ for conservative purposes. Lower values may be advisable in the ratings for cast iron and bronze if the indications of the AGMA standards are followed for s_{ty}. Cautious values may also be used for nitrided and for induction-hardened gears. The introduction of K_f in Eq. (11.80) serves again to exclude the notch effect for yielding, but this can be more easily accepted for ductile materials.

AGMA does not exclude K_s for yielding, whereas ISO does. This depends on the different concepts of the factor, but can be considered as an additional margin against yielding of big gears in the AGMA rating, provided that a K_s greater than 1 has been assumed.

11.8.3 Tooth Damage and Cumulative Gear Life

The assumption of the safety loading level Q_S is equivalent to assuming that no damage D_g occurs if $Q_H < Q_{HS}$ or $Q_F < Q_{FS}$: $D_g = 0$, that is, an unlimited life is supposed for RH or RF, respectively, although AGMA and others do not accept this view.

If, on the contrary, the loading level is $\geq Q_S$, the damage of tooth flank or fillet is:

$$D_g = \frac{L}{L_f} \qquad (11.83)$$

where L is the desired life in hours for the loading level considered and L_f is the gear life until failure, as rated by Eq. (11.78).

If x different damaging loads for RH or for RF are applied to the gear, a cumulative damage is rated

$$D_{g_c} = D_{g_1} + D_{g_2} + \ldots + D_{g_x} \qquad (11.84)$$

The allowable cumulative gear life under damaging loads is, inversely

$$L_{f_c} = \frac{L_c}{D_{g_c}} \qquad (11.85)$$

where L_c is the desired cumulative life for the considered damaging levels. If there are further, nondamaging loading levels, the total allowable gear life increases proportionally to the desired gear life.

11.8.4 Reliability

The reliability factors, C_R for RH and K_R for RF, can be easily calculated in accordance with the definitions of the standards if all cycle numbers of the damaging loads are included either in the first stretch WV of the curve (range: $N_W \leq N \leq N_V$) or in the second stretch VS (range: $N > N_V$, $Q_H > Q_{HS}$ or $Q_F > Q_{FS}$). In this case[38]

$$\text{RH: } C_R = \left(\frac{1}{D_{g_c}}\right)^{1/2A} \quad (11.86)$$

$$\text{RF: } K_R = \left(\frac{1}{D_{g_c}}\right)^{1/A} \quad (11.87)$$

(ISO symbols: S_H for C_R, S_F for K_R.)

If the cycle numbers correspond to different stretches of the curves, iterative calculations would be necessary. The assessment of reliability factors is not necessary in itself: It is given here just because the standards define them, and indeed Eqs. (11.86) and (11.87) afford the calculations even in cases where the original methods do not, that is, when more than one loading level is considered. However, the damage assessment is sufficient in itself.

A damage limitation is necessary to get reliability: $D_{g_c} < 0.3$ is usually suitable for RH; lower values for RF, especially if a gear failure is dangerous for the operators or costly because it stops a plant far more important than the gears themselves. In such cases a severe limitation of the RH damage is necessary, too, as extended wear may induce tooth breakage.

Q_{HS} and Q_{FS} must be lowered accordingly when reliability must be increased.

11.9 COPLANAR GEARS: PREVENTION OF TOOTH WEAR AND SCORING

Lubrication regimes and the associated problems of tooth wear and cold and hot scoring are amply treated in Chap. 15 and Dudley.[1] The reader is referred to it for a thorough examination of them and for a detailed calculation of the hot scoring risk. They involve not only ratings in themselves, but designer competence and suitable tooth profile definition. This chapter is limited to some first estimates.

11.9.1 Progressive Tooth Wear

Fast gears in EHD lubrication regime were examined after 40 years operation and had practically no flank wear. However, slow gears in boundary lubrication always wear. For instance, some very slow spur gears presented a large amount of metal removal in the single contact zone after 10 years—but they did operate for 10 years. In fact, wear was important if compared with cycle number, but the total cycle number was not great, due to the small number of revolutions per minute. The disturbance due to profile alteration was acceptable because of the slowness of motion.

In practice, the same limitation of the tooth pressure that is considered as a pitting prevention serves as a wear prevention, especially if an adaptation factor for tooth velocity is introduced, like ISO's Z_v, that lowers the conventional fatigue limit for slow gears. If the desired cycle numbers involve the second sloped stretch of the life curve, then a low exponent of the life Eq. (11.75) or (11.76) is necessary for slow gears. And a proper lubricant with sufficient density and viscosity is necessary. Lubricant additives are useful if they are appropriate: for example, additives that favor running-in and/or reduce the friction coefficient may threaten wear in the long term, besides their possible action in pitting prevention.

11.9.2 Scoring or Scuffing

Gears can score even at low speed (cold scoring) if inappropriate lubricants are used and if the gear design does not prevent contacts in high-sliding zones. Cold scoring is distinguished from progressive wear, not only because of a different tooth flank appearance, but because it can occur after a short operating time. "Scoring" usually means hot-scoring, that is, tooth surface destruction of fast loaded gears at high contact temperature. Many gear features can influence scoring. For instance, edge contact of gear teeth may be an important item for scoring risk,[39] and a proper tip relief can prevent it. Another important item is the tooth surface finish. A detailed rating method is given in Dudley,[1] and it is based on the concept of the "flash temperature." Important indications on suitable profile features are also given there. Other detailed methods are given by the standards.[3,4,5]

Scoring prevention does not concern common industrial gear applications, since EP lubricants have become usual, but only fast, heavily loaded gears.

This chapter gives only a simplified equation, approximately determined from Niemann's original indications for good EP oils.[24,40] For external gears with parallel shafts and $\phi_{n_s} = 20°$, $v_t < 30$ m/s, $m_n < 10$ mm, $m_{p_A} \leq \cos \psi_s$, and $m_{p_E} \leq \cos \psi_s$:

$$W_{t_{\text{scor}}} = \left[\frac{d_P}{1 + \frac{N_P}{N_G}} \right]^3 \frac{0.48 \, F \cos \psi_s}{V \, v_t \, (0.007 \, m_n^{4.5} + \sqrt{m_n})} \quad (11.88)$$

where d_P, F, and m_n are in mm, v_t is in m/s, and the scoring tangential load is determined in N. In the case of English units, it is better to translate them into metric units at first and apply the formula just as it is.

The equation emphasizes the risk of scoring for higher modules. It usually gives a safety margin, as the best EP oils may allow loads even greater by two or three times to be applied without scoring. There are many circumstances in the risk of scoring, and for any doubtful case it is advisable that proper tooth design and more detailed ratings be performed. For example, the basic gear temperature is an important item,[26,41] and higher power losses mean higher gear temperature, if all other parameters are equal.[42]

Gear pairs with higher contact ratios require special investigations, preferably experimental ones.[23]

11.10 CROSSED HELICAL GEARS

Crossed helical gears theoretically have point contact so that they can carry a very limited load for RH, while they present no problems for RF. Hertz theory can be applied to the specific case by means of tabulated factors.[1] However, such kinds of gears are often employed as small accessories for mass productions, for example, of combustion engines. Thus it is both desired and possible to perform preliminary tests.

A summary review of performances of crossed helical gears, manufactured with

TABLE 11.16 Nominal Capacity of Crossed-Helical Gears (service factor = 1)

Pinion Pitch Diam. d_p, mm (in.)	Gear Ratio m_G	Gear Pitch Diam. d_G, mm	Center Dist. C, mm	Pinion Speed n_p, r/min				
				500	1000	2000	3000	4000
				Power, kW (multiply by 1.341 for hp)				
25	1	25	25	0.008	0.015	0.025	0.033	0.041
(≈ 1)	3	75	50	0.028	0.052	0.091	0.12	0.15
	5	125	75	0.036	0.068	0.12	0.16	0.19
50	1	50	50	0.063	0.12	0.20	0.27	0.33
(≈ 2)	3	150	100	0.22	0.42	0.73	0.97	1.18
	5	250	150	0.29	0.54	0.94	1.26	1.53
75	1	75	75	0.23	0.42	0.73	0.97	1.18
(≈ 3)	3	225	150	0.81	1.51	2.64	3.51	4.26
	5	375	225	1.05	1.96	3.43	4.56	5.54

Gear data: $\phi_n = 20°$, $\psi = 45°$
Pinion in alloy steel, carburized and case hardened, 58 ÷ 62 HRC, ground teeth.
Gear in phosphorus bronze.
Short running-in.
NOTE: The table is deduced from the first edition of the *Gear Handbook*, but the helix angle is unified to 45° in order to enable grinding of the pinion teeth. The given power serves only as a rough indication. Higher performance can be achieved by special materials in tested operating conditions. Crossed helical gears should be avoided as far as possible in common industrial applications.

typical materials, is given in Table 11.16. A variant can have case-hardened pinions with ground teeth and bath-nitrided gears with shaved teeth.

11.11 HYPOID GEARS

In theory, the load capacity rating of hypoid gears might be approached in a way not too different from what is adopted for bevel gears, if one takes into account the real pinion size and corrects the curvature radii of the profiles for RH. In practice, a lot of conflicting observations can be made. Longitudinal sliding may increase the risk of scoring. However, the profile design of such gears can be very accurate at present, see Litvin and Gutman.[43] The tooth surface can be made less sensitive to axial displacements and misalignment due to mounting and to elastic deflections. They may then carry greater loads compared to bevel gears as designed by more common criteria. Finally, their load capacity must be fully exploited as the reduction of their overall size is important in their usual application fields, for example, for vehicle transmissions. All such observations lead to one conclusion: It is better to size hypoid gears according to previous field experience and test them for their specific application than to calculate their load capacity. An approximate idea of the performance of hypoid gears can be determined from the same Table 11.12 that deals with spiral bevel gears, corresponding to equal gear diameter.

LOAD RATING OF GEARS   11.57

11.12 WORM GEARING

Double-enveloping worm gears are not considered here: In fact, either they are used as precision gearing where the transmitted power is very low and so they are oversized and a capacity rating is not worthwhile—or they require specific tests, as their operating behavior depends on a lot of specific manufacturing and application conditions.

11.12.1 Cylindrical Worm Gearing

The existing standard methods for rating the capacity of worm gears are based on the Hertz theory for RH. They include some empirical factors or modifications that do not take into due consideration the surface finish of both worm threads and wheel teeth (which may be only partly improved by a running-in period) and the actual tooth bearing area. The hob that cuts the wheel teeth usually has larger diameter than the worm, and its parameters affect the contact conditions of the gear pair in an unforeseeable way, unless a specific software is used for investigating them.[44] The lubrication conditions and the lubricant types greatly influence the real performance of the gear pair. The use of synthetic oils is becoming more and more common for this kind of gear pair. Furthermore, the high sliding produces a great amount of heat that must be dissipated through the housing. Therefore, the housing design and cooling are more important for the real performances of worm gears than for common coplanar gears, especially when a continuous load is transmitted.

Thus the ratings for worm gears are less reliable than those for coplanar gears (even if the latter are not completely reliable either).

Table 11.17 gives approximate values of the allowable input power for speed-reducing worm gears manufactured by specialized firms. The values must be multiplied by the efficiency for obtaining the output power (see Chap. 12).

The table refers to service factors equal to 1 for gear durability.

Proper housing design and fan cooling ensure good heat dissipation for such speed reducers. Nevertheless, the power values must be thought of as mechanical rather than as thermal limits. The service factors usually account for the cases of continuous and prolonged loading, as well as for application conditions and required life as usual.

The table relates to

Worms: alloy-steel, case-hardened—ground threads with involute profile

Wheels: centrifugally cast special bronze—hobbing with good accuracy and surface finish

Suitable gear diameters for optimizing efficiency and durability

Suitable diameter difference between the hob (generating the wheel) and the worm, and suitable bearing zone between worm thread and wheel tooth

Either superfinishing techniques are used for both worm and wheel, or some running-in period is required.

There is a great difference between the allowable power transmission of such specialized speed reducers and that of worm gears applied, as a part of any machine, internally to a nonspecific housing. The same worm gears of said speed reducers are often employed in such cases but must be essentially derated.

The problem of designing and manufacturing a worm gear pair for a generic mechanical application is that only the best conditions of materials, manufacturing, lu-

TABLE 11.17 Nominal Capacity of Cylindrical Worm Gear Pairs for Standard Speed Reducers (service factor $C_{SF} = 1$)

Ratio	Worm Speed, r/min	Center Distance, mm (in.)									
		75	(3)	100	(4)	150	(6)	200	(8)	300	(12)
		Input Power, kW (hp)									
10	500	1.7	(2.3)	3.7	(4.9)	9.5	(13)	17	(23)	35	(47)
	710	2.1	(2.8)	4.6	(6.1)	12	(16)	22	(29)	45	(60)
	1000	2.6	(3.5)	5.7	(7.6)	15	(20)	27	(37)	57	(77)
	1420	3.2	(4.3)	7.1	(9.6)	19	(25)	35	(47)	73	(98)
	1700	3.6	(4.8)	8.0	(11)	21	(29)	39	(53)	83	(111)
20	500	1.0	(1.4)	2.3	(3.1)	5.9	(8.0)	11	(15)	23	(30)
	710	1.3	(1.7)	2.9	(3.9)	7.5	(10)	14	(19)	29	(39)
	1000	1.6	(2.1)	3.6	(4.8)	9.4	(13)	17	(23)	37	(49)
	1420	2.0	(2.7)	4.5	(6.0)	12	(16)	22	(30)	47	(63)
	1700	2.2	(3.0)	5.1	(6.8)	13	(18)	25	(33)	53	(71)
30	500	0.8	(1.1)	1.7	(2.3)	4.4	(5.9)	8.1	(11)	17	(23)
	710	1.1	(1.4)	2.2	(2.9)	5.6	(7.5)	10	(14)	22	(29)
	1000	1.3	(1.7)	2.7	(3.6)	7.0	(9.4)	13	(17)	28	(37)
	1420	1.6	(2.2)	3.4	(4.5)	8.8	(12)	16	(22)	36	(48)
	1700	1.8	(2.4)	3.8	(5.1)	9.9	(13)	19	(25)	40	(54)
40	500	0.7	(1.0)	1.4	(1.9)	3.6	(4.8)	6.6	(8.9)	14	(19)
	710	0.9	(1.2)	1.8	(2.4)	4.5	(6.1)	8.4	(11)	18	(24)
	1000	1.1	(1.5)	2.2	(3.0)	5.7	(7.6)	11	(14)	23	(31)
	1420	1.4	(1.8)	2.8	(3.7)	7.2	(9.6)	13	(18)	29	(39)
	1700	1.5	(2.0)	3.1	(4.1)	8.1	(11)	15	(20)	33	(44)
50	500	0.6	(0.8)	1.2	(1.5)	3.0	(4.1)	5.6	(7.5)	12	(16)
	710	0.7	(1.0)	1.4	(1.9)	3.8	(5.1)	7.1	(9.6)	15	(20)
	1000	0.9	(1.2)	1.8	(2.4)	4.8	(6.4)	9.0	(12)	19	(26)
	1420	1.1	(1.5)	2.3	(3.0)	6.0	(8.1)	11	(15)	25	(33)
	1700	1.3	(1.7)	2.5	(3.4)	6.8	(9.1)	13	(17)	28	(38)
60	500	0.5	(0.6)	1.1	(1.3)	2.5	(3.4)	4.7	(6.4)	10	(13)
	710	0.6	(0.8)	1.2	(1.6)	3.2	(4.3)	6.0	(8.1)	13	(17)
	1000	0.7	(0.9)	1.5	(2.0)	4.0	(5.4)	7.6	(10)	16	(22)
	1420	0.9	(1.2)	1.9	(2.5)	5.0	(6.8)	9.6	(13)	21	(28)
	1700	1.0	(1.3)	2.1	(2.8)	5.7	(7.6)	11	(15)	23	(31)

brication, and housing cooling give sufficient reliability to load capacity assessments. Each one of such items, not only derates performance, but makes it more uncertain, if its condition is not the best.

In industrial use the differences in types of thread profile, given equal surface finish and accuracy, are less important. Nevertheless, worms with so-called straight profiles may have a capacity reduction depending on the diameter of the grinding wheels, as the ground profile usually is not really straight and differs from the profile of the hob that generates the worm wheel. Such a combination may in some cases obtain a favorable relief effect, but often reduces the tooth bearing area. A prolonged running-in can usually help, unless the profile combination is such as to produce a bad operating conjugation. Some tests show an advantage for special worms with concave thread flanks.

The load capacity essentially decreases for worms with nonground threads: In this case the usual assessments are absolutely unreliable.

11.13 STATE OF THE STANDARDS FOR LOAD RATING OF GEARS

Coplanar and bevel gears. Detailed information on the state of AGMA, ISO, and DIN standards is given in Faure.[45]

Worm Gearing. The conception of a standard method is proposed in Octrue.[46]

REFERENCES

1. Dudley, D. W., *Handbook of Practical Gear Design*, McGraw-Hill, New York, 1984.
2. AGMA 218.01, "AGMA Standard for Rating the Pitting Resistance and Bending Strength of Spur and Helical Involute Gear Teeth," American Gear Manufacturers Association, 1982.
3. ANSI/AGMA 2001–B88, "Fundamental Rating Factors and Calculation Methods for Involute Spur and Helical Gear Teeth," American National Standard Institute, 1988.
4. Organisation Internationale de Normalisation, "Calcul de la capacité de charge des engrenages à denture droite et helicoïdale," Projet de Norme Internationale ISO/DIS 6336/1, ISO/DIS 6336/2, ISO/DP 6336/III, 1981–1983, and following sections (unpublished, drafting in progress, 1991).
5. Deutsches Institut für Normung, "Tragfähigkeitsberechnung von Stirnrädern," DIN 3990, 1987.
6. Niemann, G., and Winter, H., *Maschinenelemente, Bd. II*, Springer, Berlin, 1983.
7. Henriot, G., *Traité Théorique et Pratique des Engrenages, Tome 1*, Dunod, Paris, 1979.
8. Castellani, G., and Zanotti, V., *La Resistenza degli Ingranaggi*, Tecniche Nuove, Milan, 1980.
9. Castellani, G., *Programmi del gruppo IPAR e IRID-RH-RF*, Tecniche Nuove, Milan, 1984.
10. ANSI/AGMA 2003–A86, "Rating the Pitting Resistance and Bending Strength of Generated Straight Bevel, Zerol® Bevel, and Spiral Bevel Gear Teeth," American National Standard Institute, 1986.
11. Kasuba, R., and August, R., "Gear Mesh Stiffness and Load Sharing in Planetary Gearing," ASME Paper No. 84–DET–229, 1984.
12. Le Borzec, R., "Essai sur l'analyse de l'equilibrage des trains epicycloïdaux," Congrès Mondial des Engrenages, Paris, 1977.
13. Kubo, A., "Untersuchungen über das dynamische Verhalten von Hochgeschwindigkeitsgetrieben," Fakultät fur Maschinenwesen der Kyoto Universität, 1971.
14. Winter, H., and Podlesnik, B., "Zahnfederfestigkeit von Stirnradpaaren," *Antriebstechnik* 22, Nr. 3 (pp. 39–42) and Nr. 5 (pp. 51–57), 1983.
15. Castellani, G., and Meneghetti, U., "Surcharges des engrenages pendant les régimes transitoires," Congrès Mondial des Engrenages, Paris, 1977.
16. Peeken, H., Troeder, Chr., and Antony, G., "Schwingungsverhalten von Planetengetrieben," *Konstruktion* 37, Heft 11, 1985, pp. 417–421.
17. Castellani, G., "Geometrical and Load Distribution Factors of Pitting Resistance Formulae," 2nd World Congress on Gearing, Paris, 1986.
18. Castellani, G., and Moretti, G., "Computing Gear Misalignment and Setting up Tooth Thickness Tolerances," Sixth World Congress on Theory of Machines and Mechanisms, New Delhi, 1983.
19. Hösel, Th., "Flankenrichtungskorrekturen für hochbelastete Stirnradgetriebe," Second World Congress on Gearing, Paris, 1986.

20. Winter, H., and Paul, M., "Influence of Relative Displacements between Pinion and Gear on Tooth Root Stresses of Spiral Bevel Gears," ASME Paper No. 84–DET–98, 1984.

21. Henriot, G., "Accélération de l'Evolution des Méthodes de Conception et de Fabrication des Transmissions Mécaniques," 4èmes Journées d'Etude des Transmissions Mécaniques, ECAM, Lyon, 1990.

22. Castellani, G., "Improvements in Geometrical Design of Gears with Semitopping," Fifth ASME International Power Transmission and Gearing Conf., Chicago, 1989.

23. Townsend, D. P., Baber, Berl B., and Nagy, A., "Evaluation of High-Contact-Ratio Spur Gears with Profile Modification," NASA Technical Paper 1458, 1979.

24. Niemann, G., *Maschinenelemente, Zweiter Band, Getriebe,* 2nd ed., Springer, Berlin, 1965.

25. Townsend, D. P., and Zaretsky, E. V., "Effect of Shot Peening on Surface Fatigue Life of Carburized and Hardened AISI 9310 Spur Gears," NASA Technical Paper 2047, 1982.

26. Townsend, D. P., "Lubrication and Cooling for High Speed Gears," Second World Congress on Gearing, Paris, 1986.

27. Townsend, D. P., Zaretsky, E. V., and Scibbe, H. W., "Lubricant and Additive Effects on Spur Gears Fatigue Life," Second World Congress on Gearing, Paris, 1986.

28. Akin, L. S., and Townsend, D. P., "Lubricant Jet Flow Phenomena in Spur and Helical Gears with Modified Center Distance and/or Addendums—for Out-of-Mesh Conditions," ASME Paper No. 84–DET–96, 1984.

29. Castellani, G., and Parenti-Castelli, V., "Rating Gear Strength," *ASME Journal of Mechanical Design,* Vol. 103, Apr.1981, pp. 516–527.

30. Faure, L., "Analysis of Tooth Root Stresses in Internal Gears with a Boundary Integral Equation Method," International Symposium on Gearing & Power Transmissions, Tokyo, 1981.

31. Tae Hyong Chong, and Kubo, A., "Simple Stress Formulae for a Thin-Rimmed Spur Gear, Part 1, Part 2 and Part 3," ASME Papers No. 84–DET–62, 84–DET–63, and 84–DET–64, 1984.

32. Castellani, G., and Zanotti, V., "Fillet Geometry of Ground Gear-Teeth," ASME Paper No. 84–DET–181, 1984.

33. Winter, H., and Wirth, X., "Einfluss von Schleifkerben auf die Zahnfuss-dauertragfähigkeit oberflächengehärteter Zahnräder," *Antriebstechnik* 17 Nr. 1–2, 1978.

34. Aida, T., and Oda, S., "Bending Fatigue Strength of Gears," *Bulletin of JSME,* Vol. 9, No. 36, 1966, pp. 793–806.

35. Bowen, C. W., "The Practical Significance of Designing to Gear Pitting Fatigue Life Criteria," *Journal of Mechanical Design,* Vol. 100, January 1978, pp. 46–53.

36. Coy, J. J., Townsend, D. P., and Zaretsky, E. V., "Dynamic Capacity and Surface Fatigue Life for Spur and Helical Gears," *Journal of Lubrication Technology,* April 1976, pp. 267–276.

37. Dudley, D. W., "Characteristics of Regimes of Gear Lubrication," International Symposium on Gearing & Power Transmissions, Tokyo, 1981.

38. Castellani, G., "Rating Gear Life," International Symposium on Gearing & Power Transmissions, Tokyo, 1981.

39. Shotter, B. A., "Scuffing and Pitting, Recent Observations Modify the Traditional Concepts," World Symposium on Gear and Gear Transmissions, Dubrovnik, Vol. B, 1978, pp. 321–332.

40. Niemann, G., Rettig, H., and Lechner, G., "Scuffing Tests on Gear Oils in the FZG Apparatus," *ASLE Transactions* 4, 1961, pp. 71–86.

41. Townsend, D. P., and Akin, L. S., "Analytical and Experimental Spur Gear Tooth Temperature as Affected by Operating Variables," *Journal of Mechanical Design,* Vol. 103, January 1981, pp. 219–226.

42. Anderson, N. E., and Loewenthal, S. H., "Efficiency of Nonstandard and High Contact Ratio Involute Spur Gears," ASME Paper No. 84–DET–172, 1984.
43. Litvin, F. L., and Gutman, Y., "A Method of Local Synthesis of Gears Grounded on the Connections Between the Principal and Geodetic Curvatures of Surfaces," *Journal of Mechanical Design,* Vol. 103, January 1981, pp. 114–125.
44. Castellani, G., "Worm-Wheel Manufacturing for ZI Worms: Meshing Analysis," 4èmes Journées d'Etude des Transmissions Mécaniques, ECAM, Lyon, 1990.
45. Faure, L., "Point actuel sur les normes de calcul des engrenages (ISO–AFNOR–DIN–AGMA)," 4èmes Journées d'Etude des Transmissions Mécaniques, ECAM, Lyon, 1990.
46. Octrue, M., "Méthode industrielle pour le dimensionnement des engrenages à roues et vis tangentes—Vérification—Conception," 4èmes Journées d'Etude des Transmissions Mécaniques, ECAM, Lyon, 1990.

CHAPTER 12
LOADED GEARS IN ACTION

E. E. Shipley
Gear Consultant
Rochester, New Hampshire

Gear analysis and design are, in effect, engineering approximations until the gears go to work. The gears must be loaded and subjected to dynamic operating conditions to obtain proper assurance that they will perform as predicted. It is only then that the efficiency of the gear mesh, the dynamic load effects, and the overall performance of the gearbox can be measured.

Those who design gears are often surprised that some gears run better and last longer than would be expected by the design analysis, while others fail prematurely, even when operated well within the design limits of transmitted horsepower. New gear designs must be based on both theoretical and practical experience to assure that the most reliable gear sets are manufactured and installed in customer locations. Gear engineers must be able to evaluate the friction of the gear mesh, bearing and windage losses, dynamic loads, and kinds and causes of gear failures in order to design the best possible gear unit for the job. In addition, they must know the various types and kinds of gear performance testing that can be done at the factory in order to simulate and supplement field conditions.

12.1 EFFICIENCY OF GEARS

One of the difficult problems of any gear designer is that of obtaining reliable efficiency information on which to base gear designs. Most gear designers, however, realize that gears are a very efficient method of transmitting power. In fact, in many applications gears are the only practical way to perform the task. Gears have the advantage, not only of transmitting power, but also of transmitting this power at almost any speed ratio desired. Gears have efficiencies in the range of 98 percent or more. On the surface it appears that these small losses could be neglected by gear designers, and often they are. In many applications, however, these small friction losses can cause considerable concern, since the losses must be dissipated as heat throughout the gear system. Gears with nonintersecting, nonparallel axes, such as worm gears or Spiroids* have a meshing

---

*Registered trademark of the Spiroid Division of Illinois Tool Works, Chicago, Illinois.

action consisting primarily of a sliding motion and, consequently, higher friction losses than parallel axes gears such as spur or helical gears, which have a combination of a high degree of rolling motion with a somewhat lesser degree of sliding motion. It is quite evident that the efficiency of a given type of gear is tied in with the amount of sliding that takes place in the meshing action.

12.2 SLIDING VELOCITY

Often the efficiency of various types of gears is expressed in terms of a sliding velocity. This being the case, it becomes quite important to understand how to calculate the sliding velocity of different gear drives.

12.2.1 Spur Gears

In the case of conjugated gear teeth such as spur or helical gears, the sliding velocity can be determined by the following equations [1]*

$$v_s = v_1 - v_2 \qquad (12.1)$$

where v_1 is the rolling velocity of the point in question on the pinion, and v_2 is the rolling velocity of the corresponding point on the gear.

$$v_s = s(\omega_P + \omega_G) \qquad (12.2)$$

where ω_P = angular velocity of pinion, rad/s
ω_G = angular velocity of gear, rad/s
s = distance along line of action to the point in question from the pitch point, ft

Since the sliding velocity is expressed as the algebraic difference of the two rolling velocities in question, it is often more convenient to calculate v_s in terms of v_1 and v_2 as indicated in the following.

$$v_1 = \frac{n_P \pi \rho_1}{360} \qquad (12.3)$$

$$v_2 = \frac{n_G \pi \rho_2}{360} \qquad (12.4)$$

where n_P = r/min of pinion
ρ_1 = radius of curvature of pinion, in.
n_G = r/min of gear
ρ_2 = radius of curvature of gear, in.

---

*Superscript numbers refer to references at the end of the chapter.

Sometimes sliding velocities are required at specific points such as the sliding velocity at the pinion tip or the sliding velocity at the gear tip. In this case,

$$v_{sP} = \frac{n_P}{114.59} \times Q_r \, \frac{m_R + 1}{m_R} \qquad (12.5)$$

where v_{sP} is the sliding velocity at pinion tip, ft/s, and Q_r is the arc of recession of driving gears, rad.

$$v_{sG} = \frac{n_G}{114.59} \times Q_a \, \frac{m_R + 1}{m_R} \qquad (12.6)$$

where v_{sG} is the sliding velocity of gear type, ft/s, and Q_a is the arc of approach, rad.

12.2.2 Bevel Gears

As in the case of spur and helical gears, the sliding velocity of a point on a bevel gear, a distance s from the pitch point, is approximately

$$v_s = s(\omega_P^2 + \omega_G^2 + 2\omega_P\omega_G \cos \Sigma)(\sin^2 \phi + \cos^2 \psi \cos^2 \phi)^{1/2} \qquad (12.7)$$

where ω_P = angular velocity of pinion, rad/s
ω_G = angular velocity of gear rad/s
Σ = shaft angle

In the case of right-angle bevel drives, $\Sigma = 90°$ and

$$v_s = s(\omega_P \cos \gamma + \omega_G \cos \Gamma)(\sin^2 \phi + \cos^2 \psi \cos^2 \phi)^{1/2} \qquad (12.8)$$

where γ = pitch cone angle of pinion, $\gamma + \Gamma = 90°$
Γ = pitch cone angle of gear
ψ = spiral angle

12.2.3 Worm Gears

The sliding velocity, or rubbing velocity, as it is generally referred to for worm gears, can be calculated as follows:[2]

$$v_r = \frac{0.262 n_w d}{\cos \lambda} \qquad (12.9)$$

where v_r = rubbing velocity, ft/min
λ = lead angle
n_w = r/min, worm
d = pitch diameter of worm, in.

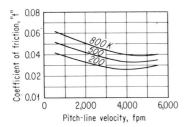
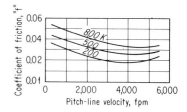

FIGURE 12.1 *(a)* Average coefficient of friction of lightweight petroleum oil. 120°F oil inlet hardened gears 45 SUS at 100°F. *(b)* Average coefficient of friction of medium-weight petroleum oil. 120°F oil inlet hardened gears 300 SUS at 100°F.

Figures 12.1*a* and 12.1*b* give values for the average coefficient of friction f versus pitch line velocity for various K-factor gear loadings based on tests run by the author at General Electric Company.[4] Additional mean or average coefficient of friction data plotted against a dimensionless ratio covering a widened viscosity range is given in Fig. 12.2. This data is the summary of work performed by Ohlendorf and was abstracted from a paper presented by Professor Winters.[15]

A short-cut method has been developed to analyze standard spur gear trains, which is quite helpful, particularly in the early design stages. The power loss of a given set

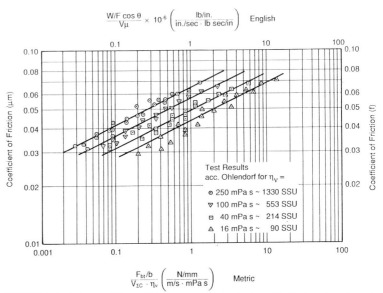

FIGURE 12.2 Mean coefficient of friction versus dimensionless ratio. Test work performed by Ohlendorf abstracted from AGMA paper by Prof. Winter.

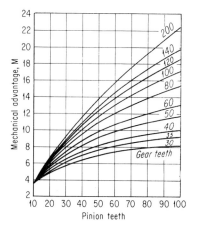

FIGURE 12.3 Mechanical advantage of 20°-pressure-angle spur gears. Percent power loss = P. Coefficient of friction = f. Mechanical advantage = M. $P = f/M \times 100$

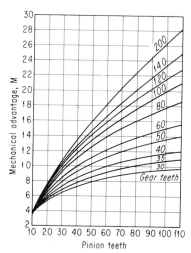

FIGURE 12.4 Mechanical advantage of 25°-pressure-angle spur gears. Percent power loss = P. Coefficient of friction = f. Mechanical advantage = M. $P = f/M \times 100$

of gears is a function of the coefficient of friction of the gear mesh and the so-called mechanical advantage of the gear mesh, that is

$$P_t = \frac{f}{M} \times 100 \qquad (12.19)$$

where P_t = percent power loss
f = coefficient of friction
M = mechanical advantage of the mesh

The mechanical advantage M, for 20- and 25°-pressure-angle standard spur gears, has been plotted for various combinations of numbers of gear and pinion teeth and can be found by referring to Figs. 12.3 and 12.4. The same coefficient of friction shown in Figs. 12.1 and 12.2 is used in the foregoing equation. The curves shown in Fig. 12.5 give the nominal efficiency of spur gears.

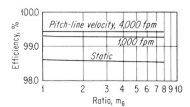

FIGURE 12.5 Nominal efficiency of spur gears. Based on a good grade of petroleum oil. Hardened and ground gears 60 Rockwell C.

12.3.2 Internal Spur Gears

The efficiency of internal spur gears is very similar in nature to that of external spur gears and can be determined by the same general equations, namely

$$E = 100 - P_t \qquad (12.13)$$

$$P_t = \frac{50f}{\cos \phi} \left(\frac{H_s^2 + H_t^2}{H_s + H_t} \right) \qquad (12.14)$$

except that H_s and H_t are determined in a slightly different manner, as specified by the following equations:

$$H_t = \frac{m_G - 1}{m_G} \left[\sqrt{\left(\frac{r_o}{r}\right)^2 - \cos^2 \phi} - \sin \phi \right] \qquad (12.20)$$

and

$$H_s = (m_G - 1) \left[-\sqrt{\left(\frac{R_i}{R}\right)^2 - \cos^2 \phi} + \sin \phi \right] \qquad (12.21)$$

where R_i is the inside radius of the internal gear.

It is quite obvious from these equations that the power loss of internal spur gears will be lower than that of external gears and, consequently, more efficient for a given coefficient of friction. Since both internal and external spur gears are very similar in nature except for the differences in specific sliding velocities, the curves of coefficient of friction versus revolutions per minute, as shown in Figs. 12.1 and 12.2, can be used for internal spur gears as well as for external spur gears.

12.3.3 Single-Helical Gears

The efficiency of helical gears is somewhat similar in nature to that of spur and internal gears. In fact, if a helical system is analyzed in terms of an equivalent spur gear, the equation for the percent power loss can be written as follows:[5]

$$P_t = \frac{50f \cos^2 \psi}{\cos \phi_n} \left(\frac{H_s^2 + H_t^2}{H_s + H_t} \right) \qquad (12.22)$$

where H_s and H_t are determined by solving Eqs. (12.15) and (12.16). It must be kept in mind, however, that

ϕ_n = normal pressure angle of helical gear
ψ = helix angle
f = coefficient of friction

The resultant efficiency is determined by the same general efficiency equation $E = 100 - P_t$.

It becomes obvious from these equations that a helical gear is somewhat more efficient than a similar spur gear. Often this is an important consideration when designing gears for minimum power loss.

Helical gears are similar in nature to spur gears; therefore, the equations of coefficient of friction as shown in Figs. 12.1 and 12.2 can be used for calculation purposes.

12.3.4 Double-Helical and Herringbone Gears

Power loss, or the efficiency, of double-helical gearing and herringbone gearing is determined in exactly the same manner as for single-helical gears. Since the power being transmitted by a set of double-helical gears splits exactly in two, the total power loss is just double that for one side. The percent-power-loss calculation need be determined for only one side. This percent loss can then be multiplied by the total transmitted horsepower of the double-helical set to obtain total power losses.

12.3.5 Straight Bevel Gears

Straight bevel gears can be likened to spur gears except, of course, that they mesh at right angles. Bevel gears are often analyzed as equivalent spur gears by using Tredgold's approximation to determine the spur gear cross sections. Fundamentally, the number of teeth in virtual spur gears can be found as follows:[6]

$$N_{v_G} = \frac{N_G}{\cos \Gamma} \quad (12.23)$$

$$n_{v_P} = \frac{N_P}{\cos \gamma} \quad (12.24)$$

where N_{v_G} = number of teeth in virtual spur gear
N_G = number of teeth in actual bevel gear
N_{v_P} = number of teeth in virtual spur pinion
N_P = number of teeth in actual bevel pinion
P = pitch cone angle of bevel gear
γ = pitch cone angle of bevel pinion

When the foregoing information is completely integrated into the power loss equations for spur gears, the following equation can be derived for determining the pecent power loss for straight bevel gears:

$$P_t = 50f \left(\frac{\cos \Gamma + \cos \gamma}{\cos \phi_n} \right) \left(\frac{H_s^2 + H_t^2}{H_s + H_t} \right) \quad (12.25)$$

H_s and H_t are determined by solving Eqs. (12.15) and (12.16). It must be kept in mind that

ϕ_n = normal pressure angle in the bevel gear
r_o = outside radius of large end of bevel pinion
r = pitch radius of large end of bevel pinion
R_o = outside radius of large end of bevel gear
R = pitch radius of large end of bevel gear

The efficiency for straight bevel gears is again determined by the same general efficiency equation

$$E = 100 - P_t$$

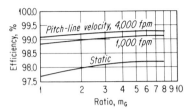

FIGURE 12.6 Nominal efficiency of straight bevel gears. Based on a good grade of petroleum oil. 33 to 40 Rockwell C.

Since bevel gears have so much in common with spur gears,[7] it is quite proper to use the curves of Figs. 12.1 and 12.2 to determine the coefficient of friction of bevel gears. The curve shown in Fig. 12.6 gives the nominal efficiency of straight bevel gears. It is interesting to note from these curves that the straight bevel gear mesh is quite efficient, being in the range of 99 percent or better.

12.3.6 Zerol* Bevel Gears

Zerol bevel gears are made in a different type of machine from the straight bevel gears.[8] The difference is that the straight bevels are cut on a bevel gear planer, which moves the cutting tool back and forth in straight lines, whereas the Zerol bevel gears are cut with a rotary cutter. This rotary motion produces a curved tooth section in the length of the tooth; however, the tooth has a 0° spiral angle and, consequently, is very similar to a straight bevel gear from an efficiency point of view. Actually, the efficiency of a Zerol bevel gear can be calculated using the information and equations for straight bevel gears as described in Sec. 12.3.5.

Figures 12.1 and 12.2 can also be used to determine the mesh efficiency of Zerol bevel gears. One of the chief advantages of Zerol bevel gears is that they can be hardened and ground quite accurately. They are often made this way and consequently can be loaded higher than straight bevel gears. This being the case, the curves in Fig. 12.6 would be slightly higher for hardened and ground Zerol bevel gears.

12.3.7 Spiral Bevel Gears

Spiral bevel gears have often been referred to as helical bevel gears.[9] In practice, however, spiral bevel gears do not have a true helical spiral. They are cut on or ground on the same machine that produces Zerol bevel gears, except that the cutting tool is set at an angle to the axis of the gear, thus producing the helical gear effect.

From an efficiency point of view, a spiral bevel gear is very good and its results can be compared with those of helical gear sets. Generally, spiral bevels are used where high load and high speeds are encountered, since they can be hardened, ground, and produced quite accurately.

The percent power loss for a spiral bevel gear can be determined by solving the following equation:

$$P_t = 50f(\cos \Gamma + \cos \gamma) \frac{\cos \psi^2}{\cos \phi_n} \left(\frac{H_s^2 + H_t^2}{H_s + H_t} \right) \quad (12.26)$$

where T = pitch cone angle of spiral bevel gear
γ = pitch cone angle of spiral bevel pinion
ϕ = spiral angle

---

*Registered trademark of the Gleason Works, Rochester, New York.

ϕ_n = normal pressure angle
f = coefficient of friction

H_s and H_t must be calculated using Eqs. (12.15) and (12.16); however, the general comment made in the discussion of straight bevel gears applies equally to spiral bevel gears. The coefficient of friction, as shown in Figs. 12.1 and 12.2, can be used for calculation purposes.

Figure 12.7 gives the nominal efficiency of spiral bevel gears versus gear ratio. Note that the efficiency has been plotted for various values of pitch line velocity. Actually, these curves have been based on full rated power being transmitted by the gear set and are only representative of what can be done. They are quite useful as a guide, but if close efficiency estimates must be made, it would be wise to calculate the design from the very beginning. By using the proper equation, the values of coefficient of friction, pressure angle, and gear geometry for a particular design can be fed into the equation, and a more precise answer will be obtained.

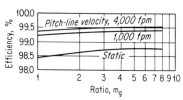

FIGURE 12.7 Nominal efficiency of spiral bevel gears. Based on a good grade of petroleum oil. Hardened and ground gears. Spiral angle = 35°.

12.4 EFFICIENCY OF NONPLANAR GEARS (NONINTERSECTING AND NONPARALLEL AXES)

Many gear applications require the features of this general class of gearing. Although coplanar gears are generally more efficient, this class of gearing can be designed with high-performance characteristics.

12.4.1 Worm Gears

The percent efficiency for worm gear drives can be determined as follows for the case in which the worm drives the worm wheel:

$$E = 100 \frac{\cos \phi_n - f \tan \lambda}{\cos \phi_n + f \cot \lambda} \quad (12.27)$$

where ϕ_n = normal pressure angle
f = coefficient of friction
λ = pinion lead angle

In many instances, not all the information pertaining to a given set of worm gears is known. In this case a simplified equation for worm gear efficiency can be used.[5]
When the worm wheel drives the worm,

$$E = 100 \frac{\tan (\lambda - \eta)}{\tan \lambda} \quad (12.28)$$

When the worm drives the worm wheel,

$$E_\omega = 100 \frac{\tan \lambda}{\tan (\lambda + \eta)} \quad (12.29)$$

where η is the friction angle, and $\tan \eta$ equals f approximately. Since, generally, the coefficient of friction has been determined with some experimental error, Eqs. (12.28) and (12.29) are fair approximations.

It is quite helpful to analyze the relationship of lead angle and the efficiency of various combinations of worm gear drives by studying Tables 12.1 and 12.2. Tables 12.1

TABLE 12.1 Efficiency of Worm Gearset, Worm Driving

Worm lead angle, deg	$f = 0.015$ eff., %	$f = 0.02$ eff., %	$f = 0.03$ eff., %	$f = 0.04$ eff., %	$f = 0.05$ eff., %	$f = 0.07$ eff., %	$f = 0.10$ eff., %
3	77.5	72.0	65.0	55.0	48.0	42.0	35.0
5	86.0	81.5	75.1	70.0	63.0	55.0	48.0
10	92.0	89.5	84.5	82.0	77.0	72.0	63.0
15	94.5	92.5	89.5	87.0	83.2	78.0	72.0
25	96.2	95.1	92.6	90.5	88.5	84.0	78.0
35	96.8	95.9	93.9	92.1	90.2	86.8	81.5
45	97.1	96.2	94.2	92.4	90.5	87.2	82.0
55*	96.8	95.8	93.8	91.8	89.8	86.0	80.5
65	96.0	94.6	92.1	89.3	87.3	82.0	75.0
75	93.9	91.9	88.1	84.0	80.0	72.0	61.0
80	91.0	88.0	82.1	75.3	71.0	59.0	40.0
85	82.5	75.3	63.0	50.3	40.0	20.0	

* Generally worms are not designed for more than 45° lead angle. An angle of 55° is usually the very upper limit for a practical design.

TABLE 12.2 Efficiency of Worm Gearset, Wormgear Driving

Worm lead angle, deg	$f = 0.015$ eff., %	$f = 0.02$ eff., %	$f = 0.03$ eff., %	$f = 0.04$ eff., %	$f = 0.05$ eff., %	$f = 0.07$ eff., %	$f = 0.10$ eff., %
3	69.0	58.0	40.0	20.0	0	20.0	
5	82.5	77.0	63.0	53.0	40.0	58.0	40.0
10	91.0	88.0	82.5	75.3	71.0	72.0	61.0
15	93.9	91.9	88.2	84.0	80.0	81.9	75.0
25	96.0	94.6	92.1	89.8	87.0	86.0	80.5
35	96.8	95.7	93.8	91.8	89.8	87.1	82.2
45	97.1	96.2	94.3	92.5	90.6	86.8	81.5
55*	96.8	96.0	93.9	92.0	90.2	84.0	78.0
65	96.3	95.1	92.6	90.7	88.5	78.0	72.0
75	94.3	92.8	89.5	87.0	83.3	72.0	64.0
80	92.0	89.5	85.3	82.0	78.0	56.0	47.0
85	86.0	81.9	81.3	70.0	63.0		

* Generally worms are not designed for more than 45° lead angle. An angle of 55° is usually the very upper limit for a practical design.

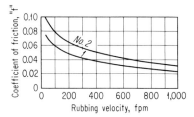

FIGURE 12.8 Coefficient of friction for worm gears. Based on a good grade of petroleum oil. No. 1 = case-carburized and ground worm and phosphor bronze gear. No. 2—cast-iron worm and worm gear.

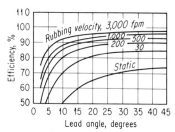

FIGURE 12.9 Nominal efficiency of worm gears. Based on case-carburized and ground worm and phosphor bronze gear.

and 12.2 were determined by assuming various values for the coefficient of friction and solving the efficiency equations. In order to use Tables 12.1 and 12.2 profitably, values of the coefficient of friction for your given conditions must be known.

Figure 12.8 gives some experimental values of f based on using a good grade of mineral oil after the gear set has been well broken-in.

Figure 12.9 shows the nominal efficiency of cylindrical worm gears as a function of lead angle and rubbing velocity. It is understood that double-enveloping worm gears also follow such a pattern as outlined in Fig. 12.9, although it has been said that double-enveloping worm gears are more efficient under starting conditions than are cylindrical worm gears.

12.4.2 Crossed Axes Helical

Crossed axes helical gears are actually nonenveloping worm gears and as such can be calculated with efficiency equations very similar in nature to equations for cylindrical worm gears.

Since crossed axes helical gears can be of the same hand or opposite hands and the spiral angle or helix angle can be different for either member, the equations for the efficiency are a little tricky and must be used with caution. For the case where the gears are of the same hand, then,[5]

$$E = 100 \frac{1 - f \tan \psi_f}{1 + f \tan \psi_d} \quad (12.30)$$

where f = coefficient of friction
ψ_d = helix angle of driver
ψ_f = helix angle of follower

When the gears are of opposite hand and ψ_d is less than ψ_f,

$$E = 100 \frac{1 - f \tan \psi_f}{1 - f \tan \psi_d} \quad (12.31)$$

When the gears are of opposite hand and ψ_d is greater than ψ_f,

$$E = 100 \frac{1 + f \tan \psi_f}{1 + f \tan \psi_d} \tag{12.32}$$

The coefficient of friction f for spiral gears is practically the same as that for worm gears; therefore, the same curves of f versus rubbing velocity can be used (see Fig. 12.8).

12.4.3 Spiroid Gears

Spiroid gears are comparable to worm gears in performance and efficiency and, in general, can be used in many applications where heretofore worm-gear drives were specified.[10] The efficiency of Spiroid gears can be determined from an equation that has much in common with the basic worm gear efficiency equation. Efficiency in percent for Spiroid gears can be expressed as follows:

$$E = 100 \frac{\cos \phi_n + f \cot \alpha_G}{\cos \phi_n + f \cot \lambda} \tag{12.33}$$

where ϕ_n = normal pressure angle
f = coefficient of friction
α_G = gear spiral angle (with respect to the plane of rotation) (*Note:* This is 90° minus the spiral angle as used for spiral bevel gears.)
λ = pinion spiral angle

$$\tan \lambda = \frac{L}{2\pi r_m} \tag{12.34}$$

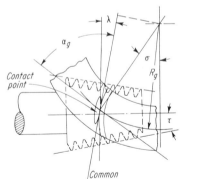

FIGURE 12.10 Contacting segment of a Spiroid gear and pinion.

where L equals the conical lead, and r_m is the mean cone pinion radius at a point halfway down the face.

In order to evaluate Eq. (12.33) fully, however, additional information is required, such as

$$\alpha_G = 90° - (\sigma - \lambda) \tag{12.35}$$

where σ is normally standard at 40° and λ equals the lead angle as indicated in the foregoing discussion.

The normal pressure angle is taken as 10° except when the gear set is run with the contact on the opposite side of the pinion teeth. Then it is driving on the high-pressure-angle side, which usually has a pressure angle of 30°. In order to understand better the geometry relationship of the foregoing equations, refer to Fig. 12.10, which shows a contacting segment of a Spiroid gear and a Spiroid pinion.

The relationship of the coefficient of friction versus sliding velocity can be seen in

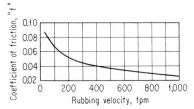

FIGURE 12.11 Coefficient of friction versus sliding velocity of Spiroids. Based on carburized pinion and gear, 60 Rockwell C, and a good grade of petroleum oil.

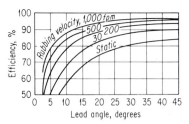

FIGURE 12.12 Nominal efficiency of Spiroid gears. Based on case-carburized pinion and gear, 60 Rockwell C, and a good grade of petroleum oil.

Fig. 12.11. Curves have been plotted for the efficiency of Spiroid gears versus pinion Spiroid angle (see Fig. 12.12). Note from Fig. 12.12 that the gears used were hardened and ground. Little difference has been observed when bronze gears have been used instead of hardened gears. At very low speeds, though, a bronze gear and hardened pinion combination does appear to be somewhat more efficient.

12.4.4 Helicon* Gears

Helicon gears are very similar in nature to Spiroid gears in that they are skew axis gears. They differ by virtue of the fact that the pinions have a 0° taper angle, whereas standard Spiroid pinions have a 5° taper angle (11). Helicon gears also are flat-faced instead of conical-faced like Spiroid gears. Other than this, the gear types are very much alike and can be calculated using exactly the same equations as outlined above for Spiroid gears. The standard σ "sigma" angle is 40°. The 10° pressure angle applies to the Helicon gear as well as the Spiroid gear. Figure 12.11 can be used to determine the coefficient of friction for Helicon gears. Figure 12.12 also depicts typical Helicon gear efficiency versus rubbing velocity. It must be kept in mind, however, that Helicon gears by design have less face width in contact and will carry less load than their brothers, the Spiroid gears. Consequently, the overall efficiency of Helicon drives will be less efficient as compared with the overall efficiency of Spiroid gear drives.

12.4.5 Hypoid Gears

Hypoid gears have much in common with plain or spiral bevel gears.[12] In fact, hypoid gears are actually offset axis bevel gears, and as such, they are subjected to a high degree of sliding as are most of the other types of offset or "nonplanar" gears. In comparison

---

*Registered trademark of Spiroid Division, Illinois Tool Works, Chicago, Illinois.

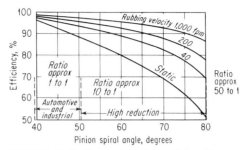

FIGURE 12.13 Nominal efficiency of hypoid gears. Based on the sum of gear and pinion spiral angles equal to 75°, and normal pressure angle of 22°30′.

with a worm gear, however, the rubbing speed of a hypoid relative to its own pitch line is not so high. The percent efficiency for hypoid gears can be determined as follows:

$$E = 100 \; \frac{\cos \phi_n + f \tan \psi_G}{\cos \phi_n + f \tan \psi_P} \qquad (12.36)$$

where ϕ_n = normal pressure angle
 f = coefficient of friction
 ψ_G = gear spiral angle
 ψ_P = pinion spiral angle

The nominal efficiency for hypoid gears can be found in Fig. 12.13. These curves were plotted for the conditions where the sum of the gear and pinion spiral angle is 75° and the normal pressure angle is 22°30′. It is quite interesting to note that at low ratios the hypoid gears are quite efficient. Generally speaking, the coefficient of friction for hypoid gears is similar to friction values of worm gears, and, consequently, the curves shown in Fig. 14.8 can be used for hypoid gear calculations.

12.4.6 Planoid* Gears

Planoid gears perform a function similar to that of hypoid gears. They are right-angle gears with offset axes. The tooth surfaces of the gears are planes and the pinions are slightly tapered with helical teeth.[13] The recommended ratio range for Planoid gears is from 15:1 to 10:1. The efficiency of Planoid gears compares quite favorably with that of hypoid gears.

The efficiency of Planoid gears can be determined from the following equation:

$$E = 100 \; \frac{\cos \phi_n + f \cot \alpha_G}{\cos \phi_n + f \cot \lambda} \qquad (12.37)$$

where ϕ_n = normal pressure angle
 α_G = gear spiral angle

---

*Registered trademark of Spiroid Division, Illinois Tool Works, Chicago, Illinois.

λ = pinion spiral angle
f = coefficient of friction

The efficiency of Planoid gears compares favorably with that of hypoid gears. Since Planoids are usually designed for low ratio, they generally are quite efficient, being in the range of 96 percent or better. In reference to Fig. 12.13, Planoid gears fit nicely in the range of industrial and automotive gears from an efficiency point of view. Like hypoid gears, the efficiency of Planoid gears can be calculated using the coefficient-of-friction data of worm gears as outlined in Fig. 12.8.

12.5 GEAR TRAINS

To analyze a complete gear train, information pertaining to all the losses of a gearbox, not just those of the gear mesh itself, must be considered. As stated previously, windage and churning losses and bearing losses are very important factors and at times contribute considerably to the overall efficiency of a complete gear train. Even the gear arrangement itself, in some cases, must be carefully analyzed to be sure that a complete loss picture is obtained. For example, a simple planetary is inherently more efficient than a simple star arrangement, even though they are generally composed of very similar gear elements. In the final analysis, it is the load transmitted and the meshing velocity, among other things, that determine the amount of power being dissipated in the mesh. Sometimes both the load and the meshing velocities are difficult to determine in complex gear arrangements.

12.5.1 Bearing Losses

For gear elements to transmit power, they must be supported in the gear casing by some sort of bearing that will permit free and accurate rotation with a minimum amount of heat generation. Generally, of course, the bearings are lubricated with an ample supply of oil, both to minimize the coefficient of friction in the bearing and to act as a cooling medium to carry away the heat that has been generated.

These bearings usually fall into two main categories; namely, rolling contact bearings such as ball and roller bearings, and hydrodynamic bearings commonly known as "sleeve"-type bearings. Usually, a given gearbox design is equipped with either one type of bearing or the other. In a few applications, however, a mixture of the two types have been used.

The horsepower loss of the hydrodynamic-type sleeve bearing can be calculated from the laws of viscosity since the power consumption of this type of bearing results from the shearing of the oil film.
If the shaft is assumed to be the center of the bearing, the losses can be calculated using the well-known Petroff equation[14] with reference to Fig. 12.14.

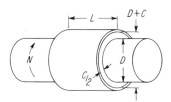

FIGURE 12.14 Elements of a sleeve bearing.

$$P_b = 3.79 \times 10^{-13} \frac{Z_2 \, n^2 \, D^3 L}{C} \qquad (12.38)$$

where P_b = horsepower loss in bearing
 Z_2 = outlet oil viscosity (absolute), cP
 n = r/min
 D = journal diameter, in.
 L = length of bearing, in.
 C = clearance diametral, in.

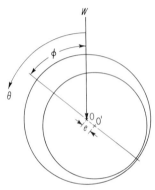

FIGURE 12.15 Definition of a loaded sleeve bearing.

This equation serves as a very good approximation for lightly loaded bearings since the shaft on lightly loaded bearings tends to ride in the center.

In practice, however, and particularly with bearings that carry an appreciable load, the shaft usually runs eccentric in the bearing. In this case, the foregoing calculated horsepower loss (P_b) must be modified by a power loss coefficient that is a numerical representation of the bearing characteristics.

$$P_b' = JP_b \qquad (12.39)$$

where J is the power loss coefficient.

The power loss coefficient can be calculated [14] by solving the following equation with reference to Fig. 12.15:

$$J = \frac{p \, Ce \, \sin \phi \, 10^9}{(4.83 \, \pi^2 Z_2 n D^2)} + \frac{1}{(2\pi)} {}_0\!\int^{2\pi} \frac{BC}{(2Lh)} d\phi \qquad (12.40)$$

where C, Z_2, D, L and n have been previously defined, and
 p = load per unit projected area, lb/in^2
 e = eccentricity of journal in bearing, in.
 θ = angle to point on bearing surface measured from load line
 B = variable width of reduced bearing oil film, in.
 h = variable film thickness, in.
 ϕ = attitude angle between line of center and load line

The foregoing equation for J has been determined for many applications, and it is plotted in Fig. 12.16 for the split cylindrical bearing. Note from Fig. 12.16 that J has been plotted for various L/D ratios versus the Sommerfeld number S.

$$S = 2.42 \times 10^{-9} \times \left(\frac{D}{C}\right)^2 \times \frac{Z_2 n}{p} \qquad (12.41)$$

In order to determine J, it is a simple matter to calculate the Sommerfeld number and the L/D ratio, then go to Fig. 12.16 and "pick off" the proper value for J.

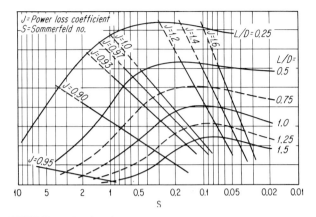

FIGURE 12.16 Plot of J split cylindrical bearings.

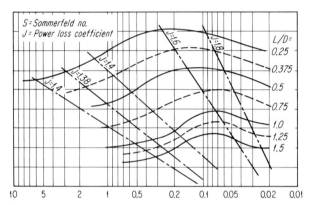

FIGURE 12.17 Plot of J elliptical bearings, ellipticity ratio of ½.

If the particular bearing under consideration is an elliptical-type sleeve bearing, then the J values found in Fig. 12.17 should be used. For applying these equations for an elliptical bearing design, the maximum clearance is used and the ratio of minimum clearance to maximum clearance is assumed ½.

Figure 12.18 can be used for determining the proper value of J for four axial-groove sleeve bearings. Again, Eqs. (12.38) and (12.39) can be used to calculate the horsepower loss.

For other journal bearing geometries, refer to CADENSE.[15]

The power loss of simple annular thrust bearings can also be determined in similar fashion by using the laws of viscous flow.

$$P_t = 3.79 \times 10^{-13} \frac{Z_2 n^2 (R_2^4 - R_1^4)}{t} \quad (12.42)$$

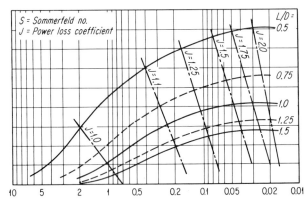

FIGURE 12.18 Plot of J four-groove bearings.

where P_t = horsepower loss of a simple thrust bearing
Z_2 = outlet viscosity, cP
t = constant oil film thickness, in.
R_1 = inside radius of thrust bearing, in.
R_2 = outside radius of thrust plate, in.

For other thrust bearing geometries refer to CADENSE 33 & 34.

Figure 12.19 gives the values of absolute viscosity in centipoises for various grades of petroleum oil versus temperature of the oil. This chart is quite useful when calculating bearing losses. Figure 12.20 gives similar information on other types and grades of oils, some of which are synthetics.

It is difficult to calculate bearing losses and have them coincide exactly with measured bearing losses. There are many sound reasons why variations between actual and predicted bearing losses exist. Some of these reasons are:

1. Bearing misalignment at assembly
2. Variation in viscosity-temperature behavior of the lubricating oil from that used in computation
3. Presence of phenomena such as oil turbulence or recirculation
4. Variation in operating variables, such as shaft speed and bearing loads, from those used in computations.

The losses of rolling-contact bearings depend to a large degree on the type and quantity of lubricant used in a given application. Exact predictions of these losses are extremely difficult. However, a considerable amount of experimental work has been done. Empirical equations were determined from these investigations. Like most experimental work, the results must be used with extreme care and understanding.

One of the simplest approaches to the bearing loss problem is to determine the torque loss in the bearing and then calculate the horsepower loss as follows:[14]

$$P_b = \frac{Tn}{63,000} \qquad (12.43)$$

LOADED GEARS IN ACTION

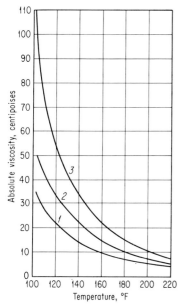

FIGURE 12.19 Viscosity of petroleum oil: (1) AGMA 1-200 SSU at 100°F, (2) AGMA 2-300 SSU at 100°F, (3) AGMA 3-600 SSU at 100°F.

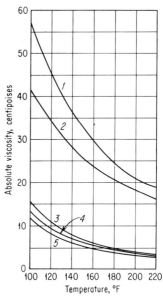

FIGURE 12.20 Viscosity of synthetic oils: (1) Dow Corning XF-258 Silicone, (2) GE Versalube F-50 (Silicone), (3) MIL-L-9236B (Celluthem 2505A), (4) MIL-7808D, (5) MIL-L-25336 (Sinclair L-743).

where P_b = horsepower loss in bearing
T = torque loss per bearing, lb-in
n = r/min of bearing

The torque loss T can be found simply by the use of a constant coefficient of friction for each type of bearing.

$$T = f \frac{D_1}{2} W \qquad (12.44)$$

where T = loss torque lb-in
D_1 = bore diameter of bearing, in.
W = load on bearing, lb

This approach offers a fast and reasonable approximation based on normal operating loads and speeds with favorable conditions.

The foregoing equation can be used for a variety of bearing types as long as the proper coefficients of friction are used[17] (see Table 12.3). It must be kept in mind, however, that, under starting conditions or if higher-viscosity lubricants are used, higher values of coefficient of friction will be obtained. Figure 12.21 can be used as a guide to determine the exact type of bearing used in a particular design.

TABLE 12.3

Type of bearing	Coefficient of friction f
Radial ball bearing (single-row deep groove)	0.0015
Self-aligning ball bearing	0.0010
Angular-contact bearing	0.0013
Thrust ball bearing	0.0013
Cylindrical roller bearing	0.0011
Spherical roller bearing	0.0018
Tapered roller bearing	0.0018

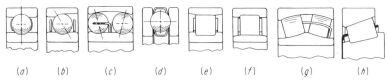

(a)　(b)　(c)　(d)　(e)　(f)　(g)　(h)

FIGURE 12.21 Types of bearings: (a) radial ball bearing, (b) angular-contact bearing, (c) self-aligning ball bearings, (d) thrust collar bearing, (e) cylindrical roller bearing (outside race free), (f) cylindrical roller bearing (inside race free), (g) spherical roller bearing, (h) tapered roller bearing.

Some gear designers prefer to determine the approximate torque loss of rolling-contact bearings by using the pitch diameter of the rolling elements for the appropriate diameter in the torque equation[18] or

$$T = f \frac{D_p}{2} W \qquad (12.45)$$

where D_p equals the pitch diameter of rolling elements.
This equation is perhaps more correct theoretically in that the rolling contact area is where the actual loss takes place, not in the bore of the bearing. This equation also permits some consideration for the differences between bearings of the same bore diameter but with larger balls or rollers. Of course, the larger ball or roller operates a larger pitch diameter. It is somewhat more difficult to obtain the operating pitch diameter of a rolling-contact bearing; however, it can be approximated by the following equation:

$$D_p = \frac{D_1 + D_2}{2} \qquad (12.46)$$

where D_p = pitch diameter of elements, approximately, in.
D_1 = bore diameter
D_2 = outside diameter

As an example of how the actual bearing loss varies for a spherical roller bearing, some #22308 bearings were tested by the author. Figure 12.22 shows these results. Note that the loss torque at slow speeds is quite high compared with that at the faster

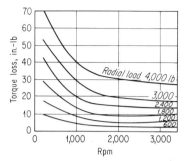

FIGURE 12.22 Losses of #22308 spherical roller bearings. Bearing torque loss versus r/min and load 40-mm bore × 33-mm width. Double-row spherical roller bearing. Light aircraft oil at 120°F inlet.

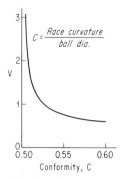

FIGURE 12.23 Plot of V versus conformity.

speeds. It is also quite interesting to note that, as the speed is increased, the losses in the bearing tend to increase, undoubtedly because of the churning effect of the oil.

A very complete work on friction torque of ball bearings was carried out by Manlio Muzzoli.[14] The equation developed for the friction torque of deep-groove ball bearings under radial load is as follows:

$$T = 0.0868KV\left[n^{0.5}(2.6d^{0.2} - 3.33)n^{0.375d-0.2}\right.$$

$$\left. + \frac{(0.454W)^m}{72n^{0.73} + 0.150N_b(d-6)^{0.055}}\right] \quad (12.47)$$

where d = ball diameter, mm
N_b = number of balls
n = r/min
W = load, lb
K = constant, lubricant
V = constant, conformity
m is determined by Eq. (12.48)

$$m = 1.65 + 0.004n^{0.054} + 0.005dN_b \quad (12.48)$$

The constant V can be evaluated by referring to Fig. 12.23. Note that V is plotted against conformity, which is the race curvature divided by ball diameter. Bearings are generally manufactured to a conformity of 52 percent, which results in a value of V equal to 1.1.

The constant K, which takes into consideration the type of lubricant, can be found in Table 12.4. Dr. Booser and Dr. Wilcock[14] state they have found that modern, good ball bearing greases will give a K value of 0.10 to 0.15, which is somewhat better than those given in Table 12.4. They also indicate, from the results of many of their tests, that the Muzzoli Eq. (12.47) is fundamentally correct.

TABLE 12.4

Type of Lubricant	K
Fluid grease	1.0
Dense grease	1.2
Very dense grease	2.0
Very light oil	0.4
Light oils	0.6

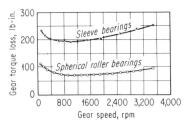

FIGURE 12.24 Comparison of gearbox torque losses (sleeve bearings versus rolling-contact bearings).

The curves as indicated in Fig. 12.24 give a good graphical comparison of the losses of a given gearbox when mounted first on sleeve-type bearings and then rolling-contact bearings. In both cases, the gearbox was loaded with exactly the same amount of torque. In many respects the curves are just what would be expected in that the sleeve bearing design has higher break-away torque. The losses of the sleeve bearing at speed are still higher in that a considerable amount of oil is required to keep the bearing functioning properly, and this generally results in a high viscosity-loss effect. However, at really high speeds and loads, these two curves tend to cross, thus making the sleeve bearing performance more comparable with the rolling-contact bearing.

12.5.2 Windage Losses

The windage losses of a gearbox are very difficult to measure or calculate with any degree of accuracy. Investigations have been carried out on this subject, but very little information has been published. In general terms, windage loss determination is still pretty much based on individual experience of the gear designer and experimental measurements on specific gearboxes in question.

Most of us that are acquainted with the operation of high-speed gears realize that, as the speed is increased, the losses from windage increase. From past test experience it has been determined that high-speed effects were not significant at pitch line velocities below 10,000 ft/min for well-designed gearboxes.

Actually the windage losses for a given gear design depend on many things such as:

1. The diameter of the rotating elements
2. The length of the rotating elements
3. The speed of rotation
4. The web or gear blank design
5. The overall casing design
6. The type of oil-feed system
7. The operating temperature and viscosity of the oil
8. The pressurization of the casing

An equation that has been used that gives an approximation of the windage losses of small-diameter gears, say up to approximately 20 in. in diameter with an L/D ratio of approximately 0.5 is as follows:

$$P_w = \frac{n^3 D^5 L^{0.7}}{100 \times 10^{15}} \qquad (12.49)$$

where P_w = horsepower loss due to windage
n = r/min

LOADED GEARS IN ACTION   12.25

D = diameter of rotating element
L = length of rotating element

Another method that may be used for windage loss approximations in horsepower is the following equation that was empirically derived for a rotating smooth body in air.

$$P_w = \frac{15}{0.746} \left(\frac{n}{1{,}000}\right)^3 \left(\frac{D}{100}\right)^4 \left(\frac{5L}{100} + \frac{D}{100}\right) \qquad (12.50)$$

The losses obtained by using this equation will undoubtedly give a minimum value of windage for a rotating body.
This equation can be broken down into its components to cover the sides of a gear blank that are smooth and the periphery of the gear blank that is also smooth before the teeth are cut. The equation for the periphery losses must be modified, however, to account for the gear teeth size and helix angle that have been machined into the gear blank.
Power loss for smooth surfaces on the sides of the gear and pinion:

$$P_w = \frac{15}{0.746} \left(\frac{N}{1000}\right)^3 \left(\frac{D}{100}\right)^4 \left(\frac{D}{100}\right) \qquad (12.51)$$

Power loss for the periphery of a gear assuming smooth surfaces, such as the gap between helices, etc.

$$P_w = \frac{15}{0.746} \left(\frac{N}{1000}\right)^3 \left(\frac{D}{100}\right)^4 \left(\frac{5L}{100}\right) \qquad (12.52)$$

Equation (12.52) must be adjusted, however, to take into consideration the modification to the smooth surface of the periphery when gear teeth are cut into the gear blank. When the gear tooth surfaces are considered the foregoing equation becomes

$$P_L = \frac{15}{0.746} \left(\frac{N}{1000}\right)^3 \left(\frac{D}{100}\right)^4 \times \left(\frac{5L}{100}\right) \left(\frac{R_f}{\sqrt{\tan \psi}}\right) \qquad (12.53)$$

R_f is a rough surface adjustment factor and is a function of the diametral pitch in the plane of rotation. Appropriate values are given in the following table. The helix angle term adjusts the losses of the rotating gear for differences in helix angle. Low-helix-angle gears produce high axial velocities.

For a double-helical gear one must solve Eq. (12.53) for the gear teeth, Eq. (12.52) for the gap between the gear teeth, and Eq. (12.51) for the sides of the gear. It is quite helpful to realize just when you should be concerned about windage losses, even if the methods of evaluating windage losses leave something to be desired. (See Table 12.5.)

TABLE 12.5 Rough Surface Adjustment Factors (R_f)

Transverse Diametral Pitch	R_f
4	7.2
6	6.7
10	6.1
16	5.0
24	3.8

Windage losses are often combined with other extraneous losses in the system. Stated in another manner, when the mesh losses and bearing losses are subtracted from the total gearbox losses, windage losses is the name given to the losses that remain.[19] Gear designers should concern themselves about windage losses when a design starts to suffer from any of the troubles indicated schematically in the following sketches, namely, Figs. 12.25, 12.27, 12.28, 12.29, 12.31, and 12.33.

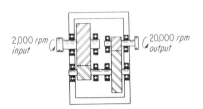

FIGURE 12.25 High pitch line velocity gearing.

Figure 12.25 indicates a high-speed gearbox with pitch line velocities higher than 10,000 ft/min. Generally, when the meshing velocity reaches 10,000 ft/min, the windage losses begin to increase to the point where an engineer should take them into consideration in the overall design.

As an indication of the effect windage losses may have on a high-speed train, refer to Fig. 12.26. This information was obtained experimentally from efficiency tests on a single-offset aircraft gearbox. It is quite interesting to note that in a well-designed gearbox little, if any, windage loss is present at the lower speeds.

Figure 12.27 indicates a tight gear casing arrangement whereby the gear casing fits tightly around the gearing. This is done because of weight and space considerations; however, it usually leads to windage troubles. Relief can often be found by shrouding the gears with oil shields. The oil-discharge port, in particular, must be protected from the "windage" within the casing in order to assure a free flow of oil from the casing.

Not only does this type of arrangement lead to excessive heat being generated in the oil because of churning, but sometimes troubles are encountered with the oil within the casing, resulting in a hydraulic braking action. If this trouble is encountered, the gear unit must be shut down immediately or the unit may suffer overheating and failure. In addition to shrouding or shielding, a unit of this nature should be scavenged with a pump that has from 2½ to 3 scavenge ratio.

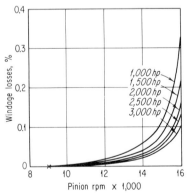

FIGURE 12.26 Windage losses—aircraft gear. Single Offset. MIL-L 7808D oil. 250°F inlet oil.

FIGURE 12.27 Tight gear casing arrangement.

Figure 12.28 illustrates a compact gear train arrangement. Usually a gear train of this sort is an accessory-drive package, whereby the gears themselves do not have excessively large face widths, but they generally are quite high-speed. Some of the gearing is not highly loaded in that often the center distance requirements dictate the size of the gearing, not the power being transmitted. Any free oil within the gearbox may be recirculated and battered about before it finally drains back to the main oil reservoir. In a gear design of this nature, it is the best policy to keep to an absolute minimum the oil flow to the gear mesh, as well as to the bearings. Generally, by carefully observing the early test results of a given design, oil can be redirected and redistributed to lubricate the complex gear train adequately without having to resort to shrouding. In some instances, however, a well-placed oil deflector can be quite beneficial. Again, removing the oil from the gear casing and preventing any buildup of surplus oil is a must; thus, a scavenge pump is usually required, but a 2:1 scavenge ratio generally is high enough.

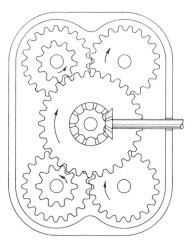

FIGURE 12.28 Compact gear train arrangement.

Figure 12.29 illustrates a poor or inadequate oil-scavenge arrangement. This type of trouble sometimes accompanies a tight gear casing arrangement or a compact gear train arrangement. It can occur in simple types of gearboxes such as the one illustrated in Fig. 12.29, primarily because of engineering oversight. When inadequate drainage and scavenge capacity are present, the gearbox will fill up with oil, causing extreme foaming and hydraulic braking. Shrouding the gears in areas where the oil has a tendency to pile up helps. The discharge ports must be shielded in order to permit a free flow of oil from the casing. Sometimes, if possible, oil strippers can be used to advantage (see Fig. 12.30). Of course, in any gear casing where excessive oil becomes a problem, it is always wise to cut down to the bare minimum the amount of oil that is initially supplied to the gear casing.

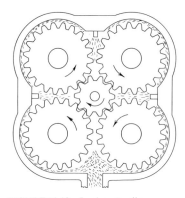

FIGURE 12.29 Inadequate oil scavenge.

Figure 12.31 shows a wide-face-width gear design. In this sketch, the L/D ratio of the gearing is quite high. Oil sometimes becomes trapped in the mesh, since the oil has a much greater distance to go before it can be expelled from the ends. Actually, there is very little that can be done to help a design of this nature, except to allow for plenty of room within the casing. If the gearing is double-helical, as most likely it would be with large-face-width gearing, the helix angle should be so arranged to force the oil toward the ends of the gear teeth. This arrangement would prevent the oil from being stacked up at the gap between helices. Considerable space should be left between the ends of the teeth and the gear casing to permit the free flow of oil to the drain. Bearing oil discharge should be channeled to the bottom of the gear casing.

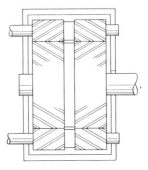

FIGURE 12.30 Oil strippers.

FIGURE 12.31 Wide-face-width gearing.

From a windage-loss point of view, the inlet oil temperature should be kept as high as is practical within good overall operational characteristics of both the oil and the equipment. An example of oil temperature effects on the no-load windage losses of a gearbox is given in Fig. 12.32. This figure shows the no-load losses of a given gearbox with respect to inlet oil temperature. These losses include churning losses of the bearings at no load and the churning losses of the gear mesh itself; the losses shown depict to a great extent the effect of oil temperature on overall windage losses.

Figure 12.33 shows a gearbox that is lubricated by the dip or splash feed system. Any time that a design of this nature is considered, the designer must be alert to the possibility of high losses due to churning of the oil. Not only must designers be aware of this problem, but they must also be aware of the tendency of the lubricant to channel, thus rendering the lubrication system ineffective. This arrangement is generally utilized for low-speed drives; in fact, many worm gear applications use this system.

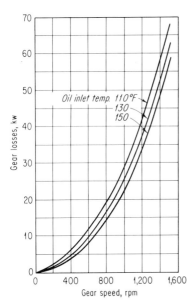

FIGURE 12.32 Oil temperature effects on windage losses. Oil inlet pressure 8 lb/in.^{2}, Navy 2190 T oil.

12.5.3 Single-Offset Gears

A single-reduction two-element gearbox is the simplest type of gear train, and consequently the determination of the overall efficiency of this arrangement is quite straightforward. First, the mesh loss is determined by solving Eq. (12.13) or (12.14) with the help of Fig. 14.1a or b. Second, the bearing losses are determined by solving Eq. (12.38) for sliding-contact bearings. This horsepower loss should then be converted to the percent loss basis, so that it may be added to the mesh loss. If rolling-contact bearings are present, then Eq. (12.43) should be used, and the loss torque determined should also be converted to a percent loss basis, and again added to the mesh loss. Third, the windage losses for the gearbox should be determined by solving Eqs. (12.49) or (12.50), (12.51), and (12.52), first for the pinion, and then for the gear. The results should be converted to a percent loss, then added to the total loss already accumulated.

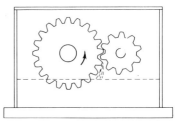

FIGURE 12.33 Dip- or splash-feed lubrication.

This new total represents the overall loss of the complete gear train. Caution should be exercised to assure that there are no bad seal rubs or oil trapping (see Figs. 12.27, 12.28, and 12.33). Factors of this nature cause additional losses and are not indicative of good design practice. Gear trains of this type are very efficient, particularly if high gear tooth loading is present. Figure 12.34 shows the results of recent tests on an aircraft single-offset gearbox with rolling-contact bearings. Note from Fig. 12.34 that the gearbox is of a relatively high-speed and high-load design.

12.5.4 Double-Reduction Gears

A double-reduction gear design is actually two single-offset gear trains in series. This being the case, the overall gearbox efficiency is determined in exactly the same way as for the single-offset gear train. Each mesh, each bearing, and each rotor is calculated for the appropriate loss. The individual losses are then totaled up and the overall gearbox efficiency is determined. Triple-reduction gear trains, etc., are also calculated in a similar manner, as well as any combination of spur, helical, bevel, or internal gears.

12.5.5 Simple Planetary Drive

A simple planetary gear system consists of three major parts: a fixed member, a driving member, and a driven member.[20] A common arrangement of this combination is shown

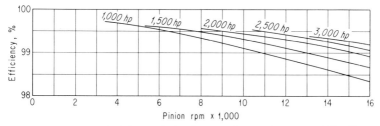

FIGURE 12.34 Overall efficiency—single offset gear. Constant horsepower; inlet-oil temperature 250°F; oil—MIL-L-7808C.

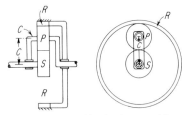

FIGURE 12.35 Simple planetary drive.

in Fig. 12.35. In this train of gears, R is the fixed member, S is the driving member, and C is the driven member. In order for the cage C to function, an idler P is required in the system. The efficiency of this drive depends on the tooth loads and the velocity of the tooth arrangements. The product of tooth load and engagement velocity gives a measure of the equivalent power being transmitted by the mesh. Generally, the equivalent power being transmitted is less than the actual power being transmitted.

Actual power transmitted can be determined as follows:

$$P_a = W_t \times v \qquad (12.54)$$

where W_t is the actual load being transmitted, and v is the pitch line velocity of actual transmitted load.

$$W_t = \frac{2T}{d_s} \qquad (12.55)$$

$$v = 0.2618 \times d_S \times n_S \qquad (12.56)$$

where T = torque transmitted
d_s = pitch diameter of sun pinion
n_s = r/min of sun pinion

The equivalent power being transmitted through the sun gear mesh is

$$P_s = W_t \times v_s \qquad (12.57)$$

where v_s is the pitch line velocity of tooth engagement in the sun pinion.

$$v_s = v + \frac{d_R}{(d_R + d_S)} \qquad (12.58)$$

where v = from Eq. (12.56)
d_R = pitch diameter of ring gear
d_s = pitch diameter of sun pinion

The equivalent power being transmitted through the ring gear mesh is

$$P_R = W_t \times v_R \qquad (12.59)$$

However $v_s = v_R$

Therefore $P_R = P_S$

The efficiency of the drive can be determined by calculating the percent loss for each mesh. The power loss for each mesh is the product of the percent loss and the equivalent

LOADED GEARS IN ACTION   12.31

power being transmitted. The overall efficiency is then found by dividing output power by the input power or

$$E = \frac{W_t \times v - \left(\dfrac{\text{percent}}{\text{loss sun}} \times W_t \times v_s + \dfrac{\text{percent}}{\text{loss ring}} \times W_t \times v_R\right)\dfrac{1}{100}}{W_t \times v} \quad (12.60)$$

Note: The percent loss of the sun mesh and the percent loss of the ring mesh can be determined by solving Eqs. (12.13) and (12.14), modified for internal gears.

To better understand the foregoing efficiency calculations, and with reference to Fig. 12.35, a sample problem is given to illustrate its usage.

Given:

10 hp
Input speed, 1000 r/min
10 diametral pitch
Sun pinion driving
$d_S = 2$ in. $d_R = 8$ in. $dp = 3$ in. $C = 3.5$ in.

Solution:

Reduction ratio $= 1 + d_R/d_S = 1 + 8/2 = 5$
r/min of driven member $= 1000/5 = 200$
Pitch line velocity of transmitted load

$$v = 0.2618 \times d_S \times n_S$$
$$= 0.2618 \times 2 \times 1000 = 523.6 \text{ ft/min}$$
$$W_t = 2 \times \frac{T}{d_S}$$
$$= 2 \times \frac{630}{2} = 630 \text{ lb}$$

Actual power transmitted

$$P_a = W_t \times v$$
$$= 630 \times 523.6 = 330{,}000 \text{ ft·lb per min}$$

Equivalent power transmitted by sun gear mesh

$$P_S = W_t \times v_s$$
$$v_s = v \times 8/(8+2) = d_R/(d_R + d_S)$$
$$= 523.6 \times 8/(8+2) = 418 \text{ ft/min}$$
$$P_S = 630 \times 418 = 263{,}000 \text{ ft·lb per min}$$

Equivalent power transmitted by ring gear mesh

$$P_R = W_t \times v_R$$
$$= W_t \times v_s = P_s$$
$$= 263{,}000 \text{ ft·lb min}$$

From Eq. (12.14) and (12.14) modified, the percent power loss for the sun gear mesh and the ring gear mesh have been determined to be 1.05 and 0.324, respectively. Then the efficiency of the overall gearbox is

$$\Sigma = \frac{330{,}000 - (1.05 \times 263{,}000 + 0.324 \times 263{,}000)\frac{1}{100}}{330{,}000}$$

$$= 98.9 \text{ percent}$$

It is quite evident, however, from studying the problem, that

$$\frac{P_R}{P_a} = \frac{P_S}{P_a} = \frac{(m_R - 1)}{m_R} \qquad (12.61)$$

where m_R is the overall ratio. Then Eq. (12.60) can be reduced to the following:

$$E = 100 - (\text{percent loss sun} + \text{percent loss ring})\frac{m_R - 1}{m_R} \qquad (12.62)$$

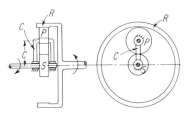

FIGURE 12.36 Simple star gear.

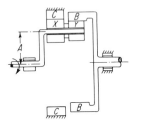

FIGURE 12.37 A simple four-gear differential.

12.5.6 Simple Star Gears

A star gear arrangement consists of the three basic parts of the simple planetary, except that the cage C is stationary and the ring gear is allowed to rotate (see Fig. 12-36). From this diagram it is obvious that the driven ring gear R will rotate in the opposite direction from the driver S. Actually, the star gear train is simply an internal gear train with an idler P, which changes the direction of rotation of the driven ring gear R. This being the case, the overall efficiency can be determined by first calculating the losses of mesh S to P and then the losses of mesh P to R. It must be remembered, however, that the bearing and windage losses of the idler P should be added into the total losses only once.

12.5.7 Differential Gears

A simple four-gear differential gear train is shown in Fig. 12.37. A is the arm through which the input torque must act, and gear B is the output gear. Generally, in practice, the fixed gear C and the rotating gear B are nearly equal in diameter, and this, of course, also applies to both pinions that must mesh with gears C and B. A differential gear train permits a very-high-speed reduction with only a few gears and in a minimum amount of space.[20] The amount of power that can be transmitted is somewhat limited by virtue of the high gear tooth loads and high engagement velocities of the mesh. In order to determine the efficiency of a differential drive, the gear tooth loads on both

LOADED GEARS IN ACTION   12.33

the fixed and rotating ring gear must be determined, as well as the meshing velocities of both members.

The power being transmitted can be calculated as follows:

$$P_a = W_t \times v \qquad (12.63)$$

where $W_t = T/A$ = load at center pinion, lb
$v = 0.2618 \times 2.1 \times n$ = velocity of center of pinion, ft/min
T = torque input, lb·in
A = torque arm, in.
n = r/min of input torque arm, A

The equivalent power on the fixed ring gear C can be determined as follows:

$$PC = v_C \times W_C \qquad (12.64)$$

where v_C = meshing velocity of ring gear C, ft/min
W_C = load applied to ring gear C, lb

$$v_C = V \frac{d_C}{d_C - d_X} \qquad (12.65)$$

$$W_C = W_t \times d_Y \frac{d_C - d_X}{d_C d_Y - d_B d_X} \qquad (12.66)$$

where d_X, d_B, d_C, and d_Y are pitch diameters of respective parts.

The equivalent power on the rotating ring gear B can be determined as follows:

$$P_B = v_B \times W_B \qquad (12.67)$$

where v_B = meshing velocity of ring gear B, ft/min
W_B = load applied to ring gear B, lb

$$v_B = v \frac{d_C d_Y}{d_X(d_B - d_Y)} \quad \text{ft/min} \qquad (12.68)$$

$$W_B = W_t d_X \frac{d_B d_Y}{d_C d_Y - d_B d_X} \quad \text{lb} \qquad (12.69)$$

The efficiency of the drive can be determined by calculating the percent loss for each mesh. The power loss for each mesh is the product of the percent loss and the equivalent power being transmitted. The overall efficiency is found by dividing output power by the input power or

$$E = \frac{W_t \times v - \left(\text{percent loss } C \times W_C \times v_C + \text{percent loss } B \; W_B v_B\right)\frac{1}{100}}{W \times v} \qquad (12.70)$$

Note: The percent loss of mesh C and the percent loss of mesh B can be calculated by using Eqs. (12.13) and (12.14).

Sample Problem. To illustrate the use of the foregoing equations for a given four-element differential, in reference to Fig. 12.37, the following information is given:

Output torque = 840,000 lb·in
Pitch diameter of fixed ring d_C = 31.2 in.
Pitch diameter of pinion d_X = 8.8 in.
Pitch diameter of rotating ring d_B = 29.2 in.
Pitch diameter of pinion d_Y = 7.6 in.
Diametral pitch = 2½
Output r/min = 1 r/min = n
Diameter of the torque arm A = 11.2

Solution. The overall ratio is found by solving the following

$$m_R = \frac{1}{1 - \left(\dfrac{d_C d_Y}{d_B d_X}\right)} \tag{12.71}$$

Since the output speed is equal to 1, this is also equal to the input speed of torque arm A.

Velocity of the center of the pinion, ft/min:

$$v = 0.2618 \times 24 \times n \tag{12.72}$$
$$= 0.2618 \times 2 \times 11.2 \times 12.9515 = 75.937 \text{ ft/min}$$

Load at the center of the pinions:

$$W_t = \frac{T}{A}$$
$$= \frac{840,000/12.9515}{11.2} = 5791 \text{ lb} \tag{12.73}$$

Input power—or power being transmitted:

$$P_a = W_t \times v$$
$$= 5791 \times 75.937 = 439,753 \text{ ft·lb per min} \tag{12.74}$$

Meshing velocity of fixed ring gear C:

$$v_C = v \frac{d_C}{d_C - d_x}$$
$$= 75.937 \frac{31.2}{31.2 - 8.8} = 105.769 \text{ ft/min} \tag{12.75}$$

Load applied to ring gear C:

$$W_C = W_t d_Y \frac{d_C - d_X}{D_c d_Y - d_B d_X}$$

$$= 5791 \times 7.6 \frac{31.2 - 8.8}{31.2 \times 7.6 - 29.2 \times 8.8} = 49{,}691 \text{ lb} \qquad (12.76)$$

The equivalent power on the fixed ring gear C:

$$P_a = W_C v_C$$

$$P_a = 105.769 \times 49{,}691 = 5{,}256{,}000 \text{ ft·lb per min} \qquad (12.77)$$

Meshing velocity of rotating ring gear B:

$$v_B = v \frac{d_C d_Y}{d_X(d_B - d_Y)}$$

$$= 75.937 \frac{31.2 \times 7.6}{8.8(29.2 - 7.6)} = 94.729 \text{ ft/min} \qquad (12.78)$$

Load applied to rotating ring gear B:

$$W_B = W_t \times d_X \frac{d_B - d_Y}{d_C d_Y - d_B d_X}$$

$$= 5791 \times 8.8 \frac{29.2 - 7.6}{31.2 \times 7.6 - 29.2 \times 8.8} = 55{,}481 \text{ lb} \qquad (12.79)$$

The equivalent power on rotating ring gear B:

$$P_B = W_b \times v_B$$

$$= 94.729 \times 55{,}481 = 5{,}256{,}000 \qquad (12.67)$$

From Eq. (12.14) modified, the percent power loss for ring mesh C and for ring gear mesh B have been calculated to be 0.503 and 0.601 percent, respectively. The efficiency for the overall gear train is

$$E = \frac{439{,}753 - (0.503 \times 5{,}256{,}000 + 0.601 \times 5{,}256{,}000)^{1/100}}{439{,}753}$$

$$= 86.8 \text{ percent}$$

It is quite evident, however, from studying the problem that

$$\frac{P_C}{P_a} = \frac{P_B}{P_a} = \frac{m_R - 1}{1} \qquad (12.80)$$

where M_R is the overall differential ratio. Equation (12.70) can be reduced to the following:

$$E = 100 - \left(\text{percent loss}_c + \text{percent loss}_B\right)(m_R - 1) \qquad (12.81)$$

An equation of this nature is quite helpful in determining the overall efficiency of a given design. It can also be used as a check on the more detailed calculations that must be carried out when a device such as this is designed for a particular application.

Often on a design problem of this nature, a quick approximation of the efficiency, which could be determined quite quickly on a slide rule, is good enough. An approximate equation for the efficiency of a four-element differential is as shown:[21]

$$E = \frac{1}{1 + \frac{1}{5}\left[\left(\frac{1}{N_X}\right) + \left(\frac{1}{N_Y}\right) - \left(\frac{1}{N_C}\right) - \left(\frac{1}{N_B}\right)\right](m_R - 1)} \quad (12.82)$$

where R = overall ratio of train
N_X = number of teeth on gear X
N_Y = number of teeth on gear Y
N_C = number of teeth on gear C
N_B = number of teeth on gear B

The efficiency calculated in this manner for the aforementioned sample problem would give an approximate efficiency of

$$E = \frac{1}{1 + \frac{1}{5}\left(\frac{1}{22} + \frac{1}{19} - \frac{1}{78} - \frac{1}{73}\right)(12.9515 - 1)}$$

$= 85.4$ percent approximately

This approximation, when compared with the efficiency obtained from Eq. (12.70) or (12.81), shows good correlations.

12.6 GEAR PERFORMANCE

To be considered satisfactory, gear drives must meet a standard of performance as interpreted by the ultimate customer. Actually, the desired performance of a given gear drive depends to a large measure on the end use of the apparatus. To marine engineers, good performance gear means extreme reliability. They envision the difficulty of cutting through the decks and compartments of ships to replace faulty gear equipment. To the navy, in many applications a good performance gear should be ultraquiet. This would give a measure of protection against detection by the enemy. Aircraft designers want a high horsepower-to-weight ratio to give them a larger payload, yet they also want reliability to prevent catastrophic failures. Controls engineers want extreme accuracy in transmitted motion, since a few thousandths in backlash may cause a malfunction. A petrochemical plant requires long-life, trouble-free equipment since gear drives are an integral part of complex processes.

LOADED GEARS IN ACTION   12.37

TABLE 12.6 Relative Performance of Gear Drives

Type of Gear Transmission	Relative Weight Per Total Job Done	Hp Per lb	Input Speed	Overall Ratio
Marine	105	0.075	6,500	64
Industrial	160	1.0	8,000	4
Industrial (lightweight)	32	2.0	19,000	16
Railway	74	0.875	2,100	4.1
Aircraft (power)	16	12–18	17,000	13
Rockets and missile	12	25–50	30,000	5

12.7 FACTORS THAT AFFECT THE PERFORMANCE OF GEAR UNITS

A good-performance gear should be described as one that operates as well as or even better than the prediction of the design engineer. From an engineering point of view, however, there are several basic fundamental criteria for judging equipment of this nature, such as the efficiency of the geared drive, the reliability as compared with the design life, or the size and weight of the drive versus the horsepower transmitted, and the overall job that the gear must do. Table 12.6 shows an interesting and revealing study of various types of gear drives. It must be remembered, however, that this table can be used only as a rough guide for ball-park comparisons.

Perhaps the ultimate in gear performance should be synonymous with the ultimate in gear capacity. The best-performing gear drive would then be the drive that could carry the most horsepower with the least weight. This being the case, in reference to Table 12.6, the aircraft and rocket geared systems would be way out front. There are several factors other than the transmitted load and the speed of operation that affect the overall performance of gear applications.

12.7.1 Power Loss

The amount of heat being dissipated within the gearbox is a good indication of how the overall gearbox is operating. This known heat loss can then be compared with the calculated power loss of the unit under consideration.

The calculated power loss should be greater than the measured power loss. If this is not the case, then the gear unit should be investigated. Often when there is an excessive heat loss within the gearbox, the unit is not scavenging properly, and the design may be of the questionable type, as indicated by Figs. 12.27, 12.28, and 12.29. Oil* that is not immediately removed from high-speed gear units will cause considerable heat loss because of churning of the excess oil. In many cases, improper scavenging will cause a hydraulic braking effect, which may overload the gear unit to the point of failure. At times, hot bearings will cause an increase in the temperature differential across the gear casing, and often this is an indication of incipient failure. When a new

---
*See Chap. 15 for detail recommendations on gear lubrication systems.

gear unit is first started up, or brought on line, it is difficult to know whether the bearings are running hot or not, unless the unit is fully instrumented. If the gear unit has been previously run, then it may be easier to determine when the temperature differential increases.

12.7.2 Vibrations

In order for a gearbox to operate satisfactorily, it must run within safe vibration limits. Normally, if the vibrations of a given gearbox run more than 2 to 3 mils peak to peak under steady-state conditions, you should suspect that something is wrong. Several things can cause undesirable vibrations in a gearbox. Dynamic balance of the rotating elements would be first on the list of suspects. All rotating elements should be checked for balanced as soon as trouble is encountered. Next, consideration should be given to connecting bodies, such as couplings and shafting. All shafting should be rechecked for critical-speed indications; this includes the gearing within the gearbox as well as the outside shafting. Often a flexible coupling will stick in an unbalanced position and then cause considerable vibration when the unit is up to speed. Runout on all bearing surfaces and the pitch line of the gears should be checked to see if they are within drawing limits. Sometimes the face of the coupling flanges will be out of square with respect to the rotating axis, causing bolted-on equipment to run somewhat out of balance. Profile errors of the gear tooth elements often can cause the mesh to run rough, which in turn can set up bad vibrations. Sometimes a bad rolling-contact bearing which has been damaged at assembly can cause a gear unit to vibrate and, of course, this usually leads to premature failures. Another possible cause of rough running is a bent or damaged gear tooth. This sometimes happens at assembly and it is usually quite difficult to detect unless the gears are pin-checked.

12.7.3 Alignment and Mounting* Improper alignment can affect the operation of a gearbox in several ways.[22] For example, if the external alignment of the gearbox to the connected load is not made carefully, the external coupling may fail because of excessive heat generation in the coupling engagement. If this happens, the coupling usually becomes quite "sticky" and then transmits bending moments back into the gearbox. If the unit continues to run, the shaft extension to which the coupling is attached often fails back near the first firm support area, which is usually the bearing just within the gearbox. Shafts have been known to fail in this manner where the shaft-extension loading would normally be considered quite conservative. Internal alignment is also very important. The gear mounting arrangement plays an important part in determining the degree of alignment that will ultimately exist within the gearbox. Sometimes at manufacturing, the casing bores are not machined true with respect to each other or parallel with respect to the set of bores in which the mating gear must run. This obvious mismatch of bores causes alignment problems in the mesh, which are compounded by additional misalignment due to internal clearance in the bearings. The alignment of the mesh is also affected by shaft deflections and gear blank deflections, as well as by casing distortions. If the pinion is wide-faced with high length-to-diameter ratio, torsional windup of the pinion can also contribute to the alignment condition of the mesh. It makes little difference what causes the misalignment in the mesh. The net result is a load concentration on one or the other end of the mesh, which usually results in premature failures of the gear or pinion teeth by pitting, scoring, breakage, or a

---

*See Chap. 10 for detail recommendations on gear mounting practice.

LOADED GEARS IN ACTION   12.39

combination of all three. The misalignment conditions within a gearbox must be eliminated if at all possible. If elimination is not possible, it must be corrected for by such means as a crowning, dubbing off the ends of the teeth, or helix-angle correction. If the mismatch is such that the ends of the teeth are being forced into tight mesh, perhaps additional backlash is required as well as some generous profile modification.

12.7.4 Lubrication

For a gearbox to operate at all, it must be supplied with an adequate lubricant. The lubricant permits the bearings, as well as the gear mesh, to operate with a minimum of power loss.[23] In addition to the job of lubricating these rubbing surfaces to cut down on friction, the lubricant usually acts as a cooling medium to keep the rubbing parts within satisfactory temperature limits.[24] In high-load, high-speed applications, the lubricant may also act as a carrying medium for antiwear and antiscoring agents. On wide-face-width gearing, it is very important that the distribution of the oil across the mesh be uniform to eliminate any uneven heating effects. The lubricant also helps in eliminating temperature differential throughout the gear casing assembly. Oil tends to wash out high-temperature areas and heat up low-temperature areas.

Normally, oil is thought of as a necessary and beneficial fluid; however, at times it can be very troublesome and detrimental. If oil is allowed to fill up within the gearbox, it will cause excessive power loss and ultimate failure due to the braking effect. Because oil is a fluid, it can cause undue concern with respect to seal leakage and proper ventilation of the gearbox. If the complete lubrication system has not been carefully designed with due regard to oil turnover and proper settling-out time, considerable trouble can be encountered with respect to oil foaming. Some antifoaming agents can be utilized in the oil; however, a healthy solution to the problem is to allow plenty of oil for the overall gear box with adequate scavenge capacity and to allow the foam to settle out in the storage tank. Oil foaming can also be a problem at reduced pressures.[25] This trouble is encountered particularly at high altitudes and it makes the use of a circulation oil system very difficult (see Fig. 12.38).[25] At high-altitude operation where a recirculation system is required, design engineers have had to pressurize the complete lube system. Because the gearbox depends on the lube oil for its very existence, the supply of oil must not be interrupted. A catastrophic gear failure may be traced back to a lube pump failure or a lube line break or even to plain carelessness on the part of an operator in failing to turn on the lube oil system at the proper time.

12.7.5 Environmental Conditions

Gears must operate in many different environments and under various adverse conditions. Gears have been designed to operate under water, with water lubrication, with poor lubrication or no lubrication, in high-temperature fields,[26] in low-pressure atmospheres,[25] in contaminated atmospheres of all kinds, and under a variety of high-speed and high-load conditions—to name a few situations. The environments in which gears must operate are important, not only from the point of view of the harm that the environment does to the gearbox, but also from the viewpoint of what the gear may do to the specific areas or atmosphere in which it is working. For example, a gear must be extremely quiet if it is located in an area where people must rest or sleep. A gear that may contaminate food will be required to operate without lubricants. Radio waves emitted by gears may interfere with various control equipment. Oil vapors escaping

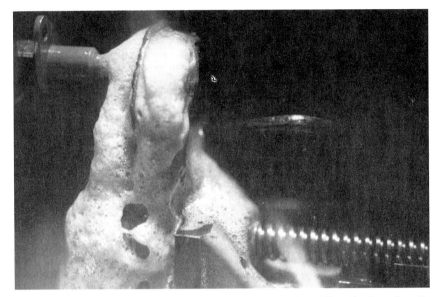

FIGURE 12.38 Oil foaming from a nozzle. *(From Hartman, M. A., "Gears for Outer Space," Machinery, August 1959, pp. 115–121.)*

from geared equipment may be objectionable to personnel who must work in the immediate vicinity.

It is not the purpose of the author to go into the details of how to cope with these many environmental conditions; however, design engineers must be aware of these conditions since, in most cases, they will affect the basic concept of whole gear designs.

12.8 GEAR FAILURES

Gears fail in service from such causes as pitting of gear teeth, breakage of the gear teeth, long-range wear, plastic deformation, scoring, and other, less common types of destructive wear. The AGMA national standard ANSI/AGMA 110.04,1980 "Nomenclature of Gear-Tooth Failure Modes," breaks down all types of gear failures into five general classes:[27]

1. Wear
2. Plastic flow
3. Surface fatigue
4. Breakage
5. Associated gear failures

Each class is then divided into several forms and types, with definitions. It is highly recommended that design engineers obtain copies of this standard for handy reference. A few photographs from this standard are used in this text.

A good gear designer needs to know how to recognize and analyze gear failures. In order for past experience to be of any benefit, failures must be properly investigated and appropriate corrective action taken.

Figure 12.39 shows a general plot of pitch line velocity versus torque capacity of a gearset.[1] The various areas indicate the type and regions where gear failures are most likely to occur. Areas 1, 2, and 3 are the failure regions associated with the wear phenomenon. In area 1, the gear is not running fast enough to develop an oil film. In area 2, the speed is fast enough to develop a film, and the gear will run for an indefinite period provided the lubricant is free of foreign material and is noncorrosive. In area 3, rapid wear or scoring will take place since the load and speed are high enough to break down the existing oil film. Ideally, of course, gear designs should be such that they operate wholly within area 2.

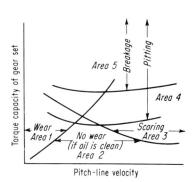

FIGURE 12.39 Regions of gear failures.

12.8.1 Wear

Wear is a very important phenomenon, particularly in high-speed gearing that must be in operation for an unlimited length of time. Wear has been defined from a gear engineer's point of view as the kind of tooth damage whereby layers of metal have been more or less uniformly removed from the surface.[1] The most common causes of gear tooth wear are metal-to-metal contact due to an inadequate oil film; abrasive particles in the oil supply; failure of the oil film in the contact area, causing rapid wear or scoring; and chemical wear due to the composition of the oil and its additives.[24]

It is a general practice in the gear trade, and a good one, to break in gearboxes and gradually build up to full load and full speed over a period of time. This procedure permits the contacting surfaces to wear off the surface asperities in a slow and controlled manner. Polishing in many cases is a very fine abrasive wear process. The amount of metal permitted to be removed by polishing should be kept well within the design life of the equipment. Figure 12.40 shows an excellent example of a well-polished spiral bevel gear. This gear has seen many cycles of operation, and, although some metal has been removed, the gear tooth surfaces are in excellent condition. Polishing wear of this nature is considered to be very fine abrasive wear, sometimes assisted by chemically active EP additives in the lubricant. Some polishing may also occur as a result of fine adhesive wear.

FIGURE 12.40 Polishing in spiral bevels.

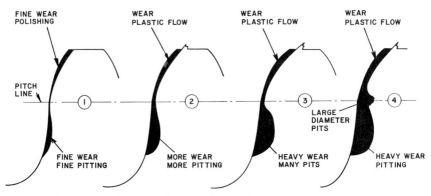

FIGURE 12.41 Progressive stages of wear (medium-hard gear).

FIGURE 12.42 Moderate wear.

The various stages of wear on gear teeth are difficult to document by photographs since the different stages have a similar appearance.

The sketches of gear teeth in Fig. 12.41 show the progressive stages of wear as they may take place on machinable-hardness gear teeth. Note in the first stage that wear and polishing are in progress, but there has been very little metal removal in the area near the pitch line. As more wear develops on the tooth surfaces, the pitch line area still shows much less wear. In the final stages, the pitch line area will carry more of the transmitted load, and pitch line pitting often results. In many applications the heavy plastic flow and feather edging in Fig. 12.41 may be minimal.

Figure 12.42 shows a helical gear with a moderate amount of wear. The pitch line is clearly visible, indicating some high or unworn metal. Unless the transmitted load is reduced or the lubrication is improved considerably, the gear will ultimately fail from pitch line pitting similar to that shown in the final stage of Fig. 12.41. Wear of this nature is often the combination of adhesive and fine abrasive wear.

12.8.2 Abrasive Wear

Abrasive wear occurs when foreign material contaminates the gear unit and lubrication system. The contamination can be generated in a variety of ways, such as machining chips, grinding residue, scale from piping, grit from cleaning processes, and a variety

LOADED GEARS IN ACTION   12.43

of other sources. A common source of abrasive wear particles is the wear debris from wearing gear tooth surfaces that often remains in the lube system.

Abrasive wear can be of considerable concern to the successful operation of a gear train. It is only logical that the abrasive particles must be kept from the lubrication system; however, this may be a difficult task to perform. Abrasive wear can be self-perpetuating in that the wear particles act as lapping compounds that tend to increase the rate of metal removal from the contacting surfaces. Closed gear units without provisions for filtering are particularly prone to abrasive wear problems. Wear particles may also pass through the filter elements. This depends, of course, on the micron size of the filter elements. Due to poor maintenance practices, the filter element may also fill up and bypass or even burst, thus contaminating the system. Many times hard wear particles may embed themselves into the wear surface of the softer member of a gear train. This embedded material will then perform as a fine abrasive hone and progressively wear away the surface of the harder gear member. A gear combination that is particularly dangerous is a nitrided pinion (with "white layer") operating against a medium-hard gear. After a time the abrasive components of the white layer material will wear away the softer gear surfaces. There is a risk, however, of loose particles of the abrasive white layer material embedding in the softer gear tooth surfaces and then wearing away the harder, nitrided pinion surface.

Another abrasive wear situation that must be avoided is the practice of hardening gear tooth surfaces that have poor surface finish. Surfaces of this nature are often induction hardened or nitrided without subsequent grinding. The rough, hard surface rubbing on a smoother, softer surface can remove considerable material in a short time.

Figure 12.43 shows a spur gear with an extreme case of abrasive wear. The oil in

FIGURE 12.43 Abrasive wear.

the closed lube system was not changed often enough, and the wear was permitted to continue. Tooth surfaces show deep radial grooves in the direction of sliding.

12.8.3 Corrosive Wear

Chemical corrosion and corrosive wear to gear units often result from the contamination of the lubrication system. Such common materials as water, salt, solvents, degreasers, and flushing compounds may lead to the contamination and corrosion of gear components. The normal ingestion of vapors from frequent start-ups and cooldowns increases the water content in gear lubrication systems. Corrosion such as rusting most often occurs in this manner. Corrosion may also result from reactions between moisture and chemicals in the lubricating oil and the base material of the components in question. Many extreme-pressure additives have chemicals such as chlorine, which may be very corrosive under certain operating conditions. Some processes in chemical plants may emit vapors or particles that can attack or combine with certain materials that are common to gear units, which may cause a chemical reaction resulting in corrosive wear.

12.8.4 Scuffing (Adhesive Wear, Scoring)

Scuffing is very rapid wear of the tooth surfaces initiated from a local failure of the lubricating film due to overheating, which permits metal-to-metal contact and adhesive wear in the form of a welding and tearing action. This removes metal rapidly from the contacting tooth surfaces. Although scuffing is usually initiated by a local failure of the lubricating oil film, there are many factors other than the oil that influence the final scuffing resistance of the gear mesh. Some of these factors are tooth surface pressure, properties of the material, surface finish, surface treatments, and surface rubbing velocities. Although scuffing failures are quite easy to recognize, they are quite hard to analyze. Scuffing can be broken down into several categories, which often helps to ferret out the real cause of the failure. The various degrees of scuffing are as follows: *initial or slight scuffing, moderate scuffing, severe scuffing, misalignment scuffing, scuffing due to local load concentrations.* [27]

Initial or light scuffing usually tends to heal itself unless the load or speed of the apparatus is increased. Many times scuffing of this nature occurs if the gear unit has not been sufficiently run-in.

Figure 12.44 shows a typical case of moderate scuffing.[27] Note that there are many areas on the tooth surfaces that have not been detrimentally affected. A gear of this nature could be put back into service if positive steps were made to improve the scoring conditions, such as reducing the transmitted load slightly, increasing the viscosity of the lubricant, decreasing the speed of operation, adding antiscoring additives to the lubricant, or reducing the inlet oil temperature.

FIGURE 12.44 Moderate scuffing.

LOADED GEARS IN ACTION   12.45

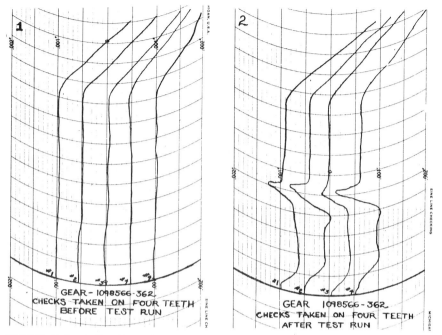

FIGURE 12.45 Scuffing failure—involute charts.

Figure 12.45 shows a chart of the involute profile of a gear before and after a scuffing failure. The amount of profile destruction is quite evident. The gear was 35 Rc hardness level. Gear tooth surface material has been scooped out of the dedendum area and built up along the pitch line of the gear teeth. This scuffing failure appeared slightly worse than that in Fig. 12.44, but not anywhere near so severe as that in Fig. 12.46, which depicts a severe case of scuffing. Observe from Fig. 12.46 the welding and tearing action that has been in progress, particularly the metal that has been dragged radially over the tips of the gear.

The transmitted load may not be high enough to cause general scuffing across the mesh; however, if the gears become internally misaligned because of deflections or manufacturing errors, the load may concentrate near the ends of the gear teeth. This concentrated load may now be high enough to break down the oil film and permit the ends of the teeth to scuff (misalignment-type scuffing). Figure 12.47 is an example of this type of failure. A failure of this nature may be helpful in that the scuffing usually shows up early, often on the test stand, indicating a misaligned condition. The engineer, realizing this condition exists, can do something about it before the gear unit is put into service. If the misaligned condition goes undetected, trouble usually shows up many hours later, maybe in the form of broken ends of the gear teeth or, at best, heavy pitting on the ends of the gear teeth.

Generally, misaligned conditions are corrected by helix-angle modifications, by relieving the ends of the teeth from contact, or, if conditions warrant, by putting a slight crown on the tooth surface in the axial direction. Crowning must be done with extreme care and careful analysis since crowning, in effect, concentrates the load near the center of the gear teeth. Crowning can cause a load-concentration failure, as indicated in Fig. 12.48.

FIGURE 12.46 Severe scuffing.

FIGURE 12.47 Misalignment scuffing.

FIGURE 12.48 Localized scuffing due to crown on pinion.

12.8.5 Plastic Flow

Gears fail from plastic flow when the contacting surfaces yield and deform under heavy load. This is the result of the rolling and sliding action of the mesh under high contact stresses. Usually cold flow occurs on soft and medium-hard material, but it can happen on hardened gear tooth surfaces as well. Generally this plastic flow is characterized by finned material overhanging the tips and ends of gear teeth as indicated in Fig. 12.46.

Failures of this nature can be reduced by lowering the applied load and by increasing the hardness of the contacting members. Increased accuracy helps by permitting the gear system to run without vibrations and dynamic loads. The hardened pinion shown in Fig. 12.49 exhibits evidence of plastic flow over the tip of the teeth and over the ends

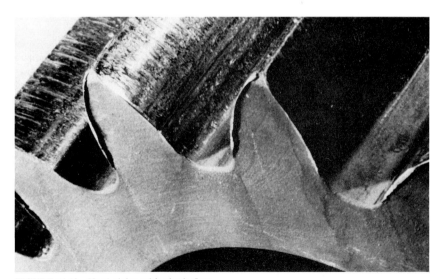

FIGURE 12.49 Hard pinion with plastic flow.

of the teeth. Hardened pinions generally require a heavy applied load to create the metal flow condition.

12.8.6 Rippling

Rippling is a failure mode associated with plastic flow. Generally, rippling occurs as a wavy formation on the contacting surface more or less at 90° to the sliding motion of the gear mesh, which may or may not lead to ultimate failure of the surfaces. Rippling is caused by high stress levels that tend to deform the surface layer of the material. Figure 12.50 shows an example of rippling on a hardened spur gear.

Rippling also appears to be associated with highly loaded gear surfaces operating at low speeds in a mixed lubrication regime. The rolling and sliding action under stress tends to foster periodic flow of the surface layer material.

Rippling can be helped by hardening the gear material if the gears are soft. Reducing contact stresses and increasing the oil viscosity are very helpful in preventing rippling. Improved surface finish and the use of oil and additives that reduce the tractive forces in the gear mesh are beneficial.

12.8.7 Ridging

Ridging is a form of plastic flow and wear characterized by the formation of ridgelike furrows in the direction of sliding along the tooth working surface. Ridging is often found on worms and worm gears and hypoid gears and pinions where the contact zone is subjected to high relative sliding velocities. The plastic flow is also associated with surface wear and polishing. Over a period of time, ridgelike markings show up and develop into a series of furrows across the tooth, as shown in Fig. 12.51.

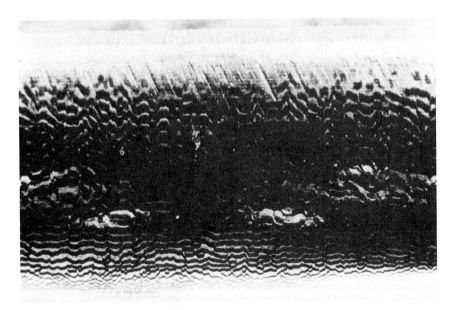

FIGURE 12.50 Rippling on spur pinion.

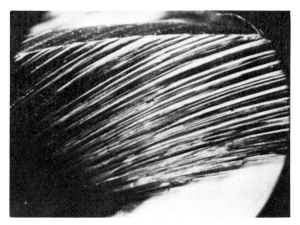

FIGURE 12.51 Ridging. (Courtesy of Socony Mobil Co., New York)

Ridging could be prevented by reducing the surface loading, as well as the relative sliding velocities of the contacting members. Improved lubrication conditions help, such as increasing the viscosity of the lubricant or by the use of improved extreme pressure agents. It is very important to keep the lubricant clean and free of fine abrasive wear particles. In some applications, the use of synthetic lubricants reduces the temperature developed in the mesh from high sliding under load, thereby maintaining a more consistent oil film.

12.8.8 Pitting (Surface Fatigue)

Surface fatigue, or pitting as it is commonly called, is a fatigue failure due to exceeding the surface capacity (or the endurance limit) of the gear material. Gears under load and in operation produce cyclic surface and subsurface stressers. If the loads are high enough and the stress cycles are repeated often enough, small pieces of material will fatigue from the surface, producing pits or cavities on the contacting surfaces.

Based on the severity of the surface damage, pitting is often categorized in three different stages: *(1) initial pitting, (2) destructive pitting, and (3) spalling.*[27]

1. Generally, pits of the initial-pitting type are quite small in diameter in the range of $\frac{1}{64}$ in. to $\frac{1}{32}$ in. in diameter. They occur in localized overstressed areas and tend to redistribute the load by progressively removing high contact spots. When the load is more evenly distributed, the pitting action is reduced and eventually goes into remission. Initial pitting of this nature is often referred to as *corrective* pitting.

2. Destructive pitting is more severe, and the pits are almost always larger in size. Destructive pitting generally occurs when the stress level is high compared with the capacity of the material (endurance limit). Figure 12.52 shows a typical pinion with destructive pitting. Pitting of this nature is almost always fatal in that the profile is destroyed beyond hope of recovery and, if the transmitted load remains the same, the pitting will get progressively worse until the gear unit runs so rough it must be taken out of service or the gear teeth crack and break off.

FIGURE 12.52 Destructive pitting.

3. Spalling is similar to destructive pitting except the term is commonly used when the pits are large in diameter and cover a considerable area. The spiral bevel pinion shown in Fig. 12.53 is an excellent example of spalling. Sometimes spalling occurs when the pits originate in case-hardened pinions and gears, for example, and fatigue away large pieces of surface material. (This should not be confused with case crushing.) Spalling can also be formed when destructive pitting voids break into each other, producing large-diameter, irregular cavities.

When gears fail from destructive pitting or spalling, the gear designer must do a considerable amount of investigation. Unless the gear mesh shows evidence of heavy misalignment, which usually can be corrected, destructive pitting and spalling is evidence that the design may not have the capacity* to carry the transmitted load.

If this is the case, then a redesign must be done to improve the load-carrying capacity. Medium-hardened gears can be fully hardened. This will increase the endurance limit of the surface material, perhaps high enough to prevent pitting. A substitutive gear material can be used, such as a high-capacity nitrided material or maybe a case-carburized material.† Either of these materials is excellent for increased pitting capacity.

Often changes in materials and heat treatments are not enough and a redesign of the gears must be made. Increasing the face width of the gear will reduce the load per inch of face and increase pitting resistance. The center distance of the gearbox could also be increased. This would reduce the transmitted load, with the net result being an improvement in surface pitting.

The classical type of pitting originates at subsurface nucleation sites where the subsurface shear stresses are highest.[28] A subsurface crack is initiated a few thousandths of an inch below the surface, and then the crack system works its way to the surface and small pieces of material are removed. The side walls of the void area are typically at right angles to the surface, as shown in Fig. 12.54. Subsurface pitting is a

---

*See Chap. 11 for methods to calculate the load-carrying capacity of a gear set.
†See Chap. 8 for detailed information on hardening gears.

LOADED GEARS IN ACTION   12.51

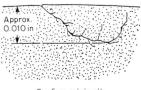

Surface origin pit

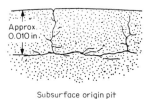

Subsurface origin pit

FIGURE 12.53 Spalling. *(Courtesy of Lubrizol Corp., Cleveland, Ohio)*

FIGURE 12.54 Surface and subsurface pitting.

function of the surface stress (and resulting subsurface shear), the number of cycles, and the cleanliness of the material. On this basis, one would obtain the maximum protection against subsurface pitting with a material such as double vacuum remelt steel.

In reality, much of the surface fatigue that occurs on gear teeth of industrial gear units is pitting that has its origin of failure on the surface. Gears that operate in the boundary and mixed lubrication regimes have considerable metal-to-metal contact. This rubbing and scrubbing action produces excellent surface-oriented crack nucleation sites. Cracks start at these surface locations and work their way into the surface material as the initial crack system is driven by repetitive stress cycles. Surface-oriented-type pits are shown schematically in Fig. 12.54. The surface-oriented pits have sloping conical sides, and as the crack system advances in depth, it branches out. When one of the branches returns to the surface, a small piece of material is removed from the surface, leaving a void or a pit.

Bronze worm gears, because of their low endurance limit in surface compression, are prone to surface pitting and spalling. The worm gear shown in Fig. 12.55 is an example of spalling where many smaller-diameter pitted areas have run together, creating a large void area.

12.8.9 Frosting (Micropitting)

Frosting is micropitting under thin oil film conditions. Patches on the tooth surface of the normal worn-in or polished surface have an etched-like finish. Under magnification the surface appears to be a field of very fine micropits less than 0.0001 in. deep. The nitrided pinion in Fig. 12.56 shows typical frosting patterns on the working flank of the pinion teeth. The damage to the pinion at this stage is minimal.[27]

The case-carburized and ground AISI 9310 pinion tooth shown in Fig. 12.57 shows considerable frosting in the dedendum region. Shallow, "arrowhead"-type pits have developed from these frosted areas. The pits are relatively shallow but rather large in area.

FIGURE 12.55 Spalling on a bronze worm gear.

FIGURE 12.56 Frosting in a nitrided pinion.

The frosted area of the gear teeth from Fig. 12.57 has been expanded 400 times to obtain a better view of the frosted surface area and is shown in Fig. 12.58. The large arrowhead pits have evolved from this bed of micropits. Once the large surface pits are formed, the life of the pinion or gear is very limited.

Case Crushing (Case-Hardened Material). Case crushing is a failure that occurs subsurface when the endurance limit of the material has been exceeded. The origin of failure is near the case-to-core junction, although the failure may not always occur at this location. Failures of this nature depend on surface contact stresses, the strength of the material at depth, the case being significantly harder than the core, and most often a large number of stress cycles.

LOADED GEARS IN ACTION 12.53

FIGURE 12.57 Frosting on a carburized and ground pinion.

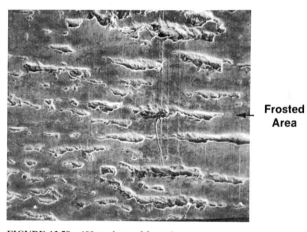

← Frosted Area

FIGURE 12.58 400× photo of frosted area.

The nitrided pinion shown in Fig. 12.59 is an excellent example of case crushing. The hard, nitrided case has failed subsurface, and most of the working case material has broken away.

A more localized case-crushing failure is shown in Fig. 12.60. Here the load has been concentrated toward the center of the teeth. A failure has started near the case-to-core junction, and pieces of the tooth have failed.

FIGURE 12.59 Case crushing on nitrided pinion.

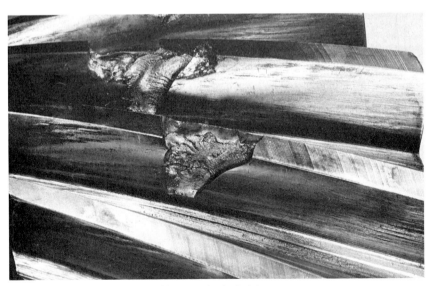

FIGURE 12.60 Localized case crushing on carburized pinion.

Gear Tooth Breakage. Gear tooth breakage is failure by breaking a whole tooth or a substantial portion of a tooth, by overload, shock, or the more common fatigue phenomena of repeatedly stressing the gear teeth above the endurance limit of the material subject to beam bending stresses. Figure 12.61 depicts a typical high-cycle fatigue failure. From this failure the focal point, or characteristic "eye," and "beach marks" of the fatigue break are clearly visible.

Failures of this nature are caused by excess loads due to design oversights, hob tears or notches in the root fillets, inclusions, heat-treatment cracks, and misalignments.[29] A fatigue failure actually takes many cycles of operation to develop. That is why the typical fatigue "eye" and "beach marks" are often evident.[1] The break initially starts with a slight crack formation. Slight motion develops some "fretting corrosion," which is usually visible in the fatigue "eye." The "beach marks" indicate crack progression and are indications of the time necessary for a failure to develop. Generally the break is quite smooth, which also indicates that a considerable amount of working of the cracked surfaces has taken place before actual failure. The final failure area generally has a fibrous, rough appearance, indicating rapid crack growth.

Overload Breakage. Overload breakage causes a stringy, fibrous-appearing break. In the harder materials this break has a more silky appearance, but the break still shows evidence of lack of fatigue and of being wrenched apart. Overload failures are caused by sudden misalignments, bearing failures that tend to freeze up the gear transmission, or large foreign particles passing through the mesh. Figure 12.62 shows a case-hardened helical pinion that has failed from overload breakage. It is quite interesting to note that several teeth have chunks broken out corresponding to the number of teeth in the

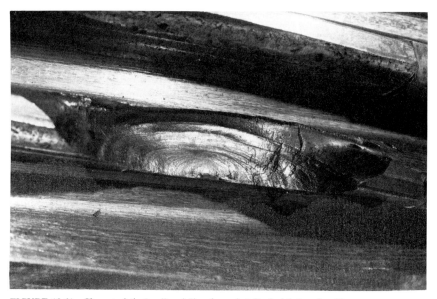

FIGURE 12.61 Characteristic "eye" and "beach marks" (typical fatigue break).

FIGURE 12.62 Overload breakage—case-hardened helical pinion.

contact zone. When gear teeth break or when foreign objects pass through the tooth engagement, quite catastrophic failure can result. Figure 12.63 shows what can happen when something wedges the gear mesh apart causing a complete mismatch of tooth engagement.

Many times gear tooth breakage is a secondary effect resulting from excessive wear or pitting failures. Figure 12.64 shows a spur gear with destructive pitting and spalling in addition to the broken tooth. This photograph clearly indicates fatigue cracks originating from stress risers set up by excessive pitting.

When gear teeth break, the investigator can often piece together the wreckage and quite accurately determine the tooth that failed first. If an end breaks off a given gear tooth, there may be some special reason. Misalignment is quite often to blame, and careful analysis of the contact markings will point this out. Sometimes gears are damaged at assembly or during manufacture. If, perchance, the end of a gear tooth is bumped, which may form upset material above the tooth surface, severe overloads will occur, and this section of the gear tooth will fail from fatigue.

12.8.10 Other Types and Causes of Gear Failure

There are many cases of gear failure other than those that fall into the categories previously outlined. A few of the widely different kinds are considered here.

FIGURE 12.63 Failed gearbox.

LOADED GEARS IN ACTION

FIGURE 12.64 Spur gear with destructive pitting and spalling.

A simple lubrication failure can cause a catastrophic failure of an entire gearbox. If, for example, a lubrication line ruptures or an operator fails to turn on the lubrication system, the complete gearbox will be starved for oil. The net result is extreme overheating, failed bearings, and badly scored and pitted gear teeth. Figure 12.65 is an example of a gear failure of this nature. The failure will show evidence of excessive heating (a deep blue color); the hardness of the material will be drawn back; and premature pitting due to lack of hardness will be under way.

The lubricant can cause other troubles that are not quite so readily detectable as a starved gearbox, such as: (1) the viscosity of the lubricant may not be high enough to produce a good oil film; (2) the "oiliness" of the lubricant may be below the level safe to prevent high mesh friction, (3) the lubricant may not have adequate "wetting" properties, (4) the lubricant may collect contaminants or break down and form harmful chemical by-products, (5) the lubricant may oxidize and sludge up the filter causing either a restriction in flow or, by forcing the filter to bypass,[30] permit unfiltered oil access to the bearings and gear mesh.

At times the location and design of oil nozzles can influence the operation of a gearbox. It is very important to deliver oil to the bearings and the gear meshes in the

FIGURE 12.65 Lubrication failure.

proper quantities to lubricate and to carry the heat away. If the nozzles are not doing a proper job of wetting down the gear and bearing surfaces, adjustments must be made. Perhaps more oil nozzles are required or a relocation of existing nozzles must be made. Often, because of the windage of the mesh, oil nozzles do not have the ability to break down this barrier. A higher back pressure must be used to create the proper jet velocity to penetrate the windage of the gear mesh.

12.8.11 Fretting Corrosion

Fretting corrosion is a form of wear caused by motion between contacting bodies that are not adequately lubricated. Usually there is sufficient load and motion to break away material from asperity interactions of contacting particles.[31] These wear particles become oxidized and turn the familiar red-rust color. The iron oxidation products can be oxidized by either air or other oxygen-containing atmospheres or oxygen-containing substances.

The products of fretting are usually quite abrasive and occur in locations where they are not readily washed away by lubricants if lubricants are used. Since much of the product of fretting remains in close approximation to where it was formed, it acts as abrasive particles and helps to promote abrasive wear.

In gear systems, the products of fretting corrosion can be found in such locations as

Bearing fits such as the inner race fit on shafting

Spline connections—both loose-fit splines and fitted splines

Bolted lightweight flange joints such as a bevel ring gear bolted to a flanged shaft under high torque loads

Pivot pads on pivoted pad bearings

Gear systems that are subjected to rapid intermittent motion

Keyed-on or shrink-fit coupling hubs and other shrink-fit parts

Bolted, splined, or keyed joints that are subjected to repeated reverse torque

About the only preventative for fretting corrosion is to stop the relative motion of the contacting parts under load. This can be accomplished by increasing shrink fits on tightly fitted parts.[31] Continually flushing the loose fits with lubricating oil will remove the products of fretting and reduce the generation of additional metal particles by abrasive wear. This is particularly helpful in such applications as splines and flexible couplings.

Fretting corrosion can be reduced but not eliminated by the use of platings and coatings that act as lubricants to deter the onset of fretting. These platings and coatings, such as silver, copper, nickel, molydisulfide, and teflon, may extend the useful life of parts that are prone to fret within acceptable limits. Fretting corrosion takes place, but the rate of rapid metal removal may be under control until the coatings wear through. The life of a flexible splined joint can be greatly extended by increasing the hardness of the contacting members. For example, 4340 or 4140 medium-hardness splined joints can be improved considerably from a fretting corrosion-wear point of view by properly nitriding the contacting members. "White layer" on nitrided surfaces must be eliminated or kept at very low thickness levels on the order of 0.0002 or lower.

The shaft pinion in Fig. 12.66 shows what happens to a gear journal when the ball bearing fit starts to fail from fretting corrosion. Many mils of material have been fretted away causing a high misalignment of the gear mesh. The by-products of fretting

FIGURE 12.66 Fretting corrosion, ball bearing fit on shaft.

corrosion are clearly visible in the spline shown in Fig. 12.67. This buildup of sludge and wear particles indicates that considerable metal has been removed from the spline fit. Continuing to operate a gear in this condition would ultimately lead to gear failures.

Another cause of gear failures can be traced to the various types of flexible couplings used to connect a given gearbox either to the prime mover or to the driven load.

Quite often, flexible couplings of the dental-tooth type are used, such as those illustrated in Fig. 12.68. This coupling has failed from the so-called worm-tracing effect as indicated by the metal that has been removed in the center of the contact zone.[32] The coupling also shows signs of heavy abrasive wear and "pickup." Generally, when couplings fail in this manner, they become sticky and seize. When this happens, bending moments are transferred into the gear casing. This causes higher bearing loads and contributes to the misalignment across the gear mesh. Couplings used for slightly

FIGURE 12.67 Fretting corrosion, spline.

FIGURE 12.68 Dental-tooth coupling.

FIGURE 12.69 Quill drive—failed.

different purposes can also cause considerable concern. Figure 12.69 shows a failed quill drive used to operate a scavenge pump. This quill failed from normal fatigue set up by stress concentration caused by hob nicks and abrupt section changes. Once a lube or scavenge pump fails, the failure of the complete gear unit is imminent.

12.9 PERFORMANCE TESTING OF GEARS

To assure the customer that the production gearbox will meet all the requirements outlined by the specifications, the gear unit should be tested in the factory. Its performance should be observed and adjustments made, if necessary.

12.9.1 Full-Power Testing

In order to measure gear performance, the gear unit must be operated under full-load and full-speed conditions. Ideally, the unit should be tested in exactly the same manner in which it is to be used in the field. It should be driven by its own prime mover and it should drive the required load. In practice, at times it is impractical to duplicate exact field conditions in the shop. The practice of transmitting full horsepower through a given gearbox into the actual field equipment or some other load device is known as *power-absorption testing*. When the speed and horsepower rating of a given gear train is low, full-power absorption testing is practical. Either the actual driven load can be used to absorb the power, or a water brake or an electric dynamometer can absorb the power.

When the horsepower rating of the gear unit is large, power absorption testing becomes more difficult. High-horsepower driving equipment is required, as well as high-horsepower absorption equipment. Not only does the equipment become expensive, but when the tests are under way, a large amount of power must be dissipated as losses in the water brake or dynamometer. To simplify and to reduce the cost of full-load testing of high-horsepower gearboxes, a technique known as *torque testing*[33] is widely used in the gear trade. Torque testing is based on the "four-square" principle whereby two gear sets are joined back to back and locked together after the full twisting couple is applied to one of the connecting shafts. Now that the gear sets are torqued together, the driving motor needs only to supply the static and dynamic friction losses of the bearings and gears as both units are brought up to speed. When the gear sets are up to speed, they are operating at an equivalent horsepower, equal to that horsepower required for a full-powered absorption test.[34] A schematic diagram of a torque-testing setup can be found in Fig. 12.70. Although torque testing does not fully duplicate exact field conditions, it does permit the factory to full-load test even the highest-horsepower gear units manufactured.

LOADED GEARS IN ACTION

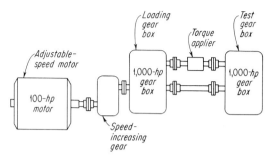

FIGURE 12.70 Torque-testing setup.

A closer view of a torque-testing setup, often referred to as a *locked torque test,* is shown in Fig. 12.71.[35]

This arrangement shows two industrial double-helical gear units tied together, that is, input shaft to input shaft and output shaft to output shaft, creating a closed loop. Torque is applied within the loop by two large torque wrenches as shown. The coupling flanges are allowed to slip relative to each other. When they have slipped the prescribed amount, the flanges are bolted together again, and a positive torque is locked within the closed gear and shafting loop. When the gear boxes are rotated at full speed, the gears and bearings are loaded similar to gears that are transmitting full power in a power absorption test.

The two industrial gear boxes, shown in Fig. 12.72, are set up in torque loop, as shown in Fig. 12.71. The torque wrenches and the load-measuring scale are clearly

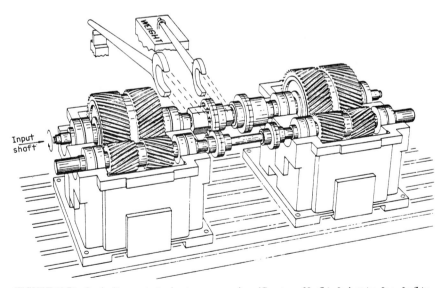

FIGURE 12.71 Locked torque test using torque wrenches. *(Courtesy of Lufkin Industries, Inc., Lufkin, Texas)*

FIGURE 12.72 Industrial gear unit in lock torque test. *(Courtesy of Lufkin Industries, Inc., Lufkin, Texas)*

visible. Once the torque has been applied and the flanges bolted together again, the torque wrenches are removed and the two gear units are ready for test.[35]

In this torque-testing setup, once the torque has been locked into the loop, the gear unit must be started up from zero speed under an equivalent full-load torque. Design engineers are often concerned that "start-up" of loaded bearings from zero speed may cause premature bearing difficulties.

The full concern of load start-up from zero speed can be eliminated by the use of a torque-applying device that can apply torque to the closed torque loop after the gear system has been brought up to speed. The schematic gear torque-testing setup shown in Fig. 12.70 is such an arrangement.

The two rather large, single-helical gear units shown in Fig. 12.73 are tied together in a four-square torque loop. The cylindrical device, located on the low-speed shaft between the two large gears, is an hydraulic, actuated torque-applying device. In this installation, the gear system can be brought up to speed with the electric motor located in the lower left-hand corner. Once the gears are up to speed, the hydraulic torque applier can be actuated and the gear system loaded up to the desired load level. Using this arrangement, neither the gear teeth nor the bearings are exposed to a high-load, low-speed start-up.

There are a few locations in this country where full-power absorption tests can be

FIGURE 12.73 Large single-helical gear in torque loop with hydraulic applier. *(Courtesy of The Cincinnati Gear Company, Cincinnati, Ohio)*

performed on high-horsepower equipment. The Naval Ship Systems Engineering Station in Philadelphia, PA, for example, can full-load test the main propulsion reduction gear of a destroyer or cruiser by absorbing output power with a high-capacity water brake.

For the most part, however, most high-horsepower gear units are full-load tested using one of the torque-testing methods described in the foregoing discussion.

12.9.2 Accessory Gearbox Testing

Accessory gears are usually designed with a single input or driver with multiple output or driven members. Accessories consist of such things as alternators, lube pumps, hydraulic pumps, power takeoffs, and governors. It becomes almost impossible to make an accessory-box torque test; therefore, most accessory gear drives are tested using the power absorption method. Since the accessories themselves are so many and varied, water brakes are usually used for the power absorption device. Generally, accessory boxes are rated in the vicinity of 200 to 400 hp. This being the case, the amount of power being absorbed and converted into losses is not prohibitive from a cost point of view. Figure 12.74 indicates an aircraft accessory gear being loaded with three water brakes to simulate various accessory loads. It is usually the practice to load accessory gears

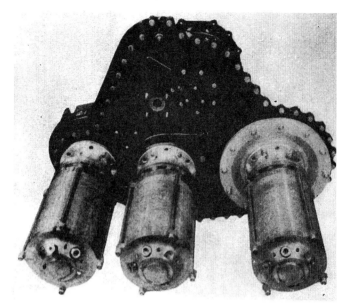

FIGURE 12.74 Accessory-box testing. *(Courtesy of General Electric Co., Lynn, Massachusetts)*

higher than the design load in order to wring out the gear tooth loadings on the test stand. Water brakes permit extra power absorption that might not be available if the actual accessories were used instead of the brakes. Many times, because of schedule considerations, accessory gearbox testing must get under way before the accessories are available. Accessory gearbox testing can also be programmed with certain prescribed load and speed cycles if water brakes are used to simulate as nearly as possible certain expected service conditions.

12.9.3 Component Testing

Although in the final analysis the production gearbox should be run under field conditions, it has been beneficial in many respects to run components of gearboxes on various types of laboratory test equipment. Component testing can be accomplished at a lower cost and, perhaps, under more controlled testing conditions. Theoretical investigations, such as a comprehensive program on the scoring resistance of various material-lubricant combinations or a study of the wear characteristics of certain types of gear steels, can be undertaken.[36]

Bearings and spline couplings can also be investigated on various component-test equipment. In fact, component testing has been an excellent source of information for design purposes. It is quite easy to try out new things on a component-test stand and observe how they operate and function. Based on this information, new equipment can be designed with a high degree of assurance for successful operation on the first production unit.

The component-test equipment shown in Fig. 12.75 is typical of the many gear

FIGURE 12.75 Component-gear tester.

testing rigs that are in operation throughout industry. This test stand uses the locked torque loop principle, shown in Fig. 12.72, whereby the narrow-faced test gears are loaded against wider-face-width, more conservatively loaded, permanent drive gears.

Equipment of this nature can be used to investigate pitting, wear, scoring, tooth breakage, lubrication, dynamic loads, and many other gear characteristics.

12.9.4 Efficiency Testing

In many gear applications, the efficiency of the gear train must be known within very close limits. Some customers, for example, request that the efficiency of a given production gearbox be measured at the factory and demand that this measured efficiency fall within the guaranteed limit. Of course, efficiencies can be calculated with a fair degree of precision,[4] however, when extreme accuracy is required, there is no substitute for efficiency testing.

If the gear train under consideration is of a design that lends itself to torque testing, then the overall efficiency of a unit may be determined on a loss-torque basis. Two identical units are set up back to back with the prescribed amount of locked-in torque. A very accurate torque meter must be installed in the drive motor shaft next to the four-square torque loop. This meter will then measure the amount of torque that is required to drive the two locked-up gear trains. Since the motor supplies only the losses to the system the inefficiency of two identical gearboxes has, in effect, been measured.

The efficiency can then be calculated on a torque basis. Namely, efficiency in percent equals

$$E = \frac{T_t - T_{L/2}}{T_t \times 100} \qquad (12.83)$$

where E = efficiency, %
  T_t = transmitted torque
  T_L = loss torque

Often a given gearbox design does not lend itself to a simple back-to-back torque test. Therefore, other means of efficiency testing must be devised. A simple heat balance on a loaded gearbox transmitting full horsepower will give the overall efficiency of the geared transmission. In effect, all the losses of a gear set are reduced to heat, which is dissipated to the atmosphere and carried away by the lubrication system. If this heat loss can be measured accurately, then the basic efficiency of the drive has been measured.

The heat carried away by the lubrication system can be measured very accurately by carefully determining the oil flow through the gear casing and observing the temperature rise of the oil as it goes into and comes out of the gearbox. The heat loss by radiation and convection is very hard to determine; however, this problem can be bypassed by "lagging" the gearbox to keep heat loss of this nature to an absolute minimum. Of course, the heat that normally would escape will now show up as an increase in oil temperature across the gearbox. The lagging job on the gear casing must be of high quality; in fact, it should be done by an expert using the very best insulating materials. The aircraft gear casing in Figs. 12.76 and 12.77 shows a before-and-after lagging job. Note from these photographs that an ample amount of lagging has been used and that a professional job of lagging has been accomplished. The efficiency of a gearbox from a heat-balance test can be determined by the following equations:

$$Q = MC_P \, \Delta T \qquad (12.84)$$

where Q = heat loss, Btu per min
  M = lb of oil per min
  C_P = specific heat of oil
  ΔT = temperature rise, °F (inlet to outlet)

$$P_L = \frac{Q}{42.4} \qquad (12.85)$$

where P_L = horsepower loss
  42.4 = Btu per hp

$$E = \frac{P_T - P_L}{P_T \times 100} \qquad (12.86)$$

where E = efficiency, %
  P_T = transmitted horsepower
  P_L = horsepower loss

FIGURE 12.76 Aircraft unit before lagging. *(Courtesy of General Electric Co., Lynn, Massachusetts)*

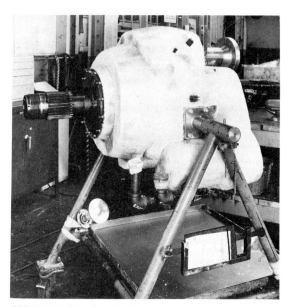

FIGURE 12.77 Aircraft unit after lagging. *(Courtesy of General Electric Co., Lynn, Massachusetts)*

12.9.5 Lubrication Testing

To get the best performance from a gearbox, a design engineer must not only have a good design, but must use the best lubricant for the application. The complete lubrication system, in fact, must meet all the performance requirements of the gearbox, or else the gear itself will never make it. Experience shows that the two major failures relating to the lubricant are scoring and normal or long-range wear.[24] To evaluate a new gear oil is a rather time-consuming and costly project. The procedure may vary somewhat, depending on the ultimate use of the oil, but, in general, the following steps should be taken: (1) All available information pertaining to the oil should be obtained from the manufacturer and the various laboratories that have been working with the oil. (2) Laboratory checks, such as viscosity, pour point, flash point, and oxidation stability should be made on an oil sample; (3) Comparison tests should be run on bench tests such as the Shell four-ball wear tester and the Timken wear tester. (4) Lubricated compression-roll tests should be run under simulated service conditions and the results compared with previous tests. (5) Scoring tests using test gears on a four-square gear testing apparatus must be run to determine the "critical scoring temperature" of the oil.

It is difficult, and often time-consuming, to run wear tests on gears since wear is very hard to measure using conventional means. Wear can be detected by very accurately weighing the parts before and after test. Pinion and gear profiles can be measured for wear on analytical inspection equipment. These checks must be taken before the parts are operated and after the tests are completed. Visual inspection and measuring surface wear by pin-checking and gear tooth vernier calipers are techniques often used to detect changes in wear.

Wear tests can be run on a radioactive gear tester to determine the wear characteristics of the new oil, or wear characteristics of the gear material using the same oil. Since wear as such is very hard to detect and measure, the radioactive technique of gear testing has been adapted for studying wear phenomena.[24] Minute quantities of wear can be detected in short periods of time, and several oils may be tested with the same sets of gears.

The schematic test setup of Fig. 12.78 gives a good flow picture of how the radioactive technique is used for testing gears and gear lubricants. The pinion to be tested is sent to the nearest atomic pile, where it is irradiated with neutrons. This makes the pinion radioactive. The pinion and gear are then mounted in a test stand. The gears are loaded and the test machine put into motion. The lubricant is circulated through the gearbox, past the counter tube, and back into the sump, as indicated in Fig. 12.78. When the active gears commence to wear, the lubricating oil transports the worn iron particles past the counter, where they are detected. It is a simple laboratory technique to convert the counting rates as recorded to milligrams of wear per unit of time.

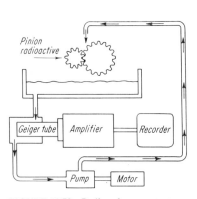

FIGURE 12.78 Radioactive wear test.

An up-to-date "takeoff" on the aforementioned radioactive technique[37] of measuring wear shows promise as a way to utilize the radioactive principle without the major inconvenience of a fully activated pinion as described in the foregoing. A small spot on a potential wear surface such as a gear or pinion is activated with high-energy ions forming a local shallow surface layer of radionuclides.

This is known as *surface layer activation* (SLA). The small spot marker technique simplifies the activation of the surface and practically eliminates the safety concerns that occur with complete part activation. The wearing away of the small spot of active material can be measured by monitoring the decrease in intensity of the gamma rays from a location outside the normal lubrication system.

12.9.6 Field Testing

Field operation is the best proving ground an engineer can find for evaluation of gear equipment. The various gear designs must operate successfully in the field, and much can be learned from obtaining and observing operational records of the gear arrangements that are in service.

In many instances, service operating conditions are much different from the original design conditions, and it is not uncommon to find a customer using the equipment far in excess of what has been recommended. The reverse is quite often true, however, and a design engineer can easily be fooled by the seemingly successful operation of gear sets in the field. The apparent successful operation may be the result of not operating the gear unit up to design conditions.

By firmly following the policy of keeping close tabs on field operations, gear engineers, not only obtain good information on which to base new designs, but also create good working relationships with customers.

REFERENCES

1. Dudley, D. W., *Practical Gear Design*, McGraw-Hill, New York, 1954, p. 141, 294, 295.
2. Philadelphia Gear Works, *Philadelphia Gears*, Philadelphia Gear Works, Philadelphia, 1958.
3. Buckingham, Earle, *Analytical Mechanics of Gears*, McGraw-Hill, New York, 1949, p. 191.
4. Shipley, E. E., "How to Predict Efficiency of Gear Trains," *Prod. Eng.*, Aug. 1958, pp. 44–45.
5. Merritt, H. E., *Gears*, 3rd ed., Pitman, New York, 1955, pp. 345, 355.
6. Candee, A. H., "Bevel Gears in Aircraft," *Trans. ASME*, May 1943, pp. 267–277.
7. Gleason Works, "20° Straight Bevel Gear System," Gleason Works, Rochester, New York, 1949.
8. Gleason Works, "Zerol Bevel Gear System," Gleason Works, Rochester, New York, 1950.
9. Gleason Works, "Spiral Bevel Gear System," Gleason Works, Rochester, New York, 1951.
10. Illinois Tool Works, "Sprioid Gears," Illinois Tool Works, Chicago, 1959.
11. Illinois Tool Works, "Helicon Gears," Illinois Tool Works, Chicago, 1959.
12. Gleason Works, "Bevel and Hypoid Gear Design," Gleason Works, Rochester, New York, 1947.
13. Illinois Tool Works, "Planoid Gears," Illinois Tool Works, Chicago, 1958.
14. Wilcock, D. F., and Booser, E. R., *Bearing Design and Application*, McGraw-Hill, New York, 1957, pp. 83, 84, 85, 192, 220.
15. CADENSE 30, "Performance of Liquid Lubricated Journal Bearings," Version 7.0, July 1983. (CADENSE is a registered trademark of Mechanical Technology Incorporated, Latham, New York 12110.)
16. CADENSE® 33 & 34, "Performance of Various Thrust Bearings."

17. Palmgren, A., *Ball and Roller Bearing Engineering,* S. J. Burbank & Co., Inc., Philadelphia, p. 21.
18. Barish, T., "Fundamentals of Ball Bearing Design," unpublished paper, March 1957.
19. Watt, H. R., "Gear Tooth Efficiencies, I, II, III," *Power Transmission,* March, April, June 1951.
20. Buckingham, Earle, *Manual of Gear Design, Sec. 2, Machinery,* 1935, p. 128, 129.
21. Tuplin, W. A., "Compound Epicyclic Gear Trains," *Machine Design,* April 1957, pp. 100–104.
22. Cleare, G. V., "Gear Tooth Failures in Highly Stressed Transmission Gears," International Conference on Gearing, The Institute of Mechanical Engineers, September 1958.
23. Tuplin, W. A., "Gear Tooth Lubrication," *Machine Design,* August 1954, pp. 123–131.
24. Shipley, E. E., "Design and Testing Considerations of Lubricants for Gear Applications," *Lubrication Eng.,* April 1958, pp. 148–152.
25. Hartman, M. A., "Gears for Outer Space," *Machinery,* August 1959, pp. 115–121.
26. Shipley, E. E., "Investigation of Factors Affecting High Temperature Gear Operation," *Lubrication Eng.,* March 1959, pp. 98–104.
27. American Gear Manufacturers Association, "Standard Nomenclature of Gear-tooth Wear and Failure," AGMA 110.04, 1980.
28. *Gear Manufacture and Performance,* American Society for Metal Material/Metal Working Technology Series, 1974, p. 125.
29. Frederick, S. H., and Newman, A. D., "Gear Failures," International Conference on Gearing, The Institution of Mechanical Engineers, September 1958.
30. Shipley, E. E., "Will Gears Operate at 600°F?" *Am. Machinist,* February 1958, pp. 101–103.
31. Derner, W. J., "Fret Wear: The Insidious Destroyer," *Power Transmission Design,* July 1978.
32. Batter, J. F., and Reed, F. E., "Wear in Heavily Loaded Lubricated Surfaces Subjected to Oscillating Motion," American Society of Lubrication Engineers, Paper 59GS-12, January 1959.
33. Shipley, Eugene E., "12 Ways to Load Test Gears," *Prod. Eng.,* January 1958, pp. 77–82.
34. Newman, A. D., "Load-carrying Tests of Admiralty Gearing," International Conference on Gearing, The Institution of Mechanical Engineers, London, September 1958.
35. Partridge, J. R., "Gears for Power Transmission Design and Application," Lufkin Industries, Inc., Lufkin, Texas.
36. Crook, A. W., "The Lubrication of Rollers," *Phil. Trans. Roy. Soc. London, A,* Vol. 250, January 1958, pp. 387–409.
37. Blatchley, C., and Sioshansi, P., "Monitoring Wear with Gamma Rays," *Machine Design,* October 25, 1990, pp. 99–102.

Other Reference Material Not Specifically Noted in Text

1. Barish, T., "Idler Gears," *Machine Design,* November 1957, pp. 121–125.
2. Barish, T., "On Hertz Contact with Sliding," American Gear Manufacturers Association, March 1960.
3. Buckingham, Earle, *Spur Gears,* McGraw-Hill, New York, 1928.
4. Crook, A. W., "Simulated Gear-tooth Contacts, Some Experiments upon Their Lubrication and Subsurface Deformations," *Proc. Inst. Mech. Eng. London,* Vol. 171, No. 5, 1957.
5. Dolan, T. J., "Analyzing Failures of Metal Components," *Metals Engineering Quarterly,* November 1972.
6. Dudley, D. W., "Modification of Gear Tooth Profiles," *Prod. Eng.,* September 1949.

7. Dudley, D. W., "b-12 Characteristics of Regimes of Gear Lubrication," Dudley Engineering Co. International Symposium on Gearing & Power Transmissions, 1981, Tokyo.

8. Editor, "Full Load Efficiency Testing of Turbine Gear Units," *Engineering,* March 1959, pp. 306–307.

9. Hirst, W., and Lanaster, J. K., "The Influence of Oxide and Lubricant Films on the Friction and Surface Damage of Metals," *Proc. Roy. Soc. London, A,* Vol. 223, 1954, pp. 324–338.

10. Ishibashi, A. and Yoshino, H., "Powder Transmission Efficiencies and Friction Coefficients at Teeth of Novikov-Wildhaber and Involute Gears," *Trans. ASME,* Vol. 107, March 1985.

11. Kron, H. O., "Gear-Tooth Sub-Surface Stress Analysis," presented at the International Symposium on Gearing, Paris, June 23, 1977.

12. Martin, K. F., "The Efficiency of Involute Spur Gears," *Trans. ASME,* Vol. 103, January 1981.

13. Moore, W. L., "Reliability of Gear Applications for Aero-space Vehicles," American Gear Manufacturers Association, March 1960.

14. Radzimovsky, E. I., "Planetary Gear Drives," *Machine Design,* June 1959, pp. 144–153.

15. Research Report, "Inert Gas Boosts Oil's Ability," *Chem. Eng. News,* October 1958, pp. 34–35.

16. Shipley, E. E., "Efficiency of Involute Spur Gears," *American Society of Lubrication Engineers,* January 1959.

17. Shipley, E. E., "Testing Can Reduce Gear Failures," *Hydrocarbon Processing,* December 1973.

18. Shipley, E. E., "Failure Analysis of Coarse-Pitch, Hardened and Ground Gears," presented at the AGMA Fall Technical Meeting, October 1982, Paper No. P229.26.

19. Tanaka, S., Ishibashi, A., Ezoe, S., "Appreciable Increases in Surface Durability of Gear Pairs with Mirror-Like Finish," *Gear Technology,* March–April 1987.

20. Terauchi, Y., Nagamura, K., Faculty of Engineering, Hiroshima University, Shitami, Saijocho, Higashihiroshima-city, 724, Japan Nohara, M., Kure Technical College, Agaminami, Kure-city, 737, Japan "Study on Rippling Failure by the Roller Test."

21. Winter, H., Michaelis, K., "Scoring Load Capacity of Gears Lubricated with EP Oils," *Gear Technology,* November 1984, AGMA Technical Paper #P219.17, October 1983.

CHAPTER 13
GEAR VIBRATION

Darle W. Dudley
President, Dudley Engineering Company, Inc.
San Diego, California

In this chapter we consider the general subject of gear vibration. This is an important technical subject. Gearboxes that have too much vibration may be rejected by the purchaser. Gear units that have operated for a few years' time with relatively low vibration may start to have high vibration. This onset of vibration is often a timely warning that a breakdown of the gearing is only a few weeks away.

Aside from the considerations just mentioned, there are other mechanical system considerations with regard to vibration. Vibration in vehicles or aircraft is unwelcome—even if the gears are quite able to run for a long time without failure. Vibrations induced in a closely connected mechanical system may cause premature failure of somewhat fragile components such as instruments, control mechanisms, small motors, or even bolts. (Critical bolted joints may suffer from bolts coming loose.)

13.1 FUNDAMENTALS OF VIBRATION

Suppose we take a gearbox with two stages of gearing. If we measure the vibration at one of the four corners of the box with sensitive instruments, we will detect a rather tiny motion that is moving back and forth at high frequency. The character of this motion tends to be like a sine wave. Figure 13.1 depicts the sine-wave character of a vibration at some point. The mathematics of the sine wave can be thought of in terms of an imaginary circle and trigonometry.

In Fig. 13.1, the radial displacement is the vector distance d. The size of this vector is determined by a point moving around a circle at a rate ω for radians per second. The vibration is both positive and negative from a no-vibration position. If d_a is the total amount of movement from the neutral position, then the diameter of the imaginary circle is $2d_a$. As a point moves around the circle, its position is specified by the angle θ. For a complete cycle of the vibration, θ goes through 360° (or 2π rad).

With the foregoing relations defined, some equations can be written.
For displacement,

$$d = d_a \sin \theta \qquad (13.1)$$

The maximum displacement, peak to peak, is D. Figure 13.1 shows that D equals $2d_a$ at the time θ equals 90°. The angle θ is taken in degrees (when taking the sine of an angle by computers or tables reading sine values versus degrees for the angle).

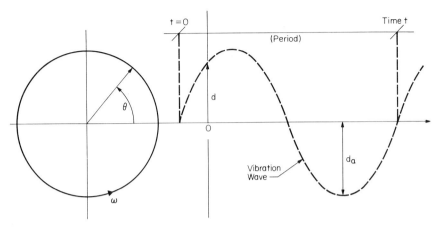

FIGURE 13.1 The vibration excursion is like a sine wave.

The *velocity* of a vibration motion is an important variable. Considering that velocity is defined as displacement divided by time, the equation for velocity becomes the first derivative of displacement with respect to time.

$$V = \omega\, d_a \cos \theta \qquad (13.2)$$

The acceleration is the second derivative of displacement with respect to time. The acceleration curve is established by

$$A = -\omega^2\, d_a \sin \theta \qquad (13.3)$$

Figure 13.2 shows schematically how the three curves for vibration behavior relate to each other from a *time* standpoint. Note that if peak displacement occurs at θ equals

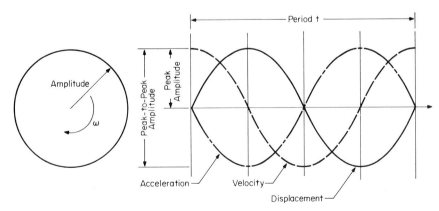

FIGURE 13.2 Wave forms for displacement, velocity, and acceleration tend to be sine waves. Note that these forms are out of phase with each other.

GEAR VIBRATION

TABLE 13.1 Relation Between Vibration Peak Values

Item	Equations* Metric	English	Eq. No.
Peak-to-peak displacement	$D = 2 \times$ peak amplitude, mm	—	(13.4)
	—	$D = 2 \times$ peak amplitude, inches	(13.5)
Peak velocity	$V = \omega \dfrac{D}{2}$, mm/s	—	(13.6)
	—	$V = \omega \dfrac{D}{2}$, in./s	(13.7)
Peak acceleration (in g values)	$A = -\omega^2 \dfrac{D}{2}$, mm/s^2	—	(13.8)
	—	$A = -\dfrac{\omega^2}{386} \dfrac{D}{2}$, in./s^2	(13.9)

*ω = radians per second = 2π Hz; Hz is cycles per second. D is in millimeters for the metric equations and in inches for the English equations.

90°, peak velocity occurs at θ equals 180°, and peak acceleration occurs at 90°—but the acceleration is negative. However, at 270° there is another acceleration peak that is positive, *and* the displacement has now become negative.

From the standpoint of vibration measurements, the common practice is to interpret vibration at double amplitude for displacement but single amplitude for velocity. Acceleration is also normally given as a single amplitude value. The units of vibration are usually in g values—where g is the acceleration of gravity.

Table 13.1 shows simplified equations for these values in both metric and English units.

Sample Problem. To illustrate the vibration relations just discussed, we take a sample problem.

Suppose that the peak-to-peak displacement at a location on a gear unit measures 0.001 in. What is the peak velocity in inches per minute and the peak acceleration in g value? The frequency is 100 Hz. (100 Hz is 6000 c/min. It is also 628.3 rad/s.) The peak displacement is 0.0005 in., which is one-half the peak-to-peak displacement.

Using Eq. (13.7), we find that the peak velocity is,

$$V = 0.314 \text{ in./s}$$

Using Eq. (13.9), the peak acceleration is,

$$A = 0.511 \ g\text{'s}$$

Figure 13.3 shows a sample of special graph paper where displacement, velocity, and acceleration can all be read—*if any one of these values has been measured.*

The Fig. 13.3 graph is based on the equations just given. Note that the point in the circle gives all the answers for the given sample problem. (Incidentally, there is a small error in printing the paper. The paper reads acceleration at 0.47 g's instead of 0.511 g's.)

13.2 MEASUREMENT OF VIBRATION

The measurement of vibration involves a sensing device that detects the vibratory motion, recording equipment that records the signals from the sensing device, and data

CHAPTER THIRTEEN

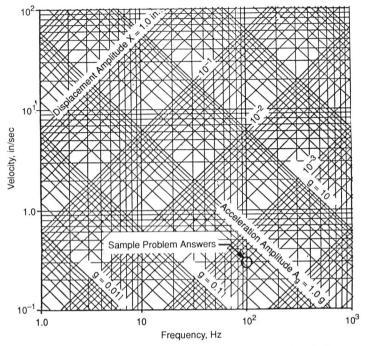

FIGURE 13.3 Special graph paper where matching values of velocity, displacement, and acceleration can all be read. Circled area shows sample problem results.

processing equipment that converts the raw data to some kind of analytical data in the form of vibration amount at different frequencies.

The device that does the sensing is generally considered to be a transducer. The output of the transducer is an electric signal that goes by wires to the data recorder. The data recorder may record these signals on magnetic tape (to be analyzed later). It is possible to have a data recorder that will instantly process incoming signals and give a readout of vibration intensity.

The processing equipment may filter the recorded data and give the maximum vibration for something like each third of an octave band. This permits plots to be made of vibration intensity versus frequency.

The processing equipment will usually have the capability of taking a signal in any one of the three vibration modes (displacement, velocity, and acceleration) and converting this signal to the other two modes. When this conversion is done, it is done on the assumption that the vibration wave is a simple harmonic motion and, therefore, the relations are the same as those given in Table 13.1.

13.2.1 Examples of Sensing Devices

We discuss three commonly used sensing devices (or transducers). These are

GEAR VIBRATION   13.5

A proximity probe
A velocity transducer
An accelerometer

Figure 13.4 shows the schematic arrangement of a proximity probe. There is an air gap between the end of the probe and a rotating shaft. The probe end is positioned about 1 mm away from the shaft. A fairly high-frequency alternating current is emitted from the probe end. Variation of probe distance causes a change in either the inductance, the capacitance, or the eddy current losses. All three types are nonlinear. They can be compensated so that over a restricted distance they will measure displacement with an accuracy that is within about 1 percent of being linear.

For good results the shaft must be homogenous, have a good surface finish, and be very accurate for roundness. The probe needs to be calibrated for the material in the shaft. The principal use of proximity probes is in measuring vibration displacement in either shafts or bearings.

A velocity transducer can be used to sense vibration velocity in a gear casing. Figure 13.5 shows a schematic of such a device.

The coil in the transducer moves along its axis in a radial magnetic field provided by a permanent magnet. The coil is attached to a probe that is spring loaded against the vibrating surface. The main mass of the magnet (and its case) is either clamped to a vibration-free base or hand held.

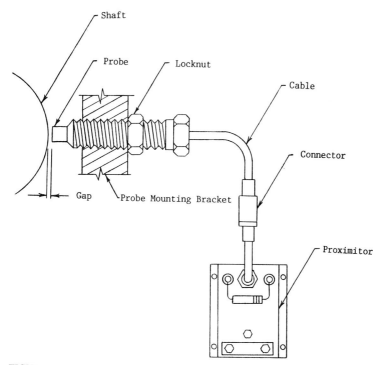

FIGURE 13.4 Schematic of a proximity probe. It is used to measure displacement.

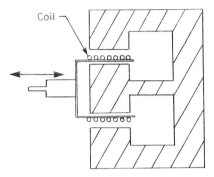

FIGURE 13.5 Schematic of a velocity transducer. It is used to measure velocity on a casing.

The coil that is moving in a magnet field in time with the vibration has relatively good linearity of electrical output with respect to vibration velocity.

Most designs of velocity transducers will work satisfactorily up to about 2000 Hz (120,000 c/min). Special miniature coil designs of velocity transducers will work satisfactorily to about 10,000 Hz.

Another commonly used sensing device for vibration measurement is an accelerometer. Figure 13.6 shows the type that is piezoelectric.

A crystal of quartz or barium titanate will produce an electrical charge proportional to the stress induced in the crystal. As vibration occurs, a mass in the accelerometer produces either shear stress or compressive stress in the crystal.

Very small accelerometers are able to measure vibration frequency up to about 100,000 Hz. The accelerometer has to be quite large to measure acceleration accurately at 0.2 Hz (12 c/min).

13.2.2 Practical Problems in Vibration Measurement

The measurement of vibration is not an easy task. With an abundance of devices on the market to sense the signal, transmit the signal, record the signal, and then process the signal, the first problem in vibration measurement is the choice of equipment to procure. If equipment is on hand, there is the related question of whether or not the equipment on hand represents the latest technology—or is appropriate to measure the kind of vibrations likely to be present in the gears involved.

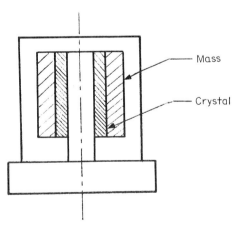

FIGURE 13.6 Schematic of a shear type of accelerometer. It is used to measure acceleration on a casing.

By one means or another, equipment is obtained to do vibration measurements. The next set of problems is apt to include one or more of the following requirements:

Obtain a calibration of the transducers to be used.

Determine how to mount each transducer so that a weakness in the mount or a vibration of the mount does not seriously affect the fidelity of the vibrations sensed.

Use calibrated wires of a proper length so that the tiny signals transmitted through the wires do not lose their accuracy from signal transmission disturbances.

Process the signal so that the results will give a valid indication of the character of the vibrations present.

Be alert to environmental things that can spoil vibration measurement accuracy. For instance, high temperatures or rather low temperatures may seriously impair the accuracy of the sensing transducer. Other things such as water vapor, oil mist, corrosive fumes, etc., may soon cause trouble.

Be sure the transducers used have the right range to cover the frequency of vibration that may be present.

In looking at the foregoing list, it might seem that all could be resolved by just reading the manufacturer's specifications for the vibration equipment on the market. Actual experience often reveals that manufacturers' specifications may not give all the needed details. Also, claims of accuracy limits may be under ideal conditions and not realizable in the actual environment of the job at hand.

13.3 SOME EXAMPLES OF VIBRATION IN GEARED UNITS

When vibrations are measured on a gear unit, several locations are chosen to take the readings. In the beginning, it is common practice to take readings in three directions. These are normally horizontal, vertical, and axial. Later on, when a unit has developed a vibration history, it may be possible to take readings at only one or two locations and only one or two directions and still know whether or not the vibration characteristics of the unit are satisfactory or unsatisfactory.

Figure 13.7 shows the full number of locations that might be used to take vibration readings.

A set of vibration readings taken by accelerometer in the horizontal direction of position A on a high-speed gear unit are shown as an example. Note that there are peaks at gear mesh frequency and at twice gear mesh frequency. There is also a peak at 9 times pinion frequency. (The pinion teeth of this unit were somewhat worn and a pattern of nine bumps was developing in the tooth spacing pattern.) See Fig. 13.8. Note that the g values for acceleration are not high, but there are three pronounced peaks. The peak at 9 times pinion frequency is quite ominous. There is no reason to have any peak at all at this frequency unless something is going wrong.

Figure 13.9 shows an example of vibration readings in velocity taken for a large, fast-running epicyclic gear unit. These readings were taken in three directions shown for positions A, B, and C. Note that the highest value of vibration in the survey was 0.20 in./s at tooth mesh frequency in the horizontal direction.

When proximity probes are used on shafts, it is possible to use two probes 90° apart. The data from these two can be so processed that the actual shaft motion can be seen on an oscilloscope.

SKETCH: Locations for housing vibration measurements "A, B, C, D"

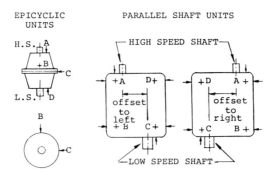

NOTE: + "means vertical reading"

FIGURE 13.7 Typical locations for taking vibration readings on gear casings.

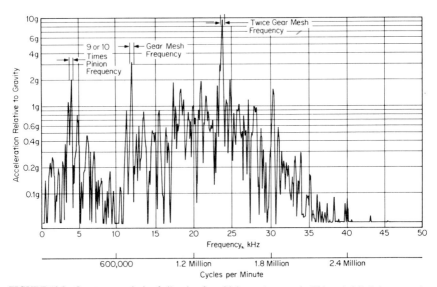

FIGURE 13.8 Spectrum analysis of vibration for a high-speed gear unit. This unit failed due to tooth wear—but the vibration amounts were not high at the time this analysis was made.

GEAR VIBRATION 13.9

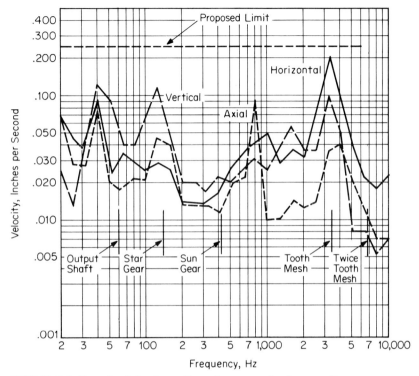

FIGURE 13.9 Example of vibration readings taken on an epicyclic gear unit.

Figure 13.10 shows the oscilloscope patterns for a "floating" sun pinion in an epicyclic gear unit running at about 60 percent of full rating. Note that a fast camera exposure can show what happens in 2½ turns of the shaft. A longer exposure shows 6 turns of the shaft. The oscilloscope data show that the orbit pattern is constantly changing due to vibrations at different frequencies.

Vibration data are particularly useful in monitoring units in service that may be prone to failure. A serious defect in geometric quality or metallurgical quality may get through even extensive and well-disciplined quality control inspections. Vibration checking of equipment in service is the "last chance" to catch a defect before a serious failure occurs.

Figure 13.11 shows an example of vibration behavior related to an impending failure. Shortly after start-up, a set of vibration readings were taken. The reading in the horizontal direction at one location was higher than the other readings. A month later this reading was rechecked and found to be still higher. In another month the reading had become substantially higher. Then, two days later, the reading was even higher. At this time the unit was taken out of service. When the unit was taken apart, serious bearing damage and gear tooth damage were found. After replacing the damaged parts, the rebuilt gear unit functioned normally for many years.

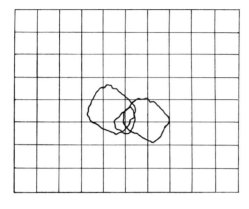

View A - Exposure 1/100 second.
Shows about 2½ loops of orbit due to 240 Hz vibration.

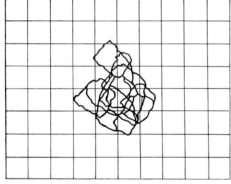

View B - Exposure 1/30 second.
Shows about 6 loops of 245 Hz vibration, 3 loops of small 92 Hz vibration and pattern for 30 Hz.

FIGURE 13.10 Examples of orbit patterns for a "floating" sun pinion in an epicyclic gear unit. Note how vibrations at different frequencies can be seen on the oscilloscope.

13.4 APPROXIMATE VIBRATION LIMITS

When we consider a wide range of gear sizes like 100 kW to over 20,000 kW, and all types of prime movers such as electric motors, turbines, and internal combustion engines, it is obvious that the vibration criteria for a well-built, good-running gearbox is bound to vary over some range of values. Some other things that make vibration vary are the kind of teeth (spur or helical), the style of manufacture (low-hardness cut teeth or high-hardness ground teeth), and the casing type (cast iron with heavy wall sections, or steel weldments with somewhat thin walls made of steel plate).

Figure 13.12 shows vibration limits based on the author's experience. Instead of showing one curve as a proposed limit, three curves are shown.

Upper curve with long dashes. This curve represents values that are on the "high" side. The statement is, the unit is *probably in trouble* if these values are exceeded.

Middle curve with solid line. This curve represents a middle-of-the-road value. It can be considered a *nominal design value*.

GEAR VIBRATION   13.11

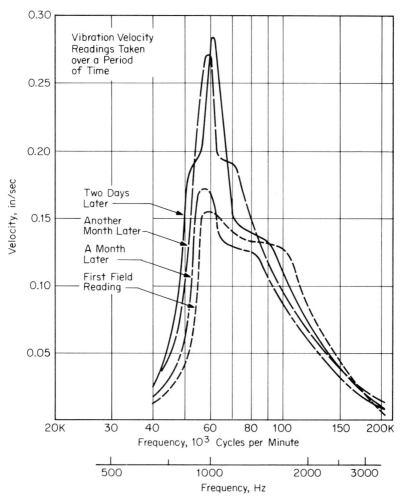

FIGURE 13.11 A turbine-driven gear unit is found to be in a failure mode and shut down before breakage of parts.

Lower curve with short dashes. This curve represents close control of vibration. Even units that are quite sensitive to mechanical failure are *probably OK* if the curve of vibration intensity versus frequency is below these values.

13.4.1 Velocity Limits

Figure 13.12 shows the three curves for casing velocity limits in part 1. Note that these curves run from 20 Hz to 1000 Hz. This is the frequency range where most velocity transducers can be expected to work well.

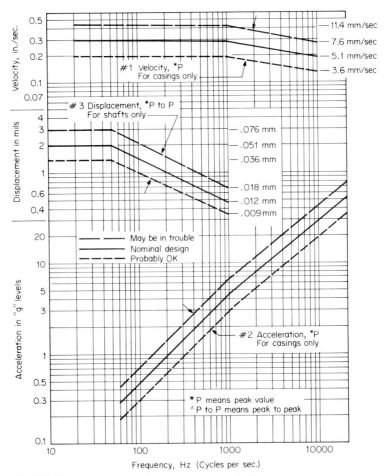

FIGURE 13.12 Range of vibration values recommended for consideration by gear designers or users.

It is rather consoling to the gear engineer that the appropriate velocity limit tends to stay fixed over a rather wide frequency range. This means, in many cases, that a measurement of vibration intensity tells the experienced engineer whether or not there is a vibration problem—without consideration for the frequency and how allowed vibration changes with frequency.

13.4.2 Acceleration Limits

Figure 13.12 in part 2 shows three curves for acceleration in g levels. These curves are plotted from 60 Hz up to 20,000 Hz. This is a range where it should be possible to use the accelerometer as the transducer to pick up casing vibration and get relatively accurate readings without undue effort. At vibration readings of more than 20,000 Hz, the problem of avoiding vibration in the accelerometer mounting device becomes quite difficult to handle.

When the vibration readings go below 60 Hz, the allowable g values become very low. This means a very low electrical current in the pickup wires and difficulty in transmitting this signal accurately back to the recording and analyzing equipment.

Quite often accelerometers are used to measure vibrations in the range of 60 Hz to 10,000 Hz, and then the readout is given in velocity. (See Table 13.1 for equations relating displacement, velocity, and acceleration.) This works fairly well. Of course, if the vibration is not close to being simple harmonic (sine wave) in pattern, then velocity converted from acceleration will not be the same as velocity read directly.

Below 60 Hz, there is apt to be trouble in getting reliable velocity readings from conversion of accelerometer readings. Considering the state of the art in 1991, the recommendation is to use velocity transducers or other means when measuring rather low-frequency vibrations.

13.4.3 Proximity Probes

These devices are generally used on rotating shafts rather than on casings. It is possible, though, to use them on casings.

The proximity probe has no particular problem working down to very low speeds. Even a speed as low as 0.01 Hz (0.6 c/min) is practical.

At high speeds, the proximity probe is usually not practical when the limit reading is very small. For instance, a displacement reading less than 0.0075 mm (0.0003 in.) tends to be in trouble due to possible surface irregularities of the shaft and small variations in composition of the shaft. If the proximity probes gives a reading of 0.0003 in., it is hard to decide whether this reading is really vibration or whether it represents shaft imperfections.

13.4.4 Displacement Limits

At frequencies below 50 Hz, one might question whether the limit values shown in part 3 of Fig. 13.12 really represent a vibration hazard. (The matching values of either velocity or acceleration get very low).

A displacement reading of 0.05 mm (0.002 in.) at 20 Hz doesn't shake a gearbox very much. Higher readings, though, tend to indicate misfit in bearings. This can spell trouble, for instance, to a low-speed output shaft in a gearbox that is turning at only 1 r/min. For good gear machinery, there is a concern to keep shaft runouts relatively low even at slow speed and low vibration frequencies.

For epicyclic gears, there is often one member "free floating." This may be the sun, the planet carrier, or the annulus gear. Displacement readings by proximity probes are quite helpful to determine whether or not the orbit pattern of the free-floating member is acceptable. The displacement limits in part 3 Fig. 13.12 do not apply to free-floating epicyclic gear elements.

13.4.5 General Vibration Tendencies

A gear designer may be working on a low-cost, short-life gear unit where weight is somewhat important. The design may use features that are bound to make the vibration levels come somewhat high. If this unit is used in an environment where noise and vibration can be tolerated, then it is quite reasonable to use somewhat high vibration limits.

In this case the key to what vibration limits should be used comes from building a number of units that meet all the design objectives—and taking enough vibration data to understand what vibration values are *normal* for this application. After the normal values are established, the gear builder and gear user can use these values as a quality control check to help determine if additional units are really as good as the first units built and found to be satisfactory.

In a similar vein, a gear designer working on a long-life, expensive gear unit to be used in a critical environment will find that the vibration values for units meeting all project objectives are rather low. This means that rather low limits must be set on vibration to maintain the desired quality in a long production run.

To help designers and users of gears understand the many factors that enter into vibration values, Table 13.2 is presented. Table 13.2 is based primarily on overall design considerations rather than just vibration logic. For instance, a turbine-driven, epicyclic gear unit made rather lightweight may show a g level of more than 40 at a frequency over 10,000 Hz. Such units at more than 1000 kW have a very large amount of kinetic energy for the mass of the parts. With good materials, good accuracy, and appropriate lubrication systems, a given design may demonstrate a service life of more than 30,000 h, even with a 40-g vibration level.

13.4.6 Trade Standard

The American Gear Manufacturers Association (AGMA) has a standard for linear vibration on gear units. This standard is ANSI/AGMA 6000-A88.[1]

Much useful data is given in this standard in regard to definitions, instrumentation, measurement, and proposed test conditions when acceptance tests are being run. The standard aims to help the gear buyer and the gear maker negotiate a reasonable contract for the vibration part of gear acceptance tests.

Those designing or purchasing gear units should study ANSI-AGMA 6000-A88 carefully.

TABLE 13.2 Factors That Raise or Lower Allowable Vibration

Factor	What Raises Vibration Limit	What Reduces Vibration Limit
Required life at or near full rating	Less than 2000 hours	Over 20,000 hours
Material hardness	38 HRC or lower (through hardened)	50 to 65 HRC (case hardened)
Kind of gear	Straight bevel gear/Spur gear	Spiral bevel gear/Helical gear
Accuracy of teeth	Medium precision (cut gears)	High precision (ground gears)
Pitch line velocity	less than 25.4 m/s (5000 ft/min)	Over 25.4 m/s (5000 ft/min)
Rated power	Less than 400 kW (500 hp)	Over 1500 kW (2000 hp)
Weight	Lightweight	Heavyweight

GEAR VIBRATION   **13.15**

Figures 13.13, 13.14, and 13.15 are extracted from the AGMA standard. They show the Class A and Class B limits for displacement, velocity, and acceleration. These AGMA limits are not binding unless specifically agreed to by the manufacturer and the purchaser.

13.5 CONTROL OF VIBRATION IN MANUFACTURING GEARS AND IN THE FIELD

The obvious starting point for vibration control is the mechanical design of the gears, shafts, bearings, and casings.

The gear teeth need to have suitable geometric accuracy. This involves tooth profile, tooth spacing, helix angle, and concentricity. It also involves tooth surface finish, profile modification, and helix modification. The teeth need to roll through mesh very smoothly with no undue shock as tips of teeth and ends of teeth enter and leave mesh.

Chapter 5 of Dudley[2] gives some general advice on gear accuracy limits that are used in a variety of gear applications. The accuracy needs for gear teeth become quite critical when one or more of the following apply:

The power transmitted across a single gear mesh is over 1000 KW, or over 1340 horsepower.

The pitch line speed is over 50 m/s, or over 10,000 ft/min.

The required life of the gearing at a high level of loading is over 10,000 h.

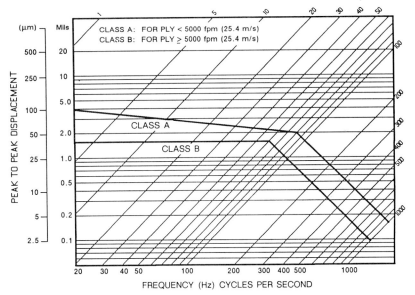

FIGURE 13.13 Displacement limits. *(Extracted from AGMA Standard 6000-A88,* Specification for Measurement of Linear Vibration on Gear Units, *with permission of the publisher, the American Gear Manufacturers Association, 1500 King Street, Suite 201, Alexandria, Virginia 22314)*

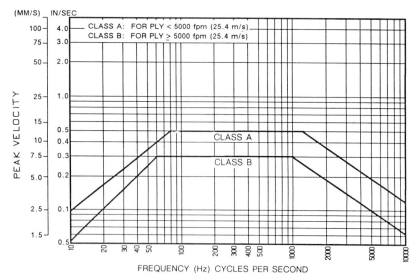

FIGURE 13.14 Velocity limits. *(Extracted from AGMA Standard 6000-A88)*

The shafts in a gear system need to so support the meshing gears as to provide good alignment even when sizable changes occur in operating speed, percent of load being carried, or operating temperature. If the shafts are rather long, the shaft windup may result in a torsional critical speed that is within the operating range of the gear unit. Large gear drives in ships or mills are often subject to torsional criticals, and therefore critical speeds must be calculated for the whole power system. If a critical speed comes too close to an operating speed, it is usually possible to shift the critical speed away from the operating speed by changing shaft diameters and masses of the rotating gears (or masses of other rotating elements).

The bearings in a gear system may be subject to "half frequency whirl," ball path excursions, or axial runout effects. Both sliding-element bearings and rolling-element bearings have their problems. The design of a critical gear unit must involve a consideration of possible vibration hazards coming from the bearings. It is usually possible to pick a kind of bearing and a mechanical design for the bearing so that the bearing does have a bad effect on the vibration behavior of the gear unit.

Large gear casings may have side panels that have critical vibration frequencies that come at some mesh frequency present in the gearbox. The attachment of the gearbox to a support structure may be such as to result in serious vibrations of the whole structure under certain conditions of operation (speed and load).

Both the mounts for a gear unit and the casing walls may need special devices to suppress vibration. To sum it up, the control of vibration starts with the mechanical design of gears, shafts, bearings, and casing structures.

13.5.1 Testing of Gear Units at the Gear Factory

Gear units for turbine drive equipment in the oil and gas industry are often given a full-speed, no-load type of test before shipment. Typically, a unit is operated for about

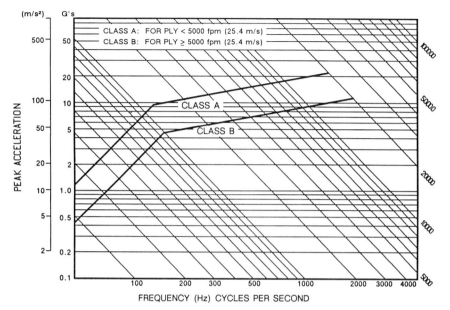

FIGURE 13.15 Acceleration limits. *(Extracted from AGMA Standard 6000-A88)*

one hour at half speed, three-quarter speed, and full speed. Vibration readings are taken. Oil flow, oil temperature rise, and critical bearing temperatures are also measured.

The full-speed, no-load test is useful to make sure that the rotating components are all balanced well enough to run without undue vibration. The oil system is also checked out. A bearing that is too tight may run hot. If an oil passage is not drilled, the oil flow will not be right.

In very critical units, the purchase contract may call for a factory test at full speed and full power. Full-power tests may be made with a dynamometer or water brake to absorb the rated power of the gear unit. Sometimes the full-power test is made with one gear unit loaded against another gear unit.

The full-power test is very useful to catch defects, such as

Too much spacing error is accumulated in the gear teeth. (Vibration is high when full torque is applied.)

Bearing fit is not right. Bearings show distress after running 20 or more hours at full torque.

The tooth fit (under load) is not right. Early signs of tooth distress show up.

Vibration shows up under torque loading at some speed in the operational range. A torsional or a lateral critical may be present. Another possibility is that some bearing may not be loaded enough to seat itself in its proper position. (Gear weight or coupling reaction causes abnormal conditions in the bearing when running at partial torque.)

13.5.2 Tests of the Assembled Power Package

After the gear unit has been tested, it is often shipped to another factory where the whole power package is put together and tested. The power package involves a prime mover, a gear unit, and a driven unit. Some examples are

Turbine, gear unit, electric generator
Turbine, gear unit, compressor
Diesel engine, gear unit, pump
Electric motor, gear unit, fan

The test of the whole power package involves starting, stopping, running at different power settings, overspeed test, high- or low-temperature tests and endurance test. During tests such as these, vibration is measured, along with many other things such as oil flow, bearing temperature, fuel rate, output energy, etc.

Supposedly, the gear unit vibration will be satisfactory—since the gear unit has already had a gear factory test. Quite often, though, the vibration behavior of the gear may be a problem. Here are some possibilities:

The gear unit vibrates due to misalignment or unbalance in couplings between the gear unit and the prime mover or between the gear unit and the driven equipment.

The heavy gear unit is within vibration limits but unbalance in the gear unit causes lightweight accessory equipment on the turbine to vibrate excessively.

The diesel engine is not timed to fire just right on all cylinders. This causes vibration to be high on the gearbox.

A centrifugal pump has a whirl mode. The pump is within limits on vibration, but it causes the gear unit to be over limits.

At the time of the power package test, it is very strategic to get a *vibration signature* of the gear unit (as well as a vibration signature of the prime mover and the driven machine).

When the power unit is installed in the field, the vibration signature can be compared with the vibration pattern at start-up. If the installation is made properly, the vibration pattern in the field should match the factory vibration signature quite closely.

13.5.3 Vibration Tests in the Field

After a power package is installed and operating in the field, vibration testing becomes most important. An increase in vibration very often means an impending failure. If it is an aircraft, it could mean that an engine would have to be removed before another flight was made. On a large ship, high gear vibration is likely to mean that gear repair work must be done before the ship can be insured to carry another load of cargo across the ocean.

In the oil and gas industry, there is always the danger of explosions or fire in an oil refinery or on an oil platform. The equipment that is running constantly is generally checked each day with some simple vibration equipment. If some gear unit or other piece of power equipment reads significantly higher one day than it did the day before, it is a matter of immediate concern. Even when vibration values are low, a change upward is an immediate warning.

When simple vibration checks show an upward shift, it is likely that more complex

GEAR VIBRATION

vibration equipment will be brought to the site. A vibration expert will make a complete analysis of the situation, and may make one of the following recommendations:

Shut down the unit immediately.

Observe the unit on a daily basis for some time to see if the vibration continues to increase.

The simple vibration check (perhaps using a hand-held instrument) was faulty—there is nothing to worry about.

In a situation such as this, a vibration signature is most helpful to the person making the decision. With the vibration signature, the whole pattern of change from a new unit running under factory conditions to a used unit running under field conditions is revealed.

In many situations, the risk of failures has led equipment operators to want to purchase gear units with installed vibration monitors. One or two strategic points are picked for a gear unit, and the vibration monitor reads continuously as the unit runs. The usual range is

Normal

Near maximum

Shut down

The continuous monitoring equipment has the obvious advantage that there can be automatic shutdown any time the vibration is dangerous.

However, there are a couple of problem areas. At start-up a unit is cold and may be out of alignment if there is a need to allow for a change in alignment when normal steady-state running conditions are reached. This can mean that a unit keeps going over the shutdown vibration limit when someone is trying to start the unit.

Another problem is that operating people assume that all is well when the vibration is low. Quite often a vibration at some location may measure no more than 0.05 in./s. If the shutdown was 0.30 in./s and normal went up to 0.25, then a change of 0.05 to 0.10 in one day and a change from 0.10 to 0.15 a day later could almost pass unnoticed. If there is a quick change of 0.10 in vibration, something is going wrong and needs to be considered—even if the vibration is still in a "normal" range. From a practical standpoint, *change in field readings is more important than the measured value.*

13.6 VIBRATION ANALYSIS TECHNIQUES

The techniques of analyzing and presenting gear unit vibration data have become somewhat numerous. Vibration specialists keep trying to develop better ways to determine what is wrong—or not wrong—with a gear unit from the vibration signals picked up on the gear unit when it is running at some output speed and at some output torque.

The concern over vibration becomes most critical when a gear unit may have a gear tooth with a developing fatigue crack—and the breakage of this tooth might lead to costly damage to the gear unit or (even worse) be a safety hazard in an oil refinery or on a passenger aircraft.

Even in a new gear unit, the risk of one or more gear teeth being accidentally damaged in installation or being malformed due to a local final tooth grinding mistake can be a matter of serious concern.

TABLE 13.3 Analysis Methods for Gear Vibration

Method	Advantages	Disadvantages
Vibration amplitude plotted against frequency (May be called "Spectrum analysis.")	A standard means of presenting vibration data. Specified in ANSI/AGMA 6000-A88. Good to determine average vibration at each shaft frequency and each mesh frequency. Instrumentation readily available.	If there are one or more bad gear teeth in the gear circumference, these are not located.
Cepstra	Can find individual bad teeth. Instrumentation is available.	Can confuse tooth defects with other shaft-related problems.
Time domain averaging	Improved vibration spectrum. Improved defect detection. Instrumention available.	Requires shaft encoder signal. Interpretation can be complex.
RMS vibration levels.	Easy to implement. Simple instrumentation.	Cannot isolate gear tooth vibration from other vibrations. High false alarm rate.
Tooth by tooth	The most effective method to find defects in all the teeth of a gear.	Instrumentation developed but not generally available in gear making shops. Requires shaft encoder. Will become better known to industry through the 1990s.

Table 13.3 shows some of the analysis methods in current use. Note that the commonly used method of plotting vibration intensity versus discrete frequencies does not have the capability to locate a single bad tooth. Of course, if there is just one bad tooth, there may be a high vibration "spike" coming in a shaft frequency—rather than gear mesh frequency.

The method that can locate any or all bad teeth in a gear is the tooth-by-tooth measuring system using encoders. (See last item in Table 13.3.)

For most gear units used in factories, aboard ships, or in oil refineries and pumping stations, the spectrum analysis shown as the first entry in Table 13.3 is quite adequate.

From a practical standpoint, many very important gear units are built and shipped without any vibration testing being done either in the gear factory or on site in the field. Although improvements and new developments in vibration analysis will be helpful, the greatest need (from the author's viewpoint) is for more gear production with a vibration control being exercised along the lines discussed earlier in this chapter.

REFERENCES

1. AGMA Standard 6000-A88, Specification for Measurement of Linear Vibration on Gear Units, American Gear Manufacturers Association, Alexandria, Virginia.
2. Dudley, D. W., *Handbook of Practical Gear Design,* McGraw-Hill, New York, 1984.

CHAPTER 14
GEAR NOISE

Professor Donald Houser
Ohio State University
Columbus, Ohio

Gear noise is an often overlooked, yet very important gear design consideration. Due to increased emphasis on noise regulation and an increase in customer awareness of noise, the importance of gear noise in design has increased significantly in recent years. Where gears are used for power transmission, gear noise is of particular concern because it is often close to a pure tone, has a high sound pressure level, and its predominant frequencies are often near the ear's most sensitive region.

The following situations are likely to cause concern over gear noise:

Legal noise limits exceeded: These situations arise when legal standards such as those developed by OSHA or the EPA are being exceeded.

Annoyance: Often gear drives do not come close to exceeding legal noise limits but yet produce sounds that are annoying to persons near the gear units. The high-pitch whine that is characteristic of many gears often results in unacceptable noises in consumer products.

Contractual limits exceeded: Gear manufacturers are often required by their customers to meet a prescribed set of noise criteria. This is often the case in industrial and marine speed reducers.

Although the applications of gears differ widely, the sources of gear noise remain much the same. Therefore, the methods used to design quiet gears in one application are typically very similar to the techniques used in totally different applications, and some general statements may be made concerning design trends that yield quiet gears.

This chapter first presents a discussion of the basic sources of gear noise and then presents different methods of measuring and displaying gear noise information. Next, a discussion of transmission error, an important gear excitation that is the major source of gear noise, is presented. The most important section of this chapter from a practical viewpoint discusses means of controlling or reducing gear noise. A section is devoted to the development of mathematical models for predicting gear noise and the subject of gear rattle is briefly discussed. While good design is desirable, many situations arise where the gears are already designed and manufactured, and it is necessary to troubleshoot a noise problem. Therefore, a short section is devoted to gear noise troubleshooting. Although most of the discussions in this chapter are directed toward spur and helical gears, the same concepts may be extended to all other types of gears.

14.1 GEAR NOISE: THE GENERAL PROBLEM

In most applications, the term *gear noise* should more aptly be called *transmission noise* since the sound that results from the meshing of gear teeth is, in reality, transmitted via forces and motions to the shafting, bearings, and transmission housing where it is then radiated to the surroundings, as depicted in Fig. 14.1. Only in open gearing applications is the gear noise transmitted directly from the gears to the listener.

The block diagram in Fig. 14.1*b* better shows the various paths by which gear noise travels to the listener. In most cases the gear sounds originating in the gear mesh are due to nonperfect (nonconjugate) action of the gears. This nonperfect action results in dynamic forces at the gear teeth, which in turn excite vibrations in the gear blanks and shafting. Gear blank vibration generates airborne noise that transmits outward to the listener in unhoused gears but is contained inside the enclosure of housed gears. The housing walls are a poor transmitter of sound, so usually little or no airborne sound from the gear blanks reaches the human ear. The more significant transmission path is structureborne, as dynamic forces in the gear mesh are transmitted through the shafting and bearings to the housing panels. The resulting panel vibrations serve as the "speakers" that propagate the gear noise heard by the listener. A third path that can be important is for forces to be transmitted through the gearbox mounts to surrounding surfaces, whose vibrations then propagate the noise.

Based on the flow of acoustic and vibration energy, several approaches may be taken to reduce gear noise:

1. Reduce the excitation at the gear mesh.
2. Reduce the dynamic force paths and the vibrations between the gear mesh and the housing.
3. Reduce the housing's acoustic radiation efficiency.
4. Modify the environment in which the gearbox is placed.

The last approach, which includes using sound barriers, room insulation, etc., is covered in a number of noise control books[1,2,3] and is not covered in detail in this chapter. Most gear noise is characterized by frequency components at the gear mesh frequency and its multiples and by modulations of mesh frequency (sidebands). Figure 14.2 provides data for a simple pair of gears as well as a depiction of the frequency spectrum components that would be expected. The primary frequencies are calculated using the following formulas:

$$f_{s1} = \frac{N_1}{60}$$

$$f_{s2} = \frac{N_2}{60} = f_{s1} \frac{Z_1}{Z_2}$$

$$f_m = f_{s1} Z_1 = f_{s2} Z_2$$

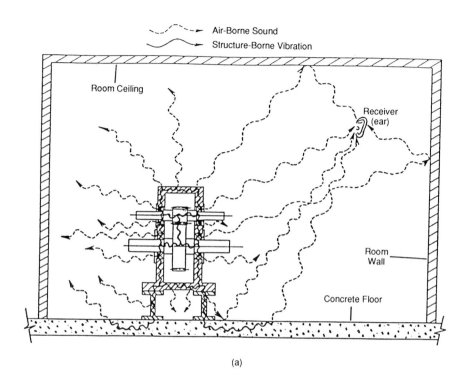

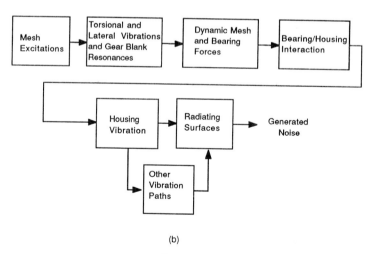

(b)

FIGURE 14.1 Gear noise transmission paths.[21]

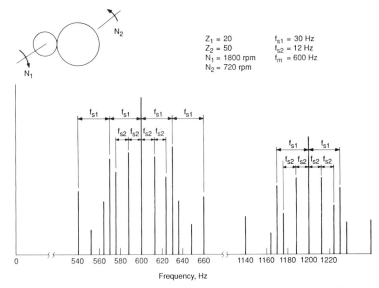

FIGURE 14.2 Components of the frequency spectrum commonly associated with mesh frequency noise.

where f_{s1} = input shaft frequency, Hz
f_{s2} = output shaft frequency, Hz
f_m = gear mesh (tooth pass) frequency, Hz
N_1 = input shaft speed, rpm
N_2 = output shaft speed, rpm
Z_1 = number of teeth on the input gear
Z_2 = number of teeth on the output gear

Harmonics occur at integer multiples of the mesh frequency, that is, the second harmonic is at $2f_m$; the third harmonic at $3f_m$, etc. Sidebands may occur about both the mesh frequency and its harmonics and are most commonly spaced at the shaft frequency. Equations for computing the sidebands are given in the following:

$$f_{sb} = f_m \pm nf_{s1}, \quad f_m \pm nf_{s1}, \quad 2f_m \pm nf_{s1}, \quad 2f_m \pm nf_{s2}, \text{ etc.} \quad (14.1)$$

where n is any integer.

The harmonics and sidebands have a strong influence on the character of the sound that is heard.

The following concepts have been proposed as excitations of gear noise that are periodic at the gear mesh frequency:

Mesh stiffness changes[4]
Transmission errors[5,6,7,8]
Gear tooth impacts at the initiation of tooth contact[4,9,10]
Dynamic mesh forces[11]

Frictional force effects [10,12,13]
Air pocketing [10,14,15,16]
Lubricant entrainment [10,17]

14.1.1 Mesh Stiffness

Mesh stiffness is the ratio of the force along the line of action to tooth deflection along the same line. The mesh stiffness varies as the gear teeth rotate through a mesh cycle. In spur gears, where contact alternates between single-tooth-pair contact and double-tooth-pair contact, each tooth pair may be thought of as a spring. The mesh stiffness alternates between the stiffness of a single spring and the stiffness of two springs in parallel, that is, the stiffness alternately changes by a factor of 2 as the gears rotate. For helical gears the mesh stiffness is roughly proportional to the sum of the lengths of the respective contact lines.[18]

14.1.2 Transmission Error

Transmission error is the single most important factor in the generation of gear noise. It is defined as "the difference between the actual position of the output gear and the position it would occupy if the gears were perfectly conjugate" and may be expressed in angular units or as a linear displacement along the line of action.[18] The respective equations for transmission error are given below:

$$\text{TE} = \theta_2 - \frac{Z_2}{Z_1} \theta_1 \text{ rad} \tag{14.2}$$

or

$$\text{TE} = R_{b2} \left(\theta_2 - \frac{Z_2}{Z_1} \theta_1 \right) \text{ in. or mm} \tag{14.3}$$

where θ_1 = angular rotation of the input gear
 θ_2 = angular rotation of the output gear
 R_{b2} = Base radius of the output gear

The foregoing definition is appropriate for both loaded and unloaded gears. When gears are unloaded, transmission errors result from manufacturing inaccuracies such as profile errors, spacing errors, and runout (a "plus" amount of material added to either a pinion tooth or gear tooth will cause a positive transmission error). When loaded, the changes in deflections due to mesh stiffness variations must be accounted for in the evaluation of transmission error (tooth deflections cause the output gear to lag behind the input gear and hence result in negative transmission errors). The time-varying component of transmission error, which is periodic at tooth mesh frequency, has been shown to be related to gear noise amplitude.[6,7] In fact, it has been shown that the transmission error of spur gears, which have large changes in mesh stiffness, can be reduced significantly by applying appropriate profile modifications.[4,14,18,18,19,20] Unfortunately, for spur gears these modifications are usually an optimum for one load level, and gears operating away from this design load will have increased noise. A later section of this chapter provides additional discussion of methods of reducing transmission error.

14.1.3 Gear Tooth Impacts

Gear tooth impacts occur when tooth deflections and/or spacing errors cause tooth contact to occur prematurely at the tooth tip. This premature contact occurs off the line of action due to a velocity mismatch normal to the tooth contact. This velocity mismatch results in an impact at the tip of the driven tooth. This impact can result in large dynamic forces, which not only can cause large mesh frequency noise levels, but can also cause significant reductions of gear tooth fatigue life. These impacts can be minimized, however, by providing adequate tip and root relief and tooth crowning. Even with proper relief, there is a sudden shear force due to the instantaneous sliding that occurs at the initiation of contact, which could be a source of noise.[5,18]

14.1.4 Dynamic Mesh Forces

Dynamic mesh forces are the result of both mesh stiffness variations and transmission errors. These forces are transmitted through the bearings to the gear housing. Although the forces are a very logical means of observing the results of mesh excitations, the basic excitations are still the mesh stiffness variations and resulting transmission errors.

14.1.5 Frictional Forces

Frictional forces due to gear tooth sliding have the potential to provide a mesh frequency excitation. The meshing action of gear teeth is characterized by a combination of rolling and sliding. As the gear teeth enter contact (approach action), sliding is at a maximum and it decreases to zero when it reaches the pitch point where pure rolling exists. The sliding direction reverses as contact progresses past the pitch point. This change in sliding direction at the pitch point causes a sudden reversal in the direction of the frictional force. This frictional force may be great enough to provide adequate excitation to affect gear noise. The occurrence of this phenomenon is conceivable for spur gears, where pitch point contact is at a distinct rotational position; but for helical gears, where contact is located along the total tooth face at any instant of time, this effect would be greatly diminished. It is generally felt that this sliding effect is of secondary importance.

14.1.6 Air Pocketing

Air pocketing is the noise generated by rapidly squeezing air through the gear teeth as they mesh. For high-speed gearing, air velocities can approach the speed of sound and can result in intense sounds at gear mesh frequency.

14.1.7 Lubricant Entrainment

Lubricant entrainment occurs when clearances are inadequate to allow the lubricant to be squeezed out of the mesh. This problem can be improved by increasing backlash and by designing gear teeth with sufficient whole depths to provide adequate clearance between the tip and root of the respective mating teeth. Also, the method of applying the lubricant, that is, whether a splash system is used or whether lubrication jets are applied at the entrance or exit to the mesh, may affect the amount of lubricant entrainment.

14.2 MEASUREMENTS AND STANDARDS

The ultimate goal of the gear designer is to meet certain noise requirements that may be specified by the customer or provided in a standard. Since the requirements of customers differ a great deal, this section outlines some of the recommended standards and discusses some of the more common procedures for performing gearbox noise measurements.

The American Gear Manufacturers Association (AGMA) has published four standards on gear noise. These standards include the three-part "Gear Sound Manual"[21] and special standards for high speed gear units,[22] enclosed drives,[23] and gearmotors.[24] The "Gear Sound Manual" presents noise fundamentals, specifications for the standardization of gear noise measurements, and a section on the control of gear noise. In addition, many foreign gear standards organizations have developed gear noise standards, and ISO is developing an international standard. Extensive listings of other acoustic standards which might apply to various gear transmission applications are available.[25,26]

Several types of measurement and analysis techniques may be performed to evaluate the noise of a gearbox:

14.2.1 Sound Pressure Level

The sound pressure level, which is usually measured with a sound level meter, provides a single measure of the sound level over the entire audible frequency range. The measurement is taken at a single point at a prescribed distance from the gearbox (usually 3 ft or 1 m) and is given in decibels.[1,2,3] Normally, A weighting, which adjusts the frequency content of the sound pressure measurement to match the human ear's sensitivity, is required by legal or health requirements (OSHA, ANSI, etc.). Other weightings such as flat (unweighted) and C weighting (less filtering at low frequencies than A weighting) are also available on most sound level meters. These different weightings allow some elementary discrimination of the frequency content of gear noise. Because sound transmission is directional, sound pressure readings will likely be different at various locations around a gearbox. Likewise, as one takes measurements at different distances from the gearbox, the sound level may also change, depending on the measurement room's acoustic characteristics.

14.2.2 Octave Band and $\frac{1}{3}$-Octave-Band Measurements

These measurements allow a coarse evaluation of the frequency content of the gearbox, but still do not adequately discriminate frequencies such as gear mesh tones from other noise sources. The A weighting is often used in specifications requiring either octave or $\frac{1}{3}$-octave-band measurements.

14.2.3 Narrow Bandwidth Analysis

Most modern spectrum analyzers perform constant bandwidth analysis where the frequency resolution is constant throughout the analysis frequency range. For instance, an analyzer having 400 output frequency lines between 0 and 2000 Hz has a resolution of 5 Hz. Additional resolution can be obtained by using frequency expansion techniques such as "zoom" analysis. When one is trying to locate the source of sounds, narrowband analysis is recommended.

Figure 14.3 presents a comparison of overall, octave, one-third-octave and narrowband spectrum analysis of a double-reduction gear set with mesh frequencies at about 290 and 1010 Hz, respectively.[21] Note that the mesh frequency and harmonics are identifiable only in the narrow $\frac{1}{10}$-octave analysis. Had a constant bandwidth analysis with zoom been applied to this gearbox, even clearer identification of the mesh frequency harmonics and sidebands would have been possible.

14.2.4 Synchronous Time Averaging

In systems with multiple reductions, synchronous time averaging provides a method of isolating the sources of sounds to individual shafts. This technique, which requires

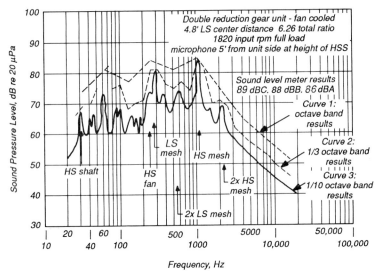

Source	Exciting frequency, Hz
HS mesh	1060
LS mesh	285
HS shaft	30.4
INT. shaft	20.4
LS shaft	4.84
HS fan	2.43

FIGURE 14.3 Comparison of octave, $\frac{1}{3}$-octave, and narrowband spectrum analysis of gear noise.[21]

an external synchronization pulse from each of the shafts to be analyzed, is available on most modern spectrum analyzers. Synchronizing the sampling of the time data to a given shaft's rotation allows only frequency peaks related to that shaft to be displayed.

14.2.5 Standard Measurements

Since gearing configurations differ a great deal, it is impossible to provide one measuring technique that satisfies all gearbox types. However, in Fig. 14.4 the AGMA shows a recommended location of the microphone for measurement of noise from a gearbox.[23]

Directivity of sound and the room surroundings are not considered in the foregoing procedure, so it is sometimes necessary to measure the overall sound power that is generated from a gearbox. Several procedures exist for measuring sound power:

1. Take measurements from an array of microphones placed around the gearbox and then integrate the results. This method normally requires the gearbox to be placed in an anechoic or semianechoic space.[1,2,3]

2. Measurement in a reverberant room.[1,2,3] This technique is difficult for gearing applications, since the prime mover and load must be isolated from the gearbox, thus making these tests extremely expensive.

3. Use of the two-microphone acoustic intensity technique.[27] This technique has the capability of measuring the sound power level by integrating microphone measurements that are taken at multiple positions in close proximity to the gearbox. If properly performed, the effect of background noises can be minimized.

A very common procedure for vehicle transmissions is to perform drive-by tests. SAE and other automotive standards organizations have prepared standard procedures for these tests.[25]

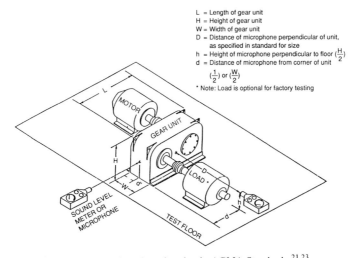

FIGURE 14.4 Single microphone location in AGMA Standards.[21,23]

14.3 POWER CONSIDERATIONS

Figure 14.5 shows measurements of the effect of transmitted power on the noise levels of numerous gearboxes of varying quality.[4] The variability in the amount of acoustic power generated is attributed to differences in the gearbox quality and a series of quality levels are superimposed on the graph. An extrapolation of the data suggests that a perfect gearbox might radiate 10^{-9} of the power transmitted. In another experimental study,[8] ground gears operating at 20 kW and 2000 r/min, achieved radiated sound power of only 24 μW which is 1.2×10^{-9} of the power transmitted. This low acoustic efficiency is possible because the dynamic forces that occur as each tooth passes through contact can be controlled so that the sum of respective tooth forces provides a smooth time-force pattern at the bearings.

Two implications are made from these data:

1. More accurate gears make less noise.
2. A doubling of power results in a 3-dB increase in gear noise.

The accuracy statement is generally true, but only if the tooth profiles and leads are designed properly (see later discussions). Increasing the power transmitted, on the other hand, can be achieved by either increasing speed or increasing torque.

Most investigators[4,9,16,17,28,29] have reported that the doubling of pitch line velocity results in a 5–7-dB increase in gear noise, whereas some investigators report increases as small as 2–3 dB and as great as 10 dB.[30] A simplistic model for predicting mesh-

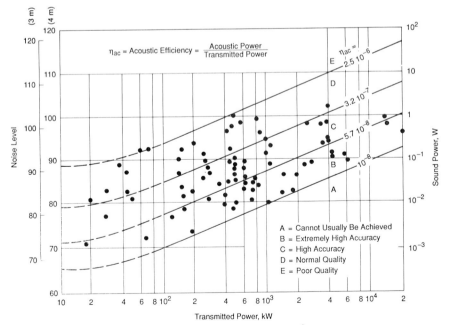

FIGURE 14.5 The effect of gear accuracy and power on gear noise.[4]

excited housing velocity (noise is usually proportional to velocity) shows a 6-dB increase in velocity for a doubling of speed.[11] When ascertaining speed effects, one should be aware that when the gear mesh frequency is near a system resonance, the noise level can be greatly enhanced, and when near an antiresonance, it will be greatly reduced.

Many investigators[4,9,15,28,29,30] have reported that doubling torque gives a sound level increase of 3 dB; yet others[16,19] report increases as high as 5–6 dB. Still others[8,14] have actually reported decreases in sound level with increased torque. This latter effect is conceivable in a properly designed gear set, as discussed in the next section.

14.4 TRANSMISSION ERROR IN GEARS

Two types of transmission error are commonly referred to in the literature. The first is the *manufactured transmission error* (MTE), which is obtained for unloaded gear sets when rotated in single flank contact, and the second is *loaded transmission error* (LTE), which is similar in principle to MTE but takes into account tooth deflections due to load.[18]

14.4.1 Manufactured Transmission Error (MTE)

The MTE is affected most by profile inaccuracies, spacing errors, and gear tooth runout, as shown in Fig. 14.6.[6,32] The frequency spectrum of a typical transmission error measurement shows the mesh frequency and its harmonics and a component at gear shaft frequency that is due to gear runout. Gears that have perfect involute profiles and no spacing or runout errors should produce a perfectly straight transmission error trace, which would result in a spectrum with no peaks at discrete frequencies. It has been shown that there is a direct relation between the manufactured transmission error and noise,[5,33,34,35] and in general it has been found that doubling the MTE gives a 6-dB increase in noise.[18]

14.4.2 Loaded Transmission Error (LTE)

When gears operating at low speed are loaded, two additional factors contribute to the transmission error:

1. A constant component due to the mean tooth compliance. This component is of major significance in choosing appropriate profile modifications, but of much less significance with regard to its contribution to mesh frequency noise.
2. A time-varying component that is a function of gear tooth geometry and mesh stiffness variation, as well as the manufactured transmission error. This component contributes heavily to mesh frequency noise.

As gears are run at higher speeds, a dynamic component that is a function of the system dynamics must be included with the aforementioned effects.[28,36]

Spur Gears. The mesh stiffness of spur gears is approximately proportional to the number of tooth pairs in contact. Since the number of tooth pairs in contact changes with rotation, the mesh stiffness and, hence, the transmission error also change with

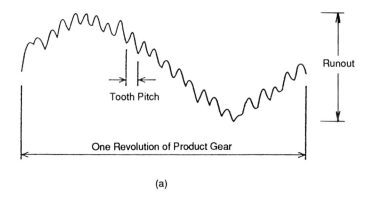

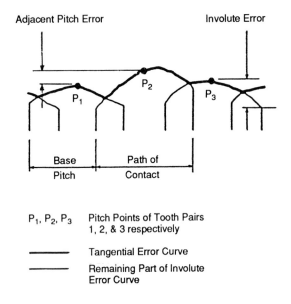

FIGURE 14.6 Manufactured transmission error measurement descriptions.[39,40]

rotation. For instance, a spur gear pair having a contact ratio of 1.60 will have two tooth pairs in contact for 60 percent of the mesh cycle, and one tooth pair in contact for 40 percent of the cycle. The mesh deflection, which is equal to the transmission error when one tooth pair is in contact, will be roughly twice the value when two tooth pairs are in contact. This alternating contact provides the analytically predicted transmission error time trace and its frequency spectrum which are shown in Fig. 14.7.[37] Changes in contact ratio will alter the time trace; and, at integer contact ratios (1.0 or 2.0), the

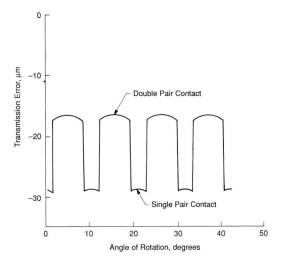

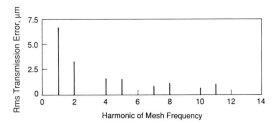

FIGURE 14.7 Prediction of transmission error of a loaded spur gear having a contact ratio of 1.60.[37]

transmission error trace will have a very small time-varying component and, in principle, provides an ideal situation with regard to gear noise.

However, because perfect profiles are highly unlikely, and because of the potential of corner impact at the initiation of contact, designers rarely design spur gears with a contact ratio of exactly 1.0 or 2.0. However, the concept of tooth modification is used so that gears may operate at "effective" contact ratios, where the design contact ratio is higher than the operating contact ratio of the gear pair.

Experimental results published as early as 1958 show that properly designed profile modifications can greatly reduce the mesh frequency components of transmission error.[38] These results, which are shown in Fig. 14.8, reveal that a given profile modification only provides a minimum transmission error at one tooth load.[20] This result is also shown in the simplified analysis presented in Fig. 14.9a.[39,40] The next sections describe a method for selecting the appropriate profile modification to minimize transmission error of spur gears.

Long Relief. A simplified analysis of the selection of the optimum tip relief is explained if we assume the mesh stiffness of a single pair of teeth in contact to be

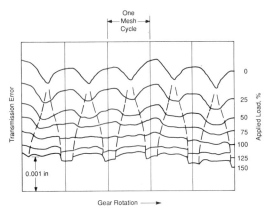

FIGURE 14.8 Measurements showing the effect of load on spur gear transmission error (4 DP, design load = 1700 lb/in of face width).[20]

exactly one-half the stiffness of two pairs of teeth in contact, and that these stiffnesses remain constant throughout the mesh cycle.[39,40] For four different loads, an unmodified, 1.6-contact-ratio spur gear pair will have the transmission error shown in Fig. 14.9a.

Long tip relief is achieved by removing material equal to the amount of the tooth deflection of a single tooth pair from both the root and tip of the gear, as shown by the solid line on the profile specification of Fig. 14.10. This deflection is roughly computed by multiplying the normal tooth load by the mesh stiffness (a commonly used mesh stiffness is 2×10^6 lbf/in/in of face width).

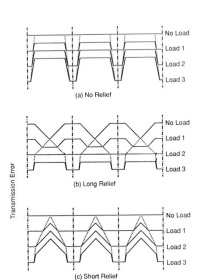

FIGURE 14.9 Effect of load on transmission error for 1.6 contact ratio spur gears with different types of profile modifications.[39,40]

The modification on each end of the tooth should start at the highest and lowest points of single tooth contact, respectively. Applying tolerance bands results in a version of the so-called K chart which is often used for tolerancing profile modifications (see Fig. 14.10 for tolerances for two AGMA classes[41]). Other methods of tolerancing profiles are discussed in Chap. 5.

Figure 14.9b shows that for load 2, the transmission error is minimized using long relief.

Refinements in the tooth stiffness model will slightly change the shape of the optimum profile modification, but will not have much effect on the observed trend. Another obvious difficulty is to accurately reproduce the profile modification. If the tolerances in the profile specifications are too broad rela-

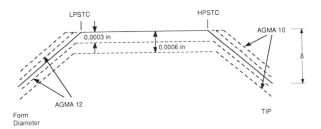

FIGURE 14.10 Profile specification chart to minimize noise for a 5 diametral pitch 6 in. diameter gear. (δ is the deflection of a single tooth pair.)

tive to the amplitude of the modification, it is possible to manufacture profile modifications that actually increase rather than decrease the transmission error.

Although the example has all of the modification on the gear, any combination of tip and root relief on both the gear and pinion that results in the net modification shown in Fig. 14.10 will yield good results. In fact, a very common procedure that keeps the relief constant even when center-distance spreading occurs under load is to provide tip relief on both the gear and pinion. Also, other modification shapes such as gentle curves (parabolas or arcs of circles) may be specified.

The long relief design procedure works well for gear drives such as pumps, compressors, blowers, conveyors, etc., which operate under relatively constant load and speed conditions. When a broad range of load and speed operating conditions are encountered, other methods such as short relief or using high contact ratio spur gears or helical gears may be necessary.

Short Relief. Short relief is used when the noise problem occurs at a load well under the gear pair's design load. In this case the relief at the tooth tip must still be large enough to avoid corner contact, so it is still selected as the single tooth pair deflection at the design load. By moving the starts of the modifications closer to the tooth tip and root, respectively, the "optimum load" may be selected to be anywhere between zero load and the design load.

Figure 14.9c shows the transmission error traces for a design that has a profile modification that provides minimum transmission error at no load and yet has adequate tip relief to avoid corner contact. In order to achieve this result, there is no modification over at least 1 base pitch along the line of action. By placing the modification starting point somewhere between the highest point of single tooth contact and the 1 base pitch point, optimum modifications for other loads may be obtained. Note, however, that the transmission error at higher loads, although improved from the baseline of Fig. 14.9a is not now near the optimum value.

In applications, such as in vehicles where the gears encounter many operating loads, an iterative procedure must be used in order to achieve an acceptable profile modification compromise for all operating conditions.

High Contact Ratio Spur Gears. Since profile contact ratios greater than 2.0 provide 33 percent less mesh stiffness variation than occurs with gear pairs having a contact ratio between 1 and 2, these "high contact ratio" gears have enjoyed much success in reducing spur gear noise. Also, the transmission error of high contact ratio spur gears is much less sensitive to load variations than are gear pairs with contact ratios less than 2.0. The major problem in implementing high contact ratio spur gears is that the required tip relief is usually very "short" and must be controlled very accurately.

Figure 14.11 shows a more complex model's[37] prediction of the transmission error load sensitivity. The start of the modification was taken at the highest point of double tooth contact. In this case, transmission errors across the total load range are significantly less than those obtained with an equivalent low contact ratio spur gear set. However, these improvements can only be achieved through proper tolerancing and accurate manufacture of the profile modification.

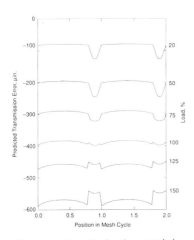

FIGURE 14.11 Predicted transmission error load sensitivity for spur gears with 2.2 contact ratio (10 DP, 1000 lb/in of face width).

Helical Gears. The mesh stiffness of helical gears is roughly proportional to the sum of the lengths of the diagonal lines of contact of all tooth pairs in contact. This has been pointed out and verified through dynamic measurements.[8] If the sum of the lines of contact is kept constant, the transmission error will be nearly constant for all loads. This is in direct contrast to spur gears for which the transmission error can only be minimized for one load, and it is the main reason that helical gears are usually much quieter than spur gears of the same accuracy. A rule of thumb in keeping the sum of the lengths of lines of contact constant is to provide integer values of profile contact ratio and/or face contact ratio.

Also, based on similar reasoning, it has been shown through analysis that the overall noise level will be minimized by maximizing the product of the profile contact ratio and the face contact ratio.[42] Many investigators have performed pertinent analyses of the noise sensitivity and transmission error of helical gears.[11,43,44] In each instance both profile and lead modifications are necessary in order to minimize transmission error, and in extreme cases topographical modifications are necessary to obtain minimum noise.[45]

Figure 14.12 shows another model's[46] predictions of the load sensitivities of the transmission error of a profile modified helical gear pair of the same face width and load used for the gears of Fig. 14.11. The transmission errors are so small for this design that each chart is expanded on the plot of Fig. 14.11b. For sake of comparison, Table 14.1 shows analytically predicted peak-to-peak transmission errors for four different spur and helical gear designs of the same pitch, face width, and loading. The profile modification of each design has an amplitude equal to the design load mesh deflection when a new tooth pair comes into contact. The table shows that increasing total contact ratio not only reduces transmission error, but also reduces the sensitivity to operating at off-design loads. Although the data in the table are for specific designs, the table is presented to illustrate trends expected from different designs.

GEAR NOISE 14.17

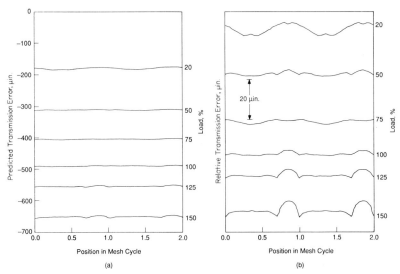

FIGURE 14.12 Predicted transmission error load sensitivity for a pair of helical gears with 2.2 profile contact ratio and 1.1 face contact ratio (10 DP, 1000 lb/in. of face width). (a) Normal scale; (b) mean value unscaled.

Bevel and Hypoid Gears. The surface geometry of bevel and hypoid gears is very dependent on the types of the manufacturing processes used, and conjugacy is usually only achieved at one shaft mounting condition. The added curvatures of bevel and hypoid gear tooth surfaces, coupled with the additional manufacturing complexities, and, until recently, the difficulties in quantifying the measured profiles and leads of these gears, has made their analysis for noise minimization very difficult. Therefore, procedures for developing quiet gears have required many iterations in developing manufacturing process parameters for each gear pair with many rules of thumb being used by designers.

TABLE 14.1 Peak-to-Peak Transmission Error Predictions in Microinches for Different Gear Designs (10 diametral pitch, design load = 1000 lbf/in., face width)

% Design Load	LCR Spur $m_p = 1.60$ $m_f = 0$	HCR Spur $m_p = 2.20$ $m_f = 0$	LCR Helical $m_p = 1.60$ $m_f = 1.10$	HCR Helical $m_p = 2.20$ $m_f = 1.10$
20	196	45	11.4	6.2
50	136	50	8.5	3.0
75	87	32	6.4	2.4
100	45	14	4.9	2.2
120	60	23	7.3	4.2
150	117	41	9.8	7.8

However, the deflection and transmission error effects discussed for spur gears are quite similar for straight bevel gears. Similarly, hypoid and spiral bevel gears behave much like helical gears except there is one more degree of curvature to contend with in order to get optimum load sharing for minimum transmission error. Experimental results have shown that minimizing loaded transmission error (LTE) of bevel and hypoid gears will provide minimum noise in most instances.

Even with the more complex geometries, procedures have been developed for analyzing the transmission error of bevel gears, both under no load[47] and under load.[48,49] Also, methods of inspecting these gears in a manner similar to spur and helical gears allows one to consider modifications using similar rationale to profile and lead modifications of parallel axis gears.[50,51]

14.4.3 Accuracy Effects on Transmission Error

The general trends of gear inaccuracies on the frequency spectrum of transmission error are very similar to their effects on gear noise. These effects are discussed in this section.

Tooth Profile Effects. Profile error is one of the main contributors to mesh frequency transmission errors and noise. The smaller, high-frequency ripple of Fig. 14.6 is principally the result of profile errors in the single tooth pair contact zone. However, as load is applied to the gears, the deflection effects discussed in the transmission error section become important and must be considered in unison with the manufactured profile errors. Normally, profile errors provide strong components in the frequency spectrum at the mesh frequency and its harmonics. It has been commonly felt that concavity in the profile chart is unacceptable[52] and that "hollows" near the pitch point are to be avoided.

Adjacent Pitch Error. Adjacent pitch error manifests itself in a step change in transmission error for each tooth. When analyzed in the frequency spectrum, these errors, which are usually random in nature, will provide frequency components at all harmonics of shaft frequency except the mesh frequency.[18,42,53] This shows that adjacent pitch error is not one of the causes of mesh frequency noise, but it may add broadband noises. In extreme cases special hobs have been used for the addition of random spacing errors, which tend to "mask" out the pure mesh frequency tones.[53]

A special case of adjacent pitch error occurs when the manufacturing process creates a "dropped" tooth. This usually occurs when there is an accumulated index error that manifests itself as a single, large spacing error at the last tooth cut. In this case a once-per-revolution transient causes a clicking sound and shows up in the frequency spectrum as a series of harmonics of shaft frequency. A similar spectrum will occur if there is a broken tooth or any other anomaly that occurs on only one tooth.

Eccentricity. Eccentricity causes an excitation frequency at once-per-shaft revolution, which is usually inaudible. More importantly, it often causes an annoying beating of the mesh frequency tone, which is evidenced in the frequency spectrum by sidebands adjacent to the mesh frequency. These sidebands are spaced at $\pm$ the shaft frequency, and, due to system dynamics, the sidebands are usually more evident in the noise spectrum than in the transmission error spectrum.

Misalignment and Lead Errors. Misalignment, shaft deflections, and lead errors each affect transmission error in a similar manner.[54] However, only the component of misalignment in the plane of the tooth normals has a significant effect. Lead crowning is usually desired to compensate for misalignment and shaft deflections so that tooth contact does not move to the edge of the tooth. Should contact shift to one side of the tooth, the effective face width is reduced, thus causing a reduction in the sum of the lengths of lines of contact, which usually increases transmission error. This effect is more pronounced on helical gears since their behavior approaches that of spur gears when severe misalignment causes a large shift of the load distribution to one end of the teeth. Providing too much lead crowning, however, has a similar effect, since both edges of the teeth could now lose contact, causing a reduction of the effective sum of the lengths of lines of contact.

Concave lead shapes usually have a harmful effect on transmission errors.[46,55] When shaft deflections are understood, lead modifications to compensate for these deflections are often used. For instance, overhung gears often require a tapered lead modification. As with profile modifications, these types of lead modifications yield minimum transmission error for only one load. It has also been found that, as load is increased, misalignment has a reduced effect on transmission error.[57]

Other Effects. Stiffness changes due to bolt holes, nonuniform rims, or spokes can result in sidebands at $\pm$ rim or spoke pass frequency.

When resonances are excited by the mesh frequency or one of its harmonics, a significant increase in the noise may occur. Also, random spacing errors have been known to excite resonances even when the resonant frequency does not match up with a mesh-related frequency. Resonances of significance include blank resonances in light-weight gearing, shaft torsional and lateral resonances, and housing resonances. Methods of improving resonant problems include applying damping to the portion of the system that is resonating, shifting the resonant frequency by altering the mass-stiffness distribution of the troublesome component, or, when possible, changing the operating speeds or the number of teeth on the gears so the excitation frequency is shifted out of the resonant region.

Inaccuracies in the cutting machine gear train have been known to cause so-called ghost tones at noninteger multiples of mesh frequency.[31] Grinding, honing, lapping, and other hard-finishing operations have often been successful in reducing these ghost tones. However, some hard-finishing operations have been known to introduce their own ghost tones.

14.5 GEAR NOISE REDUCTION AT THE DESIGN STAGE

Table 14.2 provides a summary of the improvements reported in the literature that are possible by modifying the design or manufacture of a specific gear. One should be aware, however, that these improvements are not necessarily cumulative, that is, as more improvements are made, the effect of each new improvement is significantly reduced from that shown. Also, as one factor is changed, other factors that were previously unimportant will have a greater effect on noise.

14.5.1 Design Elaboration

Based on the following table, one would assume that the optimum gear design would use extremely accurate helical gears having fine pitch, high contact ratio, and fine

TABLE 14-2 Effects of Various Design Parameters on Gear Noise

Design Improvement	dB Reduction	Comments
Contact ratio increase	0–9	Not much improvement above 3.0 unless extremely well designed and accurate profile and lead modifications are implemented.[4,13,28,30]
Damping	0–7	Only if resonance exists.[6,16,14,65]
Helical gears	0–20	Relative to spur gears, machining errors have less effect[6,9,13,19,28]
Lapping	0–10	For poor profiles; necessary for unground hypoid gears.[57]
Lead crowning	2–8	Compensates for shaft misalignment and deflection.[4]
Lower mesh frequency	0–6	Only if below 500 Hz; however, this limit is affected by gear size.[6]
Pressure angle reduction	<3	Reduces tooth stiffness, reduces runout effect, increases contact ratio.[9,21,56]
Profile modifications	5	Best for spur gears.[19,53]
Surface finish	0–7	Depends on initial finish.[16,28]
Other comments		
Backlash:		If excessive, reducing can provide improvement. At high velocities, increasing can reduce air pocketing. Too little backlash is very harmful because of interference.
Bearings:		Plain bearings reduce gear noise relative to rolling-element bearings.
Contact line length		Keep constant.[9,13,19,57,58]
Diametral pitch		Smaller teeth increase contact ratio, thus reducing noise.[56]
Helix angle (HA) incr.		Reduces noise for HA up to 40°. At higher HA, thrust loads become too severe.[28,58]
Materials		Plastic gears much quieter than steel.
Changing mesh freq.		May help in avoiding resonance conditions.
Mesh stiffness		Lowering reduces impact forces but increases loaded transmission error; achieved by lowering pressure angle.[6,59]
Phasing		May be helpful for planetaries and split-path drives that have multiple pinions, not practical for other types.[16]
Planetary gears		Proper phasing can reduce noise.[60,61] Shafts need concentricity and float.[60]
Pressure angle error		Can add 8 dB.[28]
Profile errors		S-shaped and concave profiles particularly bad, increased noise up to 18 dB.[57]
Recess action		Is favored—reduces entering contact impact and sliding.[13,19]
Roll angle		Lowest contact point should have at least 8–10° roll.[60]
Spacing errors		Increase noise up to 7 dB, but not at mesh frequency.[57]
Split paths		Splitting power paths gives smaller diameters, less noise.
Tuned absorbers		Where resonances exist, have greatly reduced noise.[59,61]

surface finish. Yet other factors in the gear design may not allow the designer to use all of these features. For example, bearing thrust forces may force the designer to use spur gears; required bending strength may lead to coarse pitch; and so forth. Also, changes in one of the variables may change any of the other variables, thus further confusing the issue.

When designing a gear set for minimum noise, one must go back to Fig. 14.1 in order to better understand the steps necessary.

14.5.2 Excitation Minimization

First, it is important to minimize the excitation, which usually means to minimize transmission error at the operating load. This is best achieved by providing adequate profile and lead modifications to compensate for deflections. This is true whether the gears are spur, helical, or bevel; however, the manufacturing process for bevel gears restricts the designer more than for parallel axis gears.

An important, but often overlooked, aspect of this procedure is to provide for adequate quality control at the design stage, since many noise problems are not due to faulty designs, but are due to poor tolerancing. It is important to note that the two main factors in causing excessive transmission errors are poor tooth profiles and shaft misalignment, both of which should be carefully controlled. This is particularly true for profile modifications, where gears cut to the extremes of the profile tolerance band can cause significant increases in transmission error. It is possible to develop combinations of lead and profile modifications that provide an allowance for shaft misalignment; however, a significant amount of modeling or experimental iterations are necessary to obtain these modifications.

When helical gear face contact ratios become large, transmission errors become very small, and other excitations such as time-varying bending moments on shafting become predominant and must be considered. Also, the larger contact ratios require very accurate modifications, since shaft deflections and misalignment are even more critical, and topographical modifications may be necessary.[45]

The effect of surface finish has been the subject of much study that has led to many inconclusive results. This is probably because baseline data have not been adequately controlled. Surface finish should not have a significant effect on transmission error but could change friction forces, which could provide an excitation at mesh frequency. Usually, improved surface finish causes a reduction in noise; however, in some instances it has been reported that shaved gears are quieter than ground and polished gears.[28] To draw a conclusion from this information would require further knowledge about the profiles and leads of the teeth manufactured by the two processes. Finally, the ghost noise that has been previously discussed is often thought to be a surface-finish-related noise, since it often cannot be ascertained from profile and lead measurements.

14.5.3 Torsional and Lateral Vibrations and Gear Blank Resonances

Any time a mesh frequency or one of its harmonics coincides with a natural frequency of transmission shafting there is the potential for increased noise levels. The coupled torsional-lateral natural frequencies of a simple gear mesh are usually in the region from 1000 Hz to 10,000 Hz, depending on gear size, and the designer should compute these frequencies in order to ascertain if they could cause problems. Many methods for making these computations are discussed in the literature.[36] These resonances are to be avoided or shifted by redesign since amplitudes of vibration at these frequencies are very difficult to reduce by adding damping.[62,63]

Gear blank resonances may also be a problem, particularly in lightweight gears such as bevel ring gearing where the blank has a large mass overhung from its shaft support. Supporting gear teeth with thin webs may also cause resonance problems. Here, however, it is possible to provide damping when a problem is encountered; however, it is still desirable to avoid operating at these resonances. Successful damping methods include free-layer damping, constrained-layer damping, frictional damping rings, and tuned absorbers.[28,62,63,64]

14.5.4 Bearing-Housing Interaction

An important, yet overlooked, aspect of quiet transmission design is the selection of the bearing configuration. By minimizing the forces that excite the gear housing, it is possible to obtain some noise reduction. For instance, plain bearings have been known to reduce gear noise by 10 dB relative to rolling-element bearings.[9,59] In other cases tilting pad bearings with compliant supports have been successful in reducing bearing force transmissibility.

There is no consistent information regarding the effects of different types of rolling-element bearings. This is probably due to the fact that the force transmissibility of rolling-element bearings may change significantly with each housing design. In fact, it has been noted that simply disassembling and reassembling a gearbox can provide 2–4-dB variability in the gear noise that is produced. However, techniques have been developed for analyzing rolling-element bearing force transmissibility.[65]

14.5.5 Housing Vibration

Reducing noise in the housing design can be highly fruitful; yet this is an extremely difficult task.[66] If resonances are a problem, shifting them may only shift the problem to a new operating speed. Damping of resonances can be very helpful. Subsectioning housing walls or adding stiffeners may reduce noise, but sometimes this method backfires, since noise is often radiated only from the edges of panels, and adding subsections adds edge area, which adds to the radiated noise, even though static stiffness is increased. Some improvements can be reached by using other housing materials such as cast iron in place of steel. In fact, a concrete housing has been used in one experiment and was found to have excellent noise reduction properties. Vibration isolation, covers, and room treatments can also be effective. Figure 14.13 shows the effects of vibration isolation and enclosure of the housing on gear noise.[21]

14.5.6 Epicyclic Gears

From a noise generation and transmission viewpoint, epicyclic gears are very complex because of the multiple meshes having the same mesh frequency and the fact that the position of each mesh rotates as the carrier rotates. This often results in heavily modulated tones which appear in the frequency spectrum as strong sidebands around mesh frequency. Phasing of the meshes is often used in an attempt to get the effects of multiple meshes to cancel one another. However, no general rules are available regarding when phasing will provide improvement, and the optimization procedure is

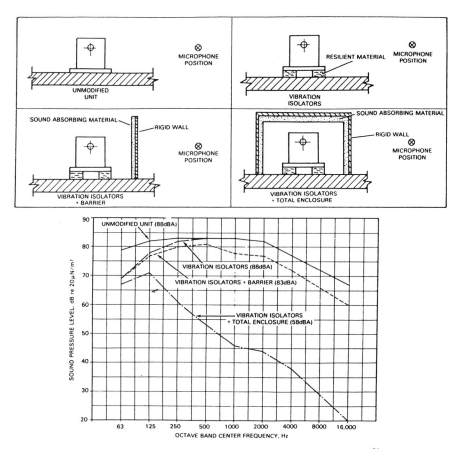

FIGURE 14.13 The effect of different housing treatments on gear noise reduction.[21]

usually of the cut-and-try nature.[62,63] The increased compliance that is usually introduced to obtain proper load sharing between meshes can result in resonant problems that must either be avoided or minimized by adding damping.

14.5.7 Unique Gear Geometries and Treatments

Many special tooth forms and treatments have been proposed over the years. For instance, gears have been proposed of either involute or near-involute design, which are purported to provide high load sharing while keeping mesh stiffness nearly constant.[67]

Schemes proposed for reducing force transmissibility include providing a large torsional compliance within the rim of the gear[61] and applying polymeric coatings to gear teeth.[68] The first scheme has potential, provided torsional compliance is added without affecting lateral compliance. The latter scheme has potential, but there is little evidence that it can be applied without adversely affecting load-carrying capacity.

The advent of active noise control techniques that achieve noise cancellation by applying controlled forces to shafting and housings so that gear forces are cancelled has aroused some interest for use in gear applications. The major hurdle will be the development of active control techniques that are successful at the relatively high mesh frequencies of many gear applications.

14.6 GEAR NOISE MODELING

An ultimate goal of the designer is to have design methods that allow the prediction of noise levels for a given transmission. Although many attempts have been made to achieve this goal, none of the methodologies used has received universal acceptance. At best, the current state of the art of noise modeling allows the comparison of different treatments but does not allow absolute prediction of noise levels to an acceptable accuracy.

Most efforts have been devoted to predicting transmission errors and then using these predictions as inputs to a dynamic model of the transmission. Time domain models have been used to predict the effects of different gear design changes (face contact ratio, misalignment, profile and lead modifications) on transmission error changes for a pair of helical gears.[46] Figures 14.14 to 14.16 show the results of changing some of these variables on transmission error. A frequency domain method has been used to predict transmission errors of wide-face-width gears.[69] Finally, transmission error has been used in conjunction with large torsional dynamics models of gear drives for the prediction of the bearing forces and housing excitations.[70,71]

As mentioned earlier, the design of housings for minimum noise is a very difficult task. Lim and Singh[66] presents a review of many of the studies in which gear housing design for minimum noise is the topic. The most common approach is to use the finite element method for modeling.[72] One such approach uses a strain energy method to reduce housing vibrations at mesh frequencies.[73] An important result of these studies is that at high frequencies (typically above 100 Hz) housing natural frequencies are closely spaced (high modal density) and the use of traditional forced-vibration analysis

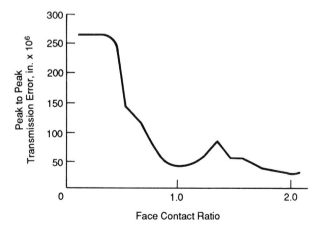

FIGURE 14.14 The effect of face contact ratio on the predicted transmission error of a helical gear pair.[46]

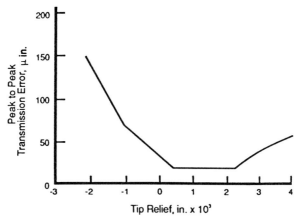

FIGURE 14.15 The effect of tip relief amplitude on the predicted transmission error of a helical gear pair (material removed is positive relief).[46]

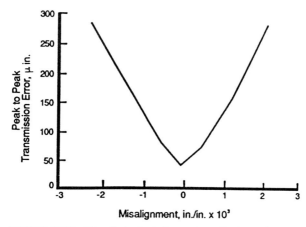

FIGURE 14.16. The effect of misalignment amplitude on the predicted transmission error of a helical gear pair.[46]

methods would be tedious at best. Therefore, there has been a trend towards using *statistical energy analysis* (SEA) for housing noise prediction.[74,75,76]

14.7 GEAR RATTLE

In lightly loaded gears and transmissions with large time-varying torsional excitations, such as automotive drives excited by an internal combustion engine, gears may often leave contact and then reengage with an impact. These impacts are repetitive in such a manner that a clatter or rattle occurs.

When rattle occurs it can be quite annoying, yet difficult to diagnose. It does not appear as a discrete peak in the frequency spectrum and usually only manifests itself in transient bursts when observing the time trace of the noise. Many analysis procedures for gear rattle have been reported,[77,78] and Fudala and Engle[79] present one of the more common experimental approaches used for gear rattle analysis.

Rattle may be reduced by proper design of dead zones in clutches, control of spacing errors, and reducing the effects of resonances excited by the tooth impacts. However, since rattle is a nonlinear systems dynamics problem, generalizations are difficult, and each problem requires its own analysis.

14.8 GEAR NOISE TROUBLESHOOTING

Since many gear noise problems occur after the gear set has been designed and is in production, this section discusses procedures for identifying and solving problems after they become evident. Since the available solutions vary with the type, size, application, and cost of the gears and the production lot size, some of the procedures outlined here might differ depending on the specific situation. Some of the factors to be included in such an analysis are presented.

14.8.1 Characterization of the Gear Noise

This step requires the identification of the objectionable or unacceptable noise and aids in the location of its source. Oftentimes transmission noise is characterized by subjective terms such as *growl, rumble, whine, moan, gravelly sound, hum,* etc. It is necessary to develop descriptions of these sounds in engineering terms so their source can be identified. In this stage, mesh frequencies should be computed as well as frequencies of all other components of the drive train (fans, pumps, bearings, etc).

14.8.2 Instrument the Gearbox and Its Attachments

Once having established that the transmission is the noise source, instrumentation must be selected to further isolate the cause. Where noise problems exist, microphones and vibration transducers are best used for measurement purposes, although strain gages, noncontact displacement probes, and torsional vibration measurement transducers are also useful. It is also very helpful to have some form of tachometer pulse available for triggering oscilloscopes or for externally sampling spectrum analyzers. Most common sources of tachometer pulses are magnetic pickups, optical pickups, and optical encoders.

14.8.3 Obtain Time Domain Traces, Frequency Spectra, and Spectral Maps (Waterfall Plots)

Each of these displays are readily available on most modern spectrum analyzers. The narrowband frequency spectrum (Figs. 14.2 to 14.3) is extremely valuable in isolating the gear mesh that is causing difficulties. Frequencies that dominate the frequency spectrum should be compared with the calculated frequencies. When frequency peaks are large around the gear mesh frequency, its harmonics and sidebands, this provides positive identification of which gear meshes are predominant noise sources. If other peaks are of comparable or greater magnitude, attempts should be made to relate them to other components in the power train. It is recommended that the speed of the tester be varied in order to determine whether the major frequency peaks are coming from components mounted on the shafts or are due to outside sources or system resonances.

Figure 14.17 shows a spectral map that demonstrates how the frequency spectrum changes in direct proportion to system rotational speed. The mesh frequency and its harmonics shift to the right as speed is increased. The peaks that remain stationary are independent of speed and are likely due to system resonances. Note how the amplitudes of the mesh frequency components increase as they pass through these resonant regions. Some of the "unidentifiable" frequencies such as the one at 3.7 times mesh frequency are called *ghost* frequencies and may be related to cutting errors coming from the final drive of the hobber, shaper, or grinder that cut the gears.

Another means of achieving similar results as the spectral map is to use a tracking filter to monitor each mesh tone as speed is changed while operating at a constant load. In addition to spectral maps it is helpful to use a similar approach to determine the effects of load on the frequency spectrum while keeping speed constant. The shape of the transmission-error-to-load-transfer function can be compared with predictions and system nonlinearities can be identified.

Synchronous time averaging can also be invaluable in isolating gear noise problems. This method allows the analysis of phenomena related to each specific shaft of a transmission and can even identify problems on individual teeth.

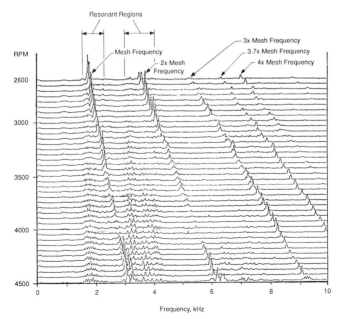

FIGURE 14.17 Spectral map for evaluating the effects of operating speed on gear noise.

Other types of noise such as gear rattle are more difficult to diagnose and are often best analyzed subjectively or through observation of the time trace of the noise or vibration signal.

14.8.4 Perform Other Special Measurements and Modeling

Once the problem has been narrowed down (mesh frequency, resonance, rattle, etc.) additional measurements may be necessary. Likewise, it will likely be necessary to perform a design and/or manufacturing audit. In high-production situations where gear noise has not previously been a problem and all of a sudden appears, the noise is most often attributed to some change or combination of changes in the manufacturing processes. When possible, gear profile, lead, and spacing charts should be performed on the gears. Single flank transmission error inspection has also proven to be a useful troubleshooting tool. Contact pattern analysis of the gear teeth often identifies severe misalignment, deflection problems, and/or interference problems. Where resonances exist, it may be necessary to perform an experimental or analytical modal analysis in order to isolate the problem.

14.8.5 Implementing a Fix

Although this is the obvious goal of problem troubleshooting, because each problem has its own unique solution, it is beyond the scope of this treatise to delve into this subject in detail.

When the gear noise problem is particularly evasive or is of a fundamental nature, it may be necessary to perform additional studies in order to obtain a better diagnosis and solution to the problem. This could include designed experiments or parametric studies, either for experiments or analytical studies.

14.9 CLOSING COMMENTS

1. Remember that gears are part of a complex dynamic system.
2. Keep transmission error to a minimum.
3. Profile accuracy is extremely important.
4. Design gears with enough tip relief to allow not only for bending, but also for the errors inherent in the manufacturing process.
5. Design for effective profile contact ratios and face contact ratios that are integers at the design load. For helical gears, design so that the sum of the face contact ratio and the profile contact ratio is 3 or greater.
6. Most production noise problems from previously quiet transmissions are usually due to a problem in the manufacturing process. Only methodical diagnosis of the manufacturing process can reduce these problems.
7. Be sure to check out resonances when a noise problem occurs. It is not uncommon that a very high harmonic of mesh frequency excites a system resonance and causes a noise problem. If you have diagnosed a resonance problem, the predominant modal features must be identified and these features must be either damped or modified to shift the resonance out of the operating range. When resonances are not the problem, damping materials will not be effective.

REFERENCES

1. Faulkner, L. L., Ed., *Handbook of Industrial Noise Control*, Industrial Press Inc., New York, 1976.
2. Harris, C. M., Ed., *Handbook of Noise Control*, McGraw-Hill, New York, 1960.
3. Beranek, L. L., *Noise and Vibration Control*, Institute of Noise Control Engineering, Cambridge, Massachusetts, 1971.
4. Opitz, H., "Noise of Gears," *Phil. Trans. of the Royal Society*, Vol. 263, December 1968, pp. 369–380.
5. Welbourn, D. B., "Forcing Frequencies Due to Gears," *Proc., Vibr. in Rotating Syst. Conf., I. Mech. E.*, 1972, pp. 25–40.
6. Smith, J. D., *Gears and Their Vibration*, Macmillan, New York, 1983.
7. Smith, R. E., "The Relationship of Measured Gear Noise to Measured Gear Transmission Errors," AGMA Paper 87FTM6, Cincinnati, 1987.
8. Niemann, G., Baethge, J., "Drehwegfehler, Zahnfederharte und Gerausch bei Stirnradern," *VDI-Z*, Vol. 2, No. 4, 1970, pp. 205–214, and No. 8, pp. 475–499.
9. Ill, M., "Some Problems of Gear Noise and Quality Control," *Vib. and Noise in Motor Veh., I. Mech. E*, 1972, pp. 82–90.
10. Moeller, K. G. F., "Gear Noise Reduction," *Noise Control*, March 1955, pp. 11–15.
11. Winter, H., Gerber, H., and Muller, R., "Investigations on the Excitation of Vibrations and Noise at Spur and Helical Gears," *Proc., 1989 ASME Intl. Power Trans. & Gearing Conf.*, Chicago, April 1989, pp. 765–779.
12. Ishida, K., and Matsada, T., "Study on Pitch Circle Impulse Noise of Gear by Simulated Gear Tooth Contact," ASME Paper, 80–C2/DET–69, San Francisco, 1977.
13. Jones, E., and Route, W. D., "Design Considerations in Gear Noise Control," AGMA Paper 299.02, June 1963.
14. Drago, R. J., "How to Design Quiet Transmissions," *Mach. Des.*, December 11, 1980, pp. 175–181.
15. Rosen, M. W., "Noises of Two Spur-Gear Transmissions," *Noise Control*, Vol. 6, November 1961, pp. 11–19.
16. Mitchell, L. D., "Gear Noise: The Purchaser's and the Manufacture's View," *Proc., Purdue Noise Control Conf.*, 1971, pp. 95–106.
17. Taggart, R., "Noise in Reduction Gears," *J. Amer. Soc. Naval Engrs.*, No. 66, 1954.
18. Welbourn, D. B., "Fundamental Knowledge of Gear Noise—A Survey," *Proc., Noise & Vib. of Engines and Transmissions., I. Mech. E.*, Cranfield, UK, July 1979, pp. 9–14.
19. Bradley, W., "How to Design the Noise out of Gears," *Machine Design*, Vol. 45, No. 30, 1973, p. 49.
20. Gregory, R. W., Harris, S. L., and Munro, R. G., "Dynamic Behaviour of Spur Gears," *Proc. I. Mech. E.*, Vol. 178, Pt. 1, No. 8, 1963-4, pp. 207–226.
21. AGMA Standard 299.01, "Gear Sound Manual," American Gear Manufacturers Association, Alexandria, Virginia, 1977.
22. AGMA Standard 295.04, "Specification for Measurement of Sound on High Speed Helical Units," American Gear Manufacturers Association, Alexandria, Virginia, 1977.
23. AGMA Standard 297.02, "Sound for Enclosed Helical, Herringbone, and Spiral Bevel Gears," American Gear Manufacturers Association, Alexandria, Virginia, 1987.
24. AGMA Standard 298.01, "Sound for Gearmotors and In-line Reducers and Increasers," American Gear Manufacturers Association, Alexandria, Virginia, 1987.
25. "Selected Standards Available in English in the Areas of Acoustics, Vibration, Human Environment, Acoustic Emission, Illumination," Bruel and Kjaer Instruments, Inc., Marlborough, Massachusetts, January 1986.

26. "Acoustic and Vibration National and International Standards," Bruel & Kjaer Instruments, Inc., Marlborough, Massachusetts, 1978
27. Gade. S., "Sound Power Determination from Sound Intensity Measurements," *Sound and Vibration,* December 1989, pp. 18–22.
28. Beuler, E., "Influences on the Noise of Transmissions," SAE Paper 680051, January 1968.
29. George, C., "Gear Noise Sources and Controls," in M. Crocker, ed., *Reduction of Machinery Noise,* Purdue University, 1974.
30. Grover, E. C., and Anderton, D., "Noise and Vibration in Transmissions," *Engineers Digest,* Vol. 32, No. 9, September 1971.
31. Greeves, C. S., "Gear Noise," *Noise Control and Vibration Reduction,* June 1974, pp. 163–165.
32. Munro, R.G., "A Review of the Single Flank Method for Testing Gears," *Annals of CIRP,* Vol. 28, No. 1, 1979.
33. Smith, J. D., "The Uses and Limitations of Transmission Error," AGMA Paper 87–FTM–5, Cincinnati, 1987.
34. Furley, A. J. D., Jeffries, J. A., and Smith, J. D., "Drive Trains in Printing Machines, Vibrations in Rotating Machinery," *I. Mech. E.,* 1980, pp. 239–245.
35. Baron, E., Favre, B., and Mairesse, P., "Analysis of Relation Between Gear Noise and Transmission Error," *Proc. Internoise,* '88, Vol. 2, pp. 611–614.
36. Ozguven, H. N., and Houser, D. R., "Mathematical Models in Gear Dynamics—a Review," *J. Sound and Vibr.,* Vol. 121, No. 3, March 1988, pp. 383–411.
37. Tavakoli, M. S., and Houser, D. R., "Optimum Profile Modifications for the Minimization of Static Transmission Errors of Spur Gears," *J. Mech, Transmissions, Automation in Design., Trans. ASME,* Vol. 108, March 1986, pp. 86–95.
38. Harris, S. L., "Dynamic Loads on the Teeth of Spur Gears," *Proc. I. Mech. E,* Vol. 172, No. 2, 1958, pp. 87–112.
39. Munro, R. G., "The D. C. Component of Gear Transmission Error," *Proc., 1989 ASME Intl. Power Trans. & Gearing Conf.,* Chicago, April 1989, pp. 467–470.
40. Munro, R. G., "A Review of the Theory and Measurement of Gear Transmission Error," *Gearbox Noise and Vibration Proc., I. Mech. E.,* Cambridge, UK, April 9–11, 1990, pp. 3–10.
41. ANSI/AGMA 2000–A88, "Gear Classification, and Inspection Handbook," American Gear Manufacturers Association, Alexandria, Virginia, 1988.
42. Mark, W. D., "Gear Noise Excitation," *Engine Noise: Excitation, Vibration, and Radiation,* Plenum, New York, 1982.
43. Umezawa, K., "The Performance Diagrams for the Vibration of Helical Gears," *Proc., 1989 Intl. Power Trans. & Gearing Conf.,* Chicago, April 1989, pp. 399–408.
44. Rouverol, W. S., and Pearce, W. J., "The Reduction of Gear Pair Transmission Error by Minimizing Mesh Stiffness Variation," AGMA Paper 88–FTM–11, New Orleans, October 1988.
45. Weck, M., and Salje, H., "Increased Efficiency in Heavily Loaded Gears through Specific Tooth Flank Correction," *Proc., 2nd World Cong. on Gearing,* Paris, March 1986, pp. 161–171.
46. Houser, D. R., "Gear Noise Sources and Their Prediction Using Mathematical Models," Chapter 16 of *Gear Design Manufacturing and Inspection Manual,* Society of Automotive Engineers, Warrendale, Pennsylvania, 1990. pp 213–222.
47. Litvin, F. L., Tsung, W., Coy, J. J., and Heine, C. "Generation of Spiral Bevel Gears with Zero Kinematical Errors and Computer Aided Tooth Contact Analysis," NASA Technical Memorandium 87273, March 1986 and *Proc., 2nd World Congress on Gearing,* Paris, 1986, pp. 399–408.
48. Krenzer, T. J., "Increasing Role for the Computer in Bevel and Hypoid Gear Manufacture During the 1980's," AGMA Paper No. P129.25, October 1982.

49. Sugimoto, M., Maruyama, N., Nakayama, A. and Hitomi, N., "Effects of Tooth Contact and Gear Dimensions on Transmission Errors of Loaded Hypoid Gears," *J. Mech. Design Trans. ASME*, Vol. 113, No. 2, June 1991, pp. 182–187.
50. Chambers, R. O., Brown, R. E., "Coordinate Measurement of Bevel Gear Teeth," SAE Paper 871645, Milwaukee, September 1987.
51. Lemanski, A. J., Frint, H. K., Glasow, W. D., "Manufacturing Perspective in the Design of Bevel Gearing," *Gears and Power Systems for Helicopter and Turboprops, AGARD Conf. Proc.*, NATO, Lisbon, October 1984, pp. 15.1–15.18.
52. Yuruzume, I., Mizutani, H., and Tsubuku, T., "Transmission Errors and Noise of Spur Gears Having Uneven Profile Errors," ASME Paper 77–DET–51, 1977.
53. Remmers, E. P., "Gear Mesh Excitation Spectra for Arbitrary Spacing Errors, Load, and Design Contact Ratio," *J. of Mech. Design, Trans. ASME*, August 1978, pp. 715–722.
54. Umezawa, K., Suzuki, T., Houjoh, H., Bagiasna, K., "Influence of Misalignment on Vibration of Helical Gear," *Proc., 2nd World Cong. on Gearing*, Paris, March 1986, pp. 615–626.
55. Kubo, A., and Kiyono, S., "Vibration Excitation of Cylindrical Involute Gears Due to Tooth Form Error," *Bull. JSME*, Vol. 23, No. 183, September 1980, pp. 1536–1543.
56. Wellauer, E. J., Schunck, R. A., "Gear Unit Sound Generation, Transmission, Control, and Specification," *TAPPI*, Vol. 56, No. 6, June 1973, pp. 69–72.
57. Opitz, H., "Gerauschuntersuchungen an Zahnradgetrieben (Noise Research on Gear Drives)," *Modern Produktionstechnik*, W. Girardet Publishing Co., Essen, Germany, 1970.
58. Battezzato, L., and Turra, S., "Possible Technological Answers to New Design Requirements for Power Transmission Systems," *Gears and Power Systems for Helicopters and Turboprops, AGARD Conf. Proc.*, NATO, Lisbon, October 1984, pp. 17.1–17.12.
59. Schlegel, R. G., King, R. J., and Mull, H. R., "How to Reduce Gear Noise," *Mach. Des.*, February 1964, pp. 134–142.
60. Route, W. D., "Gear Design for Noise Reduction," Design Practices—Passenger Car Automatic Transmissions, SAE Transmission and Drivetrain Com. No. 208E, 1960, New York, 1973.
61. Drago, R. J., "Minimizing Noise in Transmissions," *Mach. Des.*, January 1981, pp. 143–148.
62. Badgley, R. H., "Reduction of Noise and Acoustic-Frequency Vibrations in Aircraft Transmissions," *Proc., 28th Annual Natl. Forum of Amer. Hel. Society*, Preprint 661, Washington, D.C., May 1972
63. Badgley, R. H., Hartman, R. M., "Gearbox Noise Reduction: Prediction and Measurement of Mesh-Frequency Vibrations Within an Operating Helicopter Rotor-Drive Gearbox," *J. Eng. Ind. Trans.*, ASME, May 1974, pp. 567–584.
64. Moeller, K. G. F., "Means for Reducing Gear Noise," *Design News*, Vol. 14, 1959.
65. Lim, T. C., and Singh, R., "Vibration Transmission Through Rolling Element Bearings. Part 1: Bearing Stiffness Formulation," *J. Sound and Vibration*, Vol. 139, No. 2, 1990, pp. 179–199.
66. Lim, T. C., and Singh, R., "A Review of Gear Housing Dynamics and Acoustics Literature," NASA Contractor Report 185148, October 1989.
67. Watanabe, Y., Rouverol, W. S., "Maximum-Conjugacy Gearing," SAE Paper 820508, 1982.
68. Rivin, E. I., "Low Noise Power Transmission Gears," *Proc., Inter-noise 88*. Avignon, France, August 1988. pp. 627–630.
69. Mark, W. D., "Analysis of the Vibratory Excitations of Gear Systems. II: Tooth Error Representations, Approximations, and Applications," *J. Acoust. Soc. Amer.*, Vol. 66, 1979, pp. 1758–1787.
70. Laskin, I., "Prediction of Gear Noise from Design Data," AGMA Paper 299.03, October 1968
71. Kahraman, A., Ozguven, H. N., Houser, D. R., and Zakraszek, J., "Dynamic Analysis of Geared Rotors by Finite Elements," *Proc., 1989 ASME Intl. Power Trans. and Gearing Conf.*, Chicago, April 1989, pp. 375–382.

72. Lim, T. C., and Singh, R., "Modal Analysis of Gear Housing and Mounts," *IMAC Proceedings,* Las Vegas, January 30–February 2, 1989, pp. 1072–1078.
73. Royal, A. C., Drago, R. J., and Lenski, J. W., "An Analytical Approach and Selective Stiffening Technique for the Source Reduction of Noise and Vibration in Highly Loaded Mechanical Power-Transmission Systems," Fifth Eur. Rotorcraft Forum, Paper No. 66, Amsterdam, September 1979, pp. 1–15.
74. Rockwood, W. B., Lu, L., Warner, P. C., and DeJong, R. G., "Statistical Energy Analysis Applied to Structureborne Noise in Marine Structures," 1987 ASME Winter Ann. Mtg., NCA Vol. 3 Statistical Energy Analysis, pp. 73–79.
75. Mathur, G., Manning, J. E., Aubert, A. C., Bell 222 Helicopter Cabin Noise: Analytical Modeling and Flight Test Validation," *Proc., Noise-Con 88,* East Lafayette, Indiana, June 1988, pp. 573–578.
76. Yoerkie, C., Moore, J., "Statistical Energy Analysis Modeling of Helicopter Cabin Noise," *Proc., Annual Forum, Amer. Helicopter Soc.,* St. Louis, May 1983, pp. 458–471.
77. Singh, R., Xie, H., and Comparin, R., "Analysis of Automotive Neutral Gear Rattle," *J. Sound and Vibr.,* Vol. 131, No. 2, 1989, pp. 177–196.
78. Comparin, R. J., Singh, R., "An Analytical Study of Automotive Neutral Gear Rattle," *J. of Mech. Design, Trans. ASME,* Vol. 112, June 1990, pp. 237–245.
79. Fudala, G. J., Engle, T. C., "A Systems Approach to Reducing Gear Rattle," SAE Paper 870396, Detroit, February 1987.

CHAPTER 15
GEAR LUBRICATION

Dr. Aizoh Kubo
Kyoto University
Kyoto, Japan

Dennis P. Townsend
NASA Lewis Research Center
Cleveland, Ohio

Many gear system problems have been generated because of an inadequate appreciation of lubrication methods and requirements for gearing. Gears, like rolling-element bearings, require a very small amount of lubricant for lubrication; however, much more lubricant is needed for removing heat from the gears. Several methods are used to lubricate and cool gears. The most common is splash lubrication, in which one gear dips into a reservoir of oil in the bottom of the gearbox and throws the oil onto the other gear and bearings. With this method, care must be taken to assure that an adequate flow of oil is fed to the bearings for good cooling. If this is not done, bearing failure will occur because of insufficient cooling and not enough EHD oil film. The most common method used to supply oil to the bearing with the splash lubrication method is to provide a channel to catch the oil splash and feed it to the bearings. When gears operate at higher speeds and loads, the splash method of lubrication and cooling will generally not provide the cooling required for the gears. With inadequate cooling, early failure will occur. The failures may be due to scoring, wear, surface fatigue, or tooth breakage. The main cause for these failures sometimes is not recognized as overheating of the gears from inadequate cooling. The first indication of excessive temperature of the gears may be superficial pitting or gray staining caused by a reduced EHD film thickness as the oil viscosity is reduced. In many applications, the designer will use oil jet lubrication with the oil jet directed at the into-mesh position or at the out-of-mesh position. Neither of these methods provides the most effective cooling of the gears and could in many cases allow overheating of the gears. The into-mesh method of oil jet lubrication will considerably reduce the efficiency of the gears since the oil going into mesh increases oil churning losses. The into-mesh and out-of-mesh jet lubrication method gives very shallow impingement depth on the gear teeth,[1,2] and therefore, neither method provides good cooling. Much better gear tooth cooling is obtained when the oil jet is directed radially onto the gears with adequate jet velocity.

15.1 POWER LOSSES

There are four main areas of losses in high-speed gears. These losses can be controlled for the most part by careful design and construction. The losses consist of bearing losses, tooth friction losses,* oil churning losses, and gear windage losses. Bearing losses may account for about half the losses, particularly when fluid film bearings are used. However, when long life is a requirement, the fluid film bearing is the best choice. An approximate method for the friction torque in a fluid film bearing is shown in the following equation from Fuller:[3]

$$M_b = \frac{\mu A r V}{h} \tag{15.1}$$

where M_b = bearing friction torque
A = bearing area
r = radius
V = velocity
h = radial clearing
μ = viscosity

Rolling-contact bearings have considerably less friction loss when properly lubricated and will generally have a much shorter life than a fluid film bearing. However, rolling-element bearings should have sufficient life for many applications. The friction torque for rolling-element bearings may be estimated by the following equations from Harris:[4]

$$M_b = M_1 + M_\nu$$
$$M_1 = \text{friction torque}$$
$$M_\nu = \text{viscous torque}$$

The friction torque due to the applied load is

$$M_1 = f_1 \, F_\beta d_m$$

where $\quad F_\beta = 0.9 \, F_a \cot \alpha - 0.1 \, F_r \quad$ or $F_\beta = F_r$ if larger $\tag{15.2}$

$$f_1 = z \left[\frac{F_s}{C_s}\right]^y \qquad z = 0.001 \quad y = 0.33$$

where f_1 = friction factor
F = bearing load
C_S = static capacity

for 30° angle contact bearings and $f_1 = 0.0003$ for roller bearings. The viscous friction torque may be estimated by

---

*See Chap. 12, Sec. 12.2 for the equation to calculate the friction losses of spur, helical, and bevel gears.

GEAR LUBRICATION   15.3

$$M_v = 1.42 \times 10^{-5} f_o (v_o N)^{2/3} d_m^3 \quad (15.3)$$

$$M_v = 3.492 \times 10^{-3} f_o \, d_m^3 \quad v_o N \leq 2000$$

where N = speed and d_m = mean diameter. For ball bearings f_o values range from 3 to 4. Bearing computer programs have been developed that give more accurate results for bearing torque.[5,6]

The tooth friction loss is probably the lowest loss in the gear system when the gears are adequately lubricated. There is very little that can be done to reduce gear tooth friction loss once adequate lubrication has been provided. Some lubricants will have a little less friction loss than others. Some gears will have more or less tooth friction loss because of the type of design, mainly because of the different sliding conditions. For instance, high contact ratio spur gears generally have more sliding and, hence, more losses than standard contact ratio gears. However, since this is a small part of the overall loss, it generally has little effect on the total loss.

Windage losses can account for a large part of the total gearbox losses in high-speed* gears because of the high pitch line velocities. Some of this windage loss can be reduced by careful designs. For instance, it was shown in Dawson[7] that axial holes in a gear web can significantly increase the windage losses. Also, it was shown that placing a shield on the ends of the gear teeth, to prevent the air circulating into the teeth, reduced the windage losses by a large percentage. The windage losses in a gearbox with the smooth sides of the box located approximately 1 in. from the gear, and the inside diameter 0.6 in. from the teeth, reduced the windage loss to approximately one-half that for open gears. The pumping loss of the air at the entrance of the mesh accounts for some losses and can be reduced by reducing the pressure in the gearbox, which also reduces other windage losses. An equation for approximating the windage power loss in gears was given in Dawson[7] as shown in the following equation.

$$\overset{\text{Sides}\qquad\qquad\text{Periphery}}{P = N^{2.9} (0.16 \, D^{3.9} + D^{2.9} \, F^{0.75} \, M^{1.15}) \times 10^{-20} \, \phi \, \lambda} \quad (15.4)$$

where N = speed r/min
       D = root diameter
       F = face width
       M = module
       ϕ = oil mixture function: 1 = oil-free mixture
       λ = space function: 1 = free space; 0.5 = fitted gears; 0.7 = large enclosures

Churning losses are caused by the gear striking, pumping, or otherwise moving the lubricant around in the gearbox. It is very important in high-speed gearboxes to get the lubricant to perform its lubrication and cooling function and then get it out of the way. Shrouds are sometimes used in gearboxes to direct the lubricant oil away from the gears. If too much lubricant is allowed to enter the gear mesh, excessive losses will occur from oil being trapped in the gear teeth and being pumped out of the mesh in the axial direction. In spur gear applications this is more critical than for helical gears. This is one reason why most gear designers prefer helical gears for high-speed gearing. Even with helical gearing, however, considerable power loss occurs with too much oil getting into the gear mesh. Some spur gear designs have a groove in the center of the axial length of the gear to reduce pumping losses in the mesh. This is also done in some very wide helical gear applications.

---

*See Dudley[24] for the effect of helix angle on windage losses at high speeds (pp. 7.47–7.51).

15.2. LUBRICATION AND COOLING METHODS

There are many high- to moderately loaded high-speed gears operating today with oil lubrication using 30- to 50-lb/in^2 oil pressure to lubricate the gears. This type of low-pressure system does not do a good job of cooling the gears in a high-speed gear drive and will only allow the gears to operate at a moderate load. In a low-pressure oil jet system, the oil jet can penetrate only a small distance into the tooth space. This results in cooling of the tips of the gear teeth only. This causes the gear tooth temperature to be higher than that obtained with a better system, such as a high-pressure radial oil jet. When the speed is increased at this load condition and low-pressure lubrication, failure of the gears will generally occur. The gear tooth temperatures are reduced when the oil jet pressure is increased to obtain good impingement depth.[8]

15.2.1 Out-of-Mesh Jet Lubrication

A large number of gears are lubricated with low-pressure oil jets into mesh or out of mesh or both. In the out-of-mesh lubrication method, the oil jet has a very modest impingement depth. The impingement depth on the pinion can be determined by the following equation from Townsend and Akin.[9]

$$d_i = r_o - \sqrt{r^2 + L_p^2} \qquad (15.5)$$

where d_i = radial impingement depth
 r_o = outside radius pinion
 r = projected radius on line of centers to impingement point
 L_p = distance from line of centers to impingement point

In order to get the maximum impingement depth for the out-of-mesh condition, care must be exercised to get the proper oil jet location. The analysis indicates that the pinion can be completely missed by a very small change in offset distance from the intersection of the outside diameter of the gear and pinion or from a small change in the jet angle. For maximum impingement depth, in most cases the oil jet should be directed at the intersection of the two outside diameters at an angle that will intersect the pitch point of the gear and pinion.[2] For large gear ratios, it is probably better to favor the pinion to get a better cooling balance. Akin and Townsend[2] gives an analytical method for out-of-mesh jet lubrication for gears with modified center distance and/or addendums. Akin and Townsend[2] also gives the impingement depth results for various oil jet offsets and oil jet angles.

15.2.2 Into-Mesh Jet Lubrication

Into-mesh oil jet lubrication is often used as a means of getting oil to the gear tooth surfaces at a good impingement depth when the oil system is operating at a low pressure. This method is effective because it uses the gear tooth velocity moving with the oil jet velocity. See Akin and Townsend's equations[1,11] for the oil jet impingement depth for into-mesh lubrication. When the jet velocity is less than the gear velocity, the oil impinges on the backside of the teeth. When the jet velocity is greater than the gear velocity, the oil will impinge on the front of the gear tooth. The optimum impingement depth for into-mesh lubrication was shown in Akins and Townsend [1,11] to occur when the oil jet velocity and gear velocity were equal. The oil jet impingement depth at the optimum velocity is given by the following equation.

$$d_i = \frac{1}{P_d} = a \qquad V_j = \omega_g R \sec \beta \qquad (15.6)$$

where d_i = impingement depth
 P_d = diametrical pitch
 a = addendum
 V_j = oil jet velocity
 ω_g = gear rotational speed
 β = oil jet angle
 R = pitch radius

When into-mesh lubrication is used with high-speed gears, care must be taken to avoid excessive oil being trapped in the gear tooth. This trapping can cause various problems such as loss of efficiencies, high loads on the teeth, high noise, and even gear failure under some conditions. In many cases, the bulk of the cooling oil is supplied to the out-of-mesh location with only a small percentage for lubrication supplied to the into-mesh position Tucker.[10] However, there is usually sufficient oil film remaining on the gear tooth for good lubrication when adequate cooling is provided at the out-of-mesh location. In some cases where it is difficult to keep the oil out of the into-mesh zone, a circumferential groove is cut into the center of one of the gears to break up the length of the teeth, thereby reducing the trapping losses. This groove reduces the axial length required to pump the oil by at least one-half.

15.2.3 Radial Jet Lubrication

When the oil jet is directed radially inward, the best impingement depth is obtained. Since gear tooth cooling is at a maximum when the oil jet impinges on the face of the tooth, the radial oil jet offers the best method of gear lubrication and cooling. The oil pressure should be sufficient to allow the oil jet to impinge on the gear tooth a little more than halfway down the working depth of the gear tooth. The maximum cooling is obtained when the oil pressure is sufficient to cause the oil jet to reach an impingement depth equal to the full working depth of the tooth. However, adequate cooling can often be obtained with impingement depth just below the pitch line. When radial jet lubrication is used, the oil jet should be located near the out-of-mesh position with the jet directed radially at the center of the gear and pinion. In a speed reducer, the pinion will receive cooling on the loaded side of the tooth while the gear will be cool on the backface of the tooth. When the gear set is a speed increaser, the pinion will receive cooling on the backface of the tooth and the gear on the loaded side. Experiments have shown[8] that good cooling of the gear or pinion can be obtained when either the loaded flank or unloaded flank of the gar tooth is cooled. Figure 15.1 shows the effect of oil jet pressure and load on gear tooth temperature using radial oil jet cooling on the back flank of the gear tooth. The temperature was measured on the loaded flank of the gear tooth. This figure very clearly shows that good cooling is obtained when cooling the back flank of the tooth. Other data obtained during this test phase also prove that increased oil flow without good impingement depth will not provide good cooling of the teeth.

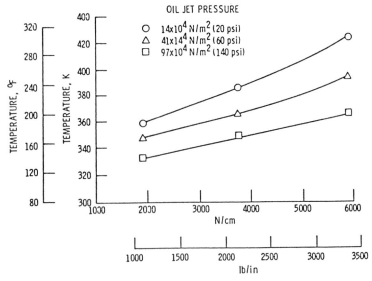

FIGURE 15.1 I.R. microscope measurement of gear average surface temperature versus load for three oil jet pressures, speed 7500 r/min, oil jet diameter 0.04 cm (0.016 in.), inlet oil temperature 308 K (95°F).

The following equations[8,12] give the impingement depth on the tooth flank for various speeds and oil jet pressures for a radially directed oil jet.

$$d_i = \frac{1.5708 + 2\tan\phi + \dfrac{B}{2}}{P_d\left(\dfrac{N_d}{2977\sqrt{\Delta p}} + \tan\phi\right)} \quad (15.7)$$

The vectorial model used to calculate the radial impingement depth is shown in Fig. 15.2. When the impingement depth for a given gear operating condition is known or desired, then the pressure required to obtain that impingement depth is given by the following equation.

$$\Delta p = \left[\frac{d_i P_d N_d}{2977\left[\dfrac{\pi}{2} + \dfrac{B}{2} + (2 - d_i P_d)\tan\phi\right]}\right]^2 \quad (15.8)$$

When the oil jet velocity equals the gear velocity, the oil jet will usually impinge to a depth approximately equal to the full working depth and will prrovide very good cooling for the gear teeth. For standard gear tooth geometry, the pressure required to obtain this velocity may be approximated by the following equation where V is the m/s (ft/s) and Δ_p is in N/cm^2 (lb/in^2)

$$\Delta_p = \frac{V_g^2}{169 \text{ English}} = \frac{V_g^2}{22.8 \text{ metric}} \quad (15.9)$$

Using the foregoing equation for gears operating at a pitch line velocity of 150 m/s (500 ft/s), the oil jet pressure required for full depth impingement would be

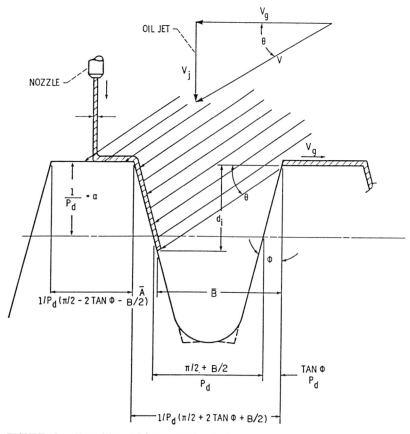

FIGURE 15.2 Vectorial model for penetration depth.

approximately 1014 N/cm^2 (1480 lb/in^2). This pressure is much higher than the oil jet pressure used in most high-speed gearboxes operating at 150 to 300 m/s (500 to 1000 ft/s). These gearboxes must operate at reduced loads because of the limited cooling available. It should be understood that the oil jet size must be reduced at these high pressures to limit the oil flow to that required for good cooling. Using the minimum orifice size of 1.02 mm (0.04 in.) specified by many gear designs would give an oil flow of approximately 1.3 g/min per orifice, which may be too much oil for most applications. However, if the orifice size is reduced to 0.5 mm (0.02 in.), then the oil flow is only 0.33 g/min or one-fourth of that for the larger orifice. In order to limit orifice plugging, the oil should be filtered through a 5- to 10-μm filter. It is much better for both the gears and bearings to filter the oil through a 5-μm or better filter.

15.2.4 Cooling Requirements

The amount of lubricating fluid required for cooling of gears and bearings may be determined by estimating the power loss for gears, bearings, etc., and then using an appropriate temperature rise in the oil to determine the required oil flow. The following equation can be used to determine the oil flow in a given gear system. Assume 2000 hp is transmitted in a gear set with one mesh and two sets of bearings. The losses per mesh in a well-designed spur or helical gearbox should be no more than 0.5 percent, for some applications, with the losses equally divided between the gears and bearings. The oil therefore must absorb 0.005 × 2000 or 10 hp, which is 424.4 Btu/min. Assuming a 50°F temperature rise, the total oil flow required would be

$$w = \frac{Q}{c_p \Delta T} = \frac{424.4}{0.6 \times 50} = 14.2 \text{ lb/min} \quad (15.10)$$

with equal flow to the gears and bearings of 7.1 lb/min. If the gear set is less efficient than the foregoing, then more oil flow would be required.

15.3 LUBRICATION OF LOADED TOOTH FLANK AND SURFACE DISTRESS

15.3.1 Contact of Smooth Surfaces

The lubrication condition between contacting bodies with rolling and sliding under loaded condition is generally understood by the concept shown in Fig. 15.3. When the lubricating oil film formation between contacting bodies is large, that is, the film thickness-to-surface-roughness ratio λ is large (λ greater than 2), there is full film

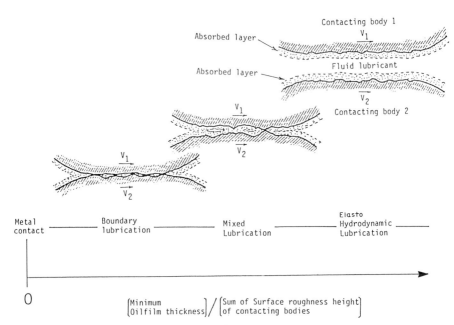

FIGURE 15.3 Concepts of the regime of lubrication.

lubrication and no metal contact occurs. This would be called full EHD film lubrication. When the oil film thickness is thinner or the surface roughness is larger (λ from 1 to 2), the asperities on the two surfaces begin to contact each other such that partial metal-to-metal contact occur. This lubrication condition is called *mixed lubrication*. When the oil film thickness-to-surface-roughness ratio is very small, that is, $\lambda < 1$, metal contact or contact via absorbed oil molecules dominate the contact zone. This condition is called *boundary lubrication*. Schematic representation of these three regimes of lubrication is shown in Fig. 15.3.

A model of contacting cylinders that have the same radii of curvature as that of the gear tooth is used to investigate the lubrication condition of gear teeth. This modeling is not accurate for the beginning or ending of tooth meshing. Lubricating oil film formation at the beginning of tooth meshing is generally poorer than expected and severe metal-to-metal contact occurs rather easily, which can cause damage at these locations.

The tooth flanks of power transmission gears contact each other with very large load and contact pressure causing local deformation of the tooth flank. The lubrication under this condition is called *elastohydrodynamic lubrication* (EHL or EHD) and many investigations have been conducted in this field.[13,14,15] Figure 15.4 shows one example of analyzed pressure distribution and shape of the oil film thickness in the contact zone between two rollers. The hydrodynamic pressure developed differs from the shape of the hertzian contact pressure at the inlet position and the outlet position. The minimum film thickness is approximately calculated[16] by the following equation.

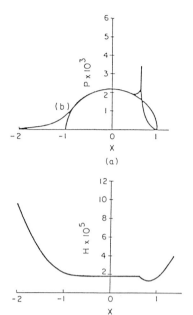

FIGURE 15.4 Example of calculated EHL oil film shape and pressure distribution (isothermal model). $W = 3 \times 10^{-5}$. $U = 10^{-11}$. *(a)* Pressure distributions. *(b)* Film shapes. $G = 2500$. [*Reproduced from the* Journal of Mechanical Engineering Science, *Vol. 2, No. 3, pp. 188–194 (1960) by kind permission of the Institution of Mechanical Engineers.*]

$$H_{min} = 2.65 \ G^{0.54} \ U^{0.7} \ W^{-0.13} \qquad (15.11)$$

where

$$H_{min} = \frac{h_{min}}{R} \qquad (15.12)$$

$$G = \alpha \ E' \qquad (15.13)$$

$$U = \frac{\eta_o \mu}{(E'R)} \qquad (15.14)$$

$$W = \frac{w}{(E'/R)} \qquad (15.15)$$

and R = effective radius of the roller pair, in. (mm)
E = effective elastic modulus of the materials of rollers, lb/in^2 (kgf/mm^2)
η_o = viscosity of oil in atmospheric pressure, lb s/in^2 (kg s/mm^2)
α = pressure-viscosity index of lubricant, in^2/lb (mm^2/kgf)
w = applied load per unit width, lb/in (kgf/mm)
u = rolling velocity, ft/5 (mm/s)
$h_{\min}$ = minimum oil film thickness

When relative sliding exists between contacting bodies under EHL contact, heat is generated in the oil film in the contact zone. Figure 15.5 shows schematically the EHL condition with consideration of thermal effect. When V_1 is not equal to V_2, frictional heat is generated in the oil film due to shearing of oil. This heat is transferred with oil mass in the flow direction, but some amount of heat is conducted across the oil film and transferred to the contacting surface.

Figure 15.6 shows one example of temperature distribution in the direction along the oil film when the inlet oil temperature is kept constant at 50°C. The temperature rises strongly with the rise of slip velocity between contacting bodies. The maximum temperature appears at the pressure spike. This instantaneous surface temperature under hertzian contact is called *flash temperature*.

The concept of flash temperature sometimes corresponds to the increment of instantaneous temperature over the inlet temperature, but sometimes to the sum of surface temperature at the inlet and the increment of instantaneous temperature. Here the former definition is incorporated for flash temperature.

Historically, the analysis of flash temperature was conducted with only one dimensional heat conduction considered. The existence of lubricating oil film, heat conduction in the oil film, and heat transfer from the oil film to the contacting body had not been considered. Blok[17] developed the following equation for the maximum value of flash temperature, assuming hertzian pressure distribution, equal surface temperature on both surfaces and no oil film.

$$T_f = \frac{1.11 \, \mu \, w \, [V_1 - V_2]}{(\sqrt{V_1} + \sqrt{V_2}) \cdot \beta \cdot 2a} \qquad (15.16)$$

where $\beta = \sqrt{\lambda \rho_s c}$; for average kind of steel β takes a value between 0.9 and 2.3, kgf/(mm·s·deg)
λ = thermal conductivity of surface material
ρ_s = density of surface material
c = specific heat of surface material

When the existence of lubricating oil film, heat conduction in the oil film and heat transfer from the oil film to the contacting body is considered, the instantaneous surface temperature in the hertzian contact zone is different from the Blok's flash temperature. Figure 15.7 shows one example of comparison between these instantaneous temperatures are not equal. The calculation of flash temperature under EHL condition is possible, but the oil viscosity is not well known under the condition of very high pressure, existence of shearing action of the oil film and very high pressure changes in a very short time. Therefore, when the Block's flash temperature, or any kind of flash temperature, is used for the estimation of scuffing resistance of gears, the calculated value of flash temperature should not be understood as the actual physical quantity, but as one reference value or guiding value for the instantaneous surface temperature. The reliability of the absolute value of the calculated flash temperature is not certain.

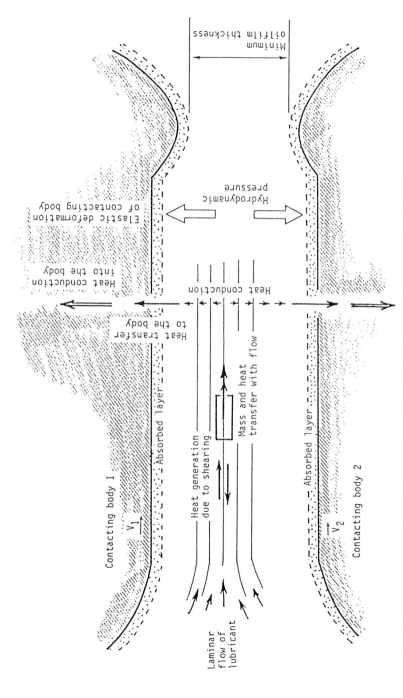

FIGURE 15.5 Model of elastohydrodynamic lubrication with consideration of the thermal effects (TEHL model).

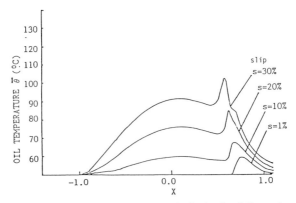

FIGURE 15.6 Average temperature distribution in oil film under TEHL contact and the effect of slip ratio.

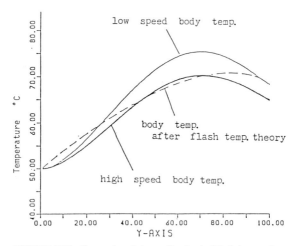

FIGURE 15.7 Comparison between the classical flash temperature after Blok with that of the model considering heat generation and heat conduction inside the oil film and heat transfer to the contacting bodies.

15.3.2 Surface Distress

The surface of actual contacting bodies is not smooth, but has various surface roughness conditions. The oil film thickness calculations consider smooth surfaces. Therefore, the calculated film thickness according to Eq. (15.11) does not show the actual condition. The effect of surface roughness on the flash temperature is not well known. The absolute value of flash temperature will be influenced by the surface roughness.

GEAR LUBRICATION   15.13

The loading condition of many of today's power transmission gears is much higher and the speed is often higher than those used in the past. In addition, many users do not like to change lubricants very often. These conditions can cause lubrication-type failures such as wear, micropitting, macropitting, and scuffing.

Wear can be caused by several conditions such as thin oil film due to high temperatures or large surface roughness, contaminated lubricant, and slow-speed, high-load condition. Some of these conditions can be improved by filters, smoother surface finish, EP lubricants, or surface coatings.

15.3.3 Micropitting and Macropitting

FIGURE 15.8 Micropitting of carburized tooth flank.

Under mixed lubricated condition in which partial metal-to-metal contact between surface asperities occurs, micropitting sometimes appears on hard surface, such as on case-carburized gear teeth, because the hard surface will not wear in. On soft surface, micropitting can also occur, when the contacting surfaces do not wear in. The observation of micropitting looks like a frosted area, and is sometimes called gray staining, Fig. 15.8. The clear form of pitting is usually not found. The depth of micropitting is about 20 μm and sometimes many microcracks are found in its area. Micropitting can increase vibration and dynamic loading of gears and can develop into severe pitting or scoring/scuffing or tooth breakage. When the contact between surface asperities is more severe, high surface shear stresses can cause macropitting. Macropitting of this kind starts on a projected part of the surface roughness, such as at the ridge of ground marks on the tooth surface. Figure 15.9 shows an example of macropitting on ground gear teeth. Macropitting of this kind is usually small and can occur on the soft and hard gears. When the loading condition is not very severe, macropitting can stay in the condition of initial pitting and not develop into destructive pitting.

Under high hertzian pressure, the subsurface shear stress is much higher. The position of the maximum subsurface shear stress approaches the surface with the existence of frictional force. Very near the surface, there can be stress concentration due to the effect of surface roughness and macropitting, and additional stress can be induced due to asperity contact. Consequently, the total stress level at or very near the contacting surface exceeds the endurance limit of the material and cracks are induced on the surface. The cracks can propagate into the surface and cause pitting. Cracks may also start below the surface with low friction. The depth of the pit is related to the material property and the position of the maximum shear stress.[18] The initial pitting can increase to larger pitting, or, on material with low fracture toughness, can cause tooth fracture.

FIGURE 15.9 Macropitting on the ridge of Maag 0° grinding mark.

15.3.4 Scuffing/Scoring

Scoring or scuffing failure is a tooth flank failure that can occur at the beginning of operation or suddenly during the operation of gear drives. Scoring often occurs without any signs and usually stops further operation of the unit. The increase in vibration and noise can cause a total breakdown of the facilities, including the gear drive unit.

Scoring or scuffing failure is a thermal damage of the tooth flank. Local welding of the contacting surface develops suddenly and the contacting surfaces are damaged in a very short time. Specialists in the field of tribology distinguish between the definition of scoring and scuffing failure, but gear engineers commonly mix both expressions. Gear researchers sometimes use the expression "hot scuffing/scoring" and "cold scuffing/scoring," which nearly corresponds to "scuffing" and "scoring," respectively in the tribologist's sense. Here the word *scuffing* is used to include all the sense of this kind of failure.

The physical state of scuffing can be roughly defined as follows:

1. Two loaded lubricated surfaces slide relative to each other with or without rolling.
2. Atoms on both surfaces approach close to each other forming a bond.
3. Enough energy for bonding the atoms of both surface materials is developed by the shearing action between the surfaces.
4. The bonding and transferring action of atoms between both surfaces and the abrading away of surface material make more energy for this reaction. This positive feedback condition accelerates itself explosively.

Scuffing can initiate when the lubrication condition is in mixed or boundary lubrication. The scuffing failure that initiates under the condition of very high contact pressure and high sliding velocity is called *hot scuffing*. The damaged surface has an appearance of melted surface as shown in Fig. 15.10. The cause is believed to be pure thermal. Shear energy dissipated in the contacting area raises the temperature of the oil film and absorbed layer on the contacting surface, melting or vaporize them so that metal-to-metal contact and local welding occur. The surface is then worn away. The real method of the initiation and development of hot scuffing is not clear, but the shear energy maximum value of flash temperature, and bulk temperature of the contacting surface are considered to have an effect on the initiation of hot scuffing.[13-38] Thermal stability of the EHD oil film[20] or surface temperature[21] is also considered to have some relation to the initiation of hot scuffing.

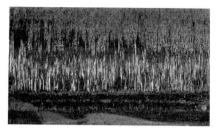

FIGURE 15.10 Typical look of hot scuffing.

Under very high loading, but very slow speed condition, a kind of scuffing failure can occur. This kind of damage is often called *cold scuffing*. Cold scuffing is initiated under a boundary lubricated condition of very poor lubrication between contacting surfaces. During the phenomena of cold scuffing, the hot area is limited to very local, contacting surface asperities. This is the biggest difference between cold scuffing and hot scuffing. An example of damaged surface due to cold scuffing is shown in Fig. 15.11: Deep scratch is observed and the material near the surface usually shows plastic flow. Cold scuffing is often found together with pitting, spalling, and plastic deformation of tooth flanks.

The phenomenon of scuffing is very complicated and there are many factors that give rise to this condition. In spite of a large amount of research on hot scuffing, no reliable method to predict the hazard of hot scuffing failure of gears has been developed. No method to predict cold scuffing failure is known.

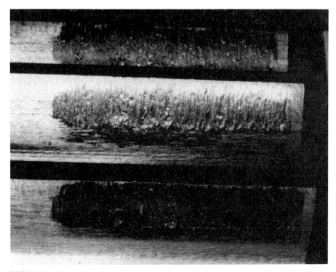

FIGURE 15.11 Typical look of cold scuffing. $PLV = 100$ ft/min

15.4 ESTIMATION OF THE INFLUENCE OF LUBRICATION ON SURFACE DISTRESS

15.4.1 Pitting

In general, the surface durability of gear teeth against pitting failure is estimated by using the maximum hertzian contact pressure as the reference value; and the influence of lubrication on durability is normally considered by a multiplication-derating factor

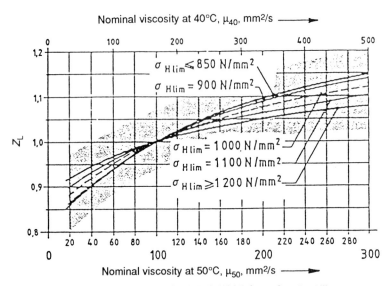

FIGURE 15.12 Lubrication factor of ISO DIS 6336/2 for surface durability.

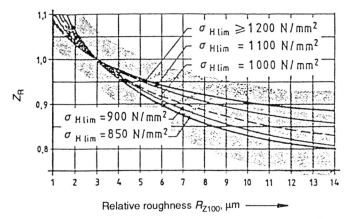

FIGURE 15.13 Roughness factor of ISO DIS 6336/2 for surface durability.

such as lubricant factor Z_L, roughness factor Z_R, and speed factor Z_V: The durability limit of the tooth flank material is calculated to increase or decrease by multiplication of these three factors $Z_L Z_R Z_V$. An example of the values of these factors is shown in Figs. 15.12, 15.13, and 15.14.[22]

Taki[18] published a remarkable investigation, which shows the influence of lubrication and surface roughness on the surface durability of gear teeth. After the publication of Dawson,[23] Taki summed up many hundreds of test results of durability of gear tooth flank in relation to the maximum hertzian contact pressure at the pitch point of gears and the D-value. The D-value is defined as the ratio of the sum of the maximum surface roughness of mating tooth flanks to the minimum oil film thickness calculated by Eq. (15.11). The result is shown in Fig. 15.15: Black marks indicate gears damaged by destructive pitting and white marks indicate no failures. It is clearly seen, that the surface durability of gear teeth deceases with increasing D-value. Between unfailed and gears damaged by destructive pitting, a definite boundary is found as functions of hertzian pressure and D-value. The boundary line is different for ground gears and for hobbed gears. He also confirmed that the results for gears cut by shaper or by pinion cutter fall in the same group as those for hobbed gears. The results for ground gears with large surface roughness and material with low limiting stress value for surface durability were also near the same group as those for hobbed gears. The difference between boundary lines for ground and hobbed gears in Fig. 15.15 seems to have a close relationship with the ratio of the wavelength of surface roughness to the hertzian contact width. Taki supposes that the reason for this difference lay in the unreliable value of the minimum oil film thickness for very rough surface, which was calculated by the EHD theory.

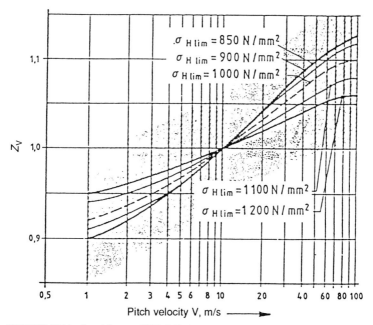

FIGURE 15.14 Speed factor of ISO DIS 6336/2 for surface durability.

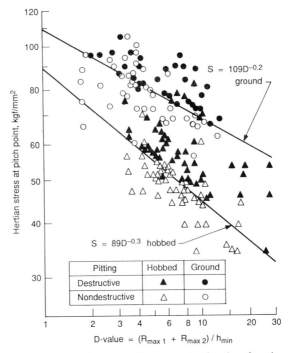

FIGURE 15.15 Surface durability of gears as function of nominal hertzian stress S and D-value. Black marks: failed gears due to destructive pitting. White marks: OK gears.

Taki proposed the following correction of the maximum hertzian contact pressure as the reference value for practical estimation of surface durability.

$$\begin{bmatrix} \text{Reference value for} \\ \text{estimation of pitting} \\ \text{load capacity} \end{bmatrix} = [\text{max hertzian stress}] \cdot D^q \quad (15.17)$$

where $\quad D = (R_{max_1} + R_{max_2})/h_{min}$
$R_{max_1}, R_{max_2} =$ maximum surface roughness of contacting bodies 1 and 2, respectively
$q = 0.2$ for smooth surface finish, for example, ground gears
$q = 0.3$ for moderate surface finish, for example, hobbed gears

Smooth surface finish means that the number of surface roughness asperities that exist within the hertzian contact width is greater than or equal to 3, and *moderate surface finish* means that the surface roughness asperity number in the hertzian contact width is less than 2.

Dudley shows also a strong influence of the lubrication condition on the surface durability of gears. He divides the EHD condition into three categories: regime I for boundary lubrication, regime II for mixed lubrication, and regime III for hydrodynamic lubrication. Guidance to determine the lubrication condition is given in Fig. 15.16.[24] The durability life of the gears is estimated by

$$\frac{L_b}{L_a} = \left(\frac{K_a}{K_b}\right)^q \quad (15.18)$$

where L_b = loading cycles for life
$L_a = 10^7$ cycles
K_a = K-factor (allowable load intensity) at L_a
K_b = K-factor at L_b
$q = 3.2$ for regime I
$q = 5.3$ for regime II
$q = 8.4$ for regime III

K-factor is defined by

$$K = \frac{W_t}{bd_1}\left[\frac{m_G + 1}{m_G}\right]$$

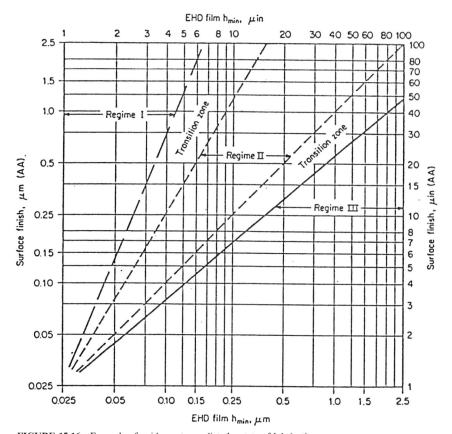

FIGURE 15.16 Example of guidance to predict the state of lubrication.

and proportional to the square of the maximum hertzian contact pressure. Here W_t denotes the tangential transmitting load; b, face width; d_1, the pitch diameter of the pinion; and m_G, the ratio of gear tooth number to pinion tooth number.

15.4.2 Scuffing

Most of today's design formulae for scuffing resistance of gears are valid only for so-called hot scuffing. The dissipating energy due to friction of the contacting tooth flanks is transformed into heat. The rate of generation is

[frictional coefficient] · [contacting pressure] · [sliding velocity]

per unit time and per unit contact area. The initiation of hot scuffing is closely related to the temperature of the contacting surface.

Reference Values to Predict Scuffing Failure

1. *PV and PVT value:* Historically, PV-value was incorporated as the first reference value to estimate scuffing resistance, where P is the maximum hetzian contact pressure and V is sliding velocity at the point at which P acts.[25]

To get a better correlation between the reference value for scuffing prediction and the actual limit of scuffing resistance for gears, the PV-value was improved to the PVT-value, where T is the distance along the line of action from the pitch point to the point at which P acts.[26] The PVT-value was found to be effective for estimating the scuffing resistance of small aircraft gears made by automobile gear manufacturers during World War II.

PVT-value for the meshing of gear teeth at the tip is calculated by the following equation:

$$(PVT)_G = \frac{\pi \cdot n_P}{360} \left(1 + \frac{N_P}{N_G}\right)(\rho_G - R \sin \phi_t)^2 \cdot P_G \qquad (15.19)$$

where n_P = pinion speed, r/min
 N_P = number of pinion teeth
 N_G = number of gear teeth
 ρ_G = radius of curvature at the gear tooth tip
 R = pitch circle radius of the gear
 ϕ_t = transverse pressure angle,°
 P_G = maximum hertzian contact pressure at the tip of gear tooth.

1,500,000 lb·ft/in·s (8,200,000 kgf/s) has been used frequently as a design limit of PVT-value for case-hardened gears of good quality, lubricated with a medium weight mineral oil for driving pitch line velocity higher that 2000 ft/min. 8,800,000 kgf/s for straight mineral oil, 11,200,000 kgf/s for gear oil and 20,100,000 kgf/s for hypoid gear oil are also proposed for a design limit on PVT-value.[27]

2. *Flash temperature:* During World War II, Blok found a strong correlation between the scuffing resistance limit of gears and the contact temperature, that is, the sum of steady bulk temperature of the surface and the flash temperature. Here the *flash temperature* is defined as the instantaneous temperature rise of the surface under contact calculated by Eq. 15.16. His proposal, using the contact temperature as the reference value for the condition of scuffing initiation, was investigated by many researchers since then. Figure 15.17 shows one result of such investigations and can be used for a survey chart of scuffing temperatures for different combinations of nonaddi-

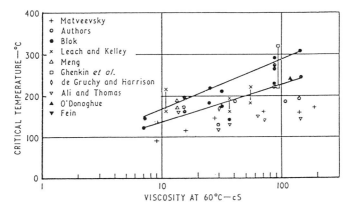

FIGURE 15.17 Survey chart of critical temperature of scuffing.

tive mineral oils and steels.[28] The sum of measured steady surface temperature of a contacting body (bulk temperature) and the instantaneous flash temperature calculated by Eq. (15.16) was incorporated for the critical temperature on the ordinate of Fig. 15.17.

Kelley modified Blok's flash temperature equation by considering the influence of surface roughness,[29] and Dudley modified Kelley's equation by considering the loading actually applied on the gear tooth flank.[30] In the AGMA 17.01 equation (1965) for contact temperature T_C, the calculation method is described in a practical form as a reference value for estimating the scuffing resistance of aerospace spur and helical gears. A part of this flash temperature is defined as a *scoring index* (SI) number and used as a reference value for scuffing prediction. The equations are:

$$T_C = T_i + \left(\frac{W_{te}}{F_e}\right)^{3/4} \cdot \frac{Z_t \cdot n_P^{1/2}}{P_d^{1/4}} \cdot \frac{50}{50 - S} \quad (15.20)$$

$$SI = \left(\frac{W_t}{F_e}\right)^{3/4} \cdot \frac{n_P^{1/2}}{P_d^{1/4}} \quad (15.21)$$

where T_i = initial temperature (inlet oil temperature), °F
W_{te} = effective tangential load, lb
F_e = effective tooth width, in.
S = surface finish, rms (after running in) in microinch
Z_t = scoring geometry factor

$$= 0.0175 \frac{\left(\sqrt{\rho_P} - \sqrt{\frac{N_P}{N_G} \cdot \rho_G}\right) \cdot P_d^{1/4}}{(\cos \phi_t)^{3/4} \left(\frac{\rho_P \cdot \rho_G}{\rho_P + \rho_G}\right)^{1/4}}$$

where n_P = pinion r/min
P_d = transverse diametrical pitch

TABLE 15.1 Maximum Design Limit of Flash Temperatures to Prevent Scoring of Spur Gears

Kind of Oil	Specification	T_f, °F
Petroleum	SAE 10	250
	SAE 30	375
	SAE 60	500
	SAE 90 (gear lubricant)	600
Diester, compounded	75 SUS at 100°F	330
Petroleum	SAE 30 plus mild EP	425

TABLE 15.2 Critical Scoring Criterion Numbers

	Blank Temperature, °F				
Kind of Oil	100°	150°	200°	250°	300°
	Critical Scoring Index Numbers				
AGMA 1	9,000	6,000	3,000	—	—
AGMA 3	11,000	8,000	5,000	2,000	—
AGMA 5	13,000	10,000	7,000	4,000	—
AGMA 7	15,000	12,000	9,000	6,000	—
AGMA 8A	17,000	14,000	11,000	8,000	—
Grade 1065, Mil-O-6082B	15,000	12,000	9,000	6,000	—
Grade 1010, Mil-O-6082B	12,000	9,000	6,000	2,000	—
Synthetic (Turbo 35)	17,000	14,000	11,000	8,000	5,000
Synthetic Mil-L-7808D	15,000	12,000	9,000	6,000	3,000

Source: This table is reproduced by permission from the *Gear Handbook*, Chap. 13.
Notes: 1. See Sec. 3.31 for more data on rating gears for scoring and the use of the scoring criterion number.
2. See Sec. 7.15 for general data on gear lubricants and the hazard of lubrication failures.
3. See AGMA 250.04 and 251.02 for general data on industrial lubricants.

ρ_P, ρ_G = radius of curvature of tooth flank at a point (e.g., tooth tip) and corresponding position on mating tooth flank, respectively, in.

Tables 15.1 and 15.2 show the design limit for contact temperature T_C and SI number respectively.[24]
The AGMA standard employs the contact temperature on the basis of the local maximum flash temperature as the reference value for rating scuffing load capacity, but, to calculate the flash temperature, Blok's original equation is modified to resemble Kelley's equation of AGMA 217.01 (1965) for considering the effect of surface roughness of tooth flank.

3. *Scoring factor:* Borsoff[31] defined a scoring factor S_f as the ratio between hertzian contact width and sliding velocity:

$$S_f = \frac{3.04 \sqrt{(W/F) \cdot \rho/E}}{(3.14 \cdot N_P/30)[\rho_P - \rho_G (n_P/n_g)]} \quad (15.22)$$

where W = normal tooth load, lb
F = tooth width
E = Young's modulus, lb/in^2
ρ_P, ρ_G = radii of curvature of contacting surfaces, in.
ρ = relative radius of curvature $\rho_P \rho_G/(\rho_P + \rho_G)$
N_P = pinion r/min
n_P, n_g = tooth numbers of pinion and gear, respectively

For teeth meshing at the tooth tip, the normal tooth flank load is supposed to be half the transmitting load, thus

$$S_f = 3.7477 \cdot 10^3 \frac{\rho^{0.5} (W/F)^{0.5}}{\rho_P - \rho_G/(n_P/n_g)N_P} \tag{15.23}$$

The limit of scoring load is given by the linear equations listed in Table 15.3.

4. Bulk temperature: Many engineers and users of gears know from experience that when the temperature of the lubricating oil or gear exceeds a certain limit, the hazard of scuffing failure increases. Niemann and Lechner[32,33,34,35] investigated various kinds of temperature as candidates for the reference value for scuffing prediction. They concluded that the steady temperature of the tooth flank has better correlation with the initiation of scuffing than does the instantaneous flash temperature. The strength of adhesion of the surface layer of lubricant, which protects the surface from scuffing, is determined by the steady temperature of the surface and not by the instantaneous temperature of very short duration.

Scuffing failures of machine parts start on the surface that makes contact periodically, and in most cases the duration of no contact is far longer than that of the contact. Hirano[36] studied the lubricating oil property change on the surface during the no-contact period and found that of the molecular construction of the base oil had a strong influence on the condition of scuffing initiation. Hirano takes the sign of the derivative of the friction coefficient to the steady temperature $d\mu/d\theta$ for the reference value of scuffing initiation. When $d\mu/d\theta > 0$, the condition of EHD contact is unstable and the surface temperature diverges to start scuffing failure. When the on/off ratio of contact time duration is small, the bulk temperature (steady surface temperature) is to be used for the temperature θ in the calculation of $d\mu/d\theta$.

Fujita[37] took as the reference value for scuffing initiation both the bulk temperature and contact temperature (sum of bulk and flash temperatures, each of which is valid when scuffing initiates in hydrodynamic–mixed lubricated condition or in boundary lubricated condition, respectively.

TABLE 15.3 Score Load—Scoring Factor Equations for Oils Tested

Author	No.	Oil	Equation
Borsoff[2]	1	1010 mineral oil	$L = 28\ S_f + 160$
	2	SAE 30 mineral oil	$L = 28.3\ S_f + 690$
	3	SAE 60 mineral oil	$L = 24.8\ S_f + 2170$
Ku and Baber[6]	4	1065 mineral oil	$L = 31.2\ S_f + 1375$
	5	A synthetic oil	$L = 43.1\ S_f + 1630$
Mansion[7]	6	A light naphthenic oil	$L = 43.4\ S_f + 1380$
	7	A blend	$L = 38.6\ S_f + 840$
	8	A light paraffinic oil	$L = 31.3\ S_f + 390$

5. Frictional power intensity and frictional power: Matveevsky[38] reports that the product of the coefficient of friction, mean contact pressure, and sliding velocity at the initiation of scuffing is constant. This product is called *frictional power intensity* (FPI).

$$\text{FPI} = \frac{0.113 W \mu \left| V_1 - V_2 \right|}{A} \text{ watts/mm}^2 \qquad (15.24)$$

where W is the specific loading (lb/unit length of contact) and A is contact area (mm^2). The Matveevsky's criterion of scuffing initiation means that every kind of lubricant has its own limit of FPI value for breakdown of the oil film.

Moorhous[39] modified this reference value for scuffing initiation to FPI/$(V_1 - V_2)^{0.8}$ to get better correlation between reference value and actual beginning of scuffing failure of gears, where V_1, V_2 are surface velocities relative to the contact zone.

Carper[40] compared frictional power; that is, the product of coefficient of friction, normal load, and sliding velocity; with critical temperature and with FPI as the candidate for the reference value for scuffing initiation. His investigation concluded that the frictional power worked best to predict scuffing initiation.

6. Ratio of flash temperature to contact stress: Gabrikov[41] proposed K_1 value, which is the ratio of flash temperature to hertzian contact stress, as the reference value for scuffing prediction. For steel material K_1, in degrees C/(kgf/cm^2)$^{0.5}$, is approximately

$$K_1 = 0.00302 \, f p_n^{0.25} \left| \sqrt{V_1} - \sqrt{V_2} \right| \cdot \rho_r^{0.25} \qquad (15.25)$$

where f = coefficient of friction
 p_n = specific normal load
 ρ_r = the relative radius of curvature

Drozdov[42] gives the limit value of K for the beginning of scuffing as $0.0053 \cdot \nu_{50}^{0.006}$, where ν_{50} is the viscosity of oil in cSt at 50°C.

7. Ratio of flash temperature to the square root of specific normal load: Drozdov[42] proposed K_2 value, which is the ratio of flash temperature to the square root of normal force per unit width, for the reference value of scuffing prediction. For steel K_2, in degrees C/(kgf/cm)$^{0.5}$, is approximately

$$K_2 = 1.88 \, f p_n^{0.25} \left| \sqrt{V_1} - \sqrt{V_2} \right| \rho_r^{-0.25} \qquad (15.26)$$

He gives the limit value of K_2 as $2.0 \, \nu_{50}^{0.06}$.

8. Mean flash temperature. Niemann and Seitzinger[43] proposed the mean temperature of the tooth flank, which is defined by the sum of oil temperature and temperature rise calculated from the power loss of gears, as the reference value for scuffing.

Winter and Michaelis[44] substituted the temperature rise of Niemann and Seitzinger's method with the average flash temperature. The average flash temperature of Winter and Michaelis' method is calculated on the tooth flank of spur gears or cylindrical involute gears in the transverse section. The sum of the oil temperature and the average flash temperature with some derating factors was called *integral temperature* and was proposed for the reference value for scuffing load calculation. This scuffing load calculation method is approved as DIN Standard 3990/4 and was proposed as one part of the international standard ISO 6336/4 for the method of calculating the scuffing load capacity of gears.

ISO Proposal for Estimation of Scuffing Load Capacity. Since the early 1960s a group of researchers (Niemann, Lechner, Seitzinger, Winter, and Michaelis) in FZG in the

GEAR LUBRICATION

T. U. Munchen have conducted extensive investigation on the scuffing resistance of gears. They concluded that the bulk temperature plus the average (mean) value of flash temperature on the tooth flank gives a better correlation with the actual scuffing failure than does the bulk temperature plus maximum value of flash temperature.[35,43,45]

The outline of the originally proposed ISO method[19] for the scuffing load capacity calculation for cylindrical involute gears using the local flash temperature criterion is as follows.

The basic formula for safety against scuffing is

$$\theta_B = \theta_M + \theta_{fl} \leq \theta_{BP} \quad (15.27)$$

where θ_B = contact temperature at any point on the line of action
θ_M = the gear bulk temperature
θ_{fl} = the flash temperature at the point considered
θ_{BP} = the allowable contact temperature

$$\theta_{fl} = \mu_{m_y} X_M X_B X_{\alpha\beta} X_\Gamma \quad (15.28)$$

$$\times \frac{w_{B_t}^{3/4} V^{1/2}}{a^{1/4}}$$

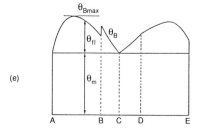

FIGURE 15.18 Assumed pattern of tooth flank load sharing and the value of X_Γ.

where μ_{m_y} = the mean local coefficient of friction
X_M = the thermal flash factor, which takes into account the moduli of elasticity and the thermal contact coefficients of the gear material. A roughly approximated value is $50.0 \, kN^{-0.75} \cdot S^{0.5} \cdot m^{-0.5} \cdot mm$
X_B = geometry factor, which takes into account the influences of the gear ratio, the radii of curvature, and sliding velocities
$X_{\alpha\beta}$ = pressure angle factor, which takes into account the influences of the pressure angle and the helix angle (For a very rough estimation, $X_{\alpha\beta}$ can be set equal to 1.)
X_Γ = load-sharing factor, which takes into account the distribution of load over adjacent tooth pairs (Value of X_Γ, for example, is shown in Fig. 15.18.)
w_{B_t} = tangential specific tooth load in N/mm
V = pitch line velocity in m/s
a = center distance in mm

$$X_B = 0.51(u+1)^{0.5} \frac{[\sqrt{1+\Gamma} - \sqrt{1-\Gamma/u}]}{[(1+\Gamma)(u-\Gamma)]^{0.25}} \quad \text{for external gears} \quad (15.29)$$

$$\Gamma = \frac{\tan \alpha_y}{\tan \alpha_t'} - 1 \quad (15.30)$$

where Γ = parameter on the line of action calculated by Eq. (15.30)
u = gear ratio
α_y = transverse pressure angle at arbitrary point
α_t' = transverse working pressure angle

$$X_{\alpha\beta} = 1.22 \frac{(\sin \alpha_t' \cos \alpha_n \cos \beta)^{0.25}}{(\cos \alpha_t' \cos \alpha_t)^{0.25}} \quad (15.31)$$

$$w_{Bt} = K_A K_{B\beta} K_{B\alpha} K_{B\gamma} \frac{F_t}{b} \quad (15.32)$$

where β = helix angle
K_A = application factor
$K_{B\beta}$ = longitudinal load distribution factor[45]
$K_{B\alpha}$ = transverse load distribution factor[45]
$K_{B\gamma}$ = helical load distribution factor[45]
F_t = nominal tangential load at reference cylinder in transverse section, N
b = smaller face width of pinion or gear, mm

The bulk temperature should be measured or determined by experience. The roughly estimated value can be given by

$$\theta_M = X_S (\theta_{\text{oil}} + 0.47 \theta_{\text{fl max}}) \quad (15.33)$$

where θ_{oil} = lubricating oil temperature before it reaches the tooth mesh
$\theta_{\text{fl max}}$ = maximum value of the flash temperature [Eq. (15.28)] along the path of contact
X_S = lubrication weight factor, which takes into account a better heat transfer for sump lubrication compared with jet lubrication (A rough guiding value of this is empirically 1.0 for dip lubrication and 1.2 for spray lubrication.)

The allowable contact temperature θ_{BP} is given by

$$\theta_{BP} = \frac{\theta_S - \theta_{\text{oil}}}{S_{B_{\min}}} \quad (15.34)$$

where $S_{B_{\min}}$ is the minimum demanded safety factor and θ_S is scuffing temperature, which should be determined by experiments using Eq. (15.35).

$$\theta_S = \theta_{M,\,T} + X_{W_{\text{rel}},\,T} \cdot \theta_{\text{fl max},\,T} \quad (15.35)$$

where the suffix T denotes the value for test gears and $X_{W_{\text{rel}},\,T} = \dfrac{X_W}{X_{W,\,T}}$ is the relative welding factor. For guiding value of X_W see Table 15.4.

TABLE 15.4 Relative Welding Factor for Scuffing Temperature Calculation in ISO DIS 6336/3

Through-hardened steel	1.00
Phosphated steel	1.25
Copper-plated steel	1.50
Bath- and gas-nitrided steel	1.50
Hardened carbonized steel	
Content of austenite less than average	1.15
Content of austenite average	1.00
Content of austenite more than average	0.85
Austenitic steel (stainless steel)	0.45

For the standard FZG gear test, it is approximated that

$$\theta_{M,\,T} = 80 + 0.23\, T_{1,\,T} \tag{15.36}$$

$$\theta_{\text{fl max},\,T} = 0.12\,(T_{1,\,T})^{1.2}\left(\frac{100}{\nu_{40}}\right)(\nu_{40}^{-0.4})$$

For the standard Ryder gear test

$$\theta_{M,\,T} = 90 + 0.0022\,\frac{F_{t,\,T}}{b_T} \tag{15.37}$$

$$\theta_{\text{fl max},\,T} = 0.0053\,\frac{F_{t,\,T}}{b_T}\left(\frac{100}{\nu_{40}}\right)^{0.17}$$

where $T_{1,\,T}$ is the torque on the pinion of the test gears, Nm, and ν_{40} is the kinematic viscosity of the lubricant at 40°C, mm²/s.

ANSI/AGMA Standard. The ANSI/AGMA standard 2001-B88 Appendix A (Sept. 1988) employs the contact temperature on the basis of the instantaneous maximum flash temperature as the reference value for rating the scuffing load capacity. To calculate the flash temperature, Eq. (15.16) is modified to resemble the second term of Eq. (15.20), that is, the flash temperature is

$$T_{fl} = \frac{0.8\,\mu\,X_r W_{Nr}\,\left|\,V_{r_1}^{0.5} - V_{r_2}^{0.5}\,\right|}{B \cdot b^{0.5}} \tag{15.38}$$

For a rough estimation, the coefficient of friction $\mu_m = 0.06\,\left(\frac{50}{50 - S}\right)$. The thermal contact coefficient B_m is about 43 lbf/in · s$^{0.5}$ · °F. The normal unit load W_{Nr} is calculated by dividing the normal operating load by the minimum length of simultaneous contact lines. The load sharing by more than one tooth pair and the tooth profile modification is accounted for by the load sharing factor X_r in Fig. 15.18. The surface roughness is taken as the average rms value of pinion and gear tooth flanks in microinches. But the term $\frac{50}{(50 - S)}$, which takes into account the effect of surface roughness, is limited to less than or equal to 3.0. The flash temperature after

TABLE 15.5 Example of Mean Critical Contact Temperature (50-percent chance of scuffing) and Standard Deviation

Lubricant	Mean Critical Total Temperature, °F	Standard Deviation, °F
MIL-L-7808	366	56.6
MIL-L-6081 (grade 1005)	264	74.4

TABLE 15.6 Example of Mean Critical Contact Temperature and Standard Deviation (15 percent of the mean value)

ISO VG	AGMA Lube No.	Mean Critical Total Temperature, °F	Standard Deviation, °F
32	—	351	53
46	1	372	56
68	2	395	59
100	3	418	63
150	4	441	66
220	5	464	70
320	6	486	73
460	7	507	76
680	8	530	80
1000	8A	553	83
1500	—	577	87

Eq. (15.38) should be calculated at several points on the pinion tooth profile and the sum of the maximum flash temperature and the gear bulk temperature, that is, contact or conjunction temperature, is employed as the reference value. The risk of scuffing is estimated by comparing this reference value with the critical value listed in Tables 15.5 and 15.6.

In this standard it is recommended, that the specific film thickness should also be calculated on the tooth flank to estimate the risk of tooth flank wear. This is used as a method to prevent cold scuffing. The same kind of explanation is given in Dudley.[24] The specific film thickness Λ is the ratio between the minimum oil film thickness h_{min} and the composite surface roughness $R'_{rms} = (R_{1rms} + R_{2rms})^{0.5}$, where R_{1rms} and R_{2rms} are rms values of surface roughness of contacting tooth flanks.

$$\Lambda = \frac{H_{min}}{R'_{rms}}$$

The minimum oil film thickness h_{min} is calculated by Eq. (15.11). Figure 15.19 is a plot of percent EHD film versus the specific film thickness, which gives an indication of the different regimes of lubrication and the expected results. The probability of wear is estimated by Fig. 15.20. The data used for Fig. 15.20 were limited-life data and should be used with extreme caution, especially for long-life applications.

GEAR LUBRICATION 15.29

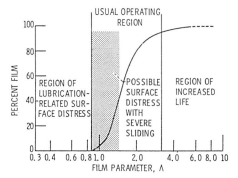

FIGURE 15.19 Percent film as a function of film parameter.

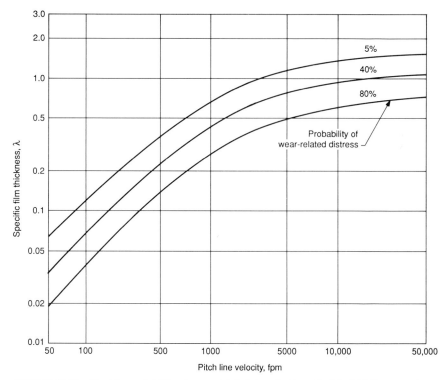

FIGURE 15.20 Chart for probability of wear-related distress (cold scuffing) in AGMA proposal.

TABLE 15.7 Lubricant Physical Properties Summary

	Lubricant					
	A Mineral Oil	C Polyol Ester Pentaerythritol	E Dibasic Acid Ester	F Synthetic Paraffinic	K Polyol Ester Pentaerythritol	L Pentaerythritol Tetraester
Kinematic viscosity, cm^2/s (cS) at -244 K (-20 °F)	35. (3500)	30. (3000)	50. (5000)	26. (2600)	30. (3000)	—
311 K (100 °F)	0.40 (40)	0.285 (28.5)	0.36 (36)	0.303 (30.3)	0.285 (28.5)	0.276 (27.6)
372 K (210 °F)	0.07 (7.2)	.053 (5.3)	0.06 (6.0)	0.055 (5.5)	0.053 (5.3)	0.052 (5.2)
Pressure viscosity coefficient, GPa^{-1} (lb/in^2) at —						
311 (100 °F)	15.4 (10.6×10^{-5})	11.6 (8.0×10^{-5})	15.5 (10.7×10^{-5})	13.4 (9.2×10^{-5})	11.4 (7.9×10^{-5})	—
372 (210 °F)	11.2 (7.7×10^{-5})	10.0 (6.9×10^{-5})	11.5 (7.9×10^{-5})	11.1 (7.7×10^{-5})	9.5 (6.5×10^{-5})	—
Flash point, K (°F)	433 (320)	527 (490)	513 (465)	508 (455)	533 (500)	527 (490)
Pour point, K (°F)	233 (−40)	211 (−80)	219 (−65)	219 (−65)	214 (−75)	213 (−76)
Specific gravity	0.862	1.005	0.932	0.829	0.983	0.982
Specific heat at 311 K (100°F), J/kg K (Btu/lb °F)	1758 (0.42)	1842 (0.44)	2847 (0.68)	2177 (0.52)	1884 (0.45)	—
Λ ratio (h/σ)	0.87	0.76	0.86	0.65	0.69	1.22

15.5 LUBRICANT AND ADDITIVE EFFECTS ON GEAR FATIGUE LIFE

Gear failure by surface pitting fatigue is affected by the physical and chemical properties of the lubricant. Knowledge of how these chemical and physical properties affect surface pitting is a useful guide both in selecting existing lubricants for mechanical power transmission applications and in developing new lubricant formulations. For transmission applications, it is important to know the effects of these lubricants and their additives on bearing and gear life and reliability.

Lubricant additives are necessary to the operation of gear systems. These additives can prevent or minimize wear and surface damage to bearings and gears whose load-carrying surfaces operate under very thin film or boundary lubrication conditions.[46,47] These antiwear extreme-pressure (EP) additives either absorb onto or react with the surfaces to form protective coatings or surface films. Although these boundary films are 1 μm (40 μin.) or less thick,[48] they can provide separation of the metal surfaces when the elastohydrodynamic (EHD) film becomes thin enough for the asperities to interact. The boundary film probably provides lubrication by microasperity-elastohydrodynamic lubrication as the asperities deform under load. The boundary film prevents contact of the asperities and, at the same time, provides low shear strength properties that prevent shearing of the metal while reducing the friction coefficient below that of the base metal.

The type of EP additive required depends on the severity of the conditions of the meshing gear teeth, such as sliding velocity, surface temperature, and contact load. The boundary films can provide lubrication at different temperature conditions, depending on the materials used.

Spur gear endurance tests were conducted with six lubricants using a single lot of CVM, AISI 9310 spur gears.[49] Properties of the lubricants are given in Table 15.7. The sixth lubricant was divided into four batches, each of which had a different additive content. Lubricants tested with a phosphorus-type load-carrying additive showed a statistically significant improvement in life over lubricants without this type of additive. The presence of sulphur-type antiwear additives in the lubricant did not appear to affect the surface fatigue life of the gears. No statistical difference in life was produced by those lubricants of different base stocks, but with similar viscosity, pressure-viscosity coefficients, and antiwear additives. Results of these tests are shown in Fig. 15.21.

REFERENCES

1. Akin, L. L., and Townsend, D. P., "Into Mesh Lubrication of Spur Gears with Arbitrary Offset Oil Jet. Part I—For Jet Velocity Less Than or Equal to Gear Velocity," *Jour. of Mech., Trans. and Automation in Des.,* Vol. 105, No. 4, December 1983, p. 713.

2. Akin, L. S., and Townsend, D. P., "Lubrication Jet Flow Phenomena in Spur and Helical Gears with Modified Center Distances and/or Addendums for Out-of-Mesh Conditions," *Jour. of Mech., Trans., and Automation in Des.,* Vol. 107, No. 1, March 1985, pp. 24–30.

3. Fuller, Dudley, D., *Theory and Practice of Lubrication for Engineers,* Wiley, 1965.

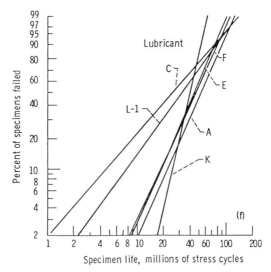

FIGURE 15.21 Surface fatigue life results of AISI 9310 at 250 KSI (1.7 GPa) maximum Hertz stress with several lubricants of same viscosity.

4. Harris, T. A., *Rolling Bearing Analysis*, Wiley, 1966.
5. Pirvics, J., and Klechner, R. J., "Predication of Ball and Roller Bearing Thermal and Kinematic Performance by Computer Analysis," *Advanced Power Transmission Technology*, G. K. Fisher, ed., NASA CP–2210, 1983, pp. 185–201.
6. Coe, H. H., "Predicted and Experimental Performance of Large-Bore High-Speed Ball and Roller Bearings," *Advanced Power Transmission Technology*, G. K. Fischer, ed., NASA CP-2210, 1983, pp. 203–220.
7. Dawson, F. H., "Windage Loss in Large High-Speed Gears," *Proc., Inst. of Mech. Eng., Part A—Power and Process Engineering*, Vol. 1984, No. 1, 1984, pp. 51–59.
8. Townsend, D., and Akin, L., "Analytical and Experimental Spur Gear Tooth Temperature as Affected by Operating Variables," *Jour. Mech. Des.*, Vol. 103, No. 1, January 1981, pp. 219–226.
9. Townsend, D. P., and Akin, L. S., "Study of Lubricant Jet Flow Phenomena in Spur Gears—Out of Mesh Condition," *Jour. Mech. Des.*, Vol. 100, No. 1, January 1978, pp. 61–68.
10. Tucker, A. I., "Bevel Gears at 203 m/s (40,000 FPM)," ASME Paper 77–DET–178, September 1977.
11. Akin, L. S., and Townsend, D. P., "Into Mesh Lubrication of Spur Gears With Arbitrary Offset Oil Jet. Part II—For Jet Velocities Equal to or Greater Than Gear Velocity," *Jour. Mech., Trans. and Automation in Des.*, Vol. 105, No. 4, December 1983, pp. 719–724.
12. Akin, L. S., Townsend, D. P., and Mross, J. J., "Study of Lubricant Jet Flow Phenomena in Spur Gears," *Jour. Lub. Tech.*, Vol. 97, No. 2, April 1975, pp. 283–288.
13. Grubin, A. N., Vinogradova, I. E., "Fundamentals of the Hydrodynamic Theory of Lubrication of Heavily Loaded Cylindrical Surfaces," Maschgiz, Moscow, 1049, D.S.I.R. Translation No. 337.
14. Petrusevich, A. J., "The Basic Conclusion from the Elasto-Hydrodynamic Theory of Lubrication," *Izvestia Akad. Nauk* SSSR(OTN), No. 2, 1951, English translation S.T.S.–138, BST–480.

15. Dowson, D., and Higginson, G. R., *Elastro-Hydrodynamic Lubrication,* Pergamon, 1966.
16. Dowson, D., "Elastro-Hydrodynamics," *Proc. AIME, Conference on Lubrication and Wear,* London, 1967, Vol. 182, 1967–68, p. 151.
17. Blok, H., "Les Temperatures de Surface Dans Conditions de Graissage Sous Extreme Pression," *Proc. Second World Petr. Congress,* Section IV, Paris, Vol. III, p. 471.
18. Taki, T., "Research on Pitting Durability of Gears," Diss. 1987, Kyoto University.
19. Proposal for ISO DIS 6336/4, "Calculation of Load Capacity of Spur and Helical Gears—Part 4: Calculation of Scuffing Resistance."
20. Blok, H., "Thermal Instability of Flow in Elastro-Hydrodynamic Films as a Cause for Cavitation, Collapse and Scuffing," *Proc. Leeds Lyon Sypm., Tribology,* 1975, Vol. 1, p. 189.
21. Kubo, A., "Fundamental on Scoring Failure (3rd report, Improvement of Scoring Model and Definition of Scoring Index)," *Bull. JSME,* Vol. 28, No. 240, 1985–86, p.1288.
22. ISO DIS 6336/2, "Calculation of Load Capacity of Spur and Helical Gears—Part 2: Calculation of Surface Durability."
23. Dawson, P. H., "Effect of Metallic Contact on the Pitting of Lubricated Roller Surfaces," *Journal of Mechanical Engineering Science,* Vol. 4–1, 1962, p. 16.
24. Dudley, D. W., *Handbook of Practical Gear Design,* McGraw-Hill, 1984.
25. Almen, J. O., "Factors Influencing the Durability of Spiral-Bevel Gears for Automobiles," *Automobile Ind.,* Vol. 73, 1935, p. 662, 692.
26. Almen, J. O., "Surface Deterioration of Gear Teeth," *Mechanical Wear,* 1950, p. 229.
27. Rettig, H., "Ermitlung der Schmierstoffeigenschaften in Sahnradtest," *Schriftreihe Antriebstech.,* Bd. 16, 1955, p. 58.
28. O'Donoghue, J. P., and Cameron, A., "Friction and Temperature in Rolling/Sliding Contact," *Trans. ASLE,* Vol. 9, 1966, p. 186.
29. Kelley, B. W., "A Newlook at the Scoring Phenomena of Gears," *SAE Transactions,* Vol. 61, 1952, p. 175.
30. Dudley, D. W., *Practical Gear Design,* McGraw-Hill, 1954.
31. Borsoff, V. N., and Godet, M. R., "A Scoring Factor for Gears," *Trans. ASLE,* Vol. 6, 1963, p. 147.
32. Niemann, G., and Lechner, G., "The Measurement of Surface Temperatures on Gear Teeth," *Trans. ASME,* Ser. D, Vol. 87, 1965, p. 641.
33. Niemann, G., and Lechner, G., "Die Erwaermung der Sahnraeder im Betrieb," *Schmiertechnik,* Vol. 14, 1967, p.13.
34. Lechner, G., "Zahnradschmierung," *Maschinenmarkt,* Vol. 74, 1968, p. 24.
35. Niemann, G., and Lechner, G., "Die 2Fress-Grenzlast bei Stirnraedern aus Stahl," *Z. 2Erdoel und Kohle,* Bd. 20, 1967, Nr. 2, p. 96.
36. Hirano, F., and Ueno, T., "Lubrication and Failure of Gears," *Journal JSME,* Vol. 79, 1976, p. 1073.
37. Fujita, K., Obata, F., and Yamaura, I., Study on the Scoring Resistance of Lubricating Oil in Two Cylinder Test (3rd Report, Seizure Behavior under Rolling-Sliding Contact Condition and the Relation between Seizure and Temperature in Non-EP Gear Oil).
38. Matveevsky, R. M., "The Critical Temperature of Oil With Point and Line Contact Machines," *Trans. ASME,* Ser. D, Vol. 87, 1965, p. 754.
39. Moorhous, P., "The Prediction of Gears in Industrial Service," *Proc. FZG-Colloquim,* April 1973, Munich, p. 61.
40. Carper, H. J., and Ku, P. M., "Thermal and Scuffing Behavior of Discs in Sliding-Rolling Contact," *Trans. ASLE,* Vol. 18, 1975, p. 39.

41. Gabrikov, Y. A., Protivozadirinaya Stoikost Tyazheponagruzhennego Zybyatogo Kontakta, Izvest. Vuz, Mash., No. 4, 1967, p. 44.
42. Drozdov, Y. N., and Gabrikov, Y. A., "Friction and Scoring Under the Conditions of Simultaneous Rolling and Sliding of Bodies," *Wear,* Vol. 11, 1968, p. 291.
43. Niemann, G., and Seitzinger, K., "Die Erwaermung Einsatzgehaerteter Zahnraeder als Kennwert fuer ihre Fresstragfaehigkeit," *VDI-Z,* bd. 113, 1971, p. 97.
44. Winter, H., and Michaelis, K., "Fresstragfaehigkeit von Stirnradgetrieben," *Antribstech.,* bd. 14, 1975, nr. 7, p. 405, 465.
45. ISO DIS 6336/1, "Calculation of Load Capacity of Spur and Helical Gears—Part I: Basic Principles—Introduction and General Influence Factors."
46. Beane, G. A., and Lawler, C. W., "Load-Carrying Capacities of Gear Lubricants of Different Chemical Classes Based on Results Obtained with WADD High-Temperature Gear Machine Used with Induction-Heated Test Gears," AFAPL–TR–65–23, Southwest Research Institute, April 1965.
47. Anderson, E. L., et al., "Gear Load-Carrying Capacities of Various Lubricant Types at High Temperatures in Air and Nitrogen Atmospheres," AFAPL–TR–67–15, Southwest Research Institute, March 1967.
48. Fein, R. S., "Chemistry in Concentrated-Conjunction Lubrication. Interdisciplinary Approach to the Lubrication of Concentrated Contacts," NASA SP–237, 1970, pp. 489–528.

CHAPTER 16
GEAR CUTTING

A. Donald Moncrieff
Gear Consultant, Moncrieff Associates
Vero Beach, Florida

The material in this chapter has been rearranged and rewritten to include new material introduced since this book was printed in 1962. The trends of today demand gears of higher speeds, greater load-carrying ability and higher accuracy. There is also much concern to produce gear parts faster and at lower costs than was possible in earlier years. With these fundamental objectives in mind, this chapter discusses the machinery currently used to cut and finish gear teeth.

16.1 GEAR MILLING

Gear teeth are produced by milling with a form-milling cutter. After each tooth space is milled, the gear blank is indexed to the next cutting position. Gear milling can be applied to the roughing and finishing of spur, helical, and straight bevel gear teeth.

16.1.1 Application of Gear Milling

Gear milling is universal in application. In practice, however, it is usually confined to producing replacement gears or small-lot gears having special tooth forms. Depending on the size of the gear and the capacity of the machine, standard milling machines are used with either manual or automatic indexing mechanisms. On a production basis, where milling is used to produce large coarse-pitch gears, special automatic indexing gear cutting machines are used. Small automatic gear cutting machines are applied to the volume production of fine-pitch gears having special tooth forms.

Gear milling finds its widest application in roughing and finishing external spur gears and in roughing external helical gears. Straight bevel gears are often roughed out by milling prior to generating; and in some cases these gears are finished by milling.

The first edition author of this chapter was Stuart J. Johnson, Chief Engineer, Small Tools Division, Barber-Colman Co., Rockford, Illinois.

In addition to gears, milling is applied to sprockets, splines, racks, ratchets, and other forms. Although it is usually applied to external tooth forms, internal tooth forms can also be produced by milling. Figure 16.1 shows a heavy-duty gear milling machine.

Milling can be used to machine almost any tooth form. The teeth can be either equally or unequally spaced and the tooth forms can be either symmetrical or unsymmetrical. In milling the small gear of a two-gear cluster or teeth adjacent to a shoulder, adequate clearance is required to allow the cutter to produce full-depth teeth for the specified face width.

Standard-type milling machines are available in various sizes for application to gear milling. The size and pitch of the gear that can be produced will largely depend on the machine capacity available.

In addition to standard milling machines, special machines are available for the primary function of milling gears and other tooth forms. They are available to cut gears of 1 diametral pitch and up to 10 ft in diameter. For fine-pitch gears, small machines with automatic features are applicable to volume production.

Since the introduction of CNC, improved indexing is available. The index gears drive the worktable through a dual-lead worm and worm gear, which is adjustable to provide minimum backlash. The index gear drive is coupled directly to a servomotor and a closed loop system that is controlled by the CNC to give 0.0001° accuracy (as reported by the manufacturer).

Whether the machine is hand-operated or automatic, the accuracy of tooth spacing depends on the indexing mechanism. Since milling gear teeth is a form-cutting operation in which the cutter tooth produces the complete gear tooth profile, fine surface

FIGURE 16.1 Large heavy-duty milling machine cutting teeth on a bevel pinion. *(Courtesy of Gould & Eberhardt Gear Machinery Corp., Webster, Massachussets)*

finish can be obtained. Depending on conditions such as gear material, milling cutter design, and milling setup, surface finishes in the range of 15 to 20 μin. can be obtained.

16.1.2 Principles of Gear Milling

Gear milling is a machining process in which the gear teeth are produced by a form cutter. In the case of spur gears, the form on the cutter teeth is reproduced on the gear. When helical gears are milled, the cutter tooth form is not reproduced on the gear. The details of the milling cutter teeth and the gear teeth being cut are shown in Fig. 16.2.

Gear cutters are not universal for a complete range of teeth. To produce theoretically correct gear teeth, the cutter tooth form must be designed for the specific number of teeth. However, where a small error in tooth form in acceptable, cutters that are designed for a range of teeth can be used. Standard cutters, which are discussed in the chapter on gear cutting tools, are available for established ranges of teeth. The form on the cutter is made correct for the lowest number of teeth in that particular range. Thus, all teeth within the range are provided with sufficient tip relief. The same form is produced on all tooth spaces within that range.

The tooth form of a standard 14½° involute gear cutter is actually a composite of the involute and cycloidal forms. The purpose of this form is to allow gears with a 14½° pressure angle and a small number of teeth to be produced by a cutter. This requires a modified involute form to allow gears and pinions to be meshed without interference. Consequently, gears cut by this type of cutter will not mesh properly with those produced by other types of standard involute gear cutting tools. They will, however, mesh properly with other gears of the same pitch that are cut by a standard gear cutter. Gear cutters with a standard involute form can be special-ordered, and gears produced by these cutters will roll properly with gears produced by other types of standard involute gear cutter tools.

For spur gears, a gear cutter will produce tooth spaces that have the same form as the cutter teeth. Since the cutter axis is at 90° to the gear axis, the axis of the cutter is contained in the transverse plane of the gear. Under these conditions, the gear tooth

FIGURE 16.2 Gear-milling cutter and spur gear.

profiles are finished when the cutter teeth pass through this plane. Therefore, the form on the cutter teeth is reproduced in the gear.

The form on helical gear tooth spaces will not be the same as the cutter tooth form. In milling helical teeth, the cutter axis is usually set at the helix angle of the gear. At this setting, the axis of the cutter is contained in the normal plane through the center of the gear tooth space. Under these conditions, only one point on the finished profile is produced in this normal plane. All other points on the finished profile are produced in different planes. Therefore, the form on the cutter teeth is not reproduced in the gear. In addition to the setting angle, the diameter of the cutter affects the gear tooth form and must be considered in designing the proper cutter.

In practice, standard cutters are sometimes used to cut helical gears. The range of the cutter to be used can be determined by finding the equivalent number of spur teeth. For small helix angles, a standard cutter for the proper range of teeth will produce a close approximation to the desired gear tooth form. However, gear cutters for gears with large helix angles are usually designed specifically to produce the desired form.

16.1.3 Milling Machine Operation

Milling gear teeth is primarily a form-milling operation to which the general techniques of milling apply. Except for the addition of an indexing mechanism, and the determination of lead gears and setup for helical teeth, machine operation is comparable with conventional milling practice.

With milling considered as a standard-type machining operation for which ample information is available, the following discussion is confined to some of the more important aspects of milling gear teeth.

In selecting the proper cutter to be used, attention should be given to the number of teeth in the gear and the tooth form specified. For spur gears, standard-range cutters, as discussed in Chap. 22, "Gear Cutting Tools," can be used to produce gears having the standard gear cutter form. It is only necessary to select the proper range of cutter for the number of teeth specified in the gear. Where the tooth form produced by a standard-range cutter is not applicable, a special cutter designed to produce the specified form is used.

Theoretically, a special cutter is required to mill a helical gear of a specified tooth form and helix angle. Determination of the equivalent number of spur teeth provides an approximation that is often used to find the standard-range cutter required. The equivalent number of spur teeth is found by dividing the actual number of gear teeth by the cosine of the helix angle cubed.

Cutters for straight bevel gears are made as standard-range cutters. Cutter selection depends on the number of teeth in both the gear and the pinion. The number of teeth for which the cutter should be selected can be found by

$$\tan C = \frac{N_G}{N_p} \qquad (16.1)$$

Number of teeth to select cutter for gear:

$$\frac{N_G}{\cos C} \qquad (16.2)$$

Number of teeth to select cutter for pinion:

$$\frac{N_P}{\sin C} \quad (16.3)$$

where N_G is the number of teeth in gear, and N_P is the number of teeth in pinion.

For miter gears, only one cutter is required. Cutting a bevel gear by milling is shown in Fig. 16.3.

In using a gear cutter, the correct gear tooth form is produced only when the cutter is set to its proper depth of cut. A reduced depth of cut increases the thickness of the gear tooth, but the form is in error. A reduction in gear tooth thickness to provide backlash is a design feature that should be incorporated into the cutter as an increase in cutter tooth thickness.

A more involved and time-consuming method of thinning the gear teeth requires that an additional cut be taken on each profile of each gear tooth space. After all the gear tooth spaces have been cut, the gear is indexed through a slight angle based on one-half the amount of backlash. A cut is then taken on one side of each tooth space. An identical operation is performed on the opposite side of each tooth space.

With a single cutter, the cutter tooth form is centered correctly with the gear axis so that a symmetrical tooth space is produced. More than one tooth space can be cut by using a gang of single cutters or a multiple-tooth cutter. Normally, a roughing and a finishing cutter are ganged, with the finishing cutter being centered with respect to the gear. Multiple-tooth cutters are usually used to mill rack teeth. Gang-type cutters and multiple-tooth cutters are designed specifically for the job. Figure 16.4 shows two cutters ganged together.

The feed for finishing gear teeth is normally in the range of 0.002 to 0.004 in. per cutter tooth. Higher feeds can be used for roughing cuts. To improve the finish, reduced feeds may be required. Other factors that affect the feed include cutter design, tooling rigidity, machine capacity, and work material.

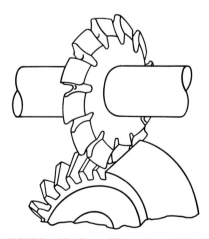
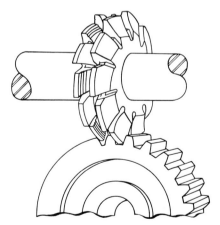

FIGURE 16.3 Gear-milling cutter and bevel gear.

FIGURE 16.4 Roughing and finishing cutters ganged.

Cutter speed is primarily dependent on the machinability of the gear blank material. Steel at 100 BHN can be milled at approximately 150 sfm with high-speed steel cutters. If the hardness is increased to 300 BHN, the speed should be reduced to approximately 50 sfm. Normally, reduced speeds are required for lower machinability rating of the work material.

16.1.4 Advantages and Limitations of Gear Milling

Gear milling can be used to advantage in producing replacement gears or a small quantity of gears on a standard milling machine, with tools that are usually less expensive than other types of gear cutting tools. Milling can be applied as a roughing or finishing operation on gears, racks, splines, and many special tooth forms. In a number of cases, milling can produce tooth forms that would not be applicable to generating methods.

Gear milling cutters are not universal for a complete range of teeth, as are standard hobs and shaper cutters. Helical gears usually require cutters designed for the specific part. Although milling can produce accurate gears, tooth spacing accuracy is limited to the accuracy of the indexing mechanism.

16.2 GEAR HOBBING

Almost any external tooth form that is uniformly spaced about a center can be hobbed. It is not necessary that the form be symmetrical about an individual axis, but each tooth group or group of teeth must be a duplicate of any other tooth or group. The kinds of work that can be hobbed include spur gears, helical gears, splines, serrations, worm gears, and special forms.

Many gears are finish-hobbed in one cut, others are hobbed with a semifinishing hob prior to shaving or grinding, and some are rough-hobbed before the finish-hobbing operation. After hardening, they may be ground, or finished by hard gear finishing. Hobbing is applicable to mass production or to jobbing operations. Automatic machines are designed for mass production of a specific part. However, the ease of setup makes hobbing machines especially adaptable to small-lot production.

16.2.1 Application of Gear Hobbing

Although the hobbing process is often associated with the manufacture of gears and splines, many other forms can be cut. Hobbing is also recognized as an economical means of producing ratchets, sprockets, and other special forms. Standard hobs have been established for some of these forms.

The versatility of hobbing makes it an economical method of cutting gears. One hob of a given pitch will cut the teeth of all involute spur and helical gears of the same normal pitch and pressure angle, including all numbers of teeth and helix angles. The size of the teeth and the size of the work are limited only by the capacity of the hobbing machine on which the part is to be cut. Hobbing machines are available for cutting any diameter up to 30 ft. However, the face width is limited to the length of hob travel that is built into the machine. Pitches from ½ DP to 350 DP have been hobbed. A helical pinion being hobbed is shown in Fig. 16.5.

GEAR CUTTING   16.7

FIGURE 16.5 Hobbing a helical pinion. *(Courtesy of Barber-Colman Co., Rockford, Illinois)*

The accuracy of parts produced by hobbing is determined by the accuracy of the hobbing machine, the rigidity of the tooling, the accuracy of the blanks, the care in mounting the hob and work, and the accuracy of the hob. With an accurate machine and proper attention to tooling, hobbing can produce AGMA quality 12 gears in the coarse pitches and AGMA quality 12 gears in the fine pitches.

As an example, 8-diametral-pitch gears of 3-in. pitch diameter and with a face width of 1.375 in. are cut with AGMA quality 12 tolerances. Runout of the pitch diameter is held within 0.0007 in., the pitch tolerance is within 0.0002 in., and the profile tolerance is held to 0.00027 in. The lead variation on this gear with 20° helix angle is held within 0.0003 in.

As an example of fine-pitch accuracy, 64-diametral-pitch spur gears with a pitch diameter of 0.750 in. and a face width of 0.093 in. are hobbed within AGMA quality 12 precision tolerances. Fine-pitch gears are normally inspected on a variable-center-distance fixture. The tolerances for this quality gear are 0.00045 in. for total composite tolerance.

The finish on the gear teeth is dependent on the accuracy of the hob, the rigidity of the tooling, and the amount of feed. With adequate tooling and a small amount of feed, a very fine finish can be obtained. Although it is difficult to inspect the finish on gear tooth profiles, it is possible to obtain a finish of 15 μin. On roughing operations, a coarse feed would be used, and the finish might be as rough as 150 μin.

Hobbing is applicable to all types of gear materials. High-speed TiN-coated steel hobs are used to cut ferrous, nonferrous, and nonmetallic materials. Carbide-tipped hobs are now used to cut steel used for automobile gears, provided the hobbing machine used is capable of the high speeds required. Steel gears can be hobbed in the hardness range of 63 to 65 RC with carbide-tipped skiving hobs. In this case the part is hobbed in the annealed state, leaving finish stock on the flanks of the teeth. After heat treatment, the part is finish-hobbed. Skiving hobs do not hob the root diameter of the part and must be used only on rigid universal hobbing machines. It may be necessary to reduce backlash between the worm and index gear to obtain the required rigidity of the machine.

Hobbing machines can also be adapted to perform special operations. Gear teeth can be chamfered with a special hob used on a standard hobbing machine. A crowned tooth can be hobbed on a machine that continually changes center distance between the hob and the work, according to the desired configuration of the crown. Taper root splines are cut on a machine that feeds the hob at an angle to the axis of the spline. Machines that have a work spindle set at an angle to the plane of the hob slide travel will cut tapered serrations. Special worm gear machines are designed to feed the hob at 90° to the work axis. Worm gears are also hobbed on machines equipped with a power feed in a radial direction. Worms or threads can be either hobbed or milled on a machine that allows the proper angular setting of the tool. Many other types of special parts can be cut on hobbing machines that are designed specifically for that part.

16.2.2 Theory of the Hobbing Process

The generation of a gear tooth is a continuous-indexing process in which both the cutting tool and the work piece rotate in a constant relationship while the hob is being fed into the work (see Fig. 16.6). As the hob is fed across the work once, all the teeth in the work are completely formed.

In most instances, the unusual feature of the hobbing process is that the form on the hob is not reproduced on the work. The largest application is the cutting of involute gears with a hob having teeth with essentially straight sides at a given pressure angle. Hobs can also be designed with other than straight sides to produce the desired form on the work.

The hob is basically a worm that has been fluted and has form-relieved teeth. The flutes provide the cutting edges. Each tooth is relieved radially to form clearance behind the cutting edge, allowing the faces of the teeth to be sharpened while retaining the original tooth profile.

Figures 16.7 and 16.8 are schematic diagrams showing the relationship between the hob and the gear. The teeth of the hob cut into the gear blank in successive order, and each in a slightly different position. Each hob tooth cuts its own profile, which in the case of an involute hob is straight-sided, but the accumulation of these straight cuts produces a curved form on the gear teeth. Therefore, instead of being formed in one profile cut as in milling, the gear teeth are formed a little at a

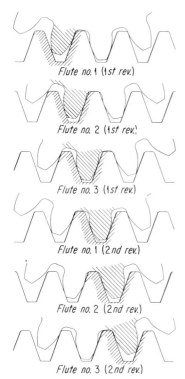

FIGURE 16.6 How different hob flutes form the gear tooth.

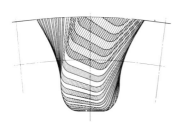 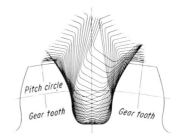

FIGURE 16.7 Chip loads cut by successive hob teeth.

FIGURE 16.8 Complete generating action of the hob.

time in a series of cuts. This is known as the *generating process* of cutting gears.

16.2.3 Hobbing Machine Relationships

The generating process requires accurate relationships between various elements of the machine in order to produce the desired results. The hob rotates in a definite relationship with the rotation of the work. For spur gears with a single-thread hob, the blank moves one tooth space while the hob rotates once. The rotation is timed by means of change gears. For helical gears, the rotation of the work is slightly retarded or advanced in relationship to the rotation of the hob, and the feed is also held in a definite relationship with the work and the hob.

Figure 16.9 is a schematic diagram of a hobbing machine, including a differential and lead gears. The differential and lead gears are shown in dotted lines to indicate the mechanism added to a nondifferential machine for cutting helical gears by the differential method. The power originating at the motor passes through speed-change gears and branches to the hob and work spindles. The differential and index-change gears are located in the branch drive to the work spindle. A branch to the feed screw is taken off between the index-change gears and the work spindle. The feed-change gears are provided for in this branch line, and drive both the feed screw and the differential. The lead-change gears are located in the drive to the differential.

Most hobbing machines in use today are equipped with a differential that holds a constant relationship between the hob, the work, and the feed even during independent movement of the hob slide. This is accomplished by holding the feed screw, the index worm, and the hob spindle in constant geared relationship at all times, even when the feed is disengaged. A differential is used for hobbing helical work when two cuts are required or when a tooth must be held in relation to a feature on a part. It can also save considerable amount of time in setting up the machine on short runs of helical parts.

The following relationships are typical for some makes of hobbing machines. To maintain the required relationship between the index and the feed, two constants are established by the gearing in the machine and are used to determine the correct change gears. The machine index constant is the number of revolutions of the hob spindle during one revolution of the work spindle when the index-change gears have a 1:1 ratio. The machine feed constant is the distance in inches that the hob slide will advance

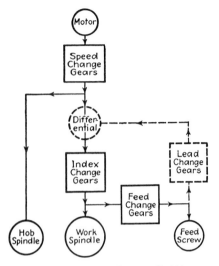

FIGURE 16.9 Diagram of a type of hobbing machine drive system.

during one revolution of the cutter when the ratio of the feed-change gears is 1:1. These constants are shown on the machine.

For spur gears, the formula for determining the index-change gear ratio is

$$\text{Index gear ratio} = \frac{KT}{N} \qquad (16.4)$$

where K = machine index constant
T = number of threads on hob
N = number of teeth on gear

The feed gear ratio for cutting spur gears is determined by dividing the feed by the machine feed constant. The formula is

$$\text{Feed gear ratio} = \frac{F}{M} \qquad (16.5)$$

where F is the feed, in. per revolution of work, and M is the machine feed constant.

For helical gears, the basic formula for figuring index-change gear is

$$\text{Index gear ratio} = \frac{KCT}{(CN) \pm 1} \qquad (16.6)$$

where K = machine index constant
C = constant
T = number of threads on hob
N = number of teeth on gear

The selection of the plus or minus sign is determined by the hand of the hob, the hand of the gear, and the feed direction. C is a different number for each different set

GEAR CUTTING   16.11

of conditions, depending on the diametral pitch, the desired feed, and the helix angle. It is actually a constant whose primary purpose is to establish a relationship between the index- and feed-change gears so that the correct helix angle will be cut on a specific gear. It can be defined as the number of revolutions of the work as the hob is fed one axial pitch of the work. Mathematically, this can be shown by the following equation:

$$C = \frac{p_n}{F \sin \psi} \qquad (16.7)$$

where p_n = normal circular pitch
F = desired feed, in. per revolution of work
ψ = helix angle of work

It is not necessary to use the theoretical value of C as determined in the foregoing, since changing its value will affect only the amount of feed. Therefore, C can be assumed to be any value that is fairly close to the theoretical value. For convenience, C is usually assumed to be a whole number that, when substituted in the basic formula, will produce a ratio that is factorable exactly into suitable numbers of gear teeth. Of course, the farther the actual C is from its theoretical value, the farther the actual feed will be from the desired feed.

Actually, neither the index nor the feed gears have to be held to an exact ratio as long as the resulting combination of these gears will produce the desired lead. Normally, however, it is easier to obtain a satisfactory setup if the index gears represent the exact ratio. The feed can then vary slightly (usually not more than 0.000015 in.) from its theoretical value and still produce a lead that is satisfactory.

In order to establish the proper relationship between the index and the feed, the exact value of C that is used in the index formula must be used in figuring the feed.

$$\text{Feed} = \frac{p_n}{C \sin \psi} \quad \text{in. per revolution} \qquad (16.8)$$

where p_n = normal circular pitch
C = exact C used in figuring index gears
ψ = helix angle of work

With the feed determined, the feed gear ratio is found as given for spur gears.

To obtain the tooth profile, cutting clearance, and tooth thickness for which the hob is designed, the hob must be accurately positioned. For spur gears, the hob swivel setting is equal to the lead angle of the hob. The actual procedure for determining the setting angle for helical gears is to add the lead angle of the hob to the helix angle of the gear if they are of opposite hands, and to subtract them if they are of the same hand.

16.2.4 Hobbing Machine Operation

As with most machining operations, a number of factors influence the optimum speed-feed combination. In hobbing, these factors include material being hobbed, finish, accuracy, machine capacity, hob specifications, and hob life. In any specific hobbing department, finish and accuracy are normally established items for the particular hobbing operation and the part to be cut. The problem is to determine a hob speed and feed for a given hob and machine capacity that will produce the established finish and accuracy at a given production rate with justifiable tool life.

Hob Speeds and Feeds. Hob speed [surface feet per minute (sfm)] is primarily a function of the machinability of the material being hobbed, the hob specifications, the machine capacity, and the hob life desired. The machinability rating of a material varies with its structure and its hardness. It is commonly compared with the machinability of AISI B-1112 cold-drawn steel, which is rated at 100 percent.

The following table provides a starting point in determining a satisfactory hob speed.

Diametral Pitch	Hob Speed (100% Machinability), Surface ft per min
1–16	180
16–32	200
32 and finer	220

With other things being equal, an increase in machinability rating would warrant an increase in hob speed. However, the hob material and hob tooth form should also be considered. Where carbide-tipped hobs have been applied to nonferrous metals and nonmetallic materials, hob speeds have been increased beyond the operating speed range for *high-speed steel* (HSS) hobs. In many cases, hob speed has been limited by the maximum machine speed available. In addition, the type of high-speed steel in the hob influences the speed that can be used and the wear that is obtained. Pointed hob teeth or forms with only a small corner radius may require reduced hob speeds to eliminate excessive wear. The top corner of a tooth on a gear hob does a great deal of work in forming the root fillet of a gear tooth. With a small portion of the cutting edge doing a big cutting job, too high a hob speed may cause premature cutting edge failure.

The amount of feed applied to a specific job varies with the finish and accuracy required and with the machine capacity and tooling employed. Fine-pitch precision gears are usually hobbed within the range of 0.005 to 0.020 in. per revolution. Gears coarser than 20 pitch are usually finish-hobbed in the range of 0.010 to 0.030 in. and are roughed or semifinished in the range of 0.040 to 0.100 in. per revolution.

For single-thread hobs and even thread-tooth ratios, the size of the feed marks or scallops increases with an increase in feed rate. With a prime thread-tooth ratio and some hob runout, the feed marks are broken up, and it is often difficult to find a feed mark pattern. In axial feeding, the feed marks follow a spiral path around the circumference of the part, and they can usually best be seen on the root surface. A high feed rate produces long, deep feed marks, resulting in a rough finish having peaks and valleys, which affects the accuracy of the part. The feed rate applied must be within the capacity of the machine and tooling used to clamp and drive the part. Lack of rigidity enhances deflections, which often cause chatter marks and gouges on the hobbed surfaces. For a preshave hobbing operation, these surface irregularities are often too deep to be removed by shaving when the amount of shaving stock allowed is in the normal range.

Although improved designs and techniques that may promote equivalent or even better results are investigated and often successfully applied, the initial hob speed and initial feed for a new job are usually based on past performance records for similar parts.

For essentially the same kind of part being produced in several different hobbing departments, past experience varies, for reasons that include the acceptance level of finish and accuracy, the conditions and capacity of the hobbing machines available, and the tooling designs employed. As a result, it is difficult to recommend a hob speed and feed for a specific job.

With all these improvements and developments in the basic hobbing machine, there was still a need, without sacrificing accuracy, for additional rigidity, higher hob speed, and a reduction in setup time. Rigidity and speed were needed to take advantage of carbide hobs as well as HSS hobs coated with titanium nitride (TiN), which came into extensive use in 1981 when they were tested and adopted for use by one of the largest auto transmission plants in the Detroit area.

Hobbing machines were designed and built to meet these requirements. In May of 1975 an NC machine built by Pfauter* was exhibited at The Paris Machine Tool Show. The first machine was equipped with rotary enciphers to control the transmission ratio between the hob spindle and the hob table. The conventional mechanical gear train was replaced by electrically controllable drive units. A rotary encoder coupled to the hob spindle generated 35,000 impulses per revolution of hob spindle, while on the index gear shaft a second rotary encoder generated 35,000 impulses. By using these impulses an electronic differential was provided in place of the standard gear differential. This closed loop system eliminated differential gears from the machine. In addition the machine was equipped with individual drives and associated incremental measuring systems for radial motion, axial motion, and tangential motion.

This machine could hob helical gears as well as produce crowns and tapers. However, there was a restriction on the maximum r/min for the work spindle when a conventional worm and worm gear were used to drive the work spindle. This was because carbide or TiN hobs used to hob pinions or parts with a low number of teeth (for automobile transmissions) performed better when higher work spindle speeds were used. Index table speeds around 500 r/min are needed. A unique answer to this problem was solved by Gleason,† who built a hobbing machine using a hypoid spindle drive with a set of spiral gears and a brake motor to eliminate backlash instead of a conventional dual-lead worm gear drive for the work spindle. Not all hobbing machines are capable of these speeds, only the most modern; perhaps those built in the last few years and certainly all new NC machines are built for these speeds. Another method, used to provide high table speed, is the use of a triple-lead worm and worm gear.

A six-axis hobbing machine offers many advantages to the user (see Fig. 16.10). The ability to calibrate axes by computer offers a means of eliminating slight errors in lead screws and index gears so that, in effect, the machine will produce parts with greater accuracy than it was built for. Then hobs can be changed automatically, permitting dull hobs to be changed or even a different hob to be mounted, along with programs suitable for the new part; and, if necessary, the work fixture can be changed. Automatic part changing, automatic probing to control size, computerized control for management information, maintenance diagnostics, optimization of machine cycles, as well as other possibilities, give the CNC user many advantages. In fact CNC hobbers are offered with as many control axes as needed to operate completely automatic hobbing centers.

The kinematics of hobbing machines was finally solved by the introduction of CNC

---

*Herman Pfauter GmbH & Co., Ludwigsburg, Federal Republic of Germany.
†The Gleason Works, Rochester, New York.

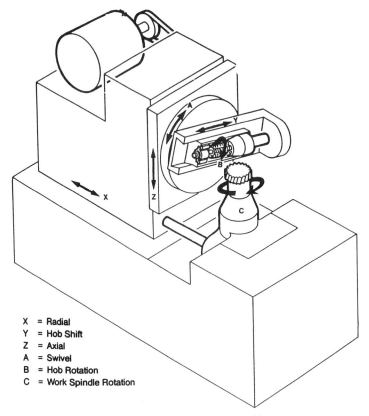

X = Radial
Y = Hob Shift
Z = Axial
A = Swivel
B = Hob Rotation
C = Work Spindle Rotation

FIGURE 16.10 Schematic of a 6-axis CNC hobbing machine.

controls. With development of electronics, CNC provided an accurate substitute for the mechanical differential. The computer numerical controller provides a control board in which data necessary to cut a gear can be entered. Once the part data are entered in the keyboard they can be assigned a reference number. The reference number remains in the computer memory and can be recalled as needed.

Carbide development has been carried on in Japan and Germany, as well as in other parts of the world. Professor Masato Ainoura of Kurume National Technical College has contributed many papers on this subject. He developed and exhaustively tested a new hobbing machine (Fig. 16.11) before CNC was available.

These tests were done in 1977 using high-yield and high-strength carbide with good physical characteristics to prevent adhesions or thermal cracks. The closest equivalent was said to be ISO-P20.

Careful attention was given to manufacturing the hob, including: use of end keys, increased number of gashes to reduce the cutting load, use of a multiple-thread hob, and careful attention to brazing and to the design of the hob tip, to provide as large a radius as the gear design permits (see Figs. 16.12 and 16.13).

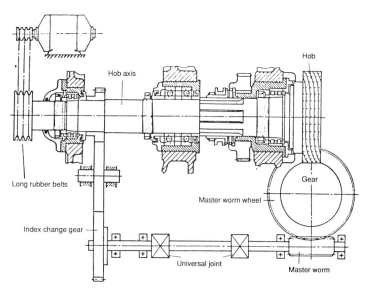

FIGURE 16.11 Test machine arrangement used by Prof. M. Ainoura in the development of carbide hobs. *(Courtesy of Kurume National Technical College, Fukuoka, Japan)*

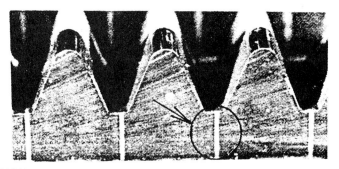

FIGURE 16.12 A good design of carbide-tipped hob, $m = 2.5$ and good brazing. *(Reprinted with permission of the publisher, ASME Journal of Mechanical Design, Vol. 100, July 1978.)*

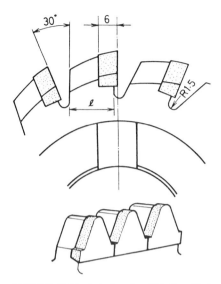

FIGURE 16.13 A design of carbide-tipped hob, $m = 2.5$. (Reprinted with permission of the publisher, ASME Journal of Mechanical Design, Vol. 100, July 1978.)

In addition all hob tolerances were held to very close limits, and special attention was paid to grinding the hob teeth, as well as to sharpening, to prevent hairline cracks and microscopic chips (see Fig. 16.14).

Single-thread, double-thread and triple-thread hobs made from 10 slightly different grades of carbide were tested. Several different size gears, from automotive pinions to differential ring gears, were tested with hobs of similar size to those mentioned in the foregoing. Test results for the differential ring gear are shown in Fig. 16.15.

The machine (Fig. 16.16) that Professor Ainoura developed was provided with a direct hydraulic drive to the hob spindle while the main drive of most hobbing machines was through the worktable. The drive to the hob spindle was through a shock-vibration absorber. The theory behind the use of a shock absorber was to absorb the shock and vibration for the fluctuation of the cutting forces when using a carbide hob.

There were several machines sold to one of the major transmission builders in the Detroit area who used Professor Ainoura's design and carbide hobs, but these machines were driven by a 1200-r/min motor through a shock absorber rather than a hydraulic motor. Carbide hobs were used successfully until they were replaced by HSS TiN-coated hobs that proved cheaper to use.

Estimating Hobbing Production. An estimate in advance of production is of prime interest in any machining operation. As applied to hobbing, the production estimate is required to provide production figures from which the cost of the part may be determined, and to ascertain the number of hobbing machines required for the desired amount of production.

GEAR CUTTING 16.17

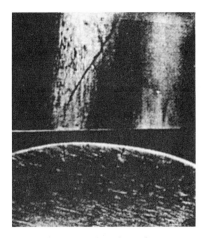

Nonused cutting edge

Cutting edge after use

FIGURE 16.14 Cracks in sharpening and chippings occurred in sharpening the carbide hob for the finish. (Hob: $m = 8$, $N = 10$, $d = 150$ mm; gear: $Z = 29$, $d = 248$ mm, $b = 70$ mm, S45c ($180H_B$), $\beta = 0°$; cutting condition: $V = 250$ m/min, $f = 2$ mm/rev.) *(Reprinted with permission of the publisher, ASME Journal of Mechanical Design, Vol. 100, July 1978.)*

16.18 CHAPTER SIXTEEN

Machine Type: KS-300HD Serial No.: D-2906	HOB WEAR	Kashifuji Works, Ltd. Date: March 10,
Workpiece data	HOB data	Cutting condition
Name: Ring Gear Pitch: DP 11.9473 No. of teeth: 86 P.A.: 17° H.A.: 23.30152431°RH Outer dia.: 206.98 Face width: 30.5 Material: SAE4320H Hardness: HBN 150~170	Hob No. 82X1-630 Type of Hob: Carbide tipped Material: AS–100 Dia.: 105 No. of starts: 3 LH No. of flutes: 16 Gash lead: ∞ Cutting depth: 5.64 Tooth length: 134(100) Maker: AZUMI	Pieces per chuck: 2 pcs. No. of cuts: 1 Hobbing method: Conv. Hob speed: 550 rpm Cutting speed: 182 m/min. Radial feed: — in/rev. Axial feed: 4.4 mm/rev. Amount of shift: 13.3 mm/40 cyc. No. of pieces hobbed: 410 pcs.

Test 4
n: 550 rpm
v: 182 m/min.
f: 4.4 mm/Trev.

0.3 mm

Cutting tooth number (32)

Cutting tooth number (44)

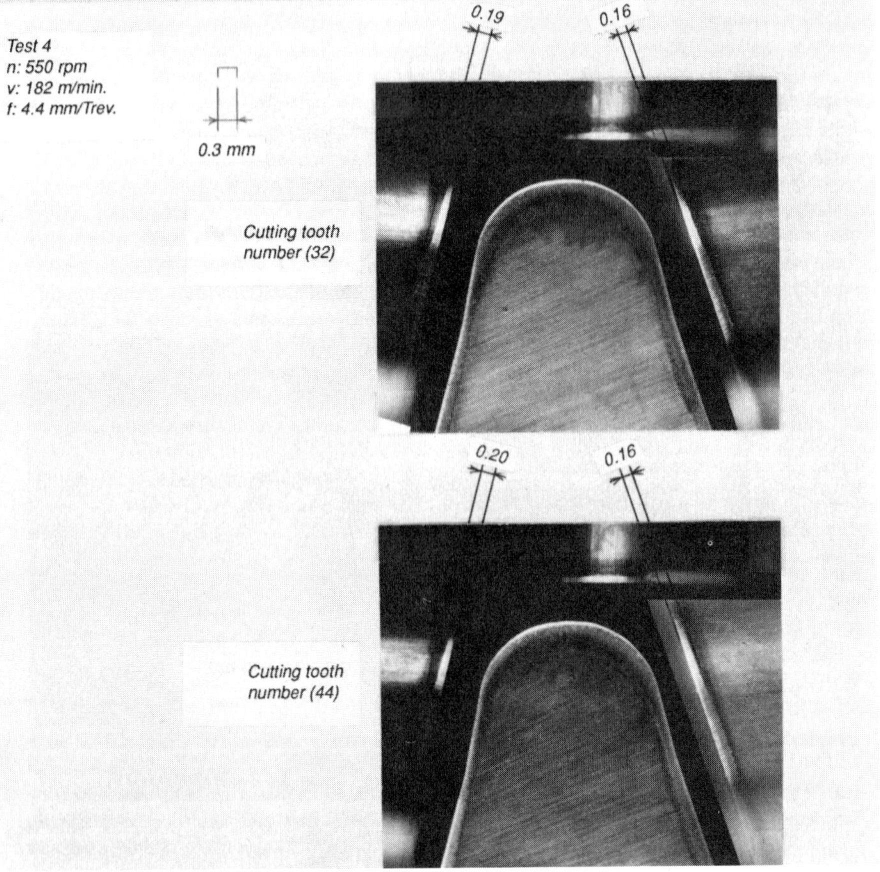

FIGURE 16.15 Test of carbide hob. *(Courtesy of Kashifuji Works, Ltd., Kyoto, Japan)*

GEAR CUTTING 16.19

FIGURE 16.16 Carbide hobbing machine. (Table diameter = 360 mm; hydraulic motor output = 40 kg.m.) *(Reprinted with permission of the publisher,* ASME Journal of Mechanical Design, *Vol. 100, July 1978.)*

The actual computation connected with finding the time required to hob a given part is relatively simple. However, many variables must be taken into consideration. If the existence of these variables is appreciated and they are properly applied, the estimate should be fairly reliable. To determine the numerical values for these variables requires broad experience tempered with careful consideration for the particular job.

Basically, the expression for determining the hobbing time is the linear distance traversed by the hob carriage on the bed ways divided by the velocity of the hob carriage. The mathematical expression is

$$\text{Hobbing time} = \frac{NL}{T \times \text{r/min} \times F} \text{ min} \tag{16.9}$$

where F = feed, in. per revolution of work
L = total travel of hob carriage, in.
N = number of teeth on gear
T = number of threads on hob
r/min = hob speed

For a reliable solution to this expression, each item should be investigated individually. Feeds and speeds should be determined by past experience on similar jobs or by a test under actual job conditions. They are discussed in the preceding article. Figure 16.16 shows an example of hob wear after cutting 410 pieces.

The one element of this equation that must be calculated is the total length of hob-carriage travel. The total length of travel to complete the generation of the part

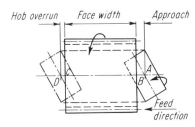

FIGURE 16.17 Total length of hob travel is from A to D.

is divided into three elements: (1) approach, (2) face width, and (3) overrun. (See Fig. 16.-17.) The face width is indicated on the part print, but approach and overrun must be calculated. When more than one blank per load are to be hobbed, the total face width of all the blanks must be considered.

The hob approach is the distance from the point of initial contact between the hob and blank to the point where the hob first cuts at full depth. This distance varies with the hob diameter, part diameter, depth of cut, and hob swivel-setting angle. These points have a geometric relationship that has been plotted in charts A and B (see Figs. 16.18 and 16.19).

Where the swivel-setting angle is small, as is often the case, chart A is sufficient. Find the approach distance by following the curve corresponding to the hob diameter until it intersects the tooth depth, and read the approach by following the horizontal line that contains that intersection point.

Setting the hob at an angle has the same effect as increasing the hob diameter since the hob has the form of an ellipse in the axial plane of the part. For the larger swivel angle settings, which usually exist when hobbing helical parts, the actual hob diameter must be increased by a certain number prior to applying chart A. Use chart B to find the number required for a specific job. Determine the point at which the sum of the hob and part diameters intersects the line representing the swivel angle setting, and read the number at the left of the chart. Add this number to the hob diameter, and use chart A. If the sum of the hob and part diameters is greater than the largest sum shown in chart B, take half the sum of the hob and part diameters, find the appropriate number on the chart, and double the answer.

Hob overrun is the linear hob-carriage travel beyond full-depth cutting that is required to complete generation of the teeth. For helical gear hobbing where the hob swivel angle settings are usually large, the overrun should be calculated and added to the approach and face width to obtain the total hob-carriage travel. A practical formula for involute forms that has been checked by test and gives safe, full-generating results is

$$\text{Overrun} = \frac{b \cos \psi \tan S}{\tan \phi_n} \quad \text{in.} \quad (16.10)$$

where b = dedendum of gear
ψ = helix angle of gear
S = swivel-setting angle
ϕ_n = normal pressure angle

The sum of the approach, face width, and overrun provides the total travel of the hob carriage, which can be substituted in the equation. The answer found from the equation gives the time required for cutting the part. Added to this cutting time should be a handling time that is influenced by the work-holding device and the factors affecting material handling. This sum gives the total time per cycle.

GEAR CUTTING

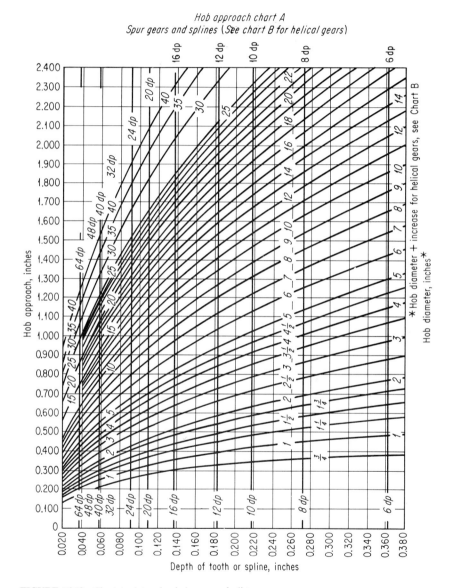

FIGURE 16.18 Chart to determine hob approach distance.

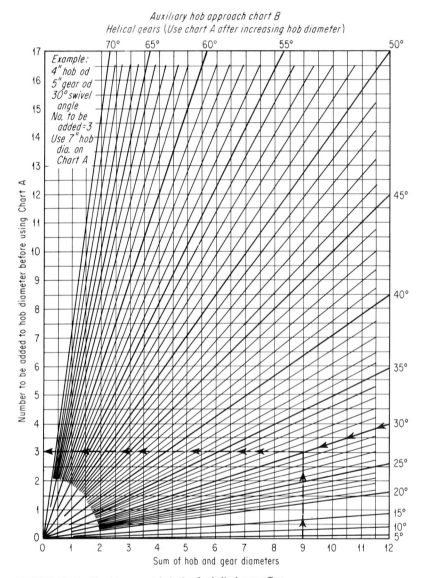

FIGURE 16.19 Chart to correct hob size for helical gear effect.

Locating the Hob. A conventional-type multiposition generating hob can be used across its entire active face width. In each location relative to the part axis, however, there must be sufficient hob teeth in the generating zone to produce completely the required gear tooth form. This restricts the location of the hob when the teeth at either end of the hob are being used. Also, roughing teeth are required when the part rotates into the end of the hob.

The blank must pass across the roughing zone prior to reaching the generating zone. Consequently, when the blank for an involute gear rotates into the end of the hob, the roughing zone should contain adequate hob teeth so that the cutting load on the first roughing tooth is not excessive. The number of roughing teeth required varies with the diameter of the blank, the hob swivel-setting angle, and the feed. A minimum setting dimension can be determined for spur and helical involute gears by the following approximate expression. In this expression, we have assumed a maximum generating length so that safe settings will result regardless of the helix angle or diameter of the blank. (See Figs. 16.20 and 16.21.)

$$A = \frac{a}{\tan \phi_n} + \frac{p_n}{2} \quad \text{in.} \quad (16.11)$$

where A = minimum setting dimension measured along hob axis
a = addendum of gear
p_n = normal circular pitch
ϕ_n = normal pressure angle

When the opposite end of the hob is used, ample roughing teeth are usually available. To ensure that generation is complete, however, the minimum end-setting dimension should be at least

$$B = \frac{a}{\tan \phi_n} \quad \text{in.} \quad (16.12)$$

where B is the minimum setting dimension measured along hob axis.

In hobbing other than involute forms, similar restrictions on the end settings of the hob are applicable. Because the geometry of the involute curve is most familiar to all users of hobbing machines, only application to the involute form is discussed. With the

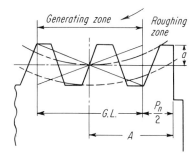

FIGURE 16.20 Distance A from hob end.

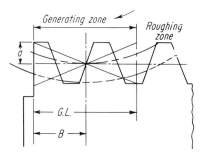

FIGURE 16.21 Distance B from hob end.

end settings restricted as indicated, the total axial hob shift is limited to its active face width minus the sum of $(A + B)$. In shifting the hob across the part, the total amount of available shift should be known.

Shifting the Hob. The purpose of hob shifting is to increase hob life by uniformly distributing the wear so that all the hob teeth have approximately the same amount of wear. The importance of proper hob shifting becomes apparent when the hob is removed from the hobbing machine for sharpening. If some of the teeth are dulled more than others, useful material will have to be ground away from the teeth that are not dulled before the badly dulled teeth are sharpened. This uneven wear can be avoided by shifting the hob regularly in small amounts.

Where the hob is shifted manually, it remains at a particular setting until the resulting wear reaches the allowed maximum. The hob is then shifted to a new setting where sharp teeth are available, and wear is allowed to reach the predetermined maximum amount before the hob is shifted to the next setting. At each setting, the maximum amount of wear can normally be found on a group of teeth located in the roughing zone adjacent to the generating zone. In shifting to a new setting so that these teeth are no longer doing any cutting, a number of teeth having less than the allowable wear are shifted past the zone where maximum wear occurs. Unless a predetermined amount of shift is made frequently, a nonuniform wear pattern is the result.

The amount of shift should be based on the maximum allowable hob wear. Primarily, the amount allowed varies depending on the pitch of the hob and whether the operation is finishing or semifinishing. After a maximum wear limit for a particular operation has been set, the amount of shift can be determined by testing various amounts of shifts. The initial shift increment can usually be found from past performance records. In arriving at the initial shift, new jobs should be compared with old jobs with consideration of such variables as machinability, numbers of teeth, length of cut, and type of tooth form. Because of the many variables that exist, no attempt has been made to recommend amounts of shift. Instead, a table of recommended amounts of maximum wear has been included.

Frequent hob sharpening pays dividends in terms of uniform accuracy, consistent finish, and longer hob life. Therefore, it is advantageous to limit the maximum amount of wear. The recommended maximum wear of Table 16.1 can be applied as a guide to establish the amount of wear that should be allowed.

The tabulated wear is the maximum amount of dullness that extends behind the tooth face. For involute gear hobs, it can usually be found on the side of the tooth near

TABLE 16.1 Recommended Maximum Hob Wear

All readings in thousandths of an inch

Diam. pitch	Operation	
	Semifinishing	Finishing
4–10	15–20	10–15
11–19	10–15	8–12
20–48	6–10	4–8
48 and finer	5–7	3–5

the top corner. The table values are representative of current practice using high-speed steel hobs.

The direction of shift is determined by whether you desire to move the hob teeth from the generating zone to the roughing zone or vice versa. Normally the former method is used. The hob teeth having sharp cutting edges are moved progressively from a light-cutting-load zone into the roughing zone where the cutting load is heaviest. Teeth that generate the part tooth form have been exposed to relatively light cutting loads. Consequently, compared with using duller teeth in the generating zone, tooth form and finish are improved. For spur gears or small hob swivel-setting angles, sharp teeth can be moved into the generating zone by shifting in a direction that is opposed to part rotation. Shifting with the rotation moves dulled teeth into the generating zone. For helical forms or large hob swivel-setting angles, a portion of the hob at one end enters the blank before the portion at the other end; consequently, it does the roughing work. Sharp teeth can be moved into the generating zone by shifting the hob toward the part.

Climb and Conventional Hobbing. For spur tooth forms, where the hob swivel-setting angle is relatively small, the type of cutting action can be either conventional or climb, depending on the rotation of the hob and the feed direction. For helical tooth forms, where the hob swivel-setting angle is usually large, the type of cutting action is determined by the rotational direction of the hob in relation to the direction of gear rotation. (See Figs. 16.22 and 16.23.)

In conventional-type cutting, when the hob swivel-setting angle is small, the hob is fed into the part in a direction that is in agreement with the tangential vector denoting the direction of hob rotation. Although the chips removed vary in size and shape, the centered hob tooth removes a chip that is comparable with those obtained in conventional milling. The chip starts out thin and becomes increasingly thicker as the hob tooth sweeps through the cut.

For parts having large helix angles, conventional-type cutting exists when a component of hob rotation is opposed to the part rotation. This holds true when the hand of the hob and the hand of the part are alike. With a large part helix angle, the hob setting angle is usually large, tending to position the axis of the hob parallel with the axis of the part. Since the direction of part rotation depends solely on the rotation and hand of the hob, the hand of the part determines whether hob rotation opposes part rotation. This kind of conventional-type cutting may be considered as tangential and independent of the axial-feed direction.

With small hob swivel-setting angles, climb-type cutting takes place when the hob

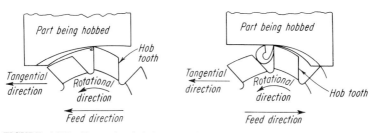

FIGURE 16.22 Conventional hobbing.

FIGURE 16.23 Climb hobbing.

is fed into the part in a direction that opposes the tangential vector denoting the direction of hob rotation. As with conventional-type cutting, the chips vary in size and shape. However, unlike conventional-type cutting, the chip starts out thick and becomes increasingly thinner as the hob tooth sweeps through the cut.

When the hand of the hob and the hand of the part are unlike for a large hob swivel-setting angle, a component of hob rotation agrees with the direction of part rotation. This combination of unlike hands produces a "roll-in" effect that is essentially climb cutting, but in a tangential direction rather than in an axial direction. Under these conditions, the hob tends to drive the part, reducing the torsional load in the index drive.

The selection of climb cutting over conventional-type cutting is usually based on a comparison of surface finish obtained from each. There is no definite rule that can be used to determine which will produce the best finish for the specific job under consideration. When conditions permit, the most satisfactory type of cutting can be determined from test runs on the specific part. In many cases, climb-type cutting has been applied to replace conventional-type cutting because a better finish was obtained, with equivalent hob life and accuracy.

16.2.5 Advantages and Disadvantages of Hobbing

Two of the primary requisites of a gear are that it must be quiet and long-wearing. Accuracy of the gear is required if it is to meet these rigid standards. With ground hobs, a machine in good condition, and reasonable care in setting up and operating the machine, hobbing will consistently produce gears to very close tolerances. Accuracy of tooth spacing, lead, and profile can be closely controlled under mass production conditions.

The constant indexing of the work makes hobbing rapid as well as accurate. Multiple-thread hobs have greatly increased production on many jobs. Carbide-tipped hobs have also increased production on phenolic, aluminum, and other nonferrous materials.

The versatility of hobbing makes it an economical method for producing gears. One hob will cut all numbers of teeth of a given pitch and pressure angle, and many different tooth forms can be cut on the same machine by changing the form on the hob.

The limitations of the hobbing process are very few. Two types of gears that are outside the range of hobbing are bevel gears and internal gears. Also, gears with an integral feature that is so close that it causes interference with the hob are not applicable to the hobbing process.

16.3 GEAR SHAPING

Like hobbing, gear shaping is a generating process. The tool used is a "pinion"-type tool while hobbing uses a "worm"-type tool. Ordinarily gear teeth are shaped on circular blanks, but the process lends itself to producing teeth on cam surfaces curved to almost any mathematical function. Shaping is used both as a jobbing operation and as a basis for automatic machines mass-producing gears. Probably the most noteworthy

GEAR CUTTING 16.27

advances in gear shaping machines and tools during the 1970s and 1980s are: (1) development of CNC machine controls, and (2) the introduction of TiN-coated shaper cutters. The cutting action processes described in the following remains the same for either CNC or conventional machines. Other important improvements are cutter spindle relief, hydrostatic lubrication of the guide and cutter spindle, and hydraulic motors and hydraulic cylinders for driving the cutter spindle stroke.

To understand the differences between a CNC and a conventional machine, see Fig. 16.24 and Fig. 16.26, showing the drive train for the conventional shaper. Then see Fig. 16.25 and Fig. 16.31, showing the drive train for a CNC shaper.

16.3.1 Application of Gear Shaping

In addition to the cutting of external and internal spur, helical, and herringbone gears, the gear shaper can also be applied to the cutting roller and silent chain sprockets, external and internal ratchets, external and internal splines, interrupted tooth gears, segment gears, elliptical gears, face gears, and racks.

There are a number of models of gear shapers, ranging in size from a machine that will cut gears as small as $\frac{1}{16}$ in. in diameter and as fine as 200 pitch to one that has the capacity of 23 ft and less than 1 diametral pitch. Machines are available with a capacity up to 60-in. face width.

As in any gear cutting process, the accuracy of gears produced by gear shapers depends on many factors, such as condition of the machine, tooling rigidity, accuracy of the cutter, the material to be cut, blank accuracy, and the skill and care of the operator.

In the coarse-pitch range, gears can be produced with AGMA quality 12 tolerances.

The material being cut has an important influence on the finish that can be obtained. For most of the commonly used alloy-steel gear materials, a finish of 40 to 60 μin is made and 20 μin can be produced as a finish cut.

Gears can be shaper-cut at comparatively high hardness. In properly heat-treated, alloy-steel forgings, such as AISI 4140 and 4340, gears have been cut at 43 Rockwell C. Steel castings of about 0.40 carbon can be cut at a maximum hardness of approximately 25 Rockwell C.

The gear shaper is suited to the addition of fixtures and attachments to increase its range of application (see Fig. 16.24). Other attachments can be added that will make them semiautomatic or fully automatic in operation.

When internal gears are cut close to a shoulder, a clearance groove or recess must be provided for the cutter to run into at the end of the cutting stroke. Also, a groove is needed when cutting herringbone gears (see Fig. 16.27).

16.3.2 Theory of Shaping Process

One gear of an interchangeable series will rotate with another gear of the same series with proper tooth action. This is the basis of the gear shaping process: The cutter is in the form of a pinion and has involute profiles and relieved flanks. The cutter is rotated in correct ratio with the gear blank. By reciprocating, the cutter metal is removed from the gear blank and teeth are developed. This is the principle on which the gear shaper

FIGURE 16.24 Conventional Gear Shaper. *(Courtesy of American Pfauter, Loves Park, Illinois, and Lorenz AG, Ettlingen, Germany)*

FIGURE 16.25 CNC Gear Shaper. *(Courtesy of American Pfauter, Loves Park, Illinois, and Lorenze AG, Ettlingen, Germany)*

GEAR CUTTING

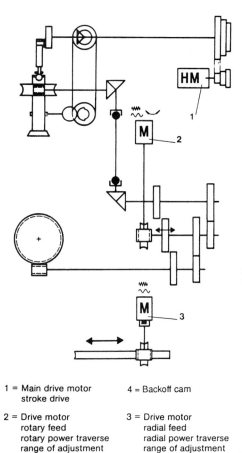

1 = Main drive motor stroke drive

2 = Drive motor rotary feed rotary power traverse range of adjustment 1:2000

4 = Backoff cam

3 = Drive motor radial feed radial power traverse range of adjustment 1:3000

FIGURE 16.26 Conventional gear shaper drive train. *(Courtesy of Maschinenfabrik Lorenz, Ettlingen, Germany)*

operates. How this shape is produced by the gear shaper is shown in Fig. 16.28 when using conventional feed technique.

When a machine with a CNC controller is used, advantage can be taken of a variable radial feed technique, sometimes called a controlled cutting process. In Fig. 16.29 notice that with conventional feed, the leading flank chip is larger, with a very small trailing flank chip; while with CCP a more uniform chip is formed from leading to trailing flank. Figure 16.30 illustrates how three gears on a five-step cluster gear can be cut in a single setup and gives an example of improved tool life and production rates, using the Lorenz* CCP method.

---

*Machinenfabrik Lorenz AG, Ettlingen, Federal Republic of Germany.

FIGURE 16.27 Herringbone gear with a groove being cut with a CNC HYDROSTROKE® gear shaper. *(Courtesy of Fellows Corporation, Springfield, Vermont)*

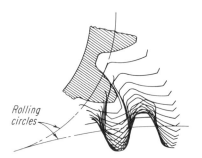

FIGURE 16.28 Generating action of shaper cutter.

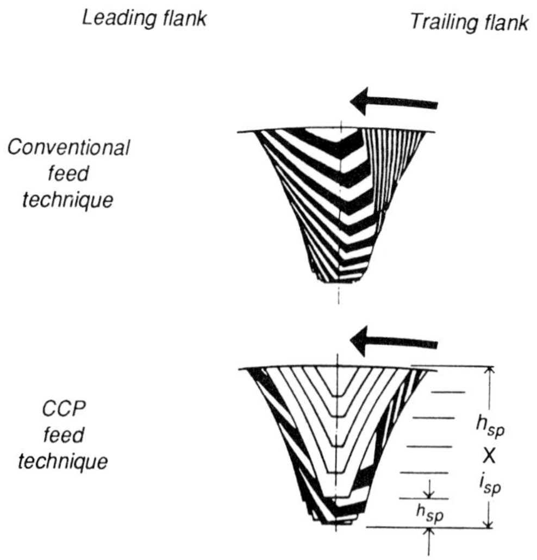

GEAR CUTTING

Leading flank   Trailing flank

Conventional feed technique

CCP feed technique

h_{sp} = radial feed per workpiece revolution
i_{sp} = number of workpiece revolutions to full depth of roughing cut

FIGURE 16.29 Comparison of feed techniques of conventional and CCP cutting with a CNC controller. *(Courtesy of Maschinenfabrik Lorenz, Ettlingen, Germany)*

16.3.3 Gear Shaper Operation

When a spur gear is being machined, the reciprocating motion of the cutter spindle is restrained to operate in a straight path by a straight guide.

If a gear having helical teeth is being machined, a helical guide is used. As the cutter spindle reciprocates, the helical guide imparts a helical motion to the cutter spindle. It is evident that the reciprocating control is independent of any other function of the machine. For external helical gears, cutters of the opposite hand are used. Internal helical gears require cutters of the same hand.

The cutter spindle is reciprocated through a crank arm mechanism. The eccentric position of the crank pin determines the length of stroke, and the crank arm is adjustable so as to position the cutter relative to the work.

We have been describing, and Fig. 16.26 relates to, a modern spindle relief disc-type cutting tool machine, with the cutter spindle and guide mounted hydrostatically and with independent drives for the stroke and the radial and traverse feeds.

CNC gear shapers provide the ability for faster setups by eliminating all the change gears and manual adjustments. A six-axis machine, and the names of the controls are given by Figs. 16.31 and 16.32.

A CNC gear shaper allows cluster to be shaped in one setup using two shaper cutters as shown in Fig. 16.33.

Although a helical guide is usually required for each helix angle, different helix angles can be shaped if the leads of the shaper cutters are the same as the lead of the guide.

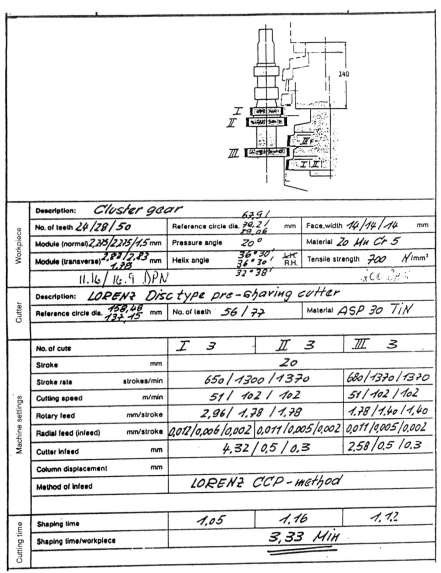

FIGURE 16.30 Shaping of three gears to preshave in one set-up with two cutters. *(Courtesy of Maschinenfabrik Lorenz, Ettlingen, Germany)*

GEAR CUTTING

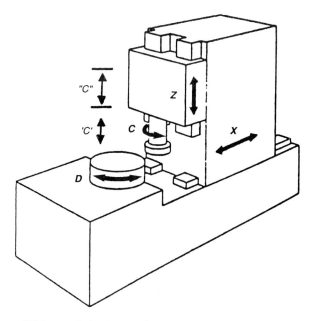

5 & 6 axes electronic control:

1. $\boxed{Z}$ Axial movement of cutterhead slide, stroke positioning
2. $\boxed{X}$ Radial movement of column slide
3. $\boxed{C}$ Rotary movement of cutter spindle
4. $\boxed{D}$ Rotary movement of work-table
5. 'C' Stroking rate per minute
6. "C" Stroke length setting

FIGURE 16.31 Axes motions on a CNC shaper. *(Courtesy of Maschinenfabrik Lorenz, Ettlingen, Germany)*

After the proper guide and cutter have been selected, determine the work change gears according to the following formula:

$$\text{Change gear ratio} = \frac{K \times N}{T} \qquad (16.13)$$

where K = machine index constant
N = number of teeth on gear
T = number of teeth on cutter

A helical guide causes the cutter to trace the same lead as the lead of the guide. For the same guide and cutter, a fixed helix angle is produced on the work regardless of the number of teeth on the work. This can be expressed in terms of lead, according to the following formula:

$$\frac{L}{T} = \frac{l}{N} \qquad (16.14)$$

where L = lead of helical guide
 T = number of teeth on cutter
 l = lead of gear
 N = number of teeth on gear

The principal factors controlling the cutting on the gear shaper are the length of the cutting stroke required, machinability, the degree of quality, and the cutter life desired. Table 16.2 indicates the cutting speeds for various materials. These speeds may be used as a guide in the selection of the initial speed. A more favorable speed can be determined by trial cuts.

General Rule: 100% machinability index = 120 sfm
  Steels of 8620 type, 145 to 227 BHN should start with 90 sfm rough and 120 sfm finish speeds and an infeed rate of 0.002 in/stroke

Increase approximately 30% for TiN-coated cutters.

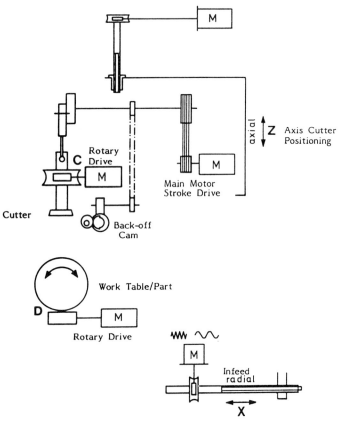

FIGURE 16.32 Kinematic drawing of drive train CNC shaper. *(Courtesy of Maschinenfabrik Lorenz, Ettlingen, Germany)*

GEAR CUTTING 16.35

FIGURE 16.33 Two gears in one set-up. *(Courtesy of American Pfauter, Loves Park, Illinois, and Lorenz AG, Ettlingen, Germany)*

When converting cutting or surface speed into strokes per minute, it is necessary to take into consideration the fact that the actual distance traveled by the cutter spindle, per revolution of the crankshaft, is twice the length of the stroke. The following formula is used for determining the number of strokes per minute:

$$s = \frac{12 \times S}{2L} = \frac{6S}{L} \qquad (16.15)$$

where s = number of strokes per minute
 L = length of cutting stroke, in.
 S = surface speed, ft/min

Based on a cutting speed of 60 ft/min and cutting steel at about 200 BHN, the following table gives the number of strokes per minute.

Gear Face Width, In.	Strokes Per Min
½	600
1	350
3	110
8	40

TABLE 16.2 Cutting Speeds for Use on Gear Shapers
(based on a maximum cutting stroke of 1 in. and the maximum degree of machinability for the analyses specified)

Material	Surface speed, fpm	Remarks
Cast iron	50–60	Ordinary gray-iron castings
Steel (mild)	75–85	0.15–0.20% carbon, no hardening alloys
Steel (high-carbon)	50–60	0.50% carbon, 1.15% manganese, 0.55% chromium
Steel (tool)	40–50	0.90–1% carbon, 0.25% manganese
Steel (chrome-nickel)	40–50	SAE N3250
Steel (stainless)	25–35	
Brass (soft)	100–110	
Bronze (naval)	45–55	
Aluminum	190–200	With coolant
Fibrous materials	75–85	

Source: From Fred H. Colvin and Frank A. Stanley, *Gear Cutting Practice*, 3d ed., McGraw-Hill Book, 1950.

TABLE 16.3 Rotary Feed of Cutter (in. per stroke)

Diametral Pitch	No. Cuts	Roughing	Semifinishing	Finishing
24–64	1		0.012	0.004
	2		0.012/0.008	0.006
	3		Not necessary	0.008
10–24	1	0.025	Not recommended	Not recommended
	2	0.030/0.020	0.030/0.015	0.025/0.012
	3	Not necessary	Not necessary	0.025/0.025/0.012
7–10	1	0.025	Not recommended	Not recommended
	2	Not recommended	0.030/0.015	0.025/0.012
	3	Not recommended	0.030/0.030/0.015	0.025/0.025/0.012
3–7	1	0.015	Not recommended	Not recommended
	2	Not recommended	0.025/0.015	0.020/0.012
	3	Not recommended	0.025/0.025/0.015	0.020/0.020/0.012

Source: Courtesy of Bourn & Koch Machine Tool Co., Rockford, Illinois.

GEAR CUTTING 16.37

TABLE 16.4 Strokes per revolution of cutter

Cutter data	No. of strokes					
	Roughing			Finishing		
	No. of gear teeth			No. of gear teeth		
	12	35	100	12	35	100
4 pitch, 16 teeth	1,000	800	500	1,200	900	600
6 pitch, 18 teeth	800	500	300	1,000	700	500
24 teeth	1,100	700	400	1,300	900	700
10 pitch, 30 teeth	800	500	300	1,000	700	500
40 teeth	1,000	700	400	1,300	900	700
20 pitch, 20 teeth	600	400	300	600	400	300
40 teeth	800	550	400	800	550	400

Because of the twisting motion imparted to a helical cutter, its actual cutting speed is greater than that of a spur cutter used at the same stroke length and number of strokes per minute. The difference depends on the helix angle of the cutter used. The actual cutting speed can be found by dividing the selected cutting speed by the cosine of the helix angle of the cutter.

The selection of the rotary feed depends largely on the finish and degree of accuracy desired. It is evident that a fine-pitch gear requires a finer feed than a coarse-pitch gear, and also that a coarser feed can be used for roughing than for finishing cuts. The most suitable combination of feed and speed to meet requirements is best determined by trial to find the combination giving the highest production with the longest cutter life.

Rotary feed is in inches per stroke of the cutter. Table 16.3 shows some representative values for the rotary feed of the cutter. These are average values that can be used for cutting good commercial gears. The rotary feed is, of course, influenced by the number of teeth on the gear being cut. Table 16.4 gives typical strokes per revolution of the cutter based on these considerations.

To compare the two tables, we consider a 35-tooth gear, 10 pitch, cut with a 30-tooth cutter. This cutter has a 3.000-in. pitch diameter and, therefore, a pitch circumference of about 9.43 in. If we take a rotary feed of 0.020 in. from Table 16.3, we get 9.43 divided by 0.020 for 471 strokes. Table 16.4 shows a nominal value of 500 strokes for this situation.

The reader is advised to consider both Tables 16.3 and 16.4 when making a preliminary estimate of how many strokes per minute to use on a new job.

The production time in minutes to cut a gear using a pinion-type cutter may be estimated by the following formula:

$$\text{Time} = \frac{N \times r \times n}{T \times s} \qquad (16.16)$$

where N = number of teeth on gear
r = strokes per revolution of cutter
n = number of cuts
T = number of teeth on cutter
s = number of strokes per minute

16.3.4 Advantages and Disadvantages of Shaping

Gear shaping is a versatile and accurate method for producing gears. It is used to cut spur and helical teeth and external and internal gears. In addition, herringbone and face gears can be produced. Shaping is the method for cutting cluster gears or gear teeth that are adjacent to a shoulder. From a production standpoint, shaping may be advantageous on gears of narrow face width because of the relatively short approach and overrun.

A limitation of gear shaping is the length of cut. Gear parts that are integral with long shafts may be difficult to mount in the machine. Another limitation is the requirement for a separate helical guide for each helix angle to be cut.

16.4 GEARS CUT BY SHEAR SPEED

Form-tool blades are used to shear-cut each space between teeth at the same time. The speed of shear cutting and the nature of the operation are quite similar to broaching. The patented machine used for this process is called the Shear-Speed* gear shaper. The Shear-Speed process is a high-production method of producing gear teeth.

16.4.1 Application of Shear-Speed Cutting.

The Shear-Speed cuts all teeth of external spur forms simultaneously at a high rate of production. Helical forms are not applicable to this method. While the majority of forms shaped by the Shear-Speed are external, there are cases where this method is most advantageous in the shaping of internal forms. There are, however, certain limitations on shaping internal forms on the Shear-Speed. Shear-Speed shaper flexibility is demonstrated by the broad range of parts that can be produced and the wide size and shape variations. The product range includes gears, splines, sprockets, ratchets, and miscellaneous external shapes. Part tooth forms may be involute, straight-sided, or contoured. The forms can be either symmetrical or unsymmetrical.

Although some clearance on both sides of the part is required for overtravel of the form-cutting tool, parts that are integral with closely spaced flanges or shoulders, such as cluster gears, can be readily cut by the Shear-Speed process.

---

*Shear-Speed is a registered trademark of the Michigan Tool Co. of Detroit, Michigan.

GEAR CUTTING 16.39

FIGURE 16.34 Shear-Speed cutter above work. *(Courtesy of Michigan Tool Co., Detroit, Michigan)*

In the Shear-Speed process, it is possible to form-cut involute spur gear teeth with practically any required tooth modification. This includes preshave root relief, semitopping, full fillet modifications, and modifications in the involute form. Normally, this method is applied to semifinishing or finishing of gear teeth.

Shear-Speed machines are designed for mass production of gears. Therefore, they are most efficiently applied to high-volume production. The Shear-Speed process can also be applied to short runs when they are repetitive at frequent intervals.

Machines are available to cut a maximum diameter of 20 in. and a maximum face width of 6 in. Gears up to 2 diametral pitch can be produced. The finest pitch recommended on these machines is 12. Gear materials that can be cut on these machines include steel, cast iron, bronze, brass, aluminum, fiber, plastic, and others.

16.4.2 Shear-Speed Principles

All the teeth in a gear are cut simultaneously in the Shear-Speed gear cutting process. The same number of gear cutting tools as there are teeth in the gear to be cut are assembled radially in the cutting tool holder mounted in the head of the machine (see Figure 16.34).

The number of tools depends on the number of spaces, slots, or forms to be cut around the periphery of the work blank. All the tools may have the same form, or the forms may vary depending on the part specifications. The tools may be equally or unequally spaced.

The form of the cutting edge of the tool is an exact duplicate of the space, slot, or form. All tools are form-relieved to maintain correct form after sharpening.

With the tool head stationary, the part is reciprocated past the cutting tools (see Fig. 16.35). Each tool is fed radially a predetermined distance with each stroke, until the

FIGURE 16.35 Shear-Speed gear shaper and stack of three gears. *(Courtesy of Michigan Tool Co., Detroit, Michigan)*

FIGURE 16.36 Progressive cutting of gear by shear process.

full tooth depth is cut. All tools are retracted on each return stroke of the gear to avoid drag and are returned to their initial position at the end of each cycle. The tool head assembly consists of three parts: a retained housing, a radially slotted member that controls tool alignment (Fig. 16.36), and a movable, double cone-shaped guiding unit (Fig 16.37) that controls the feed. All tools are fed radially into the work at the same time by use of the double cone-shaped tool head.

The cutting tools can be compared with the teeth on a broach. Since each tool must have the same form as the tooth space on the gear, each set of tools and the tool head are normally designed to produce a specific part (see Fig. 16.38).

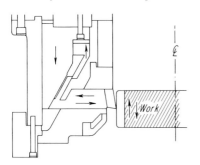

FIGURE 16.37 Feed control of shear cutters.

Gear concentricity adjustment is provided by an adjustment to the cutter head as follows (see Fig. 16.39). The Shear-Speed cutter head is held approximately in line with the centerline of the machine. It is necessary to make final adjustment, by checking the gear being produced for tooth spacing, before the machine is set to run production. This is done by adjusting the four screws that hold the

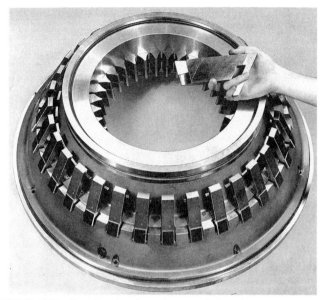

FIGURE 16.38 Shear-Speed cutting tools and toolholder. *(Courtesy of Michigan Tool Co., Detroit, Michigan)*

cutter head in position so that the cutter head is held firmly, but not tightly, and can be moved by the adjustment screws provided. To assist in this adjustment, two indicators are provided. Note that the tools, held in the cutter head on the X axis, move radially in toward the pitch diameter of the gear, while the tools on the Y axis move on a line tangent to the pitch diameter. The other tools move varying amounts depending on the angle of the cutter head slot the tool is in, to the direction of the adjustment. By this adjustment, good spacing (hence good concentricity) is obtained.

16.4.3 Machine Operation: Shear-Speed Cutting

Cutting tools are assembled in slots of the tool holder, with the cutting edges inside and facing the bottom of the tool holder. The angular projections on the cutting tools are at the outside of the tool holder with the projections pointing toward the top of the tool holder. When the tool holder is raised into position by means of jacks in the head of the machine, the angular projections fit between two cones. The outer cone controls infeed of the cutting tools on the upstroke; the inner cone controls retraction of the cutting tools on the downstroke. Thus fastening the tool holder in place in the head of the machine places the cutting tools in correct cutting position in relation to the

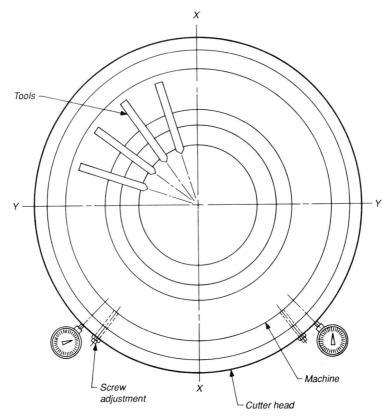

FIGURE 16.39 A shear speed cutter head can be adjusted to achieve good concentricity by adjusting the tool holder inside the cutter head (as shown).

vertically reciprocating gear blank. Provision is made for centering the gear blank in relation to the cutting tools.

The cutter head assembly is mounted in the head slide, which also carries the feed and relief cams and their drives. This slide travels on scraped ways from loading to cutting positions and is changed from one position to the other by means of a hydraulic cylinder. When the slide lowers the head into cutting position, it is automatically locked rigidly in place by two wedges, one at either side of the head assembly. These wedges are actuated by individual hydraulic cylinders. Lock action is automatic. The locking wedges are retracted automatically at the end of the cutting cycle to allow the slide to return to loading position.

The actual feed functions only when the head slide is locked in "cutting" position. Downfeed of the cones is controlled by a cam that gradually reduces the feed on each cutting stroke. The rate of feed is controlled by a ratchet mechanism that turns the cam. The pawl of this mechanism is adjustable to move the cam an amount equal to either one or two teeth on the ratchet wheel for each reciprocation of the work, thereby regulating the rate of feed of the tools. The rate of infeed decreases as the tools are fed to depth. A ratchet wheel with the correct number of teeth is selected to give the desired infeed of the tools.

The cones are raised a definite amount for each return stroke by an eccentric relief cam. As the work starts on its downstroke, this rotating cam raises the cone assembly slightly and holds it up until the downstroke is completed. This causes the inner cone to pull the tools away from the work. The cam returns the cones to a cutting position for the cutting stroke. This relief action is completely independent of the feed and acts identically on each return stroke regardless of the position of the cones during the machine cycle (see machine details in Fig. 16.40).

The gear blank is reciprocated vertically through the cutter assembly while the cutting tools are fed in a predetermined amount for each stroke of the gear, until all teeth are formed to full depth. Provision is made to retract the blades on each return stroke to avoid drag and to return the blades to starting position at the end of the cycle.

Length of reciprocating stroke of the work is adjustable. Recommended length of stroke is the face width of the gear being cut plus ¼ in. Speed of stroke is regulated by change gears on the machine. It is recommended that the change gears selected give a speed of 20 ft/min cutting stroke.

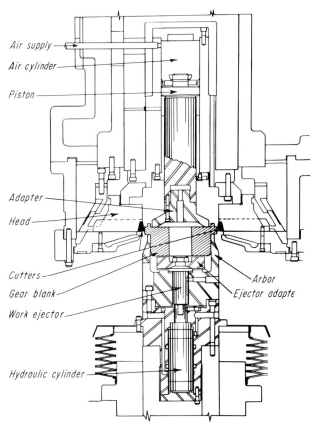

FIGURE 16.40 Shear-Speed machine. *(Courtesy of Michigan Tool Co., Detroit, Michigan)*

The work is mounted on an adapter and is clamped hydraulically or pneumatically. Mechanical clamping may also be used. Clamping and release may be either automatic or semiautomatic. Parts with shafts, as well as with holes, can be cut by the use of proper adapters. The work-holding adapter or fixture is mounted on the top of a ram.

The following examples indicate typical production data. In both cases, a finish of approximately 50 μin. can be obtained.

Example 1

Number of teeth	13 spur
Diametral pitch	12
Pressure angle	14°30'
Outside diameter	1.290"
Whole depth of tooth	0.1645"
Face width	0.786"
Strokes per minute	126
Surface feet per minute	21
Infeed per stroke, average	0.0036"
Length of stroke	1.0"
Teeth in ratchet	50
Pieces per load, manual loading	1
Pieces per hour at 80-percent efficiency	77

Example 2

Number of teeth	56 spur
Diametral pitch	12.7–16.9
Pressure angle	20°
Outside diameter	4.527"
Whole depth of tooth	0.167"
Face width	0.480"
Strokes per minute	81.1
Surface feet per minute	21.1
Infeed per stroke, average	0.0037"
Length of stroke	1.56"
Teeth in ratchet	50
Pieces per load, manual loading	2
Pieces per hour at 80-percent efficiency	102

16.4.4 Advantages and Disadvantages of Shear-Speed Cutting

Shear-Speed form cutting is applicable to cutting gears, splines, sprockets, and other spur tooth forms that may be unsymmetrical or unequally spaced. Since all the spaces between the teeth are cut at one time, gears can be cut at a high rate of production.

GEAR CUTTING 16.45

The production advantage is greater for gears with large numbers of teeth. Each machine is usually put on the production of one high-volume part.

The cutting tools are easily removed for sharpening and are sharpened all at one time on a surface grinder. The process is limited to coarse-pitch gears having spur teeth. In addition, each model is limited to a small range of diameters and pitches.

16.5 GEAR BROACHING

Broaching is a process widely used for making keyways, finishing holes, cutting splines, etc. The process is also applicable to making gear teeth. The form of the space between broached gear teeth corresponds to the form of the broach tooth. Broaching is normally a high-production process. Occasionally it may be used in fairly small-lot production to get accuracy on a critical part that cannot readily be made to a high degree of accuracy by other methods.

16.5.1 Application of Broaching

Broaching is a machining process that removes metal from a part by pulling or pushing a multiple-toothed tool called a *broach* through or along the surface of the part. It is a fast and accurate process that produces fine surface finishes. This section is limited to a discussion of broaching as it applies to gears and similar parts.

Gear broaching is usually applied to the cutting of internal spur and helical parts. Normally, all teeth are cut in one pass of the broach. In addition to gears, other parts that can be broached include splines, serrations, and other internal tooth forms. Although broaching is generally applied to internal tooth forms, it is sometimes advantageous to broach external tooth forms. When all the teeth are cut simultaneously, broaching is a high-production method of cutting gears and splines and is normally applied to high-volume production. Roughing and finishing are normally combined in one pass of the broach. However, some parts require two operations: rough broaching and finish broaching. In some special applications, broaching is done one tooth at a time.

Broaching machines can be designed to cut gears of almost any size and pitch. Each application must be considered in terms of the major and minor diameters, the length of the part, and the material. Some of the large vertical broaching machines have a capacity of 75 tons and a stroke of 72 in.

A broach can be used to machine almost any internal form. The teeth can be either equally or unequally spaced, and tooth forms can be either symmetrical or unsymmetrical. The tooth form must be uniform in the direction of the broach axis, and the wall of the part must be strong enough to withstand the broaching pressure. Gear broaching usually requires a part with a through hole. However, blind holes can be broached if there is sufficient clearance beyond the broached section and if a series of punch-type broaches are used.

Broaching can produce close tolerances on profile, spacing, lead, and size. As an example, an internal pump gear with 35 teeth, 10 diametral pitch, 24° pressure angle, and 11/16-in. face width is broached within AGMA quality 10 tolerances on profile, parallelism, and spacing. Print specifications call for involute profile errors to be within $+/-0.00025$ in. Spacing is held to 0.0005 in., lead within 0.0003 in., and lead variation within 0.0005 in. Both faces must be parallel and square with the pitch diameter, and the outside diameter must be within 0.001-in. indicator reading. In addition, the pitch diameter must be concentric with the outside diameter within 0.001 in.

Most types of ferrous and nonferrous materials can be broached. In steels, the hardness range for broaching is normally between 12 and 22 Rockwell C. However, a considerable amount of broaching is done at hardnesses up to 35 Rockwell C. In some cases, the finish on the hard gear may be superior to that on a gear made of softer material. Steels below 12 Rockwell C are difficult to broach because of their tendency to pick up metal on the sides of the broach teeth and on the lands, resulting in tears and poor surface finish. Parts harder than 35 Rockwell C can be broached, but it is seldom economical because of the high cost of broach maintenance. Examples of broached parts are shown in Fig. 16.41.

To broach work-hardening materials, it is necessary that there be sufficient step or cut per tooth so that each tooth will bite under the surface of the metal—including the first tooth. Otherwise, the material being broached may tend to break down the cutting edges of the tooth.

Steel forgings and cast iron can be broached, but the material should be of as uniform density or hardness as possible. Commercial grades of brass, bronze, aluminum, and plastic can normally be broached at high speeds.

16.5.2 Broaching Principle

Broaching is a machining method by which the cutting tool is pulled or pushed over the surface to be machined. The method of operation affects the construction of a

FIGURE 16.41 Examples of production parts with broached teeth. *(Courtesy of National Broach & Machine Co., Mt. Clemens, Michigan)*

broach. Push broaches are normally short and therefore less expensive than the longer pull broaches. However, one pull broach can sometimes remove as much stock as several push broaches. Most internal broaching is done with pull broaches. Not only do they remove more stock than push broaches, but they are also capable of taking much longer cuts.

The actual cutting action of any single broach tooth is very much like that of a single form tool. Each section of the broach has as many teeth as there are tooth spaces on the gear. The diameter of each section increases progressively to the major diameter that completes the tooth form on the part.

Assume an internal gear blank with an inside-diameter finish reamed to a 3.8000-in. diameter in which internal gear teeth are to be broached to a major diameter of 4.2314 in. The broach will produce the gear teeth by the simple method of each succeeding section of the broach increasing in diameter by increments from 3.8000 to 4.2314 in. Each broach tooth contributes some portion of the internal form. As shown in Fig. 16.42, the gear tooth space is produced by successive broach teeth whose form depth increases progressively. The broach teeth in the finishing sections complete the gear tooth form by producing the major diameter.

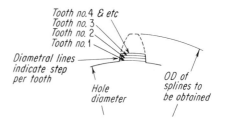

FIGURE 16.42 Cutting action of a broach.

In some cases, broaches can be designed with a removable finishing section that finishes the complete tooth form. This "side-shaving" section of the broach finishes the entire profile of each gear tooth (see Fig. 16.43, the *A* portion as shown). A full form-finishing section solves profile problems caused by off-center drift of the broach; off-center drift of the broach being due to lack of care in maintenance of machine alignment and perpendicularity or nonuniform tooth dulling.

Differences that must be considered in broaching arise more in chip disposal than in chip separation. In ordinary cutting on a lathe, milling machine, shaper, or planer, chips are removed almost as soon as they are formed. However, chips from a broach are of necessity confined in the space between the teeth for the length of the cut. Therefore, it is absolutely necessary to provide adequate chip space when designing the broach. Unless this is done, there is a chance of dulling or even breaking the tool and roughing or tearing the surface of the work being broached. Chip space is limited by the pitch of the teeth and, in small broaches, by the diameter of the broach. Broaching in general requires the use of chip breakers. Otherwise too wide a chip may be produced. Such chips

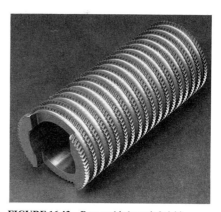

FIGURE 16.43 Removable broach finishing section. It is possible to take a light finishing cut with this kind of a section and get better profile accuracy and better concentricity than that obtainable with a conventional one-piece broach. *(Courtesy of National Broach & Machine Co., Mt Clemens, Michigan)*

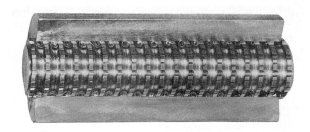

FIGURE 16.44 Broach chips revealed by cutting through a workpiece. *(Courtesy of Colonial Broach Co., Detroit, Michigan)*

are hard to control and are usually taken care of by providing staggered chip breakers on successive broach teeth. Figure 16.44 shows how a work piece may be cut to study chip formation.

16.5.3 Broaching Operation

Machine selection plays a vital part in the relative success of the broaching operation. A primary factor to be considered is the broach itself since, in good broaching practice, broach design to some extent precedes machine selection. Parts that require heavy stock removal should be applied to machines with heavy tonnage. However, lighter machines can frequently do a good job by using a finer step per tooth with more passes.

Gears are broached on both horizontal and vertical machines using pull-type broaches. Horizontal machines are generally considered to be adaptable to a wide variety of broaching operations. On some models, provisions are made for accessibility of a broach drive head to facilitate the broaching of internal helical gears. Both pull-up and pull-down vertical machines are applicable to broaching gears. The pull-down type is more convenient on heavy parts and where it is necessary to locate an internal tooth in relation to a feature on the part, such as an external gear tooth. Both types can be equipped with automatic broach-handling mechanisms, making it unnecessary for the operator to handle the broach. Other types of machines can be equipped with automatic handling devices.

In addition to the size and type of broaching machine required, there are several general considerations that must be given to the gear specifications.

Broaching of several narrow-face gears at one pass of the broach will result in an off-lead condition if the parts are not processed with all faces parallel. Most internal gear plants grind either one or both sides of the gear locating faces to ensure parallelism. With wide-face gears, a broach of excessive length would be required for a one-pass operation. Consequently, more than one broach would be applied with each one removing a portion of the stock in its pass through the gear.

Some internal gears require a pitch circle that is concentric with an external contour of the part. Since the broach has a tendency to follow the hole through which it is pulled or pushed, the internal gear teeth should be broached prior to finishing the external contour. The part can then be located on the pitch circle for finishing the external contour. When the part is to be located on the minor diameter, a dwell tooth broach, as shown in Fig. 16.45, will provide concentricity between the minor diameter and the pitch diameter of the internal gear.

Broaching of internal spur gears with the desired zero helix angle is dependent on the alignment of the puller end and the follower mechanism of the broaching machine. To achieve accuracy, it is necessary to align the puller and retriever mechanisms properly. In addition, the axis of the broach must be perpendicular to the face of the fixture that locates the gear and absorbs the broaching forces.

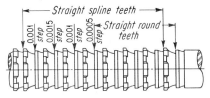

FIGURE 16.45 Dwell tooth broach.

Internal helical gears are broached in much the same manner as spur gears except that the broach must rotate as it passes through the blank. Rotation of the broach is accomplished by means of a drive head on the machine. Helical gears are produced on both vertical and horizontal broaching machines equipped with special attachments. It may not be practical to broach gears with a large helix angle. Drive head configuration is shown in Fig. 16.46.

What would be feed in other types of machining is designed into the broach and is called *chip per tooth*. In a gear broach, the teeth increase in diameter from the first roughing tooth to the first finishing tooth. This increase in diameter, or cut per tooth, need not be uniform. Normally, finishing teeth are of uniform diameter to take off the final increment of stock, completing the form to the major diameter. Cut per tooth in the broach has a bearing on finish. Usually, a smaller step helps to give a good finish

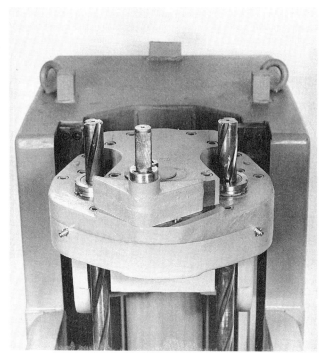

FIGURE 16.46 Drive head for helical internal broaches. *(Courtesy of Colonial Broach Co., Detroit, Michigan)*

by producing a thin chip that curls up in the chip space and keeps clear of the cutting edge. The cut per tooth must be decreased as hardness or toughness increases. A step of 0.0005 in. on broach diameter is about minimum for a clean cut.

Speed is an important consideration in broaching. Cutting speeds for gear broaching are normally in the range of 8 to 30 ft/min. Broaching speeds as low as 2 or 3 ft/min are used effectively to broach extremely hard-to-cut materials.

Cutting speed can be varied for an individual broach application. Actually, considerations of power requirements, tooth strength, broach strength, finish desired, and choice of coolant offer considerable limitations. Free-cutting material may be broached faster than tough stringy material. Other considerations, however, usually make it advisable to use a different broach for a different material.

In general, steels are broached with high-speed broaches at surface speeds ranging from 10 to 30 ft/min. Cutting speeds range from 8 ft/min for hard stainless to 20 or 25 ft/min for free-machining types.

Cast iron and malleable iron can be broached with high-speed steel broaches at maximum speeds up to 30 ft/min in normal operation and at 120 ft/min with carbide broaches. Cast iron may be cut with a greater step or cut per tooth than free-cutting steel, but break-out is a possibility and may limit the cut.

Cutting speeds up to 30 ft/min are practical in broaching commercial grades of brass and bronze. Aluminum and magnesium materials can be broached at speeds of over 30 ft/min.

It is not difficult to estimate the production possible from a broaching setup if a few simple factors are known. These include machine speed, return speed, length of stroke, and time to load and unload. The following is an example of how the production rate can be determined.

Cutting speed	30 ft/min (360 in./min)	
Return speed	60 ft/min (720 in./min)	
Stroke	48 in.	
Cutting time	$= \dfrac{48 \text{ in.} \times 60 \text{ s/min}}{360 \text{ in./min}}$	$= 8$ s
Return time	$= \dfrac{48 \text{ in.} \times 60 \text{ s/min}}{720 \text{ in. per min}}$	$= 4$ s
Starting and stopping		$= 2$ s
Loading and unloading		$= \underline{5 \text{ s}}$
Total time per piece		$= 19$ s

Due to advances in machine configuration, broaching tools, and automation systems, broaching of external tooth forms has expanded in its application.

The basic process consists of a tool with internal teeth being held in a pot being passed over a round part and producing external shapes in the forms of splines and involute gear teeth.

In the past, external gears were broached by either pushing the work piece down through a stationary pot broach or mounting the work piece on a stationary post and pushing the pot broach down over it. Problems resulted in chips packing in the chip gullets when using the push-down stationary pot broach method. Slow work piece loading, due to using a plug-type approximate locator for each part, is a problem when broaching parts with the push-down stationary work piece method.

Two more recent methods are push-up pot broach and pull-up broaching. Push-up pot broaching consists of pushing work pieces up through a stationary pot broaching tool. Chips fall by gravity away from the tool and work piece. Automatic loading is simplified due to the part being elevated by the process and using gravity to unload parts onto a conveyor. A push-up pot broach machine is shown in Fig. 16.47.

Pull-up broaching is used when the part diameter and broach length are such that the required post diameter and length do not provide sufficient strength to support the work piece rigidly. Using this method, the work piece is pulled up through a stationary pot broaching tool with a pull rod. This process permits broaching parts having deeper teeth and wider faces by using longer tools. Figure 16.48 shows a pull-up pot broach machine.

Tools used for pot broaching are either stick-type or ring-type. The stick-type design is used where work piece accuracy permits. These tools are particularly adapted to short-face-width splines and gears. The ring-type pot broaches are used for precision running gears and splines where close tooth tolerances are required.

These two types of tools are often used in combination (See Fig. 16.49). Stocks in

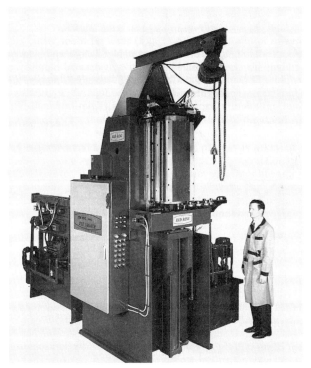

FIGURE 16.47 Push-up pot broaching machine. *(Courtesy of National Broach & Machine Co., Mt. Clemens, Michigan)*

16.52 CHAPTER SIXTEEN

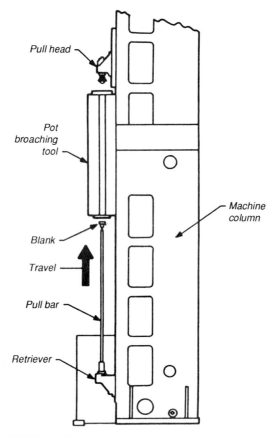

FIGURE 16.48 Sketch of broaching machine for *pull-up* pot broaching.

the pot broaching tool produce the slots on the synchronizer hub shown, and rings produce the 12 DP spur teeth.

16.5.4 Advantages and Limitations

Gear broaching is primarily applicable to high-volume jobs that require high production rates. Gears can be broached accurately, since a number of the elements are controlled by the accuracy of the broach. In addition, broaching can produce a fine finish on the gear-tooth profiles.

However, a gear broach will cut only parts that have identical gear specifications.

FIGURE 16.49 Combination pot broaching tool with stocks to produce slots in a hub and rings to produce spur gear teeth. (Courtesy of National Broach & Machine Co., Mt. Clemens, Michigan)

REFERENCES

1. Barber-Colman, *Hob Handbook,* Barber-Colman Company, Rockford, Ill.
2. Dudley, D. W., *Gear Handbook,* 1st ed., McGraw-Hill Book Company, Inc., New York, 1962.
3. Dudley, D. W., *Handbook of Practical Gear Design,* McGraw-Hill Book Company, Inc., New York, 1984.
4. Fellows, "Gear Production Equipment," The Fellows Gear Shaper Company, Springfield, Vt.
5. Fellows, "The Art of Generating with a Reciprocating Tool," The Fellows Gear Shaper Company, Springfield, Vt.
6. Fellows, "The Internal Gear," The Fellows Gear Shaper Company, Springfield, Vt.
7. Gould and Eberhardt, "Gear Cutting Machines," Bul. 250, Gould and Eberhardt, Irvington, N.J.
8. Michigan Tool, "The Michigan Shear-Speed Shaper Line," Bull. 1800–54, Michigan Tool Company, Detroit, Mich.
9. Michigan Tool, "Shear-Speed," Bull. SS-55, Michigan Tool Company, Detroit, Mich.

10. National Broach, "Broaching Practice," National Broach and Machine Company, Detroit, Mich.
11. National Broach, "Modern Methods of Gear Manufacture," 4th ed., National Broach and Machine Co., Detroit, Mich. 1972.
12. The Metal Cutting Tool Institute, "How to Use and Maintain Gear Cutting Tools," Gear Generating Tools Subdivision of the Metal Cutting Tool Institute, New York, New York.
13. Colvin, F. H., and F. A. Stanley: *Gear Cutting Practice,* 3rd ed., McGraw-Hill Book Company, Inc., New York, 1950.
14. Hartford, J. L., "Computer Numerical of Six Axes Hobbing Machine," AGMA, April 1985.
15. Lange, J. M., "CNC Gear Shaping Machine Design and Benefits," AGMA, April 1985.
16. Ainoura, M., "A Research on High Speed Hobbing with Carbide Multiple Thread Hobs," ASME, 60–C2/Det 56.
17. Moncrieff, A. D., "High Speed Hobbing with Carbide Cutters," Tooling and Production, March 1978.
18. Whitney, D. S., "Computer Aided Design for Gear Shaper Cutters," SME copyright 1986 from The Gear Processing Technology Clinic.

CHAPTER 17
GEARS MADE BY DIES

Michael J. Broglie, P.E.
President, Dudley Technical Group Incorporated
Escondido, California

Gear teeth are often made by a cutting process such as milling, hobbing, shaping, or broaching. These cutting methods all make the gear teeth by removing the material in the space between gear teeth. Collectively these processes can be thought of as "metal removal" processes. Chapter 16 discussed how cutting methods are used to make gear teeth.

In this chapter another general method of making gear teeth is discussed. Molten or powdered material can be put into a cavity that has teeth made like the spaces between gear teeth. The tool that holds the raw material to make a gear part is called a *die*.

The making of gears by dies has essentially been developed during the last century. As a method it is not nearly as old as the method of making gears by cutting.

The practice of making gears by dies has grown rapidly in the 1970s and the 1980s. All indications point to even more growth of this method in the 1990s and beyond. Some of the reasons for the fast growth in the making of gears by dies are:

New plastic materials have been developed that will stand much higher temperatures. Also, plastic materials that are much stronger are now in use.

Powdered metal gears can now be made which are quite comparable in load-carrying capacity to case-hardened steel gears.

The making of gears by dies is achieving higher accuracy.

The making of gears by dies lends itself to automation and to the control of operations by computer. In many cases the mass production of a gear part at a very low price can only be achieved through the use of dies.

17.1 DIE PROCESSES

The die process involves passing a gear material into or through a die that has, as its shape, the outside dimensions of the gear. The die may be for a segment of the gear teeth, as in the case of a die used to roll worm threads. Extrusion and cold drawing are examples of dies that form all of the teeth at once, but further machining on the work piece is needed to produce individual gears. Injection molding, the powdered metal process, and net forging are examples of processes in which the part's shape is completely formed by the die.

17.2 INJECTION MOLDING*

Plastics have been used as gear materials on a widespread basis since the 1950s, and today the majority of gears produced by dies are injection-molded gears utilizing thermoplastic materials. A thermoplastic material is one that will repeatedly soften when it is heated and harden when it is cooled.[1] Nylon, acetal, polycarbonate, and polyester are examples of thermoplastics commonly used for gearing. Thermoplastics have been used to make spur, helical, bevel, and worm gears. Often lubricants such as polytetrafluoroethylene (PTFE), silicone, molybdenum disulfide, and graphite are added to a thermoplastic material to further improve the inherent lubricity of the plastic.

It has occasionally been mentioned that molded gears are not as accurate as machined gears. This is an inaccurate statement. A molded gear held to the tolerance of AGMA quality Q8[2] is just as accurate as a machined gear made to the same quality number. It is true that gears have not yet been molded to the highest precision obtainable by machining, but the number of gears requiring such precision represents only a small percentage of all gears made. In general, molded plastic gears are usually specified AGMA quality Q6 or Q7.

Applications for molded plastic gears include a wide variety of products, among which are clocks; timers; small motors; instruments; toys of all types; small appliances, such as mixers, blenders, vacuum cleaners, and small tools; and many similar items. Today there is an extensive use of plastic gearing in business machines such as printers, plotters, fax machines, and copying machines. Each application should be thoroughly investigated, however, with respect to load, temperature, tolerances, and assembly design specifications. Figure 17.1 shows some examples of injection molded gears.

17.2.1 Injection-Molding Process

Injection molding involves forcing, by pressure, a thermoplastic that has been heated to a liquid state into a closed chamber known as the die.

A typical machine for injection molding of gears is shown in Fig. 17.2. Powdered or pelletized plastic is placed in the hopper, melted, and injected. Figure 17.3 is a sketch of the schematic cross section of a gear molding die, shown open and closed. After the molding die is closed, the melted plastic is injected under high pressure into the gear cavities. There can be as many as eight cavities in each die. The die remains closed until the plastic solidifies. This takes from several seconds to several minutes depending on the material and the geometry of the part.[3]

Once the material has solidified, the die opens and the gear or gears are knocked out of the die. If the die has cavities for more than one gear, the cavities are usually connected with feeders so that the mold will fill as quickly as possible. When the

---

*Part of the data used in this section was furnished by Raymond Paquet of Plastics Gearing Technology, Inc., Manchester, Connecticut.

GEARS MADE BY DIES 17.3

FIGURE 17.1 Some examples of injection-molded gears. *(Courtesy of Winzeler, Inc., Chicago, Illinois)*

material in the feeders cools, the parts are connected by this extra material. Under ordinary circumstances removing this material is the only work needed to finish the parts. A small amount of material is left on the gear at the feeder points. In critical cases this material must be removed with a sharp tool.

Almost any desired shape can be produced; however, care must be exercised to maintain fairly constant sections. Diametral pitches from 16 to 120 can be molded very well, and combination or clusters of gears can be made in one piece.

A molded gear can be no more accurate than the molding die that produces it. Basic considerations for gear molding dies are the same for any precision molded product. Since molded gears are usually small in relation to the size of average moldings, they lend themselves admirably to being molded in the small, high-speed automatic injection molding machines (20- to 100-ton range). This allows for compact dies, usually with one to eight cavities.

The designer of the die must consider several factors that can have an effect on the gear accuracy.

1. The increased strength of case-hardened die frames will enable them to withstand the abuses of the molding process and thus maintain their accuracy throughout extended useful life.

2. Steel surfaces in the bore eliminate the need for wear bushings. This removes the additional errors resulting from a lack of concentricity between the bore and a bushing.

3. The mold should have a balanced runner system to provide equal pressure drops in all cavities.

4. There must be adequate venting in the cavities to allow air to be displaced by the flow of plastic.

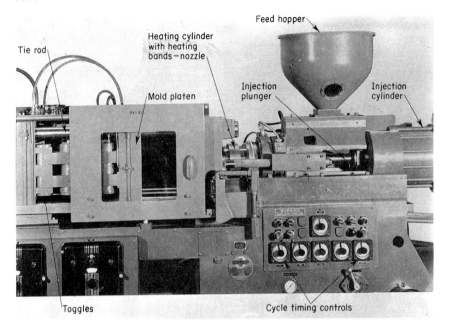

FIGURE 17.2 A machine for injection molding of gears. *(Courtesy of Fellows Gear Shaper Co.)*

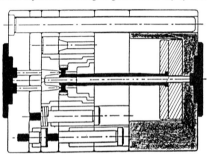

Closed position

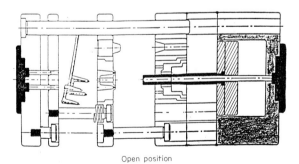

Open position

FIGURE 17.3 Schematic cross section of gear molding die shown in open and closed position.

GEARS MADE BY DIES 17.5

5. There must be an adequate ejection system to ensure minimum distortion of the product when ejected from the die. Also, the die design must provide adequate drafting.
6. There must be adequate interlocks between the die halves to remove the misalignment of running fits provided in the die system.

There are several factors in a molding die that can contribute to the gear's runout.

1. There is a 0.0001/0.0002-in. running fit required between the core pin and the stationary half of the die.
2. If sleeve ejection is required around the core pin, the solid anchor is lost and an additional 0.0001/0.0002-in. running fit will be present.
3. The gear cavities themselves are a series of concentric rings. This increases the potential for runout error.
4. The gating position on the part is usually off-center. This will result in some unevenness of plastic flow which, in turn, will introduce additional runout in the gear.

In addition to making the die accurate, the designer must consider both the size and shape changes that occur as the plastic cools. All plastics shrink when changing from a molten to a solid state. As a consequence, all mold cavities must be made larger than the product specification. For example, if a molded gear is to have an outside diameter of 1.200 in., and the plastic has mold shrinkage of 0.025 in./in.; then the outside diameter of the cavity will be required to be 1.2308 in.

In making a gear cavity, however, it is not sufficient simply to enlarge all of the gear's dimensions. The pressure angle of the cavity must also be compensated. If this is not done, the result will be a molded gear with a serious profile error. This will also result in a larger-than-acceptable tooth-to-tooth error.

For example, on the left-hand side of Fig. 17.4 is an enlargement of a 32-diametral-pitch, 20°-pressure-angle gear tooth, and, superimposed on it, is the profile of the oversize gear tooth cut with a standard 32-diametral-pitch, 20°-pressure-angle cutter. The illustration on the right side of the figure shows the standard gear, and this time, superimposed on it, is the profile of the molded tooth (after shrinkage) that would be obtained from the oversized cavity.

Note that the tooth form of the molded gear departs considerably from standard. It is thicker at the root and thinner at the tip (it has a pressure angle much in excess of 20°).

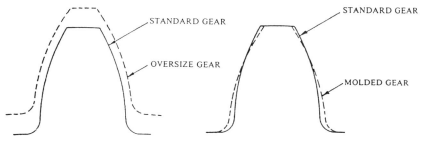

FIGURE 17.4 Standard gear tooth superimposed with oversized tooth and molded tooth. Note that the tooth of the molded gear departs considerably from standard. It is thicker at the root and thinner at the tip (it has a pressure angle much in excess of 20°).

The formula used to calculate the correct pressure angle is:

$$\cos \phi_2 = \frac{D \cos \phi_1}{D(1+S)}$$

where D = pitch circle diameter
  ϕ_1 = pressure angle of hob
  ϕ_2 = pressure angle of molded gear
  S = shrinkage

The teeth in the cavity must be carefully compensated for shrinkage so that, when the molded gear solidifies and becomes stable, the teeth will have the correct profile. This design work is further complicated in the case of a helical gear, because the axial shrinkage is usually different from the shrinkage across the diameter.

In summary, there is a wide range of materials that can be molded. Some of these have lubricants in the materials themselves. The cost of dies is moderate and the per-piece cost is low.

17.3 DIE-CAST GEARS

The die-casting process is similar to the injection-molding process but replaces the plastic with molten metal. The equipment is also similar to injection-molding equipment. The process involves forcing, under pressure, molten metal into the die cavity. The material is allowed to solidify and is then removed from the die. Though solidified, these pieces are still very hot and thermally insulated gloves are needed to handle parts.[4]

Several materials are candidates for die casting. These include zinc, brass, aluminum, and magnesium. Each material is best suited for a specific purpose. Before selecting a material, the application should be considered.

Many different types of gears can be cast, such as spur, helical, worm, clusters, and bevel gears. The process is suited to high-volume, low-cost gears that do not need to be extremely accurate. Applications for these types of gears include toys, washing machines, small appliances, hand tools, cameras, business machines, and similar equipment.

17.3.1 Die-Casting Process

There are two types of die-casting processes: hot chamber and cold chamber. The cold chamber process requires that the metal be melted in a separate furnace and ladled, either by hand or with a mechanical ladle, into the loading cylinder. The hot chamber method utilizes equipment that has the loading cylinder submerged in the molten metal. The materials that can be used in the hot chamber process are limited. Certain alloys that have a tendency to dissolve iron or steel cannot be used since they would react with the plunger and cylinder, affecting the critical clearance between these parts. Certain other materials, such as aluminum, magnesium, and brass, are also unacceptable for the hot chamber process due to the high pressures needed to get the required densities. The one material that is most commonly used in the hot chamber process since it does not have these restrictions is zinc.

From this point, the process is the same for both methods. The molten metal flows, under pressure, into the die cavities. Once solidified, the castings are removed.

As with plastic gears, the die can be machined to accommodate almost any tooth form that could be cut. The dies must be designed to compensate for the shape and size changes that occur as the material cools. It is recommended that an experienced designer be consulted before producing a die.

The shape of the die can allow for threads to be molded on the part. The accuracy is not of the same order as that to which machined threads are usually held. Gears can also be cast integral with shafts or other appendages. Though this can be an advantage from a standpoint of manufacturing, it should be noted, in general, that the tolerances in a cast part are related to the weight of the part. For instance, on parts that are 1 oz or less, a tolerance might be plus or minus 0.001 in. A 6-oz part may need a tolerance of plus or minus 0.005 in. Closer tolerances are possible, but this increases the cost of the tooling.

Several general guidelines have been developed from experience and should be followed.

Tooth proportions should follow the AGMA fine pitch standard for involute gear teeth.

The weight of the part should be considered (possibly requiring lightening holes).

Section thicknesses should be maintained as uniform as possible. Always avoid abrupt changes.

At each section intersection there should be a smooth radius. This provides better strength and helps the flow of molten metal into the die.

Long unbroken surfaces should be avoided since they may become tapered or wavy.

Holes should have as much taper as possible and, if straight holes are required, a reaming operation will be needed.

Parting line location must be carefully chosen for the sake of appearance.

The location of gates and runners should be chosen so as not to interfere with the operation of the part.

The inspection of cast gears is the same as for cut gears. Initially, a die and a process must be developed to get good gears. Once the die and process begin to produce good parts, the inspection should continue for die wear, material variances, or process changes that affect the quality of the parts.

17.4 POWDERED METAL GEARS

The *powdered metal process* (P/M) has been used since the 1930s to manufacture parts. During the 1950s designers began using P/M for gears because of the strength and precision that was obtainable for a low per-piece cost. Powdered metal gears fill a need in the manufacture of spur, straight bevel, and spiral bevel gears where the volume is high but the load-carrying capacity exceeds the capability of thermoplastic materials. The strongest and most precise gear is a machined gear, but this extra strength and precision come at a much higher cost; and in many cases the load-carrying capability of P/M gears exceeds what an application requires. Today improvements in die making have permitted closer manufacturing tolerances. While most P/M gears are designed to have tooth accuracies equal to AGMA Q7 or Q8, up to Q10 is achievable.

The powdered metal process has been used to make spur, helical, and bevel gears. An important aspect of the powdered metal process is the ability to mold intricate shapes. For instance, Fig. 17.5 shows a gear from a riding lawn mower axle, which is in high production. This gear has a pitch diameter of about 6 in. and is the final drive gear from a residential mower that is rated at more than 10 hp.

The manufacturing cost of this part has been minimized by incorporating the bore, holes, and a counterbore into the molded shape. This feature of the P/M process can also be utilized to mold cluster gears and gears with integral cams and clutch jaws or other appendages. With this process it is possible, and recommended, to include in the shape of the die, a tooth profile modification. This will allow the gear to run more smoothly.

One application that uses powdered metal gearing extensively is gear pump rotors. In this case, one set of dies, which is designed for the longest length in a pump series, is suitable for many different rotor lengths.

Fine-pitch P/M gears are commonly used for fractional horsepower electric motors, power tools, and small appliances. The repeatability that can be expected from P/M gears is helpful in these applications since there is little time for detailed inspection.

Coarse-pitch applications utilizing P/M gears include heavy appliances, automotive equipment, and agricultural equipment. In these cases extra toughness is needed and cost is very important.

FIGURE 17.5 Gear from riding lawn mower axle produced by powdered metallurgy techniques. *(Courtesy of Dudley Technical Group, Inc., San Diego, California)*

17.4.1 The Powdered Metallurgy Process

The powdered metallurgy process begins with the selection of metal or combination of metals (in powdered form) that best suit the application. For small fine-pitch gears, where the material content is small, it is wise to use the highest-strength materials since the cost of the material has only a small impact on the overall cost of the part. On larger coarse-pitch gears the amount of material is greater and has a greater impact on the overall cost of a part. In these cases cost considerations would require a study of which materials would have the required capacity for a specific application. Also, the possibility of additional heat treatment or machining after sintering must be considered when selecting the metal.

Table 17.1 shows examples of some of the materials used for P/M gears. The nickel steels are used where high strength, high impact properties, and excellent wear resistance is necessary.

Once a material has been selected, the powder is blended to insure a homogeneous mixture. Figure 17.6 shows how the powder fills the die when the bottom punch is

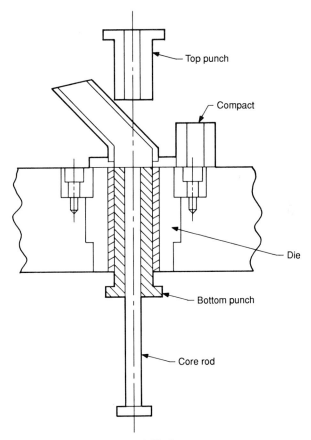

FIGURE 17.6 Powdered metal fills die.

TABLE 17.1 Typical Physical Properties for Alloyed Powdered Metal*

Nominal Composition	ASTM Specification	Dry Density, g/cc	Young's Modulus in Tension, lb/in^2	Ultimate Tensile Strength, lb/in^2	Transverse Rupture Strength, lb/in^2	Hardness (see below)
Carbon copper steel	B426-76 Type II Grade 2 Cl.C	6.0–6.4	15.0 E+6	48,000	85,000	75–80 (15T scale)
	B426-76 Type III Grade 2 Cl.C	6.4–6.8	18.6 E+6	68,000	110,000	83–88 (15T scale)
Nickel steel	B484-76 Type II Class C	6.8–7.2	18.5 E+6	55,000	105,000	71 HRB
	B484-76 Type III Grade 2 Class C	7.2 min.	23.0 E+6	160,000	245,000	37 HRC

*The top three entries pertain to materials that are sintered, but not heat treated. The last entry is for a heat-treated material.

lowered. Once the correct amount of powder fills the die it is pressed by an upper and lower die. The load pressure will vary from about 1 ton to 1500 tons, depending on the size and shape of the part. The compacted powder is then ejected from the die and is now in the "green" state and is known as a compact. The shape of the part shown is simple but multiaction (Fig. 17.7), or double die processes make it possible to produce more complicated shapes.

A compact is very brittle and has very little strength. The compact must be sintered at a temperature just below the melting temperature of the major constituent material that welds the material together. This is done in a controlled-atmosphere furnace. The amount of time in the furnace depends on the material and the shape of the gear.

Usually after sintering the part is considered complete; however, in cases where extreme physical properties are needed, coining and resintering can be done. This increases the density of the part and therefore its strength. Also, in some cases, further heat treating such as carburizing or induction hardening can be done. Additional machining may also be performed after sintering. For example, tapping of holes is one operation that must be done after sintering.

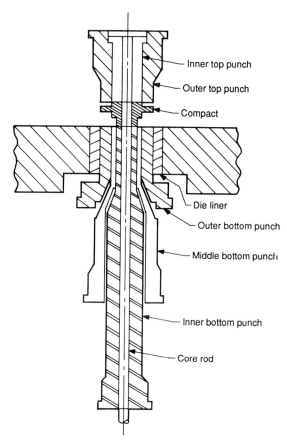

FIGURE 17.7 Multiaction die.

17.5 FORGED GEARS

Today gears are in service on which the teeth have either been finish-forged or near net forged. Finish-forged gear teeth need no machining at all. Near net forged gears leave only about 0.015 in. of material on each side of the tooth for cutting or grinding.

The type of gear that has been most successfully forged is a straight bevel gear. There is extensive use of forged and near net forged straight bevel gears at this time. Up until now the spiral bevel gears that have been forged need a small amount of machining on the teeth and are considered near net forged. This is also true for spur gears. There has been some use of net forged spiral bevel gears, but the process is still considered developmental. Figure 17.8 shows some forged gears that are in production. Current applications of net and near net forged gears in use today are highway truck axles and differentials, agricultural equipment, and marine outboard motor transmissions and stern drives. Research is planned for such extreme applications as helicopter transmissions.[5]

Forged gears have the same advantage over cut gears as other molded gears in that there is little or no material lost. This is a cost savings from the standpoint of both the cost of the material itself and, more importantly, the cost of machining. However, it turns out that forged gears may also have the advantage over cut gears of increased load-carrying capacity. This added strength comes in the form of increased bending

FIGURE 17.8 Examples of net forged straight bevel gears and near net forged spiral bevel gears in current production. *(Courtesy of Eaton Corporation, Southfield, Michigan)*

GEARS MADE BY DIES

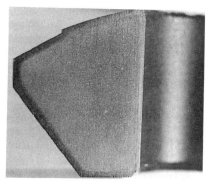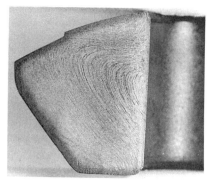

FIGURE 17.9 Difference in grain flow between (left) teeth cut from bar stock and (right) teeth that are forged. *(Courtesy of Eaton Corporation)*

strength and is due to the difference in grain flow between gears cut from bar stock and forged gears. Figure 17.9 shows the difference in grain flow between cut and forged gear teeth.

17.5.1 The Forged Gear Process

The forging process begins with a billet, cylindrical in shape, which is carefully sized and prepared for each forging. The billet is centerless ground to remove "bark" and any sharp corners. The billet is then heated to a temperature of about 2000°F. It is then forged into a die cavity that has the shape of a finished gear. Figure 17.10 is a schematic of the tooling arrangement for a bevel gear. Most gears are forged in a single blow. In some cases, because of the shape, two dies are used. The first die has an intermediate shape and the second, the finished shape.

The die is usually made by the *electrical discharge machining* (EDM) method. A master electrode, which is a "negative" of the die, is cut out of graphite or brass.[6] The electrode is brought close to, but does not touch, the die material, which is tool steel. This electrode then "burns" away the die material forming the cavity. The die cavity is the shape of the finished gear with some compensation for the shrinkage that occurs after the forging completely cools.

After being forged, the parts are allowed to air cool. Controlling the cooling is important for minimizing the distortion of the forging. An effective way to do this is on a cooling table that rotates to insure adequate flow of air around the parts. A scale forms on the outer surface during the cooling process. This scale should be removed by shot blasting before any heat-treating or subsequent machining operations are performed.

On gears where the teeth are cut, the blank usually has a bore. When the teeth are cut this bore is used to locate the part, and the gear teeth will run true with the bore. On forged gears the bore is machined to run with the teeth. This is done by holding the part with a pin plate or in a "nest," which locates the pitch diameter.

It is possible to forge gears with integral shafts. Eliminating keys and keyways in some cases further increases the strength of parts.

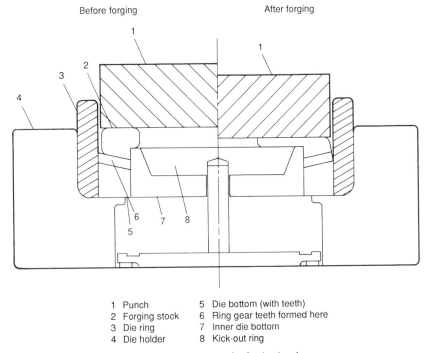

1 Punch
2 Forging stock
3 Die ring
4 Die holder
5 Die bottom (with teeth)
6 Ring gear teeth formed here
7 Inner die bottom
8 Kick-out ring

FIGURE 17.10 Schematic of tooling arrangement for forging bevel gears.

17.6 COLD DRAWING AND EXTRUSION

This process requires the least tool expenditure for mass production of spur toothed gear elements and is extremely versatile in that almost any tooth form that can be desired can be produced. As the name of the process implies, the material is passed through a series of block dies with the final die the shape of the cross section of the desired tooth element. As the material is run through these dies it is actually squeezed into the shape of the die. Since the material is displaced by pressure, the outside surface is work hardened and quite smooth.

The bars that are "blanks" for this process are usually 10 to 12 ft in length. After passing through the dies, they are known as pinion rods, and often are put into screw machines that finish the individual gears. Experience has shown that it is more economical to slice a segment off an extruded bar than to cut an individual gear. In some cases it would be impossible to produce the desired shape of pinion in any other way. Any material that has good drawing properties, such as high-carbon steels, brass, bronze, aluminum, and stainless steel, may be used for the drawn pinion rod.

Gears and pinions manufactured by this process have a large variety of applications and have been used on watches, electric clocks, spring-wound clocks, typewriters, multigraph machines, motion picture machines, carburetors, magnetos, small motors, switch apparatus, taximeters, cameras, slot machines, all types of mechanical toys, and many other parts for machinery of all kinds.

17.6.1 Cold Drawing and Extrusion Process

In this process a bar is pulled (drawn) or pushed (extruded) through a series of several dies, the last die having the final shape of the desired tooth form.

Usually from four to seven dies are made, the first-pass die being of a shape similar to that of the pinion desired but with large radii to allow easier metal displacement. After each pass the material is annealed for the next pass, and with each succeeding pass the stock assumes a shape nearer to that of the final design, until the last pass, in which the shape is exactly that of the tooth form specified.

This desired tooth form has provisions for backlash, whole depth, and corrected tooth thickness. Also, in order to compensate for wear of tools at the root, additional whole depth is provided. Figure 17.11 shows a representative tooth form for a 10-tooth, 68-pitch pinion. Note that 0.0015 in. has been added to the whole depth. Table 17.2 shows the recommended amount to be added to the whole depth for various diametral pitches. The tooth thicknesses at the pitch diameter should be reduced by the amount shown in Table 17.3.

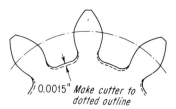

FIGURE 17.11 Representative tooth form for 10-tooth, 68-pitch drawn pinion. *(Courtesy of Tekniks Engineering Co., Elmhurst, Illinois)*

In the processes listed earlier in this chapter, very little, if any, machining was needed once the part was removed from the die. The drawing or extrusion process takes a bar or rod and puts teeth on the outside diameter. This becomes a pinion rod, which then must be further machined into individual gears.

Some of the advantageous properties of the drawn rod are the smooth finish and work-hardened surface. The hardness of the surface is several Rockwell points higher

TABLE 17.2 Recommended Whole-Depth Correction for Drawn Pinion Rod

Pitch Range	Amount to be Added, In.
16–24	0.0030
26–38	0.0025
40–62	0.0020
64–80	0.0015

TABLE 17.3 Recommended Tooth Thickness Reductions for Drawn Pinion Rod

Pitch Range	Reduction in Tooth Thickness of $1.5708/DP$, In.
100–120	-0.002
65–99	-0.003
42–64	-0.0035
20–40	-0.004

than that of the core. In cases where close tolerance holes are required, care must be exercised in drilling so that the tool will not drift because of the soft center.

Pinion rods from 16 to 100 diametral pitch can be obtained, but as the pitch becomes finer, it becomes more difficult to obtain the close tolerances that are sometimes desired on fine-pitch pinions. The tolerance of individual parts themselves depends a great deal on the screw machine operation. In this regard the concentricity depends on the accuracy of the machine, the tooling, and the setup. Concentricities of less than 0.003 in. TIR are difficult, and 0.005 in. TIR is more generally practical. A tolerance of plus or minus 0.002 in. on all pinion dimensions is considered standard, but all dimensions can be held to a total of 0.002 in., if required, without excessive difficulty. There are some so-called standard pitches, for which pinions are available without tool charges. These are 24, 32, 40, 64, and 80. Many are available in both 14½ and 20° pressure angles. It is a good idea to check with the manufacturer to ascertain the tooth number and pitches that are available.

Inspection of pinion rods usually consists of cutting a piece from the bar approximately $\frac{1}{32}$ in. thick, sanding to remove burrs, and comparing the tooth form with a master on a comparator, checking hardness, measuring with appropriate tools, and inspecting visually with a glass for defects on the surface. Figure 17.12 is a photograph of a shadow of a 10-tooth drawn pinion as it appears on a comparator window.

FIGURE 17.12 View of 10-tooth drawn pinion as it appears in comparator window. *(Courtesy of Tekniks Engineering Co., Elmhurst, Illinois)*

17.7 GEAR TEETH AND WORM THREAD ROLLING

This process, as the name implies, consists of passing bars through a set of rolls that have shapes that will produce the worm thread or gear tooth desired. The rolling process may also be a finishing operation[6] that is somewhat like shaving. In this case the parts are usually helical gears. When the entire tooth is "formed" by the rolling process, the part is usually a worm. Rolling the material between dies leaves the tooth surface smooth and burnished, free from chatter or cutter marks, and increases the hardness of the material. When a tooth is rolled, the material is reformed in a continuous pattern of the shape of the roll, and no waste is incurred because the material is displaced into the shape. The original bar is elongated by the amount of metal displaced. In practice, it is usual to make the diameter that is to be rolled somewhat smaller than the finished outside diameter. The displaced material fills the areas of increased diameter.

Any ductile material can be rolled, but straight-carbon steels, high-speed steels, nonleaded brass, and copper constitute the majority of materials used in the production of rolled worms. Many variations in rolling characteristics and finishes are encountered with the different materials. Rolled worms are used in many applications, from toys and motion picture projectors to thrust reversers in aircraft engines. In addition to worms, splines and spur gear teeth can be rolled. The automotive industry, for example, has used rolled splines extensively in vehicles. The best results for rolled splines come with 30 or 45° pressure angles.

17.7.1 Gear Tooth and Worm Thread Rolling Process

This process is used when there is mass production quantity. It is generally accepted that, if a part can be rolled as designed or redesigned slightly to be rolled, it will cost less than other methods. The resulting part will have superior strength and finish because of the material-displacement process.

Once a decision has been made to roll a part, consideration must be given to the design of the gear blank. The depth of the thread or tooth should be kept to about 18 percent of the major diameter. Greater depths are difficult to make maintaining straightness of the blank. When there are multiple threads or helical gears with several overlaps, this amount can be higher. Worm threads should not be specified between shoulders, and the runout of the worm onto the main shaft should be equal to at least two threads. The blanks should be accurate because the finished tooth accuracy depends on the blank. Since the material is designed to pass through two or more rolls that are closed together, a large blank will not allow the rolls to close properly, and the part will be oversized; also, a small blank will not fill the void area of the roll when closed, and the resulting part will be undersized. If good accuracy is desired, it is sometimes necessary to have the blank pass through a secondary operation for size control. Exact blank diameters may be found from the manufacturer; however, an approximation can be made by subtracting one whole depth of tooth from the finished diameter of the part.

Straight-carbon steels, structural alloy steels, nonleaded brasses, and copper constitute the majority of rolled toothed elements. Many variations in characteristics and finishes are obtained, and final selection usually depends on a compromise of "rollability" and machinability. Successful rolling on worms has been obtained for pitches from 48 to 8. Production rates can be quite high. The rate varies according to the material and accuracy that must be obtained.

17.8 GEAR STAMPING

The gear stamping process is unlike the other processes discussed in this chapter in that metal removal is involved. In the case of stamping, a sheet of metal is placed between the top and bottom portions of a die; the upper die is pressed into the lower section and "removes" the gear from the sheet. This is a low-cost method for producing lightweight gears for no-load to medium-duty applications.

A typical set of dies is shown in Figs. 17.13 and 17.14; an enlarged view of the punch section is shown in Fig. 17.15. The stock for punching gears usually comes in rolls, although 12-ft-long strips are occasionally used. A wide choice of materials can be used, including all the low- and medium-carbon steels; all brass alloys, which incidentally give the best finish; half- and full-hard aluminum alloys; and nonmetallic materials. When high-carbon and stainless steels are used, the material for the die and the construction are special, and each application should be considered individually.

Gears manufactured by this process have a wide variety of applications, among which are toys; clock and timer mechanisms; watches; water and electric meters; small appliances such as mixers, irons, blenders, toasters, can openers; dryers; washers; and similar devices.

FIGURE 17.13 Die for gear stamping. *(Courtesy of Sunbeam Corporation, Chicago, Illinois)*

FIGURE 17.14 Die for gear stamping. *(Courtesy of Sunbeam Corporation, Chicago, Illinois)*

GEARS MADE BY DIES   17.19

FIGURE 17.15 Enlarged view of punch section of stamping die. *(Courtesy of Sunbeam Corporation, Chicago, Illinois)*

FIGURE 17.16 Example of drawdown from stamping process. 50-tooth, 64-pitch, 0.020-in. thickness, blanked two at a time, 150 strokes per minute, No. 1010 steel. *(Courtesy of Sunbeam Corporation, Chicago, Illinois)*

17.8.1 Gear Stamping Process

This process requires the use of a press equipped with a set of tools that are the stamping dies. These dies have the form of the toothed gear element built in and will produce parts in the exact shape of the die. Aspects of the tooth form that must be considered in the design of the die are backlash, whole depth, and corrected tooth thickness. Additionally, to compensate for tool wear, it is usually good practice to provide an extra amount to the whole depth. This is done so that when the punch wears the tooth will not become shallow. The amount needed is not as much as for extrusion dies, since they wear less. In preparing the tooth form, the thickness at the pitch diameter is usually reduced by a standard amount. The recommended reduction in tooth thickness for stamped gears is shown in Table 17.4.

The stock for producing gears is usually passed through a device called a roll

TABLE 17.4 Recommended Tooth Thickness Reductions for Stamped Gears

Pitch Range	Reduction of $1.5708/DP$, In.
150–180	-0.0008
121–149	-0.0012
100–120	-0.0015
65–99	-0.0020
20–64	-0.0025

straightener, which irons out the stock and makes it flat. It then passes through the dies. As the top portion of the die comes into contact with the stock, the punches can enter, and a shearing action takes place, which produces a part to the shape of the punch that entered the die. This shearing action causes the teeth and outside periphery to be somewhat rough and rounded at one edge. This rounding is called drawdown, and a typical example is shown in Fig. 17.16. The shearing and rounding characteristic increases as the stock thickness increases, and thicknesses over $\frac{1}{16}$ in. are not recommended. If punched gears are desired with less drawdown and roughness, an additional operation can be performed, called shaving. This does exactly what the name implies in that an additional die station is provided with closer clearance, and the gear is finely trimmed or shaved to remove some of the drawdown and roughness.

Typical gears produced at a high rate from commercial quality dies are shown in Figs. 17.17 and 17.18. Generally there are no blank requirements on a stamped gear. In this case, a compound die will form the bore and teeth in one stroke. This insures good concentricity between the teeth and the bore. In cases where a lancing or special tab forming is specified as part of the gear, progressive dies are usually employed. The tolerances of the part depend on the exactness of the pilots in each die station. Typical pilot pins were shown in Fig. 17.13.

Lightweight gears are usually made from AISI 1010 steel at about 250 BHN, leaded brass, full-hard aluminum, and sometimes higher carbon steels, without any special die design requirements. However, when high-carbon steels and stainless steels are used, special carbide inserts are necessary. Stamped gears range in size from 20 through 120 diametral pitch, 0.010 to 0.125 in. thick. As the pitch becomes finer, the material specification must become thinner. Table 17.5 shows recommended stock thicknesses for various pitches that are commonly used and require no special care in die maintenance. With special setups and/or shaving, the thicknesses in Table 17.5 can be modified; however, die wear and tolerance deviation will be greater.

FIGURE 17.17 A gear in high production made by stamping. 60-tooth, 0.025-in. thickness. *(Courtesy of Sunbeam Corporation, Chicago, Illinois)*

FIGURE 17.18 A gear in high production made by stamping. 50-tooth, 60-pitch, 0.046-in. thickness, stamped two at a time, 220 pieces per minute. *(Courtesy of Sunbeam Corporation, Chicago, Illinois)*

TABLE 17.5 Recommended Stock Thicknesses for Stamped Gears

Pitch Range	Thickness, In.
20–36	0.020–0.078
38–60	0.015–0.062
62–72	0.010–0.040
74–90	0.010–0.035
92–120	0.010–0.025

Tolerances for stamped gears are good and AGMA[7] quality class 9 can be achieved without difficulty or excess cost. The surface finish usually depends on the raw material used. In running stamped gears, a first-piece inspection is usually made when setting up the part for the first time. A few parts are run, and the press is stopped until these first few pieces have been checked. This may be done by comparing the part with the tooth form on a comparator, by rolling with a master gear, by measuring over wires, or all three, depending on the accuracy desired. If the gear is satisfactory, the press will then be started. High-speed presses, operating at speeds of 200 strokes per minute, used in conjunction with a die such as the one shown in Figs. 17.13 and 17.14 (which produces 2 gears per stroke), will produce 400 gears per minute. A check should be made about every two hours to ensure that the die does not break a tooth or that excessive burrs are not developing because of some imperfections in the material or that some foreign matter has not embedded itself in the die portion.

REFERENCES

1. AGMA 141.01, "Plastics Gearing—Molded, Machined, and Other Methods," American Gear Manufacturers Association, Alexandria, Virginia, August 1984.
2. AGMA 200–A88, "Gear Classification and Inspection Handbook," American Gear Manufacturers Association, Alexandria, Virginia, March 1988.
3. Mckinlay, W., and Pierson, S., *Plastics Gearing Technical Manual,* ABA/PGT Publishing, Manchester, Connecticut, 1976.
4. Dudley, D. W., *Gear Handbook,* McGraw-Hill, New York, 1962.
5. Drago, R. J. and Lenski, J. W., "Advanced Rotorcraft Transmission Program—A Status Review," AGMA Technical Paper, FTM7, 1990.
6. Dugas, J. P., "Gear Roll-Finishing," *Gear Technology Magazine,* May–June 1987.
7. Kuhlmann, D. J., and Raghupathi, P. S., "Manufacturing of Forged and Extruded Gears," *Gear Technology Magazine,* July–August 1990.

CHAPTER 18
GEAR FINISHING BY SHAVING, ROLLING, AND HONING

John P. Dugas
National Broach and Machine Co.
Mt. Clemens, Michigan

There are several methods available for improving the quality of spur and helical gears following the standard roughing operations of hobbing or shaping. Rotary gear shaving and roll-finishing are done in the green or soft state prior to heat treating. These processes have the ability to modify the gear geometry to compensate for the distortions that occur during heat treatment. Gear honing is a particularly effective method of removing nicks and burrs from the active profiles of the teeth after heat treatment. Combined with its ability to improve surface finish and make minor form corrections, the honing process is rapidly being accepted as an operation through which many gears are processed following heat treatment.

18.1 THE ROTARY GEAR SHAVING PROCESS

Gear shaving is a free-cutting gear finishing operation that removes small amounts of metal from the working surfaces of gear teeth. Its purpose is to correct errors in index, helix angle, tooth profile, and eccentricity (Fig. 18.1). The process also improves tooth surface finish and eliminates, by crowned tooth forms, the danger of tooth end load concentrations in service. Shaving provides for profile modifications that reduce gear noise and increase a gear's load-carrying capacity, its factor of safety, and its service life. Gear finishing (shaving) is not to be confused with gear cutting (roughing). They are essentially different. Any machine designed primarily for one cannot be expected to do both with equal effectiveness or with equal economy.

Gear shaving is the logical remedy for the inaccuracies inherent in gear cutting. It is equally effective as a control for those troublesome distortions caused by heat treatment.

INVOLUTE PROFILE CHECKS

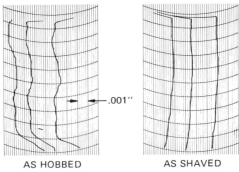

LEAD CHECKS

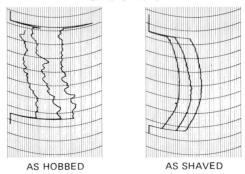

FIGURE 18.1 Improvement in profile and lead, 5.7 NPD 20°. NPA, 3.85 in. P.D., crowned shaved with stock removal of 0.011 in. over pins.

The form of the shaving cutter can be reground to make profile allowance for different heat-treatment movements due to varying heats of steel. The shaving machine can also be reset to make allowance for lead change in heat treatment.

Rotary gear shaving is a production process that utilizes a high-speed steel, hardened and ground, ultraprecision shaving cutter. The cutter is made in the form of a helical gear. It has gashes in the flanks of the teeth that act as the cutting edges.

The cutter is meshed with the work gear in crossed axes relationship (Fig. 18.2) and rotated in both directions during the work cycle while the center distance is reduced incrementally. Simultaneously, the work is traversed back and forth across the width of the cutter. The traverse path can be either parallel or diagonal to the work gear axis, depending on the type of work gear, the production rate, and finish requirements. The gear shaving process can be performed at high production rates. It removes material in the form of fine hair-like chips.

Machines are available to shave external spur and helical gears up to 5 m (200 in.) in diameter. Other machines are also available for shaving internal spur or helical gears.

FIGURE 18.2 Crossed axes meshing of shaving cutter and work gear.

For best results with shaving, the hardness of the gear teeth should not exceed 30 Rockwell C scale. If stock removal is kept to recommended limits and the gears are properly qualified, the shaving process will finish gear teeth in the 3.6- to 2.5-M (7- to 10-pitch) range to the following accuracies: involute profile, 0.005 mm (0.0002 in.); tooth-to-tooth spacing, 0.0075 mm (0.0003 in.); lead or parallelism, 0.005 mm (0.0002 in.).

In any event, it should be remembered that gear shaving can remove from 65 to 80 percent of the errors in the hobbed or shaped gear. It will make a good gear better. The quality of the shaved gear is dependent to a large degree on having good hobbed or shaped gear teeth.

Excellent surface finish is achieved with gear shaving. A value of approximately 25 μin is the normal finish achieved with production gear shaving, although much finer finishes are possible by slowing the process. In some cases, shaving cutters will finish up to 80,000 gears before they need sharpening. They may generally be sharpened from four to ten times.

To a gear designer, the shaving process offers attractive advantages in the ability to modify the tooth form. If a crowned tooth form or a tapered tooth form are desired to avoid end bearing conditions, these can be easily provided by shaving.

If modifications are desired in the involute profile, these can be made by suitable modifications in the ground cutter tooth form. If heat-treatment distortions can be controlled to a minimum, the most inexpensive way to produce an accurate, quiet, high-performance gear is to specify hobbing followed by gear shaving. The shaving process has a variety of standardized production equipment available ranging from hand loading to fully automatic loading and unloading.

18.2 BASIC PRINCIPLES

The rotary gear shaving process is based on fundamental principles. This process uses a gashed rotary cutter in the form of a helical gear having a helix angle different from that of the gear to be shaved (Fig. 18.3). The axes of cutter and gear are crossed at a predetermined angle during the shaving operation. When cutter and work gear are rotated in close mesh, the edge of each cutter gash, as it moves over the surface of a work gear tooth, shaves a fine, hair-like chip. The finer the cut, the less pressure is required between tool and work, eliminating the tendency to cold work the surface metal of the work gear teeth.

This process is performed in a shaving machine (Fig. 18.4), which has a motor-driven cutter head and a reciprocating work table. The cutter head is adjustable to obtain the desired crossed axes relationship with the work. The work carried between live centers is driven by the cutter. During the shaving cycle, the work is reciprocated parallel to its axis across the face of the cutter and up-fed an increment into the cutter with each stroke of the table. This shaving cycle (conventional) is one of several methods.

18.2.1 The Crossed Axis Principle

To visualize the crossed axis principle, consider two parallel cylinders of the same length and diameter (Fig. 18.5). When brought together under pressure, their common contact surface is a rectangle having a length of a cylinder and width that varies with contact pressure and cylinder diameter.

FIGURE 18.3 Assortment of rotary gear shaving cutters.

FIGURE 18.4 12-in. rotary gear shaving machine.

When one of these cylinders is swung around so that the angle between its axis and that of the other cylinder is increased up to 90°, their common plane remains a parallelogram, but its area steadily decreases as the axial angle increases. The same conditions prevail when, instead of the two plain cylinders, a shaving cutter and a work gear are meshed together. When the angle between their axes is from 10 to 15°, tooth surface contact is reduced and pressure required for cutting is small. As the work gear is moved away axially from the point of intersection of the axes, backlash develops. Conversely, as it is returned to the point of axial intersection, backlash decreases until the two members engage in tight mesh with the teeth of the cutter wedging between those of the work gear. Thus, each succeeding cutting edge sinks deeper into the work gear tooth until the point of axial intersection is reached.

For shaving, the cutter and work gear axes are crossed at an angle usually in the range of 10 to 15° or approximately equal to the difference in their helix angles.

Crossing of the axes produces reasonably

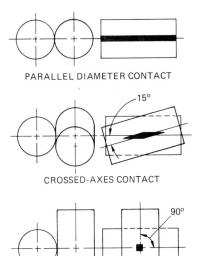

FIGURE 18.5 Contact between cylinders changes as crossed axes are varied.

uniform diagonal sliding action from the tips of the teeth to the roots. This not only compensates for the nonuniform involute action typical of gears in mesh on parallel axes, but also provides the necessary shearing action for stock removal.

18.2.2 Relationship Between Cutting and Guiding Action

Increasing the angle between cutter and work axes increases cutting action, but, as this reduces the width of the contact zone, guiding action is sacrificed. Conversely, guiding action can be increased by reducing the angle of crossed axes but at the expense of cutting action.

18.2.3 Preparation Prior to Shaving

The first consideration in manufacturing a gear is to select the locating surfaces and use them throughout the process sequence. Close relationship between the locating surface and the face of the gear itself must be held. Otherwise, when the teeth are cut and finished with tooling that necessarily contacts the gear faces, the teeth will be in an improper relationship with the locating or related surface on which the gear operates. Gears that locate on round diameters, or spline teeth must fit the work arbors closely, or these critical hole-to-face relationships will be destroyed.

Typical manufacturing tolerances for gear blanks prior to cutting of the teeth are shown in Table 18.1.

Once the gear blank has been manufactured, it is necessary to cut the gear teeth. The most common methods today for rough-cutting gear teeth are hobbing and shaper cutting. Of primary concern to the shaving cutter manufacturer is the fillet produced by the roughing operation. The tips of the shaving cutter teeth must not contact the gear root fillet during the shaving operation. If such contact does occur, excessive wear of the cutter results and the accuracy of the involute profile is affected.

The shaving cutter just finishes the gear tooth below its active profile. Thus, the height of the fillet should not exceed the lowest point of contact between the shaving cutter teeth and the teeth on the work gear.

Protuberance-type hobs and shaper cutters are often used prior to shaving to produce a slight undercut or relief near the base of the gear tooth. This method assures a smooth blending of the shaved tooth profile and the unshaved tooth fillet, as well as

TABLE 18.1 Typical gear blank tolerances

Blank Dia. In.	Face Runout In.	Hole Size In.	Hole Taper In./In.	Hole Roundness In.-Max	O.D. In.-Max.	O.D. Runout In.
Up to 1, 1-in. Thick	0.0003-0.0005	0.0003-0.0006	0.0002-0.0003	0.0002-0.0003	0.003	0.003
1 to 4, up to 1-in. Thick	0.0004-0.0008	0.0005-0.001	0.0002-0.0003	0.0003-0.0005	0.005	0.005
4 to 8	0.0006-0.0012	0.0008-0.0012	0.0002-0.0003	0.0004-0.0006	0.005	0.007
8 to 12	0.001-0.002.	0.001-0.0015	0.0002-0.0003	0.0005-0.0007	0.005	0.008

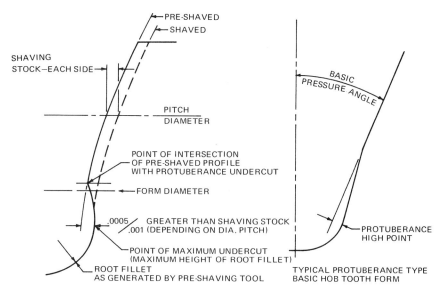

FIGURE 18.6 Undercut produced by protuberance hob and basic hob tooth form.

reduces shaving cutter tooth tip wear (Fig. 18.6). The amount of undercut produced by the protuberance-type tool should be made for the thin end of the tooth. The position of the undercut should be such that its upper margin meets the involute profile at a point below its contact diameter.

18.2.4 Shaving Stock

The amount of stock removed during the shaving process is a key to its successful application. Sufficient stock should be removed to permit correction of errors in the preshaved teeth. However, if too much stock is removed, cutter life and part accuracy are effectively reduced.

Table 18.2 shows the recommended amounts of stock to be removed during the shaving operations and the corresponding amount of undercut required.

18.2.5 Shaving Methods

There are four basic methods for rotary shaving of external spur and helical gears: (1) axial or conventional, (2) diagonal, (3) tangential or underpass, and (4) plunge. The principal difference among the various methods is the direction of reciprocation (traverse) of the work through and under the tool.

18.2.6 Axial or Conventional

Axial shaving is widely used in low- and medium-production operations (Fig. 18.7). It is the most economical method for shaving wide-face-width gears. In this method, the

TABLE 18.2 Recommended shaving stock and undercut for preshaved gears

Normal Diametral Pitch	Shaving Stock (In. per Side of Tooth)	Total Undercut (In. per Side of Tooth)
2 to 4	0.0015 to 0.0020	0.0025 to 0.0030
5 to 6	0.0012 to 0.0018	0.0023 to 0.0028
7 to 10	0.0010 to 0.0015	0.0015 to 0.0020
11 to 14	0.0008 to 0.0013	0.0012 to 0.0017
16 to 18	0.0005 to 0.0010	—
20 to 48	0.0003 to 0.0008	—
52 to 72	0.0001 to 0.0003	—

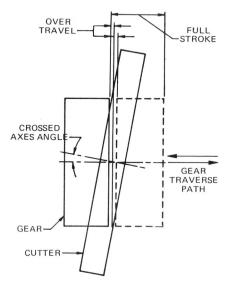

FIGURE 18.7 Axial shaving (conventional).

traverse path is along the axis of the work gear. The number of strokes may vary due to the amount of stock to be removed. The length of traverse is determined by the face width of the work. For best results, the length of traverse should be approximately 1.6 mm ($\frac{1}{16}$ in.) greater than the face width of the work, allowing minimum overtravel at each end of the work face. In axial shaving, in order to induce lead crown, it is necessary to rock the machine table by use of the built-in crowning mechanism (Fig. 18.8).

18.2.7 Diagonal

In diagonal shaving, the traverse path is at an angle to the gear axis (Fig. 18.9). Diagonal shaving is used primarily in medium- and high-production operations.

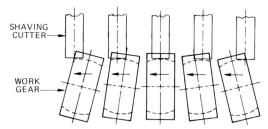

FIGURE 18.8 Rocking table action for crowning during conventional shaving.

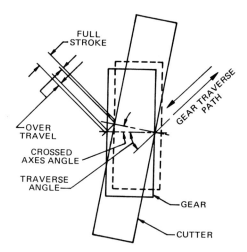

FIGURE 18.9 Diagonal shaving.

By use of this method, shaving times are reduced by as much as 50 percent. In diagonal shaving, the sum of the traverse angle and the crossed axes angle is limited to approximately 55°, unless differential-type serrations are used; otherwise, the serrations will track. Relative face width of the gear and the shaving cutter has an important relationship with the diagonal traverse angle. A wide-face-width work gear and a narrow shaving cutter restrict the diagonal traverse to a small angle. Increasing the cutter face width permits an increase in the diagonal angle. Crowning the gear teeth can be accomplished by rocking the machine table, provided the sum of the traverse angle and crossed axes angle does not exceed 55°. When using high diagonal angles, it is preferable to grind a reverse crown (hollow) in the lead of the shaving tool.

In most cases, the diagonal traverse angle will vary from 30 to 60° to obtain optimum conditions of cutting speed and work gear quality.

With diagonal traverse shaving, the centerline of crossed axes is not restricted to a single position on the cutter as it is in conventional shaving, but is migrated across the cutter face, evening out the wear. Consequently, cutter life is extended. Although conventional shaving requires a number of table strokes, each with its increment of upfeed, diagonal shaving of finer-pitch gears may be done in just two strokes with no upfeed and a fixed center distance between cutter and work. An automatic upfeed mechanism on the shaving machine materially enlarges the scope of diagonal shaving by making it available also for multistroke operations. This device feeds the work into the cutter in a series of small increments, synchronized with table reciprocation. Removing stock from the work gear in a series of small increments, instead of two large increments, further increases cutter life. It also makes the process feasible for gears requiring more stock removal than can be handled on a two-stroke cycle. When upfeed is completely automatic, there can be no danger of an error in selecting feed rates. Inasmuch as the cycle starts and stops in a position of maximum backlash, loading and unloading can be very fast.

18.2.8 Tangential or Underpass

In the tangential (underpass) method of shaving (Fig. 18.10), the traverse path of the work is perpendicular to its axis. Tangential shaving is used primarily in high-production operations and is ideally suited for shaving gears with restricting shoulders. When using this method, the serrations on the cutter must be of the differential type. Also, the face width of the cutter must be larger than that of the work gear.

18.2.9 Plunge

Plunge shaving is used in high production operations (Fig. 18.11). In this method, the work gear is fed into the shaving cutter with no table reciprocation. The shaving cutter

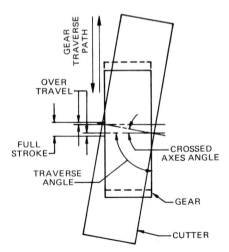

FIGURE 18.10 Tangential shaving (underpass).

must have the differential-type serrations or cutting action will be impaired. To obtain a crowned lead on the work, it is necessary to grind into the shaving cutter lead a reverse crown or hollow. In all cases of plunge shaving, the face width of the shaving tool must be greater than that of the work gear. The primary advantage of plunge shaving is a very short cycle time.

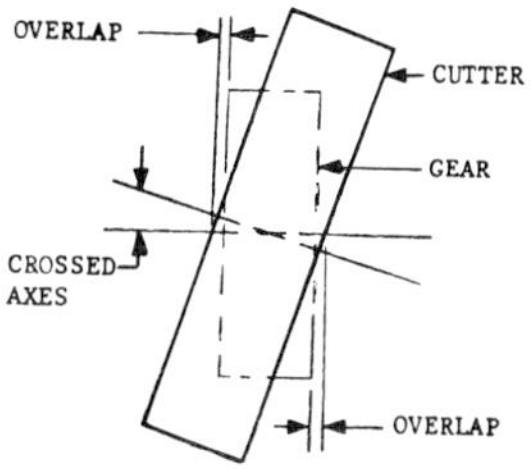

FIGURE 18.11 Plunge shaving.

18.2.10 Shaving Internal Gears

Internal gears can be shaved on special machines in which the work drives the cutter (Fig. 18.12), or by internal cutter head attachments on external shaving machines (Fig. 18.13).

Because of the crossed axes relationship between the cutter and the work gear in internal shaving, the cutter requires a slight amount of crown in the teeth to avoid interference with the work gear teeth. Crowning of the teeth on gears over 19 mm (¾ in.) wide is best achieved by a rocking action of the work head similar to the rocking table action with external gear shaving.

When internal gears are 19 mm (¾ in.) wide and under, or should interference limit the work reciprocation and crossed axes angle, a plunge shaving method can be applied. Here the cutter is provided with differential serrations and plunge-fed upward into the work. If lead crown is desired on the work gear, a reverse crowned cutter is used with the plunge feed shaving process.

FIGURE 18.12 Internal gear shaver where tool is driven by the work.

FIGURE 18.13 External shaver with internal cutter head attachment.

18.3 THE SHAVING CUTTER

Rotary shaving cutters are high-precision, hardened and ground, high-speed steel generating tools held to Class A and AA tolerances in all principle elements (Fig. 18.14). The gashes in the shaving cutter extend the full length of the tooth, terminating in a clearance space at the bottom. These clearance spaces provide unrestricted channels for a constant flow of coolant to promptly dispose of chips. They also permit uniform depth of serration penetration and increase cutter life.

The shaving cutter is rotated at high speeds up to 122 m (400 and more surface ft) per minute. Feed is fine and the tool contact zone is restricted. Cutter life depends on several factors: operating speed, feed, material and hardness of the work gear, its required tolerances, type of coolant, and the size ratio of cutter to work gear.

18.3.1 Design

Rotary gear shaving cutters are designed in much the same manner as other helical involute gears. The serrations on the tooth profiles, in conjunction with the crossing of the axes of the cutter and the work gear, make it a cutting tool. In designing rotary gear shaving cutters, the following are some of the points that must be considered:

1. Normal diametral pitch and normal pressure angle must be the same as those of the gears to be shaved.

2. Helix angle is chosen to give a desired crossed axis angle between the cutter and work. The crossed axis angle is the difference between the helix angles of the shaving cutter and work gear. The desired range is from 5 to 15°.

3. The number of teeth is chosen to give the appropriate pitch diameter required, considering helix angle and diametral pitch. Hunting tooth conditions and machine capacity are also important factors.

FIGURE 18.14 Variety of shaving cutters.

4. Tooth thickness of the cutter is selected to provide for optimum operating conditions throughout the life of the tool.
5. The addendum is always calculated so the shaving cutter will finish the gear profile slightly below the lowest point of contact with the mating gear. Tooth thickness and addendum of the cutter are not necessarily given to the theoretical pitch diameter.
6. Cutter serrations are lands and gashes in the involute profile of the tool. They extend from the top to the bottom of the tooth clearing into a relief hole at its base. The width or size is determined by the work gear to be shaved. Differential serrations with a control lead are produced on shaving cutters used for plunge shaving and diagonal with the traverse angle over 55°.
7. The involute profile of the shaving cutter tooth is not always a true involute. Very often, it must be modified to produce the desired involute form or modifications in the profile of the gears being shaved.

18.3.2 Sharpening Shaving Cutters

The shaving cutter, like other tools, dulls with use. In sharpening, minimum stock is removed on the tooth faces. With normal dullness, the resharpening operations usually reduce the tooth thickness approximately 0.74 mm (0.005 in.). An excessively dull or damaged tool must be ground until all traces of dullness or damage are removed.

The number of sharpenings varies with pitch and available depth of serrations. Usually a cutter can be sharpened until the depth of serrations has been reduced to approximately 0.15 to 0.30 mm (0.006 to 0.012 in.).

18.4 SHAVING MACHINES

Rotary gear shaving machines are manufactured in various configurations to meet the needs of the gear producing industry. Gears smaller than 25 mm (1 in.) and as large as 5.1 m (200 in.) require different approaches. Rotary gear shaving utilizes a shaving machine that has a motor-driven cutter and a reciprocating work table. The cutter head is adjustable to obtain the desired crossed axis relationship with the work. The work carried between centers is driven by the cutter. Machines are available ranging from mechanical (Fig. 18.15) to one CNC axis (Fig. 18.16) to full five CNC axes (Fig. 18.17).

During the shaving cycle, the work is reciprocated and fed incrementally into the cutter with each stroke of the table. The number of infeeds and strokes is dependent on the shaving method used and the amount of shaving stock to be removed.

FIGURE 18.15 Mechanical gear shaving machine.

FIGURE 18.16 One-axis CNC controlled shaving machine.

18.4.1 The Machine Setup

Mounting the Work Gear. The work gear should be shaved from the same locating points or surfaces used in the preshave operation. It should also be checked from these same surfaces. Locating faces must be clean, parallel, and square with the gear bore. Gears with splined bores may be located from the major diameter, pitch diameter, or minor diameter. When shaving from centers, the true center angle should be qualified and the surfaces should be free of nicks, scale, and burrs. Locating points of work arbors and fixtures should be held within a tolerance of 0.005 mm (0.0002 in.). The arbor should fit the gear hole snugly. Head and tailstock centers should run within 0.005 mm (0.0002 in.) for dependable results. Gears should be shaved from their own centers whenever possible. If this is not possible, rigid, hardened and ground arbors having large safety centers should be used (Fig. 18.18). Integral tooling is another popular method of holding the work piece, especially in high production. This consists of hardened and ground plugs, instead of centers, mounted on the head and tailstock (Fig. 18.19). These plugs are easily detached and replaced when necessary. They locate in

FIGURE 18.17 Five-axis CNC controlled shaving machine.

the bore and against the face of the gear. It is therefore essential that the gear faces be square and bore tolerances held to assure a good slip fit on the plugs.

Mounting the Cutter. Great care is required in handling the shaving cutter to avoid any accidental contacts between its teeth and other hard objects. The slightest bump may nick a tooth. Until the cutter is placed on its spindle it should lie flat and away from other objects. The cutter spindle and spacers should be thoroughly cleaned and the spindle checked before the cutter is mounted. The spindle should run within 0.005 mm (0.0002 in.) on the O.D. and 0.0025 mm (0.0001 in.) on the flange full indicator reading.

After mounting, the cutter face should be indicated to check mounting accuracy. Face runout should not exceed 0.02 mm (0.0008 in.) for a 30.5-cm (12-in.) cutter; 0.015 mm (0.0006-in.) for a 23-cm (9-in.) cutter; or 0.01 mm (0.0004 in.) for a 18-cm (7-in.) cutter.

Feeds and Speeds. Shaving cutter spindle speeds will vary with the gear material hardness, finish, and size of part. Normally, when using a 18-cm (7-in.) cutter on a 2.5-m (10-pitch) gear having a 7.6-cm (3-in.) pitch diameter, spindle speed will be approximately 200 r/min; or, using a 23-cm (9-in.) cutter, 160 r/min. This speed figured on the pitch circle is approximately 122 surface m (400 surface ft) per minute and this generally produces good results.

The following are formulas for determining cutter and gear speeds (r/min):

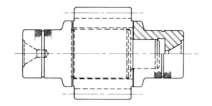

BASIC PLUG-CUP

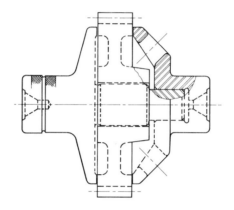

FLANGED PLUG-CUP

FIGURE 18.18 Typical hardened and ground work-holding arbors.

$$\text{Cutter r/min} = \frac{\text{desired surface speed per min}}{\text{cutter diameter} \times \pi}$$

$$\text{Gear r/min} = \frac{\text{cutter r/min} \times \text{number of teeth in cutter}}{\text{number of teeth in gear}}$$

For conventional shaving, about 0.25 mm (0.010 in.) per revolution of the gear is considered a good starting point and becomes a factor in the following formula:

Table feed rate [mm/min (in./min)] = 0.25 (0.010) × Gear r/min

For diagonal shaving, an "effective feed rate" of approximately 1.0 mm (0.040 in.) per revolution of gear is considered a good starting point. Effective feed rate of the speed at which the point of crossed axes migrates across the face of the gear and shaving cutter. The following is the formula for determining the table traverse rate (in./min) to produce a 1.0-mm (0.040-in.) effective feed rate:

$$\text{Table traverse rate [mm/min (in./min)]} = \frac{1.0 \text{ mm } (0.040 \text{ in.}) \times \text{Gear r/min}}{R_f}$$

where

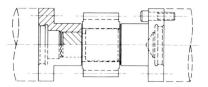

INTEGRAL PLUG-CUP

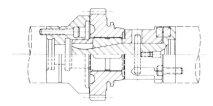

INTEGRAL EXPANDING

FIGURE 18.19 Integral work-holding arbors.

$$R_f = \frac{\text{sine traverse angle}}{\text{tangent crossed axes angle}} + \text{cosine traverse angle}$$

These suggested feed rates may be varied depending on individual operating conditions. If higher production is desired, the table feed rate can be increased, but this may result in some sacrifice of the quality of tooth finish. Where surface finish is very important, as with aviation and marine gears, table feeds are reduced below the amounts indicated. In some cases (notably, large tractor applications), feeds considerably in excess of those indicated are used.

18.5 GEAR ROLL-FINISHING

Gear roll-finishing is a fast and economical means of finishing the gear teeth of helical gears. Helical overlap is required in order to insure the smooth flow of material across the entire gear face.

Due to the speed at which gears are finished, this process is usually restricted to mass production facilities such as in the auto industry. It is not unusual for a set of rolling dies to produce over 1 million pinions before being reconditioned. Rolling dies can normally be reconditioned between three and five times before their end of life.

Gear rolling is a finishing operation requiring the teeth to be rough-cut by hobbing or shaper cutting.

In this discussion of gear roll-finishing, particular attention is called to the special tooth nomenclature resulting from the interaction between the rolling die teeth and the gear teeth. To eliminate confusion, the side of a gear tooth that is in contact with the "approach" side of a rolling die tooth is also considered to be the approach side. The same holds true for the "trail" side. Thus, the side of the gear tooth that is in contact with the trail side of a rolling die is also considered to be the trail side.

SHAVING, ROLLING, AND HONING

Gear roll-finishing is much different from gear shaving in that a flow of material is involved, rather than a removal of material. A study of gear tooth action is required to analyze the material flow in the rolling process. In Fig. 18.20 it can be seen that as a gear rolling die tooth engages the approach side of a work piece tooth, sliding action occurs along the line of action in the arc of approach in a direction from the top of the gear tooth toward the pitch point where instantaneous rolling action is achieved. As soon as the contact leaves the pitch point, sliding action occurs again, but in the opposite direction toward the pitch point in the arc of recession.

What is more interesting, however, is that the contact between the die and work gear teeth on the trail side produces exactly the opposite direction of sliding to that on the approach side (Fig. 18.21). The result of these changing directions of sliding is that material is being compressed toward the pitch point on the approach side and extended away from the pitch point on the trail side (Fig. 18.22).

This action causes a greater quantity of material to be displaced on the trail side than on the approach side by a ratio of about 3:1. On the approach side, the tendency is to trap the material rather than permit it to flow toward the top and root of the teeth as on the trail side. Thus, completely different from what occurs in a metal removal process such as gear shaving, the amount of material to be flowed during the rolling process, as well as the hardness of that material, have a significant effect on the accuracy of the produced form.

For successful roll-finishing, it appears that an undercut is desirable near the root

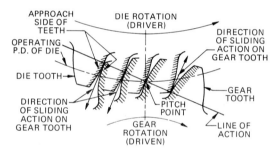

FIGURE 18.20 Contact action between one tooth of a work piece and the approach side of a rolling die tooth.

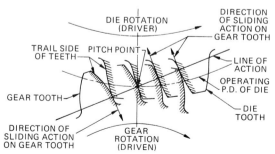

FIGURE 18.21 Contact action between the tooth of a work piece and the trail side of a rolling die tooth.

18.20 CHAPTER EIGHTEEN

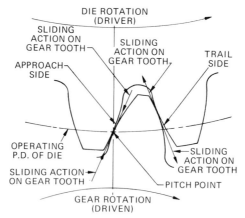

FIGURE 18.22 Differing flow directions induced by each side of a die tooth with gear roll-finishing.

section such as with conventional preshave tooth forms. Since most production gears are also provided with a tip chamfer, the material will tend to be pulled up into the chamfer on the trail side and down away from the chamfer on the approach side.

As a result, some adjustment in hobbed tooth tip chamfer depths and angle are

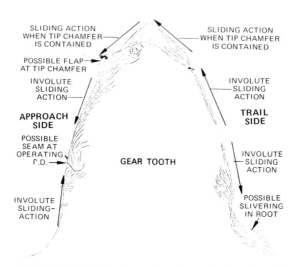

FIGURE 18.23 Tooth flow pattern that results when too much stock is left for roll-finishing, or when material hardness is excessive.

As a result, some adjustment in hobbed tooth tip chamfer depths and angle are required to balance out the opposed metal flow conditions on each tip side. These chamfer depths and angles have to be held to close tolerances. If too much stock is left for gear roll-finishing, or if the gear material tends to be too hard (above approximately 20Rc), several conditions may result. The sliding action on the approach side of the tooth may cause a "seaming" of material that builds up in the area of the pitch point. On the trail side, the flow of excess material may result in a burr on the tip of the gear tooth and a "slivering" of material into the root area. Figure 18.23 shows the condition of a roll-finished gear tooth when too much stock is flowed or high-hardness conditions are encountered.

In Fig. 18.24 photomicrographs show the conditions encountered when stock removal is excessive, material is excessive and material hardness is too high. A seam is evident in the approach side of the tooth at the left in the area of the operating pitch diameter. The trail side photomicrograph at the right in Fig. 18.24 shows slivering in the root portion with about 0.1 mm (0.004 in.) of lapped-over metal, and about 0.05-mm (0.002-in.) deep surface cold working of the material.

In contrast, photomicrographs in Fig. 18.25 show the excellent tooth structure that can be achieved with roll-finishing if stock reduction is held to a minimum and material

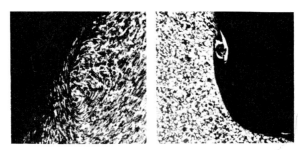

FIGURE 18.24 Photomicrographs of a gear tooth with high hardness. The approach side, left, has a seam in the area of the pitch diameter. The trial side, right, shows where excessive stock has caused cold working and a sliver near the root.

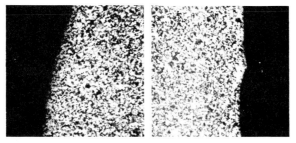

FIGURE 18.25 Photomicrographs of a properly roll-finished gear tooth. The approach side (left) has no seaming. The trail side (right) shows no slivering or cold working.

is not too hard. No evidence of cold working or seaming is seen in the approach side at the left. In the trail side at the right in Fig. 18.25, no evidence of slivering or cold working is seen.

The amount of stock reduction with roll-forming should be held to about one-half that normally associated with shaving if seaming and slivering are to be avoided. The burr condition on the tip of the trail side of the tooth can be improved by close control of the angle and location of the protective tooth chamfer generated by the hob in the tooth-generating operation.

18.5.1 Gear Rolling Dies

Since roll-finishing involves material flow rather than metal removal, it should be expected that the tooth form on the die would not be faithfully reproduced on the work piece tooth due to minute material springback and material flow conditions.

Even with gear shaving, it has been found necessary to modify the shaving cutter teeth profiles somewhat to produce a desired form on the work gear teeth. Experience to date has shown that a different type of tooth form modification is required for gear roll dies than for gear shaving cutters. The correct amount of gear rolling die tooth form modification is determined, as with gear shaving cutters, from an extensive development program. Less rigid gear roll-finishing machines usually require greater and varying die form modifications.

Gear roll dies (Fig. 18.26) are made from special fatigue- and impact-resistant high-speed steel to the tolerances shown in Table 18.3.

FIGURE 18.26 Gear roll dies.

TABLE 18.3 Standard tolerances for rolling dies

Die Specification	Tolerance-In.
Involute Profile (True Involute Form)— Active Length, tiv	
Through 0.177-in. Working Depth	0.00015
0.178 Through 0.395-in. Working Depth	0.00020
Lead—(Uniformity-tiv Per Inch of Face)	0.0003
Parallelism— (Opposite Sides of Same Tooth Alike Within)	0.0002
Helix Angle— (Deviation From True Angle—Per Inch of Face)	0.0005
Tooth Spacing— (Adjacent Teeth at Pitch Diameter)	0.00015
Circular Pitch—(Variation-tiv)	0.0002
Spacing Accumulation— (Over Three Consecutive Teeth)	0.00025
Runout—(tiv at Pitch Diameter)	0.0004
Face Runout—(tiv Below Teeth)	0.0002
Tooth Thickness	Minus 0.0010
Hole Diameter	Plus 0.0002

Note: Dies can be made in pairs alike within 0.0005-in. measured over pins, if necessary

18.5.2 Gear Rolling Machines

Several import design considerations have to be met in a roll-finishing machine. These include rigidity, strength, high-speed loading, die phasing, and independent adjustment for die axis and die positioning.

The force required to roll-finish a gear depends on its width, diametral pitch, tooth shape, cycle time, material, and hardness.

18.5.3 Double-Die Gear Rolling

The double-die gear roll machine shown in Fig. 18.27 is a vertical design with the dies mounted one above the other. It is designed to handle gears up to 10 cm (4 in.) wide and 15 cm (6 in.) in diameter. Die speeds are from 40 to 160 r/min. Dies up to 11.4 cm (4½ in. wide and 24.4 cm (9⅝ in.) in diameter can be mounted in the machine.

The upper die head is fixed and the lower die is fed upward by a hydraulic cylinder. The gears are fed into rolling position by an air-operated automatic loader (Fig. 18.28). Here they are picked up on a work arbor that is advanced by a hydraulic rotary actuator utilizing a gear rack arrangement. The work gears are advanced against a pneumatically loaded cup.

The work arbor is prerotated by a hydraulic motor at a speed slightly slower or faster than the die speed to ensure clash-free engagement. The lower die then feeds upward to a predetermined operating position to control finish-rolled gear size.

Table 18.4 illustrates the range of gearing for which gear rolling dies have been produced for finish-rolling production applications.

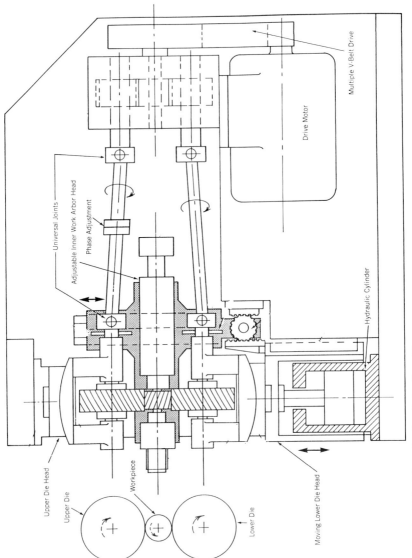

FIGURE 18.27 Schematic of vertical roll machine.

TABLE 18.4 Data on roll finished gears

No. Teeth	Pitch Diameter (in.)	Normal Diametral Pitch	Normal Pressure Angle	Helix Angle	Hand	Face Width (in.)	Material
26	4.6666	6.539	18° 28'	23° 25'	L	1.380	8620
25	3.3667	8.8709783	16° 30'	33° 10'	L	0.918	8620
14	1.0711	14	20°	21°	L	0.727	5140H
17	1.2143	15.1535	18° 35' 09"	22° 30'	L	0.758	4024
28	2.0000	15.1535	18° 35' 09"	22° 30'	R	3.04	4024
18	1.2542	15.5	17° 30'	22° 11' 30"	R	1.935	4620
16	0.9621	18	18° 30'	22° 30'	R	0.728	5130, Fine Grain (5-8)
34	2.0445	18	18° 30'	22° 30'	L	0.860	5130, Fine Grain (5-8)
20	1.1580	18.5	18°	21°	R	0.874	4027H
19	1.0549	19.3	20°	21° 03' 42"	R	0.705	4027H

FIGURE 18.28 Hinge-type auto loader on gear roll machine.

18.5.4 Single-Die Gear Rolling

Machines have been developed to finish-roll gears with a single die. This process has proven economical in low and medium production.

A single gear rolling die is mounted in a heavy-duty gear head above the work piece (Fig. 18.29). The die is driven by an electric motor to provide rotation of the work piece that meshes with it. Normally semiautomatic loading methods are utilized on single-die roll-finishing machines whose work cycles are somewhat longer than those of the fully automatic, double-die machines. The work piece is mounted on an arbor between head and tailstock (Fig. 18.30). In operation, the table supporting the head and tailstock is fed upward by a unique, air-powered, heavy-duty radial feed system. The continuous upfeed of the table provides the large force necessary to roll-finish the gear teeth.

During the work cycle, the work piece can be rotated in one direction for one part of the cycle, then reversed and rotated in the other direction for the balance of the cycle. This double rotation sequence tends to balance the metal flow action on the approach and trail sides of the work gear teeth.

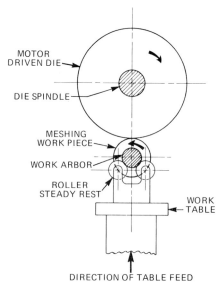

FIGURE 18.29 Operating principle of single-die gear roll finishing.

Tooth thickness size of the work piece is controlled by adjusting the height of the table with a handwheel-controlled elevating screw.

18.6 ROTARY GEAR HONING

Rotary gear honing is a hard gear finishing process that was developed to improve the sound characteristics of hardened gears by:

1. Removing nicks and burrs
2. Improving surface finish
3. Making minor corrections in tooth irregularities caused by heat-treat distortion

The process was originally developed to remove nicks and burrs that are often unavoidably encountered in production gears because of careless handling. Further development work with the process has shown that minor corrections in tooth irregularities, and surface finish quality improvement can be achieved. These latter

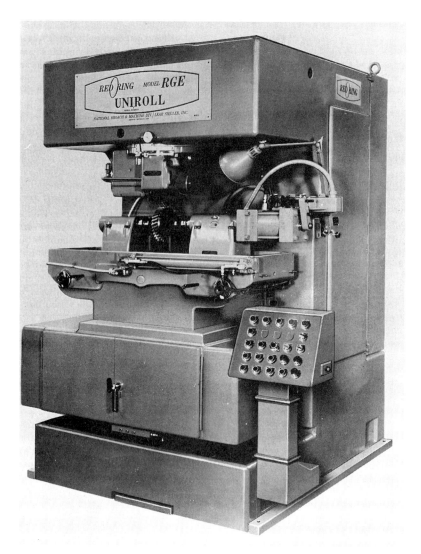

FIGURE 18.30 Single-die gear roll machine.

SHAVING, ROLLING, AND HONING   **18.29**

improvements can add significantly to the wear life and sound qualities of both shaved and ground hardened gears.

Gear honing does not raise tooth surface temperature, nor does it produce heat cracks or burned spots or reduce skin hardness. It does not cold work or alter the microstructure of the gear material, nor does it generate internal stresses.

Honing machines are available for external (Fig. 18.31) and internal (Fig. 18.32) spur and helical gears. Both taper and crown honing operations can be carried out on these machines.

18.6.1 How the Process Works

The process uses an abrasive-impregnated, helical-gear-shaped tool. This tool is generally run in tight mesh with the hardened work gear in crossed axes relationship under low, controlled center-distance pressure.

The work gear is normally driven by the honing tool at speeds of approximately 183 surface m (600 surface ft) per minute. During the work cycle, the work gear is traversed back and forth in a path parallel to the work gear axis. The work gear is rotated in both directions during the honing cycle. The process is carried out with conventional honing oil as a coolant.

The honing tool is a throw-away type that is discarded at the end of its useful life. The teeth are thinned as the tool wears. This tooth thickness reduction can continue

FIGURE 18.31 External gear honer.

FIGURE 18.32 Internal gear honer.

until root or fillet interference occurs with the work gear. Then the O.D. of the hone can be reduced to provide proper clearance.

Eventually, thinning of the hone teeth also results in root interference with the outside diameter of the work gear. When this condition occurs, the hone is generally considered to be at the end of its useful life. In some isolated cases, it has been found practical to recut the hone root diameter with a grinding wheel to provide additional hone life.

Usually the amount of stock removed from the gear tooth by honing ranges from 0.013 to 0.05 mm (0.0005 to 0.002 in.) measured over pins.

The production rate at which honing operations can be carried out depends on the pitch diameter and face width of the work. A gear 2.5 cm (1 in.) in diameter by 2.5 cm (1 in.) in width can be honed in approximately 15s. A gear 61 cm (24 in.) in diameter by 7.6 cm (3 in.) in face width will require approximately 10 min honing time. Of course, honing of salvage gears required longer cycles.

A typical external gear honing machine has the motor-driven honing tool mounted at the rear of the work spindle. The work spindle is mounted on a tilting table that can be positioned to provide four selective modes of operation.

The first mode is called loose backlash, where the hone and work gear are positioned in loose backlash operation on a fixed center distance. This method is sometimes utilized to slightly improve surface finish only, primarily on fine-pitch gears with minimum stock removal.

The second mode of operation is called zero backlash. Here the work gear is positioned in tight mesh with the honing tool. The table is locked in fixed center-distance location with a preselected hone pressure. This method is sometimes used to provide maximum gear tooth runout correction with minimum stock removal.

The third and most generally applied mode of operation is called constant pressure. The work gear is held in mesh with the honing tool at a constant pressure. This method removes nicks and burrs and provides maximum surface finish improvement in minimum time.

The fourth mode of operation is called differential pressure. A preselected low pressure is present between the hone and the low point of an eccentric gear; and a preselected increased amount of pressure is present between the hone and the high point of eccentricity. This method has all of the desirable features of the constant pressure method plus the ability to slightly correct eccentricity. The amount of eccentricity in the gears with differential pressure honing may cause the hone to wear faster than with the constant pressure method.

18.6.2 Rotary Gear Honing Tools

Standard honing tools are a mixture of plastic resins and abrasive grains such as silicon carbide, which are formed in a precision mold. They are made in a wide variety of mix numbers with grits ranging from 60 to 500, to suit special production and part requirements.

Special tools have been developed to do salvage-type honing. The tools are made from hardened steel and the active tooth surface is plated with carbide or diamond. These harder materials give the process the ability to remove an increased amount of stock and thereby make larger corrections in tooth irregularities.

18.6.3 Honing Shaved Gears

Traditionally, tooth surface finishes in the range from 15 to 40 μin. have been provided by the rotary gear shaving operation. The honing process, because it is not basically a heavy stock removal or tooth correction process, cannot substitute for gear shaving, which is performed on the soft gear. In fact, the tendency of a hone to charge a gear under 40 Rockwell C hardness with abrasive particles makes honing of soft gears a questionable application.

However, because a gear has to be heat treated, a process that usually roughens the tooth surface to a degree, the honing process tends to restore the hardened tooth surface finish to its original shaved condition and actually improves it. In all cases, the honed surface finish is better than the surface finish before honing (Fig. 18.33).

To hone production gears, economy dictates that one grit of tool and a relatively short honing cycle be used. What is produced then, in the way of surface finish, represents a compromise. First, the honing tool must remove nicks and burrs, then it should make minor tooth corrections that will improve sound level and wear life. The improvement in surface finish, which is in reality a by-product of the honing process, is a valuable adjunct that will help promote long wear life as well as improve sound characteristics.

18.6.4 Honing Ground Gears

In the aerospace industry, gears are traditionally operated at high speeds under heavy loads. They are usually cut, heat treated, and ground to provide tooth surfaces (usually

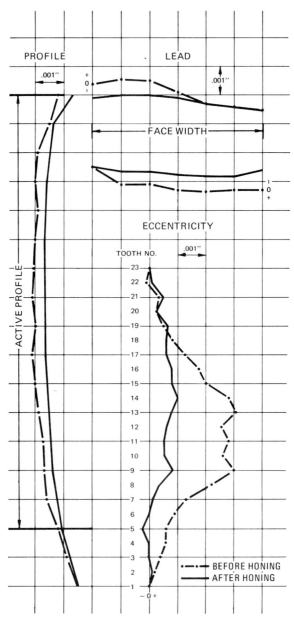

FIGURE 18.33 Improvement in tooth accuracy achieved by honing.

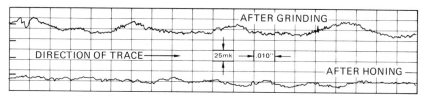

FIGURE 18.34 Surface finish improvement after honing.

of sophisticated modified forms) of the highest order of accuracy. However, tests with exotic surface measuring equipment have shown that ground surfaces have a jagged, wavy profile that will not support heavy loads or wear long unless costly break-in procedures are carried out.

Ground tooth surfaces usually have a surface finish in the 16- to 32-μin range. Honing with type AA honing tools can bring this surface finish down to the 8- to 10-μin range (Fig. 18.34). In one 39-tooth, 5-m (5-D.P.), 20° P.A., 20-cm (7.800-in.) P.D. spur helicopter drive gear, honing of the gear teeth down to 8-μin surface finish increased wear life by 1000 percent and increased load-carrying capacity by 30 percent. Other tests by the gearing industry have shown 100-percent load-carrying capacity increases by honing ground gears.

CHAPTER 19
GRINDING OF SPUR AND HELICAL GEARS

Dr. Suren B. Rao
Vice President, Engineering
National Broach and Machine Co.
Mt. Clemens, Michigan

Grinding is a technique of finish-machining, utilizing an abrasive wheel. The rotating abrasive wheel, which is generally of special shape or form, when made to bear against a cylindrical-shaped work piece, under a set of specific geometrical relationships, will produce a precision spur or helical gear. In most instances the work piece will already have gear teeth cut on it by a primary process such as hobbing or shaping. There are essentially two techniques for grinding gears: form and generation. The basic principles of these techniques, with their advantages and disadvantages, are presented in this section.

A general introduction to the basic principles of grinding process, however, precedes the discussion of gear grinding techniques. This is based on the belief that the discussion of the grinding process, in combination with the description of the gear grinding techniques, would constitute a more complete treatise on gear grinding.

19.1 REASONS FOR GRINDING

There are two primary reasons for grinding gears. In spite of several attempts to the contrary, it still remains one of the most viable techniques of machining gears once they are in a hardened state (50 Rc and above). Also, the process, in combination with highly accurate machines, is capable of gear manufacturing accuracy unmatched by other manufacturing techniques. AGMA gear quality 12 and 13 are common, and AGMA gear quality 14 and 15 are not unusual. With the advent of *cubic boron nitride* (CBN), grinding has been tried, with some success, as a primary operation on hardened material instead of hobbing or shaping before heat treatment (sometimes referred to as "direct grinding"). Obviously this is of greatest consequence only if a gear is being made from through-hardened material, a process that is not very common. It is much more common for gears to be made of case-hardened material where the economics of grinding from the solid are not as beneficial.

Gear grinding is an expensive operation and has to be justified on the basis of required gear quality in the hardened condition. The basic principles of grinding are now presented.

19.2 GRINDING PROCESS MECHANICS AND PROCESS PARAMETERS

Grinding is a metal cutting process, not unlike single- or multipoint machining such as turning, milling, hobbing, etc., but with some major dissimilarities. Grinding is characterized by the fact that the cutting tool, in this case the grinding wheel, consists of a very large number of randomly oriented cutting edges machining small amounts of material, thus resulting in extremely fine chip thicknesses. While chip thicknesses of 20 μm (0.0008 in.) or more are common in operations like turning, and chip thicknesses of 8 μm (0.0003 in.) are common in operations like milling; chip thicknesses of less than 1 μm (0.00004 in.) are the norm in grinding.

Though the abrasive particles in the grinding wheel are randomly oriented, by virtue of their shape, they generally present a large negative rake angle to the cutting velocity vector as seen in Fig. 19.1. Negative rake angles always result in higher cutting forces than do positive rake angles. Also, small chip thicknesses result in higher specific cutting forces, where *specific cutting force* is defined as the force required to cut a unit area of chip cross section (kgf/mm^2 or lbf./in.^{2}). The combination of negative rake and

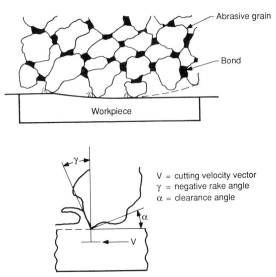

FIGURE 19.1 Grinding process schematic showing negative rake angle.

low chip thickness gives rise to high specific power requirements in grinding. Specific power is defined as power required to machine unit quantity of material in unit time (hp/mm^3/min or hp/in.^{3}/min). This is not only indicative of the low efficiency of the grinding process but also its high susceptibility to burning damage, as all the power consumed by the operation is converted into heat. The small chip thickness, however, also enables the generation of a high-quality surface and tight dimensional tolerances that make this process critical to the manufacture of high-precision gears and components.

The combination of large negative rake angle on the abrasive grain and small chip thickness also results in cutting process stiffnesses in grinding that are almost several times the cutting process stiffnesses in other machining processes such as turning and milling. *Cutting process stiffness* here is defined as the force per unit chip thickness and is generally expressed in kgf/mm^2 (lbf/in.^{2}) of chip thickness. Since the rate of reproducibility of error due to a machining process is given by the formula:

$$\delta = \frac{\mu}{1 + \mu}$$

where δ is the rate of reproducibility and μ is expressed by the formula:

$$\mu = \frac{R}{K}$$

where K is the stiffness of the machine tool and R is the cutting process stiffness.

In a typical turning operation K is several times R, and δ typically computes to less than 0.25. This signifies that only 25 percent of the initial work piece error will remain after the first turning pass. In grinding, however, since R is extremely large compared with K, values of $\delta = 0.95$ are not uncommon. This signifies that in grinding almost 95 percent of the initial work piece error may remain after the first grinding pass. It is obvious from this discussion that careful execution of all previous processes to ensure a good preground gear is essential to an economic and successful grinding operation.

Wear of the grinding wheel is an essential part of the process. As the sharp cutting edges of the abrasives wear out, cutting forces on that particular abrasive increase until either the grain fractures reveal new sharp cutting edges or the abrasive is pulled out of the bond and a new abrasive grain is exposed. In essence, if the process were in perfect harmony, the grinding wheel would be self-sharpening. But even though wheel wear is an accepted phenomenon, the rate at which it occurs is critical. For wheel wear results, not only in wheel replacement, but also in other nonproductive wheel preparatory operations such as trueing, dressing, and profiling. Therefore, the ratio of work material removed to volume of wheel lost, also called the G ratio, is a measure of grinding efficiency G ratios can range from less than one to several hundred depending on these variables.

One other distinguishing feature of the grinding process is the fractional amount of time the abrasive grain is actually creating a chip, in comparison with the total time this grain is in contact with the work piece material. Three distinct phenomena have been recognized as occurring as the abrasive grain comes in contact with and leaves contact with the work surface. These three regions have been defined as rubbing, plowing, and cutting, and the actual cutting or chip formation may be occurring for only about 30 percent of the time the abrasive and the work piece are in contact. The force at which the transition occurs from rubbing and plowing to cutting is called the *threshold force*. When the mechanisms holding the grinding wheel and the work exert a force in excess of this threshold force, grinding and metal removal will occur.

Finally, here is a word about spark-out. As a grinding process proceeds with the rotating grinding wheel being fed into the work material, cutting forces are generated. These cutting forces cause the electromechanical structure that holds the grinding wheel and the work piece to deflect away from each other. At a certain instant in the process, the infeed of the wheel and/or the work piece is stopped, resulting in a reduction of the cutting forces. This causes the strain energy stored in the structure to overcome the deflection and return the system to a state of equilibrium. As this happens, the wheel and work piece move into each other and continue to grind as the forces decay to the threshold force level, after which no more grinding occurs. This part of the grinding process, where no infeed occurs but grinding continues, is called *spark-out*. The time taken to complete spark-out is a measure of the stiffness of the structure of the machine tool–tool–work-piece system. In general, some amount of spark-out in a grind cycle will generally improve work piece quality.

19.3 ABRASIVES

Though aluminum oxide (Al_2O_3), silicon carbide (SiC), diamond (C), and cubic boron nitride (CBN) are generally considered in the category of abrasives where grinding is concerned; only aluminum oxide and cubic boron nitride are discussed further. This is because in gear grinding we are usually dealing with ferrous alloys, and diamond and silicon carbide tend to perform poorly when grinding steel. High wear rates of the diamond or silicon carbide abrasive when grinding may be due to interatomic diffusion of the carbon atoms present in these two abrasives, since steel is characterized as "carbon hungry" at the elevated temperatures that are encountered during grinding.

The characteristics of aluminum oxide and cubic boron nitride that impact their performance as abrasives are now presented in a comparative manner. It is obvious from the following that cubic boron nitride is a considerably superior abrasive, though more expensive than aluminum oxide.

19.3.1 Hardness

Figure 19.2 shows a comparative plot of diamond, cubic boron nitride, silicon carbide, and aluminum oxide hardness at elevated temperatures. It is obvious that cubic boron nitride is several times harder than aluminum oxide and even harder than diamond at temperatures higher than 1472°F (800°C). The chemical inertness of cubic boron nitride is also of significance since any chemical affinity to iron would result in increased wear rates.

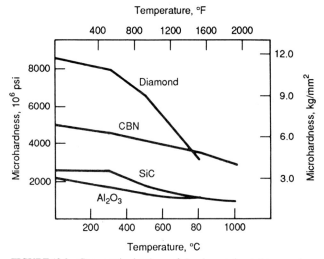

FIGURE 19.2 Comparative hardness of abrasives at elevated temperatures.

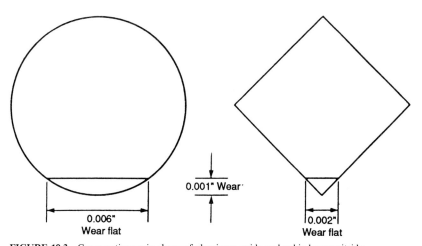

FIGURE 19.3 Comparative grain shape of aluminum oxide and cubic boron nitride.

19.3.2 Grain Shape

When comparing grain shapes of aluminum oxide and cubic boron nitride, the former is known to have a more pronounced spherical form while the latter has a block form. For a given amount of crystal wear, a spherical form exhibits a larger wear area than does a block form (Fig. 19.3). Tendency to burning has been related to wear flat area, indicating that higher degrees of burning and surface damage are possible with aluminum oxide than with cubic boron nitride.

19.3.3 Thermal Conductivity

Figure 19.4 shows a comparison of thermal conductivity of the various abrasives and some common metals. Though diamond has the highest thermal conductivity, cubic boron nitride is not far behind and considerably higher than aluminum oxide. The high thermal conductivity of cubic boron nitride allows more of the heat generated at the abrasive–work material interface to flow into the abrasive and into the wheel than into the work piece, resulting in reduced tendency for surface damage. It must be remembered, however, that it is possible to produce thermal damage with cubic boron nitride. The combination of high thermal conductivity and lower wear flat area owing to grain shape allows for much higher metal removal rates to be achieved and higher spindle powers to be utilized before thermal damage can occur. The impact of a grinding abrasive on the work piece will generally induce a compressive stress on the work surface. However, the localized heating and subsequent cooling that is more predominant when grinding with aluminum oxide overcomes the compressive stress due to mechanical impact, and the residual stresses in the uppermost layers of the work piece are highly tensile. The absence of this heating when grinding with cubic boron nitride results in residual stress that is a compressive on the work surface. Figure 19.5 shows typical residual stress profiles produced by plunge grinding with the two abrasives. Since tensile stresses are accompanied by lowered fatigue life, clearly grinding with CBN offers distinctive advantages. The only drawbacks to the application of cubic boron nitride are its costs and the need for stiffer, higher-pow-

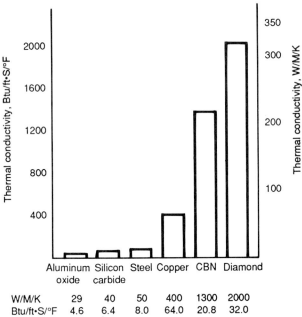

FIGURE 19.4 Comparative thermal conductivity of abrasives and other materials.

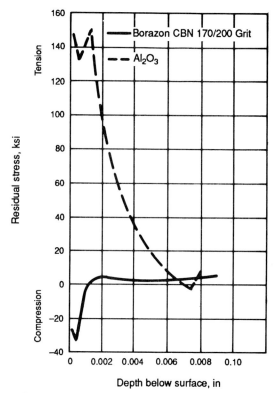

FIGURE 19.5 Comparative values of residual stress distribution.

ered machine tools to fully utilize the advantages that cubic boron nitride has to offer.

19.4 GRINDING WHEELS

Grinding wheels and their properties are only briefly discussed here, as a considerable amount of literature is in existence, especially from wheel manufacturers, that covers this in detail. All grinding wheels, except electroplated wheels, consist of an abrasive held in a bond. The physical size of the abrasive is a major determining factor in abrasive grain concentration and in the number of cutting edges engaged in the process of grinding at any given instant in time, that is, the larger the abrasive size, the fewer the number of abrasives and cutting edges, and vice versa. This in turn impacts the chip thickness as grinding proceeds and, consequently, the surface finish that is obtainable. In general, coarse abrasive grain sizes, also called grit sizes, result in rougher ground surfaces and finer grit sizes in lower surface roughness values. Surface roughness values of 0.4 μm (16 μin.) are generally possible with 60 grit size abrasives, and values better than 0.1 μm (4 μin.) are possible with abrasives of 200 grit size or finer.

Wheel hardness is another important characteristic of the wheel and is related to the amount of bond used in the manufacture of the wheel. A hard wheel has more bond, resulting in a greater abrasive retention ability. The abrasive will have to become considerably dull, in regard to its cutting edges, before sufficient forces are generated to tear it away. On the other hand, a soft wheel has less bond and consequently will lose its abrasive grains more readily. In general, soft wheels are used with hardened materials because the abrasive grain is known to dull rapidly when machining the hard materials, and fresh sharp abrasives will be required to continue grinding without the occurrence of high temperatures and surface damage. On the other hand, hard wheels are generally used with soft materials since the abrasive is expected to last longer, and abrasive grain retention is a property that is desired from an economic point of view.

Wheel structure is another important characteristic. An open structure allows greater chip clearance and is preferable in roughing operations where large quantities of material may be removed. Lack of sufficient space for chips would result in the loading of the wheel with subsequent burning of the work surface. However, open-structure wheels are also softer, because of the reduced amount of bond material.

Abrasive grain size is specified by the wire mesh size that will allow the abrasive to pass through. The smaller the number, the larger is the grain size. It must be remembered that the mesh size specified only indicates that grains larger than the specified value do not exist in the wheel, but smaller abrasive grain sizes do. In general, the grain size distribution can be assumed to follow a normal distribution. Wheel hardness or wheel grade is specified with a letter with A being the softest and Z the hardest. Wheel structure is generally specified with a number, with 1 representing a close structure and 10 representing a very open structure.

It must be remembered that there are no absolute relationships between work piece and wheel characteristics. The aforementioned facts are only guidelines, and the exact choice of a wheel for a particular work material–grinding operation combination has to be arrived at on the basis of trial and experience.

19.4.1 Wheel Preparation

In most grinding operations where a dressable wheel is used, four distinct operations may be present, singly or in combination. They are:

1. Wheel trueing
2. Wheel dressing
3. Wheel profiling
4. Wheel crushing

The purpose and procedure for these four operations are now discussed. The mechanism for accomplishing these operations is described later.

19.4.2 Wheel Trueing

In this operation wheel material is removed in order to eliminate wheel nonuniformities of shape and geometry due to wheel manufacture and mounting. The grinding wheel is mounted on its wheel holder, balanced, and then mounted on the machine spindle.

Trueing is then carried out by the motion of a diamond tool in a direction along the axis of wheel rotation as the wheel is spinning at speeds close to or at grinding speeds. After all the nonuniformity is eliminated, the grinding wheel will need to be balanced again. Wheel trueing is generally necessary only when the wheel is mounted for the first time unless nonuniform wheel wear has occurred during the grinding operation. Trueing will reduce forced vibration problems due to nonuniform wheel shape and geometry, resulting in improved surface finish.

19.4.3 Wheel Dressing

This is required to eliminate the uppermost layer of dulled abrasive grains and expose the sharp, next layer of abrasive grains in the wheel to obtain efficient cutting. On a new wheel it becomes necessary to do this when the wheel is very hard and the bond material completely encloses the abrasive grain. For softer wheels the trueing operation is generally able to expose abrasive grains, and the first wheel contact with the work piece is sufficient to break down any bond material that may still be covering the abrasive. On harder wheels the bond material may need to be pushed back with a stick of silicon carbide or naturally occurring abrasives such as corundum, etc. Too much bond removal is, however, detrimental, as abrasive grains would be unsupported and consequently lost easily, leading to loss of the wheel.

Dressing also clears the chip-loaded surface of the wheel, which may cause burning of this work piece. A loaded wheel, in combination with dull abrasive grains, will have a smooth, glazed surface. After dressing the wheel surface will be rougher to the touch.

19.4.4 Wheel Profiling

In this operation the wheel is shaped to a specific profile in order to generate the required geometry on the work piece. This is of special significance in gear grinding as the wheel is either representing a rack in some generating-grinding operations or the normal space between two adjacent teeth in form-grinding operations. Wheel trueing, dressing, and profiling can, however, be combined into one operation on a machine, especially if a medium or soft wheel is used. For hard wheels trueing and profiling can be combined, while initial dressing to push back the bond material is carried out as a separate operation.

19.4.5 Wheel Crushing

This is a technique used for rapidly removing wheel material to profile a wheel. A crushing roll, generally made of high-speed steel, with the required profile machined on it, is brought into contact under pressure with the grinding wheel, with no relative tangential velocity. Wheel speeds are generally reduced during this process to about one-fifth to one-tenth the actual grinding speeds.

Except for crushing and dressing of very hard wheels, all other wheel preparation operations are combined on most grinding machines. Profiling, trueing, and dressing can be done with a single-point diamond traversing the wheel surface in a specific relationship to generate the required profile. Where profile accuracy is influenced by the wear of the diamond point, a rotating diamond disk that represents many diamond points will improve results since the diamond wear is distributed over many points. However, the disk should run true in axial and radial directions in order to maintain profiling accuracy. For much faster profiling, in combination with dressing and trueing, formed diamond rolls can be used. These rolls are, however, expensive, and sufficient part volume may be necessary to justify the investment. With a formed diamond roll, intermittent dressing when the grinding wheel is out of the cut, or continuous dressing during the grinding operation, are possible. Continuous dressing is especially effective for high-speed, creep-feed grinding, which is discussed later.

19.5 GRINDING PROCESSES

There are two distinct grinding processes used in gear grinding, as in most grinding operations. They are as follows:

19.5.1 Conventional Grinding

Figure 19.6 illustrates the basic properties of this process, which is characterized by a wheel rotating at surface speeds of about 30 m/s (6000 ft/min), infeeds of 0.01 to 0.050 mm (0.0004 to 0.002 in.) and work velocities of 1.25m/min to 10m/min (50 to 400 in./min). The chips generated are short, due to the small arc of contact between an abrasive grain and work piece in this process, and are easily disposed of. Wheel wear rates are generally high in this process, resulting in low to medium G ratios. This is because of repeated impacts between the edge of the work piece and the wheel due to the to-and-fro oscillations of the work piece, for which reason this process is also sometimes referred to as pendulum grinding. Researchers have also found that in this mode of grinding the average force per abrasive grain is high, further contributing to rapid wheel breakdown. Coolant is generally used, though dry grinding can be done if the amount of infeed is in the 0.005-mm (0.0002-in.) range when grinding hardened steel. This is due to the fact that metal removal rates are very small, with a small fraction of the wheel surface cutting at any given instant, with low power consumption and consequently low amounts of heat generation.

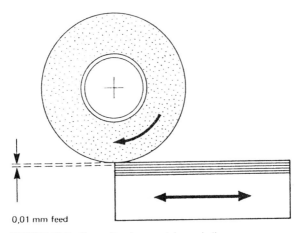

FIGURE 19.6 Conventional or pendulum grinding.

19.5.2 Creep-feed Grinding

Figure 19.7 illustrates the basic properties of this process that are characterized by large infeeds into the work in excess of 0.5 mm (0.0125 in.) and up to 10 mm (0.4 in.), depending on machine power and stiffness; but accompanied by much lower work velocities, which could be as low as 50 mm/min (2 in./min) and seldom exceeding 500 mm/min (20 in./min). Work velocity is inversely proportional to the infeed. Work velocities exceeding 250 mm/min (10 in./min) are generally accompanied by special dressing processes, such as continuous dressing, which enable the maintenance of a sharp, unclogged grinding wheel.

Since the infeeds are large, arc of contact between wheel and work is extremely large in comparison to conventional grinding. This results in each abrasive grain cutting a long chip. The wheel consequently has to have a very open structure to accommodate long chips.

The large arc of contact, which results in a large number of grains in simultaneous cutting action, requires high spindle power, which in turn results in large cutting forces and the generation of greater quantities of heat than with conventional grinding. The machine tool has to have the necessary power and stiffness to withstand the larger forces, and a copious supply of well-directed coolant to carry away the heat generated in the process.

In spite of the larger power requirement, creep-feed grinding generally enjoys a higher G ratio than does conventional grinding when grinding similar materials. The lowered wheel wear is attributed to lower forces per abrasive in creep-feed grinding and also to the fact that in conventional grinding many wheel–work piece impacts are present as the work piece oscillates from side to side about the wheel.

Any means of eliminating the long chips produced in creep-feed grinding from loading the wheel will only improve the efficiency of the process. The use of high-pressure coolants to flush the wheel has been one technique allowing higher work velocities. Another technique has been continuous dressing. Here the dressing roll, which may have the required form, is continuously fed into the wheel during the grinding process, with the grinding wheel being continuously fed into the work to compensate for reduction in wheel size. This continuous dressing keeps the wheel clean and sharp allowing higher work velocities during creep-feed grinding. Experimental

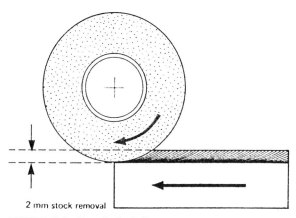

FIGURE 19.7 Creep-feed grinding.

work where work velocities were in the 1m/min (40 in./min) range and higher have been reported.

With this introduction to the various aspects of the grinding process it is now possible to discuss gear grinding as practiced by the industry. The two most common techniques are form grinding and generating grinding. The techniques are now discussed in detail.

19.5.3 Generating Grinding

There are several basic techniques of generating grinding; each technique is associated with a specific machine-tool manufacturer. These distinct techniques are now presented.

Threaded Wheel Method. The basic machine motions that generate the gear in this method are kinematically illustrated in Fig. 19.8. The similarities of the mechanics of this technique to gear hobbing are very obvious, with the threaded grinding wheel replacing the hob. The ratio of work speed and wheel speed when grinding spur gears is a simple ratio of number of teeth on the gear and number of starts on the wheel. For helical gears this has compensated (differential indexed) for the traverse of the grinding wheel along the face width of the gear.

In order to be able to carry out the grinding process, the threaded grinding wheel, unlike the hob, has to achieve surface speeds in excess of 25 m/s (5000 ft/min). The indexing mechanism has to be considerably more accurate in order to achieve the gear quality required in grinding. In the past, complex gear arrangements were normally used to obtain the simple and differential indexing requirements between the grinding wheel and work piece. However, *electronic gear boxes* (EGBs) are now commercially available to maintain the kinematic relationships. Figure 19.9 shows a typical machine with an EGB for generating grinding.

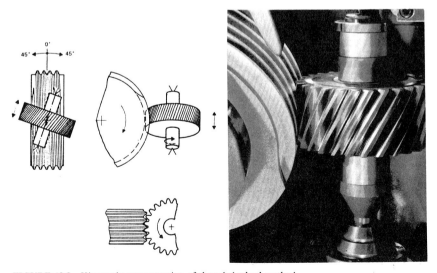

FIGURE 19.8 Kinematic representation of threaded wheel method.

GRINDING OF SPUR AND HELICAL GEARS 19.13

FIGURE 19.9 Electronic threaded wheel–type gear grinder.

The quality of the gear being ground is also significantly affected by the rack-type profile of the grinding wheel. It must first be introduced on a cylindrical wheel and then maintained through the grinding process as the wheel breaks down due to wear.

Introduction of the rack-type profile on a cylindrical wheel is done in two steps. A rough rack-type profile is crushed into the wheel using a steel crushing roll. This can then be finished by a variety of techniques using diamond tools such as a single-point dresser or coated dressing disk. Profile modifications are introduced in this second step as required. With single-point dressing tools, profile modifications are made using special cams. If coated disks are being used, the modifications are lapped into the disk by the machine manufacturer.

Saucer Wheels Method. This is another generating technique where two saucer-shaped wheels are used as shown in Fig. 19.10. The grinding surfaces of the two wheels represent the rack, and the involute profile is generated by the gear rolling relative to and in contact with the two grinding wheels. The wheels may be set parallel to each other or at an angle up to 10°. The work piece is reciprocated in the axial direction to provide the feed motion as two flanks of two different teeth are ground in one pass. At the end of the pass the entire gear is indexed using mechanical index heads so that two flanks of two more teeth are then ground by the wheel. The depth of cut is determined by the infeed of the two grinding wheels toward each other.

For spur gears the axes of the grinding wheels are perpendicular to the axis of the gear and only simple motion, to simulate the rolling of the gear on the rack represented by the grinding wheel, is needed to generate the involute. This generating motion is produced by steel tapes fixed to a stationary tape stand, the other end of which is wound over a rolling block that is generally the same diameter as the base-circle diameter of the profile being ground. This is illustrated in Fig. 19.11. When grinding a helical gear, the rolling motion that is necessary to generate the involute has to be compensated for

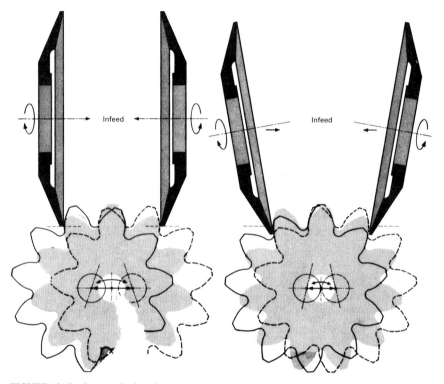

FIGURE 19.10 Saucer wheel method.

the helix angle as the grinding wheels move along the face width of the gear. This is accomplished by a helix guide mechanism attached to the tape stand that is used to generate the rolling motion. The helix guide is set to the base helix angle, and, as the gear moves along its axis, additional motion is imparted to it to produce the helix along with the involute.

On older machines of this type, changeover from one gear to another required the change of the rolling block for each change in base-circle diameter. On modern machines, mechanisms have been developed that allow a range of base-circle diameters that can be ground with the same rolling block.

The contact between the saucer-shaped wheel and the tooth flank is generally restricted to a very small area at any given time. This generally makes this technique of gear grinding time consuming and slow. However, it also enables point-by-point profile and lead modification along the flank of the tooth technique, termed *topological modification*. Consequently, the profile of the gear tooth can be different along the entire face width of the gear, a feature that none of the other gear grinding techniques, form or generating, can duplicate. Use of computers to control this topological grinding feature allows an infinite variety of tooth forms to be ground. It must be remembered, however, that this feature is at the price of slower cycle time, and tradeoffs have to be examined before a decision is made to use this technique. Also, at the present time,

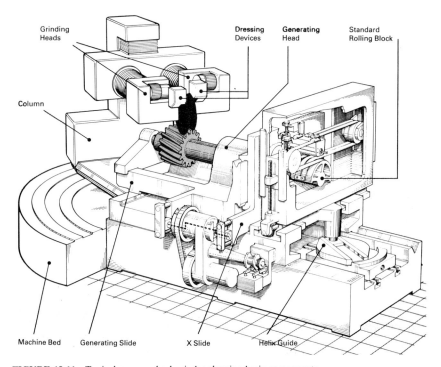

FIGURE 19.11 Typical saucer wheel grinder showing basic components.

apart from a few gear tool applications, no other applications of topological grinding have been applied.

Vitrified aluminum oxide wheels are most commonly used in this method of gear grinding, and the grinding process is generally done "dry." Dressing is carried out with wheel compensation, using single-point diamonds. The purpose of dressing is to ensure that the rotating surface of the wheel represents a straight tooth of a generating rack. Single-layer, cubic boron nitride–plated wheels have also been tried as they eliminate the need for dressing, and wheel life is high for reasons already explained in the text.

Conical Wheel Grinder. This is another version of generating grinding using a grinding wheel that represents a single tooth of the rack, as shown in Fig. 19.12. The sides of the wheel correspond to the pressure angle of the gear being ground. The work gear rotates and translates linearly to generate the rolling action required to generate the involute profile.

The simultaneous rolling and linear motion is generally obtained by having a master gear with the same number of teeth mounted on the work spindle rolling

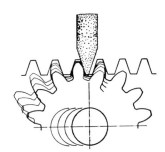

FIGURE 19.12 Basic concept of a conical wheel grinder.

on a stationary master rack. The master gear and rack must correspond to the gear being ground in terms of number of teeth, pressure angle, diametral pitch, etc. Electronic means of varying the base roll diameter to correspond to the gear being processed are, however, now available. In this process, as with the previously discussed saucer wheels method, two flanks of two different teeth are finished before the gear is indexed to grind two more flanks of two more teeth.

Helical gears can also be generated by this technique, though helical master gears and racks are required. If the wheel needs to be dressed, diamond points operating at the specific pressure angle are required. Since simple straight line forms need to be dressed, the dressing mechanism is relatively simple. Tooth profile modifications are produced by the modified grinding wheel. For lead modifications, the tool slide with the grinding wheel is radially advanced in synchronism with the stroking motion of the grinding slide, controlled through a tracer roll following the slope of the templet. A moderate-sized conical wheel grinder, grinding a double helical gear, is shown in Fig. 19.13.

FIGURE 19.13 Conical wheel grinder finishing a double-helical gear in one setup.

19.5.4 Form Grinding

In this technique the abrasive grinding wheel is profiled to represent the space between two adjacent teeth on a gear. The wheel is then passed through the space while grinding occurs on the two adjacent teeth flanks and the root, if required, as shown in Fig. 19.14. This is one of the primary advantages of form grinding in that various simple and compound root forms can be produced. Form grinding also enables the grinding of internal gears and external gears positioned against a shoulder.

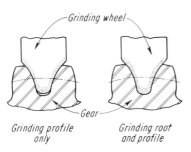

Grinding profile only   *Grinding root and profile*

FIGURE 19.14 Basic concept of form gear grinding.

When a spur gear is being ground, the wheel is simply moved along the axis of the gear. When a helical gear is required, the axial motion of the wheel is combined with a motion of the gear about its axis in order to produce the lead. The various axes of motion required to manufacture a gear on a horizontal axis grinder are illustrated in Fig. 19.15. The A axis provides the tooth-to-tooth index and, when interpolated with the X axis, generates the lead. The Y axis provides size control and, in combination with the X axis, provides lead modifications. The B axis allows the wheel to be set to the helix angle of the part. On a horizontal axis machine, two more axes

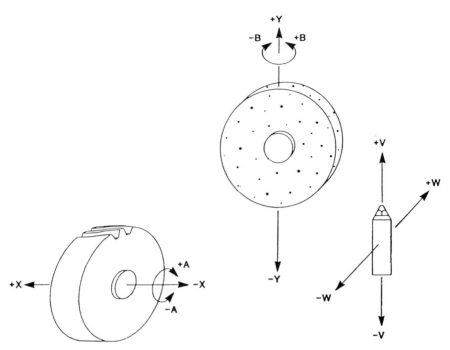

FIGURE 19.15 Axes of motion for form grinding.

are generally required to dress the profile on the grinding wheel. These are marked as the V and W axes.

An analysis of current gear grinding equipment indicates that form grinders are ahead of the generating machines in the application of computer control to gear grinding. In most generating machines it was found that only some aspects of the process were under computer control, while other aspects used mechanical control devices such as index plates, sine bars, or cams. However, contemporary form grinders appear to have completely abandoned mechanical devices in favor of computer control and appear to be doing as well or better than the older form-grinding machines. Also, computer control enables these form grinders to be more flexible and require less setup time than their generating counterparts.

Since the wheel profile is constant, modifying the lead by Y axis motion, which results in a change in center distance between the grinding wheel and the gear, will result in slight distortions to the profile. Lead modifications through change in the interpolation relationships between the A and X axes are also possible.

It is also important to note that, when grinding a helical gear, the normal tooth space that is represented by the grinding wheel has to be modified to account for the interference that occurs between the wheel and the helical groove, commonly termed *heel-toe action*. This heel-toe action is a function of the wheel diameter. Consequently, this has to be compensated for wheel diameter reduction during dressing to avoid errors in profiles.

In the past, mechanical devices such as sine bars, index plates, and cams were used to generate the helix, index, and profile, respectively. On modern computer-controlled machines, such as the one shown in Fig. 19.16, software generates and controls the relationships. Consequently, compensating for the involute as the wheel diameter changes due to dressing can be done just as easily as speeding up the spindle is carried out for changes in wheel diameter in order to maintain constant wheel surface speeds. The only necessity is that the machine constitute a set of necessary accurate linear and rotary axes.

Wheel trueing, profiling, and dressing are accomplished by the dressing mechanism. A diamond disk or single-point diamonds can be used. If production volumes can justify, a diamond preformed dressing roll can be used to reduce dressing times and increase productivity. Alternatively, electroplated preformed cubic boron nitride wheels can be used, and dressing times can be completely eliminated (keeping in mind that preformed wheels may cost up to 100 times the cost of a dressable aluminum oxide wheel). Vitrified, dressable cubic boron nitride wheels can also be used. These need to be dressed, but not as often as aluminum oxide wheels, and are generally cheaper or about the same cost as a plated wheel.

The coordinates describing the profile that is used to control the dressing device are also generated by software. Two basic approaches are evident. One is a more fundamental approach based on solid geometry where the grinding wheel and work piece are considered as two cylinders intersecting each other at a present distance and angle between the two axes. The shape of the intersecting surface on one of the cylinders, that is, the work piece, is defined by the specified profile. Consequently the shape of the wheel surface can be computed. The profile may be an involute with modifications or a noninvolute if required. The other basic approach is heuristic or data-based in which

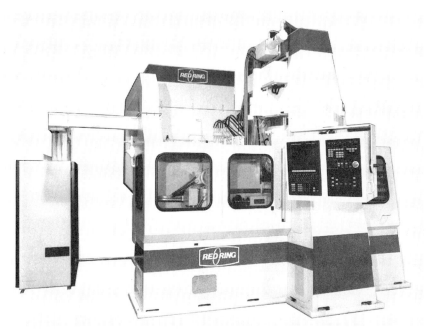

FIGURE 19.16 CNC form gear grinder.

GRINDING OF SPUR AND HELICAL GEARS   19.19

profile coordinates corresponding to different pressure angles, modules (diametral pitch), base-circle diameters, and helix angles at P.D. are stored. Interpolated values of coordinates for other profiles can then be obtained. This approach is more limited in scope and may need a few trials to arrive at the right profile.

The ability of form grinding to produce noninvolute forms cannot be overstressed. Generating grinding is limited in this area as the gear profile is due to the rolling action of the work against the wheel. Form grinding is, on the other hand, limited only by the type of forms that can be generated on the wheel.

Since the accuracy of profile obtained in form grinding is directly impacted by wheel wear, any technique that could reduce wheel wear is obviously of benefit to the economics of the operation. Plated cubic boron nitride wheels, where wear of the wheel is almost nonexistent, represent one approach, providing it can be cost-justified. Creep-feed grinding, with its accompanying reductions in wheel wear, is another. In order to accommodate the creep-feed grinding process, current machines have been designed and built with high spindle power and high static stiffness to utilize the power, low table speeds, and large coolant flows. All these features have enabled the application of advanced processes to the technique of form gear grinding.

19.6 CYCLE TIME ESTIMATES

These are essential in job shops for quoting purposes before a job of grinding a gear can be started. Keeping in mind the variety of gear grinding techniques available and the variety of grinding processes that could be utilized, as discussed in the preceding chapter, development of specific formulas to suit each process and technique was considered futile. Instead, a more general approach is now presented: an approach that can be modified to suit each technique or process as necessary.

The total time required to grind a gear is given by the expression:

Total time = grind cycle time + work handling time + setup time per gear

Since the work handling time and setup time per gear are functions of sophistication and type of work-handling equipment and machine tool and the skill of the operator, further discussion is restrained to the grind cycle time only.

The *grind cycle time* is given by the generalized expression

Grind cycle time = grind time + index time + wheel dress time + reset time

Further, *grind time* is given by the expression

$$\text{Grind time} = \frac{\text{gear face width} + \text{overtravel}}{\text{work traverse velocity}} \times \text{number of traverses} \\ \times \text{number of gear teeth}$$

Gear face width and number of teeth can be obtained from a part print; the amount of overtravel is a value necessary to clear the part for the purposes of indexing; and the work traverse velocity is a parameter that is dependent on a variety of factors, including the process and the type of machine tool being utilized. The number of traverses required for grinding is given by the formula

$$\text{Number of traverses} = \frac{\text{rough stock}}{\text{rough infeed}} + \frac{\text{semifinish stock}}{\text{semifinish infeed}}$$

CHAPTER NINETEEN

$$+ \frac{\text{finish stock}}{\text{finish infeed}} + \text{number of spark-out traverses}$$

All these aforementioned parameters are part- and process-dependent variables. The number of spark-out traverses is also dependent on the incoming quality of the gear, the required outgoing quality, and the stiffness of the machine tool being utilized. If the work traverse velocity in the expression for grind time changes during the rough, semifinish, finish and spark-out, different grind times have to be calculated for each part of the process and summed to get total grind time.

The *index time*, which does not exist in the case of generation grinding with a threaded wheel, is given by the expression

$$\text{Index time} = \text{time per index} \times \text{number of gear teeth}$$
$$\times \text{number of index traverses}$$

The time per index, usually a few seconds, is dependent on the type of machine tool and the number of teeth and is a part-dependent parameter. The number of index traverses is based on the processes and can be computed from the expression

$$\text{number of index traverses} = \frac{\text{number of rough traverses}}{\text{number of traverses/index}}$$
$$+ \frac{\text{number of semifinish traverses}}{\text{number of traverses/index}} + \ldots$$

This combination is due to the possibility of a number of traverses with grinding infeed on the same tooth before indexing, a technique that is used occasionally while roughing on a form grinder or generating grinder using saucer-shaped or conical grinding wheels. If the grinding process being used requires indexing after every traverse, then the number of index traverses is the same as the number of traverses.

The *dress time* is given by the expression,

$$\text{Dress time} = \text{time per dress} \times \text{number of dresses per gear}$$

In some types of generating grinding a number of gears may be finished between dresses and so a fractional value will have to be used for the parameter "number of dresses per gear." Also, the time per dress during roughing may be different from the time per dress during semifinishing, finishing, or spark-out in which case the formula may have to be expanded to account for all these variables. Since a certain amount of time is usually required to bring the dressing mechanism into action at the start of each dressing cycle, this time should be added for a more accurate estimate of dress time.

The gear being ground, in almost all instances, is held between two elements on the machine during the grinding process, for example, a headstock and a tailstock for an external gear. Of the two, one is generally the stiffer member, and consequently it is preferable to grind against this member in what is generally characterized as unidirectional grinding. (This is not to say that bidirectional grinding cannot be done, though this is restricted to roughing passes only.) The *reset time* is the idle time lost to reset the machine to do unidirectional grinding and given by the formula

$$\text{Reset time} = \frac{\text{face width} + \text{overtravel}}{\text{wheel return speed}} \times \text{number of traverses}$$
$$\times \text{number of gear teeth}$$

The wheel return speed is generally the rapid traverse rate on the machine, though depending on the amount of travel, the table of the machine may never reach that speed. A lower rate should generally be used to account for acceleration and deceleration. If a combination of bidirectional and unidirectional grinding is used, the formulas have to be modified to suit the requirements.

As stated earlier, only the general approach to estimation of cycle time is presented here. These have to be modified to suit the grinding technique and process selected. Above all, process parameters such as feed rates and infeeds have to be valid because the quality of the cycle time estimate is vitally dependent on it.

REFERENCES

1. Boothroyd, G., *Fundamentals of Metal Machining and Machine Tools,* McGraw-Hill, New York, 1975.
2. Dodd, H., and Kumar, D.V., "Technological Fundamentals of CBN Bevel Gear Finish Grinding," *Gear Technology,* Vol. 2, No. 6, November–December 1985.
3. Dudley, D. W., *Gear Handbook,* 1st ed., McGraw-Hill, New York, 1962.
4. Schwartz, R. W., and Rao, S. B., "A Wheel Selection Technique for Form Gear Grinding," AGMA Paper No. 85FTM7, 1985.
5. Technology of Machine Tools, Vol. 3 Machine Tool Mechanics, UCRL–52960–3, LLL, Livermore, California, 1980.
6. Tlusty, J., and Koenigsberger, F., "Specifications and Tests of Metal-Cutting Machine Tools," *UMIST,* Manchester, England, 1970.

CHAPTER 20
BEVEL AND HYPOID GEAR MANUFACTURING

Robert G. Hotchkiss
Director, Gear Technology, Gleason Works
Rochester, New York

William R. McVea
Manager, Application Engineering, Gleason Works
Rochester, New York

Richard L. Kitchen
Senior Research Project Engineer, Gleason Works
Rochester, New York

The various types of bevel and hypoid gears are described in Chap. 2, and the theory for producing conjugate pairs is outlined in Chap. 1. Information concerning tooth proportions and the design of bevel and hypoid gears is given in other chapters. The material in this chapter deals primarily with the machinery and equipment used to produce the teeth of such gears—their basic principles of operation and application. It also briefly touches on other aspects of the manufacturing process that directly affect the production of the teeth, including processing of the gear blank and heat treating.

The tools, methods, and machines used for producing bevel and hypoid gears are quite different from those used for other types of gears. As is the case with all other gear machines, there are various models to accommodate differing requirements of size, range, cutting method, and productivity. To understand these specific machines, it is first necessary to understand the basic principle of a generating machine, and also of a nongenerating machine. In order to understand the application of the differing machines, it is important to know of the different processes used to produce the teeth in gear sets for various requirements.

20.1 THE BASIC PROCESS

Generation can be called the basic process in bevel and hypoid gear making because at least one member of every pair must be generated; this is usually the member having the lesser number of teeth, called the pinion. The theory of generation, as applied to these gears, involves the following concepts:

1. An imaginary nondeformable bevel or hypoid gear called the generating gear. (This may be conceived of as a crown gear, a mating gear, or some other bevel or hypoid gear.)
2. A gear blank or work piece made of plastic, deformable, nonelastic material on which teeth are to be produced.
3. The positioning of generating gear and work piece so that the teeth of the generating gear are in mesh with the teeth of the work piece.
4. The turning of generating gear and work piece on their respective axes according to a prescribed motion. (With bevel gears the turning motion is defined by the imaginary nonslip rolling of the pitch surfaces of generating gear and work piece.)
5. The envelopment of teeth of the workpiece by the teeth of the generating gear.

20.2 THE BASIC GENERATOR

The actual production of gear teeth by the generating process requires:

1. That at least one tooth of the generating gear be described by the motion of a cutting tool or a grinding wheel
2. That the workpiece be positioned in relation to the cutting tool or grinding wheel so as to mesh the teeth of the work piece with the teeth of the generating gear as they are represented by cutter or tool
3. That the cutting tool or grinding wheel be carried on a rotatable machine member called a cradle, the axis of which is identical with the axis of the generating gear
4. That cradle and work piece rotate on their respective axes, or "roll together," exactly as would the work piece and the imaginary generating gear
5. That a means of indexing be provided

Thus, the rotating cradle carrying a cutter, tool, or grinding wheel represents a generating gear rolling with the work piece, according to a prescribed relative motion. The motion of the tools or rotation of the cutter or wheel must sweep out, or describe, the teeth of the generating gear. As the tool, cutter, or wheel is carried about the cradle axis, teeth are enveloped on the rotating work piece.

The concept of the imaginary generating gear is the real key to understanding the generating process. Figure 20.1 illustrates the representation of a generating gear by a face-mill cutter on a Gleason hypoid gear generator. Fundamental design differences that exist among different types of gear generators have originated because of different concepts of the imaginary generating gear. For example:

1. The generating gear might be one of a pair of complementary gears that fit one another like a mold casting. If such gears have 90° pitch angle they are referred to as complementary crown gears. A series of interchangeable bevel gears could be based on these generating gears. Straight bevel and some spiral bevel gears are derived from generating gears having pitch angles close to 90°.
2. The generating gear for a spiral bevel pinion could be either a spiral bevel or a hypoid gear.

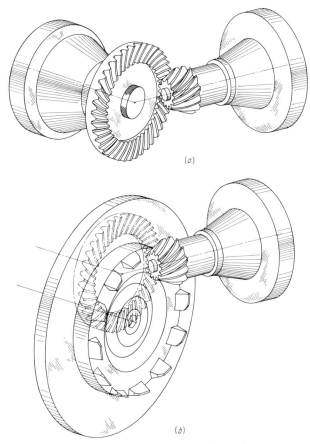

FIGURE 20.1 How a face-mill cutter represents a generating gear: *(a)* pinion in mesh with mating gear; *(b)* face-mill cutter representing generating gear.

3. The generating gear for a hypoid pinion could be a hypoid gear having a pitch angle and offset different from the mating gear.
4. The generating gear for a spiral bevel or hypoid pinion could be a gear having a helicoidal instead of a conical pitch surface; this is used for the generation of two tooth sides simultaneously.

Tooth surfaces of the aforementioned generated gears are described by tools, cutters, or wheels that are adjustable to some degree relative to the axis of the generating gear or cradle. Most of the machines with which such generating gears are represented use face-mill cutters, face-hob cutters, or cup-type grinding wheels. Spindles of cutter or wheel are adjustable radially from the cradle axis; they can also be swiveled about a point in the cutter or wheel axis. This permits the representation of a variety of generating gear teeth. Machines having tiltable cutter axes (known as tilting spindle machines) usually operate so as to maintain a constant ratio of turning between cradle and work during the generating roll.

Some of the larger generators and generating grinders have wheel or cutter spindles that are not tiltable; that is, they are fixed in direction parallel to the cradle axis but may be adjusted radially with respect to it. With this type of generator the ratio angular velocities between cradle and work spindle may be varied during generation. The ratio of turning velocity between work spindle and cradle is called the *ratio of roll,* and methods employing a nonconstant ratio of roll are called *modified roll methods.* Machines on which the cutter or wheel spindle is fixed in direction parallel to the cradle axis have provision for modified roll and are referred to as *modified roll machines.*

The two basic methods of generating spiral bevel or hypoid gears are face milling and face hobbing.

The face-milling method utilizes a circular face-mill type of cutter and the cradle and work axes rotate in a timed relationship. The rotational speed of the cutter is independent and is selected based on the desired surface feet per minute. The method employs an intermittent index. It is a very flexible method and offers the gear manufacturer a wide variety of cutting cycles. The face-milling method may employ both tapered- and uniform-depth teeth although normally tapered-depth teeth are used. The method may be applied for producing gears from the solid in one operation or by taking multiple cuts on each member.

The face-hobbing method utilizes a circular face-hob type of cutter and the cradle, work, and cutter axes must rotate in a timed relationship. During the cutting process the work has continuous rotation and the cutter rotates in a timed relationship with successive cutter blade groups engaging successive tooth slots as the gear is being cut. The face-hobbing method requires uniform-depth teeth and parts must be produced from the solid in one operation.

For the actual gear cutting process, the generating gear must be replaced by a rotatable member that carries some type of cutting tool for sweeping out one or more of its tooth surfaces. This member is known as the cradle (see Fig. 20.2). The gear to be cut is held by a rotatable spindle that is part of a work head. The work head is adjustable axially so that gears of different mounting distance can be accommodated. For the simplest generating methods, the work head is positioned so that the pitch cone apex of the work piece falls on the cradle axis.

The work head is supported by a swinging base that is rotationally adjustable about a vertical axis. The swinging feature of this base is required in order to accommodate gears of different pitch angles. Beneath the swinging base is the sliding base. This member is movable and adjustable in a direction parallel to the cradle axis. Adjustment is made in the position of this base in order to establish the correct depthwise relationship between the cutting tool and the work piece. It can also be moved back from the cutting position in order to obtain clearance for mounting and dismounting the work piece. The sliding base is supported by the main frame of the machine, to which is attached a housing that supports the cradle member.

The sequence of operation of a typical face-milling bevel gear generator is as follows: The sliding base moves forward to bring the work into engagement with the cutter, and the generating motion of the cradle and work begins (Figs. 20.3a and b). After the operation is completed on a tooth space, the sliding base is moved back (Fig. 20.3c) and the cradle and work spindle reverse their rotation. The cradle returns to its original position, but the work is indexed one tooth during its return roll (Fig. 20.3d). The sliding base then moves forward for the start of the next cycle.

The sequence of operation of a typical face-hobbing bevel gear generator is as follows: The sliding base moves forward to bring the work into engagement with the cutter. The work is continually rotating as the cutter is fed to full depth. Once full depth

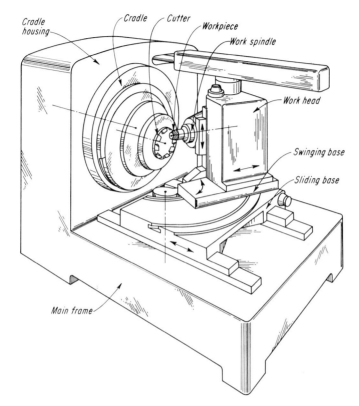

FIGURE 20.2 Arrangement of basic generator.

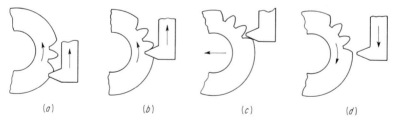

FIGURE 20.3 Four stages of generating cycle: *(a)* start of roll; *(b)* center of roll; *(c)* withdraw; *(d)* return.

is reached, the generating motion of the cradle begins. At the completion of the generating roll, the sliding base is moved back and the cutting operation is complete.

Computer numerical control (CNC) has been introduced in bevel and hypoid gear producing equipment, and the most modern machines all utilize this technology to replace gear-driven generating trains, index mechanisms, feed- and speed-change gears, and setup axes. This was made possible with the development of the *electronic gearbox*

(EGB), which eliminates the mechanical connection between axes used during the generating process and most of the generating gear train. The EGB (Fig. 20.4) controls the cutter, cradle, and work spindle axes while the sliding base is controlled by the CNC.

For two-axis intermittent face milling, the relationship between the cradle and work rotation is controlled by the EGB while cutter speed is independently controlled.

For three-axis continuous face hobbing, the relationship between the cradle, work spindle, and cutter rotation are all controlled by the EGB during cutting.

The introduction of CNC and the EGB has allowed both the face-milling and face-hobbing cutting operations to be performed on the same machine. A typical generating train for this universal generator is shown in Fig. 20.5.

More recently, full CNC machines have been developed that operate on the principle that any bevel or hypoid gear tooth geometry can be produced with only three rectilinear and three rotational axes. These six-axis CNC machines (Fig. 20.6) eliminate the entire cradle mechanism while retaining the capability of producing gears by all previ-

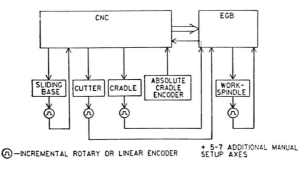

FIGURE 20.4

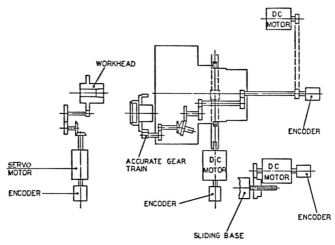

FIGURE 20.5 Two-axis, three-axis combination generator with EGB.

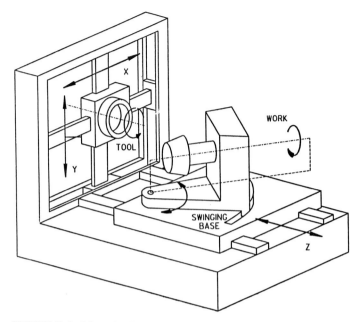

FIGURE 20.6 Schematic of universal six-axis CNC bevel/hypoid gear generator.

ous methods. Other advantages include improved accuracy and repeatability of setup and reduced operator involvement.

20.3 THE BASIC NONGENERATOR

To achieve fastest production rates, the gear member of spiral bevel pairs (approximately 3:1 ratio and higher) and of hypoid pairs (approximately 2.25:1 and higher) can be made with nongenerated tooth surfaces, which are straight in the normal section. These tooth surfaces, formed by a single sweep of cutting tool, can be produced much more rapidly than the tooth surfaces of generated gears. (Generated profiles, because they are curved, require a series of generated cuts to envelop the profile shape.) Theoretically, a generating machine can be used to produce these nongenerated gears. The only requirements are a means of locking the generating train in a fixed position and of feeding the gear into the cutting tool. Some current models of generators can be furnished with these provisions, so that they can be used for producing both generated and nongenerated parts.

The nongenerating machines are relatively simple compared with generating machines: Fewer machine components are required, since, except for the indexing of the work piece, the only motion involved is that of the cutting tool. Nongenerated gears, which include both Formate* and Helixform* gears, are cut with a conical face-mill type of cutter. For Formate gears a Single Cycle* cutter makes the finishing cut on both sides of one tooth space with each revolution. The inside and outside cutting blades are

---

*Formate, Helixform, and Single Cycle are registered trademarks of Gleason Works, Rochester, New York.

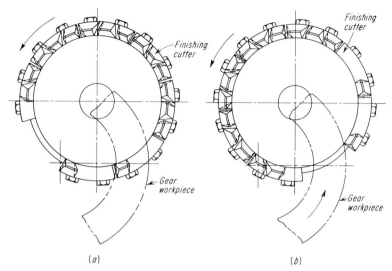

FIGURE 20.7 Cutting and indexing sequence on nongenerator: (a) cutter rotation with blank stationary; (b) blank rotation for indexing.

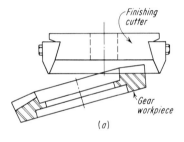

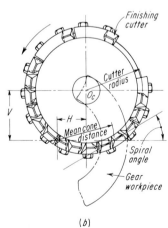

FIGURE 20.8 Relative position of single-cycle cutter and formate gear: (a) plan view; (b) elevation view.

stepped radially, causing the cutter to act like a broach. The last two blades are an inside and an outside finishing blade, which are spaced so that they cut one at a time. After they pass through the tooth space, the work is indexed quickly while a gap in the cutter passes the work. This sequence is indicated in Fig. 20.7.

The position of the Single Cycle cutter relative to the nongenerated gear is indicated in Fig. 20.8. The cutter axis is located with respect to the gear apex by the distances H and V, which are functions of the mean cone distance, the spiral angle, and the cutter radius. Figure 20.9 is a schematic picture of a machine for cutting these gears. The cutter is mounted in a spindle that is adjustable upward for the V setting and across for the H setting. The work spindle is adjustable for root angle and also in the direction of the cutter axis for depth setting.

Nongenerated gears can also be produced with a face-hob type of cutter. The motions required are the continuous rotation of the work piece, rotation of the cutting tool, and feed in a direction parallel to the cutter axis.

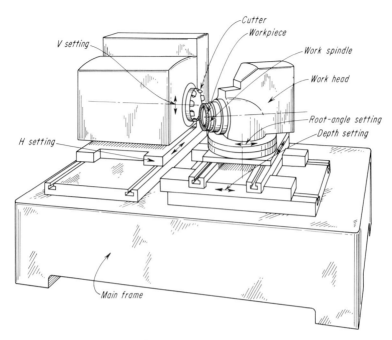

FIGURE 20.9 Arrangement of basic nongenerator.

20.4 GEAR BLANK REQUIREMENTS PRIOR TO TOOTH PROCESSING

The sequence of operations and the type of equipment used for production of bevel gear blanks vary widely, depending on size, quantity, quality, and shape. In general, the machining processes and tooling used are similar to those used in the manufacture of any other accurate machine part. The purpose of this section is not to specify in detail methods of blank preparation or tolerances, but to list the general requirements that must be met so that the gear machines can produce parts of desired quality.

The primary requirement is that the blank, as delivered to the gear machine, must be in proper condition for producing the teeth in accurate relation to certain important surfaces on the blank. The most important of these surfaces are those used for locating the gear radially and axially on the gear machine. When the size and shape of the gear permit, it is preferable to use these same surfaces for cutting, testing, die quenching, and grinding to avoid an accumulation of errors and to eliminate the need for machining numerous surfaces to close tolerances. Equally important, in the case of hardened gears, are the reference or locating surfaces used for the aforementioned operations that follow the tooth cutting.

Wherever possible, the surfaces used for locating the blank in the gear machine should be the same surfaces used for mounting the gears in the final assembly. Figure 20.10 shows the various surfaces and dimensions that should carry specific tolerances, in the case of a bevel gear blank of conventional bored-type design.

As a general rule in small and medium sizes of blanks for industrial gears of good

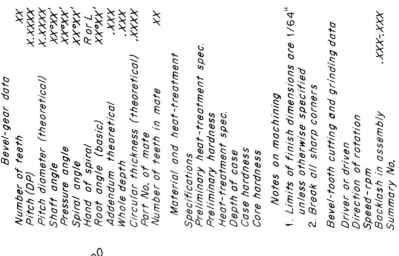

FIGURE 20.10 Toleranced bevel gear—blank dimensions.

quality, locating bores or shanks are held to a total tolerance of 0.0013 cm (0.0005 in.). This is reduced to 0.0005 cm (0.0002 in.) for precision gears, and to a "wring fit" on work-holding equipment if extreme precision is required.

The axial locating surface must be flat and square with the locating bore or shank within a total axial runout of 0.0013 cm (0.0005 in.) for good-quality industrial gears, and a smaller tolerance may be required for precision gears.

The axial and radial location of the crown point, as well as the face angle, must be controlled within specific limits, which depend generally on diametral pitch and gear size. Otherwise tooth depth and operating clearance will vary from gear to gear because the gear machine produces teeth in fixed relative position to the blank locating surfaces.

20.5 BEVEL AND HYPOID GEAR MANUFACTURING PROCEDURES

Tables 20.1 and 20.2 show the sequence of processes directly related to tooth formation of gears for various types of applications.

20.6 BEVEL GEAR WORK-HOLDING EQUIPMENT

In the cutting, testing, lapping, and grinding of bevel and hypoid gears, the chucking equipment plays a vital part. The important points to consider in connection with chucking equipment are rigidity; concentricity; accuracy of size; and an ample, uniform force for holding the gear in place. Excessive overhang, loose bushings, a multiplicity of fits, and equipment that must be trued up on the ma-

TABLE 20.1 Straight-Toothed Bevel Gears: Applications and Manufacturing Procedures

Type	Precision, Fine-Pitch (Unhardened)	Precision Load-Carrying (Hardened)	Commercial	Commercial Coarse-Pitch, Also Gears with Hub or Flange Projecting Above Root Plane
Typical application	Instruments	Instruments and aircraft accessory drives	General industrial, automotive differential	General industrial
Procedure	Cut teeth in one operation, test	Cut teeth, soft test, harden, finish-grind teeth, hard test	Cut teeth in one operation, soft test, harden, hard test	Rough-cut teeth, finish-cut teeth, soft test, harden, hard test

TABLE 20.2 Spiral Bevel, Zerol Bevel, and Hypoid Gears: Applications and Manufacturing Procedures

Type	Precision, Fine-Pitch (Unhardened)	Precision Load-Carrying (Hardened)	Commercial	Automotive Axle
Typical application	Instruments	Instruments and aircraft accessory drives	General industrial	Passenger cars, trucks, and buses
Procedure	Cut teeth in one operation, test	Cut teeth, soft test,* harden,* finish-grind teeth, hard test	Rough and finish-cut teeth, (or cut teeth in one operation) soft test, harden, lap, hard test	Rough and finish-cut teeth, (or cut teeth in one operation) soft test, harden, lap in pairs, hard test

*May be omitted on gears of through-hardening material approximately 20 diametral pitch and finer.

chine should be avoided. Chucks and arbors should, in general, be maintained within 0.0005 cm (0.0002 in.) of the correct size at all times. When mounted in the cutting or testing machine, the radial and axial locating surfaces should run true within 0.0005 cm (0.0002 in.) of the full indicator reading. Even closer tolerances are required for gears of extreme precision.

Bevel gear cutting and testing machines are usually equipped with a work spindle having a taper bore that receives the taper shank of the arbor. In manufacture, the large end of the bore is held very accurately to size, so that the shoulder of a correctly made arbor will seat on the face of the work spindle when the arbor is pulled snugly into the taper bore.

To seat properly in the taper bore and on the face of the spindle, the arbor must be made with the correct draw. The *draw* is the amount that the shoulder of the arbor is wrung into the spindle by hand. This amount should be from 0.010 to 0.030 cm (0.004 to 0.012 in.), depending on the size of the taper.

All modern bevel gear cutting machines are equipped with hydraulic chucks to which nearly all types of work-holding equipment may be readily adapted. Some typical methods of chucking blanks for gear cutting are shown in Figs. 20.11*a*, *b*, and *c*.

20.7 STRAIGHT BEVEL GEAR MACHINES

This section deals with the four basic types of straight bevel gear cutting machines. The subjects covered include description of machine operating cycles, required setup data, and tooling requirements of the various machines.

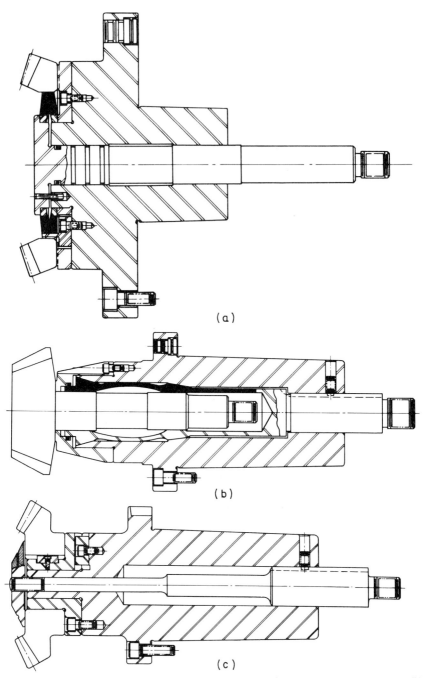

FIGURE 20.11 Typical work-holding equipment: *(a)* production-type arbor for cutting gear; *(b)* production-type chuck for cutting pinion; *(c)* adaptable-type arbor for bored gear.

20.7.1 Two-Tool Generators and Roughers

As the name implies, these machines employ two reciprocating tools, each of which cuts on opposite sides of a tooth. These are the most universal of the straight bevel machines, and initial tooling cost is the least expensive, but production rates are appreciably slower than on other types of straight bevel generators. The two-tool machines are used for: (1) work beyond the size range of interlocking cutting machines (Sec. 20.7.2) and the Revacycle* machine (Sec. 20.7.3); (2) parts that have integral hubs or flanges projecting above the root line, precluding use of the aforementioned machines; and (3) installations for minimum quantity requirements over a wide range of work, which cannot be accommodated by one model of the other types of straight bevel machines.

Roughing. Gears that are finish-cut on these machines must be rough-cut. Where the production requirements warrant, the roughing is done on separate machines that are similar in appearance to the generators but have no provision for generating roll. Otherwise the rough-cutting operation is performed on the generators. On both the roughing machines and the generators, the roughing cut is made by feeding the blank straight into the tool without roll.

A method of roughing that is used to good advantage on flat gears having a relatively large number of teeth and a face width not exceeding one-quarter of the cone distance, is the "double-index" method. By means of this method, two slots are roughed during each cutting cycle, one roughing tool operating in each of the tooth slots. The index mechanism is arranged to index the blank for half the number of teeth in the gear being roughed.

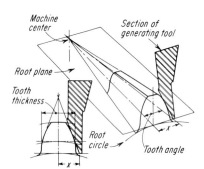

FIGURE 20.12 Straight bevel gear tooth angle.

Required Setup Data. In order to make the machine setups for producing a pair of right-angle gears, the operator must be provided with the gear data and one calculated machine setting called the tooth angle (Fig. 20.12). The remainder of the setup data are taken from tables furnished with the machine. When the shaft angle of the pair is not 90°, the ratio-of-roll gears and data for checking the roll must be calculated, in addition to the usual setup information.

Tools. Figure 20.13 shows typical roughing and finishing tools that are specified by diametral pitch and pressure angle. These are mounted in holders that are attached to the front of reciprocating slides on the face of the cradle. Each tool can be adjusted relative to the holder so that the end of the tool is in the gear root plane and so that the tool edge travels along a line that intersects the cradle axis. This tool positioning is checked by means of gauging fixtures that are set on a proof block. The position of the tool holder on the face of the tool slide is adjusted to accommodate the cone distance of the work, and the length of stroke is adjusted so that the tool overtravels the face width by a small amount.

Several types of tools are available for the roughing operation, the simplest being

---

*Revacycle is a registered trademark of Gleason Works, Rochester, New York.

BEVEL AND HYPOID GEAR MANUFACTURING 20.15

FIGURE 20.13   Roughing tools (left) and finishing tools.

straight-sided V-shaped tools used for gears of 8 m (3 diametral pitch) and finer, as in Fig. 20.13. Another type of roughing-tool set, Tanruf,* which is used for heavier pitches [usually 8 m (3 diametral pitch) and coarser] consists of a slotting tool, followed by a side-cutting tool. The first tool cuts only on the bottom of the tooth space while the second tool cuts only on the side. The second tool is usually made with a straight cutting edge but can be furnished with a curved profile for closer roughing.

Coniflex† Provision. Coniflex gears are Gleason straight bevel gears whose teeth are crowned lengthwise, to localize the tooth contact. In order to produce such gears, the generators have two angularly adjustable guides on the back of each slide, which ride on a pair of fixed rollers. When these guides are in line, the tool is stroked along a straight line. When the guides are set out of line, as in Fig. 20.14, the tool is stroked along a curved path. The amount and position of the curvature can be varied by the setting of the guides. A table furnished with the

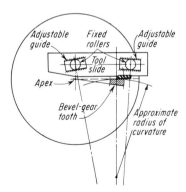

FIGURE 20.14   Coniflex provision on two-tool generators.

---

*Tanruf is a registered trademark of Gleason Works, Rochester, New York.
†Coniflex is a registered trademark of Gleason Works, Rochester, New York.

machine lists the guide settings that are used to make the tooth bearing length approximately one-half the face width. The machine operator can readily vary these settings to make the tooth contact longer or shorter.

Finishing Cycle. The completely automatic cutting cycle of the finishing operation consists of three phases. The first is a semifinishing operation that is done with the cradle turning clockwise and the sliding base held back from the full-depth position by a small amount. Next, the sliding base is automatically advanced to full depth and the cradle and work rotations are reversed to make the finishing cut. At the end of the generation, the work is withdrawn and indexed to the next tooth. Semifinishing the tooth surface prior to finishing leaves a uniform thickness of material for the finishing, and thereby ensures the best possible surface finish.

20.7.2 Interlocking Cutter Machines

A more productive type of straight bevel generator is that which employs a pair of interlocking disk-mill cutters having a multiplicity of blades that cut on the two sides of a tooth space. Figure 20.15 is a close-up view of a finished gear and the interlocking cutters. These cutters are not stroked in the direction of the teeth; consequently slightly concave root land surfaces are produced.

In nearly all cases, gears cut on these machines are completed from the solid blank in one operation that combines the roughing and finishing cuts. The size range of each of these machines is not so great as that of a two-tool generator, but the productivity is as much as four to five times greater. These machines are well suited for the economical production of straight bevel gears in widely varying quantities.

Coniflex Provision. As seen in Fig. 20.15, the disk cutters rotate on axes that are inclined to the face of the cradle. The cutting edges do not lie in the planes of rotation of the cutters but instead sweep out slightly internal conical surfaces. These dished

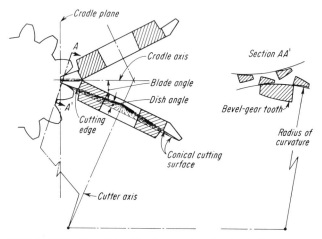

FIGURE 20.15 How disk-mill cutters produce Coniflex teeth.

cutter surfaces produce teeth that are slightly crowned in the lengthwise direction. The amount of crowning, or tooth bearing localization, varies with the dish angle, which is determined by the blade angle. The cutter blade angle is selected according to the length of contact desired. The pressure angle of the gears being cut is obtained by means of machine settings.

Cutters. The cutters for the smallest machine are available with one of three different blade angles, but one angle, 20°, is satisfactory for all nearly all gears. Seven pairs of these cutters with different blade point widths cover the the pitch range from 1.6 m to 0.4 m (16 to 64 diametral pitch). The cutters for the larger machines are made with five different blade angles. A series of nine tooth bearing lengths can be obtained with these cutters by using different blade angles (in some cases, for the two members of a gear set). Normally a selection can be made from two or three bearing lengths that are between 35 and 65 percent of the face width.

The theoretical blade point width is calculated directly from the gear tooth proportions. In actual use, the next smaller blade point is selected from the series of standard widths furnished by the manufacturer.

The blades of Coniflex cutters are relief ground when manufactured and thus require sharpening only on the front face. Blade-to-blade spacing and cutter diameter, as well as angle and surface finish of the front face, must be closely controlled in sharpening. Special-purpose sharpeners, offered by the gear machine manufacturer, are recommended to assure consistency in sharpening, as well as longest cutter life.

Machine Cycles. The small generator uses a continuous-index type of cycle in which the work spindle turns continuously in one direction at a constant rate, while the cradle is oscillated in timed relationship to produce the generating roll. The cradle rotation is produced by a generating cam that actuates the cradle through an adjustable linkage. A series of cams is furnished to cover the required range of rotation.

On the larger machines, the generating train is driven by a cam that permits the use of the type of cycle illustrated in Fig. 20.16. This cycle begins with the work and cradle stationary—the cutters situated somewhat above the center of the generating roll. A cam feeds the gear blank into the cutters, stopping just short of full depth. After this infeed roughing, the work and cradle roll down, rough-generating the tooth space. At the bottom of the roll, the base sets in to the full-depth position and the tooth space is finish-generated during the uproll. At the top of the roll, the work is withdrawn and the cradle and work roll down to the infeed roughing position. During this brief downroll, the blank is indexed to the next tooth space.

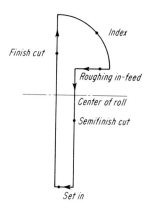

FIGURE 20.16 Completing cycle diagram for Coniflex generator.

Required Setup Data. In order to make the machine setups for producing a pair of gears, the operator must be provided with the gear data, a summary of the machine settings and cutter specifications, and the necessary work-holding equipment and cutters. The machine settings can be obtained from the machine manufacturer or can be determined readily by means of a relatively short calculation procedure.

20.7.3 Revacycle Machine

The Revacycle process is the fastest method of cutting straight bevel gears of commercial quality. It is used primarily for large-volume production of gears for such applications as automotive-type differentials and agricultural implements. The Revacycle machine is usually furnished with an automatic loading and unloading unit. Although initial tooling cost is greater than for other types of straight bevel gear machines, the extremely fast production rate results in lowest overall cost for large-scale production. Revacycle-cut gears are truly conjugate but are not interchangeable with gears produced by other methods. Revacycle tooth proportions differ from those of other straight bevel gears, and thus gear blanks must be designed specifically for this process.

Cutter and Completing Machine Cycle. Most Revacycle-cut gears are produced with one completing operation, using a large-diameter [53- or 64-cm (21- or 25-in.)] cutter, continuously rotating in a horizontal plane at a uniform rate. The cutter blades, which extend radially outward from the cutter head, have concave edges that produce convex profiles on the gear teeth. During the cutting operation the work piece is held motionless while the cutter is moved by means of a cam in a straight line across the face of the gear and substantially parallel to its root line.

This motion makes it possible to produce a straight tooth bottom, while the desired tooth shape is produced by the combined effect of the motion of the cutter and the shapes of the cutter blades. There is no depthwise feed of the cutter into the work, the effective feed being obtained by making each cutter blade progressively longer than the one before it.

The completing cutter contains three kinds of blades: roughing, semifinishing, and finishing. One revolution of the cutter completes each tooth space, and the work is indexed in the gap between the last finishing blade and first roughing blade.

Figure 20.17a shows the Revacycle cutter in position at the beginning of the cut. As the cutter rotates counterclockwise, blades of gradually increasing length contact the work piece until the root line of the tooth is reached. Figure 20.17b shows a transverse view of the tooth space as roughed to full depth. Each chip extends the full width of the slot, except for the amount of stock left for finishing. Figure 20.17c is an axial section of the same tooth showing that the roughing chips extend the full length of the tooth. During the first part of the roughing, the cutter moves from A to B (Fig. 20.17a). It then remains stationary at B until full depth is reached.

The roughed tooth does not have the proper taper. It is substantially correct along the diagonal line in Fig. 20-17c, but the portions of the tooth surface lying to the right of this line toward the heel of the tooth still have considerable stock that must be removed before finishing. This is accomplished by the semifinishing blades, which cut while the cutter center travels from B to C (Fig. 20.17a).

Finishing is done while the cutter center returns from C to D at a uniform rate. The finishing blades are given the proper shape to produce the correct tooth taper and proper profile at every point along the tooth. The blades cutting at either end of the tooth space are made slightly wider than necessary for correct taper in order to ease off the ends of the tooth and produce a crowning, which results in localized tooth bearing.

The resulting finished tooth surface is generated, being made up of a series of inclined cuts similar to those shown in Fig. 20-17d. In general, a different cutter is required for each different gear specification.

BEVEL AND HYPOID GEAR MANUFACTURING 20.19

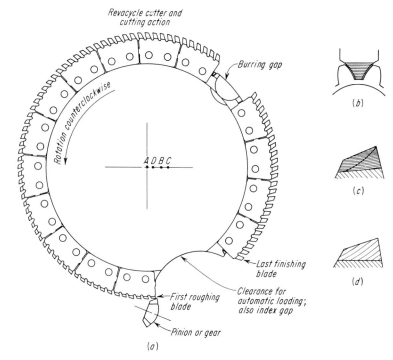

FIGURE 20.17 Revacycle cutter and tooth generation.

Two-Cut Operation. For Revacycle work that is too deep for completing in one cut, separate roughing and finishing operations are used. Separate cutters and separate machine setups are required for each operation. The cutters and machine cycle are similar to those described for completing, with some slight differences. The roughing cutter has no semifinishing or finishing blades, and there is no translation of the cutter in the roughing operation. The finishing cutter, however, translates as in the completing operation, with semifinishing blades cutting on the first translation and finishing blades cutting during the return stroke.

Required Setup Data. In order to set up for Revacycle cutting, the operator requires a summary, or list, of cutter and cam specifications and machine settings.

Automatic Deburring. A deburring attachment is provided so that burrs may be removed during the actual cutting of the gear teeth. The attachment is designed to remove the profile and root burrs from the heel of the teeth. A gap in the cutter between the semifinishing and finishing blades (see Fig. 20.17a) provides the space for mounting the burring tools. The shape of the burring tool is determined by the design of the gear, and, generally, the shape of the tool must be changed for each different gear specification.

Cutter Maintenance. Revacycle cutter blades are relief ground when manufactured and thus require sharpening on the front face only. Blade-to-blade spacing, the angle of the plane of the front face, and surface finish of the front face must all be closely controlled in sharpening. In addition, when new segments are assembled in heads, cleanliness of assembly, accuracy of position of segment-locating keys, and close control of segment-holding bolt tension are all necessities for proper cutting results. The gear machine manufacturer recommends use of a special cutter assembly stand and a special-purpose, automatic wet-type sharpener for all Revacycle production units.

20.8 FACE-MILL AND FACE-HOB TYPES OF CUTTING MACHINES AND GRINDERS—SPIRAL BEVEL, ZEROL BEVEL, AND HYPOID GEARS

With the exception of the large gears cut on planing generators, spiral bevel, Zerol* bevel, and hypoid gears are cut with face-mill or face-hob cutters or ground with cup-type wheels. A wide variety of: (1) cutters, (2) methods, and (3) machines is available to cover the entire range of this work—from the smallest instrument-type gears up to approximately 254-cm (100-in.) diameter.

This section explains these three factors in detail, so that the best choice can be made for producing a specific part at a given production level.

After the type of cutter, cutting or grinding method, and machines have been chosen, it is necessary to provide the operator with a summary, which lists in detail the cutter or wheel specifications, machines to be used, and their settings. This summary can be obtained from the machine manufacturer, or can be prepared by the user's engineering department following detailed calculating instructions furnished by the manufacturer. Computer timesharing services are available to the gear manufacturer for making the required calculations using the latest calculation techniques.

20.8.1 Face-Mill Cutters; Types and Uses

Face-mill cutters have three general types of blades: integral, segmental, and inserted. They are used for rough-, finish-, and completing cutting.

Integral blade or solid cutters are made from a single piece of tool steel. That is, the blades are part of the cutter head. Usually these cutters are used for fine-pitch work and are 15 cm (6 in.) or less in diameter.

Segmental cutters are made up of segments that have two, three, or four blades per segment. The segments are bolted to the cutter head around the periphery.

Inserted blade cutters have individual blades bolted to slotted heads.

The three types of face-mill cutters are shown in Fig. 20.18. Inserted blade cutters usually have parallels for changing diameter and adjusting wedges for trueing as shown. Some designs, however, such as the Helixform cutter, have neither parallels nor wedges, while other designs, such as the Triplex† rougher, have only parallels.

For finishing, each of the three types of cutters can be furnished with all outside-cutting blades, all inside-cutting blades, or alternate outside and inside blades. Roughing cutters may have either alternate inside and outside blades or may be of the Triplex

---

*Zerol is a registered trademark of Gleason Works, Rochester, New York.
†Triplex is a registered trademark of Gleason Works, Rochester, New York.

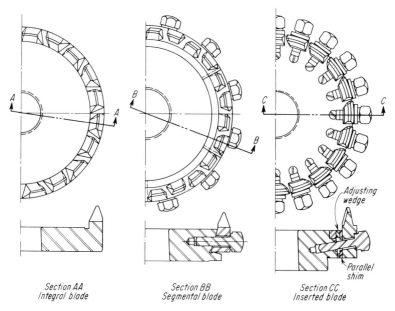

FIGURE 20.18 Three general types of face-mill cutter.

type, which has end- or bottom-cutting blades alternately spaced with inside and outside blades.

Table 20.3 shows the various categories and uses of face-mill cutters of the segmental and inserted blade types. Some of the preceding categories merit special explanation.

Cutters Used for Nongenerated Gears. Helixform, Single Cycle, and Cyclex* cutters are used for nongenerated gears. The gears produced are referred to as nongenerated because the teeth are not enveloped by successive positions of the cutter blades (see Sec. 20.3, "The Basic Nongenerator," and Sec. 20.8.6).

Helixform Gear Finishing Cutter. The inserted alternate blade cutter has eight blades spaced at 36/ intervals with the remaining space being used as the index gap. The first two inside and first two outside blades shown in Fig. 20.19 are semifinishing blades, while the last four blades are finishing blades. The blades are stepped in both a radial and axial direction to take a precalculated chip load, and one cutter revolution finishes each tooth slot. The semifinishing blades remove from 0.010 to 0.018 cm (0.004 to 0.007 in.) total stock depending on the design and tooth size of the particular gear. The finishing blades remove a total of 0.008 cm (0.003 in.) stock in steps of 0.005 and 0.003 cm (0.002 and 0.001 in.), respectively. Each blade of the Helixform cutter is positioned so that only one blade is cutting at any given time. Helixform cutters can also be used for Single Cycle gear finishing in place of the segmental alternate blade cutter.

---

*Cyclex is a registered trademarks of Gleason Works, Rochester, New York.

TABLE 20.3 Types and Uses of Face-Mill Cutters

Type of Cutter	Type of Blade	Kind of Blade Inside	Kind of Blade Outside	Adjusting Wedges for Trueing	Parallels for Diameter Changes	Used for
Helixform	Inserted	Both		No	No	Nongenerated gear finishing
Single Cycle	Segmental	Both		No	No	Nongenerated gear finishing
Cyclex	Segmental	Both		No	No	Nongenerated gear completing
Cyclex	Segmental	Both		No	No	Nongenerated gear completing
Triplex	Inserted	Both plus bottom cutting blades		No	Yes	Gear roughing
Inside blade	Inserted	All	None	Yes	Yes	Pinion finishing
Outside blade	Inserted	All	None	Yes	Yes	Pinion finishing
Alternate blade	Inserted	Both		Yes	Yes	Gear and pinion finishing and roughing, also jobbing and Duplex* completing processes

Remarks: Alternate blade, inside blade, and outside blade Gleason finishing cutters having special hardened heads are called Hardac cutters. Alternate blade cutters designed for roughing, or Triplex cutters, using special hardened heads, are referred to by the trade name Roughac. Both are registered trademarks of the Gleason Works, Rochester, NY.
*Fine-pitch gears, completed by the Duplex method, use integral blade cutters.

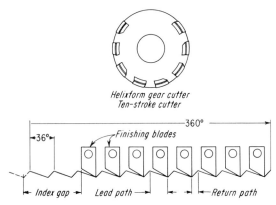

FIGURE 20.19 Helixform gear cutter diagram.

Single Cycle Gear Finishing Cutter. This cutter is a segmental alternate blade cutter with from 10 to 16 blades, depending on the cutter size. A gap between the last finishing blade and first semifinishing blade provides the space for indexing the work (see Fig. 20.7). Blades are stepped in both a radial and axial direction to take a precalculated chip load. Figure 20.7 shows the blade arrangement for a 23-cm (9.0-in.) mean diameter segmental Single Cycle cutter. Between 0.038 and 0.076 cm (0.015 and 0.030 in.) total finishing stock is removed by Single Cycle cutters, depending on the tooth size of the gear. The finishing blades are spaced so that there is never more than one blade cutting in the tooth slot at any given time.

Cyclex Completing Cutter. The Cyclex cutter is an alternate blade segmental completing cutter that roughs and finishes both sides of a tooth slot in one operation. The cutter is composed of two sections, roughing and finishing, with the two finishing blades set lower than the highest roughing blade so that they do not engage the work during the roughing portion of the cutting cycle.

There are two blade arrangements for the roughing section of the cutter. The Cyclex cutter for use on the Formate machine has all the roughing blades set at the same height, and corresponding points on the cutting edges are at the same radius from the center of the cutter. The Cyclex cutter for use on a generator specially arranged for Formate cutting differs from that described in the foregoing discussion in that the roughing blades are stepped in an axial direction to take a precalculated chip load.

The two finishing blades on these cutters are set at the same height but with space between them so that they are not in engagement with the work piece at the same time. They are also separated from the roughing blades so that a roughing and finishing blade will not engage the gear blank at the same time.

Triplex Gear Roughing Cutter. The Triplex cutter has bottom- or end-cutting blades alternately spaced with inside- and outside-cutting blades. The end-cutting blades are generally set 0.025 cm (0.010 in.) higher than the side-cutting blades to relieve the amount of end cutting necessary in the finishing operation. Triplex cutters are most generally used when the slot width of the gear tooth in the root plane is 0.165 cm (0.065 in.) or greater.

Hardac Finishing Cutter. These finishing cutters may be one of two types: cutters with alternate inside and outside blades, or cutters with all inside or all outside blades. The blades of these cutters are mounted in hardened and ground heads that have adjusting wedges for trueing. Hardac cutters of the alternate inside and outside blade type can also be used for the duplex completing process.

Hardac Carbide Finishing Cutter. These finishing cutters may be one of two types: cutters with alternate inside and outside blades, or cutters with all inside or all outside blades. These cutters are used to finish cut gears that have been rough-cut, hardened and process ground. The cutting process is a corrective one, eliminating the distortion produced by the hardening operation.

The blade and head design are essentially the same as for the previously described Hardac finishing cutter with the addition of carbide material being brazed to the front face of the blades.

These cutters can be used for either the Duplex spread blade method of cutting or the single-side fixed-setting method of cutting. In either case the gear member must be generated.

Maintenance. Most face-mill cutter blades are sharpened on the front face so that the primary cutting edge lies in a plane containing the cutter axis. The blades are relieved in such a manner as to maintain pressure angle and diameter with successive sharpenings. Blade-to-blade spacing and angle of front face must be closely controlled in sharpening, specific tolerances varying with type and size of cutter. The finish of the sharpened surface should be within a limit of 10 μin.

The gear machine manufacturer offers special-purpose sharpeners to maintain the required tolerances. Several models are available, varying as to size, range of cutters accommodated, and degree of automatic operation.

A high degree of precision is essential in the assembly of segments, or blades, in heads, for purposes of either refilling used cutters or changing cutter specifications. Duplication of blade angles and concentricity of blades must be held to close tolerances—the latter within a fraction of a ten-thousandth of an inch on some finishing cutters, for example. This necessitates care, insistence on cleanliness, and use of torque wrenches to maintain uniform cutter bolt tension.

Stand-type fixtures are offered by the gear machine manufacturer to facilitate cutter reassembly. These fixtures include the equipment necessary for measuring trueness of angle and concentricity of blades.

20.8.2 Face-Mill Cutters: Specifications

The important cutter specifications that directly affect the tooth form of the gears are the diameter, pressure angle, and point width. Hand of cutter must also be specified, depending on cutting conditions.

Diameter. The diameter referred to in a cutter having alternate inside- and outside-cutting blades is the average diameter (see Fig. 20.20). Blade point diameter is specified for cutters having all inside or all outside blades.

The cutter diameter for a particular pair of gears is chosen from the 12 available standard diameters, by considering cone distance, face width, whole depth of tooth, and spiral angle. Table 20.4 is given as a general guide for cutter selection. A good rule is to select a cutter diameter equal to or slightly less than twice the cone distance. Generally, if there is a choice between two diameters, the smaller is preferred.

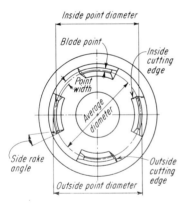

FIGURE 20.20 Face-mill cutter specifications.

Pressure Angle. Generally the nominal pressure angle of the cutter blade is the same as the normal pressure angle of the gear to be cut. The standard is 20°, although 16, 22½, and 25° are also commonly used. Since the cutter blade tips travel along the root cone instead of the pitch cone, the inside and outside blade angles must be made, respectively, greater and less than the normal pressure angle of the gear teeth. This difference between cutter blade angle and normal pressure angle is denoted by a system of cutter numbers.

Point Width. For cutters having alternate inside and outside blades, the point width indicates the distance between the inside and outside cutting edges at the tip of the blades as shown in Fig. 20.20. For cutters having all outside or all inside blades, the

BEVEL AND HYPOID GEAR MANUFACTURING   20.25

TABLE 20.4 Range Covered by Available Face-mill Cutter Diameter

Cutter Diameter	Number of Blades	Approximate Mean Cone Distance, In.	Approximately Maximum Face Width, In.	Maximum Standard Whole Depth, *In.		
1.1	8	0.5–0.75	0.25			0.125
1.5	12	0.75–1.0	0.3125			0.1875
2.0	16	1.0–1.5	0.4375			0.1875
2.75	20	1.25–1.75	0.5625			0.250
3.5	20	1.5–2.75	0.750			0.350
4.5	20	2.5–3.0	1.0			0.375
6.0	16, 20	2.75–3.5	1.25			0.375
7.5	16, 20	3.5–4.0	1.5			0.50
9.0	16, 20, 24	4.0–5.25	1.875			0.5625
12.0	12, 20, 28, 32	5.25–7.5	2.5			0.750
16.0	24, 40	7.5–15.0	3.5			0.750
18.0	28	7.5–15.0	4.0			0.750
18.0	24	7.5–15.0	4.0	14½°		1.3125
				17½°		1.125
				20°		1.000

*Can be exceeded by use of special-length blades.

blade point indicates the distance between the cutting edge and clearance side at the tip of the blade. Alternate blade cutters must have blade points less than the cutter point width in order to provide clearance for chips.

Toprem* Feature. Lapping a pair of gears above 2:1 ratio may produce interference between the flank of the pinion tooth and the top of the gear tooth. Toprem blades in the pinion-finishing cutters prevent this interference.

Near the tip of a Toprem blade there is a small straight segment of the blade profile where the pressure angle is slightly less than the pressure angle of the rest of the blade. The amount of this modification can be selected from the following general rule: depth of modification equals one-third the gear addendum plus the clearance between the top of the gear tooth and the bottom of the pinion tooth space at the large end.

Hand of Cutter. Another specification for cutters relates to the direction in which the cutter rotates when it is cutting. A right-hand cutter is one that rotates clockwise when viewed from the back, while a left-hand cutter rotates counterclockwise.

When the hand of cutter is the same as the hand of the spiral of the gear or pinion, the direction of cut is from the small end to the large end of the teeth. Therefore, flat gears (greater than 3:1 ratio) that are roughed without generation use cutters having the same hand of spiral as the gear, so that the thrust of the cut forces the work against the chuck.

On generating cuts, however, another factor must be taken into account. When the machine cradle is rotating counterclockwise as viewed from the front of the machine, the machine is said to be on the uproll; clockwise rotation is called downroll. Right-

---

*Toprem is a registered trademark of Gleason Works, Rochester, New York.

hand cutters climb cut on the uproll, whereas left-hand cutters climb cut on the downroll. Better finish and cutter life are obtained by climb cutting during both the roughing and finishing operations, and the hand of cutter is usually specified accordingly.

RSR* Completing Cutter. These duplex completing cutters are used for producing spiral bevel and hypoid gears by the face-milling method of cutting. These cutters have alternate inside and outside blades. The cutters utilize tool-bit-style blades whose relieved surfaces are ground to sharpen the cutting edges. The front face or chip-tool interface need not be reground as on the more conventional-type cutters. See Fig. 20.21. By avoiding the necessity of grinding the front face, the grinding time is reduced, and this allows the front face to be TiN coated. This system also provides users with the ability to manufacture cutter blades in their own plants.

The cutter body consists of an inner portion that contains precision-ground blade slots. An outer clamp ring is attached and contains the blade clamp screws and clamp blocks (see Fig. 20.22). A series of parallels are used to move the blades inward or outward radially in large increments. For smaller radial moves the profile is shifted on the blade while the relieved surfaces are being ground.

Blades are assembled and advanced axially to the same gauge point after each sharpening, eliminating the need for any compensating machine adjustments. The blades are then trued by adjusting axially for radial trueing.

This cutter system can also be used for gear and pinion roughing when employing the fixed setting method of cutting bevel gears.

*RSR is a registered trademark of Gleason Works, Rochester, New York.

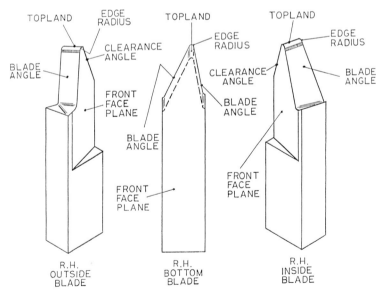

FIGURE 20.21 RSR blades.

FIGURE 20.22 RSR completing cutter.

Tan-Tru Roughing and Finishing Cutter. These roughing and finishing cutters are used to produce very-coarse-pitch spiral bevel gears to a maximum of 42 m (0.6 diametral pitch), 9.5-cm (3.750-in.) whole depth, and 254-cm (100-in.) diameter. These cutters are used for face-milling type of cutting.

The blade design is similar to other types of face-mill designs with the front face of the blade being sharpened. There are two types of blades: side-cutting and bottom-cutting. The bottom-cutting blades are symmetrical and can be used for either left-hand (L.H.) or right-hand (R.H.) cutting. There are two types of side cutting blades. One type is used as a L.H. outside or R.H. inside blade, and the other type is used as a R.H. outside or L.H. inside blade.

The cutter bodies are either four-sided or six-sided, depending on diameter and size of gear being cut. Each side of the body contains a tool block into which a side-cutting blade and a bottom-cutting blade can be mounted. A series of parallels and wedges control both cutter diameter and pressure angle. The side-cutting blades are used for both roughing and finishing with the bottom cutting blades used only for roughing. The tool blocks can be reversed to change hand of cutter.

20.8.3 Tri-Ac* Completing Cutter

These completing cutters are used for producing spiral and hypoid bevel gears by the continuous face-hobbing method of cutting. These cutters have alternate inside and outside blades and are very similar in design to RSR completing cutters. The cutters utilize tool-bit-style blades with relieved surfaces being ground to sharpen the cutting

---

*Tri-Ac is a registered trademark of Gleason Works, Rochester, New York.

edges. The feature of providing users with the ability to manufacture their own blades applies here also.

The cutter body design is also very similar to the RSR cutter body design, consisting of an inner body containing blade slots and an outer clamp ring containing clamp screws and clamp blocks (see Fig. 20.23). The Tri-Ac cutters do not use parallels. The amount of radial move of blade profile needed to change from one gear specification to another is small enough so that it can be accommodated within the width of the blade blank.

Blades are assembled and advanced axially to the same gauge point after each sharpening, eliminating the need for compensating machine adjustments. Trueing is not required for face-hobbing cutters.

20.8.4 General Classification of Methods

There are four basic cutting or grinding methods; namely, spread blade, single side, fixed setting, and single setting. The rotating cutting edges of the cutter generate an imaginary cutter surface; when grinding is the finishing operation, these cutter surfaces may be described by grinding wheels. In general, the methods defined below apply to both cutting and grinding.

Spread-Blade Method. This refers to the process where the member being produced is generated with a cutter with alternate inside and outside blades that cut the tooth

FIGURE 20.23 Tri-Ac completing cutter.

surfaces on both sides of a tooth space simultaneously. The width of the tooth space is controlled by the point width of the cutter. Uses of the spread-blade method are:

1. For the gear member of Formate or Helixform pairs
2. With the jobbing system—for the gear member
3. For either or both members of gear pairs completed from the solid
4. For roughing

With gear pairs completed from the solid in one chucking, using either a face-milling or face-hobbing cutter, the process employs the basic spread-blade method.

Single-Side Method. This is the process where the member being produced is finished by a circular face-mill cutter with alternate inside and outside blades that cut the tooth surface on each side of a tooth space in separate operations with independent machine settings. The cutter point width is considerably less than the space width of the finished gear. The width of the tooth space is controlled by rotating the blank about its axis through a measured distance, or by means of a stock-dividing gauge if sufficient production is involved. This method is used for pinion finishing and sometimes for the gear member of generated pairs.

Fixed-Setting Method. This refers to the process where the member being produced is finished by two circular face-mill cutters—one with inside blades only for cutting the convex side of the tooth and one with outside blades only for cutting the concave side of the tooth. The two sides of the tooth are produced separately with two entirely different machine setups. For volume production, pairs of machines are used—one machine of the pair for one side of the tooth, the other machine for the other side of the tooth. Width of the tooth space is usually controlled with the stock-dividing gauge. The term *fixed setting* is applied to the finish-cutting of pinions only, which may be pinions of a generated pair or Formate or Helixform pinions.

Single Setting Method. This is a variation of the spread-blade method and is used when available cutters have smaller point widths than required for spread-blade cutting. Both sides are cut with the same machine settings, and the blank is rotated about its axis in either or both directions to remove the amount of stock necessary to produce the correct tooth thickness. After the first cut, only one blade at a time is cutting in the tooth slot. This method is used with the Unitool* system.

Combinations of cutting methods, calculating methods, and special machining procedures make up the various gear cutting processes that are applicable to Helixform, Formate, and generated pairs. In general, all processes are theoretically adaptable to grinding.

20.8.5 Rough-Cutting

The first roughing cut on all work is a spread-blade operation. For most work, this is the only roughing operation. However, it is occasionally desirable to perform a second roughing cut on a pinion, in order to conform more closely to the finished taper of the tooth. If this is done, the second cut is performed with the same cutter, but with a set-over of the machine settings. This second cut is therefore a single-side operation. Current generating-type roughing machines include provisions for automatically

---

*Unitool is a registered trademark of Gleason Works, Rochester, New York.

changing these settings between uproll and downroll of the cradle, so that the two cuts are performed in one machine operation.

Generating roll is omitted when the pitch angle of the gear to be roughed is approximately 65°, and all pinions are roughed with generating roll.

20.8.6 Processes for Nongenerated Gears

The gear member of spiral bevel, Zerol bevel, or hypoid pairs has relatively little profile curvature when the ratio is high, and a substantial gain in productivity can be made by cutting or grinding the gear without generating roll. Such gears are referred to as nongenerated gears, the classification including both Helixform and Formate gears. In volume production, generation of the gear is usually omitted for spiral bevel pairs of ratio 3:1 and higher, and for hypoid pairs having ratios of 2.25:1 and higher. Helixform and Formate "pairs" consist of a nongenerated gear produced by the named method, and a pinion with teeth generated to match. Most automotive rear-axle-drive gears are made by these processes.

Figure 20.24 illustrates the basic difference between tooth profile formation by generation and nongeneration. In the former type of operation, the cutter contacts the tooth profile at only one point at any one time; whereas in the latter, the cutter makes contact over the entire profile at once.

Helixform. Gears called by this trade name are of a nongenerated type having helicoidal tooth surfaces cut by the spread-blade method. Helixform pinions are generated, generally using the fixed-setting method, to match the Helixform gear. The process can be applied to spiral bevel, hypoid, and Zerol gears.

Helixform gears, like Formate gears, are straight-sided in the normal section. As each blade of the Helixform cutter passes through a tooth space, the cutter is advanced axially and then returned before entry of the subsequent blade. The cutter thus has combined rotational and reciprocating motion, and the work is indexed in a gap between the blades.

In Fig. 20.25 the gear-cutter axis has been placed perpendicular to the gear face

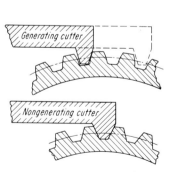

FIGURE 20.24 Difference between generating and nongenerating.

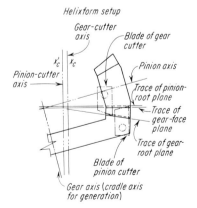

FIGURE 20.25 Relationships of cutters and work pieces in Helixform cutting.

plane. This is possible because, as the cutter rotates, it is given an axial advance so that the path of the cutter blade tips is tangent to the gear root plane. Since gears are designed with constant clearance, the gear face plane and pinion root plane are parallel. If the cutter blade angles are made equal to the pinion root plane pressure angles, the gear-cutter axis, which is perpendicular to both the gear face plane and pinion root plane, is parallel to the pinion-cutter axis.

Helixform gears are roughed on a companion machine to the Helixform gear finishing machine by a method similar to Formate roughing. The cutter is fed in the direction of the cutter axis until the full-depth position is reached. The difference between the conical surface of the roughed gear and the helical surface of the finished gear is small enough to be corrected by the semifinishing blades of the Helixform gear finishing cutter.

Formate. The gears to which this trademark is applied are nongenerated gears produced with face-mill cutters. They have tooth surfaces that are surfaces of revolution and are usually straight in the normal section. The Formate process can be applied to hypoid, spiral bevel, and Zerol gears.

Because of tooth design considerations, it can be used only on flat gears having pitch angles greater than approximately 70°.

All Formate gears are cut by the spread-blade method, except Unitool gears, which are cut by the single-setting method.

Formate pinions are generated to mate with Formate gears. For most Formate pinions, the fixed-setting method is used; the Unitool process, however, uses the single-side method, while duplex-helical and face-hobbed pinions are cut by the spread-blade method.

Figure 20.26 shows the positions for the gear- and pinion-cutter axes in the Formate setup. If the cone angles of the gear cutter are made equal to the root plane pressure angles of the gear, the gear-cutter axis x_C is perpendicular to the gear root plane. In this position, the tips of the gear-cutter blades always lie in the gear root plane as the cutter rotates. The conical surfaces of the gear teeth are thus defined by the position of the gear cutter and its cone angles.

The pinion root plane is inclined to the gear root plane by the angles. To produce teeth having depthwise taper, the peripheral direction of the pinion-cutter blade tips must be tangent to the pinion root plane near the middle of the face, as shown.

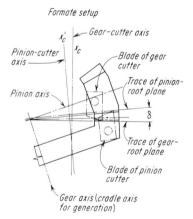

FIGURE 20.26 Relationships of cutters and work pieces in Formate cutting.

Use of Single Cycle Cutter (Sec. 20.3). The use of this cutter requires that gears be roughed without generation before finishing by feeding the cutter in an axial direction into the full-depth position. The conventional alternate blade face-mill roughing cutter may be used or, more commonly, a Triplex roughing cutter (Sec. 20.8.1), which contains alternate end- and side-cutting blades.

Gear finishing is accomplished with a single revolutioon of the Single Cycle cutter in each tooth space (Fig. 20.7).

Use of Cyclex Cutter (Sec. 20.8.1). This cutter provides a means of completing both roughing and finishing in one cut from the solid blank. It is used where the quantity of production does not warrant separate roughing and Single Cycle finishing equipment. The major advantages of Formate gears with regard to accuracy of tooth spacing, quality of finish, and fast production rates are closely maintained by the Cyclex method.

During the roughing portion of the cutting cycle, the cutter makes a number of turns, and feed is provided by a cam that delivers uniform motion in the direction of the cutter axis.

After the last roughing blade has passed through the tooth slot, and as full depth is reached, the cutter is rapidly advanced axially by the cutter feed cam to a fixed position. The cutter is held in this advanced position until both finishing blades pass through the tooth slot, finishing the tooth shape. The cutter is then rapidly withdrawn to prevent the roughing blades from contacting the work. Additional withdraw of the cutter provides the necessary clearance to index to the next tooth.

Use of Alternate Blade Cutters. Fine-pitch gears of sets produced by the Formate helical Duplex process may be cut by the spread-blade method with simple alternate blade cutters on generators having tilting cutter spindles. A special setup places the cutter in a position for correct contact when the gear and cradle are coaxial. The cradle and gear are rotated together as one unit during infeed to whole depth. Details and advantages of the Duplex helical method are given below (Sec. 20.8.7).

Formate gears cut by the Unitool process using alternate blade cutters depend on the single-setting method. The gear member is cut on a generator arranged for Formate cutting, that is, with both the cradle and work spindle held fixed by clamps during the cut. The work spindle clamp is released automatically for indexing. The Unitool cutter is fed to whole depth with no relative generating motion so that the surface on the tooth is a counterpart of the surface of revolution of the cutter.

20.8.7 Processes Used for Either Generation or Nongeneration

There are three processes that can be used for either generation or nongeneration; Duplex spread blade, face hobbing, and Unitool.

Duplex Spread-Blade Method. This is a method in which both the gear and pinion are cut spread-blade. The gear may or may not be generated, while the pinion must always be generated. This method is suited to large- or small-scale production of spiral bevel, Zerol bevel, and hypoid gears.

The setup for generating the pinion member of Formate duplex cut pairs represents an imaginary generating gear having a helicoidal pitch surface. This requires relative feed of the cutter along the cradle axis during the generating roll, which is referred to as *helical generation.* When semifinishing is required prior to Duplex helical grinding, it is usually done with helical generation.

Duplex grinding on machines without tilting spindles or helical movement is done with wheels having surfaces that are spherical or which are curved surfaces of revolution. Semifinishing prior to grinding on this type of machine must be done by a comparable process without cutter spindle tilt.

In order to use the Duplex spread-blade method or face-hobbing method, special tooth proportions are required. The blanks must be designed for the particular method of cutting to be employed, following calculating instructions available form the machine manufacturer.

Although control over the tooth bearing is not quite so complete with this method as with the fixed-setting method, control over tooth size, concentricity between the two sides of the teeth, and smoothness of the fillets and tooth bottoms is excellent.

Unitool Method. The purpose of the Unitool method is to make practical and economical the production of good-quality spiral bevel, Zerol bevel, and hypoid gears in very small quantities. It is not intended for mass production. It is slower than the jobbing system (Sec. 20.8.8) because both members of the pair are cut single side. In brief, the Unitool method covers the following:

1. Spiral bevel and Zerol bevel gears of any ratio
2. Hypoid gears of 3:1 ratio and higher
3. All reasonable tooth numbers, pitches, spiral angles, and pressure angles
4. All reasonable tooth designs including equal-addendum teeth, as well as standard long- and short-addendum designs, stub teeth, and special root-line direction

It features are the following:

1. The same cutter is used for both members of the pair.
2. Each part is completed before removing from the cutting machine.
3. One cutter covers a wide range of work.
4. Six cutters cut practically all jobs within the range of the machines used by this method.
5. Roughing is generally performed with the finishing cutter.
6. Calculations for machine settings are comparatively short.

20.8.8 Processes Used Only for Generation

Gleason 20 Spiral Bevel Gear Jobbing System. The purpose of the jobbing system is to make practical and economical the production of high-quality spiral bevel gears in moderate quantities. It is not intended for mass production. This system is applicable to the cutting of a wide range of spiral bevel gear ratios:

1. With mean cone distances between 6.4 and 18 cm (2.5 and 7.0 in.)
2. With 35° spiral angle
3. With 20° pressure angle

This system features a simplified method for obtaining tooth proportions, blank dimensions, and machine settings. Only a minimum of calculation work is necessary, together with a minimum of cutters.

Cutting of the gear is accomplished by the spread-blade method, while the mating pinion is finished by the single-side method. Both members are roughed prior to finishing. All cutters used with this system contain alternate inside- and outside-cutting blades.

Quality of the tooth bearing produced by this method is good, since direction and width of contact are controlled, and a choice in the length of the tooth bearing is obtained by selection of the proper pinion cutter. Set-over from one side of the pinion tooth to the other requires a change in only four machine settings.

A few roughing cutters are provided, but roughing can also be done with the finishing cutters for small quantities.

Special Jobbing. This is similar to the 20 jobbing system explained in the foregoing but can be used for cutting other spiral angles using the 20 jobbing cutters. This process is also applicable to cutting spiral bevel gears with other pressure angles provided the necessary cutters are used.

Versacut.* This is a process that requires only a few cutters to accommodate a wide variety of spiral bevel gears on the largest face-mill generator, covering the following range:

	Versacut	Versacut II
Outer cone distance, cm (in.)	7.6–33 (3–13)	11.2–38.9 (4.4–15.3)
Module (diametral pitch)	2.5–15.7 (10–1.6)	4.8–12.7 (5.3–2.0)
Pressure angle	20°	20°
Spiral angle	0–40°	0–40°
Ratios	1:1–10:1	1:1–10:1

The cutters can be used for both roughing and finishing, and each gear can be completed before being removed from the machine. Where production quantities merit, however, a roughing cutter can be used for a separate roughing operation. The machine settings for roughing are also the finishing settings for one side of the tooth. Each side of each member is finish-cut separately.

Arco.† This is a special process of cutting and grinding angular spiral bevel gears having cone distances that are beyond machine limits for conventional generation. Gears made by this method are usually designed with constant depth, and both members are generated. Modified roll is used when generating the pinion.

LSVM.‡ This term has been applied to a special process for generating gears of long cone distance. With this process the work piece has motion of translation with respect to the cradle axis in addition to the generating roll. The method is used only on machines with linear vertical motion of the work head during generation.

20.8.9 Grinding Considerations

The requirement for ground bevel and hypoid gears for certain types of power transmission has increased greatly in recent years. A knowledge of the merits of ground bevel and hypoid gears, of the machines employed, and of the machine cycles used is needed before the decision to grind is made.

Purpose. The primary reason for grinding bevel and hypoid gears is to obtain optimum accuracy in the teeth of gears that require a heat-treatment or hardening operation. To obtain maximum life, sometimes under heavily loaded conditions, the use of hardened gears is mandatory. These gears are normally cut in a gear cutting machine prior to the heat treatment operation. Some distortion of the teeth is unavoidable in the heat-treatment process, and this distortion destroys the accuracy of shape and dimensions originally produced in the cut member.

---

*Versacut is a trademark of Gleason Works, Rochester, New York.
†Arco is a trademark of Gleason Works, Rochester, New York.
‡LSVM is a trademark of Gleason Works, Rochester, New York.

BEVEL AND HYPOID GEAR MANUFACTURING   20.35

There are procedures other than grinding that are used to minimize or partially correct the heat-treatment distortion. Generally, quenching dies are employed in the heat-treatment operation to decrease the amount of distortion. A subsequent lapping operation, which is a partially corrective measure only, often provides gears that are quite satisfactory for their particular application. However, when optimum smoothness in motion transmission is required because of high speed, heavy loads, or accuracy requirements, a grinding operation must be used on hardened gears in lieu of lapping to assure satisfactory results.

Typical Applications Requiring Ground Gears.   It is the application of the gear, rather than the type of gear, that dictates the need for grinding. The following outline indicates the general fields of power transmission in which ground bevel and hypoid gears are employed, and the predominating reasons for the requirement of ground teeth.

1. *Aircraft:* Aircraft gears are designed to provide maximum load capacity with minimum weight, necessitating the use of case-hardened steel. The same design restrictions usually result in considerable section variations that induce excessive heat-treatment distortions. Grinding the teeth to remove hardening distortion and to improve accuracy of the gears has permitted higher operating speeds, longer operation, and heavier loading.
2. *Machine Tools:* The requirement for accuracy (gear trains, dividing heads, etc.) and long life under heavily loaded conditions has increased the demand for case-hardened ground bevel and hypoid gears.
3. *Instrument:* The prime requirement is accuracy. Many fine-pitch instrument gears are ground from the solid in a single operation.
4. *Marine propulsion:* Long life under heavily loaded conditions requires the use of hardened and ground gears.
5. *High-speed hand tools:* The high-speed application requires the use of ground gears that can operate at speeds considered impractical with unground gears.
6. *High-speed mills and presses:* Hardened and ground gears have proved advantageous for the requirements of high-speed smooth power transmission under heavily loaded conditions.
7. *Automotive, truck, and tractor:* Improved accuracy and higher residual compressive stress obtained by CBN grinding provides more load-carrying capacity for these applications.

Accuracy of Ground Bevel and Hypoid Gears.   The following values are indicative of the accuracy and quality which can be maintained in production when using current gear grinding machines: tooth-to-tooth spacing as close as 0.0005 cm (0.0002 in.); eccentricity of the teeth relative to gear bore or shaft within 0.00064 cm (0.00025 in.) if proper regard is given the gear blank processing and the work-holding equipment; tooth size (measured in terms of backlash) to within 0.0025 to 0.005 cm (0.001 to 0.002 in.) between mating members; and surface finish within a range of 0.38 to 0.76 μm (15 to 30 μin.). This uniformity of quality and dimensional accuracy, as provided by grinding machines, can be obtained by no other process on hardened bevel and hypoid gears.

Portions of Teeth That Are Ground.   Bevel and hypoid gear grinding machines grind the entire configuration of the tooth slot—that is, the profile, fillet radius, and root or bottom land. It is primarily due to this fact, combined with the accuracy and uniformity of ground teeth, that longer life is obtained with ground bevel gears even under operation at high speeds or in heavily loaded applications.

The grinding wheel is dressed so that the fillet radius, top land, and profile are all tangent at their junction points, creating a smooth blend without an abrupt change in curvature. The rate of dress is controlled so as to permit a uniform dressing action at all points of the wheel. This is particularly important in the area of the edge radius where the proper dressing speed helps to prevent burning in the fillet.

Recommended Semifinishing Practices and Grind Stock Allowances. Minimum amount and uniform distribution of stock left by the semifinishing operation are extremely important. These are the bases for economical production in the grinding operation, and a necessity for obtaining case-hardened gears with high load-carrying capacity. Table 20.5 lists the recommended semifinishing practices and stock allowances for spiral bevel, Zerol bevel, and hypoid gears of various diametral pitches. (Similar values are applicable when grinding straight bevel gears.)

Gears of diametral pitch coarser than 20 should be semifinished before grinding. When the Duplex method of grinding is used, a single Duplex semifinishing cut from the solid will suffice, but the same calculating method must be used for both cutting

TABLE 20.5 Recommended SemiFinishing Practices and Grind Stock Allowances for Spiral Bevel, Zerol Bevel, and Hypoid Gears

Diameter Pitch	Stock Allowance Per Tooth Side, In.	Stock Allowance in Tooth Bottom, In.	Remarks
20 and finer	—	—	All gears 20 diam. pitch and finer can be ground from the solid by the Duplex method.
10–20	0.003–0.004	0.003	1. Gears of 12 to 20 diam. pitch should be semifinished using conical (straight-profile) cutters in all cases.
Coarser than 10	0.004–0.005	0.003	2. Gears coarser than 12 diam. pitch should be semifinished with cutter shapes similar to shape employed in grinding wheel; i.e., if spherical (curved-profile) grinding wheels are to be used, spherical (curved-profile) semifinishing cutters are required. If conical (straight-profile) grinding wheels are to be used, conical (straight-profile) semifinishing cutters are required.

Notes:
1. Exceeding the above limits for stock allowance will: *(a)* reduce the load-carrying capacity of case-hardened gears, *(b)* increase the grinding time in order to eliminate the risk of burning or checking, and *(c)* increase wheel and diamond wear.
2. To maintain the recommended minimum stock allowance, it is essential that related control surfaces and dimensions of the gear blank be held to close tolerances.
3. It is essential that hardening distortion be kept to a minimum by means of the necessary quenching dies and presses.

and grinding. This is important for the following reason: since there are two types of grinding machines—one having a grinding wheel axis fixed parallel to the cradle axis, and the other permitting inclination of the wheel axis with respect to the cradle axis—there are, perforce, two different methods of representing the generating gear. With modified roll machines, the angle between work axis and cradle axis is always 90° plus the root angle; with the tilting type of machines, the angle between cradle axis and work axis is not so restricted, and imaginary generating gears of many different pitch angles may be represented by tilting the grinding wheel axis and work axis so that the surface of the generating gear tooth is tangibly represented. Gears, and particularly pinions, have substantially different root surfaces when manufactured on different types of machines.

In order to hold the stock allowance to the limit established in Table 20.5, semifinishing and grinding must be based on the same calculation method.

The foregoing requirements in method, which are based on uniform stock removal, do not preclude interchangeability, to a practical degree, of gears ground on both types of machines. For optimum interchangeability, however, it is best to follow the recommendation of the machine manufacturer.

Nital Etch Inspection. Nital inspection can reveal grinding burns that may occur during grinding. Therefore, it should be especially noted that production procedures should include a periodic nital etch inspection to check for burning. Consult the grinding machine manufacturer for details.

Coolants. Preferably, bevel gear grinding is done with an oil coolant and without water, a typical oil coolant having specifications as follows:

Viscosity: 150 to 180 SSU, 32 to 39 cSt

Mineral base: 2-percent sulfur contained

Water-soluble oil coolants can be used, but when grinding case-hardened steel, results will be less satisfactory. Provision should be made for keeping the coolant clean.

The Grinding Cycle. Prior to production grinding, the most efficient grinding cycle for a particular gear is established. This depends on the material to be ground. The grinding cycle, which is fully automatic, consists of sequential operations of dressing, wheel feed, and grinding. Repetition of identical operations for every gear of a production lot ensures uniformity of the product.

20.8.10 Machines for Nongenerated Gears

A Helixform machine can be used to cut both Helixform and Formate gears, whereas Formate gear finishers can be used only for Formate gears.

Helixform Machines. Helixform gear finishing machines have a stroke cam that transmits axial motion to the cutter spindle by means of an adjustable pivoted arm. A no-stroke cam is substituted for the Helixform stroke cam in order to cut Formate gears.

The stroke cam is designed with two lead and return paths and rotates five times for each revolution of the cutter. The proportion of lead to return path is approximately 3:1, which means that 27° is the maximum angular face width of the gear that can be cut. Since the stroke cam has a timed relationship with the cutter, one lead path is always used to cut the coast side of the gear and the other lead path the drive side. A range of modified-lead-path cams permits independent drive and coast-side tooth bearing developments to suit individual assembly deflections and hardening distortions.

The work is mounted on an arbor (Fig. 20.11a), which is actuated hydraulically. The work spindle in which the arbor is mounted is locked in place during cutting and automatically released for indexing. In dechucking action, the gear is pushed off the centering portion of the centering ring so that it needs only to be lifted off the arbor.

Rapid indexing is accomplished with an index-plate mechanism, or an electronic index mechanism.

On the gear roughing machine, the cutter is advanced toward the work in the direction of its axis by a cam motion. A linkage adjustment permits from 60 to 100 percent of the cam throw to be delivered to the spindle. Work-holding, index, and cutter speed and feed are controlled the same as on the finisher.

The gear is chamfered automatically on the outside diameter by a tool that cuts on both the bottom and concave side of the tooth slot.

Formate Cutting Machines. The Formate process can utilize either the face-milling or the face-hobbing method. In the case of face milling, the rotation of the cutter on its axis is the only motion used when finishing a gear. Work-holding equipment and indexing are essentially the same as with the Helixform machine.

Formate roughing and face-mill-type completing utilize a depth feed mechanism that advances the work or cutter in the direction of the cutter axis until the full depth position is reached. Additionally, face-hobbing machines doing Formate completing require a work rotation timed to the cutter.

Automatic chamfering may be used on either the roughing or completing machines. They chamfer the acute angles (that is, heel and toe edges) of the gear teeth after or, in the case of face hobbing, while the tooth is being cut.

Formate Grinding Machines. There are two basically different Formate grinding machines.

The first type uses a flared cup-type grinding wheel that is positioned so as to have line contact with the sides of the tooth and is translated from end to end of the tooth while grinding. Both sides of the tooth space are ground simultaneously. Figure 20.27 shows the position of the grinding wheel axis and the equivalent cutter axis.

The grinding wheel axis is inclined at an angle with the cutter axis or axis of translation. The normal radius of curvature of the grinding wheel must remain smaller than the radius of the equivalent cutter on the concave side of the tooth and larger on the convex side of the tooth. This makes it possible at any given moment to have the grinding wheel in contact at only one section b-b' along the tooth as shown in section a-a' of Fig. 20.27.

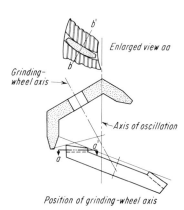

FIGURE 20.27 Relationship of wheel and work piece on Formate grinder.

Wheels are usually CBN-plated wheels. Some older-style machines use conventional aluminum oxide wheels for this method.

The second type of machine uses a cup-shaped wheel similar in shape to a face-mill cutter. It is usually an aluminum oxide–dressable type of wheel. The operation is similar to face-mill cutting. An oscillation of the wheel about a slightly eccentric center relieves pressure between the wheel and the work, permitting coolant access to the work-wheel interface to decrease likelihood of thermal damage to the work.

20.8.11 Machines for Generated Gears

Generators using face-mill cutters, and grinding machines using cup-type wheels, are made in two types—tilting spindle machines and modified roll machines. Machines utilizing face-hobbing cutters can be used to produce generated gears. The basic differences and relation to the means of representing the imaginary generating gear are explained in Sec. 20.2, "The Basic Generator."

Cutting Machine Cycles. Operation on both types of machines is entirely automatic. On a typical face-milling cycle, as soon as the work piece is chucked, the work head and gear move into cutting position and generating roll begins. At the completion of one tooth space, the cradle and work spindle reverse direction, the work is withdrawn, and indexing takes place. The work head moves forward and the sequence of motion is repeated until all teeth are cut; the machine then stops automatically.

With a typical face-hobbing cycle, after the work piece is chucked, the work head and gear move into the cutting position, and generating roll timed to a continuous rotation of the work begins. At the end of the generating roll, the gear is complete and the work is withdrawn. The machine then stops automatically.

Tilting Spindle Cutting Machines. This is the type of machine used extensively for automotive pinion production. It is made in a series of sizes ranging from the smallest machine for fine-pitch work, which mounts face-mill cutters of from 2.8- to 8.9-cm (1.1- to 3.5-in.) diameter, up to the largest sizes, which mount cutters up to 50-cm (19.7-in.) diameter. Some tilting spindle machines are optionally equipped with modified roll.

Modified Roll Cutting Machine. The larger machines, mounting face-mill cutters of 23 to 46 cm (9 to 18 in.) in diameter are made without tilt adjustment of the cutter spindle. That is, the direction of the cutter axis always remains parallel to the cradle axis. As explained in Sec. 20.2, the ratio of angular velocity between cradle and work varies during generation.

Grinding Machine Cycles. Although the generating cycles of the two types of grinders are different, as explained in the following, the complete grinding sequence is automatically controlled in similar manner on both types. The number of grinding passes and dresses is determined prior to production grinding, depending on the material and amount of stock to be removed. The complete grinding cycle consists of the operations of wheel feed, dressing, grinding, and work advance in predetermined sequences. Once the machine cycle is started, the grinder automatically proceeds through all the aforementioned operations in sequence until final size is reached.

Grinding Wheels. Cup-type wheels for generated grinding are generally operated between 1158 and 1372 m/min (3800 and 4500 ft/min). Typical wheel specifications are A-60H-7V and A60M-B.

Tilting Spindle Grinding Machines. These machines use a cup-shaped grinding wheel instead of a face-mill cutter, and, in general, all cutting methods and processes not requiring modified roll are equally applicable to grinding. The generating process has been described in Sec. 20.2.

Grinding takes place while the cradle and the work turn about their respective axes far enough to complete the generation of the side of a tooth or the adjacent sides of a tooth space, depending on the method used. When generation is complete, the work is withdrawn from engagement with the wheel, and indexing of the work piece takes place while cradle and work spindle are returning to their starting positions.

Modified Roll Grinding Machines. These machines use cup-shaped wheels in place of face-mill cutters and, like some of the larger cutting machines, have wheel spindles that are always parallel to the cradle axis. Operation is different from the larger cutting machines, however, in that the work spindle turns continuously at a constant rate, but when generation of a tooth side or sides is complete, the wheel is withdrawn from engagement with the work. During the time that the wheel is disengaged, the work continues to turn while the cradle motion reverses and the cradle returns to its starting position. The wheel then moves forward, and engagement with the work is resumed in a different tooth space.

Wheel Dressing. There are three different types of dressers used for bevel gear and hypoid grinders. They are the single-point diamond dresser, the form dresser, and the rotary diamond dresser.

The single-point diamond dresser uses three natural dressing diamonds mounted in rotatable nibs that are actuated either mechanically with cams or hydraulically through linkages. Shape of the grinding wheel is controlled by changing settings, cams, and/or dresser arms.

The form dresser uses dresser rolls, and the final form of the wheel is controlled by the form on the rolls and related feed between the rolls and the wheel. The dresser is actuated either by a mechanical linkage or by an electrical or hydraulic mechanism.

The rotary diamond dresser is usually numerically controlled and the shape of the grinding wheel is generated by a contouring motion. This type of dresser is usually associated with CNC grinders.

20.9 SPIRAL BEVEL AND HYPOID GEAR LAPPING

Lapping is a polishing operation used to refine the tooth surface and to improve the tooth bearing. The mating gear and pinion are run together under a controlled light brake load while a mixture of abrasive compound and suitable vehicle is pumped onto the pair.

Hardened spiral bevel and hypoid gears, which are required to transmit appreciable loads smoothly and quietly, are nearly always lapped, unless the teeth are to be ground after hardening. Zerol bevel and straight bevel gears are not lapped because they do not have the progressive lengthwise tooth contact required for effective lapping action.

After a gear and pinion have been lapped together, they must remain a pair; they cannot be used as interchangeable single pieces.

Lapping is a routine production operation and should not be considered a salvage operation for the correction of excessive hardening distortion, improper gear cutting, or runout.

To ensure best results and to minimize lapping time, it is necessary to make allowances, in the soft tooth bearing development, for heat-treatment distortion. Therefore, control of both cutting and heat treatment is essential for production lapping. Variations in either of these processes will result in changes in tooth bearing, which require special lapping machine settings or extra lapping time, thus losing the production features of the lapping operation.

20.9.1 Lapping Machine Cycles and Rates

The pair to be lapped is mounted in the lapping machine in much the same manner as for testing (Sec. 23.4). Setup fixtures may be used to position the machine heads at correct mounting distances, and the desired backlash is automatically established by a machine setting. As the pinion drives the gear during the lapping operation, it is necessary that their relative position be changed, so that contact will move progressively from one end of the tooth to the other, thus lapping the entire working surface of the tooth. This is done automatically by the lapping machine.

After the start button is pushed, to start the lapping compound pump and driving spindle rotation, the gears are run for a specific time cycle on one side. This cycle includes a predetermined number of passes during each of which the point of lapping action is moved to "wipe out" entirely the desired portion of the tooth surface. Spindle rotation then momentarily stops, settings are automatically changed to predetermined independent positions for the other side, and the second side is lapped. At the conclusion of the complete lapping cycle, the compound pump shuts off before the gear stops rotating, so that they are free of excessive compound before handling.

Average lapping time for conventional automotive passenger car hypoid gears varies from 4 to 6 min per pair.

20.9.2 Lapping Machine Types

There are two general types of lapping machines, the primary difference between them being in the motions that they impart to the mating gear and pinion during the lapping cycle.

Machines with All-Linear Movements. Machines of this type have a lapping cycle in which the position of the gear is changed automatically in three linear directions, as shown in Fig. 20.28. The movements are accomplished by a cam-controlled primary vertical motion of the gear spindle, and any required change of throw in any of the three directions is accomplished by a simple adjustment of dial-operated controls. In order to determine the movements required, a vertical and horizontal (V and H) check of tooth contact is made for both forward and reverse sides of the gear. The vertical movement is determined directly from the vertical check thus obtained. Horizontal motion is controlled by a dial setting and backlash control is obtained by an automatic axial movement of the gear spindle. The number of passes is set by a counter, and the speed of the movement can be varied between the two ends of the teeth.

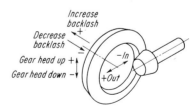

FIGURE 20.28 Gear movements on lapping machines with all-linear motions.

When the machine automatically changes from reverse to forward side lapping, a set-over mechanism provides a means of automatically shifting the relative position and movement of gear and pinion to obtain the correct lapping positions for each side of the tooth, independent of the other.

An adjustable dynamometer-type brake on the driven spindle provides the load applied during the lapping operation.

Machines with Swinging and Linear Movements. Machines of this type use three motions in a horizontal plane to accomplish the lapping action: a swinging motion of the pinion head, a motion in the direction of the pinion axes, and a motion in the direction of the gear axes. The swinging motion is obtained by rotating the pinion head perpendicular to both the gear and pinion axes, and passes through the teeth of the two mating members at a central point of mesh. This is the command motion that causes pinion and gear head slides to reciprocate simultaneously and proportionately.

All motions are independently controlled so machine setup is simplified. The gears are mounted in the machine at standard mounting distances, and, by means of dial settings, trial amounts of the three motions are activated. When the desired extremity of lapping position is reached, the corresponding dial settings are "locked." The procedure is followed to determine motion limits for toe and heel position on both drive and reverse sides.

During the lapping cycle, the machine automatically moves to the aforementioned predetermined limits, thus lapping the entire desired tooth area. Movements and length of time for each side are independently set, and the lapping rate can be made either uniform or variable from one end of the tooth to the other.

An automatic backlash-control mechanism assures that each pair of a production lot is lapped with the same amount of backlash throughout the entire lapping cycle.

Load is provided by an adjustable dynamometer-type brake on the drive spindle.

20.9.3 Lapping Compounds

The compounds used in lapping machines should consist of a mixture of oil and abrasive. The compound must flow readily and still be heavy enough to hold the abrasive grains in suspension. For lapping spiral bevel gears, a compound such as Carborundum C–280–V20S (RI medium) or its equivalent is suitable. Because of the sliding action present, a finer grade, such as Carborundum C–400–V20S (RI fine), is used for lapping hypoid teeth.

20.9.4 Lapping by Manual Positioning

This method of lapping is necessarily slow and is generally used only on large gears beyond the range of the lapping machines, for which production quantities are small. It is often referred to as *brush lapping in fixed position.* The gears are lapped in central, toe, and heel positions on both sides, each position requiring an independent vertical and horizontal setting, as in testing. A lapping paste, which can be obtained with several different grades of abrasive, is applied to the gear teeth with a brush.

20.9.5 Other Spiral Bevel Gear Systems

Dennis Townsend
NASA, Cleveland, Ohio

The Cyclo-Palloid Method for Manufacture of Spiral Bevel Gears. There are two other manufacturing machines used for cutting spiral bevel gears. The Klingelnberg and Oerlikon systems are the same type system and are nearly identical in their method used for manufacturing spiral bevel gears. A description of the cyclo-palloid method is given in the following discussion.

BEVEL AND HYPOID GEAR MANUFACTURING   20.43

During face hobbing, the cutter head that rotates synchronously with the work piece being produced, together with the cutting edges of its cutter blades, embodies an imaginary crown wheel, which meshes and rotates continuously with the gear (Fig. 20.29). The flanks of this imaginary crown wheel represent epicycloids (Fig. 20.30).

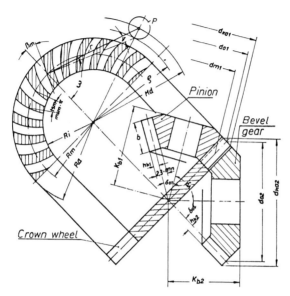

FIGURE 20.29 Pair of cyclo-palloid bevel gears with superimposed crown wheel.

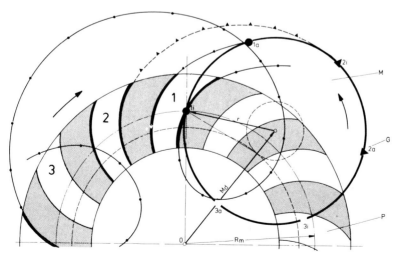

FIGURE 20.30 Development of flank traces that follow the course of extended epicycloids of the imaginary crown wheel during face hobbing. (For clarity, the layout is based on a three-start cutter head.)

From a kinematic point of view, the hobbing feed motion is a supplementary rotary movement of the imaginary crown wheel, which is compensated in the work piece rotation by the CNC control. By comparison, in the forming (cyclomet) method, the components of one pair are not generated. In this case, the teeth of the crown wheels are cut by time-saving plunge cutting.

However, this method can be used only on gears with flat crown wheels (transmission ratios exceeding 1:2.5), and is preferably used for series production.

The spiral angles of the gears being generated can be freely chosen within customary limits: Gears with spiral angles from zero (Zerol gears) up to customary maximum values can be cut.

In addition to spiral and hypoid bevel gears, this method can also be used for the precise and economic production of self-centering couplings for positive connection of coaxial drive shafts.

Infinitely Variable Longitudinal Tooth Crowning. A characteristic of the cyclo-palloid and cyclomet methods is the use of solid and two-part cutter heads, whereby the latter consists of two cutter heads that are enmeshed with each other (Fig. 20.31), with one carrying the inner and the other, the outer flank cutting blades.

Both parts are driven synchronously and in the correct angular position, while rotating about different axes. The distance between these axes (the eccentricity) can be infinitely varied between customary limits. As a consequence, the effective pitch circle radius of the outer cutter head is greater by the amount of the eccentricity than the pitch circle radius of the inner cutter head.

During gear cutting, this produces variable curvatures between convex and concave flanks and thus a limited longitudinal tooth bearing on the flanks when the pinion and crown wheel mesh.

Through this infinitely variable eccentricity setting, it is possible to optimally adapt the amount of longitudinal tooth bearing and thus the length of the contact pattern to suit specific operating conditions. It also permits separate contact pattern modifications for each flank.

Although both gears of a set are usually produced with a longitudinal crowning when universal cutter heads and cutter blades are employed for single-piece and small-batch production, this is by no means an absolute necessity. The same effect can be achieved when only the pinion or the crown wheel have longitudinally crowned teeth. However, in such cases, the longitudinal crowning of the gear or pinion flanks must then be increased accordingly. Consequently, solid cutter heads may be used in conjunction with two-part cutter heads for the cyclo-palloid method, as has been customary with the cyclomet method for years.

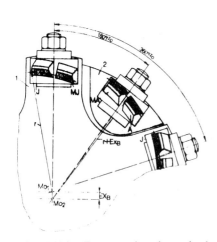

FIGURE 20.31 Two-part universal cutter head (section): 1 = inner cutter head; 2 = outer cutter head; z_o = cutter head starts; r = pitch circle radius; EX_B = crowning eccentricity; J = inner cutter blade; MJ = inner center cutter blade; MA = outer center cutter blade; A = outer cutter blade; MO_1 = inner cutter head axis; MO_2 = outer cutter head axis.

Predeterminable Contact Pattern Formation Without Contact Pattern Developments. The length and position of contact patterns can be precalculated very reliably for the cyclo-palloid and the cyclomet methods, so that the first pair of spiral bevel gears of a layout can be expected to have a good contact pattern, thus eliminating the need for lengthy contact pattern developments.

The precalculated machine setting data required for contact pattern formation are set up easily and quickly on the machine and can similarly be altered quickly, where required.

With a contact pattern produced in this manner, with a uniform tooth depth and a considerably longitudinal tooth curvature, due to appropriately small pitch circle radii of the cutter heads, the tooth system guarantees not only good insensitivity to shifting under load, but also very good strength properties.

Uniform Tooth Depth Along the Entire Tooth Width. The most prominent characteristic of spiral bevel gears produced with multistart cutter heads by the cyclo-palloid or cyclomet method, is a tooth of parallel depth along its entire width. Together with the predeterminable contact pattern formation, this makes the gears insensitive to mounting dimension deviations, a characteristic that is so typical of all cyclo-palloid gears. Deformation of the drive housing or bearings can thus be virtually fully absorbed without causing edge bearing, which would adversely affect the service life of the drive.

20.10 TOOLING SYSTEM

20.10.1 Universal Tools for Single-Piece and Small-Batch Production

The characteristic two-part universal cutter heads for the cyclo-palloid method for soft machining (Fig. 20.32) are suitable for producing pinions and crown wheels: The cutter

FIGURE 20.32 Two-part, five-start universal cutter head.

heads are not for a specific spiral direction and generally have five starts, that is they have five cutter blade groups with inner and outer cutting, profile relief-ground universal blades, which are arranged in a regular sequence. However, by comparison to the cutter heads, these are dependent on the spiral direction, so that cutter blades are required for left-hand and right-hand spiral work pieces.

The universal blades are universally suitable for a specific module range: Differing tooth thicknesses of the crown wheel and pinion can be machined simply by altering the cutter head or the cutter blade setting. However, the tooth number of the gears to be machined should not be a whole number multiple of the number of starts of the cutter head, so as to prevent periodic pitch errors.

20.10.2 Hard Finishing

The cyclo-palloid gear can be hard finished by skiving on an HPG-S machine to produce a gear with finishes approaching ground gears. This process can be used to remove distortions caused by heat treating. The HPG-S tools for hard precision machining are similarly designed: Here again, the two-part cutter heads are mainly independent of the spiral direction. The blades are held in the cutter heads directly or by clamps. They have a brazed cutter strip on their cutting face, consisting of a 0.7-mm-thick layer of polycrystalline cubic boron nitride, which is applied to a carbide carrier. This extremely hard and wear-resistant cutter strip, together with the actual cutter, also forms the tool clearance (Fig. 20.33). All HPG-S blades can be used independent of the spiral direction of the work pieces.

Universal and HPG/HPG-S cutter heads available from conventional cyclo-palloid machines of the AMK series can be easily used on KNC machines, provided the cutter heads have the same connection dimensions. This also applies for soft and hard precision blades.

For finish-machining following case hardening, as an alternative to lapping and hard precision cutting by means of HPG-S gear cutting on KNC machines, the CNC-controlled spiral bevel gear grinding machine W800 CNC—System Wiener (Fig. 20.34) can be used. This machine is used preferably where precision bevel gears with given profile correction such as defined tooth depth crowning, tip and root relief, or root fillets have to be produced.

FIGURE 20.33 HPG-S clamp cutter blade.

20.10.3 Grinding of Constant-Depth, Face-Hobbed Spiral Bevel Gears

It has been stated and assumed in the past that the face-hobbed spiral bevel gears could not be ground because of the epicycloid produced when the gears are hobbed.

FIGURE 20.34 CNC-controlled spiral bevel gear grinding machine W 800 CNC—System Weiner.

However, Weiner[13] states that these gears can now be ground because the radius of an epicycloid can now be determined according to Euler-Savary. The deviations of this radius from the epicycyloid are smaller than the grinding stock, which allows grinding of these spiral bevel gears on the Klingelnberg Wiener system machine shown in (Fig. 20.34).

20.10.4 Description of the Klingelnberg KNC Spiral Bevel Gear Cutting Machine

The Klingelnberg KNC range is based on full utilization of all the possibilities offered by presently available CNC technology. The kinematic motions and setting functions of the conventional cyclo-palloid spiral bevel gear cutting machines have been retained wherever this seemed appropriate with regard to optimum machine rigidity and dynamics. The requirement for "increased flexibility and greater economy" necessitated optimum design adaptation of the basic concept to the possibilities of latest CNC technology. This machine is shown in Fig. 20.35.

The operating movements during the production of spiral bevel gears in a continuous hobbing process consist of three associated rotational movements: both the rotation of the tool and work piece in the ratio of the number of teeth of the work piece to the number of starts of the tool (number of cutter groups in the cutter head), and the feed movement, which can consist of a plunge feed, a generating feed, or a combination of both.

The generating feed consists of a swiveling of the cutter head mounted on the face plate of the machine around its axis (generating rotation axis); the plunge feed takes

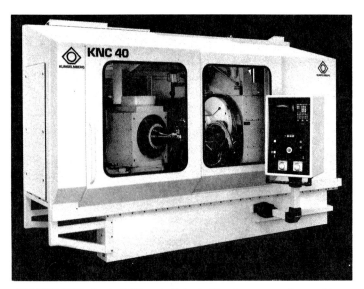

FIGURE 20.35 8-axis, CNC-controlled spiral bevel gear generating machine.

place by moving the cutter head toward the tooth depth. A complex gear mechanism with various change gear ratios was required for these operating movements on conventional spiral bevel gear hobbing machines. This gear mechanism has been completely elimated on the KNC machines. Instead, all operating axes have their own drive units, which are interlinked by means of extremely fast control electronics within the CNC control according to the characteristics of the gear being generated.

The heart of the KNC machines is thus the electronic coupling of the three rotational movements involved in the generating process.

Through program recall, the coupling movements can be adapted to the changing gear data within seconds. The required machine setting is just as fast; with the exception of the offset, all setting and positioning axes are motor-operated and program-controlled for automatic setting, so as to ensure absolute repeat accuracy.

A summary of the individual axis diagram (operating, setting, and positioning axes) is given in (Fig. 20.36)

20.10.5 Life Data for Face-Milled and Face-Hobbed Spiral Bevel Gears

A study was conducted on the difference between the face-milled (variable depth) and face-hobbed (constant depth) spiral bevel gears.[7] Fatigue tests were conducted using three sets of each type of gears. The test results show a life for the face-hobbed spiral bevel gears of nearly two times that for the face-milled gears. While this data is somewhat limited (only three sets each and nonoptimum loading), it does indicate that the face-hobbed spiral bevel gears should give much better life than the face-milled spiral bevel gears.

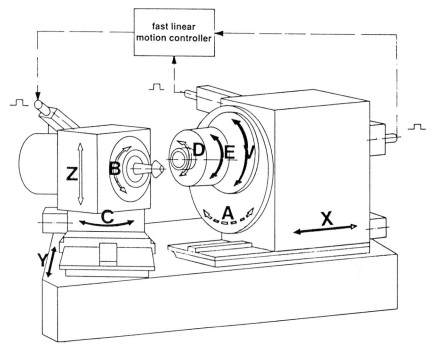

FIGURE 20.36 Axis diagram of KNC machines: D = tool (cutter head) rotation axis; B = work piece rotation axis; A = rolling rotation (feed) axis; X = plunge feed resp. cutting depth setting axis; Y = work piece positioning axis; C = work piece swiveling axis; E = tool positioning axis; V = machine distance setting axis; Z = axial offset setting.

REFERENCES

1. Candee, A. H., "Large Spiral Bevel and Hypoid Gears," ASME Paper MSP–51–9, 1929.
2. Gleason Works, "How to Test Bevel Gears," Gleason Works, Rochester, New York.
3. Gleason Works, "Gleason Bevel Gear Manufacture," Gleason Works, Rochester, New York.
4. Hart, H. J., King, C.B., Pederson, H., and Spear, G. M., Chap. 20, "Bevel and Hypoid Gear Manufacturing," in D. Dudley, ed., *Gear Handbook*, 1st ed., McGraw-Hill, New York, 1962.
5. Kimmet, Gary J., "CBN Finish Grinding of Hardened Spiral Bevel and Hypoid Gears," AGMA Paper 84 FTM6, 19xx.
6. King, C. B., and Spear, G. M., "Helixform: A Recent Development in Spiral and Hypoid Gears," Institution of Mechanical Engineers, Paper 11, London, September 1958.
7. Krenzer, T. J., "Face Milling or Face Hobbing," AGMA technical paper number 90 FTM 13, October 1990.
8. McVea, W. R., "Carbide Hard Finishing Generated Spiral Bevel and Hypoid Gears," SME Paper MR81–993, November 17, 1981.
9. Spear, G., King, C. B., and Baxter, M. I., Jr., "Helixform Bevel and Hypoid Gears, *J. Eng. for Industry, ASME,* August 1960.

10. Wildhaber, E., "Precision Bevel Gears Cut Quickly," *Am. Machinist,* June 7, 25, 1945.
11. Wildhaber, E., "Basic Relationship of Hypoid Gears," *Am. Machinist,* 1946.
12. Wildhaber, E., "Relationships of Bevel Gears," *Am. Machinist,* August 30, September 27, October 11, 25, 1945.
13. Wiener, D., "Grinding of Spiral Bevel Gears in Small Lot Sizes with Different Geometry," presented at the 17th Annual Gear Manufacturing Symposium, Cincinati, Ohio, April 9–10–11, 1989 by AGMA

CHAPTER 21
HIGH RATIO RIGHT-ANGLE GEARING MANUFACTURING

Walter L. Shoulders
Reliance Electric
Columbus, Indiana

In the marketplace dominated by worm gear drives, the reduction ratio of a stage of gearing varies from about 5:1 to about 70:1. This will be our high ratio. All commercial units, consisting of freestanding reducers, NEMA C face-coupled reducers, and integral gear motors, are right-angle units. While there are skew axis worm gears built into custom units, they are beyond the scope of this chapter. Right-angle gearing opens the door to considering Helicon,* high-reduction hypoid, Spiroid,* and the like, in addition to cylindrical worm gearing, and double-enveloping worm gearing.

Rolled worm threads, which are applicable to both cylindrical worm gearing and Helicon gearing, are discussed in this chapter. Molded powdered metal processing, applicable to gearing having teeth on an end of the gear, is also discussed. Hard gear finishing is discussed as an alternate to grinding. Considerations are to cover the effects of tool design and wear on the generating process as it affects fit between conjugate gear surfaces.

As in the first edition of the *Gear Handbook,* the material in Chap. 16 on hobbing and milling, the material in Chap. 22 on hobs and cutters, and the material in Chap. 23 on worm gear inspection is applicable; therefore those chapters are recommended references. Chapter 11 covers rating methods pertinent to high ratio right-angle gearing. Chapter 4 gives general geometric data on cylindrical and double-enveloping worm gearing, and on Helicon and Spiroid gearing.

---

*Helicon and Spiroid are trademarks registered by the Illinois Tool Works, Chicago, Illinois.

21.1 THE REARRANGEMENT OF MATERIAL

21.1.1 The Removal of Material

Machining chips, to remove that which is not to be the worm, worm gear, pinion, gear, or whatever toothed element, is the most common method of manufacturing gears of all sorts. Because of the importance and long standing of chip machining and grinding, extensive study of machines, tools, methods, and techniques has been made. Results have been good and continued study is recommended. Cost of a gear is a function of the speed with which the chips are removed. In recent years there has been an accelerating tendency to make machine tools more rigid and more powerful with tooling capable of taking heavier cuts, or running faster, or lasting longer between sharpenings. *Computer numerical controlled* (CNC) machines result in the elimination of long gear and shaft trains to control timing between the cutting tool and the work piece, thereby improving the precision and finish of the work piece.

21.1.2 The Displacement of Material

Worm rolling, a cold displacement of material in the thread section of cylindrical worms, has become widely used in the last 20 years. It is similar to, and is a logical sequence to, thread rolling on bolts.

There are two methods in use. The through-feed method uses two dies on fixed centers with the stock feeding through guide bushings, producing a continuously threaded rod stock. This may be further processed into parts. The infeed method uses two dies on variable centers, usually with the worm blank held between centers in a plane common to the die centerlines. The infeed method is more commonly used on worm blanks. The great speed at which worms can be rolled, and the as-rolled finish obtainable (compared with cutting operations) recommend it for consideration.

21.1.3 The Compacting of Material

Compacting of powdered metal particles in closed forming dies, and subsequent sintering, is applicable to the manufacture of parts that are columnar in shape and near-uniform in thickness. Uniformity of thickness promotes uniformity in density, therefore, uniformity of strength within the part. Higher compacting pressure also promotes higher density and strength. Closer communication between the die maker, molder, heat treater, designer, and manufacturing people is necessary compared with older established processes. This is because of the wide variety of materials, heat treatments, shrinkage rates during sintering, compacting pressure, and combinations thereof.

21.2 THE THREAD-MILLING PROCESS

A worm thread milling machine is basically a screw-cutting lathe and is designed specially for the production of machine components having thread forms. Its basic elements are a work spindle, cutter head, lead screw, tailstock, and change-gear train. The lead screw actuates the cutter head and provides the latter with a controlled linear motion, parallel to the axis of the worm blank. The gear train provides the proper time relation between the rotation of the work spindle and the linear motion of the cutter head. The cutter head is geared, through the change gears, in the proper time relation with the rotation of the worm blank so that, for each revolution of the work, the cutter advances one lead parallel to the axis of the worm.

The cutter head is mounted on a rotating member so that its axis may be set to any desired angle with respect to the axis of the worm blank. The angle between the axis of the cutter and the axis of the worm is usually equal to the lead angle of the worm being cut at the mean working depth of the thread. Saari[1] shows how the profile of the worm thread can be changed by setting the cutter head angle differently than the mean lead angle. This is useful for changing the contact pattern.

21.2.1 Milling Single-Thread Worms

The foregoing discussion is the procedure to follow for the production of single-thread worms. After the finished worm has been removed from the machine, the cutter is returned to the starting position. A new blank is mounted in place and the cutting cycle is started.

21.2.2 Milling Multiple-Thread Worms

When milling multiple-thread worms, the added operations are those that are made for purposes of indexing. After the first thread is cut, the cutter is retracted from the work so that the cutter tip is moved slightly beyond the outside diameter of the worm. With the gear trains still engaged, the cutter head is brought back to the starting position. A handwheel is provided for this purpose. After the cutter head is in starting position, the indexing device is actuated to rotate the work spindle to the next thread. The indexing mechanism is then secured and the cutter is again set to the full depth of thread. Another complete pass is made through the work. This process is continued until all threads have been cut.

The working area of a thread miller is shown in Fig. 21.1.

FIGURE 21.1 Thread milling. *(Courtesy Reliance Electric Co., Columbus, Indiana)*

FIGURE 21.2 Worm-hobbing operation. *(Courtesy of Barber-Colman Co., Rockford, Illinois)*

21.3 THREAD-HOBBING PROCESS

Hobbing worm threads is usually done when there are three or more threads in the worm. Under certain conditions double-thread worms may be cut in this manner. Hobbing worms requires setting up the hobbing machine the same way as would be done for a helical gear except that a 90° hob swivel is used in most cases because of the low lead angle. If, as in some instances, the number of threads is such that the lead angle permits the use of a standard hob swivel, then, of course, this may be used.
The working area of a thread-hobbing operation is shown in Fig. 21.2.

21.4 DETERMINATION OF CHIP LOAD

Because of the many variables involved in the removal of metal when producing gears of various kinds, the amount of chip load to use for any particular job is a matter of judgment based on long experience in the field. The following definition of this term may be helpful to understand its true meaning.

Chip load is the amount of material each cutter tooth removes per revolution of the cutter.

The chip load used helps to determine the surface finish obtained during cutting. It would be rather difficult to lay down rigid rules for determining the chip load required for different materials. As stated in the foregoing, one's judgment and experience must be used, as there is no other way to establish these data. However, the logical approach to this problem, and one that is used in many instances, is to start with a chip load of 0.025 mm (0.001 in.) and keep making adjustments in speeds and feeds until the results needed are obtained.

For milling worms, the chip loads used vary from 0.013 to 0.075 mm (0.0005 to 0.0030 in.).

For hobbing worms, the chip load used may vary from 0.013 to 0.03 mm (0.0005 to 0.0020 in.).

On thread-milling machines the chip load is varied by changing the rate at which the cutter passes through the work. The speed of rotation of the cutter is independent of the rate of cutter head traverse. Therefore, the higher the rate of traverse, the greater will be the chip load. Experience will dictate what may be used under given circumstances.

When milling threads on a hobbing machine, the problem of determining and varying the chip load is somewhat different. Here we have a machine that was designed, originally, without any thought of using it to mill worm threads. In hobbing machines the rate of rotation of the worm blank is determined by the gears that make up the index gear train.

The equation that may be used to determine the make-up of the index gear train to give the required chip load is as follows:

$$N = \frac{D}{TC_T}$$

where N = number of teeth to index for
D = mean diameter of worm, mm (in.)
T = number of flutes in cutter
C_T = chip load, mm (in.)

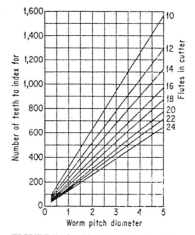

FIGURE 21.3 Chart to determine index for proper chip load.

A chart may be constructed to simplify the procedure. The chart shown in Fig. 21.3 may be used as a first choice. The chart itself is based on a chip load of 0.025 mm (0.001 in.) per cutter tooth per revolution of cutter.

If a finer or coarser chip load is required, the number of teeth, as represented by the ordinate, must be multiplied or divided by the required multiple of 0.025 mm (0.001 in.). In other words, if a chip load of 0.050 mm (0.002 in.) was required, the number of teeth would be divided by 2. If a finer chip load of, say 0.013 mm (0.0005 in.) was required, the number of teeth would be multiplied by 2.

21.5 SPEEDS AND FEEDS

In the manufacture of any product *time* is an important element. The time it takes to produce a worm or gear is a function of the rate of removal of material in a unit interval of time. The rate at which material is removed from a blank is a function of the *speed* of the cutter in revolutions per minute and the *feed* of the cutter in millimeters (inches) per minute or in millimeters (inches) per revolution of the work.

Many different kinds of material are used, and each shop has its own peculiarities that affect the speeds and feeds that may be used. The kind and age of the available equipment must be considered, as well as the ingenuity and experience of the available working force. Whether *climb* or *conventional* milling or hobbing is used may well affect the time taken to produce a part. The type of work-holding fixtures used may be another factor.

The determination of speeds and feeds used in a machine shop should be established with care and by those well grounded in the various processes and materials used for producing their products. The hob and cutter manufacturers can be valuable advisors in this regard. However, each shop should experiment with various combinations of speeds and feeds and determine what is best for their purposes. These experiments need not be formal laboratory tests. They may be effectively carried out during the initial run of the parts. The important point, however, is to keep and maintain proper records of such tests for future reference and use.

The capacity of the machine to deliver power at the point of cutting and the rigidity of the equipment are of vital importance and must be taken into account. Other factors that must be considered are as follows:

1. Depth of cut
2. Material and its cutting properties
3. Hardness of the material
4. Surface finish required
5. Accuracy required
6. Type of coolant used
7. Type of cutter used
8. Material from which cutter is made

21.6 CALCULATE THREAD-MILLING TIME: SINGLE CUTTERS

The following procedure may be used for computing the floor-to-floor time, in minutes, when using single cutters. It may also be used for computing the floor-to-floor time for multistart worms when using a single-type cutter.

The feed rate is usually between 0.013 and 0.076 mm (0.0005 and 0.0030 in.) chip load per cutter tooth. The cutter speeds usually run between 18 and 60 m/min (60 and 200 ft/min).

The following equation will be found very useful for these computations:

$$t = \frac{\pi\,DN}{nCG\,\cos\lambda}\left(\frac{B}{L} + 1.2\right) + F + I(N - 1)$$

where t = floor-to-floor time, min
B = length of full thread, mm (in.)
D = outside diameter of worm, mm (in.)
L = lead of worm, mm (in.)
C = chip load per cutter tooth, mm (in.)
G = number of teeth in cutter
n = speed of cutter, r/min
λ = lead angle of thread, degrees
N = number of threads in worm
F = loading plus unloading time, min
I = indexing time, min

21.7 CALCULATING HOBBING TIME FOR WORMS

The worm-hobbing process has been discussed in Sec. 21.3. The following procedure may be used to determine the floor-to-floor hobbing time when using this process. The following equations may be used in the order given.

$$A = \frac{\sqrt{h_t(D - h_t)}}{\tan H} + \frac{W}{2 \sin H}$$

$$C = \frac{a}{\tan \phi_n}$$

$$t = \frac{N(A + B + C)}{nfT} + F$$

where t = floor-to-floor time, min
 A = hob approach, mm (in.)
 B = length of worm, mm (in.)
 C = extra cut for full generation, mm (in.)
 D = worm outside diameter, mm (in.)
 a = addendum of worm, mm (in.)
 n = speed of hob, r/min
 N = number of threads in worm
 f = feed, mm (in.) of hob advance per revolution of worm
 T = number of threads in hob
 F = loading plus unloading time, min
 H = setting angle of hob swivel, degrees
 W = width of hob face, mm (in.)
 h_t = whole depth of worm thread, mm (in.)
 ϕ_n = normal pressure angle of worm, degrees

21.8 HOBBING WORM GEARS

The hobbing of worm gears may be accomplished by either one of two convenient methods: the *radial* or *infeed method* or the *tangential feed method*. Figure 21.4 shows substantially the principle of each of these methods. Worm gears produced by either of these methods have "throated" teeth, which immediately establish the location of the axis of the mating worm. Also established is the location of the axis of the hob used to produce these worm gears.

The hob must not travel across the face of the worm gear blank as is done in the case of spurs, spirals, and helicals. The hob slide must therefore be fixed in one position in order to obtain the results required. The hob itself may travel radially or tangentially but not parallel to the axis of the gear being cut.

Locating the proper position of the axis of the hob with respect to the throat of the gear being cut is important in the hobbing of worm gears. The procedures used to find

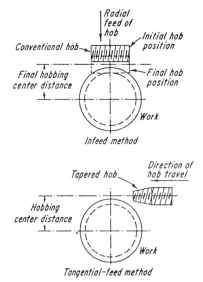

FIGURE 21.4 Illustrating infeed and tangential feed methods of cutting worm gears.

the proper relationship are applicable to both methods described—namely, infeed and tangential hobbing.

The hob setting requires special attention because of the nature of the contact between the profiles of the worm gear and its mating worm. The profiles of both these elements are warped surfaces, and care must be exercised in their production. Because of the throated feature of the worm gear, it would be assumed that the hobs used to produce them should have the exact specifications of the mating worm, and the hob axis should be set to the same angle as the worm is to the gear when operating.

If this is done, the hob would lose its usefulness after the first sharpening because it would then be smaller than the worm. Hobs made to such specifications are not practical, nor are they economical from the standpoint of unit cost per gear. The ideal contact between the worm and its mating gear is line contact across the face of the gear. This would be obtained if the hob had the same specifications as those of the worm. As already noted, on the first sharpening the hob would be smaller than the worm, resulting in reduced radius of curvature of the throat, and contact would then be at the ends of the gear teeth with none in the center.

For those who want to control the relationships of oversize and lead angle between the worm and hob, and the hob axis deviation from the gear centerline, the method described by Wildhaber[2,3] is recommended. Some of the characteristics of this method are:

1. Base pitch of the worm and hob are equal; therefore the lead of an oversize hob is shorter than that of the worm.
2. The ease-off desired at the sides of the gear, with the contact at the gear centerline, varies with the transverse pressure angle; hence, with ratio and, to a lesser degree, with normal pressure angle. When the transverse pressure angle is below 45°, then the ease-off will increase with decreasing oversize. This unusual condition can be found in some 5:1 ratio designs. When the transverse pressure angle is at 45°, then the ease-off will remain constant with changing oversize, and a small ease-off value can be maintained to accommodate other errors. This desirable condition occurs around 6:1 ratio. When the transverse pressure angle is above 45°, then the ease-off will decrease with decreasing oversize, which should start at 0.064 to 0.050 mm (0.0025 to 0.0020 in.) with a new hob, and wear to 0.013 to 0.0 mm (0.0005 to 0.0000 in.), at which time the hob should be replaced. This condition occurs from a ratio of about 8:1 upward.
3. On two similar designs the same hob oversize on the larger specimen will produce the same ease-off as on the smaller specimen.

4. The hob deviation angle Δ for setting up the hobber will vary throughout the life of the hob.

$$\sin \lambda_{P_h} \times D_{P_h} = \sin \lambda_{P_w} \times D_{P_w}$$

where λ_{P_h} = lead angle of hob at pitch diameter at design oversize
D_{P_h} = pitch diameter of hob at design oversize
λ_{P_w} = lead angle of worm at pitch diameter
D_{P_w} = pitch diameter of worm

Where design oversize of the hob describes the size at which the pressure angles of the hob and worm match at the pitch diameter.

When matching axial pitch on the worm to circular pitch on worm gear, then

$$L_H = L_w \times \frac{\cos \lambda_{P_w}}{\cos \lambda_{P_h}}$$

where L_H equals the lead of hob and L_w the lead of worm. At any value of hob oversize

$$\lambda_{o_h} = \tan^{-1} \frac{L_H}{\pi \times D_{P_w} + O_H}$$

where λ_{oh} equals the pitch lead angle of hob at current oversize, and O_h the current oversize of hob, then

$$\Delta = \lambda_{P_w} - \lambda_{o_h}$$

where Δ equals the hob setup angle as shown in Fig. 21.5.

One other source of error occurs in the pressure angle generated on the gear, when the hob teeth are formed by a constant shape back-off tool, rather than being generated from the hob base circle. Figure 21.6 shows the instant radii of curvature generated both ways. Displacing a tool tooth curve radially changes the pressure angle that it generated on the cut part. This produces an increase in pressure angle when the pitch radius is reduced. These characteristics for one 10:1 ratio design follow:

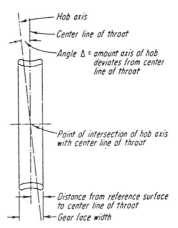

FIGURE 21.5 Showing relationship between hob axis and worm gear when hob is larger than mating worm.

Worm Lead 20.951 mm (0.83485 in.)
Hob lead 20.862 mm (0.82132 in.)

Hob Oversize	Ease-Off, Both Sides	Hob Setup Angle	Gear Pres. Angle
New size 1.45 mm	0.050 mm	1°2′4″	22°20′12″
(0.057 in.)	(0.0020 in.)		
Design size 1.14 mm	0.046 mm	0°50′25″	22°28′8″*
(0.045 in.)	(0.0018 in.)		
Discard size −0.63 mm	0.008 mm	0°−23′−38″	23°22′21″
(−0.025 in.)	(0.0003 in.)		

*This is the worm pressure angle produced by a 22°30′ cone angle grinding wheel of 390 mm (15.5 in.) by Saari's method.

Some manufacturers have the opportunity of changing the center distance, worm pressure angle, diameter, and lead, and are therefore able to use a hob a long time. One such hob that was made in 1917 and is still in use had about 80 percent of the tooth thickness remaining.

21.9 INFEED HOBBING OF WORM GEARS

For this infeed method of hobbing, the index gears are selected from a chart that accompanies the machine. The change gears used to provide the radial feed are also

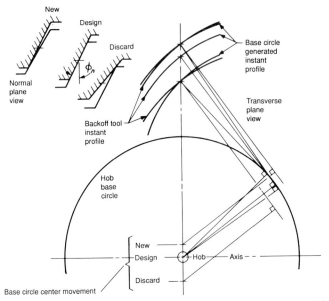

FIGURE 21.6 Worm gear pressure angle θ_n generated from hob teeth cut with backoff tool.

HIGH RATIO RIGHT-ANGLE GEARING MANUFACTURING   21.11

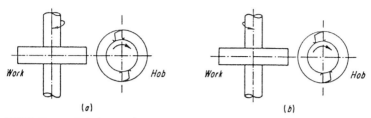

FIGURE 21.7 Relative rotations between work and hob: *(a)* right-hand hob, right-hand work; *(b)* left-hand hob, left-hand work.

selected from a chart provided for the purpose. The change gears for the infeed motion should be selected to give the proper chip load per hob tooth. This will be dictated by the material being cut. The hob speed will also be governed by the material being cut and the finish required. This center distance will be the computed standard plus half the current distance at the hob oversize.

The diagrams in Fig. 21.7 show the proper direction of rotation for both right- and left-hand hobs and work.

The hob is fed in until the outside diameter is very slightly scored by the tips of the hob teeth showing how many teeth are in the gear.

The hob is then fed in to full depth (or to the correct center distance between hob and work) and allowed to make at least one full revolution to complete the generation of the gear teeth. Inspection then follows to determine correct size and tooth contact pattern.

FIGURE 21.8 Cylindrical worm gear hobbing. *(Courtesy of Reliance Electric Co., Columbus, Indiana)*

The infeed method of hobbing is usually used only for relatively low lead angles, particularly for worm gears that may be considered to be of commercial grades. Worm gear hobs have relatively few cutting flutes because their sizes are restricted to be near the size of the worm. Therefore, the worm gear tooth profiles consist of a series of flats or facets, not the smooth continuous curve desired for tooth profiles. An infeed-hobbed worm gear is shown in Fig. 21.8. When the lead angle is high, and where accuracy is required, some arrangement must be made to develop a smooth continuous tooth profile for the gear teeth. The tangential method described in the following section provides the means for achieving this end.

21.10 TANGENTIAL HOBBING OF WORM GEARS

FIGURE 21.9 Tapered hob for tangential hobbing of worm gears. *(Courtesy of Barber-Colman Co., Rockford, Illinois)*

When hobbing worm gears using the *infeed method,* the design of the hob is substantially cylindrical in form and similar to hobs used to produce spur gears. When using the tangential method, the hob is designed as shown in Fig. 21.9. This hob has a tapered section and a cylindrical section. The tapered section provides cutting teeth for roughing out the worm gear teeth, and most of the material is removed by this portion of the hob. The cylindrical section provides cutting teeth for finishing the worm gear teeth.

When this hob is used, the hob slide is fixed in position, as was discussed in Sec. 21.8. However, in order to perform the complete cutting cycle, the hob travels along its axis and tangent to the worm gear blank. The combined rotary motion of the hob about its axis, and its longitudinal motion along its axis, creates a condition similar to that of a hob having a great many cutting flutes. This action develops worm gear teeth with a great many extremely small flats, which results in a tooth profile of high accuracy. The quality of the tooth profile is a function of the feed of the hob along its axis.

Tangential hobbing is not as common as infeed hobbing. The cost of producing worm gears by tangential hobbing is greater than by infeed hobbing, and the deciding factors are quality required and unit cost.

To use this method, a hobbing machine equipped with *differential change gears* or a CNC hobbing machine is required.

The center distance between the work and the hob must be determined by calculation and based on the specifications of both. The final center distance and hob-setting angle will be determined when the proper contact pattern is obtained. The specific details for any particular tangential hobbing machine may be obtained from the operator's manual.

A two-axis computer numerical controlled hobbing machine can be used to tangentially hob worm gears. The increased inherent stiffness produces greater precision than is available using mechanical drive trains.

21.11 FLY CUTTING

A hob may be designed to do a particular job and then one finishing tooth is made in a tool room and mounted on a suitable arbor. This one tooth is then traversed through the work, tangentially, as shown in Fig. 21.10. This type of tool is known as a *fly tool*.

A fly tool has some advantages, in addition to cost, over a complete worm gear hob. It has no lead errors, no spacing errors, and no errors due to profile variation (from tooth to tooth). All these errors are to some degree inherent in a worm gear hob and are reflected in the gear being cut. The fly tool, properly designed, fabricated, and mounted, could have decided advantages over a hob in terms of accuracy.

The fly cutter will rough and finish a worm gear in a single pass, unless the ratio between the number of teeth in the gear and the threads in worm is an integral number. Under such circumstances the fly tool will have to be indexed for each thread in the worm until the gear has been completed.

Another advantage in using a fly tool is in the constancy of the cutting center distance. The fly-cutter specifications can be chosen with precision relative to those of the worm so that the fit between the worm and worm gear teeth can be controlled without sharpening considerations.

FIGURE 21.10 Tangential hobber fly-tooling a steel-mill worm gear of 23 teeth and 3.5 in. C.P. *(Courtesy of Gould & Eberhardt Div., Norton Co., Worcester, Massachusetts)*

21.12 FEEDS FOR USE IN CUTTING WORM GEARS

21.12.1 Infeed Method

When using this method, the feeds used must, of necessity, be less than those used when hobbing spur or helical gears because fewer teeth are doing the work. Depending on the conditions, as outlined in Secs. 21.4 and 21.5, the feeds may vary from 0.075 to 0.50 mm (0.003 to 0.020 in.) per revolution of the work. The material used for the worm gear blank will be a deciding factor.

21.12.2 Tangential Feed Method

The feeds used when this method is being employed may be somewhat greater than for the infeed method. The tapered hob has substantial roughing teeth and the finishing teeth have relatively little metal to remove. Furthermore, the action inherent in the process presents considerably more teeth to the cutting path, which distributes the chip load accordingly.

Certain types of equipment have built into their controls a sequence that permits a rapid approach of the hob to the work, then a coarse feed during the roughing phase and a fine feed for the finishing cut. They may also have a rapid and coarse infeed motion, and then, when this has been completed, the tangential motion commences at a finer feed to complete the work.

The roughing-phase feed rates will vary from 0.075 to 0.50 mm (0.003 to 0.020 in.) per revolution of the work to 0.25 to 1.0 mm (0.010 to 0.040 in.) per revolution of the work for the finishing cut.

21.12.3 Fly-tool Feed Method

Hobbing by this method must be necessarily slower, by far, than the methods using a complete hob. The reason is that one tooth is removing all the material from the spaces. For gears that run with single-thread worms the feed rate may vary from 0.1 to 0.13 mm (0.004 to 0.005 in.) per revolution of the work. Multiple-thread work, particularly that which has a prime or nonintegral ratio, needs finer feeds still.

A single-thread fly tool will pass through each consecutive tooth space during the cutting action. If the feed rate is 0.1 mm (0.004 in.) per revolution of the work, then in one revolution of the gear blank the tool will have advanced 0.1 mm (0.004 in.).

For a double-thread job and 51 teeth in the gear, the blank would have to make two revolutions before the fly tool would again pass through the first space. This means that the tool would have advanced 0.2 mm (0.008 in.) as compared with the aforementioned single-thread job, which had 0.1 mm (0.004 in.) per revolution. For the double-thread job the feed must be reduced to 0.05 mm (0.002 in.) per revolution so that each time the tool entered the first space the tool advance would be 0.1 mm (0.004 in.).

In other words, for prime- or nonintegral-ratio jobs the feed rate for multiple-thread work should be equal to the feed rate for single-thread jobs divided by the number of threads in the mating worm.

The feed rate for multiple-thread jobs having an integral ratio will be the same as for single-thread work. For this kind of work the fly tool *must* be indexed after each complete cycle to the next thread. The action is therefore exactly like a single-thread job and must be handled in the same way.

21.13 TIME REQUIRED TO CUT WORM GEARS

The time required to hob a worm gear may be determined by the following equation. It may be used for either infeed or tangential hobbing. It may also be used for work cut with fly tools.

$$t = \frac{N_G L_t}{nfT} + F$$

where t = floor-to-floor time, min
N_G = number of teeth in worm gear being cut
L_t = total length of travel of hob, mm (in.)
n = speed of hob or fly tool, r/min
f = Feed of hob fly-tool advance, per revolution of work, mm (in.)
T = number of threads in hob or fly tool
F = loading plus unloading time, min

21.14 DOUBLE-ENVELOPING WORMS AND WORM GEARS

The discussions covered so far in this chapter, with regard to worms and worm gears, involved gear sets having one throated or enveloping member. In this section the manufacture of double-enveloping worm gear sets will be of prime concern. With drives of this kind, both members are throated and therefore envelop the mate, providing greater load-carrying capacity. This means that both members must be accurately located with regard to center distance; shaft angle; and, in the axial position, with respect to the mate. Cutting positions, therefore, between the cutting tools and the member being cut must also be accurately located in the hobbing machine.

The hob used to cut the worm gear is as shown in Fig. 21.11, and it will be noticed that the hob is throated the same as the mating worm. The hob is made the same as the mating worm except the teeth are made somewhat thinner. The infeed method is used exclusively for this process.

The hob and the blank are at spread machine center distance so that both may rotate freely without interference. The hob is then radially fed into the blank until the worm and worm gear center distance has been reached. It is only at this final center distance that the correct helix angle is produced. If the hob was made to produce the gear with the correct tooth thickness, parts of the gear teeth would be removed during the infeed phase of the cutting cycle; therefore the hob teeth are made thinner.

After the hob reaches the position of the designed center distance of the gear set, the gear blank is driven first in one direction and then slowed down to take up the machine backlash, which permits the hob to side-cut the other side of the gear teeth to the required thickness. This permits proper generation of the gear tooth form at the correct operating center distance.

It is also important that the proper relationship between the end and side positions of the hob is maintained, and the side-cutting procedure as outlined permits this

FIGURE 21.11 Hob used to cut double-enveloping worm gear. *(Courtesy of Cone-Drive Division, Michigan Tool Co., Detroit, Michigan)*

relationship to be maintained throughout the cutting cycle and at the same time produces the correct tooth form.

21.14.1 Cutting the Throated Worm

Cutting the throated worm is an operation performed on the same machine used to produce its mating gear. The procedure is similar to that used for cutting the worm gear, and the setup is shown in Fig. 21.12. It will be noticed, however, that the worm blank is mounted on the spindle that previously carried the hob. The cutter, similar in construction to a gear shaper, is mounted on the work table.

The cutter used to cut the worm also has thin teeth for the reasons previously explained. The cutting cycle is also the same, consisting of an infeed phase and a side-cutting phase to complete the cycle.

One other operation must be performed to ensure that no improper contact takes place at the ends of the worm. The drawing shown in Fig. 21.13 shows that portion of the worm thread which must be removed to avoid the undesirable contact at that point. Removing the portion of the worm thread to avoid undesirable contact is called relieving.

That portion of the worm thread which contacts the gear tooth profile while entering or leaving engagement cannot extend beyond a distance equal in length to the radius of the base circle and measured in both directions from the common centerline and in an axial direction with respect to the worm. As a general rule, relief should begin at a

FIGURE 21.12 Cutting a throated double-enveloping worm. *(Courtesy of Cone-Drive Division, Michigan Tool Co., Detroit, Michigan)*

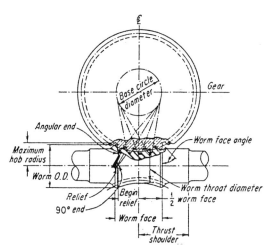

FIGURE 21.13 Portion of worm thread that is removed to prevent undesirable contact.

distance equal to the radius of the base circle minus 0.01 times the center distance. Because of the special nature of this type of drive, it is suggested that the design details of the gears and their respective cutting tools be reviewed or established by the machine-tool and cutting tools manufacturer for the final specifications and method of processing.

21.15 CUTTING TIME FOR DOUBLE-ENVELOPING GEAR SETS

It would be rather difficult to quote speeds and feeds for gears of this type because of the many variables involved and because of the special nature of the components themselves. In general, the average cutting speed used for the gear blank is about 61 m/min (200 ft/min) for hobbing time and about 15.2 m/min (50 ft/min) when cutting the worm. These figures are for gears that operate on center distances of about 6.000 in. These speeds would be increased for the smaller sizes and decreased for the larger sizes. This is due to the variation in the removal of material per revolution, which is related to the circular pitch.

How the cutting time varies can best be shown in terms of typical time required to produce two gear sets of different size. The following values are a good measure of the time, in-house, that it takes to produce such gears:

50.8-mm (2.000-in.) Center Distance Gear Set	Hours
Worm:	
Blank	0.36
Generate and relieve	0.56
Gear:	
Blank	0.11
Hob, chamfer, and match	0.57
Total	1.60
152.4-mm (6.000-in.) Center Distance Gear Set	
Worm:	
Blank	1.28
Generate and relieve	3.53
Gear:	
Blank	0.20
Hob, chamfer, and match	2.49
Total	7.50

These figures cover complete manufacturing time from raw material to final inspection—ready for shipment.

21.16 WORM-THREAD GRINDING

The grinding of worm threads is accomplished by using the same basic principle as described in Sec. 21.2, the difference being the use of a grinding wheel in place of a thread-milling cutter. A grinding machine, used for grinding worm threads, is shown in Fig. 21.14.

The usual procedure is to rough-mill the worm threads, harden, and then finish-grind the threads to the final and desired specifications. In some instances, the worm is ground from the solid when the worm is through-hardened. Case-hardened worms, however, must be rough-milled first, then ground after the case-hardening operation in order to maintain the desired case on the active surfaces of the worm threads.

In many instances worms of high precision are required, and this precision can be maintained by finishing with a grinding wheel or a cubic boron nitride (CBN) layered wheel. The CBN wheel produces very fine chips rather than the swarf associated with grinding.

Be aware of the difference in worm profile produced by a small-radius mill and a large-radius grinding wheel, both with the same cone angle, as can be determined by the method in Saari.[1]

Some of the methods used in thread milling may be used in thread grinding. For the manufacture of fine-pitch worms having a lead less than 5 mm (0.200 in.), a grinding wheel with multiple ribs may be used for high production.

FIGURE 21.14 Grinding worm threads. *(Courtesy of Reliance Electric Co., Columbus, Indiana)*

When designing worms that are to be ground, the blank must be made to allow clearance for the grinding wheel to run out and avoid any damage to the wheel. A commonly used wheel size is 500-mm (20-in.) diameter.

The grinding wheels may be dressed to their proper dimensions and forms using either *diamond dressers* or by using the *crush-dressing method.*

The accuracy and finish obtained by using a single ribbed grinding wheel is greater than when multiribbed wheels are used.

Cubic boron nitride layered wheels do not need to be dressed because the accuracy

TABLE 21.1 Feed Rates Normally Used in Worm Thread Grinding Using Vitreous Wheels

	Feed Rate, In. Per Min	
Linear Pitch	10-In. Wheel	20-In. Wheel
0.250	25	40
0.500	20	35
0.750	15	30
1.000	12	25
1.250	10	20

TABLE 21.2 Suggested Starting Feed Rates for Worm Thread Grinding Using CBN Wheels

Stage	Cutter Speed	Stock Removal	Feed Rate	Finish
Semifinish	2.5 m/s (480 ft/min)	0.13 mm (0.005 in.)	400 mm/min (16 in./min)	0.6 μm (25 μm)
Finish	2.5 (480)	0.013 (0.0005)	1500 (59)	0.6 μm (25 μm)

is put into the wheel form. The CBN-plated layer is uniform in thickness. The life of the CBN layer, in length of cut, equals 10,000 times the wheel diameter, after which it can be stripped and replated.

The limited data in Table 21.2 are drawn from helical gear grinding using CBN (cubic boron nitride) wheels. The rate of doing work is the same in finishing a pinion or a worm of the same tooth depth and stock thickness; therefore it is believed that data transfer is feasible. Due to the early state of development in the art, experimenting may prove beneficial in improving productivity. Circular pitch on the teeth was 13 to 20 mm (0.52 to 0.78 in).

21.16.1 Production Time

The following equation may be used to estimate the actual grinding time. To this must be added the time required for loading and unloading.

$$T = \frac{n}{C_T}\left(I + \frac{L_d}{R_f}\right) N$$

where T = actual grinding time, min
n = number of threads
C_T = threads per cut
I = indexing time, min
L_d = developed length of worm, mm (in.)
R_f = feed rate, mm/min (in./min)
N = number of cuts

The length of the thread L_d, as developed, may be computed as follows:

$$L_d = \frac{\pi \, dF}{L \cos \lambda}$$

where L_d = developed length of worm, in.
d = diameter of worm, in.
F = actual length of worm face, in.
L = lead of worm, in.
λ = lead angle of worm, degrees

21.17 WORM ROLLING

Parameters that affect the design of the rolled worm, the rolled worm blank, the rolling dies, and the cost are:

Whole thread depth/rolled worm outside diameter, ratio
For single thread worm, ratio = $\frac{1}{6}$ max.
For even number thread worm, ratio = $\frac{1}{4}$ max.
For odd number thread worm, ratio = $\frac{1}{5}$ max.

Length of fully formed threads needs to run from the first contact on the approach action boundary to beyond the last contact on the recess action boundary. How to determine the approach and recess action boundaries is explained in Buckingham and Ryffel.[4] Unless the drive in which this worm is used is unidirectional, the thread length should be the greater of: (1) twice the distance to the approach boundary, or (2) twice the distance to the recess boundary.

Refer to Fig. 21.15. The diameter of the worm blank, *BD*, in the fully threaded section is about mean thread diameter. This varies with designs, particularly when the worm thread thickness is not half pitch. See an equation for a trial blank diameter in Fig. 21.15. To get tightly rolled thread crests and hold pin size on the worm usually

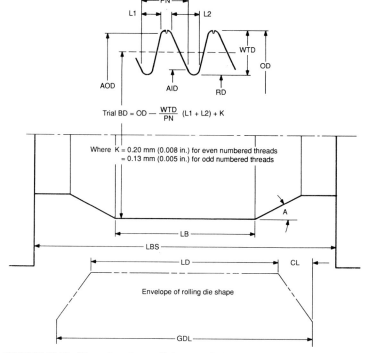

FIGURE 21.15 Dimensions for a rolled worm blank.

requires experimenting with this diameter, with the final adjustment about 0.025 mm (0.001 in.).

The length of the worm blank diameter LB, above, should be 1.5–2.0 pitches longer than the length of the fully formed threads.

When the worm body diameter beyond the threaded section is slightly less than the root diameter, the taper angle A in the transition sections should not exceed 30°; however a 20° angle will lessen stress on the die thread and reduce bending forces on the worms and should be used where room permits.

Full diameter die length LD should be one thread pitch longer than the length of the worm blank diameter LB to allow for axial feed during penetration. This is a rough allowance and may be reduced on any specific design.

Gross die length GDL is the full diameter die length LD plus two chamfer lengths CL.

Length between shoulders LBS to clear the die during plunge feeding, is the gross die length plus one axial pitch, or whatever was decided to accommodate axial feed.

All tolerances should be outside these limits.

To produce structurally sound worms, maximize die life, and minimize cost, the following parameters are recommended:

Use a 20°–25° pressure angle to facilitate metal flow during forming and to produce a shorter-length worm requirements.

Use a full fillet root on the worm thread.

Use a generous radius on the worm tooth tip (thread roll dedendum) for die strength and good thread hat forming. Note that this superstructure is above the working thread form, does not contribute to capacity, should not interfere with the root of the worm gear, and need not be removed just to make the worm look conventional.

When the worm blank is unrelieved, then the die threads must taper beyond the working length LD at great expense. This option is not recommended.

The profile produced by the dies on the rolled worm should be controlled to fit the profile produced on the worm gear by the hob used. There are many profile generating methods available, substantially equal in result, but it is necessary that both mating parts be of one family.

21.17.1 Unsymmetrical Worms

Rolling is feasible for a uniform-pitch, cylindrical-form worm with different pressure angles on opposite sides. This form is used in Helicon gearing. By keeping the average pressure angle in the 20°–25° range, material flows well. The position of the crease in the hat on unsymmetrical threads may be moved by changing direction of rotation of rolling.

Material that rolls well has uniformity, absence of MnS or other hard inclusions and has at least 12-percent elongation in finer pitches to 27 percent in coarser pitches.

The larger the pitch, the lower the Brinell hardness number (BHN) allowable. Recent data recommends as an upper limit, 6.35 module (4 normal diametral pitch) for infeed rolling and 3.18 module (8 normal diametral pitch) for through-feed rolling. Check with the vendor since these data change rapidly.

Time to infeed roll a 19-mm (0.75-in.) outside diameter worm runs about 12 s per part floor to floor, and a 38-mm (1.50-in.) worm runs about 20 s, using manual loading and unloading.

A worm being rolled is shown in Fig. 21.16.

FIGURE 21.16 Worm rolling. *(Courtesy of Reliance Electric Co., Columbus, Indiana)*

21.18 CARBURIZING AND OPTIONAL FINISHING

Most cylindrical worms and many other forms of right-angle gearing are carburized to provide a hard and stable surface with which to wear in the mating part. The case depth need be only half to two thirds that of carburized parallel gearing due to the lower working stress.

When great accuracy, low noise, or low backlash is needed, it may be necessary to post–heat treat finish the threads, as described in Sec. 21.16.

21.19 POWDERED METAL COMPACTS

The material must be carefully chosen in order to successfully manufacture gearing with high sliding velocities between mating parts and low rolling velocities perpendicular to the contact lines. The lower the lead angle, with conventional cylindrical worm gearing, the more nearly horizontal are the contact lines. This gives low oil film thickness, in the neighborhood of 0.05–0.1 μm (0.000002–0.000004 in.). Asperities on the worm alone are several times this high. In addition, the gear is brought to only approximate shape by the hobbing process. It is estimated that about 1–4 million cycles are required to break in a worm gear set. Bronze is a good worm gear material because it does not weld easily to steel. Higher ratio worm gears tend to wear, whereas the surfaces of lower ratio worm gears tend to pit when they fail. This is because the oil film thickness is greater on the lower ratio gearing due to an increase in the inclination of the contact lines. By carrying this contact line situation to its logical conclusion, which is exhibited in the hourglass worm drive and in Helicon gearing, the contact lines are nearly perpendicular to the direction of sliding between the mating parts. This produces a greater oil film thickness, which makes steel gearing feasible. The break-in problem is still present; therefore, a phosphate coating should be used to last through this period.

One of the desired attributes of the compact is that it be to finished dimension. It is therefore desirable to minimize growth or shrinkage variation from part to part and from run to run. It is particularly desirable to establish a stable base pitch on the gear through control of the processing sequence before making worm rolling dies.

21.19.1 Classification of Compact Shapes and the Die Requirements

Class I parts are single-level parts. These can be made with a female die and a single punch. Because of the tendency for density of the compact to decrease with distance from the punch, significant variation begins at about 8-mm (0.31-in.) thickness. When gears are molded in this arrangement, the teeth are in the lowest density part of the compact.

Class II parts are single-level parts of any thickness pressed from both ends. These require at least two punches, a die, and usually a core rod. Of the first three parts, at least two have to move to allow compacting action from both ends. This produces symmetry of density with the lowest density near the central plane between the ends. This would make a toothed section only of a gear with end teeth, such as Helicon.

When this set of dies is operated with the bottom punch and die held and only the top punch moves, then Class I parts would be produced. The bottom punch is often used to eject the work from the die.

Class III parts are two-level parts of any length pressed from both ends. These require a punch for each level, usually on the bottom end; at least one punch for the top end; a die; and usually a core rod.

This could make a complete gear, but the hub and web inside the toothed portion would have to be one length. This is not common in gearing.

Class IV parts are multilevel parts of any thickness pressed from both ends. Required are individual bottom punches for each level to control powder fill and density, at least one top punch, a die, and usually a core rod. This allows selective density and thickness in the toothed portion, web, and hub section; and is capable of using minimum material overall.

Figure 21.17 shows some of these possibilities.

The press size available can be influential on the compact configuration. With a compacting pressure in the neighborhood of 550 MPa (40 tons/in.^{2}), a 7.2-MN (800-ton) press could process only a 12,900-mm^2 (20-in^2) part. Three size categories emerge: (1) smaller parts, which in total are within press capacity; (2) intermediate parts, of which only the toothed area can be molded; and (3) parts too large for the press, which would be chip machined. Powder forming presses to the 22-MN (2500-ton) range are available. There are presses of the same wire-wound construction, used in plate forming operations, to 535 MN (60,000 ton).

A thorough discussion of processes and equipment can be found in the *Metals Handbook.*[7]

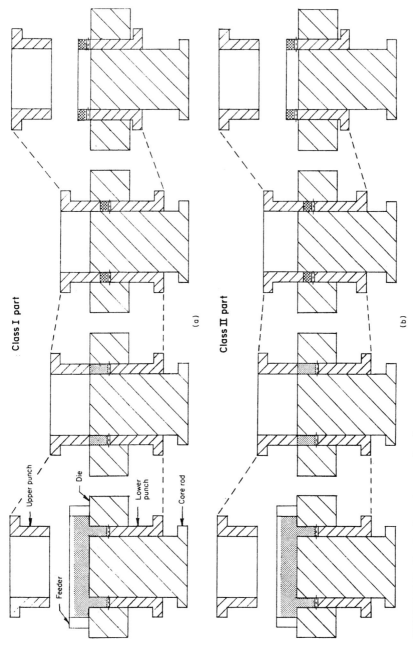

FIGURE 21.17 Powdered metal compacting dies.

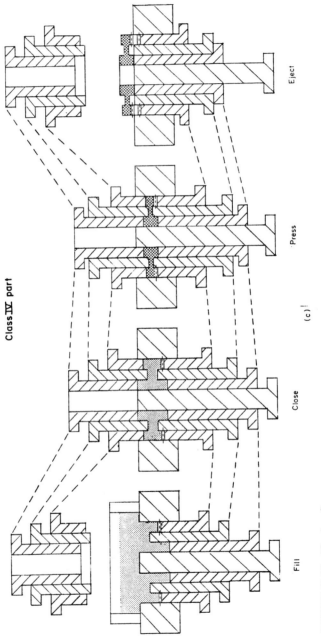

FIGURE 21.17 (*Continued*)

Published test data for Hertz contact stress versus life for powdered metal materials is extremely scarce. Therefore, these data should be obtained for life analysis. One way of obtaining these data is by use of the roll test machine. Figure 21.18 shows a failed specimen in place on one of these machines. United Shoe Machinery built and used this type of machine from the late 1920s, and Caterpillar Tractor Company built a heavier-duty machine later, the design of which was turned over to Mechanical Technology Inc., who currently markets them. Several are in use and some joint industry studies have been made utilizing them, a few of which were sponsored by the American Society of Mechanical Engineers.

FIGURE 21.18 Failed bronze roll test specimen. *(Courtesy of Reliance Electric Co., Columbus, Indiana)*

Some users of these machines are not confident in the absolute accuracy of results. This leads to the ultimate selection by making and testing parts in the machine in which they will be used. There is value in testing rolls for screening materials, lubricants, heat-treating process, and combinations thereof. Regression analysis of six data points, two specimen at each of three loads, has proven to constitute sufficient data in accordance with the method of Hill.[8]

It may be of interest to look at the variation in oil film thickness on the 10:1 ratio worm gear described earlier in the chapter. By using an elastohydrodynamic lubrication equation similar to those in Dowson and Higginson[9] and Ku,[10] oil film thicknesses near both ends and the center of one continuous contact line for four conditions of size and speed are shown in Table 21.3.

Due to the high sliding velocity, and the necessity to break in the parts, the following recommendations are made for selecting materials:

Highest yield strength consistent with other requirements.

Some malleability is required, suggesting a matrix hardness in the low to middle 30s Rc. This can be obtained by drawing heat-treated parts.

Materials suggested for trial are medium-carbon iron and copper mixtures, low alloys containing little nickel, and copper infiltrated steel such as the following from Metal Powder Industries Federation Standard 35:

FC–0205–80 HT + Draw to 30–35 Rc
FL–4205–100 HT + Draw to 30–35 Rc
FX–1005–110 HT + Draw to 30–35 Rc

It is best to avoid much nickel, particularly in the free state, in a high sliding gear design since it is one of the most weldable of metals.

Powdered metal parts are associated with long production runs, whereas worm gear production runs tend to be short because there are so many ratios and sizes in a typical industry product line. The following will improve this situation for production, by adding in lieu of a coupling a stage of spur gears between the motor and worm shaft, as shown in Fig. 21.19, whereby the following spectrum can be made:

Number of Teeth On Motor Shaft Gear	Number of Teeth On Worm Shaft Gear	High Ratio Right-Angle Stage	Overall Ratio	ISO 10/10 Series
34	19	9:1	5.03	5.01
31	22	9:1	6.39	6.31
28	25	9:1	8.04	7.94
25	28	9:1	10.08	10.00
22	31	9:1	12.7	12.6
19	34	9:1	16.1	15.8
34	19	36:1	20.1	20.0
31	22	36:1	25.5	25.1
28	25	36:1	32.1	31.6
25	28	36:1	40.3	39.8
22	31	36:1	50.7	50.1
19	34	36:1	64.4	63.1

TABLE 21.3 Oil Film Thickness Study

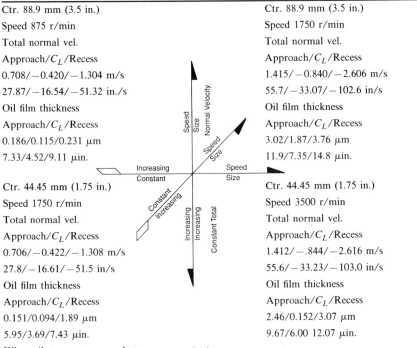

Ctr. 88.9 mm (3.5 in.)
Speed 875 r/min
Total normal vel.
Approach/C_L/Recess
0.708/−0.420/−1.304 m/s
27.87/−16.54/−51.32 in./s
Oil film thickness
Approach/C_L/Recess
0.186/0.115/0.231 μm
7.33/4.52/9.11 μin.

Ctr. 88.9 mm (3.5 in.)
Speed 1750 r/min
Total normal vel.
Approach/C_L/Recess
1.415/−0.840/−2.606 m/s
55.7/−33.07/−102.6 in/s
Oil film thickness
Approach/C_L/Recess
3.02/1.87/3.76 μm
11.9/7.35/14.8 μin.

Ctr. 44.45 mm (1.75 in.)
Speed 1750 r/min
Total normal vel.
Approach/C_L/Recess
0.706/−0.422/−1.308 m/s
27.8/−16.61/−51.5 in/s
Oil film thickness
Approach/C_L/Recess
0.151/0.094/1.89 μm
5.95/3.69/7.43 μin.

Ctr. 44.45 mm (1.75 in.)
Speed 3500 r/min
Total normal vel.
Approach/C_L/Recess
1.412/−.844/−2.616 m/s
55.6/−33.23/−103.0 in/s
Oil film thickness
Approach/C_L/Recess
2.46/0.152/3.07 μm
9.67/6.00 12.07 μin.

When oil, temperature, and stress are constant,
then: At constant r/min, oil film thickness $\propto$ (size ratio)$^{1.0}$
At constant size, oil film thickness $\propto$ (r/min ratio)$^{0.7}$
At constant total normal velocity, oil film thickness $\propto$ (size ratio)$^{0.3}$

HIGH RATIO RIGHT-ANGLE GEARING MANUFACTURING   21.29

Thus, with two ratios of high ratio right-angle stages and six input gears, 12 overall ratios that approximate the ISO $^{10}\sqrt{10}$ preferred number series can be obtained. This makes for better interchangeability with parallel gear units than does the arithmetic series current in American worm gear practice.

To benefit from compacting and sintering processes versus chip machining, some economy must result. Continuous cast-phosphor bronze tubing is about $2.00 per pound, 8620 H steel is about $0.40 per pound, and iron powder is about $0.40–0.50

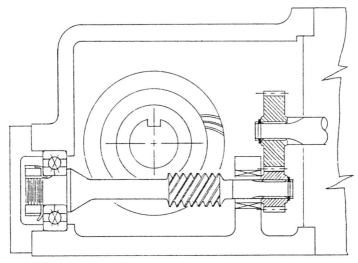

FIGURE 21.19   Spur gears (shaded) acting as a coupling.

TABLE 21.4   Sequence of Events and Time Required to Compact and Sinter Powdered Metal Gear

Event	Time	Remarks
Compact powder in die	900–3000 parts per hour	
Sintering at 1120°C (2050°F)	30 min, mesh belt conveyor loading density 49 kg/m² (10 lb/ft²)	Cost can be evaluated from time and tonnage
Infiltrating	Copper or alloy wafer, simultaneous with sintering	
Heat treating		Same as wrought steel
Sizing	120–240 parts per hour, manual load and unload	Recommended for improved accuracy tooth to tooth
Repressing		Alternate to sizing plus 3–5% densification by using greater force
Resin Impregnation		When not infiltrated
Phosphate Coated		To aid break-in

per pound. Yield on machined parts may run about 50 percent, thereby increasing effective cost per pound of the finished part. Copper infiltrating will add about 30 percent to P/M material cost.

REFERENCES

1. Saari, O., "Nomograph Aids Solution of Worm-Thread profiles," *American Machinist*, July 5, 1954, pp. 113–116.
2. Wildhaber, E., "A New Look at Wormgear Hobbing," AGMA Paper No. 129.10, June 1954.
3. "Discussion on the Wildhaber Paper," AGMA Paper No. 129.13, December 1954.
4. Buckingham, Earle, and Ryffel H., *Worm and Spiral Gears*, The Industrial Press, New York, 1960, pp. 270–288.
5. Schwartz, R. W., and Rao S. B., "A Wheel Selection Technique for Form Gear Grinding," AGMA Paper No. 85FTM7, October 1985.
6. MPIF Standard 35, "Material Standards for P/M Structural Parts," Metal Powder Industries Federation, Princeton, New Jersey.
7. *Metals Handbook*, 9th ed., Vol. 7, Powder Metallurgy, American Society for Metals, Metals Park, Ohio.
8. Hill, W. J., "Regression Analysis Digs Out Equations from Scattered Data," *Product Engineering*, July 20, 1959, pp. 62–64.
9. Dowson, D., and Higginson G. R., *Elasto-Hydrodynamic Lubrication*, Elmsford, New York, Pergamon, 1966, pp. 190.
10. Ku, P. M., ed., "Interdisciplinary Approach to the Lubrication of Concentrated Contacts," NASA SP-237, 1970, pp. 48.
11. Dudley, D. W., ed., *Gear Handbook*, 1st ed. McGraw-Hill, New York, 1962.

CHAPTER 22
GEAR CUTTING TOOLS

A. Donald Moncrieff
Gear Consultant, Moncrieff Associates
Vero Beach, Florida

Gear cutting tools are complex and expensive. This chapter is written to acquaint you with the basics, as well as some of the improvements, developed for the manufacture of gear tools made during the last 30 years. There have been important developments in gear tools as well as the machines that use them. One of these developments, which is applicable to all gear cutting tools, is the PVD method used to coat HHS tools with a thin coating of titanium nitride (TiN), which appreciably increases tool life. Another development is the use of cubic boron nitride (CBN). Either plated tools with a single layer of CBN or bonded tools with a thicker layer of CBN crystals are used to finish gear teeth. This is called hard gear finishing and is done on a hard gear finishing machine, which is usually controlled by CNC. This chapter describes these developments and continues with a description of the tools used.

22.1 HOBS FOR SPUR AND HELICAL GEARS

The first patents on a hob were issued to Joseph Whitworth in 1835. Since then there has been relatively little change in the appearance of the hob, except for the addition of proof diameters and other refinements. Although the first hobs would not be considered precision tools by today's standards (which produce such accurate gears that the finest equipment is needed to inspect them), they provided the most efficient ways of making gears in those days.

22.1.1 Principles of Hobbing

The generation of gear teeth is a continuous-indexing process in which the hob and the gear rotate in a timed relationship, while the hob is fed through the work. The hob is basically a worm that has been fluted, and each tooth has been form-relieved. The flutes

---

The author of this chapter in the first edition was Richard S. Hildreth, Chief Engineer, Michigan Tool Company, Detroit, Michigan.

provide the cutting edges. Each tooth is relieved radially to form clearance behind the cutting edges, to allow sharpening while retaining the original tooth form.

The hob is used to cut teeth in external spur or helical gears, splines, serrations, worm gears, worms, and special forms. The largest application of hobbing is the cutting of involute gears, with the hob having teeth with straight sides at a given pressure angle. Each hob tooth cuts its own profile, but the accumulation of these cuts produces a curved form on the gear tooth.

Types of Hobs by Mounting. Arbor-type or shell-type hobs have straight or tapered holes for mounting on an arbor. They usually have a keyway in the bore, or sometimes in the end face, for driving purposes.

Shank-type hobs are equipped with straight or tapered shanks to fit the hobbing machine spindle or adapter. It is usual practice to have a pilot diameter or center to support the end opposite the shank end.

Hand of Hob. This defines the thread helix of the hob teeth. A right-hand helix has a thread helix corresponding to a right-hand screw. A left-hand helix has a thread helix corresponding to a left-hand screw. It is sometimes necessary to qualify the hand specification further by specifying "top coming" or "bottom coming." This terminology is used when a hob is equipped with a special feature and it is necessary to define its location.

Method of Profile Generation. The hob threads may be excluded as a factor in understanding the manner in which a hob generates. As opposed to a milling cutter, a hob has a conjugate form to produce the part form. A milling cutter has a counterpart form. In determining the conjugate form, a simple rack profile can be utilized. The straight-line elements of the rack are combined to produce involute profiles, the pitch line of the rack representing the hob pitch line. For hobbing noninvolute forms on a part, the rack pitch line is the basis for determining the conjugate hob form. From the engineering standpoint, the rack pitch line comes into tangent contact with the selected generating diameter of the noninvolute form. The rack contour is then back-generated from a layout of the part to be formed by hobbing. This back-generated form becomes the radial form of the hob cutting edge.

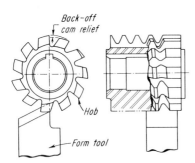

FIGURE 22.1 "Backing-off" hob with form tool.

Manufacturing. Once the necessary hob form for generating the part is known, this form is transferred to a form tool. The form tool is identical to a lathe tool and is used in almost the same manner. The tool is accurately ground with a form equivalent to the space of the hob space that was generated (see Fig. 22.1). The tool is then used on a relieving lathe to cut the hob form. The relieving lathe is equipped with a cam-controlled saddle that plunges the tool deeper into the hob during the tooth forming. The hob has been previously gashed axially with a desired number of flutes. The thread lead has usually been roughed in on a thread mill. The relieving lathe is geared to feed the cam-controlled saddle parallel to the hob axis on the desired thread lead. The saddle is also geared for the same number of relieving cycles as there are flutes in the hob. The form tool then meets the proposed cutting edge of the hob and feeds in towards the hob axis in a precise amount in time with the hob rotation. Upon clearing the heel of the hob tooth, the tool is in the gash space. It then retracts to meet the next cutting section, and the cycle is cam-controlled. The cam used is selected to produce the desired clearance and relief at the outside diameter of the hob. The process is referred to as *backing off* a hob. At this point, the hob is ready for heat treatment and subsequent operations. If the hob is to be unground, it will not have a grinding operation on the tooth profile. The bore, hubs, and proof diameter will be ground after heat treatment. After sharpening, if the hob meets classification tolerances, it is ready for use. Heat-treat growth is allowed for at "back-off time" by using a shorter thread lead during forming. The hob manufacturing process is reliant on the consistency of steel and heat-treat practices.

22.1.2 Hob Nomenclature

Figure 22.2 shows the basic elements of an arbor-type hob. Figures 22.3 and 22.4 show relative hob features and the resulting gear contour after hobbing. Features such as semitopping and protuberance should be considered for the tooth-number range designated by the hob manufacturer. For example, if a semitopping chamfer is designed into a hob for a 16-tooth pinion, a greatly increased chamfer would be produced on a 60-tooth gear. This results in decreased contact of the gear's outside diameter in the hob space.

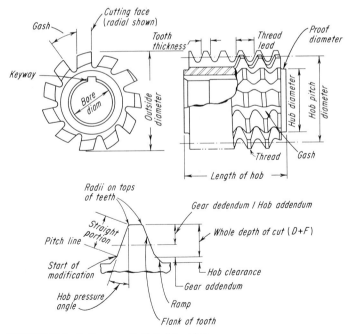

FIGURE 22.2 Basic elements of shell-type hob.

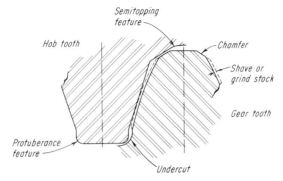

FIGURE 22.3 Semitopping hob with protuberance.

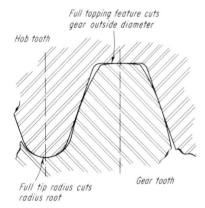

FIGURE 22.4 Full-topping hob with full tip radius.

22.1.3 Hob Classification by Threads

Hobs are classified as single- or multiple-thread hobs. The single-thread hob, when used in conjunction with a gear, is interconnected for rotation by means of a positive system. For example, when a single-thread hob is used on a 30-tooth gear, each revolution of the hob is the equivalent of a one-tooth index of the gear being hobbed. Therefore, 30 revolutions of the hob would be required while the 30-tooth gear is making one revolution. A double-thread hob in one revolution would be in timed relation to a two-tooth index of the gear. A double-thread hob would be making 15 revolutions while the gear is making one revolution. As may be noted, there are some advantages of a multiple-thread hob in terms of production time. This production time is not proportionate to the increase in the number of threads and could not be so unless the number of flutes were increased in proportion to the number of threads.

The gashes or flutes can be straight and parallel with the hob axis or can be of the spiral type on a lead, such that the spiral would be normal to the thread angle of the hob. The spiral gash is usually incorporated in hob design when the thread angle is in excess of 4°, as caused by a coarse pitch or multiple threads.

The greatest accuracy is obtained by the single-thread design as opposed to the multiple-thread design. This results from the "hunting-tooth" condition of a single-thread hob. It is very desirable to avoid the use of multiple-thread hobs when the number of threads in the hob is divisible into the number of teeth in the gear being hobbed. Inherent accuracy of the hobbing process is negated if this condition exists.

Another consideration in the use of multiple-thread hobs is the fact that the speed of the indexing worm gear increases when multiple-thread hobs are used. In the case of low numbers of teeth, it is necessary to check the speed of the index worm gear to ascertain that it does not exceed the maximum speed recommended by the hobbing machine manufacturer.

It is noteworthy that improved methods of manufacturing have increased the tolerance quality of the class C accurate unground hobs, enough so that they are frequently compared with class B hobs in terms of tolerance and are frequently used for finishing operations if the accuracy of the part permits this application.

The selection of the number of flutes to be introduced into a hob depends on tooth depth and hob diameter. It is pretty well standardized with the cutting tool manufacturers in order to give maximum tool life for any particular size of hob. Increasing the number of flutes in a hob has the tendency of reducing the amount of available sharpening life on the hob.

Many features can be added into a hob, such as protuberance, semitopping, and special involute control, including modified involute, or full-fillet radii, etc.

22.1.4 Classification by Tolerance

Hobs are generally classified into four common areas:

1. Class A, precision
2. Class B, commercial ground
3. Class C, accurate unground
4. Class D, commercial unground

The class A hob is a ground precision-type hob used most often as a finishing hob where no subsequent operations, such as shaving or grinding, would be involved.

The class B commercial ground hob is used for both finishing and semifinishing preparatory to grinding or shaving.

The class C accurate unground hob is most commonly used as a semifinishing hob, preparatory to grinding or shaving as the finishing operation.

The class C hob is an unground hob, formed by "backing off" in an annealed state and then heat treating. Except for the grinding of the bore, proof diameter, and hub faces, no further refinement of the tooth form, spacing, or lead is made.

With the class A and B ground hobs, the subsequent grinding operation is for the refinement of the limits on profile, lead, and spacing.

Class D hobs are of commercial unground tolerances, generally used as roughing-type hobs.

Ultraprecision, or class AA, hobs are available from the ground hob manufacturers. Because of manufacturing problems, it is advisable to consider ultraprecision hobs as single-thread hobs only.

22.1.5 When and How to Select a Hob

Of primary importance are the physical characteristics of the part to be hobbed. Will it be arbor-mounted or shank-mounted? In either case, common-sense consideration in providing rigid support will be rewarding. The part to be hobbed must not contain interfering adjacent shoulders or diameters. In some cases it is permissible to hob into adjacent bearing diameters. In this event, the hob depth is increased, as shown in Fig. 22.5.

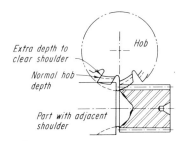

FIGURE 22.5 Hob designed with extra depth to avoid complete mutilation of shoulder.

Practically all profiles that are equally spaced on a cylindrical base surface can be hobbed. The profiles need not be symmetrical, such as with involute gear teeth or straight-sided splines. Nonsymmetrical part profiles such as ratchet teeth can be readily hobbed. It is desirable that all profiles to be generated by hobbing have greater than a radial form.

When it is impossible to hob by conventional multiposition design of hobs, a single-position nongenerating hob may be used. A single-position nongenerating hob allows cutting of contours that exceed the capabilities of a generating-type hob. Small fillets and radii that are impossible to produce by generating can be produced by single-position hobs.

Hob Size. The sizes of hobs used to generate involute spur and helical gears have been standardized. Table 22.1 shows the standard sizes for hobs in the coarse-pitch series.

TABLE 22.1a Single-Thread Coarse-Pitch Hob Sizes for Ground and Unground Hobs

(1–19.99 Normal Diametral Pitch) For Spur and Helical Gears

Normal Diametral Pitch	Nominal Hole Diameter	Outside Diameter	Overall Length
1	2½	10¾	15
1¼	2	8¾	12
1½	2	8	10
1¾	2	7¼	9
2.00–2.24	1½	5¾	8
2.25–2.49	1½	5½	7½
2.50–2.74	1½	5	7
2.75–2.99	1½	5	6
3.00–3.49	1¼	4½	5
3.50–3.99	1¼	4¼	4¾
4.00–4.99	1¼	4	4
5.00–6.99	1¼	3½	3½
7.00–7.99	1¼	3¼	3¼
8.00–11.99	1¼	3	3
12.00–13.99	1¼	2¾	2¾
	¾	2	2
14.00–19.99	1¼	2½	2½
	¾	1⅞	1⅞

Source: Extracted from AGMA 120.01 Standard, *Gear-cutting Tools, Fine- and Coarse-Pitch Hobs,* with the permission of the publisher, American Gear Manufacturers Association, 1500 King Street, Suite 201, Alexandria, Virginia 22314.
All dimensions in inches.

GEAR CUTTING TOOLS 22.7

TABLE 22.1b Multiple-Thread Coarse-Pitch Hob Sizes for Ground and Unground Hobs
(2–19.99 Normal Diametral Pitch) For Spur and Helical Gears

Normal Diametral Pitch	Number of Threads	Nominal Hole Diameter	Outside Diameter	Overall Length
2–2.99	2	1½	6½	8
3–3.99	2	1½	5½	5½
4–4.99	2 or 3	1½	5½	5½
5–6.99	2, 3, or 4	1½	5	5
7–7.99	2, 3, or 4	1¼	4	4
8–8.99	2, 3, or 4	1¼	3¾	3¾
9–11.99	2, 3, or 4	1¼	3½	3½
12–13.99	2, 3, or 4	1¼	3¼	3¼
14–15.99	2, 3, or 4	1¼	3	3
16–19.99	2, 3, or 4	1¼	2¾	2¾

Source: Extracted from AGMA 120.01 Standard, *Gear-cutting Tools, Fine- and Coarse-Pitch Hobs*, with the permission of the publisher, American Gear Manufacturers Association, 1500 King Street, Suite 201, Alexandria, Virginia 22314. All dimensions in inches.

Table 22.2 shows standard-size fine-pitch hobs. These tables are based on standards of the American Gear Manufacturers Association.[2,3]*

Figure 22.6 shows hob feeds plotted against tooth numbers for single- and multiple-thread hobs. These are suggested feeds within hob capacity based on a 12-flute hob for 3- to 12-pitch gears. The suggested feeds apply to steel only, using a sulfur-base coolant. The feeds do not necessarily apply to cast iron but are safe starting points. The feed rate is based on 0.0002-in. advance of hob per flute. In other words, since feed is expressed in inches per revolution of part being hobbed, the total number of flutes exposed to the cut in one revolution of part is reduced to the share of feed for one flute.

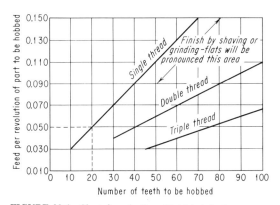

FIGURE 22.6 Chart for selection of trial hob feed.

---

*Superscript numbers refer to references at the end of the chapter.

TABLE 22.2a Single-Thread Fine-Pitch Hob Sizes for Ground and Unground Hobs

(20 Normal Diametral Pitch and Finer) For Spur and Helical Gears

Normal Diametral Pitch	Nominal Hole Diameter	Outside Diameter	Overall Length
20–21.99	1¼	2½	2½
	¾	1⅞	1⅞
22–23.99	1¼	2½	2
	¾	1⅞	1⅞
24–29.99	1¼	2½	2
	¾	1⅞	1⅞
	½	1¼	1¼
30–55.99	¾	1⅞	1½
	½	1⅛	1⅛
	0.3937 (10 mm)	1⅛	¾
	0.315 (8 mm)	¾	½
56–85.99	¾	1⅝	1½
	½	1⅛	1⅛
	0.3937 (10 mm)	1⅛	¾
	0.315 (8 mm)	¾	½
86–130.99	¾	1⅝	1½*
	½	1⅛	⅞*
	0.3937 (10 mm)	1⅛	¾
	0.315 (8 mm)	¾	½
131–200	¾	1⅝	1½†
	½	1⅛	⅞†
	0.3937 (10 mm)	1⅛	¾†
	0.315 (8 mm)	¾	½

*¾ Active face width
†⅝ Active face width
All dimensions in inches.
Source: Extracted from AGMA 120.01 Standard, *Gear-cutting Tools, Fine- and Coarse-Pitch Hobs*, with the permission of the publisher, American Gear Manufacturers Association, 1500 King Street, Suite 201, Alexandria, Virginia 22314.

TABLE 22.2b Multiple-Thread Fine-Pitch Hob Sizes For Ground and Unground Hobs

(20 Normal Diametral Pitch and Finer) for Spur and Helical Gears

Diametral Pitch	Number of Threads	Nominal Hole Diameter	Outside Diameter	Overall Length
20–29.99	2, 3 or 4	1¼	2½	2½
30–50	2, 3 or 4	¾	1⅞	1½

Source: Extracted from AGMA 120.01 Standard, *Gear-cutting Tools, Fine- and Coarse-Pitch Hobs*, with the permission of the publisher, American Gear Manufacturers Association, 1500 King Street, Suite 201, Alexandria, Virginia 22314.

Example. Hob 20-tooth pinion with single-thread hob of 12 flutes. Twenty revolutions of hob occur in one revolution of pinion. Assuming 0.050 feed per revolution of pinion, 240 flutes are exposed to cut. Share of feed for one flute is 0.050 in. divided by 240 flutes, or 0.0002 in. per flute.

There are fallacies involved in this approach when large numbers of teeth are considered, as can be shown by calculating the graph; nevertheless, it can be used as a starting point to determine the most satisfactory feed rate compatible with the machine, material, and hob wear. As indicated on the graph, the flats from generating will be very pronounced in the upper feed area. Also, the machine may be incapable of these feeds except as a second cut following a roughing operation.

Hobbing Speeds. The speed of hobbing should be based on machinability ratings in conjunction with the following table:

Diametral Pitch	*sfm* (100% *Machinable*)
1–4	160–180
5–16	180–200
17–32	200–220
32 and finer	220–240

Machinability is based on AISI B.1112 cold-drawn steel. If the machinability rating of the steel to be hobbed is known, adjust the surface feed per minute accordingly. For example, if hobbing 8620 with a 60-percent rating, the sfm for a 10-pitch gear would be 110 to 120.

The necessary hob r/min can be determined in the formula

$$\text{r/min} = \frac{\text{sfm} \times 12}{\pi \times \text{hob diameter}} \qquad (22.1)$$

Hob Sharpening. Users of hobs should recognize limitations in their own plant for hob sharpening. Usually, for spiral gashed hobs, a hob sharpener is required in order to maintain the original accuracy of sharpening. Most hob manufacturers furnish straight gashed (in line with hob axis) hobs when the thread angle is less than 4°. After 4°, a spiral gash is introduced to maintain the cutting face normal to the thread angle. Failure to follow the original sharpening detracts from hob accuracy and causes physical changes in the form of the hobbed tooth.

Attention to rake angle of the cutting face is important, especially so in topping hobs. Most hobs are sharpened radially. Positive or negative rake angles are used as predicated by their need, depending on the characteristics of the material being hobbed.

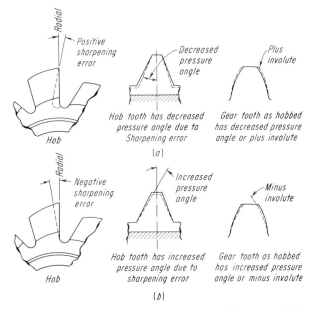

FIGURE 22.7 Effect of hob-sharpening errors: *(a)* hob sharpened with positive error; *(b)* hob sharpened with negative error.

Assuming a hob was originally sharpened with a radial face, the error of sharpening reflects the physical changes listed:

Positive Sharpening Error	Negative Sharpening Error
a. Decreases hobbed pressure angle	*a.* Increases hobbed pressure angle
b. Increases whole depth of hob	*b.* Decreases whole depth of hob

These changes are the major ones occurring and result from the error of sharpening and its effect on the hob's tooth relief. Sketches in Fig. 22.7 show various effects of sharpening errors.

22.1.6 Hob Materials and TiN Coating

Users of hobs are seldom concerned with the high-speed steel used in their hobs; however, in the case of hob failure, it is natural to blame the hob as the source of trouble. Most often the fault lies in the setup, coolant, and variations in material being hobbed. Hob manufacturers have an extremely tight control on chemical composition of their high-speed steel and its heat treatment. Because of the control needed in manufacturing, all steel, as furnished by suppliers, is certified with a report of chemical analysis. This is monitored by the metallurgical department of the hob manufacturer.

GEAR CUTTING TOOLS   22.11

The high-speed steel used in hobs must have several cutting properties and combinations of properties. From the hob manufacturer's viewpoint, consistent machinability in the annealed state is important. When a hob form is produced by the "back-off" method for accurate unground hobs, the material must cut freely without tearing out on the tooth flanks. The material must separate cleanly at the heel of the tooth without flaking out. Grindability is essential to hob production. As the percentage of vanadium in the molybdenum-class steels increases, the grindability decreases.

From the hob user's viewpoint, cutting ability is paramount. Cutting ability is dependent on properties including edge toughness, red hardness, and resistance to abrasion. Hardness and abrasion resistance are improved by increased amounts of carbon and vanadium. As mentioned before, increased vanadium goes hand in hand with increased difficulty in sharpening. The specification of the part material to be hobbed determines the most desirable hob properties for maximum hob performance.

The majority of today's hobs are made of M-2 high-speed steel. Certain applications require the use of M-3 and M-4 steels, and occasional use of the tungsten class of steels. The hob manufacturer will offer suggestions on hob steel when advised of the need for special consideration based on prior machine operations. The behavior of tools in turning and boring operations is a good guide to the need for special consideration in selection of hob material.

To coat precision high-speed hobs successfully with titanium nitride, the use of a low temperature range of 530 to 890°F, as used by the PVD process, is necessary to prevent annealing of HSS. When this low temperature is used there is no appreciable annealing of the tool. PVD is a process in which ions of titanium react with ions of nitrogen at the substrata surface, which results in a compound of TiN that is physically deposited on the tool surface. This coating is 80 RC, which results in a significant improvement in tool life for hardened HSS. The coating is extremely thin, only 0.00012 in., which does not affect the accuracy of the tool.

22.2 GEAR SHAPER CUTTERS

Two major processes are used for cutting gears: hobbing and shaping. Because of its universal application and its high performance, hobbing is usually preferred. Shaping is used due to the form of the parts to be cut, that is, internal gears, shoulder gears, herringbone gears of solid configuration, clutch gears with thick and thin teeth, and other shapes that can be obtained with a specially designed cutter.

22.2.1 The Molding Generation Process

It is evident that if one gear (shaper cutter) is provided with suitable cutting clearances on the teeth and is hardened, it can be used as a generating tool. If this tool is rotated in the correct ratio with a gear blank, and at the same time reciprocated, it will generate teeth in the gear blank as shown in Fig. 22.8.

Down and Up Shaping. Figure 22.8 illustrates the shaper cutter mounted on the cutter spindle for *down shaping*. Either the tool or the work is relieved so that it does not drag on the return stroke. The cutter is mounted with the teeth facing upward for *up shaping,*

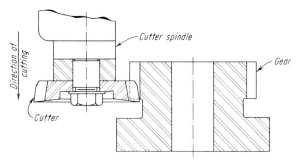

FIGURE 22.8 Shaper cutter set up for push stroke.

and the tool or work piece is relieved on the downstroke. This entails considerable setup time to change the setup when it is required.

Cutter Clearance Angles. The side relief angle of a shaper cutter generally ranges between 1½ to 2½°. This is a relatively low amount for a metal cutting tool. Any substantial increase in this angle can cause drastic changes in cutter geometry and cause a reduction in tool life.

The cutter outside angle is a function of cutter gear geometry, amount of side relief, and pressure angle. It ranges from 3.5° for 30°-pressure-angle gears to 8.5° for 14.5°-pressure-angle gears. The outside angle should be slightly curved, but is universally accepted as a straight line because of the small change in tooth depth it causes.

Occasionally it is necessary to shaper-cut a part maintaining a close relationship between tooth thickness and root diameter. If this is the case, the top angle is usually provided with a lesser angle than theoretically required. With each sharpening of the cutter, the top angle can be resized, giving an accurate cutter addendum for holding the proper depth in the part.

In general, a 5° positive sharpening angle is incorporated in the design of shaper cutters; however, in special cases, as little as 0° and as much as 10° is used.

It is important to remember that the sharpening angle is to be maintained throughout the life of the tool, as any deviation will result in a change in the involute form. A further explanation of this is contained in the section covering the sharpening of shaper cutters.

TABLE 22.3 Shaper Cutter Top Angles

If Pressure Angle Is	Top Angle Will Be
14½°	7°56'
20°	5°49'
25°	4°43'
27½°	4°20'
30°	4°

22.2.2 Application to CNC Gear Shaper

Full CNC shaping machines (Fig. 22.9) feature separate drives for the following:

Stroke motion with dead center position
Rotation of tool
Rotation of work piece
Radial motion
Stroke position
Stroke length
Offset (size) of work piece
Relief angle

The identification of the drives shown in Fig. 22.9 is as follows:

Stroke motion with depth centre positioning (S axis)
Rotation of tool (D axis)
Rotation of workpiece (C axis)
Radial motion (X axis)
Stroke length (V axis)
Offset cutter stroke workpiece (Y axis)
Relief angle (taper) (B axis)

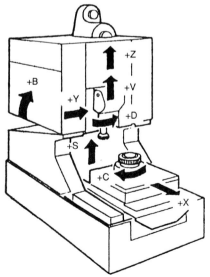

FIGURE 22.9 Schematic of a CNC gear shaping machine. *(Courtesy of Liebherr-Verzahntechnik GMBH, Kempten, Germany)*

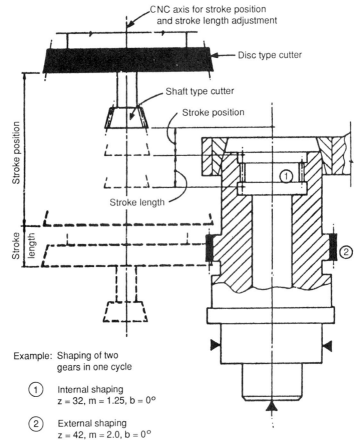

FIGURE 22.10 With a CNC type of machine two different gears can be cut in the same setup. *(Courtesy of Liebherr-Verzahntechnik GMBH, Kempten, Germany)*

Figure 22.10 illustrates one of many time-saving steps possible with a full CNC machine. When the face width, pitch, and helix angle are entered into the program, the stroke length itself is calculated and set by the control system.

22.2.3 Machine Mounting of Cutter Types

The most commonly used types of shaper cutters are shown in Fig. 22.11. Shaper cutters are made in various types to suit the special needs of the part or the machine being used. Special care must be taken when using shank-type cutters, including those used with taper or straight shank adaptors, to indicate the proof diameter before using.

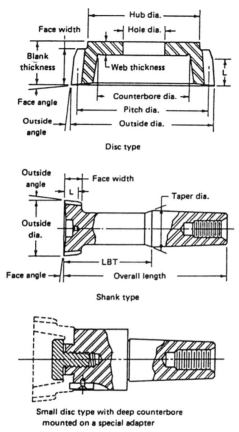

FIGURE 22.11 Typical types of shaper cutters and their nomenclature. *(Courtesy of Fellows Corp., Emhart Machinery Group, Springfield, Vermont)*

Helical Shaper Cutters. The helical shaper cutter can be made in all the previously mentioned types; namely, disk, deep-counterbore, hub, and shank. When cutting right-hand external gears and left-hand internal gears, a left-hand helical shaper cutter is used. If a left-hand external gear or right-hand internal gear is to be cut, a right-hand helical shaper cutter is used.

A helical guide must be used in conjunction with the helical shaper cutter. This guide has the same lead and hand as the shaper cutter. It is possible, then, with the use of a left- and right-hand helical shaper cutter and corresponding guides, to produce an entire range of tooth numbers of mating parts having the same helix angles. An exception to the foregoing is when extremely small numbers of teeth are to be cut. Here specially designed tools may be required to prevent excessive undercuts or possible flank interference. Figure 22.12 illustrates a typical helical shaper cutter.

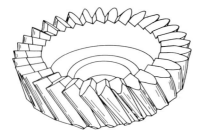

FIGURE 22.12 Typical helical shaper cutter.

Disk-type cutters are used in matched pairs of one right-hand and one left-hand cutter to generate true herringbone gears with solid continuous teeth. These cutters are known as Sykes cutters in reference to the firm that originated this gear cutting process. The cutter geometry requires that the cutter face be flat, and a special groove and lip sharpening be used in order to have equal shear faces do the cutting. Special herringbone machines are used to cut herringbone gears.

Tool Clearance. It is important to remember in the design of gears having a close shoulder interference that a sufficient clearance groove should be incorporated. Figure 22.13 illustrates the clearance groove required for spur and helical shaper cutters.
For a spur gear, the minimum clearance groove can be calculated as follows:

(Whole depth of gear × tan sharpening angle) + $\tfrac{1}{16}$ in. = groove width   (22.2)

When calculating the clearance groove for a helical gear, the following formula should be used:

(Whole depth of gear × tan sharpening angle)
+ (normal circular pitch × sin helix angle)
+ $\tfrac{1}{16}$ in. = groove width   (22.3)

In cases where the clearance groove must be held to an absolute minimum on a helical gear, an off-normal sharpening can be incorporated.
The off-normal sharpening can vary from slightly less than the helix angle to a spur-type sharpening as used on herringbone gear shaper cutters. Any changes from a normal sharpening will result in a change in the involute form. For this reason, if an off-normal sharpening is desired, the tool manufacturer should be so advised.

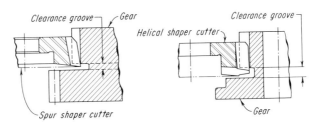

FIGURE 22.13 Clearance groove for shaping.

22.2.4 Shaper Cutter Tooth Modifications

There are many tooth modifications that can be incorporated in the design of shaper cutters. These are ordinarily described as "special features" by the manufacturers and are priced slightly higher than tools without modification. In most cases, the modified tool is tailor-made for one specific gear and is not applicable over a wide range of gears having different tooth numbers. The most common modifications are protuberance, semitopping, approach type, topping cutters, and full tip radius. The following is a description of each of the aforementioned features, along with the resultant change in the gear tooth profile.

Protuberance. When a gear is to be finished by shaving or grinding, it is advantageous to have an undercut in the prefinished gear. This allows the shaving cutter, or grinding wheel, to blend smoothly with the fillet radius (Fig. 22.14).

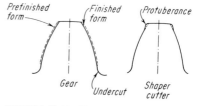

FIGURE 22.14 Undercut in gear as cut by protuberance of shaper cutter.

The amount and position of the undercut are very important. Too little undercut will leave a step after the finishing operation. This step would then be in a very critical area on highly stressed gears and would possibly be the starting point for fatigue cracks. It should also be pointed out that any step presents a point of interference for the shaving cutter and can cause an erratic gear profile. As a general rule, on gears of 18 pitch and finer, the step is not large enough to interfere with the shaving operation, and the protuberance feature is not used.

If too much protuberance is designed into the tool, the undercut may extend too high on the tooth profile and result in the finishing tool not cleaning up above the start of active profile. This is detrimental in two ways: (1) It reduces the involute contact ratio between mating parts, and (2) it reduces the involute contact ratio between the shaving cutter and the preshave gear. In the first case, this can cause a higher noise level and/or an earlier fatigue breakdown between the mating gears. In the second case, the loss of involute overlap can cause loss of involute control between the shaving cutter and the gear, and result in undesirable profile errors.

Semitopping. The semitopping feature is used to produce a chamfer at the point where the tooth profile and the outside diameter intersect.

One of the prime problems of gear manufacturers is the nicks that occur in the handling of gears after the teeth have been finished and prior to heat treatment. This is particularly true on shaved gears where a slightly dull cutter will push up a small burr at the outside diameter. If the gear does not have a chamfer and comes in contact with another object, a nick will usually appear at point *A* in Fig. 22.15. This nick then interferes with the rolling action between the mat-

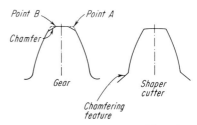

FIGURE 22.15 Shaper cutter with semitopping feature.

ing parts and results in a noisy gear set. If the gear has a chamfering feature and is mishandled in any way, the nick will usually appear at point B in Fig. 22.15. A nick at point B will not affect the rolling action of the gears, nor will it increase or affect the noise level.

Approach.   The purpose of the approach feature is to provide a tip relief on the gear (Fig. 22.16). The tip relief is used by gear manufacturers to provide a smoother rolling action on certain gear sets. There are two types of approach features: The first is called the flank type, or the tangent approach, where the amount of tip relief will vary as the tool is sharpened back; the second is called the H type, or the constant-approach type. Here the tool is designed in such a manner that the tip relief remains nearly constant throughout the life of the tool.

Full-Topping Shaper Cutters.   The full-topping shaper cutter, in addition to producing the tooth form, sizes the outside diameter, thereby establishing a relationship of pitch diameter to outside diameter (Fig. 22.17). This feature is seldom used because most shaper cutter applications do not require it. There is also a problem with a continuous chip that retards the chip flow, sometimes results in a poor finish, and can also decrease the accuracy of the gear.

Full Tip Radius.   When a full fillet or round root is desired on the gear, a full tip radius is incorporated on the shaper cutter (Fig. 22.18).

Wafer Shaper Cutter.   The Wafer Shaper Cutter, introduced by Pfauter-Maag in 1985, is a cutter assembly (Fig. 22.19) designed to replace a conventional shaper cutter. The

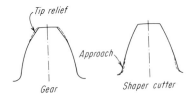

FIGURE 22.16   Shaper cutter with approach feature for tip relief.

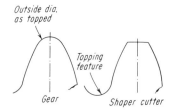

FIGURE 22.17   Shaper cutter with full-topping feature.

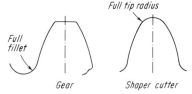

FIGURE 22.18   Shaper cutter with full-radius tip.

GEAR CUTTING TOOLS

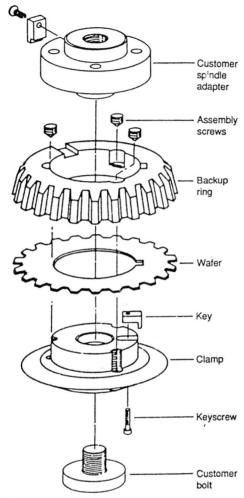

FIGURE 22.19 Exploded view of wafer cutter assembly. *(Courtesy of Pfauter-Maag Cutting Tools, Loves Park, Illinois)*

first high-production installation at the plant of a major automotive supplier, where a line of 20 gear shaper machines are producing timing chain sprockets, has proven very successful. The main feature of the cutter is that it does not have to be resharpened. The wafer is made from M2 HHS, hardened to 64–66 RC and coated with TiN. After prolonged testing, the makers of the tool claim that the new tool will reduce tool cost 30 percent, reduce scrap 70 percent, reduce manpower 60 percent, and save several hours a day by the elimination of sharpening.

Several Wafer Shaper Cutters are working successfully in a wide range of applications up to 5 DP for both external and internal gears and splines. Helical applications with a helix angle of 20° and above are being developed and are under test in selected production applications.

22.2.5 Sharpening Shaper Cutters

The importance of the sharpening of shaper cutters cannot be overemphasized, for an incorrectly sharpened tool can cause involute profile errors, pitch-diameter runout, and spacing errors. In most cases, the face angle of the sharpening angle is 5°. Any deviation from this is indicated on the marking of the tool by the manufacturer.

Because the side clearance angle causes the tooth thickness to become progressively thinner at any given diameter, a sharpening angle larger than the tool is designed for will produce a plus involute form on the gear. When the sharpening angle is smaller than the tool design angle, the effect is in the opposite direction. The involute on the gear is a minus form.

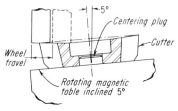

FIGURE 22.20 Sharpening of spur shaper cutter.

In the sharpening of spur shaper cutters, the tool is mounted on a rotary surface grinder (see Fig. 22.20) which is tipped to the correct sharpening angle.

It is important that a rigid machine be used, because of the varying area of contact as the tooth and space alternately pass under the grinding wheel. It is also important that the shaper cutter be properly centralized, because any misalignment will cause the tool to produce pitch line runout.

Rough-Sharpening. It is also recommended that the tool be "rough"-sharpened using a vitrified grinding wheel with a 12- to 14-in. diameter of 60 grain size, grade H. The surface speed of the grinding wheel should be 5,000 to 6,000 ft/min, traveling at about 45 strokes per minute, removing approximately 0.001 in. per stroke. The amount of stock left for finishing should be from 0.001 to 0.004 in., depending on the pitch of the tool.

Finish-Sharpening. The "finish" grind should be done with a shellac or resinoid-bond silicon-carbon grinding wheel of about 400 grain size in a soft grade with free-cutting properties.

It is possible under optimum conditions to sharpen shaper cutters in a single grinding operation. For maximum tool life, the cutting edges must be free from burning or heat checks and have a 10 microfinish or less.

In the sharpening of shank- or hub-type shaper cutters having small diameters, the diameter of the grinding wheel must be reduced. If too large a wheel is used, the sharpening angle deviates from a straight line, becomes concave, and as a result produces an involute error on the part.

Sharpening Helical Shaper Cutters. The sharpening of helical shaper cutters requires a hand-indexed grinding fixture as illustrated in Fig. 22.21.

Automatic fixtures are also available. Because the teeth are ground in a fixed position, one at a time, it is not possible to obtain a high-quality finish. For coarse-pitch tools, free-cutting 90-grain I or J grade aluminum oxide wheels are recommended. On

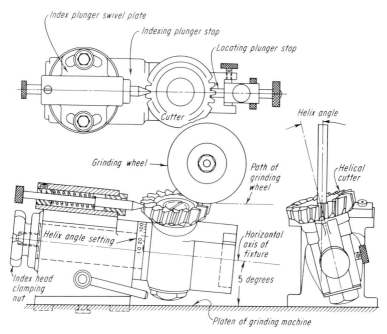

FIGURE 22.21 Sharpening of helical shaper cutter.

finer pitches, 150 grain in J or K grades is usually used. It is very important that all teeth be ground to the same height, as any deviation will result in tooth spacing errors. In most cases, the face-sharpening angle is 5°. Any deviation from this, and also any change from a "normal" sharpening, will be marked on the tool. If the correct angles are not adhered to, an error in the involute profile will result.

22.3 GEAR MILLING CUTTERS

There hasn't been a lot written about TiN-coated gear milling cutters. However, with the success shown with hobs as well as with shaper cutters, there are reasons to believe that TiN will work equally as well for gear milling. One of the most important characteristics of TiN is its extremely high hardness. The hardness of titanium nitride at Rockwell C 85 is much greater than that of the high-speed steel on which it would be deposited. The high hardness of TiN proves excellent wear and abrasion to heavy and prolonged cutting loads. TiN rubbing against steel has a considerably lower coefficient of friction than does hardened high-speed steel. Tests have proven that, due to lower friction, machine loads and resultant power required have been reduced.

The majority of present-day uses of gear milling cutters are on coarse-pitch applications. A common application on coarse-pitch gears has been a roughing operation, leaving stock for finish-hobbing or shaving.

22.3.1 Types of Gear Milling Cutters in Common Use

The standard involute gear cutter (see Fig. 22.22) will cut a range of gear teeth. For each pitch, eight cutters are made as follows:

Cutter No.	Will Cut Gears From
1	135 teeth to rack
2	55–134 teeth
3	35–54 teeth
4	26–34 teeth
5	21–25 teeth
6	17–20 teeth
7	14–16 teeth
8	12–13 teeth

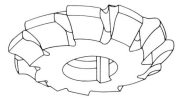

FIGURE 22.22 Standard involute-gear cutter.

The cutters are made in either 14½ or 20° pressure angles, radial face sharpening, and a "backed-off" unground form. The cutter is designed for the lowest number of teeth in the range.

As higher accuracy of tooth form is frequently required, the "half number" gear milling cutters are used. These provide a better tooth form for the high end of each tooth range.

Cutter No.	Will Cut Gears From
1½	80–134 teeth
2½	42–54 teeth
3½	30–34 teeth
4½	23–25 teeth
5½	19–20 teeth
6½	15–16 teeth
7½	13 teeth

Involute Gear Stocking Cutters. These are roughing cutters designed for high-speed cutting. They are used singly and in duplex or triplex setups. The stocking cutters are made in four styles, as shown in Fig. 22.23.

Type A cutters have steps in both sides of all teeth and will produce steps in the

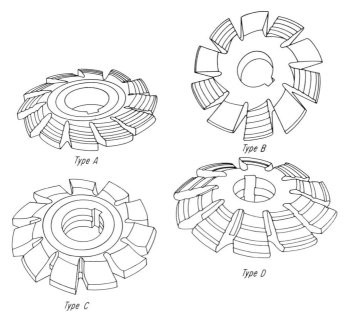

FIGURE 22.23 Involute-gear stocking cutters.

gear tooth space. These steps cut only on the diameter, giving clearance for rapid removal of stock. Type A is used for 4 pitch and coarser.

Type B cutters are stepped on one side of each tooth alternately. The stepped teeth are in plane with the smooth tooth profile, giving the advantage of chip breakers while cutting a smooth tooth surface. Type B is used for 4- to 10-pitch gears.

Type C cutters are used for 10 pitch and finer. They are made with smooth teeth on both sides.

Type D cutters have three steps on one side and two on the other, alternating every other tooth. These are generally used on the coarse-pitch applications. Stocking cutters are usually furnished with a positive rake angle.

Multiple-Type Involute Gear Cutters (see Fig. 22.24). These are made in sets to cut two or three teeth at one pass. These are usually single-purpose cutters used only for cutting the number of teeth for which they are designed. They are used in duplex and triplex in one of the following three combinations:

1. Roughing: with cutters centered correctly in relation to the gear.
2. Roughing and finishing: the finishing cutter is set on the gear centerline.
3. Finishing: The centerline setting is the same as for roughing.

Figure 22.25 shows a duplex and triplex arrangement generally used for roughing.

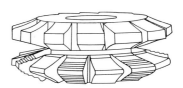

FIGURE 22.24 Multiple-type involute-gear cutters.

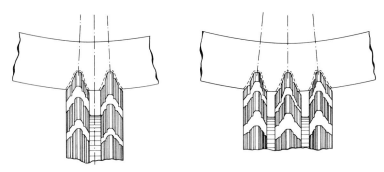

FIGURE 22.25 Duplex and triplex arrangement of cutters for roughing *or* finishing.

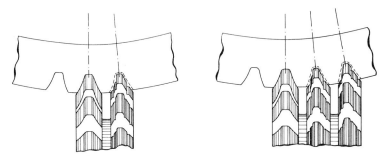

FIGURE 22.26 Duplex and triplex arrangement of cutters for roughing *and* finishing.

Finishing cutters are set up in the same arrangement. When roughing and finishing, the duplex and triplex arrangement as in Fig. 22.26 is generally used.

22.3.2 Milling Cutter Design Practices

As mentioned in Sec. 22.3.1, the cutter is designed for the lowest number of teeth in the range. The reason for this is apparent when the tooth-space configuration of a pinion is compared with the space configuration of a large gear or a rack. A No. 8 cutter designed for 12 teeth and used on 13 teeth produces a tooth curvature of predominantly minus involute at the tip of the 13-tooth gear. This is generally desirable since a minus-involute curvature rolls smoothly in conjugate action. If the design practice were reversed and the cutter were designed for the highest number of teeth, then plus-involute curvature would result in the low range. Conjugate rolling action under these conditions would be rough and probably noisy.

The cutter is a formed type, as opposed to a generating cutter, such as a hob. Therefore, size or tooth thickness of the gear is an important factor. If the gear is milled oversize, it is likely that tooth curvature will be of plus-involute type. If undersize from milling too deep, the tooth curvature will be of minus-involute type.

When generating gears with a hob, the involute curvature is unchanged, regardless of size.

Allowances are made in the design of standard-type cutters to accommodate normal limits of tooth thickness so that, even when milling at the high side of tooth thickness tolerance, the tooth curvature will be compatible for good conjugate rolling action.

The cutters vary in size from 8½-in. diameter with a 2-in. bore to 1¾-in. diameter with a ⅞-in. bore. Cutters are normally unground but can be obtained with ground forms.

22.4 SHAVING CUTTERS

Since the introduction of CNC-controlled gear grinding machines, and CNC gear checking machines, the accuracy of gear shaving cutters has improved. In addition, a six-axis CNC-controlled gear shaving machine is using these cutters to shave gears up to 18 in. in diameter, to quality levels of $+/-0.00004$ in., as reported by Cima Kanzaki. While these achievements are remarkable, they are made on relatively low-hardness gears.

A versatile, five-axis CNC-controlled machine, introduced recently by National Broach and Machine Co., permits the user to shave, hard hone, roll, CBN-generate and form grind, or hard finish gears on their model GF-300 Gear Finisher.

The addition of CNC control to the gear finishing machine provides a controlled rotary relationship between the shaving cutter spindle and the work spindle, which allows the correction of gear errors not possible without CNC control; while the ability to finish hardened gears with a CBN tool eliminates errors due to heat-treat distortion.

22.4.1 Basic Principles of Shaving Cutters

The rotary or rack-type shaving cutter uses serrated tooth flanks (Fig. 22.27), with the teeth having a helix angle plus or minus 10 to 15° different from that of the part being shaved, except when shaving close shoulder gears. Then the difference in helix angle is limited to about 3 to 5°. The shaving cutter is rotated in close mesh with the gear being shaved. As the teeth of the cutter move across the gear teeth, the serrations shave fine chips from the gear teeth. During the shaving cycle the work spindle and the tool spindle are driven in timed relation to each other by CNC-controlled motors. When a non-CNC machine is used, the rotation of the work spindle is not timed with that of the cutter spindle. Other motions controlled in a CNC machine are: the vertical slide, the tangential slide, adjustment of the crossed axis angle, and crowning adjustment.

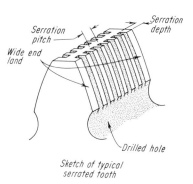

FIGURE 22.27 Typical serrated tooth.

The Crossed Axis Angle. The crossed axis angle is quite easy to visualize. By holding two straight arbors together with their axes parallel, you will note that they contact along a path parallel to their axes. When the arbors are swung to 90° of each other, they have a small contact; however, if they are placed at approximately 10° with one another, the contact will increase to the approximate crossed axis angle used between a shaving cutter and a gear. The crossed axis angle between the cutter and work produces a diagonal sliding path from the tip of the gear tooth to its root. When they are rotated together under pressure, this causes the serrations in the cutter to remove fine chips from the profiles of the gear teeth.

Design of Tool. The rotary shaving cutter is usually made with a prime number of teeth. This "hunting-tooth" condition further refines spacing errors and increases the accuracy of the shaved part. Average recommended size is an 8-in.-pitch diameter tool for most external gears. A 9-in.-pitch-diameter tool is frequently used for low tooth numbers in the part. The choice of diameter is predicated on the usually noticeable control of the shaved involute profiles. Face widths of the shaving tools vary from 1 to 4 in., depending on the requirements and type of application. Fine-pitch shaving cutters are furnished with face widths ⅜ to ⅝ in. and with a diameter of 3 to 4 in. The shaving tools are priced according to pitch, pitch diameter, and face width. In some cases of extreme tolerances required in the tool, and special modifications, the price of tools is adjusted as needed. The tooth flank serrations are cut serrations and are spaced according to the pitch of part and type of shaving to be done. The end lands are designed to be wider than the inner serrating lands. Depending on the type of shaving, whether transverse, modified underpass, or full underpass, the serrations will be annular or with a staggered pattern. The annular serration is used for transverse or modified underpass methods of shaving. The staggered type of serration for full underpass shaving compensates for the reduced axial sliding and prevents the tracking of serration lands on the tooth flank of the shaved part. The dedendum area of the shaving tool is supplemented with a drilled hole to form an undercut to allow the serrating tool to clear itself during manufacturing.

Transverse Shaving. This method is desirable for extreme face widths of gears. A standard 1-in. face width of cutter may be used. With use of the shaving machine feed slide, the tool can transfer shaving contact from one edge of the gear to the other edge. It is customary design to move the work past the cutter by means of the feed slide. On some of the larger shaving machines, for obvious reasons, the shaving cutter is mounted on a feed slide. In this discussion, the work piece is the moving member. The table supporting the part being shaved can be rocked about a pivot point in timed relation with the feed. This method of shaving, while feeding and rocking, will produce a crowned lead on the part being shaved. Straight shaving of the lead with combinations of crown, if desired, can be done by the transverse method. Production time is the slowest by this method, since multiple feed passes are required to remove stock and properly size the gear. In transverse shaving, the pivot point of the cutter must pass both edges of the gear to obtain lead control as described in Fig. 22.28.

Modified Underpass Shaving (Diagonal Shaving). This is the most widely used method of shaving. It permits shaving gears with face widths slightly in excess of the tool width (see Fig. 22.29). Its production time in shaving is high because of the short stroke of the feed slide. One shaving cycle is usually sufficient to complete the shaving. The stock removal method is efficient because of the higher cross axis angle possible when compared with underpass shaving. Crowning can be accomplished by use of

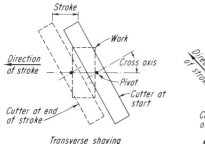

FIGURE 22.28 Transverse shaving method.

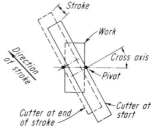

FIGURE 22.29 Modified underpass shaving method.

either a crowning attachment, as with transverse shaving, or by use of a tool ground with a concave lead. The angle of feed recommended is from 40 to 70°, measured from a normal to the work axis. When crown shaving with a tool ground with a concave lead, the amount of crown can be varied by altering the feed angle. The greatest amount of crown is obtained by feeding on an angle approaching the full underpass condition. Crown can be decreased by feeding on an angle approaching the transverse condition. As mentioned before, the tool will have annular serrations. Excluding the ground-concave-lead feature in a shaving tool, the same tool may be used for either transverse or modified underpass shaving, if the gear to be shaved permits.

Underpass Shaving. This method is usually dictated when shaving cluster gears or gears with close adjacent shoulders. The cross axis angle is usually in the area of 4 to 6° in order to avoid interference of the tool with the shoulder. In all cases of underpass shaving, the tool must be wider in face width than the gear to be shaved. This is obvious by study of the diagram in Fig. 22.30 showing the direction of feed. Crowning of the gear is feasible by the underpass method, but only when the shaving tool is ground with a concave lead. The underpass method of shaving is usually the fastest method of shaving, because of the short feed stroke and the one cycle required. The low cross axis angle has its disadvantages in that the helical side sliding, promoting stock removal, is reduced. To compensate for this, the amount of shaving stock should be maintained at a minimum, but still allow for a full cleanup on the part tooth flanks. In this method of shaving, it may be expected that the number of pieces per sharpening of the tool will be lower than for transverse or modified underpass shaving. This is a result of the lower cross axis angle and the

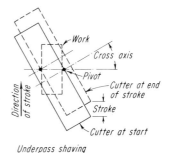

FIGURE 22.30 Underpass shaving method.

consequent lower helical side sliding. It is common to find that, in spite of good intentions in maintaining control of the preshave gear, the amount of stock left for shaving will become critical because of negligence. The amount of stock removal is critical whenever crown shaving is introduced to this method. The amount of crown per side of tooth at the gear faces must be considered as shave stock along with the normal preshave stock. The serrations must be the type with staggered pattern.

Plunge-Feed Shaving. This method is not commonly used but is worth mention as a possible method. It is an adaptation of the underpass method, applied to both internal and external gears, usually of narrow face widths. The staggered pattern of serrations is used on the tool. Instead of feed-slide motion, as with the other types of shaving, the plunge-feed method uses strictly an infeed to obtain size. As would be expected, the cutting pressures are high and work holding becomes important to obtain rigidity.

Parallel Axis Shaving. In this application, the tool and work are rotated in mesh with the axes parallel. Without benefit of the cross axis angle, there is no helical side sliding. Therefore, a reciprocating motion must be induced at a comparatively high frequency. This requires a special machine. The tool cuts only on its face, as it is rotated in one direction and then the other, as the reciprocation is taking place. It is generally applied to internal shaving. This method is not commonly used.

The Pivot Point. It is impossible to design tools or make intelligent setups without giving consideration to the pivot point. Each shaving machine is built with a swivel rotatable for cross axis setting between the tool and work. In the construction of the shaving machine, the pivot point of this swivel is so located that its axis, if projected, would intersect both the work and shaving tool axis. Its position of intersection with the tool axis is a known distance from the mounting shoulder of the shaving cutter spindle. Usually, the distance of pivot location is 1 in. from the mounting face. If a 1-in.-face-width shaving cutter was to be set up with the pivot point in the center of the tool, a ½-in. spacer should be mounted between the cutter face and the mounting shoulder. By substituting spacers of different widths, the pivot point can be transferred to various positions within the face width of the tool. This procedure is valuable to the person doing the setup for any of the three methods of shaving, in terms of the stroke of the feed slide. The pivot point must pass beyond both faces of the gear being shaved in order to control the lead of the tooth. Shaving is not complete unless the pivot point passes the gear faces (see Fig. 22.31). Since the pivot point defines the shortest distance between the part and the tool axis, complete sizing is impossible in any other position. In troubleshooting, failure to observe the fundamental of the pivot point in setting up the machine and determining stroke has been the major cause of downtime. The operator's manual for each machine will describe the location of the pivot for each machine model.

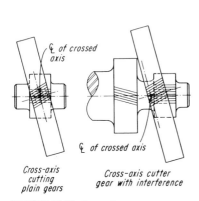

FIGURE 22.31 In cutting plain or shoulder gears, the pivot point must pass beyond both faces of the gear.

Direction of Feed. In shaving either spur or helical gears by the modified underpass shaving method, there is a choice of direction of feed and rotation. Refer to Fig. 22.32a

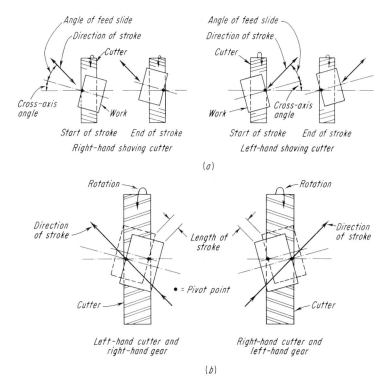

FIGURE 22.32 Direction of feed and rotation for modified underpass shaving: *(a)* spur gears, *(b)* helical gears.

for shaving spur gears, with either a right- or left-hand shaving cutter. The direction of the feed slide should be set to feed in the general direction of the cutter helix. Rotation of the cutter should be in a direction opposing the feed.

Refer to Fig. 22.32*b* for shaving helical gears. The direction of the feed slide should be set to feed in the general direction opposing the helix of the cutter. Rotation of the cutter should be in a direction opposing the feed. Occasionally, a machine setup is made contrary to these recommendations with satisfactory results. However, these setups are in a minority.

As a rule of thumb for spur gears:

1. Feed in direction of cutter helix.
2. Rotate cutter against direction of feed.

As a rule of thumb for helical gears:

1. Feed against helix of cutter.
2. Rotate cutter against direction of feed.

As a matter of note, in underpass shaving, always rotate against direction of feed.

Tool Geometry. The purpose of comment here is for general coverage of operating pressure angle and basic principles. Either the gear or the tool is the driving member in the shaving operation. The model of the machine predicates which is the driving member. Since the other member depends on the driver for rotation, special consideration must be given to the tool geometry, the most important consideration being the choice of the operating pressure angle to obtain the most consistent results.

The operating pressure angle is not usually the same as the theoretical pitch line pressure angle. The tool is designed to operate and shave at a diameter of rolling other than the pitch diameter—thus the special operating pressure angle. The angle used is a calculated element, selected to contribute to optimum conditions of tooth contact. Failure to provide optimum conditions results in lack of involute control of the shaved part. One of the symptoms of lack of involute control is the hollow involute at the pitch line area of the part. Other factors such as improper preshave can contribute to hollow irregular involutes. The tool engineer with sufficient experience can differentiate as to which factor is the cause of trouble. To eliminate tool design as a factor narrows the area of tool development. The tool throughout its life is designed to maintain the same "true form diameter" on the part. This is accomplished by retaining the correct proportions of tooth addendum and tooth thickness. The tooth thickness of the tool is decreased with each resharpening of the tool. This causes a reduction in operating center distance and the consequent change in operating pressure angle. The addendum must be decreased to contact at the original *true form diameter* (TIF) diameter. This does not change in the same proportion as the change in center distance. It is a common misunderstanding to believe the base diameter of the tool changes with each sharpening. The base diameter is constant and remains so throughout the life of the tool. The operating pressure angle is the element changing as the center distance changes. The cross axis angle decreases slightly with each resharpening. This is a function of the operating pitch diameter, decreasing with each sharpening as the lead of the tool remains constant. This is obvious from examination of the formula for lead:

$$\text{Lead} = \frac{\pi \times \text{diameter (operating pitch diameter)}}{\tan \text{helix angle}} \quad (22.4)$$

Since lead and π are constants, the helix angle decreases along with the operating pitch diameter with each sharpening. Only in rare cases is the involute of the shaving cutter altered during the entire life of the shaving tool. This, when it is required, is a result of experience and tool development, based on case history and records of the tool's past performance. This practice is discouraged, since the development is costly and sometimes is a result of inadequate tool design. The entire life of the shaving tool is carefully planned by the designer. Having full knowledge of the part to be shaved, its mating part, and the preshave tool, it is possible to predict performance of the tool in terms of expected regrinds, involute characteristics, and sources of deviations that may result from preshave. The basic geometry of the tool will parallel the design of the mating gear, except for cross axis angle and the no-backlash operating condition. With shaving tools, troubles generally stem from "action and reaction." If a fillet interference exists, as a result of an improper hob job, a crowding of contact will result, and a "reaction" one base pitch away on the part can be observed. This is the reason that shaving tool manufacturers press for satisfactory preshave conditions. The tool is capable of a great deal of correction, but only if the preshave part will allow it to do so. Shaving, therefore, is not a "cure all" but a means of making a good gear better.

22.4.2 Preshave Gear

The preparation of the gear blank is of utmost importance in obtaining the most economical shaving operation. Concentricity of gear pitch line and bore, squareness and parallelism of the gear faces, and the bore itself are important factors. An oversize bore, for example, can completely mislead tool development for production. Profiles will be erratic, making it next to impossible to make adjustments in the grind of the shaving tool; or, if the tool has been previously developed, the erratic profile may cause a needless redevelopment.

Gear faces that are not square with the bore and parallel with each other will produce leads that are misleading. The profile and lead in the shaved gears are the important elements. The damaging factors mentioned are the most common to shaving trouble. The depth of the preshave gear should be the recommended amount of 2.35/pitch.

It is extremely important that an undercut in the fillet area or at base of the part tooth be provided for the blending of the shaved profile as shown in Fig. 22.33. A protuberance-type hob or shaper cutter is required for this purpose (see Fig. 22.34).

The stock allowance for shaving varies between users of shaving equipment. Table 22.4 is intended only as a guide. The amounts are based on circular tooth thickness and may be measured by balls, rolls, or blocks, or a master gear on a rolling fixture. This table covers a pitch range from marine gears to instrument and control gears, as observed in the field. One manufacturer, on a production basis, is controlling stock to 0.001-in. allowance on 16 pitch. Another manufacturer, on an identical gear, finds it difficult to clean up the tooth, even with a 0.003-in. stock allowance. The answer is in

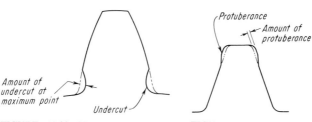

FIGURE 22.33 Undercut on a gear tooth.

FIGURE 22.34 Protuberance on a gear-hob tooth.

TABLE 22.4 Stock Allowance for Shaving

Diametral Pitch	Min.	Max.
2	0.003	0.004
6	0.002	0.004
10	0.002	0.003
16	0.001	0.003
20	0.0005	0.002
48	0.0003	0.001

the preshave operation throughout. The condition of the machines, the skill of the operators, and the desire to keep tool costs down determine how much stock allowance the individual plants can tolerate. A well-known tractor manufacturer reduced transmission rejections from an average of four a day to an equal number per working week simply by requalifying his hobbing machine spacers. The hobs, as mounted, were weaving; the shaving tool could not correct the lead variations; inspection had to compromise to maintain production; and, consequently, the entire operation suffered by failure to observe a simple fundamental of good gear cutting.

22.4.3 Resharpening

The tool, upon dulling, will react in a manner to warn that a tool change is needed. One of the symptoms will be failure to maintain size. Upon making size adjustments, the response of the tool is inadequate to the change made. If monitoring with an involute checker, a common symptom is a plus involute near the tip of the tooth. This will grow steadily worse. A rolling fixture utilizing a master gear will signal rough rolling action. Deterioration of the surface finish is usually visible. When it has been determined that the tool is dull, the tool should be returned to the manufacturer for resharpening. Continued operation of a dull tool will cause damage that will seriously detract from the life of the tool. Serrations may flake out or excessive stock may have to be removed to recondition the cutting edges. Most users keep a record of pieces shaved per grind with each cutter in order to avoid overdulling. A frequent complaint about shaving tools is that the resharpened tool does not give so many shaved pieces as a new tool. Contrary to this, it is not uncommon to hear that the resharpened tool produced more pieces than the new tool. There are so many factors that influence tool life that at this point a careful study is necessary to determine the reason. The reason is usually found in the usage, operator technique, change in material, contaminated coolant, stock removal, or any number of things or combinations. The greatest step forward has been the clinics, or educational programs, within the user's plants.

The amount of sharpening life depends primarily on the depth of the serration. This varies with the pitch. A coarse-pitch tool can be designed with a deep serration and provide many resharpenings. A fine-pitch tool will have a shallow serration and a limited number of usages. Table 22.5 is representative of serration depth, and the amount a tooth may be reduced in tooth thickness by successive sharpenings. A shaving tool is not resharpened when its serration depth has been reduced to approximately 0.012 in. In the case of the 6-pitch gear shaving tool, the 0.045-in. depth of serration provides a possible reduction of tooth thickness of 0.090 in. The serrating tool requires radius corners on its cutting edge and reduces the possible reduction of tooth thickness from 0.090 to 0.066 in. By observing the 0.012-in. minimum, the user can be

TABLE 22.5 Amount a Shaving Cutter Tooth May Be Reduced in Sharpening

Pitch	Depth of Serration	Amount Tooth Thickness May Be Reduced
6	0.045	0.066
10	0.030	0.036
20	0.020	0.020

assured that the shaving tool will present the proper cutting edge to the part to be shaved.

The chip from a shaving tool is a thin hairlike chip. This is flushed out of the serration by the coolant and removed by a magnetic chip separator. Failure to clean the coolant by chip separating can cause early tool dulling. Evidence of this can be found by studying the shaved tooth flank, watching for chip impressions in the shaved profile.

Even though the tool is dull, take the precaution of handling as a new tool in storage and in shipping. Before shipping out for resharpening, study the tool serrations with a glass, watching for chipped serrations or impacted teeth. These are clues to whether overdulling is occurring, if the gear is being deburred properly, or whether the coolant is doing its job. Improper coolant and flaked-out serrations go hand in hand. Broken serrations, resulting from impacted serration slots, are indicative of lack of deburring. The impacted metal will bear no resemblance to the fine hairlike chips from shaving; instead they will be larger masses that can come only from burrs left in hobbing or shaping.

Coolants. Tests have proved that a coolant must always be used, even when shaving cast iron. Shaving dry, or with a water-soluble coolant, results in flaking of the cutting edges. When shaving cast iron, coolants become a problem. The heavier fatty oils hold the cast-iron particles in suspension and abrade the serration edges. The best coolant is kerosene, if safety regulations do not forbid its use. Mineral spirits are frequently used for cast iron, but again the flash point is low. For shaving steel, sulfurized mineral oils are recommended. When mixed production on the same shaving machine is necessary, compromises that suit both cast iron and steel must be worked out. Otherwise it is necessary to change coolants as the material changes. Manufacturers of tools can assist in recommending the proper coolant when they have been made aware of the requirements.

22.5 GEAR TOOTH BROACHES

Broaches are made for both internal and external spur or helical gears. The general principles of cutting gear teeth by broaching were covered in Sec. 16.5.

TiN coating is beneficial to broaches. The PVD process physically deposits a thin (0.000012-in.) coating of titanium nitride, with a hardness of 85 RC and a low coefficient of friction. There are several other benefits that together result in longer tool life. Due to their length, not many broaches can be coated with TiN, however. Broaches with renewable or removable finishing shells may be coated by one of several sources. Figure 22.35 shows a typical full-length broach.

22.5.1 Types of Broaches

Broaches are basically classified as follows:

1. By their method—push or pull broaches
2. By operation—internal or external

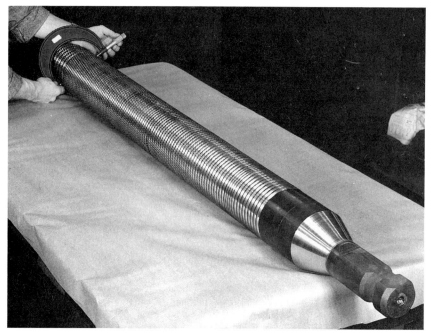

FIGURE 22.35 Full form finishing broach. *(Courtesy of National Broach & Machine Co., Mt. Clemens, Michigan)*

3. By construction—solid, built-up, progressive, circular, inserted tooth, and overlapping tooth

Pull Broaches. Pull broaches are used for most internal operations. They are capable of more stock removal than push broaches and are more suitable for longer cuts on internal work. The work-clamping pressure and broaching thrust are of importance, especially on work having thin walls. In most of these cases, a pull broach is preferable. A push broach is necessarily shorter and less expensive than the pull broach; however, based on stock removal, a pull broach can do the work that would require several push broaches. A typical pull broach is shown in Fig. 22.36.

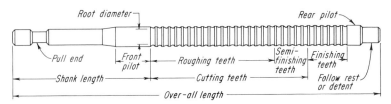

FIGURE 22.36 Typical internal pull broach.

22.5.2 Alternate Round and Spline Broaches

Frequently it is desirable to maintain concentricity between pitch line of a splined hole and pitch line of an integral external involute gear. A broach having alternate round and splined teeth at the finishing end is used. This serves to maintain concentricity between minor diameter, pitch diameter, and major diameter of the internal spline. To facilitate processing of subsequent operations, the minor diameter may be used with a plain round arbor for locating to cut the external teeth, ensuring the concentricity desired. This eliminates the need of a splined arbor and using the major diameter to accomplish the same qualified location for subsequent external gear cutting (see Fig. 22.37).

22.5.3 Design Modifications

In broaching splines utilizing a major-diameter fit in final assembly, a modification of involute form of the broach is made to avoid interference with the mating part (as shown in Fig. 22.38). The profile of the broach at its major diameter is altered with a tip protuberance. This provides an undercut at the major diameter of the internal spline to eliminate corner-radius interference of the mating members. By making this modification in the broached spline and not on the mating external part, the top land of the external part is preserved for location.

As a note of warning to the designer, be certain when using a major-diameter fit that tooth proportions of male and female splines are such that a pitch-line fit cannot overcome the major-diameter fit. This can be accomplished by providing sufficient backlash in excess of a loose effective fit.

Broaches of the pull type can be furnished with many types of pull ends. Figure

FIGURE 22.37 Alternate round and spline broach sizes minor diameter and maintains concentricity of minor, pitch, and major diameters.

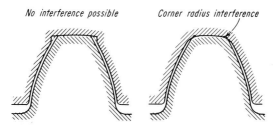

FIGURE 22.38 Means of avoiding corner radius interference.

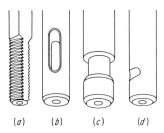

(a)   (b)   (c)   (d)

FIGURE 22.39 Four most common pull ends.

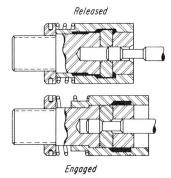

FIGURE 22.40 Automatic-sleeve-type puller.

22.39 shows four of the most common: *(a)* threaded type, *(b)* key type, *(c)* automatic sleeve type, and *(d)* pin type.

The most common type for spline production is the automatic-sleeve type for pull-up and pull-down vertical machines and on horizontal machines. Figure 22.40 shows the method of engaging and releasing the broach end. This type of puller can also be furnished with a locating flat at the pull end.

22.5.4 Spiral Involute Splines and Internal Spiral Gears

The broaching process has made huge gains in the area of production of internal spiral gears. Usually broach manufacturers recommend that the helix angle of the part not exceed 60° measured from the axis of the part. A lead bar is usually employed when the helix angle exceeds 15°. The part is clamped against rotation when a lead bar is used.

A particular area of interest is the planetary ring gears of automatic transmissions for the automotive industry (as shown in Fig. 22.41). The broaches are pulled through the proper helical path by guiding with a lead bar. The lead bar turns the broach as

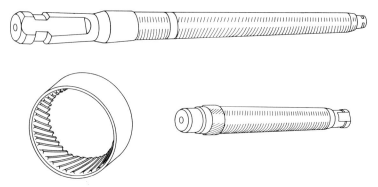

FIGURE 22.41 Ring gear broached in two passes.

it is pulled through the clamped work. Finishing broaches and preshave broaches have been made and put into production with results difficult to duplicate by other methods of production. Excellent lead and involute control have been achieved, providing all the elements of a good finished part or a good preshave part. The parts invariably reflect the original high-accuracy components developed into the broach.

Once the broach and lead bar have been developed, allowing for heat-treat distortion, high production with good duplication can be expected. The broach user must observe good practices in sharpening, blank preparation, machine maintenance, and consistent steel characteristics in order to retain production repeatability. This is paramount in any method of gear production to obtain tool efficiency.

22.5.5 Broaches for External Gears

The broaching application for external gears is feasible but limited to production quantities and size of part. The broaches have been made by series of broach rings through which the work is pushed. The broach rings are stacked, in proper sequence, held in a female adapter. The ram containing the part and piloting in bushings carries the part through the stacked broach rings.

Recently one broach manufacturer succeeded in constructing a solid broach for external gears. The results were very satisfactory, and broaches of this type have been put in production. There is at present a limitation as to diameter for the ring-type broaches. Diameter must be sufficiently large to permit introduction of a grinding wheel and grinding wheel support internally.

22.5.6 Design Elements

The length of the broach, along with the diameter of the broach, governs the price. Special features, of course, have an influence on price, such as profile deviations and alternate round and splined teeth.

Broach Pitch. The pitch, or tooth spacing, determines the construction and strength of the tooth, the allowable chip space, and the number of teeth in contact at one time. At least two broach teeth, preferably more, should be in contact at one time. In cases of inadequate machine capacity, the step per tooth is reduced rather than the teeth in contact.

The formula for pitch frequently used is

Pitch = 0.35 × length of cut   (22.5)

For cast iron a shorter pitch may be used.

The chip has no place to go while the tooth is in contact with the part. Therefore, the chip space must be designed to curl the chip and contain it until the tooth is clear of the work and can unload the curled chip (see Fig. 22.42).

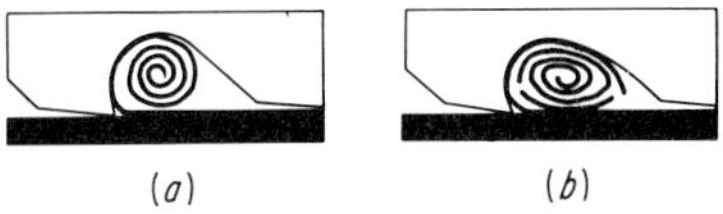

(a)   (b)

FIGURE 22.42 Tooth space must be designed to control chip: (a) curled chip in well designed tooth space; (b) chips in improperly designed tooth space.

Chip Breakers. Chip breakers are ground into round broach teeth to prevent ring formations of teeth that cannot be readily removed. Usually they are staggered in a pattern on the roughing and semifinishing teeth. Improper chip breaking can result in production of oversize splines, as well as dangerous tearing and galling in the work piece.

Chip Load. The cut per tooth in free-cutting metal is usually 0.004 to 0.006 in. measured on diameter for splines. Characteristics of the application and material being broached have considerable influence on the values of cut per tooth. Too heavy a cut will overload a broach and often result in a complete tool failure. Then, too, an insufficient cut per tooth will drag and burnish, causing heat, cutting edge wear, and galling. A practical minimum cut is in the area of 0.0007 in. on diameter.

Material Hardness. The broach manufacturers advise that part hardness be within 10 to 38 Rockwell C. Higher-hardness material can be broached, but in these cases, the broach manufacturer should be consulted. It is likely that the manufacturer's case histories will provide life expectancy of the broach and recommendations on processing, as well as the possibility of a premium-steel broach.

Most broach manufacturers are using molybdenum high-speed steels of the M-2 designation for 2-percent vanadium. The molybdenum steels have proved to be fully adequate for shock loads encountered. They respond well to surface treatment for increasing hardness of cutting edges. Special applications of broaching in ductile iron, stainless steel, and other high-hardness materials have required broaches made of M-3 steel. The M-3 steel is receptive to heat treatment to obtain a high hardness and a tough cutting edge that is resistant to abrasive wear and early dulling by part hardness.

Face Angle of Broaches. The face angle of the broach is determined by the material being broached. Hardness of part, toughness, and ductility are the influencing factors.

A rough guide for face angles is as follows:

Material	Face Angle, Deg
Cast iron	6–8
Hard steel	8–12
Soft steel	15–20
Aluminum	10 and plus
Brittle brass	−5 to +5

These are general values and are subject to varying, as needed, to accommodate a specific material.

The Broach Pilot. The internal broach contains a front and rear pilot for the purpose implied. The front pilot is designed to extend completely through the work to guide the broach in its cutting stroke. It also functions as a protective measure to avoid overloading the first roughing tooth. If the front pilot does not pass through the unbroached hole of the part, it is an indication that the part is not properly prepared for broaching. Never grind down the front pilot to fit unbroached holes. The hole should be altered; otherwise an excessive cutting load will be placed on the first roughing tooth.

The rear pilot supports the broach as the last finish teeth pass through the work part.

Without proper piloting, the rear end of the broach may whip, causing an unnatural form in the finished work piece.

Broach Tooth Side Clearance. Depth of spline tooth being broached determines whether or not broach should be relieved on the tooth flanks. For spline depths exceeding 0.100 in., side relief is generally incorporated to avoid galling or pickup of material. Because of the manufacturing limitations on shallow-depth broaches, it is difficult to side-relieve a broach cutting less than 0.100-in. depth. Occasionally, side relief is omitted to prevent broach drift when close limits are required in the broached part. The amount of side relief varies to suit the application and should therefore be left to the discretion of the broach manufacturer.

22.5.7 Tooling Tips

If the broach becomes stuck in work piece. Before a broach completes its cutting stroke, it may become stuck in the part. *Never* try to back the broach out by reversing the machine. Free the broach and work from the machine and fixture. Try to remove the part by tapping lightly on all sides, trying to work the part toward the starting end of the broach. Failing in this, the part may be reduced by putting the broach and part in a lathe. Another means is by sawing the part in two, if practical. Several factors may have caused sticking, such as dull teeth, power failure, or insufficient machine capacity.

Broach Cutting Fluids. The fluid lubricates chip-tooth flow, reducing frictional heat and abrasion. The chips are washed out of chip spaces. Heat generated by normal broaching is dissipated.

Internal broaching requires a heavy cutting fluid, as compared with surface broaching. A sulfur-base oil is generally preferred for steel. Cast iron may be broached dry; however, finish is improved by use of a very light coolant, such as mineral spirits or kerosene.

The broach manufacturer may have specific recommendations for coolants, depending on the application.

Broach Care. As with other tools used in gear production, the broach is a high-precision cutting tool, subject to damage and loss of efficiency by careless handling and improper sharpening.

Upon receipt of a new broach, check the last finishing tooth for size, also check the first cutting tooth to be sure the broach will not be overloaded. Measure and record, for sharpening purposes, the face angle, tooth depth, and root radius. Try the front pilot with a work piece. Do not drive the pilot into the work piece hole.

In storing or shipping the broaches, a little common sense will be rewarding. Store the broach in racks with partitions so that no broach can contact another. If stored for future use, or if returning for sharpening, always coat with a rustproofing lubricant. Never ship a broach with any other steel item loose in the case or carton, such as a sample part.

Sharpening. Overdulling of a broach will reduce its life at a high rate. If possible, keep a record of number of pieces per broach sharpening versus amount of grind-back on cutting face. In a short time, an adequate history can be obtained that will profit by extending tool life. The broach manufacturer can make excellent recommendations based on experience. The best results, however, are obtained by actual case history and records taken from the production floor.

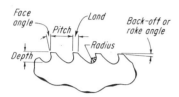

FIGURE 22.43 Broach-tooth form.

For internal broaches, grind the front face only. When sharpening, make certain that the ground face blends with the chip-space contour (see Fig. 22.43). An interrupted surface or rough surface adds resistance to the chip curl and disposal.

It is not necessary to sharpen all finishing teeth with each resharpening. Sharpen only the first finishing tooth, or the two initial finishing teeth as required. As they become undersize from subsequent sharpenings, treat the second or third or fourth tooth in sharpening as the first or second teeth were treated. The undersize finishing teeth can have chip breakers added to avoid wide chips.

Proper setup precautions should be observed in resharpening so as not to permit runout when the broach is between centers. It is advisable to contact the broach manufacturer for sharpening recommendations as to wheel grit, size of wheel, and whether to grind wet or dry. Normally, dry grinding is employed, but care must be exercised to avoid burning of the cutting edges. On long broaches, a steady rest is recommended to avoid deflection or sag of the broach.

Today's broach manufacturers have developed a great deal of know-how. This is reflected in the high precision achieved in both straight and spiral splines being cut in production. They have kept abreast of the precision of the easier-to-make external spur and helical involute splines produced by hobbing, cold forming, or grinding. It is recommended that, for new applications and chronic broaching problems in production, a broach engineer be consulted. Such services invariably will be rewarding in terms of instruction and application advice.

Information Required in Ordering Broaches.   The following list shows the vital information needed by the broach manufacturer.

1. Print of the part, giving dimensions and range of limits, if possible.
2. Material specifications, casting, forging, etc.
3. Prior machining operations on the part; also whether the machined surface is to be used for locating during the broaching operation.
4. Prior heat treatment of the part. State hardness.
5. Subsequent heat treatment of the part, which may affect the accuracy of the broached spline.
6. Production rate and quantity desired.
7. Finish required.
8. Broaching machine to be used, and information as to special fixtures or guides that are to be used.
9. Specify type of puller and whether rear support is used.

10. Radius or chamfer, if required on the corner of splines at either inside or outside diameter.
11. Is the broach to size the inside diameter as well as cut the splines?
12. Concentricity requirements and relationship to reference diameters.
13. Type of fit desired.
14. Mating part information, if available.

22.6 HARD GEAR FINISHING

Hard gear finishing, a recently introduced new technique for finishing gears after heat treatment, has made impressive inroads for gear manufacturing. Gears have been finished after heat treatment by actually cutting the teeth smaller as with Skive hobbing, or by the crossed axis principle using a carbide tool similar to a shaper cutter. However, this text relates to electroplated *hard gear finishing* tools (HGF) that have a single layer of CBN, or a resin or metal-bond CBN tool that requires dressing.

22.6.1 About CBN Crystals

In 1961 Cubic Boron Nitride (CBN) was introduced by General Electric. CBN crystals have sharp, long-lasting cutting edges that produce cleanly cut chips. The crystals are hard; next to diamonds, they are the hardest material known. They are more than twice as hard as aluminum oxide.

CBN wheels have an advantage due to their hardness and thermal stability. The temperatures encountered while grinding hard material are quite high. It can be seen in Fig. 22.44 that the hardness of CBN has an advantage over aluminum oxide.

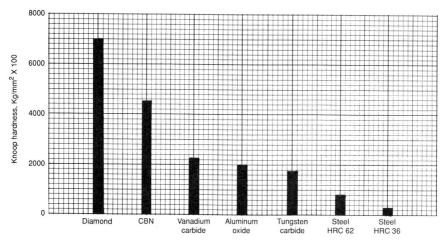

FIGURE 22.44 Comparison of the hardness of CBN crystals with the hardness of aluminum oxide and the hardness of certain other materials.

In 1981, General Electric introduced microcrystalline-structured crystals, as compared to single-structure crystals. Thus, when crystals become dull, a small microfracture occurs that makes the crystal self-sharpening throughout its life.

Thermal conductivity of CBN is high. This results in much of the heat generated by grinding going into the wheel, so that it is very important to flood the wheel and work piece with coolant to carry away chips and to lower the wheel temperature.

CBN grinding wheels allow a much greater rate of stock removal of hardened parts without producing grinding burns that are prevalent with vitreous grinding.

Tool life is estimated to be 10,000 times the wheel diameter for each pass across the face of a 1-in. gear. As CBN wheels grind, most of the heat generation is carried away by the wheel and by the chips, eliminating metallurgical damage and distortion to the gear being ground. Gears tested for aircraft use proved to have slightly better surface fatigue life than vitreous ground gears, while the subsurface residual stress remained approximately the same.

Figure 22.45 shows an historical review of gear grinding machine techniques developed over many years. With an appropriate HGF tool, all of these can be adapted to hard finishing. In the last few years, special machines for hard finishing have been developed by Liebherr and Pfauter-Kapp companies in Germany, Karatsu in Japan, and National Broach and Machine Company in the United States.

A stiff, rigid machine must be used to successfully apply CBN techniques. The machine should be made with ball-screw drives for feed slides, and materials with low coefficients of friction and high dampening qualities. The grinding spindle must be both a precision running spindle and of rigid design. It goes without saying, CNC control is almost a must.

22.6.2 Cutters for Hard Gear Finishing

The crystals used for hard gear finishing tools range in grit size from 90 to 170. The finish not only depends on crystal size, but also on the machine condition, the feeds, and the speeds at which it is operated. The wheel r/min works best at the highest speed possible. Care must be taken to have the wheel balanced and running as true as possible to avoid vibration. Wheels that are not mounted with care will cause chatter and vibration, making good surface finish impossible. With the wheel properly mounted, a finish of 2 to 3 μm* peak to valley is expected, using a work-type wheel, 120-grit crystal, 6.3 in. in diameter, running at a speed of 3300 r/min.

Form-Wheel-Type Cutters. A single layer of CBN crystals is applied electrolytically to a hardened and precision-ground steel base using a nickel bond. The form of the grinding wheel represents the form of the gear tooth. It can be modified as desired for optimum results. This wheel is shaped for a specified number of teeth, diametral pitch, pressure angle, and tooth modifications. If any one of these parameters changes, a different form wheel is required. This type of wheel is used on Pfauter-Kapp form-grinding machines, for grinding either internal or external gears. The crystals can be stripped and replaced several times.

Worm-Wheel-Type Cutters. A standard or modified gear rack is formed as a thread to the outside diameter of the grinding wheel. The grinding wheel and the work piece

---

*A micron (μm) is 40 μin.

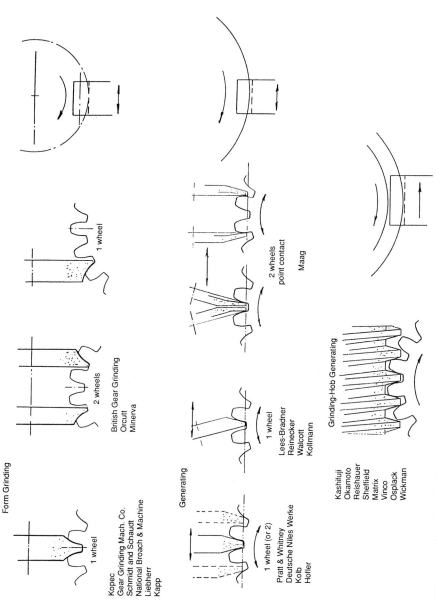

FIGURE 22.45 Schematic diagrams of gear tooth grinding methods. The names shown are companies who have produced grinding machines of these types. Several of these companies are no longer in business.

engage one another like a worm driving a worm wheel. Each revolution of a single-thread worm will rotate the gear one tooth in a continuous motion. The grinding wheel is rotated up to 6000 r/min as it is fed across the face of the gear.

A typical use of this wheel is by Kashifuji Gear Work, whose hard gear finishing machine was used to hard-gear-finish pinions for several months. A tangential feed of 0.140 in. was used for three roughing passes, with an infeed of 0.0004 in. each pass and a finishing pass using 0.040-in. feed rate. These pinions were finished at a rate of 81 pieces per hour.

The HGF worm used was 6.300-in. diameter by 3.540 in. long, single thread. The pinion the HGF tool was made for had 16 teeth, 15.58 NDP, 20° NPA, 18° HA (LH).

More than 228,000 pieces were run, 6000 for each of 38 settings across the active face width of the HGF tool. The first piece run in one of the 38 settings, as well as the six-thousandth piece, were checked and compared. The quality level checked the same—no deterioration in quality. Every fifth piece was nital etched, looking for surface damage. This test showed that no surface damage occurred during the test run.

Lange[14] and Moncrieff[18] may be consulted for considerable additional material on hard finishing.

REFERENCES

1. AGMA 112.03, "Terms and Definitions," American Gear Manufacturers Association, 1954.
2. AGMA 120.01, "Gear-Cutting Tools Fine- and Coarse-Pitch Hobs," American Gear Manufacturers Association, 1975.
3. ANSI/AGMA 1012-F90, "Gear Nomenclature, Definitions of Terms with Symbols," American Gear Manufacturers Association, 1990.
4. Buckingham, Earle, "Analytical Mechanics of Gears," Dover, 1963.
5. Dodd, H. and Kumar, D. V., "Technological Fundamentals of CBN Bevel Gear Finish Grinding," *Gear Technology*, November–December 1985, p. 30.
6. Drago, R. J., "Comparative Load Capacity Evaluation of CBN-Finished Gears," *Gear Technology*, May–June, 1990, p. 8.
7. Dudley, D. W., *Gear Handbook*, 1st ed., McGraw-Hill, New York, 1962.
8. Dudley, D. W., *Handbook of Practical Gear Design*, McGraw-Hill, New York, 1984.
9. General Electric, "Borazon CBN Superabrasives," General Electric Company, Lynn, Massachusetts, 1981.
10. Haug, E., "The Wafer Shaper Cutter," *Gear Technology*, March–April, 1989, p. 26.
11. Hildreth, R. S., "Gear Cutting Savvy," *American Machinist*, 1953.
12. Janninck, W. L., "Shaper Cutters-Design and Application, Part 1," *Gear Technology*, March–April, 1990, p. 35.
13. Kelley, P. W., "Advantages of Titanium Nitride Coated Gear Tools," *Gear Technology*, May–June 1984, p. 12.
14. Lange, J. M.: "Gear Grinding Techniques Parallel Axes Gears," *Gear Technology*, March–April 1985, p. 34.
15. Merritt, H. E., *Gear Engineering*, Pitman, New York, 1971.
16. Metal Cutting Tool Institute, *Metal Cutting Tool Handbook*, MCTI Publishing, 1959.
17. Moncrieff, A. D., "Pre-shave Gear Cutting Tools," AGMA 129.07, 1949.
18. Moncrieff, A. D., "Hard Gear Finishing," *Gear Technology*, March–April 1988, p. 7.

19. National Broach and Machine Division, "Modern Methods of Gear Manufacture," National Broach & Machine Div., Lear Siegler, Inc., Detroit, 1972.
20. Staub, C. R., "Be Kind to Your Hobs," *Machinery,* 1952.
21. Sulzer, Dr. G., "CNC Gear Shaping," *Gear Technology,* March–April 1986, p. 16.
22. Vogel, W. F., "Involutometry and Trigonometry," Michigan Tool Co., Detroit, 1945.
23. Whitney, D. S., "Computer Aided Design for Gear Shaper Cutters," *Gear Technology,* November–December 1987, p. 11.

CHAPTER 23
GEAR INSPECTION DEVICES AND PROCEDURES

Dennis P. Townsend
NASA Lewis Research Center
Cleveland, Ohio

Michael J. Broglie, P.E.
Dudley Technical Group, Inc.
San Diego, California

Danny F. Smith
Dudley Technical Group, Inc.
San Diego, California

There has been considerable development in recent years on gear inspection devices and this development is expected to continue at a fairly rapid pace for several years. A wide variety of gear inspection equipment and technical data for inspection are available today. Recent developments in gear inspection equipment have made it possible to make measurements with an accuracy of less than 1 μm (40 millionths of an inch), and, in addition, some inspection machines can automatically measure several aspects of a part on a single machine. Also many different types of gears can be measured on just one machine, provided the software is available. These developments have made checking faster and easier, making it possible to make gear measurements on a greater number of parts.

During the manufacture of gears, a certain amount of error will be present. The amount and type of error produced may vary, depending on how the part was manufactured. The error must be determined so that the manufacturing machines can be adjusted to correct the errors and ultimately determine the quality of the gear.

A wide variety of equipment and techniques are used to inspect the gear teeth. The question of how much tolerance to allow on various gear elements was covered in Chap. 9. This chapter presents detailed information on the equipment used in gear inspection.

The late Albert S. Beam of the Vinco Corp., Detroit, Michigan, was the first edition author of this chapter. Theodore F. Klem of the Gleason Works, Rochester, New York, contributed some material on bevel gear inspection and Darle W. Dudley of the General Electric Co., Lynn, Massachusetts contributed material on worm gear inspection.

Gear inspection involves two broad categories. The analytical inspection involves dimensional elements such as profile, lead, spacing, tooth thickness, and concentricity. Functional inspection involves rolling with a master or mating production gear to determine composite variations, transmission error, or vibration spectrum. The gear must also be inspected for other dimensions, which include the bore, hub, length, etc. Surface finish has become an important parameter in high-performance gears and hardness of case and core may be a required inspection parameter.

This chapter deals primarily with inspection of gear teeth. It gives information on how each kind of gear is checked and what is done to handle very large gears or very small gears, as well as how to handle normal-size gearing.

23.1 ANALYTICAL GEAR CHECKING

Analytical gear inspection comprises the various methods that check dimensional characteristics of a gear. These include involute, lead, tooth thickness, etc. When the gear is manufactured, the specifications may call for limits on the various deviations from the standard gear tooth form which result from the manufacturing process. The purpose of gear inspection is to determine what these errors are and identify when it is necessary to make corrections to the manufacturing process to maintain the required tooth accuracy. With some of the automatic inspection equipment available today, it is possible to have a direct link between inspection and manufacturing so that manufacturing machine settings are adjusted automatically based on inspection results.

23.1.1 Involute Measurement

The involute chart is a measure of the gear tooth profile. Figure 23.1 shows how the involute chart is developed by rolling the gear on the base circle, producing contact traces of the profile. All profile variations such as tip and diameter modifications show up as deviations from a straight line. Other gear errors, such as runout and gear wobble,

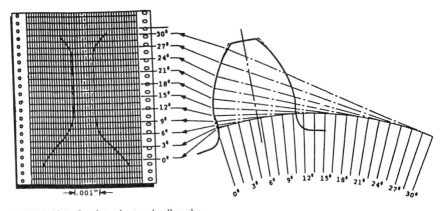

FIGURE 23.1 Involute chart and roll angles.

also effect the involute profile trace. The involute profile may change along the length of the tooth which, of course, contributes to measured error, so it is necessary to take measurements of the profile at several axial locations.

Involute checking equipment may have a fixed base diameter or a variable base diameter. Newer equipment usually has the variable base diameter. For large gears, where it is impractical to mount the gear, a portable involute checker is used. A typical profile chart is shown in Fig. 23.2.

23.1.2 Lead Measurement

The lead measurement determines how accurate the gear tooth profile is in the axial direction. For helical gears, the lead is a measure of the helix trueness of the teeth. Often it is necessary for a designer to specify a lead shape that deviates from a true line. This may be done to compensate for such things as bending and twisting of the gear under load, or to compensate for thermal effects that vary across the face width. Other modifications, such as crowning and end easement, may also be needed to avoid concentrating loads on the tooth ends under certain misalignment conditions. Data recorded on a lead chart can be affected by gear wobble, and on helical gears the runout will have some effect. The lead measurement must be corrected for these other causes, and, since the lead may vary along the profile, it is good practice to require checks at more than one profile location. A typical lead chart for a gear is shown in Fig. 23.3.

23.1.3 Tooth Spacing and Index Measurement

The tooth pitch is the theoretical distance between corresponding points on adjacent teeth while spacing is the measured distance. The index is the theoretical angular position of teeth about an axis or, in the case of a rack, the theoretical linear position. The index variation is the displacement from the theoretical angular or linear position. Figure 23.4 is a comparison of theoretical and actual index measurements on the same gear.

When a gear is measured for tooth spacing and index, all the teeth and several other things are measured and usually recorded and computerized with several items of information about the gear. The computed output from these measurements gives adjusted readings that are plotted to show the different spacing and index variations. The spacing measurements are affected by runout, tooth thickness, or backlash, as well as the pitch, index, and spacing variations. Output from the adjusted readings is sometimes called true index variation and includes the tooth spacing index error, the tooth pitch error, the backlash, and pitch line runout. A printout of maximum values of these errors may also be included. Figure 23.5 is a printout of some of these variables with some charts showing index errors, tooth thickness errors, and runout errors.

23.1.4 Combination-Type Analytical Checking Equipment

There are many inspection instruments on the market that perform a variety of analytical measurements on one instrument. These may be gear checking instruments or three-dimensional coordinate measuring instruments with software to provide for the

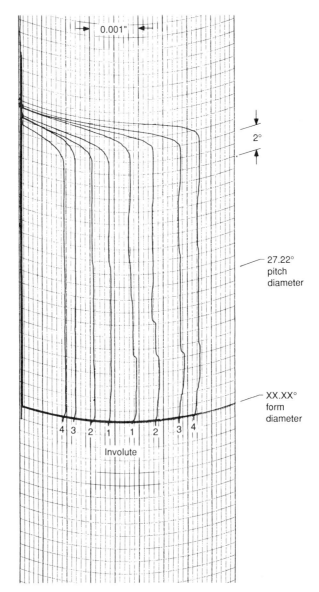

FIGURE 23.2 Typical profile chart. *(Courtesy of Dudley Technical Group Inc., San Diego, California)*

inspection of gears. Some of these inspection centers are capable of checking the profile, lead, tooth spacing, and runout on just one machine. It is also possible to check cylindrical gears, worms and worm gears, and bevel gears, as well as hobs and shaper and shaving cutters on one machine. Not all machines in this class are capable of checking all types of gears or making all the types of measurements possible. Machine-

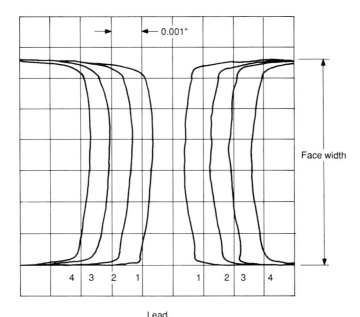

FIGURE 23.3 Typical lead chart. *(Courtesy of Dudley Technical Group Inc., San Diego, California)*

tool builders have designed machines with different capabilities to meet user requirements.

The combination instruments come in a variety of sizes and are usually more expensive than a comparable set of single-purpose instruments (involute, lead, spacing). They are usually operated with a computer system that completely automates the measurement of the various gear parameters and stores these measurements for later use or sends the information to the manufacturing machine for expedition of manufacturing process. The operator sets the gear in the instrument and, upon initial setup, locates the measuring probe relative to the gear; automatic controls then finish the measurements and plot or print the data on the charts.

With the automated measuring instrument it is possible to make a three-dimensional map of the gear tooth lead and involute shape as shown in Fig. 23.6. This type of printout gives a topological map of the tooth, which can be very useful for finding errors or checking gear tooth modification that would not show up on a one- or two-dimensional trace plot.

This three-dimensional trace of the tooth flank can also be mated with a master gear or other gear to determine a contact pattern, as shown in Fig. 23.7.

By not having to move the gear being inspected to different devices, time is saved by eliminating the need to true up the part for separate devices. Also, because the machine is computer-controlled, it can be programmed to automatically make a predetermined series of checks on a part. Once programmed, the process can be stored in the machine's memory and can be recalled at a later date to run the same checks on another lot of the same parts. The savings in time is an important advantage of this type of machine over traditional devices.

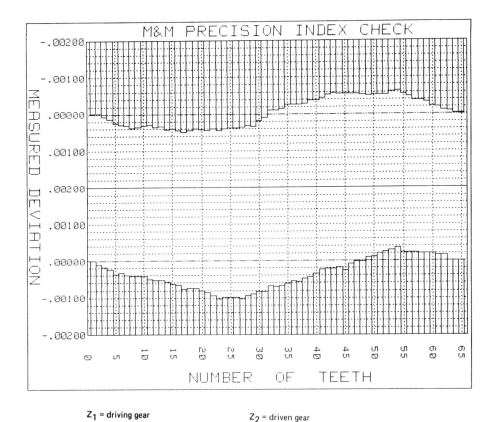

Z_1 = driving gear Z_2 = driven gear

FIGURE 23.4 Index measurement chart. *(Courtesy of M&M Precision Systems Inc., Dayton, Ohio)*

23.2 FUNCTIONAL GEAR CHECKING

Functional gear testing consists of testing gears by the way they function or roll. The earliest methods for checking were functional testers. Until recently the major advances in gear checking equipment have been analytical checking machines. Recently, advances in functional checking devices have shown the benefit of the data produced by this type of checker.

23.2.1 Gear Roller or Double Flank Tester

A schematic diagram of a gear rolling fixture is shown in Fig. 23.8. Using rollers, the gear being checked is meshed with a master gear and the trace is recorded through a full revolution of the test gear. The output is a measure of variations of the test gear plus the variation of the master gear. Figure 23.9 is a typical chart obtained when using a gear roller. The data presented in this chart can be interpreted to determine total composite error, tooth-to-tooth composite error, and deviations such as tooth-to-tooth error.

GEAR INSPECTION DEVICES AND PROCEDURES

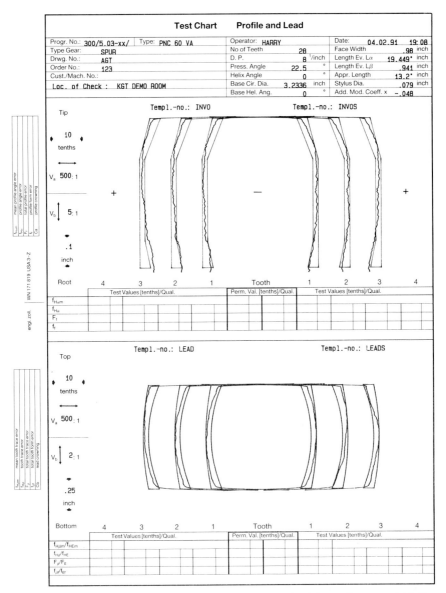

FIGURE 23.5 Printout of multiple parameters. *(Courtesy of Klingelnberg Sohne, Remscheid, Germany)*

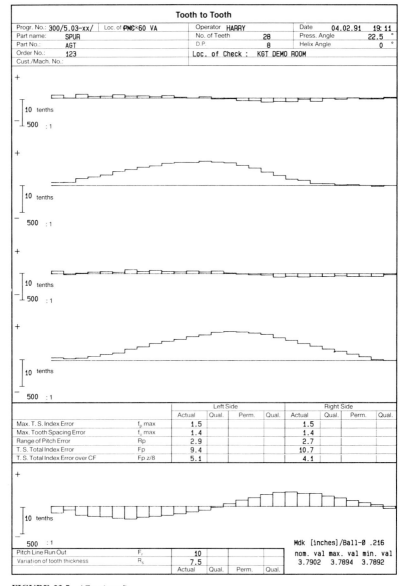

FIGURE 23.5 (*Continued*)

Some rollers are equipped with a special sensing head called a *gimbal head*, which allows the master gear to pivot in two planes. This allows for indication of lead variations.

Gear rolling machines come in a variety of types. Some are hand-operated, some are semiautomatic, and some are fully automatic for use on production when several

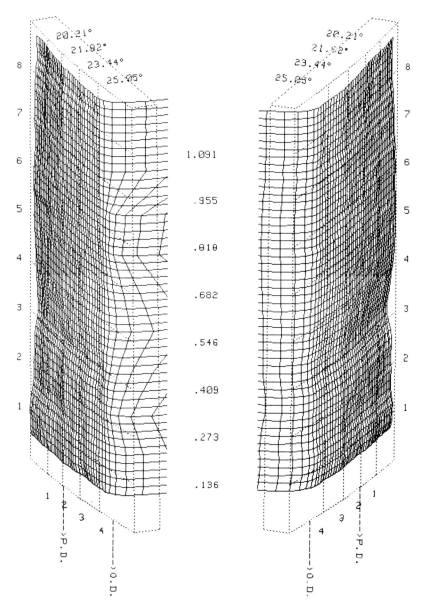

FIGURE 23.6 Topological plot from an automated measuring machine. *(Courtesy of M&M Precision Systems Inc., Dayton, Ohio)*

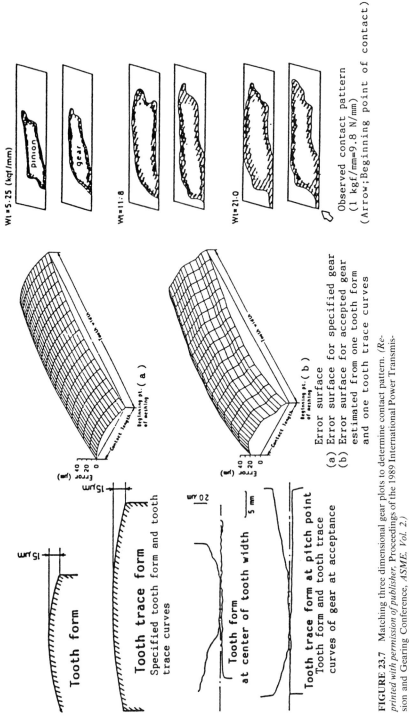

FIGURE 23.7 Matching three dimensional gear plots to determine contact pattern. *(Reprinted with permission of publisher, Proceedings of the 1989 International Power Transmission and Gearing Conference, ASME. Vol. 2.)*

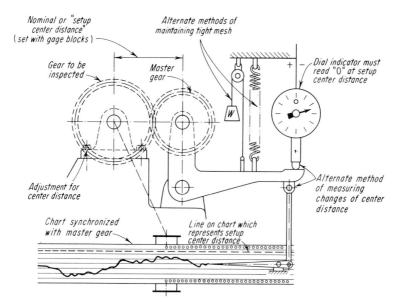

FIGURE 23.8 Schematic of gear rolling device.

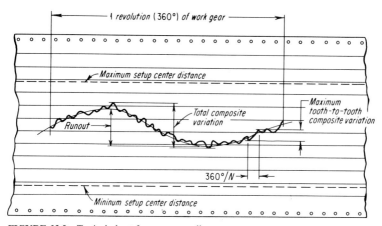

FIGURE 23.9 Typical chart from a gear roller.

hundred gears are measured per hour. The fully automatic unit outputs an electrical signal which can be used to sort gears according to different types of errors.

The gear rolling instrument can determine various errors such as rough tooth action, lead errors, and pitch diameter size. However, it cannot tell whether action, etc., is bad because of involute errors, lead errors, errors of profile modification, or some combination of these.

The drawback of this method is that the gears roll in tight mesh with no backlash. This means the production gear runs on a different pitch diameter with the master than it would with its mate.

23.2.2 Single Flank Gear Testers

The single flank gear tester is used to measure several types of gear parameters such as those determined in a gear roller, except in this case there is backlash and the gear rolls on its operating pitch diameter. It will also measure transmission error, which is sometimes called variable velocity. Transmission error is defined as deviation of the position of the driven gear, for a given angular position of the driving gear, from the position the driven gear would occupy if the gears were geometrically perfect.

The single flank gear tester measures transmission error by rolling with a master gear with backlash. One shaft is motor-driven at a constant angular velocity, and the relative motion of the driving shaft to that of the driven shaft is measured by optical encoders located on each shaft. Figure 23.10 is a schematic of a single flank measuring

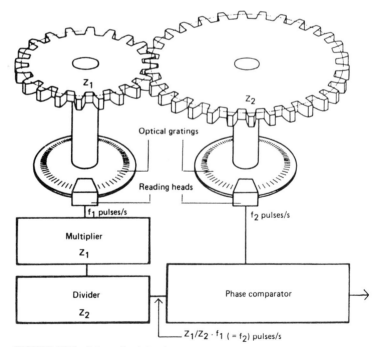

FIGURE 23.10 Schematic of the single flank instrument.

system. The output of the single flank instrument is the transmission error in the form of a fairly regular once-per-tooth pattern, superimposed on a large once-per-revolution pattern. Most single flank testers operate with a light load on the gear pair and are called unloaded single flank testers. Some single flank testers are being made that can load the gear, yielding loaded transmission error, which is more closely related to the actual condition. The profile modification on gear teeth will give different results depending on whether a loaded or unloaded single flank tester is used. Transmission errors of less than a second of an arc can be detected with single flank instruments. Single flank testers can be made for all types of gears, such as spur, helical, spiral bevel, worm, etc. Figure 23.11 is a picture of a single flank tester for spur and helical gears, and Fig. 23.12 is a picture of a single flank tester for hypoid gears.

23.3 CHECKING SPUR AND HELICAL GEARS

The majority of gears made throughout the world are spur and helical gears, and these primarily have an involute tooth form. Because of the extensive use of spur and helical gears there are many different types of devices to inspect them. The complexity and precision of the measurements used depends on the criticality of the gear application.

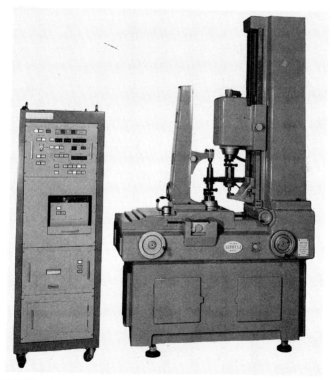

FIGURE 23.11 Single flank tester for spur and helical gears. *(Courtesy of Osaka Seimitsu Kikai Co., Ltd.)*

FIGURE 23.12 Single flank tester for hypoid gears. *(Courtesy of Gleason Works, Rochester, New York)*

23.3.1 Involute Profile Measurement

An involute can be checked with equipment as simple as a dividing head and height gage. This type of inspection is known as incremental checking since the gear is rotated in increments; the increments are controlled by the dividing head. At each increment a height measurement is taken on the tooth at a point on a vertical line that is tangent to the base circle. Figure 23.13 is a schematic of how this is done. Some calculation work is then needed to convert the height measurements into deviation from true involute. This is a very accurate method, but is quite slow.

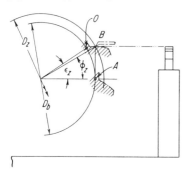

FIGURE 23.13 Involute-profile inspection by increments.

Most involute checks are made using involute checking machines. The basic principle of an involute checking machine is to indicate the amount of variation from true involute without any further calculation work. Since an involute is generated by rolling on the gear's base circle, the action of an involute machine can work on the same principle. Some machines do this with base circle disks, a master disk with compensators or linkages and cams that simulate base circle action. Other machines use a master involute cam and linkages to trace the involute curve.

Involute checking machines generally plot the results on a strip chart. A true involute would appear as a straight line on the strip chart. In most cases the desired involute shape is not a true involute, but is modified in the addendum to improve the running conditions. The profile shape may vary across the tooth face. To determine variations, checks can be made at several locations across the face width.

The latest designs of checking machines have incorporated computers into the machines. Due to the integration of the computer, there is a lot of flexibility in the way results are presented. For example, Fig. 23.14 shows how several checks can be combined and plotted to show the topological condition of a tooth. Another format is shown in Fig. 23.15, which shows a combination of plotted and tabulated data for an involute check.

Involute Errors. As mentioned above, if a gear is cut with a pure, unmodified involute and there are no errors, the plot off of the involute checking machine will be a straight line. If, on an actual involute plot, the chart indicates a negative material condition, the error is said to be minus. If the error is in the direction of extra material, the error is said to be plus. Most power transmission gears are modified so there is a minus involute in the area of the tooth tip. A plus tip can cause the gears to run noisily and may also result in premature wear.

Involute errors are the result of either errors in the cutter, cutting machine, or in the cutting process. Initial checks are made on the first piece cut on a machine after a new setup. This is to establish whether the cutter has the right shape to generate the proper involute profile and the machine has been set up correctly. Once it has been established that the cutter and the setup are correct, it is necessary to continue to make periodic checks in order to determine when it is time to sharpen the cutter. It is necessary to check the first piece cut after the tool has been sharpened to verify that the sharpening has been done correctly. Sharpening errors can cause the profile shape to be very different from what may be required.

23.3.2 Lead Measurement

Also known as helix measurement, lead measurement refers to the alignment of the tooth with its proper helix angle (spur gears have a 0° helix angle). The lead is measured normal to the tooth surface along the face width. Figure 23.16 illustrates how lead checks are made on gears.

A sine bar is the most basic way to the check helix angle. However, since the sine bar measurement is based on the points where balls touch the tooth, this method can give an inaccurate measurement because it does not account for local undulations. The more accepted method is to use a lead checking machine such as the one shown in Fig. 23.17.

Helix Checking Machines. There are a variety of helix checking machines on the market. These machines mount the gear to be checked between centers. The check is made by rotating the gear and moving a carriage in an axial direction at the right rate of lead advance per degree of rotation for the gear to be checked. The carriage carries an indicating means that is set to read deviations normal to the tooth surface. The deviation reading may be made by dial indicator, or an electrical pickup may be used and the variation recorded on chart paper just as involute checkers chart involute variation.

Figure 23.18 shows a typical helix checking machine. Note gear A, the carriage B that moves axially, and the electronic pickup C that measures the lead variation.

Helix checking machines for a gear of low helix angle (long lead) generally have a disk that rolls against a straight edge. The tangential movement of the straightedge is translated into an axial movement by sine bar or similar device. In some cases these

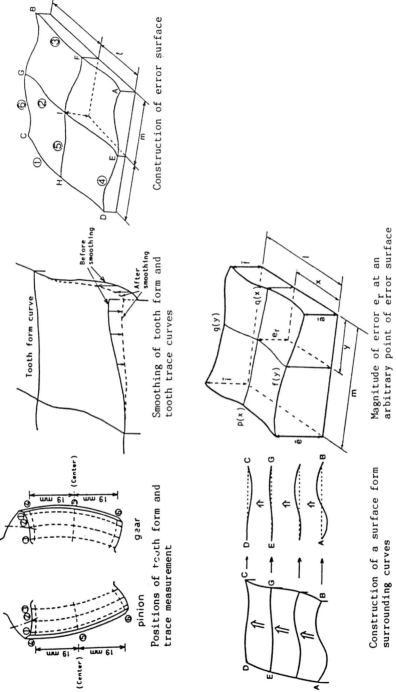

FIGURE 23.14 Combination of checks used to produce topological map of tooth surface. (Reprinted with permission of publisher, Proceedings of 1989 International Power Transmission and Gearing Conference, ASME, Vol. 2.)

GEAR INSPECTION DEVICES AND PROCEDURES   23.17

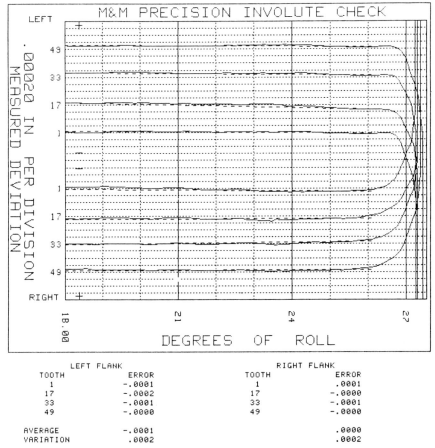

FIGURE 23.15 Involute check showing plotted and tabulated data. *(Courtesy of M&M Precision Systems Inc., Dayton, Ohio)*

machines are built in combination with involute checkers, and the roll of the base-circle disk is used to create both an involute curve for profile checking and a helix for lead checking. Other machines employ a master lead bar and follower to obtain the helical motion.

Contact Checking. Any given pair of gears that are to run together may have their effective lead accuracy determined by a simple contact check of the two parts. This may be done in a checking stand where adjustments can be made to get the two reference axes exactly parallel to each other. The check may be made in the casing that the gears are to run in. If it is made in the casing, errors in location of casing bores will show up in the contact check; a poor contact may be due to gear cutting error or due to error in machining the casing.

Where interchangeability is required, a gear may be contacted with a master in a

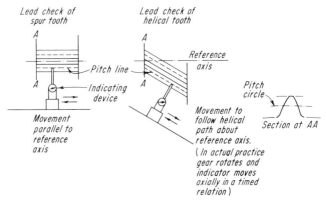

FIGURE 23.16 Schematic method for making a lead check of a gear tooth.

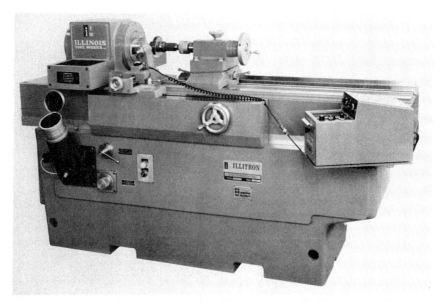

FIGURE 23.17 Modern lead-checking machine. *(Courtesy of Illinois Tool Works, Inc., Chicago, Illinois)*

checking stand. Pairs of gears that properly fit their respective masters can be assumed to be capable of fitting each other (provided that the casing is good).

Contact checks are often made by putting a thin layer of "red lead" on one member and a thin layer of "prussian blue" on the other member. The parts are then rolled together and the blue makes a clear marking on the red. The thickness of the marking paste enters directly into the check. If the paste is thinned down with solvent to an extreme, a lack of contact marking may happen with a gap between the surfaces of as little as 0.0001 in.! On the other hand, thick paste can be used, and contact may be

FIGURE 23.18 Lead-checking machine. *(Courtesy of Michigan Tool Co., Detroit, Michigan)*

indicated when a separation of as much as 0.0005 in. occurs. A proper application of the colors should show contact at 0.0002 in. and a lack of contact at 0.0003 in.

Some manufacturers will put a flash of copper plate or a flash of silver on one member. Then the gears are run under power for 15 min to ½ hr in their casing. This will show up as a clear contact marking. This kind of test gives very clear-cut indications and is easier to control than red and blue colors. It has the hazard, though, that a bad pinion might have a "wobble" type of lead error. After a few minutes of running, the wobbling pinion might mark up the gear all across its face width and give the illusion that the gear was fitting the pinion all right. If the gear had a little spacing error, it might in turn mark the pinion so that a wobble type of contact on the pinion could not be seen.

Contact checks on gears require the least equipment. Gears of low cost and low accuracy are frequently checked for lead accuracy by this method.

The contact check has the disadvantage that the gap between the teeth when no contact exists is not measured. Some manufacturers have developed unique and proprietary techniques to measure the tiny gap that may exist between contacting gear teeth. Measurement of a gap as little as 0.0001 in. can be made.

23.3.3 Measurement of Gear Tooth Spacing

A variety of devices are used to measure gear tooth spacing. To understand the different kinds of equipment used to measure spacing, it is first necessary to understand the different ways in which tooth spacing may be evaluated. These are:

1. *Pitch variation:* The difference between one circular pitch and the pitch immediately before or after the pitch in question is called *pitch variation*. The maximum allowable pitch variation for a gear is the greatest variation between any two adjacent pitches. Pitch variation is widely used as a measure of gear accuracy.

2. *Pitch variation over a span of teeth:* Sometimes the accuracy of a gear is specified by a tolerance on pitch variation over a span of so many teeth, for instance, over four teeth or five teeth. The variation is determined by comparing the span length over the specified number of teeth with the span over a like number of teeth anywhere on the gear. This measurement is not used as often as the pitch measurement.
3. *Total pitch variation:* This is the difference between the *longest* and *shortest* circular pitch in the whole gear. In the case of power gears; for instance, aircraft power gears; this concept, which was originally proposed by Darle W. Dudley, seems more significant than the pitch variation since it establishes control over the worst possible pitch difference between meshing pairs of adjacent teeth in two mating gears.
4. *Base pitch variation:* This is the variation in pitch taken along the line of action (see Fig. 23.19). Like pitch variation, base pitch variation is the difference between a base pitch and an adjacent base pitch. Base pitch variation is not used much because it tends to reveal only the base pitch error of the cutting tool and not the error in spacing produced by the generating machine.
5. *Index variation:* This is the "accumulated" spacing error over any number of teeth. Total index variation may be defined as the greatest out of position of any tooth with respect to any other tooth in the whole gear. Total index variation is important in any case where absolute trueness of angular motion is required. Radar gears, index gears for machine tools, gears of planetary systems, and high-speed power gears must be held to a high order of index accuracy.

A spacing check may be taken from any point on the surface of one tooth to a corresponding point on the surface of another tooth. It is customary, though, to make the spacing check near the center of the face width and approximately at the pitch line.

Since pitch variation and index variation are the two most important of the different kinds of spacing variation, the discussion of spacing checking devices centers around these two kinds of measurements.

Measurement of Pitch Variation by Machine. Several kinds of devices or machines are on the market which will check the pitch of a gear. Figure 23.20 shows the basic elements of a pitch checking machine. Note that the checking head has a fixed finger, a movable finger attached to an indicating device, a spring device to hold the fingers in contact with the teeth, and a stop to hold the checking head at the same position each time a reading is taken.

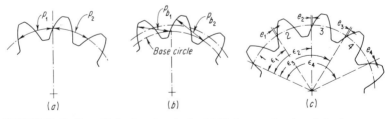

FIGURE 23.19 Some kinds of spacing checks. (*a*) Pitch error of p_1 is variation between p_1 and p_2. (*b*) Base pitch error of p_{b_1} and p_{b_2}. (*c*) N = number of teeth. Angle $\epsilon_1 = 360°/N$, $\epsilon_2 = 360°/N \times 2$, $\epsilon_3 = 360°/N \times 3$. e_1 = index error of tooth 1, e_2 = index error of tooth 2. Total index error of gear is difference between largest and smallest value of e. If the smallest e is *negative*, take the *algebraic* difference.

A gear to be checked by a spacing checking machine is mounted in the machine on a rotatable axis. Preferably this axis should be the *reference* axis of the gear. If the part is not mounted on its reference axis, the runout of its reference axis with respect to the checking axis must be noted and allowed for in interpreting the spacing check. The gear is checked by moving the head in against the stop, taking a reading, withdrawing the head and rotating to the next set of teeth, moving the head in again, and taking another reading.

Figure 23.21 shows a typical pitch checking machine. The machine has a headstock and a tailstock for mounting the gears between centers. The gaging head is mounted on a cross slide and contacts corresponding sides of adjacent teeth with one fixed contact member A and one movable contact member B. The contact is maintained by a third, spring-actuated finger C. Helical gears are checked by swiveling the barrel D, positioning the indicating mechanism to the correct helix angle. When checking the circular pitch of spur, helical, or worm gears, the gaging head is adjusted by means of handwheel E so that the gaging fingers contact the gear approximately at the pitch line. When checking base pitch, both fingers must be brought into a vertical tangent plane of the base circle. This is accomplished by gage-block settings. The indicating assembly is carried in a slide F at right angle to the barrel axis of the gaging head to permit adjustment normal to the teeth of spur or helical gears. A hand lever H permits withdrawal of the gaging

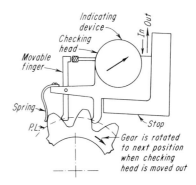

FIGURE 23.20 Schematic arrangement of a pitch-checking device.

FIGURE 23.21 Machine for making pitch checks or base-pitch checks on spur or helical gears. *(Courtesy of Michigan Tool Co., Detroit, Michigan)*

FIGURE 23.22 Spacing-checking machine for internal gears. *(Courtesy of National Broach & Machine Co., Detroit, Michigan)*

head by a preset distance, which causes an index pawl I to rotate the gear through one angular pitch for the check on the next tooth.

Figure 23.22 shows a spacing checking machine for internal gears. Besides checking spacing, this machine has equipment to permit runout checks and checks tooth parallelism on spur gears. Figure 23.23 shows an automatic tooth spacing checker. This machine is a motor-driven instrument that mounts an indexing and sensing finger on a slide that performs a cyclic variable-speed motion along an oval path in a plane normal to the gear axis. Through an electrical recording system, pitch variations are plotted on a chart. The chart is read by noting the peak points for each tooth.

Automatic-type machines like the one just mentioned are coming into widespread use in high-production gear shops such as shops making automobile transmissions. The manually operated machines are used widely in jobbing gear shops doing high-precision work.

The most popular size of gear checking machine on the market will handle gears up to 12 in. diameter. Some kinds of checkers are built in 18-in. diameter size and 24-in. diameter size. Above 24-in. size there is very little available except that "custom" designed machines can be obtained for larger work. The larger gears pose problems in tending to be heavy and not produced in quantity like smaller gears.

Most of the manually operated machines can be obtained with either dial indicators or electrical recording equipment to measure the variation. The more popular amplification ratios for electrical recorders are 200:1, 400:1, and 800:1.

Advanced Tooth Spacing Checking Machines. With new computer-controlled checking machines, inspection results can be analyzed by the computer. This type of machine, such as the one shown in Fig. 23.24, will analyze concentricity and separate these values from spacing and index errors. These types of machines will also identify the particular teeth that have the highest errors.

23.4 BEVEL AND HYPOID GEAR INSPECTION

The curvatures of bevel teeth are more complex than those of spur or helical teeth. Checks for profile and lead are made somewhat indirectly by means of contact checks between mating gears or checks between a gear and a master. Bevel teeth may be checked for spacing by equipment somewhat similar to that used for spur gears. This section covers the inspection procedures and equipment that are peculiar to the bevel and hypoid gear field.

23.4.1 Testing Procedures and Equipment

Smooth, quiet-running gears must have uniform tooth spacing, teeth concentric with axis of rotation, and proper tooth shape. Inaccuracies in any of these can lead to rapid wear, tooth breakage, or a nonuniformity in the transmission of angular velocity.

Even well-constructed bevel gear mountings will have some deflection. These deflections tend to concentrate the load at one end of the teeth with resultant rapid wear or breakage. Bevel gear testing machines make it possible to make allowances for reasonable deflections and obtain a satisfactory tooth contact under full-load conditions.

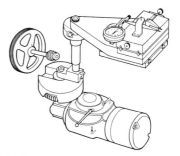

FIGURE 23.23 Automatic tooth-spacing comparator. *(Courtesy of Illinois Tool Works, Inc., Chicago, Illinois)*

Running Test. Since the most conclusive test of bevel gear operation is a *running* test, bevel inspection machines (usually called *bevel gear testers*) are designed to provide a test simulating running conditions. In these testing machines, the gears are rigidly mounted in the correct operating position and run under power and a light brake load.

The smoothness and quietness of operation, the tooth bearing, the tooth size, the surface finish, and appreciable runout and tooth-to-tooth spacing errors are all checked by the running test. These tests are made both when the gears are in the soft state and after they have been hardened.

Tooth Bearing. The tooth bearing is an indication of correct tooth shape, both up and down the tooth profile and lengthwise on the tooth. It is that portion of the gear tooth surface that actually makes contact with its mate and can readily be observed by painting the teeth with a marking compound and running the gears for a few seconds.

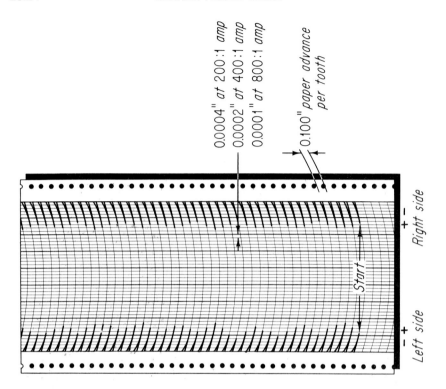

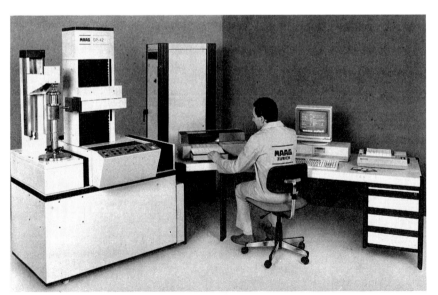

FIGURE 23.24 Computerized tooth space checking machine. *(Courtesy of American Pfauter and Pfauter-Maag, Loves Park, Illinois)*

Testing Machines. Bevel gear testing machines are provided with axial adjustments for both the gear and pinion, and usually with a vertical offset adjustment for raising and lowering the axis of one member vertically with respect to the other for the testing of hypoid gears. In addition, the *axial adjustment of the pinion* is used to produce the same effect on the tooth bearing as a change in pressure angle. The *axial adjustment of the gear* is used to control backlash. The *vertical offset of the axes* is used to produce the same effect on the tooth bearing as a change in spiral angle.

With these adjustments, it is possible to obtain the equivalent position of the gears in the actual mounting and to measure the changes that must be applied to the cutting machines for thus locating the tooth bearing correctly.

The checking of involute tooth profiles, which is applied so commonly in the case of spur and helical gears, is not carried out on bevel gears for two principal reasons. First, the shape of a bevel gear tooth varies at every position from end to end, instead of being constant, as in a spur gear. Hence a different reading is obtained at every checking position along the tooth. Second, even if this check were practical for bevel gears, it would not show the true tooth contact along the entire length of the tooth as shown in a bevel gear testing machine.

Bevel gear testing machines are considered to be superior to specially built test fixtures. Not only do fixtures fail to indicate how the gears will operate in service, but they lack the adjustment necessary for determining slight changes in setup of the cutting machine for tooth-bearing control. Furthermore, fixtures soon become an investment problem, as well as a space problem, every new design requiring a different fixture. However, in some special instances, a good design of a special fixture has been used advantageously.

General Testing Procedure. When a new bevel gear ratio is produced, it is first necessary to *develop* the pair—that is, to secure the desired location and shape of the tooth bearing to ensure a smooth- and quiet-running pair of gears under final operating conditions—before the gears can be cut in production. The gear and pinion are first cut to the calculated machine setting, as recorded on a summary. Then a check is made by running the two together on the testing machine under power, to determine what changes must be made on the pinion in order to make proper allowances for deformation in heat treatment and for mounting deflections. It is customary to make all corrections on the pinion, since changes may be made more rapidly and satisfactorily on this member. The required changes, as determined on the test machine, are then made on the pinion-generating machine, and the pinion is recut and again run with the gear. When approved tooth bearing and desired backlash are obtained, the production gears are cut.

Similarly, by means of this development procedure on the testing machine, it is possible to cut gears to match those cut at some time previously. This is particularly important where duplication is required. Once the development is completed, several sets of *control gears* should be made which duplicate exactly the newly developed pair. These control gears are used on the gear testing machine for testing the remaining gears as they are cut, to assure uniform quality and interchangeability.

A *final test* is made on each pair of gears when they are completed. This final test serves to find the best running position of the gears and provides the final mounting distances for assembly. This in turn eliminates the need of *fitting* or *tearing down* and rebuilding of an assembly to reposition the gear.

Operation of Testing Machines. There are primarily two types of bevel gear testing machines, namely, the *right-angle tester* for 90°-shaft-angle work and the *universal tester* for gears operating on both 90° and other-than-90° shaft angles.

The Right-Angle Testing Machine. The right-angle testers (Fig. 23.25) are equipped with adjusting screws for raising or lowering the gear head vertically, for displacing the gear head horizontally at right angles to the gear axis (that is, in the direction of the pinion axis), and for displacing the pinion head horizontally at right angles to the pinion axis (that is, in the direction of the gear axis).

The Universal Testing Machine. This machine (Fig. 23.26) has all the movements described for the right-angle tester with one additional feature; that is, the angle between the gear and pinion axes may be adjusted by swinging the gear head on a slide to the desired shaft angle.

Development and Testing Procedure. Throughout the development, three things should be borne in mind:

1. A smooth- and quiet-running pair of gears is required.
2. Some adjustability to prevent load concentration along the tooth edges must be provided.
3. The size, shape, and location of the tooth bearing must be consistent with the application.

To obtain smoothness and quietness, the gear teeth must have as large an area of contact as possible. This will also increase surface durability. However, with the increase in contact area, there will be a decrease in adjustability. For these reasons, the

FIGURE 23.25   Hypoid tester for testing bevel gears at 90° shaft angle. *(Courtesy of Gleason Works, Rochester, New York)*

FIGURE 23.26 Universal bevel-gear tester for testing bevel or hypoid gears at any shaft angle. *(Courtesy of Gleason Works, Rochester, New York)*

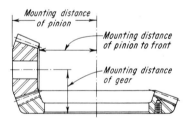

FIGURE 23.27 Mounting distance of bevel gears.

balance between smoothness and adjustability must be determined by the requirements of the specific job.

Before the gears to be tested can be run together in the testing machine, it is necessary to locate them in the same relative position as in their final mountings. This is done by positioning the gear and pinion to their respective *mounting distances* as specified on the drawings. The work heads on the testing machines are positioned so that the gears, when mounted, will be located at their correct mounting distances (see Figs. 23.27 to 23.31).

During the development of the tooth bearing, the gears are usually set to a greater mounting distance than the final one, because the teeth are not cut to the correct size until a satisfactory development has been obtained.

Checking Tooth Thickness. On medium- and coarse-pitch gears, the measurement of tooth thickness is usually made with a gear tooth vernier. This vernier has one setting for the (normal) chordal addendum and a second setting for the (normal) chordal thickness. When the correct settings have been made on the verniers, the caliper is drawn along the gear tooth from the inner and toward the outer end of the tooth. The caliper should touch both sides of the tooth and the top land at the outer end when the tooth is the correct size (see Fig. 23.32). In order to use this vernier, the outside diameter and face angle of the gear blanks must be held to recommended tolerances. It will usually be found necessary to reduce the tooth thickness slightly from the theoretical value, in order to mount the gears at their correct mounting distances with the desired backlash. Generally, the gear member is cut to the theoretical size (as measured by the tooth vernier), and the pinion is then cut to give desired backlash. This method is applied only to the first gears cut. Production checking methods are covered in the next section.

Checking Backlash. Bevel gears are cut to have a definite amount of backlash, which varies according to the pitch and operating conditions. In production testing, backlash measurement is used as a gauge of tooth thickness. Backlash should be measured at the tightest point, with the gears mounted on their correct centers. In order to make this measurement on the testing machine, the pinion is held solidly against rotation and a rigidly mounted dial indicator is placed against the tooth at the extreme heel perpendicular to the surface (Fig. 23.33). The backlash is determined by moving the gear back and forth.

Backlash variation is measured by locating the points of maximum and minimum backlash in the pair of gears and obtaining the difference. For precision gears, this variation in general should not exceed 0.001 in.

After checking backlash by the foregoing method on the finally approved pair of gears, the production control of backlash may be checked more easily by the following method.

The approved pair of gears (control gears) are mounted in the testing machine at their correct mounting distance. The gear is then moved axially into metal-to-metal contact with the pinion.

The amount of movement of the head from its initial position is observed. Then, when testing the production gears, the head should move past the correct mounting distance by this same amount, allowance being made for the specified backlash tolerance.

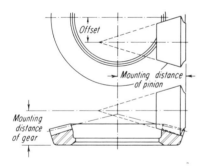

FIGURE 23.28 The mounting distance of a hypoid gear is the linear dimension from its mounting surface to the common perpendicular from its axis to the axis of its mate.

Allowances for Heat-Treat Changes and Mounting Deflections. Gear teeth that are hardened usually undergo minor changes during the heat-treating process. The degree of these changes depends on the shape and size of the gear and the method of hardening. In addition, as a result of deflections in the gearbox or carrier, the gears do not remain in their theoretical mounting position when the load is applied. For these reasons, in the shop development of bevel and hypoid gears, the tooth bearing is located with an allowance for heat-treat changes and known mounting deflections.

FIGURE 23.29 Illustration of use of setup gages for precisely locating testing machine workheads. These gages are specially made for each job.

FIGURE 23.30 Use of gage blocks with proof bar (or mating arbor in many cases) to position testing machine workheads correctly. Gage blocks are employed only where the quantities of gears are too small and the variety is too large to warrant the purchase of special setup gages for each job.

Heat-Treat Changes. In the heat treating of curved-tooth bevel gears, the pinion teeth usually change more than the gear teeth in such a direction as to reduce the spiral angle. This change in spiral angle results in a shift of the bearing length-wise on the tooth, toward the heel on the convex side of the pinion tooth and toward the toe on the concave side of the pinion tooth. Accordingly, heat-treatment allowance on the right-angle testing machine settings is made by raising the gear axis on a left-hand pinion and by lowering the gear axis on a right-hand pinion from the standard position. (By standard position is meant the relative position of the gear with respect to the pinion when the two members are assembled in their specified operating positions.) The amount of displacement of the gear or pinion axis from standard

FIGURE 23.31 The same setup as Fig. 23.30 with the gears in place.

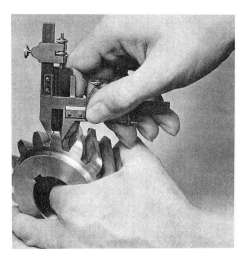

FIGURE 23.32 Use of gear tooth vernier to check tooth thickness of a spiral bevel pinion. Note that the vernier is held perpendicular to the top and sides of the tooth.

usually varies from 0.000 to 0.015 in. The exact value for any particular job can be determined only by trial. That is, the gears are developed to have the tooth bearing in the desired location (i.e., central toe) when tested with these modified settings. Hence, during heat treatment, the tooth bearing will move into the proper position, so that when the hardened gears are given a final test on correct centers, they will run with the correct tooth contact.

GEAR INSPECTION DEVICES AND PROCEDURES

FIGURE 23.33 Measurement of backlash in a pair of gears by means of a dial indicator.

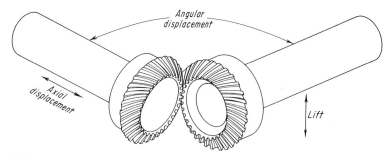

FIGURE 23.34 This drawing illustrates the displacements to which bevel gears are subject in their mountings.

In the case of straight bevel gears, these allowances are unnecessary, since there is no spiral angle. This, however, does not mean that there is no change in straight bevel gears as a result of the heat treatment. On the contrary, the change that does occur is not readily predictable and thus is difficult to allow for in cutting.

Mounting Deflections. Bevel gears running in their mountings are subject to axial displacements, lift, and angular displacements (see Fig. 23.34). The *axial displacement* is movement in the direction of the gear or pinion axis; the *lift* is the relative linear movement between the gear and pinion axes, and at right angles to both axes; and the *angular displacement* is a change in the shaft angle. Values of the aforementioned displacements will vary according to the size of the gear, the rigidity of the mountings, and the accuracy of machining. In general, the bearing will shift toward the heel (large end of the tooth) under load. To compensate for this, the tooth bearing under light load is placed slightly nearer the toe (small end of the tooth). The resulting bearing is called a *central toe bearing*.

Mounting deflection cannot readily be predetermined, unless experience has been gained from the effect on the same size gears of deflections in similar mountings. Two methods are open for determining the actual deflections after a preliminary set of gears has been made. The first of these is the deflection test, in which the gears are mounted in their case and the whole unit is subjected to dynamic loading (Fig. 23.35). Indicators are placed in the required positions to record relative displacement of the gear and pinion.

The second, and more common method, involves a running test of the gears under actual load in their final mountings and observation of the tooth bearing, which usually moves to the heel under load.

These gears are then disassembled and mounted in the testing machine. The testing machine settings are adjusted to duplicate the position of the tooth bearing observed on the gears in their final mountings under load. These new settings on the testing machine are the new operating positions for the gears and are equivalent to the final mounting position of the gears under load. The gears should be redeveloped to have the desired tooth bearings (that is, central) in these new operating positions. Then, when the gears are installed in the final assembly, they will run with a full-length tooth contact under load, allowance having been made for deflection.

When the bearing has been located in the proper position on the tooth, it should be possible to move the pinion axially in or out a slight amount, approximately 0.003 in., without causing a heavy bearing along the top edge or deep in the bottom of the tooth.

Vertical and Horizontal Movements. The readings for all dials on the testing machine may be considered as *zero* readings when the gears are mounted in their specified mounting positions. All horizontal and vertical movements are measured from these *zero* positions. The following rules will determine the correct signs to be used with these movements:

1. Horizontal movements of the gear on the right-angle testing machines away from the pinion head (out) = (+) plus.
2. Horizontal movements of the gear head on the right-angle testing machines toward the pinion head (in) = (−) minus.
3. Vertical movements of the gear head up or the pinion head down = (+) plus.
4. Vertical movements of the gear head down or the pinion head up = (−) minus.

As will be seen in the next section, it is occasionally very desirable to determine the vertical and horizontal settings necessary to place the tooth bearing in the center of the tooth profile at the extreme toe and extreme heel. The total movement between the toe and heel readings is obtained by subtracting the heel reading from the toe reading algebraically. To obtain the average reading between toe and the heel readings, add the values algebraically and divide by 2.

GEAR INSPECTION DEVICES AND PROCEDURES   23.33

FIGURE 23.35 Deflection test of main drive gears in an automotive rear axle.

Example

	Reading at Toe	Reading at Heel	Total Movement Toe to Heel	Average Reading
Vertical movement	+0.010	−0.018	0.028	−0.004
Horizontal movement	−0.014	+0.020	0.034	+0.003

$$\text{Total vertical movement} = (+0.010) - (-0.018) = +0.028$$
$$\text{Total horizontal movement} = (-0.014) - (+0.020) = -0.034$$

The algebraic signs of these totals are ignored, since the magnitude of these quantities is the item of interest.

$$\text{Average vertical setting} = \frac{(+0.010) + (-0.018)}{2} = -0.004$$

$$\text{Average horizontal setting} = \frac{(-0.014) + (+0.020)}{2} = +0.003$$

When obtaining the average readings, both the magnitude and the direction (sign) are important. The average reading is used to place the tooth bearing in the center of the tooth to observe its appearance or to compare it with the average reading for a master set of gears for the same job.

V and H Check. The V and H check is a method for measuring the amount and direction of the vertical and axial displacements of the pinion from its standard position, to obtain a tooth bearing in the middle of the tooth profile at the extreme toe and at the extreme heel of the tooth. In the preceding example, the first three columns entitled, "Reading at Toe," "Reading at Heel," and "Total Movement Toe to Heel" constitute the V and H check.

It is customary to specify a tooth bearing approximately equal to one-half the tooth length under light load. This, in general, will allow adequate compensation for mounting deflections under load and sufficient adjustability in assembly. However, as stated above, the V and H check is a more accurate method of determining the required length of tooth bearing.

Referring to the example just given, the line entitled "Vertical movement" gives the relative displacement from the central position to the toe (in example, +0.010 in.), from the central position to the heel (in example, −0.018 in.), and from toe to the heel (in example, 0.028 in.). This last value (total movement from toe to heel) is a measure of the bearing length in terms of displacement and is sometimes referred to as a ". . . -length bearing" (in example, 28-length bearing). When the total vertical movement of the V and H check is too large, it indicates that the tooth bearing is too short and that the load will therefore concentrate on too small an area of the tooth surface, causing danger of excessive wear. When the total vertical movement is too small, it indicates that the tooth bearing is too long, and hence the gears will lack sufficient adjustability to compensate for mounting deflection, which may lead to load concentration at the ends of the teeth.

To determine whether the tooth bearing is correctly placed on the tooth, when a central toe bearing is desired after hardening:

1. On the concave side of the pinion tooth, the toe vertical reading should equal the allowance in vertical offset of the testing machine for hardening, *plus* approximately one-quarter of the total vertical movement from toe to heel.

2. On the convex side of the pinion tooth, the toe vertical reading should equal the allowance in the vertical offset on the testing machine for hardening *minus* approximately one-quarter of the total vertical movement from toe to heel (see Fig. 23.36).

A bias tooth bearing tends to cross the tooth on a slant (see Fig. 23.37) and is undesirable principally because of the danger of load concentration at the corners of the teeth. However, it is sometimes necessary on spiral bevel and hypoid gears to have a slight amount of *bias out* in soft gears because, when gears are hardened, bias will be introduced.

The visual method of determining bias in a tooth bearing by observing the bearing when placed in the central position for its diagonal direction on the tooth is, of course, helpful but rather inaccurate. In determining bias by the V and H check, the amount of vertical movement on the testing machine that is necessary to shift the tooth bearing from toe to heel is compared with the corresponding amount of horizontal movement necessary to keep the bearing in the middle of the tooth profile.

For spiral bevels with a 90° shaft angle and a spiral angle between 35 and 40°, the ratio of total vertical to total horizontal movement from toe bearing to heel bearing will usually be about 1:1 on the concave side of the pinion tooth, and about 3:4 on the convex side of the pinion tooth, to produce a tooth bearing with freedom from bias. If the ratio is greater than specified in the foregoing, it denotes *bias in,* and if the ratio is less, it denotes *bias out.* As the spiral angle decreases, the ratio of total vertical to total

GEAR INSPECTION DEVICES AND PROCEDURES

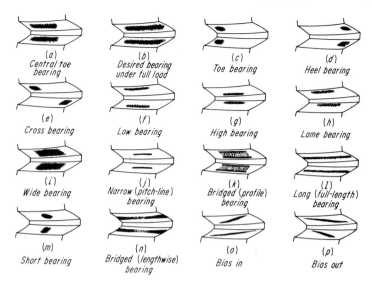

FIGURE 23.36 Tooth bearing on the pinion tooth. Although a left-hand pinion is used throughout, the bearings are representative of those of a right-hand pinion or a straight bevel pinion as well. *(a)* A central toe bearing. Note that the bearing extends along approximately one-half the tooth length and that it is nearer the toe of the tooth than the heel. In addition, the bearing is relieved slightly along the face and flank of the tooth. Under the light loads the tooth bearing should be in this position on the tooth. *(b)* The same tooth with a bearing as it should be under full load. It should show slight relief at the ends and along the face and flank of the teeth. There should be no load concentration at the extreme edges of the teeth. *(c, d, e)* Differences in spiral angle between the gears tested. *(f, g, h)* Differences in pressure angle between the gears tested. *(i, j, k)* Width of tooth bearing. *(l, m, n)* Length of tooth bearing. *(o, p)* Bias bearings. Regardless of the hand of spiral on the pinion, "bias in" will always run from the flank to the toe to the top at the heel on the convex side and from the top at the toe to the flank at the heel on the concave side.

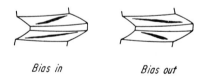

FIGURE 23.37 Bias tooth bearing can cause dangerous concentration of load at corners of teeth.

horizontal movement should increase until, in the case of a Zerol,* the horizontal movement becomes zero and the ratio becomes infinite. All the movement will be vertical. As the shaft angle decreases from 90°, the ratio of total vertical to total horizontal movement should decrease until, in the case of very low shaft angles, the vertical movement becomes zero.

---

*Registered trademark of Gleason Works, Rochester, New York.

23.36 CHAPTER TWENTY-THREE

A few representative V and H checks for gears operating at 90° shaft angles are included to serve as a guide. The examples given show both the soft and corresponding hard (after lapping) check, with the exception of the aircraft gears, which are for gears after the teeth have been ground. References to concave and convex side of the tooth apply to the pinion. The values given are average, and it may therefore be necessary to alter them considerably, depending on the distortion from heat treatment and on the rigidity of the gear mountings (see Tables 23.1 and 23.2).

23.4.2 Control Gears

A bevel or hypoid control gear is a hardened gear made to the necessary specifications of size, mounting distance, and tooth bearing to ensure proper tooth action in final

TABLE 23.1 Representative V and H Checks, Spiral Bevel and Hypoid Gears

Concave	Toe	Heel	Total	Convex	Toe	Heel	Total
Passenger car: 8″ gear							
Soft:							
V	+8	−12	20	V	−3	+17	20
H	−8	+16	24	H	+4	−23	27
Hard:							
V	+3	−9	12	V	−3	+9	12
H	−3	+8	11	H	+3	−11	14
Passenger car: 9½″ gear							
Soft:							
V	+8	−16	24	V	−3	+19	22
H	−11	+22	33	H	+2	−26	28
Hard:							
V	+4	−9	13	V	−4	+10	14
H	−4	+8	12	H	+3	−13	16
Truck: 10″ gear							
Soft:							
V	+10	−15	25	V	−8	+22	30
H	−10	+19	29	H	+7	−34	41
Hard:							
V	+4	−12	16	V	−5	+13	18
H	−4	+12	16	H	+4	−16	20
Truck: 13″ gear							
Soft:							
V	+7	−30	37	V	−6	+32	38
H	−8	+32	40	H	+7	−37	44
Hard:							
V	+2	−36	38	V	−3	+38	41
H	−3	+37	40	H	+4	−40	44

TABLE 23.2 Representative V and H checks, Zerol bevel gears

Concave	Toe	Heel	Total	Convex	Toe	Heel	Total
Tractor: 13″ gear							
Soft:							
V	+4	−24	28	V	−3	+25	28
H	0	0	0	H	0	0	0
Hard:							
V	+4	−20	24	V	−4	+24	28
H	0	0	0	H	0	0	0
Tractor: 19″ gear							
Soft:							
V	+15	−55	70	V	−12	+56	68
H	0	0	0	H	0	0	0
Hard:							
V	+5	−55	60	V	−6	+55	61
H	0	0	0	H	0	0	0
Aircraft planetary 10″ gear							
Ground:							
V	+3	−10	13	V	−2	+12	14
H	0	0	0	H	0	0	0
Aircraft 2 to 6″ gears							
Ground:							
V	+3	−10	13	V	−2	+12	14
H	0	0	0	H	0	0	0
Low-spiral-angle propeller gears 5 to 9″ gear							
Ground:							
V	+3	−10	13	V	−3	+11	14
H	0	0	0	H	0	0	0

assembly under load. Its purpose is to serve as a working model for the duplication of gears in regular service.

Gear teeth that are hardened usually undergo minor changes during the heat-treating process. The degree of these changes depends on the shape and size of the gear and the method of hardening. In addition, as a result of deflections in the housing and bearings under load, the gears do not remain in their theoretical mounting position. For these reasons, in the shop development of bevel and hypoid gears, the tooth bearing is located with an allowance for heat-treat changes and known mounting deflections. These allowances are made in the manufacture of the control gears.

Types and Uses

1. *Master control gears:* Used for checking *inspection control gears* only.
2. *Inspection control gears:* Used for checking finished production gears.
3. *Green production control gears:* Used for checking gears in the *green* or *unhardened* state. In the manufacture of these gears, compensation must be made for the effects of the final mounting and for the changes in the production gears during hardening as determined by actual trial.
   Production control gears, for checking production gears that are to have their teeth hardened and ground, are made thin by the amount of grinding stock allowed on the teeth of mating production gears. These production control gears must also allow for heat-treat changes and final mounting conditions.
4. *Pregrind production control gears:* Used for checking after the production gears have been *hardened,* but before the teeth are ground. They are also used during the grinding development. These control gears are made thin by the amount of grinding stock on the mating production gears. The tooth bearing is made the same as on the master control gears. The primary purpose of these gears is to check the uniformity of stock distribution on the sides of the gear teeth prior to grinding.

Manufacture of Control Gears. In the manufacture of these control gears, it is imperative that the effects of the final mountings be known and allowed for in the control gears.

A minimum of two control gears are made for each item, one set being suitably marked and held as a *master control gear,* and used only for checking the *inspection control gear* periodically and for reference when ordering replacement inspection control gears. The other set is used as an inspection control gear.

Production control gears are similar to the master control gears, with the exception that the size, mounting distance, and tooth bearing have been altered, as explained in the following to serve as working models during various stages of the gear manufacture.

The master control gear and inspection control gear, as well as the green and pre-grind production control gears for all bevel gears, should be selected gears that have been made in the same plant and under the same conditions as the production gears. If the production gears are not hardened, the control gears in general should be made either of Nitralloy N,* nitrided to a case depth of 0.005 to 0.010 in., or of a surface-hardening steel with flame-hardened teeth.

Certain special cases, such as control gears for gears that operate with little or no backlash on exact mounting distances, should be made with a minimum of backlash on exact mounting distances.

23.5 WORM GEAR INSPECTION

The inspection of worms and worm gears has some similarity to both helical gear inspection and bevel gear inspection. Profile and lead checks can be taken on the worm member even though the profile is generally not involute. Contact checks are frequently used between mating members or between a worm part and a master.

---

*Registered trademark of the Nitralloy Corp., New York.

23.5.1 Cylindrical Worm Gear Inspection

The worm gear member has a complex curvature that is generated from a hob or other tool that duplicates the mating worm. The resulting warped surface does not lend itself to either profile or lead checks since any section through the tooth varies from a neighboring section.

Worm gears are normally checked by meshing them with their mate—or a *master worm*—in a checking fixture. The contact check with a worm reveals the following things:

1. Width, length, and location of contact pattern
2. Amount of backlash and variation in backlash
3. Height setting of worm with respect to gear to make the worm lie in the throated-out cup of the gear teeth

Figure 23.38 shows a contacting setup between a worm and worm gear. Red lead is put on the worm threads and prussian blue on the gear teeth. The worm is raised to the correct height setting. The worm is rotated to drive the gear, and this marks the contact all the way around the gear. The contact pattern is then judged for length, width, and location. Design requirements may call for contact across two-thirds of the face width. Generally it is desirable that contact does not start right where the worm enters mesh with the gear tooth. A heavy contact here serves to scrape the lubricant off the worm threads and impair lubrication. In the middle of the face width, the contact should spread out to a fairly full profile—say two-thirds of the tooth height.

Worm gears wear in more than most gear types; so it is quite likely that the design requirements will permit a more limited contact pattern on a new set than would be the case for other gear types where the wearing-in of the tooth surfaces is much more

FIGURE 23.38 Contact check between a cylindrical worm and wormgear. *(Courtesy of General Electric Co., Lynn, Massachusetts)*

limited. The designer will specify appropriate contact pattern limits based on the material of the worm gear, the severity of the tooth loading, and the job requirements of speed, reliability, and smoothness.

The tooth thickness of worm gear teeth cannot be measured conveniently with calipers or pins (because of the warped shape). Measurement of backlash with a worm of known thread thickness serves indirectly to measure the gear tooth thickness. Variation in backlash indicates eccentricity or tooth spacing variations.

If satisfactory tooth contact is not obtained at the specified height setting of the worm, a contact check may be repeated at a different height setting. If satisfactory contact is found at some other height setting, the gear has a *variation from design height* setting. It may be permissible to stamp the observed height setting on the gear and pass the gear as acceptable.

Worm gears may be checked for tooth spacing by means similar to those described in Sec. 23.3.3. Highly precise worm wheels used for index gears may require the most precise measurement of pitch and index-spacing variations.

Cylindrical Worm Inspection. The worm may be measured by direct checks. Its lead and spacing may be measured by equipment similar to that used for helical pinions (see Secs. 23.3.2 and 23.3.3). Its profile is usually not involute; so involute checkers are not used. The thread form is usually straight-sided—or almost straightsided. A checking machine that can traverse the profile at the right angle of inclination for the normal pressure angle can measure the profile by finding the deviation from a straight line. Figure 23.39 shows a general-purpose worm and hob checking machine. This machine can measure worm profile, worm lead, and (with an index plate) thread-to-thread spacing. The thread thickness of worms is determined by caliper checks or by diameter over rolls.

23.5.2 Double-Enveloping Worm Gear Inspection

Most double-enveloping[18] worm gears are checked for profile, lead, and tooth thickness by the indirect means of checking with a mating worm or a master worm. The general technique is similar to that used on cylindrical worm gears.

FIGURE 23.39 Checking a worm on a universal measuring machine. *(Courtesy of General Electric Co., Lynn, Massachusetts)*

A contact check of a double-enveloping worm gear has two important differences over that for a cylindrical worm. These are:

1. The worm must be set for both *height* and *axial position* before it will fit into the double throating of both the worm and the gear.
2. The contact pattern of a good set is somewhat different. The lengthwise contact of a good set tends to be somewhat shorter than that of a comparably good cylindrical worm gear. The shape of the contact pattern is somewhat different.

All double-enveloping gears are generated from the unmarked side face to an exact side-position dimension shown on the part drawing. The double-enveloping worm is generated from a fixed end-position dimension to a bearing shoulder. Double-enveloping worm gear sets should always be assembled to these closely held dimensions.

Double-enveloping worm gear sets generally have more teeth in simultaneous contact than do similar cylindrical worm gear sets. This means that a very high load-carrying capacity can be obtained even when the contact of new parts is a fairly short band. These gears have excellent wearing-in capacity; so that contact pattern will broaden rapidly in service. Figure 23.40 shows a machine used to check the spacing and pressure angle of double-enveloping worm gears. The Cone-Drive* gear has a relatively straight-sided profile in an axial plane. The indicator can be traversed across the gear tooth in the center of the face width at an inclination corresponding to the pressure angle to check the profile. By the use of a fixed finger and a movable finger, pitch checks can also be made.

---

*Cone-Drive is a registered trademark of Cone-Drive Div., Michigan Tool Co., Detroit.

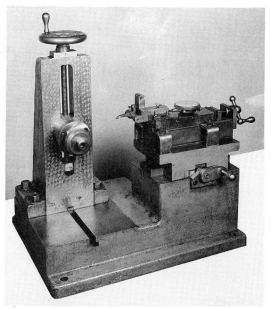

FIGURE 23.40 Machine to check double-enveloping wormgears. *(Courtesy of Cone-Drive Division, Michigan Tool Co., Detroit, Michigan)*

Double-Enveloping Worm Inspection. The worm member may be checked for lead, thread spacing, and profile by a machine like the one shown in Fig. 23.41.

By the use of the two index plates shown, the lead may be checked in any desired increments. If it is desired to check the lead for each 90° rotation of the worm, a four-tooth index plate may be used on the worm spindle and a plate having four times the number of gear teeth may be used on the gear spindle. By continuously indexing each plate one tooth at a time, the worm may be checked (by increments) for its entire length.

This system also permits a check of thread spacing. By setting the indicator slide to the axial pressure angle of the worm and traversing the indicator in and out, it is possible to check the profile of the thread. The worm thread is *straight-sided in the axial section*. Its slope is defined by a tangent line to the *base circle* of the mating gear.

Checkers of the type shown in Fig. 23.41 make it possible to give a Cone-Drive worm a complete check for lead, spacing, and profile.

23.6 CHECKING FOR GEAR WEAR

Usually the only time gears are inspected is when they are manufactured. Occasionally there is the need to inspect gears after they have run. The need most often arises when there is wear on the tooth and the user is trying to determine how much life is left in the gear.

FIGURE 23.41 Machine to check double-enveloping wormgears. *(Courtesy of Cone-Drive Division, Michigan Tool Co., Detroit, Michigan)*

GEAR INSPECTION DEVICES AND PROCEDURES

23.6.1 Measurement of Gear Wear

A common method often used to determine wear in the field would be to measure the diameter over (or under) balls or pins. This is a rough estimate of wear. Though this method is easy to do, it is not very precise. Since the tooth generally does not wear evenly, either up and down the involute or across the face, the point where the pin touches could underestimate the wear. Wear across the face could vary if there is misalignment. Wear along the involute will depend on the geometry of the gear and its mate, and will vary simply because of the mechanics of meshing teeth tend to cause the dedendum to wear more than other places on the tooth.

Sometimes it is possible to bring the gearing back into the shop and inspect it on an analytical or functional gear tester. Figure 23.42 is the plot made on an involute checking machine. For the most part, the gear was loaded on only one side of the tooth. The unloaded side would then be representative of the as-manufactured condition of the profile. The plot of the worn side of the tooth is a good example of the way wear can vary along the profile.

Similar checks can be made on the helix of this worn gear. Checking a gear this way is a lot more revealing than simply checking wire size. The problem is that the gears must be removed from service to get this much detail.

When it is not practical to remove the gear from service (such as when the gear is

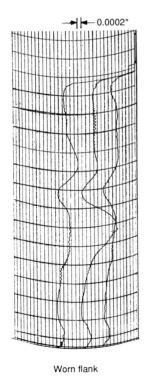 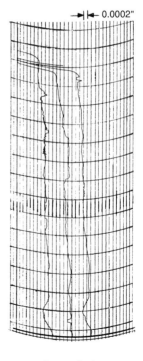

Worn flank   Unworn flank

FIGURE 23.42 Involute checks of worn gear.

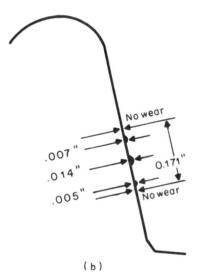

(b)

FIGURE 23.43 Mold taken to determine wear. *(Courtesy of Dudley Technical Group Inc., San Diego, California)*

in the engine room below the waterline of a large ship), a technique has been developed to take molds of the gear teeth. A range of materials are used to make molds. When the teeth are spur or helical (if the helical windup is not too great), molding material such as plaster of Paris or epoxy, which becomes hard, can be used. If the helical windup is large or there is undercutting, the molding material needs to be pliable to remove the mold from the gear. The fast-setting silicone impression materials, like those dentists use on human teeth for bridgework, are excellent materials when the mold must be pliable.

Figure 23.43 shows how a silicone mold has been sectioned and an enlarged photograph taken to determine wear on a pinion tooth.

23.7 CHECKING EXTRA LARGE AND EXTRA SMALL GEARS

Even though it is important to know the accuracy of a part before it gets out of the shop and into a critical application, it is not always possible to inspect gears on the usual types of checking equipment. One problem comes when the gear is so large that it does not physically fit on any existing inspection machine. Another problem comes when the weight of the gear exceeds the capacity of worktable of the checking machine. The other extreme occurs when the gear tooth is so small that the probe on a conventional checking machine cannot reach into the tooth slot. This section discusses techniques and equipment to handle these difficult circumstances.

23.7.1 Checking Extra Large Gears

Many gears are made that are so large that there are no checking machines large enough to check them. However, because the accuracy of the gear can have a significant affect on its ability to perform the function it has been designed to do, it is important to develop a technique to check large gears.

The most important thing that must be checked is tooth thickness. If the teeth are too thick there may not be enough backlash and they might bind. Of course it can also be a problem if there is too much backlash, especially when the gears must run loaded in both directions. On smaller gears the usual way to check tooth thickness is to measure the diameter over measuring wires. On very large gears this is not practical. For large gears a span micrometer is used, which checks the distance over several teeth. Another way to measure tooth thickness is with a tooth caliper, as shown in Fig. 23.44.

FIGURE 23.44 Tooth caliper. *(Courtesy of General Electric Co., Lynn, Massachusetts)*

This instrument need only be large enough to handle one tooth, no matter how large the gear's diameter.

Without expensive equipment, another important check that can be made is for runout. Runout can be checked while the gear is still on the cutting machine by attaching a dial indicator to the machine and measuring the variation in position of a measuring pin placed in each tooth slot as it is rotated past the indicator. Though this is a slow method, if carefully done, the accuracy will be very good.

This method can be improved on by attaching a portable fixture to the cutting machine on which a spring loaded master is mounted. A recording device can be added and the result is a double flank tester. An example of how this is done on a large marine gear is shown in Fig. 23.45.

At times it is also desirable to check the profile on large gears. Figure 23.46 shows a profile checking device designed by the General Electric Company to check large gears.

The instrument compares the transverse profile of the gear teeth with that of an accurate template A, which may be mounted adjustably at either side of the instrument for the inspection of either side of the gear teeth. The instrument rests on the tooth spaces of the gear with three ball-point legs B, the centers of which form a right triangle with sides of adjustable length. The length of the side that is parallel to the gear axis is optional on spur gears and equal to a suitable multiple of axial pitches on helical gears. The distance of the two front legs is adjusted so that both ball points are seated in tooth spaces at equal distance from the tooth to be checked. This adjustment varies from gear to gear with the tooth thickness and also the accuracy of the tooth spacing. The horizontal micrometer slide C is provided with opposite template mounting pads

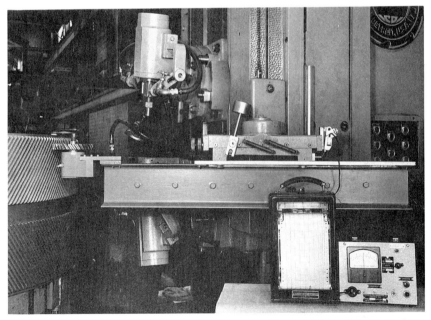

FIGURE 23.45 Checking of large gear by rolling with master. *(Courtesy of General Electric Co., Lynn, Massachusetts)*

GEAR INSPECTION DEVICES AND PROCEDURES

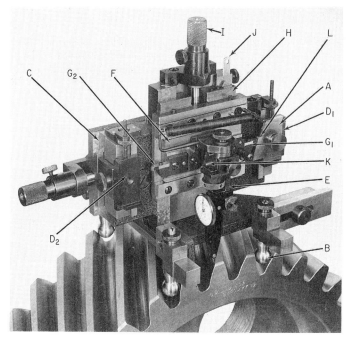

FIGURE 23.46 Portable involute checker. *(Courtesy of General Electric Co., Lynn, Massachusetts)*

D_1 and D_2 and can be adjusted to centralize the indicating assembly E, which swivels about a vertical axis for zeroing in at either side of the central tooth. The indicating assembly E is attached to a horizontal ball bearing slide F, which also mounts the two followers G_1 and G_2, which are urged by reversible spring pressure against the profile of the template located on one or the other mounting pad. Slide H, operated by the graduated hand knob I, is mounted on the micrometer slide C and serves as a base of the horizontal slide F. It provides the vertical component of the follower motion registered on scale J and on the graduations of hand knob I. The distance of the indicating stylus from the swivel axis of the indicating assembly may be adjusted to the gear tooth thickness by means of screw K.

The instrument requires one master template for each gear to be checked. The profile of the template must be of the highest order of accuracy, since its deviations from the required condition are reflected directly in the readings of the indicator. On the other hand, the tooth profiles of large gears have little change in curvature and can be approximated to a great degree of accuracy by simple circular arcs. It is also possible to incorporate desired profile modifications directly into the template. Care must be taken to attain the correct correlation between the starting position of the follower on the template and that of the indicator on the gear tooth. A flat surface L, which represents the theoretical outside diameter of the gear, is provided for this purpose on the template. The template mounting pad is vertically adjusted until the flat bottom of the follower G_1 or G_2 rests on surface L when the indicating stylus contacts the tip of the gear tooth. Further adjustment may be required for the measured outside diameter of the gear unless there is little change in the angularity of the tooth profile within the range of the outside-diameter variation. These adjustments may have to be repeated for each gear because of the effect of tooth thickness variations on the relation between the master template and the gear tooth.

23.7.2 Fine-Pitch Gear Inspection

In general, fine-pitch gearing consists of all gearing with diametral pitches of 20 or greater (modules of 1.3 or less). Although the practical limits on fine-pitch gearing are, at present, between 200 and 500 diametral pitch, pitches of as fine as 720 have been made on occasion.

In recent years the demand for high-accuracy fine-pitch gearing has increased and the demand is expected to increase well into the future as the trend toward miniaturization continues and as the need for small and highly accurate instrumentation in the scientific, industrial, and domestic arenas continues. In proportion to the increase in demand for such gearing, the demand for easier and better inspection techniques has also grown.

Inspection of fine-pitch gearing, however, has some difficulties, brought on by the fact that tooth dimensions are very small when compared with gearing of more conventional pitch. Whereas a 5-pitch gear has a standard whole depth of about 0.48 in. and a tooth space width of about 0.31 in., a 500-pitch gear has a standard whole depth of about 0.005 in. and a tooth space width of about 0.003 in. One can easily see that analytical inspections that require use of a mechanical stylus can be quite impossible.

Despite these limitations and shortcomings, a number of good inspection techniques are available to the fine-pitch gear manufacturer. This section describes some of the techniques that are used.

When thinking of fine-pitch gears it is convenient to divide such gears into two ranges according to size. These are:

Normal fine-pitch gears (We consider these gears to be in a size range from 20 to 96 pitch. This corresponds to approximately 1.3 to 0.25 module.)

Ultrafine-pitch gears (These gears are in the range of 100 to 500 pitch or, in terms of module, from 0.25 to 0.050.)

Analytical Checking. Over the years there has been a large-scale use of "normal fine-pitch gears" in motion control devices where the power transmitted is relatively negligible and the chief function of the gearing is to transmit angular motion. In recent years, though, fine-pitch gears have often been used in mechanical power devices where the principal objective is the transmission of mechanical power. For instance, a small turbine drive may run somewhere in the range of 50,000 to 100,000 r/min. The transmitted power may be somewhere in the range of 5 to 25 hp. If the output speed is 12,000 r/min or 6000 r/min, it is obvious that gearing is needed. It will turn out that the gears used will be rather small and will probably fall in the range of 96 to 20 pitch. In other words, normal fine-pitch gears can easily find themselves needed for very critical, high-speed power devices. When this situation occurs, very-high-accuracy gears are needed. The accuracy needs to be controlled by involute, spacing, and helix-type measurements taken by analytical equipment.

Such analytical checking is relatively practical to do providing the stylus used to take measurements is quite small in size. Also, the checking machines used must be relatively small. The parts of the checking machine, with regard to bearings, ways, and so forth, must be small enough and precise enough to have relatively little friction as the checking machine goes through its geometric measurements. To sum it up, it is relatively practical to check normal-size fine-pitch gears by presently available analytical checking machines.

An exception to what has been said occurs when the fine-pitch gears are not made from hardened steel. If the tooth loading is not unduly high, medium-hard steel, aluminum, bronze, or nonmetallic materials can be used. With the lower hard-

GEAR INSPECTION DEVICES AND PROCEDURES   23.49

ness it becomes impractical to use a relatively sharp-pointed stylus—even if the teeth are large enough to get a stylus between them. The small, somewhat sharp-pointed stylus, made from hardened steel or even a diamond, tends to cut into the surface of the fine-pitch gear being checked. This damages the part and tends to give erroneous checking results. When this situation occurs, the answer is to section the tooth and photograph the teeth under a microscope. Malformed teeth that could easily fail in service or bind so as to create a gear lockup can readily be found by appropriate microphoto techniques.

Magnified photographs of small, fine-pitch gear teeth serve as useful indicators of substantial tooth form error. Although this method is more qualitative than quantitative, profile errors that are "invisible" to the naked eye can often be seen using this technique.

Figure 23.47 shows two photos of several gear teeth from a fine-pitch, plastic gear. One of the errors that is immediately seen is the existence of extra material at the profile tips. The photograph at higher magnification shows that there are profile protuberances on both flanks of one tooth at this position. These errors, though quite large, are impossible to see unless the gear teeth are viewed under magnification.

Although not technically an error in tooth form, another interesting condition that can readily be seen using this technique is objectionable undercut. Figure 23.48 is a photograph of a mounted steel gear of fine pitch. Clearly shown is the fact that the undercut condition is severe. Though undercut is not always a consideration with respect to bending strength for lightly loaded, fine-pitch gears, the undercut condition can lead to unfavorable wear in areas of the root fillet and lower dedendum and this can lead to gear binding. For this particular gear, a gear used in a critical medical application, binding is intolerable and, thus, so is the undercut condition.

In addition to the qualitative uses of magnified profile photographs, there is also the aforementioned use of comparison of the magnified view with an enlarged layout of

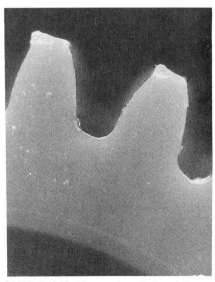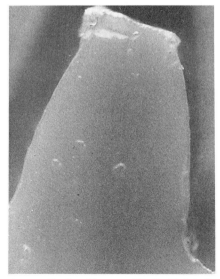

FIGURE 23.47 Photos taken of a fine pitch gear under magnification. *(Courtesy of Dudley Technical Group Inc., San Diego, California)*

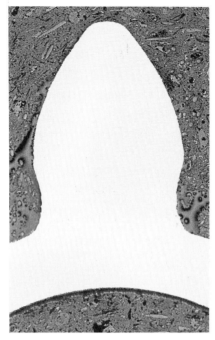

FIGURE 23.48 Magnified photograph of mounted, fine pitch gear shows undercut condition. *(Courtesy of Dudley Technical Group Inc., San Diego, California)*

tooth form. It is imperative that centers for base-circle diameters coincide for the enlarged view and the enlarged layout and that the magnification is the same.

It is also sometimes possible to check helix error on fine-pitch gearing given gears with sufficient face width. If this is the case, then one can mount the master gear on a gimbal with its first point of contact at the end of one side of the face width of the test gear. As mentioned in the foregoing, the master gear is then moved parallel to the test gear and the angular movement of the master gear is recorded.

Functional Checking. Both the single flank and double flank tests are available for fine-pitch gearing. However, double flank testing is more widely used in the United States at present. As with medium-pitch gearing, rolling testers are often set in series with analyzing computers so that the various quantities that describe gear accuracy can be computed in a fast and reliable fashion. Figure 23.49 is a picture of one such system, a double flank tester that is coupled with an analyzing computer. The particular machine shown in this figure, like many of the other machines on the market today, offers a variety of modularized fixtures that allow for testing of spur and helical external and internal gears, bevel gears, and worms and worm gears.

The normal-size fine-pitch gears are usually checked with master gears in a double-flank-type checking machine setup. Tooth-to-tooth composite error and total composite error can be measured. Usually recording machines are used and the chart will show at least one turn of the gear being checked. The composite errors just mentioned can be seen on the chart.

FIGURE 23.49 Double flank tester and analyzing computer. *(Courtesy of Hommel America, New Britain, Connecticut)*

When the size of the gear teeth is in the range of 100 to 500 pitch (ultrafine-pitch gears), double flank testing can still be carried out advantageously. With ultrafine-pitch gears the technique of sectioning the gear and then photographing microscope images of the gear teeth becomes most important. Defects in the teeth that cannot be seen by ordinary eyesight become quite visible at high magnification.

Single flank testing machines are also very useful on ultrafine-pitch gear teeth. The rolling action of the gear teeth in a single flank tester does reveal, very clearly, errors in tooth form, errors in tooth spacing, and eccentric conditions. Although it is somewhat difficult, even the most critical ultrafine-pitch gears can be adequately checked by a combination of double flank, single flank, and magnified-section-type inspections.

REFERENCES

1. AGMA 390.03a, "Handbook—Gear Classification, Materials and Measuring Methods for Bevel, Hypoid, Fine Pitch Wormgearing and Racks Only as Unassembled Gears," American Gear Manufacturers Association, Alexandria, Virginia, 1980.
2. ANSI/AGMA 2000-A88, "Gear Classification and Inspection Handbook—Tolerances and Measuring Methods for Unassembled Spur and Helical Gears (including Metric Equivalents)," American Gear Manufacturers Association, Alexandria, Virginia, 1988.
3. Dudley, D. W., *Gear Handbook,* 1st ed., McGraw-Hill, New York, 1962.
4. Dudley, D. W., *Handbook of Practical Gear Design,* McGraw-Hill, New York, 1984.
5. Illinois Tool Works, "Spiroid Gearing," Design Manual No. 6 Spiroid Division, Illinois Tool Works, Chicago, 1986.

6. Munro, Dr. R. G., "The Interpretation of Results from Gear Measuring Instruments," presented at the Half-Yearly Technical Meeting of the British Gear Manufacturers Association, May 14, 1968.
7. SAE, "Gear Design, Manufacture and Inspection Manual, AE-15," Society of Automotive Engineers, Inc., Warrendale, Pennsylvania, 1990.
8. Smith, R. E., "What Single Flank Measurement Can Do For You," presented at the Fall Technical Meeting of the American Gear Manufacturers Association, 1984.

CHAPTER 24
NUMERICAL DATA TABLES

Darle W. Dudley
President, Dudley Engineering Company, Inc.
San Diego, California

Certain tables are used very frequently in gear tooth calculations. These tables and the directions for using them are given in this chapter.

The material in this chapter is quite similar to the material presented in the first edition of the *Gear Handbook*.[1]* Some additional material is given plus instructions to facilitate the use of this data with metric dimensions. Some material has been left out due to the widespread use of computers—and therefore less need for looking up data in tables.

24.1 GEAR MEASURING TABLES

The following gear measuring tables are used extensively by gear manufacturers and users and provide a quick means of checking the tooth thickness of gears at the pitch diameter. A measurement over wires is made and compared with the measurement for a perfect gear, which is obtainable from the tables. The values in the tables are accurate to 0.0001 in. for diametral pitch gears. (This is equivalent to 0.0025 mm for metric measurements.) After the value is selected in the proper table for the specified pressure angle, number of teeth, and wire size to be used, this value is divided by the diametral pitch of the gear being measured. If the actual measurement is greater than the theoretical value, the teeth are thick; if it is less, the teeth are thin. The change factors K_m, which permit easy conversion of oversize or undersize readings into the approximate amount the teeth are thick or thin, are tabulated alongside the measurement values.

When doing a metric calculation the value of the M in Tables 24.3, 24.4, etc. is the *right value* in millimeters for a 1 module tooth size—providing that the measuring wire size is the right value in millimeters for the module. (For instance the 1.728 wire for 1 module is 1.728 mm. See Tables 24.1 and 24.2.)

In this chapter the measurement tables (except for Table 24.7 covering 25° gears) are presented so that, for each pressure angle measurement, values are available based

---

*In the first edition, this chapter was prepared by Robert T. Parsons, President of the Van Keuren Company, Boston, Massachusetts.

TABLE 24.1 Guide to Wire Selection

Series		Use
English, in.	Metric, mm	
(1) $\dfrac{1.92''}{P_d}$	$1.92 \times m$	(a) Enlarged pinions. Wires project above OD. (b) Alternate for 14½- and 20°-pressure angle, standard-addendum external gears
(2) $\dfrac{1.728''}{P_d}$	$1.728 \times m$	(a) External standard-addendum spur and helical gears. Wires project above OD
(3) $\dfrac{1.68''}{P_d}$	$1.68 \times m$	(a) Suitable for 14½° internal gears above 31 teeth and 20° internal gears above 29 teeth. Wires project below ID (b) Alternate for standard-addendum external gears
(4) $\dfrac{1.44''}{P_d}$	$1.44 \times m$	(a) Internal standard-addendum gears. Wires make good contact on gear teeth but do not project below ID

P_d is the diametral pitch of a spur gear. m is the module of a spur gear. mm is millimeters.

TABLE 24.2 Gear Wire Sizes

	English					Metric			
P_d	1.92 in./P_d	1.728 in./P_d	1.68 in./P_d	1.44 in./P_d	m	1.92 m	1.728 m	1.68 m	1.44 m
1	1.92	1.728	1.68	1.44	1	1.920	1.728	1.680	1.440
2	0.960	0.864	0.840	0.720	2	3.840	3.456	3.360	2.880
2.5	0.768	0.6912	0.672	0.576	2.5	4.800	4.320	4.200	3.600
3	0.640	0.576	0.560	0.480	3	5.760	5.184	5.040	4.320
4	0.480	0.432	0.420	0.360	4	7.680	6.912	6.720	5.760
5	0.384	0.3456	0.336	0.288	5	9.600	8.640	8.400	7.200
6	0.320	0.288	0.280	0.240	6	11.520	10.368	10.080	8.640
7	0.27428	0.24686	0.240	0.20571	7	13.440	12.096	11.760	10.080
8	0.240	0.216	0.210	0.180	8	15.360	13.824	13.440	11.520
9	0.21333	0.192	0.18666	0.160	9	17.280	15.552	15.120	12.960
10	0.192	0.1728	0.168	0.144	10	19.200	17.280	16.800	14.400
11	0.17454	0.15709	0.15273	0.13091	11	21.120	19.008	18.480	15.840
12	0.160	0.144	0.140	0.120	12	23.040	20.736	20.160	17.280
14	0.13714	0.12343	0.120	0.10286	14	26.880	24.192	23.520	20.160
16	0.120	0.108	0.105	0.090	16	30.720	27.648	26.880	23.040
18	0.10667	0.096	0.09333	0.080	18	34.560	31.104	30.240	25.920
20	0.096	0.0864	0.084	0.072	20	38.400	34.560	33.600	28.800
22	0.08727	0.07855	0.07636	0.06545	22	42.240	38.016	36.960	31.680
24	0.080	0.072	0.070	0.060	24	46.080	41.472	40.320	34.560
28	0.06857	0.06171	0.060	0.05143	28	53.760	48.384	47.040	40.320
32	0.060	0.054	0.0525	0.045	32	61.440	55.296	53.760	46.080
36	0.05333	0.048	0.04667	0.040	36	64.120	62.208	60.480	51.840
40	0.048	0.0432	0.042	0.036	40	76.800	69.120	67.200	57.600
48	0.040	0.036	0.035	0.030	48	92.160	82.944	80.640	69.120
64	0.030	0.027	0.02625	0.0225	64	122.880	110.592	107.520	92.160

TABLE 24.3 External Gears—20° Pressure Angle

Measurements over wires and change factors for 1 diametral pitch external standard-addendum 20° pressure angle spur gears with 1.728, 1.92, and 1.68 diameter wires. For any other diametral pitch divide measurements given in table by the diametral pitch. Change factors are the same for all diametral pitches.

No. of teeth	1.728 wire dia.		Alternate 1.92 wire dia.		Alternate 1.68 wire dia.	
	M	K_m	M	K_m	M	K_m
5	7.0153	1.67	7.5271	1.48	6.8800	1.75
6	8.3032	1.83	8.8449	1.62	8.1600	1.91
7	9.1260	1.88	9.6702	1.67	8.9822	1.96
8	10.3271	1.94	10.8907	1.72	10.1783	2.01
9	11.1905	1.98	11.7573	1.77	11.0410	2.06
10	12.3445	2.01	12.9252	1.81	12.1914	2.10
11	13.2332	2.05	13.8173	1.84	13.0795	2.13
12	14.3578	2.09	14.9525	1.87	14.2013	2.16
13	15.2639	2.12	15.8618	1.91	15.1068	2.19
14	16.3683	2.14	16.9748	1.93	16.2091	2.21
15	17.2871	2.17	17.8964	1.96	17.1273	2.24
16	18.3768	2.19	18.9934	1.98	18.2154	2.26
17	19.3053	2.21	19.9244	2.01	19.1432	2.28
18	20.3840	2.23	21.0091	2.03	20.2205	2.29
19	21.3200	2.25	21.9475	2.05	21.1561	2.31
20	22.3900	2.26	23.0227	2.07	22.2249	2.33
21	23.3321	2.28	23.9670	2.08	23.1665	2.34
22	24.3952	2.29	25.0346	2.10	24.2286	2.35
23	25.3423	2.30	25.9837	2.12	25.1754	2.36
24	26.3997	2.32	27.0450	2.13	26.2318	2.38
25	27.3511	2.33	27.9982	2.15	27.1828	2.39
26	28.4036	2.34	29.0543	2.16	28.2346	2.40
27	29.3586	2.35	30.0109	2.17	29.1892	2.41
28	30.4071	2.36	31.0626	2.19	30.2371	2.42
29	31.3652	2.37	32.0222	2.20	31.1948	2.43
30	32.4102	2.38	33.0701	2.21	32.2392	2.43
31	33.3710	2.39	34.0323	2.22	33.1997	2.44
32	34.4130	2.40	35.0768	2.23	34.2412	2.45
33	35.3761	2.41	36.0413	2.24	35.2041	2.45
34	36.4155	2.41	37.0830	2.25	36.2430	2.46
35	37.3807	2.42	38.0495	2.26	37.2079	2.47
36	38.4178	2.43	39.0886	2.27	38.2445	2.48
37	39.3849	2.43	40.0569	2.28	39.2115	2.48
38	40.4198	2.44	41.0938	2.29	40.2460	2.49
39	41.3886	2.45	42.0636	2.29	41.2147	2.49
40	42.4217	2.45	43.0986	2.30	42.2473	2.50
41	43.3920	2.46	44.0699	2.31	43.2174	2.50
42	44.4234	2.46	45.1030	2.32	44.2485	2.51
43	45.3951	2.47	46.0756	2.32	45.2200	2.51
44	46.4250	2.47	47.1071	2.33	46.2496	2.52
45	47.3980	2.48	48.0809	2.34	47.2224	2.52
46	48.4265	2.48	49.1109	2.34	48.2506	2.53
47	49.4007	2.49	50.0858	2.35	49.2246	2.53
48	50.4279	2.49	51.1144	2.35	50.2516	2.53
49	51.4031	2.50	52.0903	2.36	51.2266	2.54
50	52.4292	2.50	53.1177	2.37	52.2525	2.54

TABLE 24.3 External Gears—20° Pressure Angle *(Continued)*

No. of teeth	1.728 wire dia.		Alternate 1.92 wire dia.		Alternate 1.68 wire dia.	
	M	K_m	M	K_m	M	K_m
51	53.4053	2.50	54.0945	2.37	53.2284	2.54
52	54.4304	2.51	55.1208	2.38	54.2533	2.55
53	55.4074	2.51	56.0985	2.38	55.2302	2.55
54	56.4315	2.52	57.1237	2.39	56.2541	2.55
55	57.4093	2.52	58.1022	2.39	57.2318	2.56
56	58.4325	2.52	59.1265	2.40	58.2548	2.56
57	59.4111	2.53	60.1057	2.40	59.2333	2.56
58	60.4335	2.53	61.1291	2.41	60.2555	2.56
59	61.4128	2.53	62.1090	2.41	61.2347	2.57
60	62.4344	2.54	63.1315	2.41	62.2561	2.57
61	63.4144	2.54	64.1121	2.42	63.2360	2.57
62	64.4352	2.54	65.1338	2.42	64.2567	2.57
63	65.4159	2.54	66.1150	2.43	65.2372	2.58
64	66.4361	2.55	67.1360	2.43	66.2572	2.58
65	67.4173	2.55	68.1177	2.43	67.2383	2.58
66	68.4369	2.55	69.1381	2.44	68.2577	2.58
67	69.4186	2.55	70.1203	2.44	69.2394	2.59
68	70.4376	2.56	71.1401	2.44	70.2582	2.59
69	71.4198	2.56	72.1228	2.45	71.2405	2.59
70	72.4383	2.56	73.1420	2.45	72.2587	2.59
71	73.4210	2.56	74.1252	2.45	73.2414	2.59
72	74.4390	2.57	75.1438	2.46	74.2592	2.60
73	75.4221	2.57	76.1274	2.46	75.2423	2.60
74	76.4396	2.57	77.1455	2.46	76.2596	2.60
75	77.4232	2.57	78.1295	2.47	77.2432	2.60
76	78.4402	2.57	79.1471	2.47	78.2600	2.60
77	79.4242	2.58	80.1316	2.47	79.2440	2.60
78	80.4408	2.58	81.1486	2.48	80.2604	2.61
79	81.4252	2.58	82.1336	2.48	81.2448	2.61
80	82.4413	2.58	83.1501	2.48	82.2607	2.61
81	83.4262	2.58	84.1354	2.48	83.2456	2.61
82	84.4418	2.58	85.1516	2.49	84.2611	2.61
83	85.4271	2.59	86.1372	2.49	85.2463	2.61
84	86.4423	2.59	87.1529	2.49	86.2614	2.61
85	87.4279	2.59	88.1389	2.49	87.2470	2.62
86	88.4428	2.59	89.1542	2.50	88.2617	2.62
87	89.4287	2.59	90.1405	2.50	89.2476	2.62
88	90.4433	2.59	91.1555	2.50	90.2620	2.62
89	91.4295	2.60	92.1421	2.50	91.2482	2.62
90	92.4437	2.60	93.1567	2.50	92.2624	2.62
91	93.4303	2.60	94.1436	2.51	93.2489	2.62
92	94.4441	2.60	95.1579	2.51	94.2626	2.63
93	95.4310	2.60	96.1450	2.51	95.2494	2.63
94	96.4445	2.60	97.1590	2.51	96.2629	2.63
95	97.4317	2.60	98.1464	2.52	97.2500	2.63
96	98.4449	2.61	99.1601	2.52	98.2632	2.63
97	99.4323	2.61	100.1477	2.52	99.2506	2.63
98	100.4453	2.61	101.1611	2.52	100.2635	2.63
99	101.4329	2.61	102.1490	2.53	101.2511	2.63
100	102.4456	2.61	103.1621	2.53	102.2638	2.63

TABLE 24.3 External Gears—20° Pressure Angle *(Continued)*

No. of teeth	1.728 wire dia.		Alternate 1.92 wire dia.		Alternate 1.68 wire dia.	
	M	K_m	M	K_m	M	K_m
101	103.4335	2.61	104.1502	2.53	103.2516	2.64
102	104.4460	2.61	105.1631	2.53	104.2640	2.64
103	105.4341	2.61	106.1514	2.53	105.2520	2.64
104	106.4463	2.62	107.1640	2.53	106.2642	2.64
105	107.4346	2.62	108.1526	2.53	107.2525	2.64
106	108.4466	2.62	109.1649	2.54	108.2644	2.64
107	109.4352	2.62	110.1537	2.54	109.2529	2.64
108	110.4469	2.62	111.1658	2.54	110.2645	2.64
109	111.4357	2.62	112.1548	2.54	111.2533	2.64
110	112.4472	2.62	113.1666	2.54	112.2647	2.64
111	113.4362	2.62	114.1558	2.54	113.2537	2.65
112	114.4475	2.62	115.1675	2.55	114.2649	2.65
113	115.4367	2.63	116.1568	2.55	115.2541	2.65
114	116.4478	2.63	117.1683	2.55	116.2651	2.65
115	117.4372	2.63	118.1578	2.55	117.2544	2.65
116	118.4481	2.63	119.1690	2.55	118.2653	2.65
117	119.4376	2.63	120.1587	2.55	119.2548	2.65
118	120.4484	2.63	121.1698	2.55	120.2655	2.65
119	121.4380	2.63	122.1597	2.56	121.2552	2.65
120	122.4486	2.63	123.1705	2.56	122.2656	2.65
121	123.4384	2.63	124.1605	2.56	123.2555	2.65
122	124.4489	2.63	125.1712	2.56	124.2658	2.65
123	125.4388	2.63	126.1614	2.56	125.2558	2.65
124	126.4491	2.64	127.1719	2.56	126.2660	2.65
125	127.4392	2.64	128.1622	2.56	127.2562	2.66
126	128.4493	2.64	129.1725	2.56	128.2661	2.66
127	129.4396	2.64	130.1630	2.57	129.2565	2.66
128	130.4496	2.64	131.1732	2.57	130.2663	2.66
129	131.4400	2.64	132.1638	2.57	131.2568	2.66
130	132.4498	2.64	133.1738	2.57	132.2664	2.66
131	133.4404	2.64	134.1646	2.57	133.2571	2.66
132	134.4500	2.64	135.1744	2.57	134.2666	2.66
133	135.4408	2.64	136.1653	2.57	135.2574	2.66
134	136.4502	2.64	137.1750	2.57	136.2667	2.66
135	137.4411	2.64	138.1661	2.58	137.2577	2.66
136	138.4504	2.64	139.1756	2.58	138.2669	2.66
137	139.4414	2.64	140.1668	2.58	139.2579	2.66
138	140.4506	2.65	141.1761	2.58	140.2670	2.66
139	141.4418	2.65	142.1674	2.58	141.2582	2.66
140	142.4508	2.65	143.1767	2.58	142.2671	2.66
141	143.4421	2.65	144.1681	2.58	143.2584	2.67
142	144.4510	2.65	145.1772	2.58	144.2672	2.67
143	145.4424	2.65	146.1688	2.58	145.2587	2.67
144	146.4512	2.65	147.1777	2.58	146.2674	2.67
145	147.4427	2.65	148.1694	2.59	147.2589	2.67
146	148.4513	2.65	149.1782	2.59	148.2675	2.67
147	149.4430	2.65	150.1700	2.59	149.2591	2.67
148	150.4515	2.65	151.1787	2.59	150.2676	2.67
149	151.4433	2.65	152.1706	2.59	151.2594	2.67
150	152.4516	2.65	153.1792	2.59	152.2677	2.67

For 1 diametral pitch, arc tooth thickness is 1.57080 in.
For 1 module, arc tooth thickness is 1.57080 mm.

TABLE 24.3 External Gears—20° Pressure Angle *(Continued)*

No. of teeth	1.728 wire dia.		Alternate 1.92 wire dia.		Alternate 1.68 wire dia.	
	M	K_m	M	K_m	M	K_m
151	153.4435	2.65	154.1712	2.59	153.2596	2.67
152	154.4518	2.65	155.1797	2.59	154.2678	2.67
153	155.4438	2.65	156.1718	2.59	155.2598	2.67
154	156.4520	2.66	157.1801	2.59	156.2679	2.67
155	157.4440	2.66	158.1723	2.60	157.2600	2.67
156	158.4521	2.66	159.1806	2.60	158.2680	2.67
157	159.4443	2.66	160.1729	2.60	159.2602	2.67
158	160.4523	2.66	161.1810	2.60	160.2681	2.67
159	161.4445	2.66	162.1734	2.60	161.2604	2.67
160	162.4524	2.66	163.1814	2.60	162.2682	2.67
161	163.4448	2.66	164.1739	2.60	163.2606	2.68
162	164.4526	2.66	165.1818	2.60	164.2683	2.68
163	165.4450	2.66	166.1744	2.60	165.2608	2.68
164	166.4527	2.66	167.1822	2.60	166.2684	2.68
165	167.4453	2.66	168.1749	2.60	167.2610	2.68
166	168.4528	2.66	169.1826	2.60	168.2685	2.68
167	169.4455	2.66	170.1754	2.60	169.2611	2.68
168	170.4529	2.66	171.1830	2.61	170.2686	2.68
169	171.4457	2.66	172.1759	2.61	171.2613	2.68
170	172.4531	2.66	173.1834	2.61	172.2687	2.68
171	173.4459	2.66	174.1763	2.61	173.2615	2.68
172	174.4532	2.66	175.1838	2.61	174.2688	2.68
173	175.4461	2.66	176.1768	2.61	175.2617	2.68
174	176.4533	2.67	177.1842	2.61	176.2689	2.68
175	177.4463	2.67	178.1773	2.61	177.2619	2.68
176	178.4535	2.67	179.1846	2.61	178.2690	2.68
177	179.4465	2.67	180.1777	2.61	179.2621	2.68
178	180.4536	2.67	181.1849	2.61	180.2691	2.68
179	181.4467	2.67	182.1781	2.61	181.2622	2.68
180	182.4537	2.67	183.1852	2.61	182.2691	2.68
181	183.4469	2.67	184.1785	2.61	183.2623	2.68
182	184.4538	2.67	185.1856	2.62	184.2692	2.68
183	185.4471	2.67	186.1789	2.62	185.2625	2.68
184	186.4539	2.67	187.1859	2.62	186.2693	2.68
185	187.4473	2.67	188.1793	2.62	187.2627	2.68
186	188.4540	2.67	189.1862	2.62	188.2694	2.68
187	189.4474	2.67	190.1797	2.62	189.2628	2.68
188	190.4541	2.67	191.1865	2.62	190.2694	2.68
189	191.4476	2.67	192.1801	2.62	191.2629	2.69
190	192.4542	2.67	193.1868	2.62	192.2694	2.69
200	202.4548	2.68	203.1883	2.63	202.2698	2.69
201	203.4487	2.68	204.1822	2.63	203.2636	2.69
300	302.4579	2.70	303.1977	2.66	302.2719	2.71
301	303.4538	2.70	304.1937	2.66	303.2678	2.71
400	402.4596	2.71	403.2026	2.68	402.2730	2.72
401	403.4565	2.71	404.1996	2.68	403.2699	2.72
500	502.4606	2.72	503.2056	2.70	502.2736	2.72
501	503.4581	2.72	504.2032	2.70	503.2711	2.72
∞	$(N+2).4646$	2.75	$(N+3).2180$	2.75	$(N+2).2762$	2.75

TABLE 24.4 External Gears—14½° Pressure Angle

Measurements over wires and change factors for 1 diametral pitch external standard-addendum 14½° pressure angle spur gears with 1.728, 1.92, and 1.68 diameter wires. For any other diametral pitch divide measurements given in table by the diametral pitch. Change factors are the same for all diametral pitches.

No. of teeth	1.728 wire dia.		Alternate 1.92 wire dia.		Alternate 1.68 wire dia.	
	M	K_m	M	K_m	M	K_m
5	6.9936	1.90	7.5296	1.61	6.8485	2.03
6	8.2846	2.09	8.8551	1.77	8.1298	2.23
7	9.1116	2.17	9.6871	1.84	8.9555	2.31
8	10.3160	2.24	10.9147	1.90	10.1535	2.39
9	11.1829	2.31	11.7872	1.96	11.0189	2.45
10	12.3399	2.36	12.9617	2.01	12.1712	2.51
11	13.2317	2.42	13.8590	2.06	13.0615	2.57
12	14.3590	2.47	15.0002	2.11	14.1851	2.62
13	15.2677	2.51	15.9143	2.15	15.0925	2.66
14	16.3746	2.55	17.0327	2.19	16.1964	2.70
15	17.2957	2.59	17.9588	2.23	17.1163	2.74
16	18.3877	2.63	19.0606	2.26	18.2058	2.78
17	19.3182	2.66	19.9958	2.29	19.1351	2.81
18	20.3989	2.70	21.0850	2.33	20.2137	2.84
19	21.3368	2.73	22.0272	2.36	21.1505	2.87
20	22.4087	2.76	23.1065	2.39	22.2205	2.90
21	23.3524	2.78	24.0543	2.41	23.1634	2.93
22	24.4172	2.81	25.1257	2.44	24.2265	2.95
23	25.3658	2.83	26.0781	2.46	25.1743	2.98
24	26.4247	2.86	27.1430	2.49	26.2317	3.00
25	27.3774	2.88	28.0992	2.51	27.1836	3.02
26	28.4314	2.90	29.1586	2.53	28.2363	3.04
27	29.3876	2.92	30.1181	2.56	29.1918	3.06
28	30.4374	2.94	31.1729	2.58	30.2404	3.08
29	31.3966	2.96	32.1351	2.60	31.1990	3.09
30	32.4429	2.98	33.1859	2.62	32.2441	3.11
31	33.4047	2.99	34.1506	2.63	33.2053	3.13
32	34.4478	3.01	35.1979	2.65	34.2475	3.14
33	35.4119	3.03	36.1647	2.67	35.2110	3.16
34	36.4523	3.04	37.2090	2.69	36.2505	3.17
35	37.4185	3.06	38.1777	2.70	37.2161	3.18
36	38.4565	3.07	39.2193	2.72	38.2533	3.19
37	39.4245	3.08	40.1896	2.73	39.2208	3.21
38	40.4603	3.10	41.2288	2.75	40.2558	3.22
39	41.4299	3.11	42.2007	2.76	41.2249	3.23
40	42.4638	3.12	43.2378	2.78	42.2582	3.24
41	43.4348	3.13	44.2110	2.79	43.2287	3.25
42	44.4671	3.14	45.2461	2.80	44.2604	3.26
43	45.4394	3.15	46.2205	2.82	45.2323	3.27
44	46.4701	3.17	47.2539	2.83	46.2624	3.28
45	47.4437	3.18	48.2294	2.84	47.2355	3.29
46	48.4729	3.19	49.2613	2.85	48.2642	3.30
47	49.4477	3.20	50.2378	2.87	49.2385	3.31
48	50.4756	3.21	51.2682	2.88	50.2660	3.32
49	51.4514	3.21	52.2457	2.89	51.2413	3.33
50	52.4781	3.22	53.2748	2.90	52.2676	3.33

TABLE 24.4 External Gears—14½° Pressure Angle *(Continued)*

No. of teeth	1.728 wire dia.		Alternate 1.92 wire dia.		Alternate 1.68 wire dia.	
	M	K_m	M	K_m	M	K_m
51	53.4547	3.23	54.2531	2.91	53.2439	3.34
52	54.4804	3.24	55.2810	2.92	54.2691	3.35
53	55.4579	3.24	56.2601	2.93	55.2463	3.35
54	56.4826	3.25	57.2868	2.94	56.2705	3.36
55	57.4609	3.26	58.2667	2.95	57.2485	3.37
56	58.4847	3.27	59.2924	2.96	58.2719	3.37
57	59.4637	3.28	60.2729	2.97	59.2506	3.38
58	60.4866	3.28	61.2977	2.98	60.2731	3.39
59	61.4664	3.29	62.2788	2.99	61.2526	3.39
60	62.4884	3.30	63.3027	3.00	62.2743	3.40
61	63.4689	3.31	64.2844	3.01	63.2545	3.40
62	64.4902	3.31	65.3075	3.01	64.2755	3.41
63	65.4712	3.32	66.2898	3.02	65.2562	3.42
64	66.4918	3.32	67.3121	3.03	66.2765	3.42
65	67.4734	3.33	68.2949	3.04	67.2579	3.43
66	68.4933	3.34	69.3165	3.05	68.2775	3.43
67	69.4755	3.34	70.2998	3.05	69.2594	3.44
68	70.4948	3.35	71.3206	3.06	70.2785	3.44
69	71.4775	3.35	72.3044	3.07	71.2609	3.45
70	72.4963	3.36	73.3247	3.07	72.2794	3.45
71	73.4795	3.37	74.3089	3.08	73.2623	3.46
72	74.4977	3.37	75.3285	3.09	74.2803	3.46
73	75.4813	3.38	76.3131	3.10	75.2636	3.46
74	76.4990	3.38	77.3322	3.10	76.2811	3.47
75	77.4830	3.39	78.3172	3.11	77.2649	3.47
76	78.5002	3.39	79.3357	3.12	78.2819	3.48
77	79.4847	3.40	80.3211	3.12	79.2661	3.48
78	80.5014	3.40	81.3391	3.13	80.2827	3.49
79	81.4863	3.41	82.3249	3.13	81.2673	3.49
80	82.5026	3.41	83.3423	3.14	82.2834	3.50
81	83.4877	3.42	84.3285	3.15	83.2684	3.50
82	84.5037	3.42	85.3455	3.15	84.2841	3.50
83	85.4892	3.42	86.3319	3.16	85.2694	3.51
84	86.5047	3.43	87.3485	3.16	86.2847	3.51
85	87.4906	3.43	88.3352	3.17	87.2704	3.51
86	88.5057	3.44	89.3515	3.17	88.2854	3.52
87	89.4919	3.44	90.3384	3.18	89.2714	3.52
88	90.5067	3.44	91.3543	3.18	90.2860	3.52
89	91.4932	3.45	92.3415	3.19	91.2723	3.52
90	92.5076	3.45	93.3570	3.19	92.2866	3.53
91	93.4944	3.45	94.3445	3.20	93.2732	3.53
92	94.5085	3.45	95.3596	3.20	94.2872	3.54
93	95.4956	3.46	96.3474	3.21	95.2741	3.54
94	96.5094	3.46	97.3621	3.21	96.2877	3.54
95	97.4967	3.46	98.3502	3.22	97.2749	3.54
96	98.5102	3.46	99.3646	3.22	98.2882	3.55
97	99.4978	3.47	100.3529	3.23	99.2757	3.55
98	100.5110	3.47	101.3670	3.23	100.2887	3.55
99	101.4988	3.47	102.3555	3.24	101.2764	3.55
100	102.5118	3.48	103.3693	3.24	102.2892	3.56

TABLE 24.4 External Gears—14½° Pressure Angle *(Continued)*

No. of teeth	1.728 wire dia.		Alternate 1.92 wire dia.		Alternate 1.68 wire dia.	
	M	K_m	M	K_m	M	K_m
101	103.4998	3.49	104.3580	3.24	103.2771	3.56
102	104.5125	3.49	105.3715	3.25	104.2897	3.56
103	105.5008	3.49	106.3604	3.25	105.2778	3.57
104	106.5132	3.50	107.3737	3.26	106.2901	3.57
105	107.5017	3.50	108.3628	3.26	107.2785	3.57
106	108.5139	3.50	109.3758	3.27	108.2905	3.57
107	109.5026	3.50	110.3651	3.27	109.2791	3.58
108	110.5146	3.51	111.3778	3.27	110.2910	3.58
109	111.5035	3.51	112.3673	3.28	111.2798	3.58
110	112.5152	3.51	113.3798	3.28	112.2914	3.58
111	113.5044	3.52	114.3695	3.29	113.2804	3.59
112	114.5159	3.52	115.3817	3.29	114.2918	3.59
113	115.5052	3.52	116.3716	3.29	115.2809	3.59
114	116.5165	3.53	117.3835	3.30	116.2921	3.59
115	117.5060	3.53	118.3736	3.30	117.2815	3.59
116	118.5171	3.53	119.3853	3.30	118.2925	3.60
117	119.5068	3.53	120.3756	3.31	119.2821	3.60
118	120.5177	3.53	121.3871	3.31	120.2929	3.60
119	121.5075	3.53	122.3775	3.31	121.2826	3.60
120	122.5182	3.54	123.3888	3.32	122.2932	3.60
121	123.5082	3.54	124.3793	3.32	123.2831	3.61
122	124.5188	3.54	125.3905	3.32	124.2936	3.61
123	125.5089	3.54	126.3811	3.33	125.2836	3.61
124	126.5193	3.55	127.3921	3.33	126.2939	3.61
125	127.5096	3.55	128.3829	3.33	127.2841	3.61
126	128.5198	3.55	129.3937	3.34	128.2941	3.61
127	129.5103	3.55	130.3846	3.34	129.2846	3.62
128	130.5203	3.55	131.3952	3.34	130.2945	3.62
129	131.5109	3.56	132.3863	3.34	131.2851	3.62
130	132.5208	3.56	133.3967	3.35	132.2948	3.62
131	133.5115	3.56	134.3879	3.35	133.2855	3.62
132	134.5213	3.56	135.3982	3.35	134.2951	3.62
133	135.5121	3.57	136.3895	3.36	135.2859	3.63
134	136.5217	3.57	137.3996	3.36	136.2954	3.63
135	137.5127	3.57	138.3911	3.36	137.2863	3.63
136	138.5221	3.57	139.4010	3.37	138.2957	3.63
137	139.5133	3.57	140.3926	3.37	139.2867	3.63
138	140.5226	3.57	141.4023	3.37	140.2960	3.63
139	141.5139	3.58	142.3940	3.37	141.2871	3.64
140	142.5230	3.58	143.4037	3.38	142.2962	3.64
141	143.5144	3.58	144.3955	3.38	143.2875	3.64
142	144.5234	3.58	145.4050	3.38	144.2965	3.64
143	145.5149	3.58	146.3969	3.39	145.2879	3.64
144	146.5238	3.58	147.4062	3.39	146.2967	3.64
145	147.5154	3.59	148.3982	3.39	147.2883	3.64
146	148.5242	3.59	149.4074	3.39	148.2970	3.64
147	149.5159	3.59	150.3996	3.40	149.2887	3.65
148	150.5246	3.59	151.4087	3.40	150.2972	3.65
149	151.5164	3.59	152.4009	3.40	151.2890	3.65
150	152.5250	3.59	153.4098	3.40	152.2974	3.65

TABLE 24.4 External Gears—14½° Pressure Angle *(Continued)*

No. of teeth	1.728 wire dia.		Alternate 1.92 wire dia.		Alternate 1.68 wire dia.	
	M	K_m	M	K_m	M	K_m
151	153.5169	3.60	154.4022	3.41	153.2893	3.65
152	154.5254	3.60	155.4110	3.41	154.2977	3.65
153	155.5174	3.60	156.4034	3.41	155.2897	3.65
154	156.5257	3.60	157.4121	3.41	156.2979	3.65
155	157.5179	3.60	158.4046	3.41	157.2900	3.66
156	158.5261	3.60	159.4132	3.42	158.2981	3.66
157	159.5183	3.60	160.4058	3.42	159.2903	3.66
158	160.5264	3.61	161.4143	3.42	160.2983	3.66
159	161.5188	3.61	162.4070	3.42	161.2906	3.66
160	162.5267	3.61	163.4153	3.43	162.2985	3.66
161	163.5192	3.61	164.4081	3.43	163.2909	3.66
162	164.5270	3.61	165.4164	3.43	164.2987	3.66
163	165.5196	3.61	166.4092	3.43	165.2912	3.67
164	166.5273	3.61	167.4174	3.43	166.2989	3.67
165	167.5200	3.62	168.4103	3.44	167.2915	3.67
166	168.5276	3.62	169.4184	3.44	168.2990	3.67
167	169.5204	3.62	170.4114	3.44	169.2917	3.67
168	170.5279	3.62	171.4194	3.44	170.2992	3.67
169	171.5208	3.62	172.4125	3.44	171.2920	3.67
170	172.5282	3.62	173.4203	3.45	172.2994	3.67
171	173.5212	3.62	174.4135	3.45	173.2922	3.67
172	174.5286	3.62	175.4212	3.45	174.2996	3.68
173	175.5215	3.63	176.4145	3.45	175.2926	3.68
174	176.5288	3.63	177.4221	3.45	176.2998	3.68
175	177.5219	3.63	178.4155	3.45	177.2928	3.68
176	178.5291	3.63	179.4230	3.46	178.3000	3.68
177	179.5223	3.63	180.4165	3.46	179.2930	3.68
178	180.5294	3.63	181.4239	3.46	180.3002	3.68
179	181.5226	3.63	182.4174	3.46	181.2933	3.68
180	182.5297	3.63	183.4248	3.46	182.3003	3.68
181	183.5230	3.63	184.4183	3.47	183.2936	3.68
182	184.5299	3.64	185.4256	3.47	184.3005	3.68
183	185.5233	3.64	186.4193	3.47	185.2938	3.69
184	186.5302	3.64	187.4265	3.47	186.3006	3.69
185	187.5236	3.64	188.4202	3.47	187.2940	3.69
186	188.5304	3.64	189.4273	3.47	188.3008	3.69
187	189.5239	3.64	190.4210	3.48	189.2942	3.69
188	190.5307	3.64	191.4281	3.48	190.3010	3.69
189	191.5243	3.64	192.4219	3.48	191.2944	3.69
190	192.5310	3.64	193.4289	3.48	192.3011	3.69
200	202.5321	3.66	203.4326	3.50	202.3018	3.70
201	203.5260	3.66	204.4267	3.50	203.2957	3.70
300	302.5395	3.72	303.4573	3.60	302.3063	3.75
301	303.5355	3.72	304.4534	3.61	303.3022	3.75
400	402.5434	3.75	403.4706	3.66	402.3087	3.78
401	403.5404	3.75	404.4676	3.66	403.3056	3.78
500	502.5458	3.78	503.4788	3.70	502.3101	3.80
501	503.5433	3.78	504.4763	3.70	503.3076	3.80
∞	$(N+2).5558$	3.87	$(N+3).5145$	3.87	$(N+2).3159$	3.87

For 1 diametral pitch, arc tooth thickness is 1.57080 in.
For 1 module, arc tooth thickness is 1.57080 mm.

on two or more wire diameters. In general, it is recommended that wires from the 1.728-in. series be used for measuring external gears and that wires from the 1.44-in. series be used for internals. The tabulated values for the other wire diameters, however, are available for checking involute profiles and to permit greater freedom in wire selection. Table 24.1 will serve as a guide to the selection of a wire diameter for a particular application.

When spur gears are made to even pitches in the metric system, the pitch is given in terms of *module* values rather than *diametral pitch* values. The tables given in this chapter can be used for checking the tooth thickness of even-module metric gears providing that wires are on hand that are made to appropriate metric sizes in millimeters. (If appropriate wires are not on hand, it is not difficult to precision grind pairs of wires to the metric sizes needed.)

Table 24.2 shows standard wire sizes for commonly used pitches in both the English system of measurement and the metric system.

24.1.1 Numerical Examples

1. An even-tooth external gear of 26 teeth, 10 diametral pitch, 14½° pressure angle is measured with 0.1728-in. diameter wires. From Table 24.4, column 2, the measurement for 1 P_d is 28.4314 in. For 10 P_d the measurement over wires is $28.4314/10 = 2.8431$ in. If the teeth are to be cut 0.003 in. thin, the reduced measurement over wires can be determined by using the change factor K_m. From column 3, Table 24.4, the change factor is 2.90. Thus, the reduction in measurement for teeth 0.003 in. thin would be 0.003 in. $\times$ 2.90 or 0.0087 in. The measurement over wires with teeth 0.003 in. thin would be $2.8431 - 0.0087$, or 2.8344 in.

As an example of metric dimensions, we will take a 2.500 module gear. We will assume an even-tooth external gear of 26 teeth, 14½° pressure angle measured with two wires, measured with basic 1.728-mm wires (for one module). At 2.500 module the wire size is 4.320 mm. From Table 24.4 for 26 teeth we read a measurement value of 28.4314 in. This value is for 1 module. We need to interpret this value in millimeters and then change the millimeter value from 1 module to 2.500 module. The calculation procedure is

$$\frac{M}{0.0393701 \text{ in/mm}} \times \frac{2.500 \text{ module}}{25.400} = M \text{ at } 2.500 \text{ module}$$

$$\frac{M \times 2.500 \text{ module}}{1.000000} = 28.4314 \times 2.500 = 71.0785 \text{ mm} \quad (24.1)$$

If the teeth are to be cut thin by 0.075 mm, the change in M is

$$M \text{ in millimeters} = 0.075 \text{ mm} \times K_m$$

$$= 0.075 \times 2.90 = -0.2175 \quad (24.2)$$

The measurement over two pins, 4.320 mm in diameter is then $71.0785 - 0.2175$, or 70.8610 mm.

2. An even-tooth external gear of 20 teeth, 20 diametral pitch, 20° pressure angle with teeth to be cut 0.004 in. thin is measured with three sizes of wires 0.096 in., 0.0864 in., and 0.084 in. Using Tables 24.2 and 24.3, the following simple computations are made:

Wire size 1 P_d, in.	1.92	1.728	1.68
Wire size 20 P_d, in.	0.096	0.0864	0.084
Measurement 1 P_d, in.	23.0227	22.3900	22.2249
Measurement 20 P_d, in.	1.1511	1.1195	1.1112
Change factor K_m, in.	2.07	2.26	2.33
Teeth to be thin, in.	0.004	0.004	0.004
Reduction in measurement for thin teeth 0.004 in. $\times$ K_m, in.	0.0083	0.0090	0.0093
Measurement with teeth 0.004 in. thin, in.	1.1428	1.1105	1.1019

In using three sizes of wires in this case it is possible to measure the tooth thickness at the pitch diameter and also to obtain a check on the involute profile. (If all three wire measurements indicate the *same* pitch line tooth thickness, the involute curve going through the points touched by the wires must be a *true* involute.)

3. An odd-tooth internal gear of 51 teeth, 8 diametral pitch, 20° pressure angle is measured with 0.180-in. diameter wires, which, when placed in the tooth spaces of the gear are not diametrically opposite by one-half tooth interval. From Table 24.5, column 2, the measurement between wires for 1 P_d is 49.6404 in. For 8 P_d the measurement is 49.6404 in./8, or 6.20505 in. The change factor K_m from Table 24.5 is 2.71, and this, when multiplied by the amount the teeth are to be thin, will indicate the amount of increase in measurement between wires.

24.1.2 Interpretation of Results

The approximate amount the teeth are thick or thin may be determined by the change factors K_m and the amount the actual measurement differs from the computed measurement. To determine the variations in tooth thickness the following relationship is noted:

$$\Delta t = \frac{\Delta M}{K_m} \qquad (24.3)$$

where Δt = the amount the teeth are thick or thin
ΔM = the difference between computed and actual measurement
K_m = the change factor taken from the appropriate table

Tables 24.3 to 24.7 are based on the arc tooth thickness being exactly half the circular pitch. For 1 diametral pitch this value is 1.57080 in. For other diametral pitches the value is 1.57080 *divided by the diametral pitch.*

When metric dimensioning is used, the arc tooth thickness value for 1 module is 1.57080 mm. For other modules the arc tooth thickness is *1.57080 multiplied by the module.*

When teeth are "thinned" to get backlash, the actual design tooth thickness is the thickness calculated (for the diametral pitch or the module) *minus* the amount of tooth thinning used in Eq. (24.3) for external teeth—or *plus* the amount of tooth thinning used in Eq. (24.3) for internal teeth. The change in the M value is subtractive for external teeth and additive for internal teeth.

TABLE 24.5 Internal Gears—20° Pressure Angle

Measurements between wires and change factors for 1 diametral pitch internal standard-addendum 20° pressure angle spur gears with 1.44 and 1.68 diameter wires.
For any other diametral pitch divide measurements given in table by the diametral pitch.
Change factors are the same for all diametral pitches.

No. of teeth	1.44 wire dia.		Alternate 1.68 wire dia.		No. of teeth	1.44 wire dia.		Alternate 1.68 wire dia.	
	M	K_m	M	K_m		M	K_m	M	K_m
5	3.4090	2.30			56	54.6647	2.71	53.6965	3.02
6	4.6595	2.46			57	55.6431	2.71	54.6756	3.01
7	5.4823	2.43			58	56.6648	2.71	55.6975	3.01
8	6.6608	2.52	1.68 wires too		59	57.6438	2.71	56.6774	3.00
9	7.5230	2.50	large below		60	58.6648	2.71	57.6985	3.00
10	8.6617	2.56	15 teeth						
11	9.5490	2.55			61	59.6445	2.71	58.6789	2.99
12	10.6623	2.59			62	60.6648	2.71	59.6994	2.99
13	11.5669	2.58			63	61.6452	2.71	60.6805	2.98
14	12.6627	2.61			64	62.6648	2.71	61.7003	2.98
15	13.5801	2.60	12.4222	8.58	65	63.6458	2.71	62.6819	2.97
16	14.6630	2.63	13.5459	5.95	66	64.6649	2.72	63.7011	2.97
17	15.5902	2.62	14.5026	5.09	67	65.6464	2.72	64.6832	2.96
18	16.6633	2.64	15.5901	4.65	68	66.6649	2.72	65.7018	2.96
19	17.5981	2.63	16.5414	4.35	69	67.6469	2.72	66.6844	2.96
20	18.6635	2.65	17.6147	4.15	70	68.6649	2.72	67.7025	2.95
21	19.6045	2.64	18.5668	3.98	71	69.6475	2.72	68.6856	2.95
22	20.6636	2.66	19.6311	3.87	72	70.6649	2.72	69.7032	2.95
23	21.6099	2.65	20.5854	3.76	73	71.6480	2.72	70.6867	2.94
24	22.6638	2.67	21.6429	3.69	74	72.6649	2.72	71.7038	2.94
25	23.6143	2.66	22.5997	3.61	75	73.6484	2.72	72.6878	2.94
26	24.6639	2.67	23.6520	3.56	76	74.6649	2.72	73.7044	2.94
27	25.6181	2.67	24.6112	3.50	77	75.6489	2.72	74.6888	2.93
28	26.6640	2.67	25.6591	3.46	78	76.6649	2.72	75.7049	2.93
29	27.6214	2.67	26.6206	3.42	79	77.6493	2.72	76.6897	2.93
30	28.6641	2.68	27.6649	3.38	80	78.6649	2.72	77.7054	2.92
31	29.6242	2.68	28.6285	3.35	81	79.6497	2.72	78.6905	2.92
32	30.6642	2.68	29.6699	3.32	82	80.6649	2.72	79.7059	2.92
33	31.6267	2.68	30.6353	3.29	83	81.6501	2.72	80.6914	2.92
34	32.6642	2.69	31.6739	3.27	84	82.6649	2.72	81.7064	2.91
35	33.6289	2.69	32.6411	3.25	85	83.6505	2.72	82.6922	2.91
36	34.6643	2.69	33.6773	3.23	86	84.6650	2.72	83.7068	2.91
37	35.6310	2.69	34.6462	3.21	87	85.6508	2.72	84.6929	2.91
38	36.6643	2.69	35.6804	3.20	88	86.6650	2.72	85.7072	2.91
39	37.6327	2.69	36.6507	3.18	89	87.6511	2.72	86.6936	2.90
40	38.6644	2.70	37.6831	3.16	90	88.6650	2.72	87.7076	2.90
41	39.6343	2.69	38.6547	3.15	91	89.6514	2.72	88.6943	2.90
42	40.6644	2.70	39.6855	3.14	92	90.6650	2.72	89.7080	2.90
43	41.6357	2.70	40.6582	3.13	93	91.6517	2.72	90.6950	2.90
44	42.6645	2.70	41.6875	3.11	94	92.6650	2.72	91.7084	2.89
45	43.6371	2.70	42.6614	3.10	95	93.6520	2.72	92.6956	2.89
46	44.6645	2.70	43.6893	3.09	96	94.6650	2.73	93.7087	2.89
47	45.6383	2.70	44.6644	3.08	97	95.6523	2.73	94.6962	2.89
48	46.6646	2.70	45.6910	3.08	98	96.6650	2.73	95.7090	2.89
49	47.6394	2.70	46.6670	3.07	99	97.6526	2.73	96.6968	2.89
50	48.6646	2.71	47.6926	3.06	100	98.6650	2.73	97.7093	2.88
51	49.6404	2.71	48.6694	3.05	101	99.6528	2.73	98.6974	2.88
52	50.6646	2.71	49.6940	3.04	102	100.6650	2.73	99.7096	2.88
53	51.6414	2.71	50.6716	3.04	103	101.6531	2.73	100.6979	2.88
54	52.6647	2.71	51.6953	3.03	104	102.6650	2.73	101.7099	2.88
55	53.6422	2.71	52.6737	3.02	105	103.6533	2.73	102.6984	2.88

TABLE 24.5 Internal Gears—20° Pressure Angle *(Continued)*

No. of teeth	1.44 wire dia.		Alternate 1.68 wire dia.		No. of teeth	1.44 wire dia.		Alternate 1.68 wire dia.	
	M	K_m	M	K_m		M	K_m	M	K_m
106	104.6650	2.73	103.7102	2.88	156	154.6652	2.73	153.7148	2.83
107	105.6535	2.73	104.6989	2.88	157	155.6573	2.73	154.7070	2.83
108	106.6650	2.73	105.7105	2.87	158	156.6652	2.73	155.7149	2.83
109	107.6537	2.73	106.6994	2.87	159	157.6574	2.73	156.7072	2.83
110	108.6651	2.73	107.7107	2.87	160	158.6652	2.73	157.7150	2.83
111	109.6539	2.73	108.6998	2.87	161	159.6575	2.73	158.7074	2.83
112	110.6651	2.73	109.7110	2.87	162	160.6652	2.73	159.7151	2.83
113	111.6541	2.73	110.7002	2.87	163	161.6576	2.73	160.7076	2.83
114	112.6651	2.73	111.7112	2.87	164	162.6652	2.73	161.7152	2.83
115	113.6543	2.73	112.7006	2.87	165	163.6577	2.73	162.7078	2.83
116	114.6651	2.73	113.7114	2.86	166	164.6652	2.73	163.7153	2.83
117	115.6545	2.73	114.7010	2.86	167	165.6578	2.73	164.7080	2.83
118	116.6651	2.73	115.7117	2.86	168	166.6652	2.73	165.7154	2.83
119	117.6547	2.73	116.7014	2.86	169	167.6579	2.73	166.7082	2.83
120	118.6651	2.73	117.7119	2.86	170	168.6652	2.73	167.7156	2.82
121	119.6548	2.73	118.7018	2.86	171	169.6580	2.73	168.7084	2.82
122	120.6651	2.73	119.7121	2.86	172	170.6652	2.73	169.7157	2.82
123	121.6550	2.73	120.7022	2.86	173	171.6581	2.73	170.7086	2.82
124	122.6651	2.73	121.7123	2.86	174	172.6652	2.73	171.7158	2.82
125	123.6552	2.73	122.7026	2.86	175	173.6582	2.73	172.7087	2.82
126	124.6651	2.73	123.7125	2.85	176	174.6652	2.73	173.7158	2.82
127	125.6554	2.73	124.7029	2.85	177	175.6583	2.73	174.7089	2.82
128	126.6651	2.73	125.7127	2.85	178	176.6652	2.73	175.7159	2.82
129	127.6556	2.73	126.7032	2.85	179	177.6584	2.73	176.7090	2.82
130	128.6652	2.73	127.7129	2.85	180	178.6652	2.73	177.7160	2.82
131	129.6557	2.73	128.7036	2.85	181	179.6584	2.73	178.7092	2.82
132	130.6652	2.73	129.7130	2.85	182	180.6652	2.73	179.7161	2.82
133	131.6559	2.73	130.7039	2.85	183	181.6585	2.73	180.7094	2.82
134	132.6652	2.73	131.7132	2.85	184	182.6652	2.73	181.7162	2.82
135	133.6560	2.73	132.7042	2.85	185	183.6586	2.73	182.7095	2.82
136	134.6652	2.73	133.7134	2.85	186	184.6652	2.73	183.7162	2.82
137	135.6561	2.73	134.7045	2.84	187	185.6587	2.73	184.7097	2.82
138	136.6652	2.73	135.7135	2.84	188	186.6652	2.73	185.7163	2.82
139	137.6563	2.73	136.7047	2.84	189	187.6588	2.73	186.7098	2.82
140	138.6652	2.73	137.7137	2.84	190	188.6652	2.73	187.7164	2.82
141	139.6564	2.73	138.7050	2.84	200	198.6652	2.74	197.7168	2.81
142	140.6652	2.73	139.7139	2.84	201	199.6591	2.74	198.7107	2.81
143	141.6565	2.73	140.7053	2.84	300	298.6654	2.74	297.7192	2.79
144	142.6652	2.73	141.7140	2.84	301	299.6612	2.74	298.7151	2.79
145	143.6566	2.73	142.7055	2.84					
146	144.6652	2.73	143.7142	2.84	400	398.6654	2.74	397.7203	2.78
147	145.6568	2.73	144.7058	2.84	401	399.6623	2.74	398.7172	2.78
148	146.6652	2.73	145.7143	2.84	500	498.6654	2.74	497.7210	2.77
149	147.6569	2.73	146.7061	2.84	501	499.6629	2.74	498.7185	2.77
150	148.6652	2.73	147.7144	2.84					
151	149.6570	2.73	148.7063	2.84	∞	$(N-2).6655$	2.75	$(N-3).7238$	2.75
152	150.6652	2.73	149.7145	2.83					
153	151.6571	2.73	150.7065	2.83					
154	152.6652	2.73	151.7146	2.83					
155	153.6572	2.73	152.7068	2.83					

For 1 diametral pitch, arc tooth thickness is 1.57080 in.
For 1 module, arc tooth thickness is 1.57080 mm.

TABLE 24.6 Internal Gears—14½° Pressure Angle

Measurement between wires and change factors for 1 diametral pitch internal standard-addendum 14½° pressure angle spur gears with 1.44 and 1.68 diameter wires. For any other diametral pitch divide measurements given in table by the diametral pitch. Change factors are the same for all diametral pitches.

No. of teeth	1.44 wire dia. M	K_m	Alternate 1.68 wire dia. M	K_m	No. of teeth	1.44 wire dia. M	K_m	Alternate 1.68 wire dia. M	K_m
5	3.5517	2.38							
6	4.8157	2.61			56	54.8701	3.58	53.6109	4.93
7	5.6393	2.63			57	55.8486	3.59	54.5913	4.90
8	6.8262	2.77			58	56.8705	3.59	55.6144	4.87
9	7.6894	2.79			59	57.8497	3.60	56.5953	4.85
10	8.8337	2.89			60	58.8709	3.60	57.6175	4.82
11	9.7219	2.91			61	59.8508	3.60	58.5990	4.80
12	10.8394	2.99			62	60.8712	3.61	59.6204	4.77
13	11.7449	3.01			63	61.8517	3.61	60.6024	4.75
14	12.8438	3.07			64	62.8715	3.61	61.6230	4.73
15	13.7620	3.09			65	63.8526	3.62	62.6055	4.71
16	14.8474	3.13	1.68 wires too large		66	64.8718	3.62	63.6254	4.69
17	15.7752	3.15			67	65.8535	3.62	64.6083	4.67
18	16.8504	3.19			68	66.8721	3.63	65.6277	4.65
19	17.7858	3.20			69	67.8543	3.63	66.6110	4.64
20	18.8529	3.24			70	68.8724	3.63	67.6297	4.62
21	19.7945	3.25			71	69.8551	3.63	68.6135	4.60
22	20.8550	3.27			72	70.8727	3.64	69.6316	4.59
23	21.8017	3.29			73	71.8558	3.64	70.6158	4.58
24	22.8569	3.31			74	72.8729	3.65	71.6334	4.56
25	23.8078	3.32			75	73.8565	3.65	72.6179	4.55
26	24.8585	3.34			76	74.8731	3.65	73.6351	4.54
27	25.8130	3.35			77	75.8572	3.65	74.6199	4.52
28	26.8599	3.37			78	76.8734	3.65	75.6366	4.51
29	27.8176	3.38			79	77.8578	3.66	76.6218	4.50
30	28.8612	3.40			80	78.8736	3.66	77.6381	4.49
31	29.8216	3.40			81	79.8584	3.66	78.6236	4.48
32	30.8623	3.42	29.4774	9.11	82	80.8738	3.66	79.6395	4.47
33	31.8251	3.43	30.4589	8.18	83	81.8590	3.67	80.6253	4.46
34	32.8633	3.44	31.5095	7.57	84	82.8740	3.67	81.6408	4.45
35	33.8282	3.45	32.4867	7.14	85	83.8595	3.67	82.6270	4.44
36	34.8642	3.46	33.5314	6.81	86	84.8742	3.67	83.6420	4.43
37	35.8311	3.46	34.5071	6.55	87	85.8600	3.67	84.6285	4.42
38	36.8650	3.47	35.5472	6.34	88	86.8743	3.68	85.6431	4.41
39	37.8336	3.48	36.5229	6.15	89	87.8605	3.68	86.6299	4.41
40	38.8658	3.49	37.5599	6.00	90	88.8745	3.68	87.6442	4.40
41	39.8359	3.50	38.5357	5.86	91	89.8610	3.68	88.6313	4.39
42	40.8665	3.51	39.5702	5.75	92	90.8747	3.68	89.6452	4.38
43	41.8380	3.51	40.5463	5.65	93	91.8614	3.69	90.6326	4.38
44	42.8672	3.52	41.5788	5.56	94	92.8749	3.69	91.6462	4.37
45	43.8399	3.52	42.5556	5.48	95	93.8619	3.69	92.6338	4.36
46	44.8678	3.53	43.5861	5.41	96	94.8750	3.69	93.6472	4.36
47	45.8416	3.54	44.5635	5.34	97	95.8623	3.69	94.6350	4.35
48	46.8683	3.54	45.5924	5.28	98	96.8752	3.69	95.6481	4.34
49	47.8432	3.55	46.5704	5.22	99	97.8627	3.70	96.6361	4.34
50	48.8688	3.56	47.5979	5.17	100	98.8753	3.70	97.6489	4.33
51	49.8447	3.56	48.5765	5.13	101	99.8631	3.70	98.6372	4.33
52	50.8692	3.57	49.6027	5.08	102	100.8754	3.70	99.6497	4.32
53	51.8461	3.57	50.5820	5.04	103	101.8635	3.70	100.6382	4.31
54	52.8697	3.58	51.6070	5.00	104	102.8756	3.70	101.6505	4.31
55	53.8474	3.58	52.5869	4.97	105	103.8638	3.70	102.6392	4.30

TABLE 24.6 Internal Gears—14½° Pressure Angle *(Continued)*

No. of teeth	1.44 wire dia. M	K_m	Alternate 1.68 wire dia. M	K_m	No. of teeth	1.44 wire dia. M	K_m	Alternate 1.68 wire dia. M	K_m
106	104.8757	3.71	103.6512	4.30	156	154.8777	3.75	153.6627	4.14
107	105.8642	3.71	104.6401	4.29	157	155.8699	3.75	154.6551	4.14
108	106.8758	3.71	105.6519	4.29	158	156.8778	3.75	155.6630	4.13
109	107.8645	3.71	106.6410	4.28	159	157.8701	3.75	156.6555	4.13
110	108.8759	3.71	107.6526	4.28	160	158.8778	3.75	157.6633	4.13
111	109.8648	3.71	108.6419	4.27	161	159.8702	3.76	158.6559	4.13
112	110.8760	3.71	109.6532	4.27	162	160.8779	3.76	159.6636	4.13
113	111.8651	3.71	110.6427	4.27	163	161.8704	3.76	160.6563	4.12
114	112.8761	3.72	111.6538	4.26	164	162.8779	3.76	161.6639	4.12
115	113.8654	3.72	112.6434	4.26	165	163.8705	3.76	162.6567	4.12
116	114.8762	3.72	113.6544	4.25	166	164.8780	3.76	163.6642	4.12
117	115.8657	3.72	114.6442	4.25	167	165.8707	3.76	164.6570	4.12
118	116.8763	3.72	115.6550	4.24	168	166.8780	3.76	165.6645	4.12
119	117.8660	3.72	116.6450	4.24	169	167.8708	3.76	166.6573	4.11
120	118.8764	3.72	117.6556	4.24	170	168.8781	3.76	167.6647	4.11
121	119.8662	3.72	118.6457	4.23	171	169.8710	3.76	168.6576	4.11
122	120.8765	3.73	119.6561	4.23	172	170.8782	3.76	169.6649	4.11
123	121.8663	3.73	120.6464	4.23	173	171.8711	3.77	170.6579	4.11
124	122.8766	3.73	121.6566	4.22	174	172.8782	3.77	171.6651	4.11
125	123.8668	3.73	122.6470	4.22	175	173.8712	3.77	172.6582	4.10
126	124.8767	3.73	123.6571	4.22	176	174.8783	3.77	173.6653	4.10
127	125.8670	3.73	124.6477	4.21	177	175.8713	3.77	174.6585	4.10
128	126.8768	3.73	125.6575	4.21	178	176.8783	3.77	175.6655	4.10
129	127.8672	3.73	126.6483	4.21	179	177.8715	3.77	176.6588	4.10
130	128.8769	3.73	127.6579	4.20	180	178.8784	3.77	177.6658	4.10
131	129.8675	3.73	128.6489	4.20	181	179.8717	3.77	178.6591	4.10
132	130.8769	3.74	129.6583	4.20	182	180.8784	3.77	179.6659	4.10
133	131.8677	3.74	130.6495	4.19	183	181.8717	3.77	180.6593	4.09
134	132.8770	3.74	131.6588	4.19	184	182.8785	3.77	181.6662	4.09
135	133.8679	3.74	132.6500	4.19	185	183.8718	3.77	182.6596	4.09
136	134.8771	3.74	133.6592	4.19	186	184.8785	3.77	183.6665	4.09
137	135.8681	3.74	134.6505	4.18	187	185.8719	3.77	184.6599	4.09
138	136.8772	3.74	135.6596	4.18	188	186.8786	3.77	185.6667	4.09
139	137.8683	3.74	136.6510	4.18	189	187.8721	3.77	186.6601	4.09
140	138.8773	3.74	137.6600	4.17	190	188.8786	3.77	187.6668	4.09
141	139.8685	3.74	138.6515	4.17	200	198.8788	3.78	197.6678	4.07
142	140.8773	3.74	139.6604	4.17	201	199.8727	3.78	198.6618	4.07
143	141.8687	3.74	140.6520	4.17	300	298.8795	3.81	297.6725	4.00
144	142.8774	3.75	141.6608	4.16	301	299.8759	3.81	298.6694	4.00
145	143.8689	3.75	142.6525	4.16	400	398.8803	3.82	397.6752	3.96
146	144.8774	3.75	143.6612	4.16	401	399.8776	3.82	398.6731	3.96
147	145.8691	3.75	144.6530	4.16	500	498.8810	3.83	497.6767	3.94
148	146.8775	3.75	145.6615	4.16	501	499.8786	3.83	498.6753	3.94
149	147.8693	3.75	146.6534	4.15	∞	$(N - 2).8827$	3.87	$(N - 3).6841$	3.87
150	148.8775	3.75	147.6618	4.15					
151	149.8694	3.75	148.6538	4.15					
152	150.8776	3.75	149.6621	4.15					
153	151.8696	3.75	150.6543	4.14					
154	152.8776	3.75	151.6624	4.14					
155	153.8698	3.75	152.6547	4.14					

For 1 diametral pitch, arc tooth thickness is 1.57080 in.
For 1 module, arc tooth thickness is 1.57080 mm.

TABLE 24.7 External Gears—25° Pressure Angle

Measurements over wires and change factors for 1 diametral pitch external standard-addendum 25° pressure angle spur gears with 1.728 and 1.68 diameter wires. For any other diametral pitch divide measurements given in table by the diametral pitch. Change factors are the same for all diametral pitches.

No. of teeth	1.728 wire dia. M	K_m	Alternate 1.68 wire dia. M	K_m	No. of teeth	1.728 wire dia. M	K_m	Alternate 1.68 wire dia. M	K_m
5	7.0472	1.47	6.9202	1.59	56	58.4287	2.03	58.2726	2.05
6	8.3340	1.60	8.2003	1.64	57	59.4071	2.03	59.2509	2.05
7	9.1536	1.64	9.0199	1.68	58	60.4293	2.04	60.2730	2.05
8	10.3533	1.67	10.2155	1.72	59	61.4084	2.04	61.2521	2.05
9	11.2142	1.70	11.0762	1.75	60	62.4299	2.04	62.2735	2.05
10	12.3667	1.73	12.2260	1.77					
11	13.2536	1.75	13.1126	1.79	61	63.4097	2.04	63.2532	2.06
12	14.3768	1.77	14.2338	1.81	62	64.4304	2.04	64.2739	2.06
13	15.2814	1.79	15.1381	1.83	63	65.4109	2.04	65.2543	2.06
14	16.3846	1.81	16.2397	1.85	64	66.4309	2.05	66.2742	2.06
15	17.3021	1.83	17.1570	1.86	65	67.4120	2.05	67.2553	2.06
16	18.3908	1.84	18.2445	1.88	66	68.4314	2.05	68.2746	2.06
17	19.3181	1.85	19.1716	1.89	67	69.4130	2.05	69.2562	2.06
18	20.3959	1.86	20.2483	1.90	68	70.4319	2.05	70.2749	2.06
19	21.3310	1.88	21.1832	1.91	69	71.4140	2.05	71.2571	2.06
20	22.4002	1.88	22.2515	1.92	70	72.4323	2.05	72.2752	2.07
21	23.3415	1.89	23.1926	1.92	71	73.4150	2.05	73.2579	2.07
22	24.4038	1.90	24.2542	1.93	72	74.4327	2.05	74.2755	2.07
23	25.3502	1.91	25.2005	1.94	73	75.4159	2.06	75.2586	2.07
24	26.4069	1.92	26.2566	1.95	74	76.4331	2.06	76.2758	2.07
25	27.3576	1.93	27.2071	1.95	75	77.4167	2.06	77.2594	2.07
26	28.4096	1.93	28.2586	1.96	76	78.4335	2.06	78.2761	2.07
27	29.3640	1.94	29.2128	1.96	77	79.4175	2.06	79.2601	2.07
28	30.4120	1.94	30.2603	1.97	78	80.4339	2.06	80.2763	2.07
29	31.3695	1.95	31.2177	1.97	79	81.4183	2.06	81.2607	2.07
30	32.4141	1.95	32.2619	1.98	80	82.4342	2.06	82.2766	2.08
31	33.3743	1.96	33.2220	1.98	81	83.4190	2.06	83.2614	2.08
32	34.4159	1.96	34.2632	1.99	82	84.4345	2.07	84.2768	2.08
33	35.3786	1.97	35.2258	1.99	83	85.4196	2.07	85.2620	2.08
34	36.4176	1.97	36.2664	2.00	84	86.4348	2.07	86.2771	2.08
35	37.3824	1.98	37.2292	2.00	85	87.4203	2.07	87.2625	2.08
36	38.4191	1.98	38.2655	2.00	86	88.4351	2.07	88.2773	2.08
37	39.3858	1.98	39.2323	2.01	87	89.4209	2.07	89.2631	2.08
38	40.4205	1.99	40.2666	2.01	88	90.4354	2.07	90.2775	2.08
39	41.3889	1.99	41.2349	2.01	89	91.4215	2.07	91.2636	2.08
40	42.4217	2.00	42.2675	2.02	90	92.4357	2.07	92.2777	2.08
41	43.3917	2.00	43.2374	2.02	91	93.4221	2.07	93.2641	2.08
42	44.4228	2.00	44.2683	2.02	92	94.4359	2.07	94.2779	2.08
43	45.3942	2.00	45.2396	2.02	93	95.4227	2.07	95.2646	2.08
44	46.4239	2.01	46.2690	2.02	94	96.4362	2.07	96.2780	2.08
45	47.3965	2.01	47.2417	2.02	95	97.4232	2.08	97.2650	2.09
46	48.4248	2.01	48.2697	2.03	100	102.4369	2.08	102.2785	2.09
47	49.3986	2.01	49.2435	2.03	101	103.4247	2.08	103.2663	2.09
48	50.4257	2.02	50.2704	2.03	200	202.4424	2.11	202.2825	2.11
49	51.4006	2.02	51.2452	2.03	201	203.4363	2.11	203.2764	2.11
50	52.4265	2.02	52.2710	2.04	300	302.4443	2.12	302.2839	2.12
51	53.4024	2.02	53.2468	2.04	301	303.4402	2.12	303.2798	2.12
52	54.4273	2.03	54.2716	2.04	400	402.4453	2.13	402.2845	2.13
53	55.4041	2.03	55.2483	2.04	401	403.4422	2.13	403.2815	2.13
54	56.4280	2.03	56.2721	2.05	500	502.4458	2.13	502.2850	2.13
55	57.4056	2.03	57.2497	2.05	∞	$(N+2).4482$	2.14	$(N+2).2866$	2.14

For 1 diametral pitch, arc tooth thickness is 1.57080 in.
For 1 module, arc tooth thickness is 1.57080 mm.

24.1.3 Odd-Tooth Gears

It should be noted in Tables 24.3 to 24.7 that the measurement values for odd numbers of teeth represent dimensions over two wires that are not diametrically opposite to each other by one-half tooth interval. Thus, to determine the dimension from the top of one wire to the *center of the gear* the following computation should be used:

$$\text{Dimension over wires} - \text{wire diameter} = \text{dimension between wire centers} \quad (24.4)$$

$$\frac{\text{Dimension between wire centers}}{\cos(90°/N)} \times 0.5 + \frac{\text{wire diameter}}{2} =$$

$$\text{radius from top of one wire to center of gear} \quad (24.5)$$

24.1.4 Special Pressure Angles and Number of Teeth

Because of the relative simplicity of the exact formulas (see Sec. 5.8), it is not recommended that interpolation be made from the tables in the case of special pressure angles. Where the number of teeth is greater than is to be found in the appropriate table, an interpolation may easily be made. Thus from Table 24.3 the measurement over 1.728-in. diameter wires for a 1-diametral-pitch, 335-tooth, 20° external gear would be

$$303.4538 + \left(\frac{34}{100} \times 100.0027\right) = 337.4547 \quad (24.6)$$

24.1.5 Measurement of Racks

For 1.728-in. wires the decimal values of Tables 24.3, 24.4, and 24.7 show the amount the measurement over wires exceeds the outside diameter of a standard even-tooth gear. Therefore, if the decimal value given for infinity (∞) is divided by 2, the amount the 1.728-in. wire projects above the top of the 1-pitch rack is obtained. These values are:

14½°	20°	25°
0.2779"	0.2323"	0.2241"

Since the addendum of a 1-pitch gear is 1.0000 in., the amount the wire projects above the pitch line is 1.0000 plus the above values.

For any other diametral pitch P_d, divide the values by P_d and use wires 1.728-in./P_d in diameter.

24.1.6 Index to Tables

Pressure Angle	Type	Wire Diameters Standard Addendum	Table	Page
20°	External	1.728, 1.92, 1.68	24.3	24.3
14½°	External	1.728, 1.92, 1.68	24.4	24.7
20°	Internal	1.68, 1.44	24.5	24.13
14½°	Internal	1.68, 1.44	24.6	24.15
25°	External	1.728, 1.68	24.7	24.17

Measurement values in tables should be in inches for the English system of measurement and in millimeters for the metric system of measurement. (See details in the text of this section.) For numerical examples see Secs. 24.1.1 and 24.1.7. For basic formulas and equations used to compute values in the tables (or outside the tables) see Secs. 5.5 and 5.8.

24.1.7 Relation between Depth of Cut and Tooth Thickness

If the generating gear cutter* is fed into the blank an amount exactly equal to the whole depth of tooth, the gear is cut† without backlash. To provide backlash the teeth must be made thinner by an equal amount Δt (increment of t). The formulas for finding the excess depth of cut E from Δt and vice versa are as given in Table 24.8. The relationships are exact for hobs or rack-shaped tools. They are approximate for gear shaper cutters.

Example. A total backlash of 0.016 in. is desired between a 30-tooth, 14½° pinion and a 100-tooth 14½° gear; and 0.008-in. backlash will be taken on the pinion and 0.008 in. on the gear. From Table 24.4, the wire measurement for the pinion should be 2.98 $\times$ 0.008 in. = 0.0238 in. under the computed value, and for the gear 3.48 $\times$ 0.008 in. = 0.0278 in. undersize. If after a trial cut, the wire measurement on the gear is only 0.010 in. under the computed value, the backlash provided is 0.010/3.48 = 0.0029 in. This is 0.008 in. $-$ 0.0029 in. = 0.0051 in.—too little, and, from Table 24.8, the cutter must be fed in deeper by 1.93 $\times$ 00.051 in. = 0.0098 in.

24.1.8 Helical Gears

The pin measurement of helical gears may be computed as outlined in Chap. 5. A simple but approximate calculation may also be made which utilizes the spur gear measuring tables (see Table 24.9).

Even-tooth helical gears should be measured with two wires. Odd-tooth helical gears should not be measured with two wires but may be measured with three wires under

---

*Backlash in gears milled with a form-milling cutter is usually produced by side-cutting a desired amount to thin the teeth that have been cut to standard depth.

†This rule applies to common trade practices with small gear teeth. Large gear teeth are often cut with a hob that produces thin teeth with the right backlash at design whole depth.

TABLE 24.8 Change in Depth of Cut versus Change in Tooth Thickness

Pressure angle of gear, deg	E, extra depth of cut to thin teeth (Δt)	Δt, amount teeth are cut too thin due to excess depth of cut E
14½	$E = 1.93 \times \Delta t$	$\Delta t = \dfrac{E}{1.93} = 0.52E$
17½	$E = 1.59 \times \Delta t$	$\Delta t = \dfrac{E}{1.59} = 0.63E$
20	$E = 1.37 \times \Delta t$	$\Delta t = \dfrac{E}{1.37} = 0.73E$
25	$E = 1.07 \times \Delta t$	$\Delta t = \dfrac{E}{1.07} = 0.93E$
30	$E = 0.87 \times \Delta t$	$\Delta t = \dfrac{E}{0.87} = 1.15E$

certain conditions, or indirectly with one wire. Three wires may be used, provided the face width and helix angle of the gear will permit the arrangement of two wires in adjacent tooth spaces on one side of the gear and a third wire in a tooth space on the opposite side, so that all three wires will automatically adjust themselves to line contact with the micrometer's measuring surfaces. In cases where three wires cannot be used, the gear should be mounted on an arbor and a radial measurement made from the top of one wire to the axis of the gear. See Table 24.10.

It should be emphasized that the practice of using two wires for measuring odd-tooth helical gears and correcting the results by the application of a cos 90°/N factor is not valid and may result in serious errors.

24.2 TRIGONOMETRIC FUNCTIONS

Table 24.11 lists some trigonometric functions for general reference. Those doing an extensive amount of gear calculating will find it necessary to have a whole book of seven-place trigonometric tables. With modern computer equipment for engineering work, trigonometric values are immediately available from the computer.

24.3 INVOLUTE FUNCTIONS

The involute function is widely used in gear calculations. Chapter 1 in Sec. 1.1 defines the involute function. Chapter 5 in Sec. 5.8 discusses the involute and gives a table of involute functions from 1 to 90°. A short table of involute functions for general reference is given in this section. The involute function of an angle is called "inv ϕ" or simply "θ." Inv ϕ equals θ in radians. See Table 24.12.

TABLE 24.9 Wire Measurement of Standard Helical Gear
For use when normal pressure angle is 14½, 20, or 25°

For: A-Z Gear Co.
    Data from Drawing

Date: Aug. 12, 1954
Drawing No.: N90-1536
Functions of ψ (psi)

1. No. of teeth $N = 14$ $\cos 40° = 0.7660444$
2. Diametral pitch in plane of rotation $P_d = 18.3850656$ $\dfrac{1}{\cos 40°} = 1.305407$
3. Pitch diameter $D = 0.7615''$
4. Helix angle (psi) with axis ($\tan \psi = \pi D/L$) $\psi = 40°$ $\dfrac{1}{(\cos 40°)^3} = 2.224529$
5. Normal pressure angle (phi sub n) $\phi_n = 14.5$
6. Normal tooth thickness $T_n = 0.0654''$

Computations

7. Find normal diametral pitch (may show on drawing) $P_n = P_d \times \dfrac{1}{\cos \psi} = 18.38 \times 1.305407 = 24$
8. Find wire diameter $G = \dfrac{1.728''}{P_n} = \dfrac{1.728''}{24} = 0.072''$
9. Find number of teeth in equivalent spur gear $N_e = N \times \dfrac{1}{(\cos \psi)^3} = 14 \times 2.224529 = 31.143406$

Data from Tables (see Table 24-4)

10. Refer to proper table and using *even-tooth values* only interpolate for values (M-D) for N_e teeth (for 1 diametral pitch gears, $D = N$):

1 Diametral Pitch Values

	N	D	M	(M-D)	Diff.
Next lower (even)	(a) 30	(c) 30''	(d) 32.4429''	(e) 2.4429''	(f) 0.0049''
Next higher (even)	(b) 32	32''	34.4478''	2.4478''	

N_e value (9) = 31.1434
N value (a) = 30
Diff. (g) = 1.1434

$(M\text{-}D) = 2.4429'' + \dfrac{1.1434}{2} \times 0.0049'' = 2.4457''$
        (e)   (g)   (f)

11. Find (M-D) for specified normal diametral pitch $(M\text{-}D)_1 = \dfrac{(M\text{-}D)}{P_n} = \dfrac{2.4457''}{24} = 0.1019''$
12. Find measurement over wires $M_1 = (M\text{-}D)_1 + D = 0.1019'' + 0.7615'' = 0.8634''$
13. Measurement for odd-tooth gear $\dfrac{M_1}{2} =$ (odd-tooth helical gears should be mounted on an arbor and a radial measurement should be made from one wire to the center of the gear)
14. Reduction in measurement over wires to thin teeth 0.002'' in normal plane (Table 24-10) $\Delta M = \dfrac{K \times \Delta t_n}{\cos \psi} = \dfrac{2.33 \times 0.002''}{0.7660444} = 0.0061''$
15. Measurement over wires for teeth 0.002'' thin in normal plane $M_2 = M_1 - \Delta M = 0.8634'' - 0.0061'' = 0.8573''$

TABLE 24.10 Change Factors K for 1.728-in. Wires Used on External Helical Gears

$$K = \frac{\cos \phi}{\sin \phi_w}$$

Reduction in measurement = $K \times$ amount teeth are to be thin (in normal plane) $\div \cos \psi$
or Reduction in measurement = $K \times$ amount teeth are to be thin (in plane of rotation of the gear)

No. of teeth	Normal pressure angle = $14\frac{1}{2}°$								Normal pressure angle = $20°$						
	$\psi=15°$	$\psi=20°$	$\psi=25°$	$\psi=30°$	$\psi=35°$	$\psi=40°$	$\psi=45°$	$\psi=15°$	$\psi=20°$	$\psi=25°$	$\psi=30°$	$\psi=35°$	$\psi=40°$	$\psi=45°$
6	2.07	2.06	2.05	2.03	2.00	1.96	1.91	1.81	1.79	1.77	1.73	1.69	1.64	1.58
10	2.34	2.33	2.30	2.27	2.23	2.17	2.10	2.02	2.00	1.96	1.92	1.85	1.78	1.69
15	2.56	2.56	2.54	2.49	2.42	2.37	2.25	2.14	2.12	2.07	2.01	1.93	1.84	1.75
20	2.72	2.69	2.65	2.59	2.52	2.44	2.32	2.22	2.18	2.13	2.07	1.99	1.90	1.79
25	2.81	2.77	2.73	2.66	2.58	2.49	2.35	2.28	2.24	2.18	2.11	2.03	1.93	1.81
30	2.89	2.84	2.80	2.71	2.63	2.53	2.39	2.33	2.28	2.22	2.14	2.05	1.95	1.83
35	2.95	2.90	2.86	2.76	2.68	2.56	2.42	2.36	2.31	2.25	2.17	2.08	1.97	1.84
40	3.01	2.96	2.91	2.81	2.72	2.59	2.44	2.40	2.34	2.28	2.20	2.10	1.98	1.85
45	3.07	3.01	2.96	2.85	2.75	2.62	2.46	2.42	2.36	2.30	2.21	2.12	2.00	1.86
50	3.12	3.06	3.00	2.88	2.78	2.64	2.48	2.44	2.38	2.32	2.23	2.13	2.01	1.87
60	3.20	3.13	3.07	2.95	2.83	2.69	2.52	2.47	2.41	2.34	2.25	2.15	2.03	1.89
70	3.27	3.20	3.12	3.00	2.88	2.72	2.55	2.49	2.43	2.36	2.27	2.16	2.04	1.90
80	3.32	3.25	3.16	3.05	2.92	2.76	2.58	2.51	2.44	2.38	2.28	2.17	2.05	1.90
90	3.37	3.29	3.20	3.08	2.95	2.79	2.61	2.53	2.46	2.39	2.29	2.18	2.06	1.91
100	3.40	3.33	3.23	3.12	2.98	2.81	2.63	2.54	2.47	2.40	2.30	2.19	2.06	1.91
120	3.45	3.37	3.26	3.16	3.02	2.84	2.65	2.56	2.49	2.42	2.31	2.20	2.07	1.91
140	3.48	3.40	3.29	3.18	3.04	2.86	2.67	2.57	2.50	2.42	2.32	2.21	2.07	1.91
160	3.51	3.42	3.31	3.20	3.06	2.87	2.68	2.58	2.51	2.43	2.32	2.22	2.07	1.91
180	3.53	3.45	3.33	3.22	3.07	2.88	2.69	2.59	2.52	2.44	2.33	2.23	2.07	1.91
200	3.55	3.46	3.35	3.23	3.08	2.89	2.69	2.60	2.52	2.44	2.34	2.23	2.07	1.92
250	3.59	3.50	3.38	3.25	3.09	2.90	2.70	2.61	2.53	2.45	2.35	2.23	2.08	1.92
300	3.61	3.52	3.40	3.26	3.10	2.91	2.70	2.62	2.54	2.46	2.37	2.23	2.09	1.93
350	3.63	3.53	3.41	3.27	3.11	2.92	2.70	2.62	2.54	2.46	2.37	2.23	2.09	1.93
400	3.64	3.54	3.42	3.28	3.11	2.92	2.71	2.63	2.55	2.47	2.37	2.24	2.09	1.94
450	3.65	3.55	3.43	3.29	3.12	2.93	2.71	2.64	2.56	2.47	2.37	2.24	2.09	1.94
500	3.66	3.56	3.44	3.30	3.12	2.93	2.71	2.64	2.56	2.47	2.37	2.24	2.09	1.94

NOTE: The amount of reduction is the same for all diametral pitches for a given tooth-thickness reduction.

TABLE 24.11 Trigonometric Functions

$\phi°$	arc ϕ	sin ϕ	tan ϕ	cot ϕ	cos ϕ	sec ϕ	cosec ϕ
1°	0.01745	0.01745	0.01746	57.290	0.99985	1.0002	57.299
2	0.03491	0.03490	0.03492	28.636	0.99939	1.0006	28.654
3	0.05236	0.05234	0.05241	19.081	0.99863	1.0014	19.107
4	0.06981	0.06976	0.06993	14.301	0.99756	1.0024	14.336
5	0.08727	0.08716	0.08749	11.430	0.99619	1.0038	11.474
6°	0.10472	0.10453	0.10510	9.5144	0.99452	1.0055	9.5668
7	0.12217	0.12187	0.12278	8.1444	0.99255	1.0075	8.2055
8	0.13963	0.13917	0.14054	7.1154	0.99027	1.0098	7.1853
9	0.15708	0.15643	0.15838	6.3138	0.98769	1.0125	6.3925
10	0.17453	0.17365	0.17633	5.6713	0.98481	1.0154	5.7588
11°	0.19199	0.19081	0.19438	5.1446	0.98163	1.0187	5.2408
12	0.20944	0.20791	0.21256	4.7046	0.97815	1.0223	4.8097
13	0.22689	0.22495	0.23087	4.3315	0.97437	1.0263	4.4454
14	0.24435	0.24192	0.24933	4.0108	0.97030	1.0306	4.1336
15	0.26180	0.25882	0.26795	3.7321	0.96593	1.0353	3.8637
16°	0.27925	0.27564	0.28675	3.4874	0.96126	1.0403	3.6280
17	0.29671	0.29237	0.30573	3.2709	0.95630	1.0457	3.4203
18	0.31416	0.30902	0.32492	3.0777	0.95106	1.0515	3.2361
19	0.33161	0.32557	0.34433	2.9042	0.94552	1.0576	3.0716
20	0.34907	0.34202	0.36397	2.7475	0.93969	1.0642	2.9238
21°	0.36652	0.35837	0.38386	2.6051	0.93358	1.0711	2.7904
22	0.38397	0.37461	0.40403	2.4751	0.92718	1.0785	2.6695
23	0.40143	0.39073	0.42447	2.3559	0.92050	1.0864	2.5593
24	0.41888	0.40674	0.44523	2.2460	0.91355	1.0946	2.4586
25	0.43633	0.42262	0.46631	2.1445	0.90631	1.1034	2.3662
26°	0.45379	0.43837	0.48773	2.0503	0.89879	1.1126	2.2812
27	0.47124	0.45399	0.50953	1.9626	0.89101	1.1223	2.2027
28	0.48869	0.46947	0.53171	1.8807	0.88295	1.1326	2.1301
29	0.50615	0.48481	0.55431	1.8040	0.87462	1.1434	2.0627
30	0.52360	0.50000	0.57735	1.7321	0.86603	1.1547	2.0000
31°	0.54105	0.51504	0.60086	1.6643	0.85717	1.1666	1.9416
32	0.55851	0.52992	0.62487	1.6003	0.84805	1.1792	1.8871
33	0.57596	0.54464	0.64941	1.5399	0.83867	1.1924	1.8361
34	0.59341	0.55919	0.67451	1.4826	0.82904	1.2062	1.7883
35	0.61087	0.57358	0.70021	1.4281	0.81915	1.2208	1.7434
36°	0.62832	0.58779	0.72654	1.3764	0.80902	1.2361	1.7013
37	0.64577	0.60182	0.75355	1.3270	0.79864	1.2521	1.6616
38	0.66323	0.61566	0.78129	1.2799	0.78801	1.2690	1.6243
39	0.68068	0.62932	0.80978	1.2349	0.77715	1.2868	1.5890
40	0.69813	0.64279	0.83910	1.1918	0.76604	1.3054	1.5557
41°	0.71559	0.65606	0.86929	1.1504	0.75471	1.3250	1.5243
42	0.73304	0.66913	0.90040	1.1106	0.74314	1.3456	1.4945
43	0.75049	0.68200	0.93252	1.0724	0.73135	1.3673	1.4663
44	0.76794	0.69466	0.96569	1.0355	0.71934	1.3902	1.4396
45	0.78540	0.70711	1.00000	1.0000	0.70711	1.4142	1.4142
46°	0.80285	0.71934	1.0355	0.96569	0.69466	1.4396	1.3902
47	0.82030	0.73135	1.0724	0.93252	0.68200	1.4663	1.3673
48	0.83776	0.74314	1.1106	0.90040	0.66913	1.4945	1.3456
49	0.85521	0.75471	1.1504	0.86929	0.65606	1.5243	1.3250
50	0.87266	0.76604	1.1918	0.83910	0.64279	1.5557	1.3054

TABLE 24.12 Involute Functions (inv ϕ = tan ϕ - arc ϕ)

ϕ	inv ϕ	Diff.	ϕ	inv ϕ	Diff.	ϕ	inv ϕ	Diff.	ϕ	inv ϕ	Diff.
0.0°	.0000000	0	5.0°	.0002222	136	10.0°	.0017941	548	15.0°	.0061498	1262
0.1	00000	0	5.1	02358	142	10.1	18489	559	15.1	62760	1279
0.2	00000	0	5.2	02500	147	10.2	19048	571	15.2	64039	1298
0.3	00000	1	5.3	02647	154	10.3	19619	582	15.3	65337	1315
0.4	00001	1	5.4	02801	158	10.4	20201	594	15.4	66652	1333
0.5	00002	2	5.5	02959	165	10.5	20795	605	15.5	67985	1352
0.6	00004	2	5.6	03124	171	10.6	21400	617	15.6	69337	1369
0.7	00006	3	5.7	03295	177	10.7	22017	629	15.7	70706	1389
0.8	00009	4	5.8	03472	183	10.8	22646	642	15.8	72095	1406
0.9	00013	5	5.9	03655	190	10.9	23288	653	15.9	73501	1426
1.0°	.0000018	6	6.0°	.0003845	196	11.0°	.0023941	666	16.0°	.0074927	1445
1.1	00024	7	6.1	04041	203	11.1	24607	678	16.1	76372	1463
1.2	00031	8	6.2	04244	209	11.2	25285	690	16.2	77835	1483
1.3	00039	10	6.3	04453	216	11.3	25975	703	16.3	79318	1502
1.4	00049	11	6.4	04669	223	11.4	26678	716	16.4	80820	1522
1.5	00060	13	6.5	04892	230	11.5	27394	729	16.5	82342	1541
1.6	00073	14	6.6	05122	237	11.6	28123	742	16.6	83883	1561
1.7	00087	16	6.7	05359	245	11.7	28865	755	16.7	85444	1581
1.8	00103	19	6.8	05604	252	11.8	29620	769	16.8	87025	1601
1.9	00122	20	6.9	05856	259	11.9	30389	782	16.9	88626	1621
2.0°	.0000142	22	7.0°	.0006115	267	12.0°	.0031171	795	17.0°	.0090247	1642
2.1	00164	25	7.1	06382	275	12.1	31966	809	17.1	91889	1662
2.2	00189	27	7.2	06657	282	12.2	32775	823	17.2	93551	1683
2.3	00216	29	7.3	06939	291	12.3	33598	836	17.3	95234	1703
2.4	00245	32	7.4	07230	298	12.4	34434	851	17.4	96937	1725
2.5	00277	35	7.5	07528	307	12.5	35285	865	17.5	98662	1745
2.6	00312	37	7.6	07835	315	12.6	36150	879	17.6	.0100407	1767
2.7	00349	40	7.7	08150	323	12.7	37029	894	17.7	102174	1789
2.8	00389	44	7.8	08473	332	12.8	37923	908	17.8	103963	1810
2.9	00433	46	7.9	08805	340	12.9	38831	923	17.9	105773	1831
3.0°	.0000479	50	8.0°	.0009145	349	13.0°	.0039754	938	18.0°	.0107604	1854
3.1	00529	52	8.1	09494	358	13.1	40692	952	18.1	109458	1875
3.2	00581	57	8.2	09852	367	13.2	41644	968	18.2	111333	1898
3.3	00638	60	8.3	10219	376	13.3	42612	983	18.3	113231	1920
3.4	00698	63	8.4	10595	385	13.4	43595	998	18.4	115151	1943
3.5	00761	67	8.5	10980	395	13.5	44593	1014	18.5	117094	1965
3.6	00828	71	8.6	11375	404	13.6	45607	1029	18.6	119059	1989
3.7	00899	75	8.7	11779	413	13.7	46636	1045	18.7	121048	2011
3.8	00974	79	8.8	12192	423	13.8	47681	1061	18.8	123059	2034
3.9	01053	83	8.9	12615	433	13.9	48742	1077	18.9	125093	2058
4.0°	.0001136	88	9.0°	.0013048	443	14.0°	.0049819	1093	19.0°	.0127151	2081
4.1	01224	92	9.1	13491	453	14.1	50912	1109	19.1	129232	2104
4.2	01316	96	9.2	13944	463	14.2	52021	1126	19.2	131336	2129
4.3	01412	101	9.3	14407	473	14.3	53147	1142	19.3	133465	2152
4.4	01513	106	9.4	14880	483	14.4	54289	1159	19.4	135617	2177
4.5	01619	110	9.5	15363	494	14.5	55448	1176	19.5	137794	2200
4.6	01729	116	9.6	15857	505	14.6	56624	1193	19.6	139994	2226
4.7	01845	120	9.7	16362	515	14.7	57817	1210	19.7	142220	2250
4.8	01965	126	9.8	16877	526	14.8	59027	1227	19.8	144470	2274
4.9	02091	131	9.9	17403	538	14.9	60254	1244	19.9	146744	2300
5.0°	.0002222		10.0°	.0017941		15.0°	.0061498		20.0°	.0149044	

TABLE 24.12 Involute Functions (inv ϕ = tan ϕ − arc ϕ) *(Continued)*

ϕ	inv ϕ	Diff.	ϕ	inv ϕ	Diff.	ϕ	inv ϕ	Diff.	ϕ	inv ϕ	Diff.
20.0°	.0149044	2325	25.0°	.0299753	3813	30.0°	.0537515	5841	35.0°	.0893423	8589
20.1	151369	2350	25.1	303566	3847	30.1	543356	5889	35.1	902012	8653
20.2	153719	2375	25.2	307413	3882	30.2	549245	5936	35.2	910665	8717
20.3	156094	2401	25.3	311295	3918	30.3	555181	5983	35.3	919382	8783
20.4	158495	2427	25.4	315213	3953	30.4	561164	6032	35.4	928165	8847
20.5	160922	2453	25.5	319166	3988	30.5	567196	6080	35.5	937012	8913
20.6	163375	2479	25.6	323154	4025	30.6	573276	6129	35.6	945925	8979
20.7	165854	2505	25.7	327179	4060	30.7	579405	6177	35.7	954904	9045
20.8	168359	2532	25.8	331239	4097	30.8	585582	6227	35.8	963949	9112
20.9	170891	2558	25.9	335336	4134	30.9	591809	6277	35.9	973061	9179
21.0°	.0173449	2585	26.0°	.0339470	4170	31.0°	.0598086	6326	36.0°	.0982240	9247
21.1	176034	2612	26.1	343640	4207	31.1	604412	6376	36.1	991487	9315
21.2	178646	2640	26.2	347847	4245	31.2	610788	6427	36.2	1000802	9383
21.3	181286	2667	26.3	352092	4282	31.3	617215	6477	36.3	1010185	9452
21.4	183953	2694	26.4	356374	4320	31.4	623692	6529	36.4	1019637	9522
21.5	186647	2722	26.5	360694	4357	31.5	630221	6580	36.5	1029159	9591
21.6	189369	2750	26.6	365051	4396	31.6	636801	6631	36.6	1038750	9662
21.7	192119	2778	26.7	369447	4434	31.7	643432	6684	36.7	1048412	9732
21.8	194897	2806	26.8	373881	4473	31.8	650116	6735	36.8	1058144	9803
21.9	197703	2835	26.9	378354	4512	31.9	656851	6789	36.9	1067947	9875
22.0°	.0200538	2863	27.0°	.0382866	4550	32.0°	.0663640	6841	37.0°	.1077822	9947
22.1	203401	2892	27.1	387416	4590	32.1	670481	6895	37.1	1087769	10019
22.2	206293	2922	27.2	392006	4630	32.2	677376	6948	37.2	1097788	10092
22.3	209215	2950	27.3	396636	4670	32.3	684324	7002	37.3	1107880	10166
22.4	212165	2980	27.4	401306	4709	32.4	691326	7057	37.4	1118046	10239
22.5	215145	3009	27.5	406015	4750	32.5	698383	7110	37.5	1128285	10314
22.6	218154	3039	27.6	410765	4790	32.6	705493	7166	37.6	1138599	10388
22.7	221193	3069	27.7	415555	4832	32.7	712659	7221	37.7	1148987	10464
22.8	224262	3099	27.8	420387	4872	32.8	719880	7277	37.8	1159451	10539
22.9	227361	3130	27.9	425259	4913	32.9	727157	7332	37.9	1169990	10615
23.0°	.0230491	3160	28.0°	.0430172	4956	33.0°	.0734489	7389	38.0	.1180605	10692
23.1	233651	3191	28.1	435128	4996	33.1	741878	7446	38.1	1191297	10769
23.2	236842	3221	28.2	440124	5039	33.2	749324	7502	38.2	1202066	10847
23.3	240063	3253	28.3	445163	5082	33.3	756826	7559	38.3	1212913	10925
23.4	243316	3284	28.4	450245	5124	33.4	764385	7618	38.4	1223838	11004
23.5	246600	3316	28.5	455369	5166	33.5	772003	7675	38.5	1234842	11082
23.6	249916	3347	28.6	460535	5210	33.6	779678	7733	38.6	1245924	11163
23.7	253264	3378	28.7	465745	5253	33.7	787411	7793	38.7	1257087	11242
23.8	256642	3411	28.8	470998	5297	33.8	795204	7851	38.8	1268329	11323
23.9	260053	3444	28.9	476295	5341	33.9	803055	7911	38.9	1279652	11404
24.0°	.0263497	3476	29.0°	.0481636	5384	34.0°	.0810966	7970	39.0°	.1291056	11486
24.1	266973	3508	29.1	487020	5430	34.1	818936	8031	39.1	1302542	11568
24.2	270481	3542	29.2	492450	5474	34.2	826967	8091	39.2	1314110	11651
24.3	274023	3575	29.3	497924	5518	34.3	835058	8152	39.3	1325761	11734
24.4	277598	3608	29.4	503442	5564	34.4	843210	8214	39.4	1337495	11818
24.5	281206	3642	29.5	509006	5610	34.5	851424	8275	39.5	1349313	11903
24.6	284848	3675	29.6	514616	5655	34.6	859699	8337	39.6	1361216	11987
24.7	288523	3709	29.7	520271	5702	34.7	868036	8399	39.7	1373203	12072
24.8	292232	3744	29.8	525973	5748	34.8	876435	8463	39.8	1385275	12159
24.9	295976	3777	29.9	531721	5794	34.9	884898	8525	39.9	1397434	12245
25.0°	.0299753		30.0°	.0537515		35.0°	.0893423		40.0°	.1409679	

NOTE: For any value of ϕ not given in table, special working tables may be easily constructed. Thus:

inv 46.6 = tan 46.6 − arc 46.6 = 1.0574704 − .8133234 = .2441470
inv 46.7 = tan 46.7 − arc 46.7 = 1.0611742 − .8150688 = .2461054
inv 46.8 = tan 46.8 − arc 46.8 = 1.0648918 − .8168141 = .2480777

TABLE 24.13 Chordal Tooth Thickness and Chordal Addendum for Gears of 1 Diametral Pitch—Divide Values by Diametral Pitch for Other Pitches

(for notation, see illustration)

Notation
Chordal addendum—a_c
Chordal thickness —t_c
Rise of arc   —h_c
Addendum   —a
Pitch diameter —d

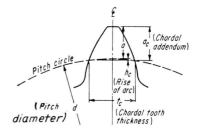

No. of teeth N	Chordal thickness, t_c	Chordal addendum, a_c	Rise of arc, h_c	No. of teeth N	Chordal thickness, t_c	Chordal addendum, a_c	Rise of, arc, h_c
8	1.56072	1.07686	0.07686	43	1.57045	1.01435	0.01435
9	1.56283	1.06836	0.06836	44	1.57046	1.01401	0.01401
10	1.56434	1.06158	0.06158	45	1.57048	1.01371	0.01371
11	1.56546	1.05597	0.05597	46	1.57049	1.01318	0.01318
12	1.56631	1.05133	0.05133	47	1.57050	1.01312	0.01312
13	1.56697	1.04733	0.04733	48	1.57051	1.01285	0.01285
14	1.56750	1.04401	0.04401	49	1.57052	1.01259	0.01259
15	1.56793	1.04109	0.04109	50	1.57053	1.01234	0.01234
16	1.56827	1.03852	0.03852	51	1.57053	1.01209	0.01209
17	1.56856	1.03623	0.03623	52	1.57055	1.01186	0.01186
18	1.56880	1.03425	0.03425	53	1.57056	1.01164	0.01164
19	1.56901	1.03245	0.03245	54	1.57057	1.01142	0.01142
20	1.56918	1.03083	0.03083	55	1.57058	1.01121	0.01121
21	1.56933	1.02936	0.02936	56	1.57059	1.01102	0.01102
22	1.56946	1.02803	0.02803	57	1.57059	1.01082	0.01082
23	1.56957	1.02681	0.02681	58	1.57060	1.01063	0.01063
24	1.56967	1.02569	0.02569	59	1.57061	1.01045	0.01045
25	1.56976	1.02466	0.02466	60	1.57062	1.01028	0.01028
26	1.56984	1.02372	0.02372	61	1.57062	1.01011	0.01011
27	1.56991	1.02284	0.02284	62	1.57062	1.00995	0.00995
28	1.56997	1.02202	0.02202	63	1.57063	1.00979	0.00979
29	1.57003	1.02126	0.02126	64	1.57064	1.00964	0.00964
30	1.57008	1.02056	0.02056	65	1.57064	1.00949	0.00949
31	1.57012	1.01989	0.01989	66	1.57064	1.00934	0.00934
32	1.57016	1.01927	0.01927	67	1.57065	1.00921	0.00921
33	1.57020	1.01869	0.01869	68	1.57065	1.00907	0.00907
34	1.57024	1.01814	0.01814	69	1.57065	1.00884	0.00884
35	1.57027	1.01762	0.01762	70	1.57066	1.00881	0.00881
36	1.57030	1.01713	0.01713	71	1.57066	1.00869	0.00869
37	1.57032	1.01667	0.01667	72	1.57066	1.00857	0.00857
38	1.57035	1.01623	0.01623	73	1.57067	1.00845	0.00845
39	1.57037	1.01581	0.01581	74	1.57067	1.00834	0.00834
40	1.57039	1.01541	0.01541	75	1.57068	1.00822	0.00822
41	1.57041	1.01504	0.01504	76	1.57068	1.00812	0.00812
42	1.57043	1.01468	0.01468	77	1.57068	1.00801	0.00801

NOTE: Table values reprinted from "The Involute Curve and Involute Gearing," with permission of the publishers, The Fellows Gear Shaper Co., Springfield, Vt.

TABLE 24.13 *(Continued)*

No. of teeth N	Chordal thickness, t_c	Chordal addendum, a_c	Rise of arc, h_c	No. of teeth N	Chordal thickness, t_c	Chordal addendum, a_c	Rise of arc, h_c
78	1.57068	1.00791	0.00791	93	1.57071	1.00664	0.00664
79	1.57068	1.00781	0.00781	94	1.57071	1.00657	0.00657
80	1.57068	1.00771	0.00771	95	1.57072	1.00649	0.00649
81	1.57068	1.00762	0.00762	96	1.57072	1.00641	0.00641
82	1.57069	1.00752	0.00752	97	1.57072	1.00634	0.00634
83	1.57069	1.00743	0.00743	98	1.57072	1.00631	0.00631
84	1.57069	1.00734	0.00734	99	1.57072	1.00623	0.00623
85	1.57069	1.00725	0.00725	100	1.57073	1.00617	0.00617
86	1.57070	1.00717	0.00717	110	1.57074	1.00561	0.00561
87	1.57070	1.00713	0.00713	120	1.57075	1.00514	0.00514
88	1.57070	1.00701	0.00701	Rack	1.57080	1.00000	0.00000
89	1.57070	1.00693	0.00693				
90	1.57070	1.00685	0.00685				
91	1.57071	1.00678	0.00678				
92	1.57071	1.00671	0.00671				

24.4 ARC AND CHORD DATA

Gear teeth are often measured by vernier calipers. The addendum setting for the calipers has to be made a slight amount larger than the gear addendum. Conversely, the chordal tooth thickness value is a slight amount smaller than the arc tooth thickness of the part. Table 24.13 illustrates the "rise of arc" h_c and the "chordal thickness" t_c. Tabulated values are given for 1 diametral pitch. The chordal thicknesses given have no allowance for backlash. The values shown should be divided by the pitch and then given a proper allowance for backlash and an appropriate tolerance.

The values given in Table 24.13 may be read in millimeters for 1 module. For other sizes of teeth, *multiply* the table value by the module.

24.5 HARDNESS TESTING DATA

Almost all drawings for gear parts will need a hardness testing specification. The method of hardness testing will depend on the material and the size and shape of piece. Table 24.14 gives general recommendations for the more commonly used hardness test methods. Table 24.15 shows the approximate conversion between the different hardness test scales.

TABLE 24.14 Hardness Testing Apparatus and Application for Gears

Instrument	Shape and Type of Indenter	Loading	Recommended Use
Brinell	10-mm steel or tungsten carbide ball	3000 kg	For large gears and shafts in range of hardness from 100 to 400 Brinell. (If gears have a hard case, the case depth must be sufficient and the core strength adequate to support the area under test. Rockwell C tests may be used to reveal any error from such irregular conditions.) Brinell tests, when fairly representative of the general hardness, are a good measure of the ultimate tensile strength of the material.
Rockwell C	Diamond Brale penetrator (120° diamond cone)	150 kg	For gears medium to large in size. Range approximately 25 to 68 Rockwell C
Rockwell A	Diamond Brale penetrator (120° diamond cone)	60 kg	For small gears and tips of large gear teeth. Range 62 to 85 Rockwell A
Rockwell 30-N	Diamond Brale penetrator (120° diamond cone, special indenter)	30 kg	For small parts and shallow case-hardened parts
Rockwell 15-N	Diamond Brale penetrator (120° diamond cone, special indenter)	15 kg	Lightest Rockwell load. For testing very small parts and checking the working sides of teeth. Check for very thinly case-hardened parts. Rockwell 15-N is used 72-93.
Vickers, pyramid	136° diamond pyramid	50 kg	All applications where piece will not be too heavy for machine. For use in testing hardness of shallow cases, etc.
Scleroscope	Does not indent surface. Diamond-tipped tup bounced on specimen		For applications permitting no damage or indentation to surfaces (results not always comparable with indentation hardness tests)
Tukon	Knoop indenter	500–1000 g for practical use	A laboratory instrument used only for finding the hardness of material on pieces of cross sections, e.g., hardness from surface

TABLE 24.14 Hardness Testing Apparatus and Application for Gears *(Continued)*

Instrument	Shape and Type of Indenter	Loading	Recommended Use
			inward, every 0.05 mm (0.002 in.), or hardness of individual microconstituents. All test areas must have flat mirror-polished surfaces. An extremely delicate and precise test. Used for any hardness

TABLE 24.15 Approximate Relation between Hardness Test Scales

Brinell 3000 kg, 10 mm	Rockwell				Vickers Pyramid	Scleroscope (Shore)	Tukon (Knoop)
	C	A	30-N	15-N			
	70	86.5	86.0	94.0	1076		
	65	84.0	82.0	92.0	820	90	846
	63	83.0	80.0	91.5	763	87	799
614	60	81.0	77.5	90.0	695	81	732
587	58	80.0	75.5	89.3	655	79	690
547	55	78.5	73.0	88.0	598	74	630
522	53	77.5	71.0	87.0	562	71	594
484	50	76.0	68.5	85.5	513	67	542
460	48	74.5	66.5	84.5	485	64	510
426	45	73.0	64.0	83.0	446	61	466
393	42	71.5	61.5	81.5	413	56	426
352	38	69.5	57.5	79.5	373	51	380
305	33	67.0	53.0	76.5	323	45	334
250	24	62.5	45.0	71.5	257	37.5	
230	20	60.5	41.5	69.5	236	34	
	B	Rockwell					
			30-T	15-T			
200	93		78.0	91.0	210	30	
180	89		75.5	89.5	189	28	
150	80		70.0	86.5	158	24	
100	56		54.0	79.0	105		
80*	47		47.7	75.7			
70*	34		38.5	71.5			

*Based on 500-kg load and 10-mm ball.

REFERENCES

1. Dudley, D. W., *Gear Handbook,* McGraw-Hill, New York, 1962.
2. Buckingham, Earle, *Manual of Gear Design,* Sec. 1, Mathematical Tables, The Industrial Press, New York, 1935.
3. Dudley, D. W., *Handbook of Practical Gear Design,* McGraw-Hill, New York, 1984.
4. *The Van Keuren Handbook 36,* The Van Keuren Co., Watertown, Massachusetts.
5. Vogel, W. F., *Involutometry and Trigonometry,* Michigan Tool Co., Detroit, 1954.

Index

Abrasive wear, **15.**13
Acceleration limits, **13.**12, **13.**17
Accuracy effects on noise, **14.**10
Adaptation factor, **11.**30, **11.**43
Addendum:
  avoiding under cut, **6.**11
  for balanced strength, **4.**39
  for balanced fatigue life, **6.**12
  for balanced flash temperature, **6.**12
  basic geometry, **6.**8, **6.**9
  for bevel gears, **4.**58
  for face gears, **4.**65
  for helical gears, **4.**48
  internal gears, **4.**50, **4.**51
  modification calculations, **6.**8–**6.**15
  nonstandard, **5.**16
  (*See also* Tooth proportions)
  for small tooth numbers, **4.**38
  tip shortening, **6.**12–**6.**14
  tooth thinning for backlash, **6.**12, **6.**14
  for zero backlash, **6.**10
Aircraft gear, quality, **7.**34
Aircraft unit, **12.**67
AISI steels, **8.**12, **8.**13
Alignment of gears, **12.**38
Aluminum bronze, **8.**30
Aluminum gears, die-cast, **8.**33–**8.**36
  general practice, **8.**33, **8.**34
ANSI/AGMA standard 1001-B88, **15.**27
Antiwear additives, **15.**31
Apex, angle, **1.**29
Approximate vibration limits, **13.**10–**13.**15
Arc and chord data, **24.**27
Arrangements:
  articulated gear, **3.**9, **3.**12
  automatic transmissions, **3.**10, **3.**24
  bevel gears, **3.**25–**3.**28
  contrarotating gears, **3.**23
  counter shaft, **3.**9, **3.**10

Arrangements (*Cont.*):
  differential gears, **3.**26–**3.**28
  epicyclic gears, **3.**14–**3.**24
  (*See also* Epicyclic gears)
  gear-shift, **3.**28–**3.**32
  harmonic drive, **3.**33, **3.**34
  high-ratio units, **3.**31, **3.**32
  MDT-type gear, **3.**9, **3.**11, **3.**12
  parallel-axis, **3.**6, **3.**7, **3.**14
  reverse gear units, **3.**26
  reverted, **3.**9
  right-angle gears, **3.**24–**3.**26
  split torque, **3.**26, **3.**27
Articulated gear, **3.**9, **3.**11, **3.**12
Assembly:
  bores and shafts, **7.**40
  center distance tolerances, **7.**44
  dimensioning of center distance, **7.**45
Automatic honing, **18.**27
Automatic shaving, **18.**14
CNC, **16.**14, **16.**30, **18.**14, **20.**5
Axis, of screw rotation, **1.**17, **1.**18
Axodes, **1.**8, **1.**13, **1.**15

Backlash:
  for assembled gears, **5.**8
  for bevel gears, **4.**59
  design considerations, **4.**27, **4.**28
  for power gears, **4.**39
  for worm gears, **4.**63
Base cylinder, **1.**9
Basic concepts, helical gears, **1.**8
Basic kinematic relations, **1.**32
Basic law of meshing, **1.**30, **1.**31
Bearing contact, **1.**38, **1.**39
Bearing losses:
  of ball and roller bearings, **12.**17–**12.**24
  of sleeve bearings, **12.**17, **12.**18
  of thrust bearings, **12.**19–**12.**21

Bearings, **15.2**
  design recommendations, **10.16**
  load calculations, for bevels, **10.23–10.46**
    for hypoids, **10.23–10.46**
    for Spiroid gears, **10.52–10.54**
    for worm gears, **10.46–10.51**
  reactions on, **10.4, 10.5**
  separating load chart for bevels, **10.25**
  thrust reaction chart for bevels, **10.24**
Bending stress calculations, **6.19–6.22, 11.47**
Bevel and hypoid gears transmission error, **14.17**
Bevel and hypoid manufacturing procedure, **20.11**
  grinding:
    accuracy, **20.36**
    aircraft gears, **20.35**
    nital etch, **20.37**
    stock allowance, **20.36**
  lapping:
    compounds, **20.42**
    machines, **20.41**
    by manual position, **20.42**
  spiral bevel, zerol bevel, and hypoid, **20.11, 20.12**
  straight toothed, **20.11**
  work-holding equipment, **20.12, 20.13**
Bevel cutters, **20.8, 20.14, 20.17, 20.18, 20.21–20.28, 20.31, 20.44–20.46**
Bevel gear blanks, requirements, **20.9, 20.10**
Bevel gear inspection:
  backlash, **23.28**
  blank tolerances, **20.10**
  machines for, **23.25**
  procedures for, **23.23–23.25**
  tooth thickness, **23.28**
Bevel gear mounting, heat-treat allowances, **23.28, 23.29**
  load deflections, **23.31–23.34**
  V and H checks, **23.34–23.38**
Bevel gears:
  assembly procedure, **10.33, 10.34**
  axial thrust chart, **10.24**
  axodes, **1.13**
  blank design, **10.31, 10.32**
  calculation of bearing loads, **10.34–10.46**
  calculation procedure, **4.52–4.58**
  cutting (*see* Bevel generating)
  definition, **2.10**
  drawing practice, **9.14–9.18**

Bevel gears (*Cont.*):
  efficiency, **12.4–12.17**
  formulas, **2.39**
  grinding (*see* Bevel grinding)
  lapping, **20.40–20.42**
  manufacturing procedures, **20.11**
  nomenclature, **2.38**
  pitch cones, **1.14**
  scoring criterion, **11.55**
  separating load chart, **10.25**
  spiral (*see* Spiral bevel gears)
  straight (*see* Straight bevel gears)
  strength rating, **11.40–11.47**
  surface durability rating, **11.21–11.40**
  work-holding devices, **20.13**
Bevel generating, basic generator, **20.2, 20.5**
  basic process, **20.1**
  CNC computer numeric control, **20.5, 20.6**
  universal 6 axis machine, **20.7, 20.47**
  cutting methods, **20.28**
  cutters and types:
    cyclex completing, **20.23**
    diameter of, **20.25**
    hardac finishing, **20.23**
    Helixform completing, **20.23**
    maintenance, **20.23**
    nongenerating, **20.21**
    RSR completing, **20.26**
    single cycle finishing, **20.23**
    specifications, **20.24–20.26**
    tan-tru roughing and finishing, **20.26, 20.27**
    Tri-Ac completing, **20.27, 20.28**
    use of, **20.22**
  face hobbing, **20.4, 20.20, 20.43**
    cutters, **20.45, 20.46**
    life data for gears, **20.48**
  face milling, **20.4–20.20**
    life data for gears, **20.48**
    machines for, **20.39, 20.40**
    process, **20.34**
    rough cutting, **20.29**
Bevel generating:
  blank requirements, **20.9–20.11**
  CNC, **20.5–20.7**
  cutting methods, **20.16, 20.17**
  face hobbed, **20.3, 20.4**
  face milled, **20.3, 20.4**
  general information, **20.1–20.7**
  straight bevel machines, **20.7–20.20**

Bevel generating (*Cont.*):
  work-holding devices, **20**.13
  (*See also* Bevel nongenerator)
Bevel generating and nongenerating
    processes, **20**.32–**20**.34
  semifinishing practices, **20**.33
  stock allowance, **20**.33
Bevel grinding:
  accuracy, **20**.36
  aircraft gears, **20**.35
  consideration, **20**.36
  general information, **20**.35–**20**.37
  nital etch inspection, **20**.37
  semifinishing practice, **20**.36
  stock allowance, **20**.36
  work holding devices, **20**.13
  (*See* also Bevel nongenerator)
Bevel nongenerator:
  basic nongenerator, **20**.7
  correlation of machines, **20**.17, **20**.32
  gear blank requirement, **20**.9
    dimensions, **20**.10
    tolerance, **20**.10
  general information, **20**.7, **20**.8
  machines for, **20**.34, **20**.37–**20**.39
    format, **20**.38
  processes, **20**.30–**20**.34
    helixform, **20**.30
    format, **20**.31
  Revacycle machine, **20**.18
Beveloid gears, **2**.16
Block, **15**.10, **15**.21, **15**.22
Borsoff, **15**.22
Boundary lubrication, **15**.8, **15**.9
Brass die castings, **8**.37
Brinell hardness, **24**.18, **24**.29
Broaches:
  care and use, **22**.39, **22**.40
  design elements, **22**.37–**22**.39
  design modifications, **22**.35
  for external gears, **22**.37
  general information, **22**.33–**22**.35
  pull, **22**.34
  for spiral splines or gears, **22**.36, **22**.37
  tooling tips, **22**.39–**22**.41
  types, **22**.33–**22**.35
Broaching:
  advantages and limitations, **16**.52
  application of, **16**.45, **16**.46
  cutting speeds, **16**.5
  general information, **16**.45–**16**.53
  operation, **16**.48–**16**.52

Broaching (*Cont.*):
  principles, **15**.46–**15**.48
Bronze:
  aluminum, **8**.30, **8**.32
  gear bronzes, **8**.29, **8**.30, **8**.32
  manganese, **8**.30–**8**.32
  phosphor, **8**.30, **8**.32
  silicon, **8**.30, **8**.32
  strength, **8**.32
  typical uses, **8**.33

Capacity, estimating (*see* Rating)
Carburizing:
  case depth, **8**.19
  case hardness, **8**.21
  cycles, **8**.20
  general practice, **8**.18–**8**.21
  time vs. depth, **8**.19
Carper, **15**.24
Case hardness, **8**.21
Cast iron:
  ductile, **8**.28
  general properties, **8**.27
  gray, **8**.27
  Meehanite, **8**.28
Center distance:
  adjustment for backlash, **5**.8, **7**.40
  backlash control, **4**.90
  dimensioning, **7**.45
  effect of mounting distance, **4**.92, **4**.93
  general equations, **4**.73–**4**.76
  nonstandard, **4**.17
    calculations, **4**.82, **4**.83
    equations, **4**.81, **4**.82
  operating, **4**.80
  special cases, **4**.77, **4**.78
  standard, **4**.78, **4**.79
  temperature effects, **6**.90, **6**.91
  tolerance effects, **6**.83–**6**.90
Centrodes, **1**.2–**1**.6
Chordal addendum, tables, **24**.26, **24**.27
Chordal thickness:
  calculation, **5**.6, **6**.2
  tables, **24**.26, **24**.27
Churning losses, **15**.3
Circular pitch, **1**.3
Climb hobbing, **16**.25, **16**.26
Clockwork tooth profiles, **4**.3
CNC bevel gear machines, **20**.5–**20**.7, **20**.47
Coefficient of friction:
  for ball bearings, **12**.22
  for Spiroid gears, **12**.14

Coefficient of friction (*Cont.*):
  for spur gears, **12.**6
  for worm gears, **12.**13
Cold drawing (*see* Extrusion)
Cold scuffing, **15.**14, **15.**15
Compound helical, **3.**9
  planetary, **3.**15, **3.**18
Composite checking, **23.**6
Composite tolerances:
  definition, **7.**15
  total, **7.**15
Compound:
  epicyclic gear index, **5.**16
  gear train, **3.**9
  planetary, **3.**15
Computer-aided drawings (CAD), **9.**22–**9.**27
Computer program:
  for fillet coordinates, **6.**3, **6.**7
  for inverse of involute, **6.**6, **6.**9
  use of, **6.**1
Cone drive gears, sample drawing, **9.**20, **9.**21
  (*See also* Worm gears)
Conical involute, **2.**16
Coniflex cutters, **20.**17
Coniflex gears, **20.**15, **20.**16
Coniflex provisions, **20.**15, **20.**16
Conjugate action, **4.**8
Contact ellipse, **1.**38, **1.**39
Contact length, **5.**4
Contact lines, **1.**31
Contact patterns for bevel gears, **23.**35
Contact ratio, **4.**7, **5.**4
Continuity of action, **4.**5
Contrarotating gears, **3.**21, **3.**23
Control gears, **23.**36–**23.**38
Control of vibration, **13.**15
Copper base die castings, **8.**29
Critical scoring index for oils, **15.**22
Critical temperature, **15.**21
Creep-fed grinding, **19.**11
Crossed axis helical gears:
  description, **2.**18
  estimated power capacity, **11.**55
  formula, **2.**36
  geometry, **1.**14
  nomenclature, **2.**36
Crowned teeth:
  drawing notes, **9.**11
  by shaving, **18.**9
Curvature coefficient, **11.**23
Cutting methods, bevel gears, **20.**28

Cyclo palloid bevel gear system, **20.**42–**20.**45
Cycloidal tooth profiles, **4.**30

Damping noise, **14.**22, **14.**23
Delrin gears, **8.**40
Diameter over pins (*see* Measurement over wires)
Diametral pitch, **1.**3
  recommended values, **4.**35
Die-cast materials, **8.**31–**8.**37
Die casting:
  general information, **17.**1
  production practice, **17.**7
Die processes:
  casting, **17.**6
  cold drawing or extrusion, **17.**14–**17.**16
  general information, **17.**1–**17.**2
  injection molding, **17.**2–**17.**6
  rolling or forming, **17.**16–**17.**17
  sintering powdered metals, **17.**11
  stamping gears, **17.**17–**17.**21
Differential gears:
  efficiency, **12.**32–**12.**36
  fixed differentials, **3.**27–**3.**30
  free differentials, **3.**26
  hypoid differentials, **3.**25
Displacement limits, **13.**13, **13.**15
Displacement probes, **13.**5
Double-enveloping worms and worm gears, cutting practice, **20.**15–**20.**18
Double helical gears, **2.**7, **2.**8
Drafting (*see* Drawings)
Drawings:
  approximating involute profile, **9.**4
  bevel and hypoid gears, **9.**14–**9.**18
  CAD systems, **9.**22–**9.**27
  face gears, **9.**22, **9.**24
  gear sets, **9.**8
  general practice, **9.**1, **9.**2
  helical gears, **9.**11–**9.**14
  Spiroid gears, **9.**22, **9.**23
  spur gear practice, **9.**7–**9.**11
  surface finish, **9.**6
  tolerances on tooth accuracy, **9.**9
  tooth profile, **9.**9–**9.**11
  worm and wormgears, **9.**18–**9.**19
Ductile iron, **8.**28
Dudley, Darle W.:
  internal gear practice, **4.**49, **15.**21
  total pitch variation, **23.**30
Duplex spread-blade method, **20.**28

Durability (*see* Surface durability)
Duration of contact, **5.**4
Dynamic factor K_v, **11.**14, **11.**15

Efficiency:
  approximate losses, **10.**8
  arc of approach and recess, **4.**20, **4.**21
  of crossed-axis helical gears, **12.**13
  of differential gears, **12.**32–**12.**36
  general considerations, **12.**1
  helical gears, **12.**8, **12.**9
  of Helicon gears, **12.**15
  of hypoid gears, **12.**16
  of internal gears, **12.**7, **12.**8
  of planetary gears, **12.**28–**12.**32
  of planoid gears, **12.**16
  of spiral bevel gears, **12.**10
  of Spiroid gears, **12.**14, **12.**15
  of spur gears, **12.**5–**12.**7
  of star gears, **12.**32
  of straight bevel gears, **12.**9
  typical, **3.**4
  of worm gears, **12.**11–**12.**13
  of Zerol bevel gears, **12.**10
Elastic coefficient, **11.**29
Elastohydrodynamic lubrication (EHL EHD), **15.**9, **15.**19
Elliptical gears, **2.**28
Envelope of contact lines on generating surface, **1.**35
Environment conditions, **12.**39
Epicyclic gears, **3.**14–**3.**24
  compound epicyclic trains, **3.**18, **3.**19
  coupled epicyclic trains, **3.**19
  efficiency, **12.**29–**12.**32
  planetary, **3.**20
  simple epicyclic trains, **3.**14–**3.**16
Epicyclic gears (noise), **14.**22
Epicyclic gears:
  solar, **3.**17
  star, **3.**16
Equation of meshing, **1.**30
Error in action, **1.**40–**1.**42
Euler-Savary equation, **20.**47
Extreme pressure additive, **15.**31
Extrusion:
  design data, **17.**15, **17.**16
  general information, **17.**15, **17.**16
  production practice, **17.**16

Face contact ratio, m_F, **11.**23
Face contact ratio modeling, **14.**24

Face gears:
  description, **2.**15–**2.**16
  design, **4.**63–**4.**67
  formulas, **2.**47
  nomenclature, **2.**47
  sample drawings, **9.**24
Face hobbing bevel gears, **20.**4, **20.**20, **20.**43
Face-mill cutters, **20.**3
Face milling bevel gears, **20.**4–**20.**20
Failures of gears:
  general considerations, **12.**40, **12.**41
  by pitting, **12.**48–**12.**51
  by plastic flow, **12.**47
  by tooth breakage, **12.**55, **12.**56
  by wear, **12.**41–**12.**44
Fatigue stress, **11.**46
Fellows 20° stub tooth system, **4.**30
Ferrous gear materials, **8.**1
  (*See also* Materials)
Field testing, **12.**69
Fillet radius calculations, **6.**3–**6.**7
Finish (*see* Surface finish)
Finishing gears (*see* Grinding; Honing; Lapping; Shaving)
Flash temperature, **15.**10, **15.**11, **15.**21
  calculations, **6.**12
Fly cutting wormgears, **21.**13
Forced feed lubrication, **15.**4–**15.**8
Forged gears, **17.**12, **17.**13
  process, **17.**13
Form diameter, definition, **4.**3
Form grinding:
  estimating time, **19.**19, **19.**20
  general information, **19.**16, **19.**17
Formate gears, **20.**31, **20.**38
Formate tooth form, **4.**55
Formulas:
  circular pitch, **2.**33
  for crossed helical gears, **2.**36, **2.**37
  diametral pitch, **2.**33
  for double-enveloping worm gears, **2.**46
  for face gears, **2.**47
  general relationships, **2.**30, **2.**33
  for helical gears, **2.**33, **2.**34
  for internal gears, **2.**35, **2.**36
  for spiral bevel gears, **2.**41
  for Spiroid gears, **2.**49
  for spur gears, **2.**33
  for straight bevel gears, **2.**37–**2.**39
  for worm gears, **2.**44, **2.**46
  for Zerol bevel gears, **2.**42
Frequency spectrum, **14.**4

Fretting corrosion, **12**.58
Friction coefficient, **15**.23, **15**.24
  of spur gears, **12**.6
  of worm gears, **12**.12
Frictional forces, effect on noise, **14**.26
Fujita, **15**.23
FZG tests, **15**.26

Gabrikov, **15**.24
Gear arrangements, **3**.1
Gear checking:
  analytical, **23**.2–**23**.6
    automatic machines, **23**.3
    fine pitch gears, **23**.48–**23**.51
    index, **23**.3
    involute, **23**.2
    load, **23**.3
    spacing, **23**.3
    topological plots, **23**.9
  bevel and hypoid, **23**.23–**23**.38
  contact patterns, **23**.10, **23**.35
  extra large gears, **23**.45–**23**.48
  fine pitch gears, **23**.48–**23**.51
  functional, **23**.6–**23**.13
    double flank tester, **23**.6–**23**.12
    gear roller, **23**.1
  helical, **23**.13–**23**.22
  spur, **23**.13–**23**.22
  wear, **23**.42–**23**.44
  worm gears, **23**.38–**23**.42
Gear definition, **2**.32
Gear grinding (*see* Grinding)
Gear hobbing (*see* Hobbing)
Gear-milling cutters (*see* Milling cutters)
Gear noise characteristics, **14**.1–**14**.28
Gear ratio, **1**.2
Gear rolling (*see* Rolling)
Gear shaving (*see* Shaving)
Gear teeth (*see* Tooth proportions; Tooth thickness)
Gear teeth impact, **14**.6
Gear tooth surface modification, **1**.39–**1**.42
Gear types, **2**.1–**2**.3
  special, **2**.27–**2**.30
Generated surfaces, **1**.31
Generating bevel gears (*see* Bevel generating)
Generating grinding:
  conical wheel, **19**.15
  disk wheel, **19**.8, **19**.9
  saucer wheel, **19**.14

Generating grinding (*Cont.*):
  (*See also* Bevel grinding)
Geometry factor:
  I, **11**.25, **11**.26
  J, **11**.40, **11**.42, **11**.44
Ghost frequencies, **14**.27
Gray cast iron, **8**.27
Grinding bevel gears, **20**.35–**20**.37, **20**.46
Grinding:
  abrasives, **19**.4–**19**.7
    aluminum oxide, **19**.4–**19**.6
    cubic boron nitride (CBN), **19**.4–**19**.6
    diamond, **19**.4–**19**.6
    grain shape, **19**.5
    hardness, **19**.4, **19**.5
    silicon nitride, **19**.4–**19**.6
    thermal conductivity, **19**.6
  form, **19**.16–**19**.20
  general information, **19**.16, **19**.17
  generating methods, **19**.12–**19**.19
  mechanics, **19**.2, **19**.3
  process control, **19**.10, **19**.11
  reasons for, **19**.1, **19**.2
  thread, worm, **21**.18–**21**.20
  threaded wheel method, **19**.12, **19**.13
  (*See also* Bevel grinding)
Grinding wheels, **19**.7–**19**.15
  general information, **19**.7, **19**.8
  types of wheels, **19**.3, **19**.12–**19**.16
  wheel preparation, **19**.8, **19**.9

Hand of spiral, **10**.34
Hard gear finishing, **20**.46, **22**.41–**22**.44
  CBN crystals, **22**.41, **22**.42
  cutters for, **22**.42–**22**.44
Hardness:
  effect of carbon content, **8**.2, **8**.4
  hardenability of steel, **8**.7
  of steel and iron, **8**.8
Hardness scales, **24**.29
Hardness testing data, **24**.28, **24**.29
Harmonic drive:
  applications, **3**.34
  arrangements, **3**.33
  special properties, **3**.34
Harmonics of gears, **14**.21
  wave generator, **3**.33
Heat treatment of steels, **8**.3
  data, **8**.16
  glossary, **8**.8–**8**.12
Helical gears:
  basic concepts, **1**.8

Helical gears (*Cont.*):
  basic formulas, **2**.33
  bearing reactions, **10**.18–**10**.23
  calculation procedure, **4**.47
  description, **2**.6, **2**.7
  design practice, **4**.43–**4**.49
  drawing practice, **9**.11–**9**.14
  estimating power capacity, **11**.37, **11**.55
  face width, **4**.46
  hobbing, **16**.7, **16**.13, **16**.22
  honing, **18**.27–**18**.33
  nomenclature, **2**.33
  quality classes, **7**.15–**7**.23
  rack cutters, **1**.10
  scoring criterion, **11**.35
  screw parameter, **1**.10
  shaping, **16**.35
  shaving, **18**.1–**18**.18
  splash lubrication, **15**.1
  strength rating, **11**.40–**11**.47
  surface durability rating, **11**.21–**11**.40
  threaded wheel grinding, **19**.12
Helical pinion and rack, **3**.8
Helical rack, **1**.10
Helicon gears:
  description, **2**.22, **2**.26
  design, **4**.67–**4**.72
  efficiency, **12**.15
Helix angle, **1**.10
  design, **4**.46
Helix base cylinder, **1**.12
Helix measurement (*see* Lead measurement)
Helix tangent, **1**.11
Helixform cutter, **20**.21
Helixform gear, **20**.21
Herringbone gears, shaping, **16**.30, **22**.11
Hertzian pressure, **15**.13
Hertz stress, **11**.33
  derivation, **6**.15–**6**.19
  for external gears, **5**.7
  for internal gears, **5**.7
High contact ratio gears, **5**.15
Hirano, **15**.23
Hob approach distance, **16**.20, **16**.21
Hob overrun, **16**.20
Hob shifting, **16**.24
Hobbing:
  application, **16**.6
  climb, **16**.25, **16**.26
  CNC, **16**.14
  conventional, **16**.25, **16**.26

Hobbing (*Cont.*):
  of double-enveloping worm gears, **21**.15, **21**.18
  estimating production, **16**.16–**16**.21
  feeds and speeds, **16**.12–**16**.16
  general information, **16**.8–**16**.11
  locating of hob, **16**.23, **16**.24
  machines, **16**.9–**16**.12
  steel, **22**.9
  theory, **16**.8
  wear of hob, **16**.24
  of wormgears, **21**.7–**21**.13
  of worm threads, **21**.4–**21**.7
Hobs:
  accuracy, **22**.5
  carbide tipped, **16**.15
  classification, **22**.4, **22**.5
  design, **16**.16
  feeds and speeds, **22**.7, **22**.9
  general principles, **22**.1–**22**.4
  material, **22**.11, **22**.12
  nomenclature, **22**.3
  selection, **22**.5–**22**.9
  sharpening, **22**.9, **22**.10
  for spur and helical, **22**.1
  standard sizes, **22**.6–**22**.8
  TiN coating, **22**.10, **22**.11
  tolerance, **22**.5
  wear, **16**.19, **16**.24
Honing:
  accuracy, **18**.32
  automatic cycle, **18**.29
  finish, **18**.33
  general information, **18**.27–**18**.33
  ground gears, **18**.31
  machines, **18**.29, **18**.30
  process, **18**.29
  shaved gears, **18**.31
  tools, **18**.31
Hot scuffing, **15**.14, **15**.15
Housing treatment for noise, **14**.23
Housing vibration, **14**.22
Hyperboloids of revolution, **1**.17, **1**.18
Hypoid gears:
  calculation of bearing loads, **10**.34–**10**.46
  cone distances, **1**.23
  description, **1**.23–**1**.26
  efficiency, **12**.15
  estimating power capacity, **11**.56
  four arrangements, **10**.35
  geometry, **1**.23–**1**.26

Hypoid gears (*Cont.*):
  lapping machines and processes, 20.40–20.42
  manufacturing procedures, 20.11, 20.12
  operating pitch surfaces, 1.23
  pitch plane, 1.24
  pitch surfaces, 1.23
  sample drawings, 9.18
  velocity polygon, 1.25

Idler bearing reaction, 10.16
Impingement depth, 15.4–15.7
Index tolerance, 7.2, 7.38
Index variation, 23.3, 23.20
Induction hardening:
  depth of case, 8.25
  frequencies required, 8.25
  general information, 8.22–8.26
Infeed hobbing, 21.8, 21.10, 21.11
Injection molding:
  design data, 17.5, 17.6
  general information, 17.2–17.6
  process, 17.2–17.5
Inspection of gears (*see* Gear checking)
Installation (*see* Assembly)
Instantaneous axis of rotation, 1.13
Instantaneous center of rotation, 1.13
Integral temperature, 15.24
Interchangeability, design considerations, 4.21
Interchangeable gear system, 3.1, 3.2
Interchangeable gears, definition of, 3.2
Interference:
  addendum modification coefficient, 6.27
  calculation, 5.13–5.15
  cycloidal overcut, 6.24
  description, 4.22
  involute interference, 6.22
  radial assembly, 6.26
Interference limit, 5.13
Internal gears:
  bearing reactions, 10.15
  broaching, 16.45–16.51
  description, 2.8, 2.9
  design, 4.49–4.52
  efficiency, 12.7, 12.8
  form grinding, 19.16–19.19
  formulas, 3.46
  interference, 5.13
  measurement over rolls, 5.10
  nomenclature, 2.36
  shaping, 16.27

Internal gears (*Cont.*):
  shaving, 18.11–18.14
  space checking, 23.22
Interval of contact, 5.4–5.7
Into-mesh oil jet, 15.4
Involute:
  approximate equation, 6.8
  coordinate equations, 6.1–6.3
  curve, 1.28
  definition, 6.6
  exact equation, 6.8
  error, 23.15
  functions, 5.3, 5.4, 24.24, 24.25
  inverse of, 6.6–6.9
  measurement, 23.2, 23.14
  pointed teeth, 4.13–4.15
  study of tooth development, 4.12
  zone of action, 4.10–4.13
Involute charts, 9.10
Involute checking machines (*see* Profile measurement)
Involute function, equation, 5.3
  polar angle, 5.3
  tables, 24.24, 24.25
Involute profile measurement (*see* Profile measurements)
Involute roll angles, 23.2, 23.17

Jominy end quench, 8.7
Journal failure, 12.59

Kelley, 15.21, 15.22
K factor, 11.14, 11.21
Kinematic viscosity, 15.27
Ku and Baber, 15.23

Lambda, 15.8, 15.9, 15.28, 15.29
Lapping:
  of bevel gears, 20.40–20.42
  capacity of bevel and hypoid machines, 20.41
  compounds, 20.42
  machines and processes, 20.41, 20.42
Lay of finish, 7.41
Lead angle, 1.11
Lead measurement:
  by contact check, 23.17
  definitions, 23.3, 23.15
  by helix checking machines, 23.15, 23.16
Length of contact, 5.4
Lexan gears, 8.40
Life factors, general data, 11.48

INDEX

Limit contact normal, **1.35–1.37**
Limit diameter, definition, **4.3**
Limits of accuracy (*see* Tolerances)
Line of action, **5.4**
  definition, **1.6**
  effective, **1.6**
Line of contact, helical gears, **1.12**
Load distribution factor, C_m K_m, **11.19, 11.26**
Load rating (*see* Rating)
Loading factors, **11.48**
Long addendum design, **4.16, 4.17, 4.39**
Low friction gearing, **4.19–4.21**
Lubricant entrainment, **14.6**
Lubricant functions, **12.39**
Lubricant properties, **15.30**
Lubrication:
  ANSI/AGMA standard 20001-B88, **15.27**
  Borsoff, **15.22**
  bulk temperature θ_m, **15.23, 15.26**
  Carper, **15.24**
  critical scoring index for oils, **15.22**
  critical temperature, **15.21**
  Dudley, **15.21**
  estimation of influence on surface distress, **15.16–15.30**
  estimation of scuffing load capacity, **15.24**
    equation, **15.25**
  flash temperature and X_r, **15.25**
  flash temperature to contact stress ratio (equation), **15.24**
  flash temperature to square root of specific normal load, **15.24**
    equation, **15.24**
  forced feed, **15.4–15.7**
  friction coefficient, **15.23, 15.24**
  friction power (equation), **15.23**
  Fujita, **15.23**
  FZG gear test, **15.26**
  Gabrikov, **15.24**
  geometry factor X_B, **15.25**
  guidance to determine lubrication condition, **15.18**
  Hirano, **15.23**
  integral temperature, **15.24**
  intensity, **15.23**
  Kelley, **15.21, 15.22**
  kinematic viscosity, **15.27**
  Ku and Baber, **15.23**
  limit of scoring load (table and equation), **15.23**
  load sharing factor X_r, **15.25**

Lubrication (*Cont.*):
  lubrication factor for pitting Z_L, **15.16**
  lubricant properties, **15.30**
  Matveevsky, **15.23, 15.24**
  mean critical contact temperature, **15.28**
  mean flash temperature, **15.24**
  Moorhous, **15.24**
  Niemann and Lechner, **15.23**
  Niemann and Seitzinger, **15.24**
  pitting, **15.16–15.20**
  pitting versus Hertzian stress, **15.18**
  practical estimation of pitting (equation), **15.18**
  pressure angle factor $X_{\alpha\beta}$, **15.25**
  probability of wear, **15.27**
  PVT equation, **15.20**
  roughness factor or pitting Z_R, **15.16**
  Ryder gear tests, **15.27**
  scoring factor (equation), **15.22**
  scoring index, **15.21**
  scuffing, **15.20**
  scuffing initiation, **15.24**
  smooth surface finish, **15.18**
  speed factor for pitting Z_v, **15.17**
  surface distress, **15.8, 15.9, 15.12, 15.29**
  surface finish, **15.21**
  Taki, **15.17**
  thermal flash factor X_m, **15.25**
  wear failures, **15.29**
  wear-related distress, **15.29**
  welding factor, table, **15.27**
  Winter and Michaelis, **15.24**
Lubrication, additive effects on pitting life, **15.31, 15.32**
  antiwear additives, **15.31**
  extreme pressure additives, **15.31**
  surface fatigue life of, **15.32**
Lubrication and cooling methods, **15.4–15.8**
  cooling requirements, **15.7, 15.8**
  impingement depth, **15.4–15.7**
  into-mesh jet, **15.4**
  out-of-mesh jet, **15.4**
  radial jet, **15.5**
  vectorial model, **15.6, 15.7**
Lubrication and surface distress:
  asperity contact, **15.13**
  Block, **15.10, 15.21, 15.22**
  boundary lubrication, **15.8, 15.9**
  cold scuffing, **15.14, 15.15**
  elastohydrodynamic lubrication EHL
    EHD, **15.9, 15.19**

Lubrication and surface distress (*Cont.*):
  flash temperature, **15.**10, **15.**11, **15.**21
    calculation, **15.**10, **15.**21
    maximum design life for oils, **15.**22
  Hertzian pressure, **15.**13
  hot scuffing, **15.**14, **15.**15
  lambda, **15.**8, **15.**9, **15.**28, **15.**29
  macropitting, **15.**13, **15.**14
  micropitting, **15.**13
  minimum film thickness H_{min}, **15.**9
  mixed lubrication, **15.**8, **15.**9
  regimes of lubrication, **15.**9, **15.**18, **15.**19
  scuffing/scoring, **15.**14
  subsurface shear stress, **15.**13
  surface distress, **15.**12, **15.**13
  temperature in oil film, **15.**12
  thermal EHL model (TEHL), **15.**11
  thermal stability, **15.**15
  wear, **15.**13
Lubrication factor, **11.**32
Lubrication, power losses, **15.**2, **15.**3
  bearings, **15.**2
  churning, **15.**3
  tooth friction, **15.**3, **15.**23, **15.**24
  windage, **15.**3

Machinability of steel and iron, **8.**4
Magnesium-base die castings, **8.**31–**8.**37
Master gear, **23.**11, **23.**12
  drawing specifications, **9.**9
Material fatigue limits, **11.**30
Materials:
  ferrous, **8.**1
  nonferrous, **8.**28
  (*See also* Steels; other specific materials)
Matveevsky, **15.**23, **15.**24
MDT-type gear, **3.**11, **3.**12
Mean flash temperature, **15.**24
Measurement of gears:
  over rolls or balls, **5.**9–**5.**13
  by vernier calipers, **23.**30
Measurement over wires, data:
  for 14 1/2° pinions, **24.**22
  for 20° pinions, **24.**22
  extra depth allowance, **24.**20
  for helical gears, **24.**21–**24.**22
  numerical examples, **24.**18
  table, for 14 1/2° external gears, **24.**7–**24.**10
  for 14 1/2° internal gears, **24.**15, **24.**16
  for 20° external gears, **24.**3–**24.**6
  for 20° internal gears, **24.**13, **24.**14
  for 25° external gears, **24.**17

Measurement over wires, data (*Cont.*):
  wire size, **24.**2
Mechanics of gear reactions (*see* Reactions)
Meehanite, **8.**28
Mesh, region of, **4.**11–**4.**14
Milling application, **16.**1–**16.**3
  advantages, **16.**6
  general information, **16.**1–**16.**6
  machine operation, **16.**4, **16.**5
  principles, **16.**3, **16.**4
  rate of cutting, **16.**5
  thread, wormgear, **21.**2–**21.**3, **21.**6
Milling cutters:
  design practices, **22.**24, **22.**25
  general information, **22.**21
  standard cutter numbers, **22.**22
  types used, **22.**22–**22.**24
Misalignment effects on noise, **14.**25
Mixed lubrication, **15.**8, **15.**9
Module, definition of, **2.**31
Molded tooth, **17.**5
Molding die, **17.**4
Moorhous, **15.**24
Mounting distance, relation to center distance, **4.**91, **4.**92
Mounting:
  of bevel gear blanks, **10.**30–**10.**33
  of bevel gears, **10.**25–**10.**30
  of gears on shaft, **10.**7, **10.**11
  general recommendations, **10.**9, **10.**10
  general types, **10.**7, **10.**8
  housing design, **10.**11
  overhung, **10.**28, **10.**29
  straddle, **10.**27, **10.**28
  of worm gears, **10.**49

Narrow band width analysis, **14.**8
Niemann and Lechner, **15.**23
Niemann and Seitzinger, **15.**24
Nital etch, **20.**37
Nitralloy (*see* Nitriding)
Nitriding:
  case depth, **8.**24
  general practice, **8.**22
  materials for, **8.**23
  process data, **8.**23
  time vs. depth, **8.**23
Nodular iron, **8.**28
Noise, **14.**1–**14.**32
  causes of, **14.**5–**14.**7
  air pocketing, **14.**6

Noise, causes of (*Cont.*):
    dynamic mesh forces, **14.**6
    frictional forces, **14.**6
    gear impact, **14.**6
    lubricant entrainment, **14.**6
    mesh stiffness, **14.**5
    transmission error, **14.**5
  comments, **14.**28
  effect of gear accuracy, **14.**10
  gear rattle, **14.**25
  general, **14.**1–**14.**5
    excitations, **14.**4
    frequency spectrum, **14.**4
    harmonics, **14.**4
    limits, **14.**1
    transmission paths, **14.**3
  measurement, **14.**7–**14.**9
    narrow band width analysis, **14.**8
    octave band, **14.**7
    sound pressure level, **14.**7
    time synchronous averaging, **14.**8
  modeling, **14.**24, **14.**25
    face contact ratio, **14.**24
    misalignment, **14.**25
    time domain models, **14.**24
    tip relief, **14.**25
  power consideration, **14.**10
  reduction, **14.**19
    bearing housing interaction, **14.**22
    damping, **14.**22, **14.**23
    design parameters, **14.**20
    epicyclic gears, **14.**22
    excitation minimization, **14.**21
    gear blank resonances, **14.**22
    housing treatment, **14.**23
    housing vibration, **14.**22
    torsional and lateral vibration, **14.**21
    unique gear geometries, **14.**23
  standards, **14.**7
  trouble shooting, **14.**26–**14.**28
    characterization of gear noise, **14.**26
    frequency plots, **14.**26, **14.**27
    ghost frequencies, **14.**27
    implementing a fix, **14.**28
    instrumentation, **14.**26
    special measurement and modeling, **14.**28
    time synchronous averaging, **14.**28
    tolerance considerations, **7.**38
    waterfall plots, **14.**26, **14.**27
Nomenclature:
  elements of teeth, **2.**32, **2.**34

Nomenclature (*Cont.*):
  general information, **2.**30, **2.**31
  metallurgy, **8.**8–**8.**12
  tooth calculations, **5.**1, **5.**6
  (*See also* Symbols)
Nonferrous gear materials, **8.**28–**8.**41
Nonstandard addendum, **5.**16
Nonstandard gears, **1.**2, **5.**16
Nonundercutting conditions, **1.**33–**1.**35
Normal operating pressure angle, **1.**20
Normal section, calculation of, **8.**3
Number of teeth:
  for bevel gears, **4.**56
  for face gears, **4.**64–**4.**66
  pinion, to avoid undercut, **4.**33
    general data, **4.**32
    for power gears, **4.**35
  for worm gears, **4.**62
Nylon gears:
  material, **8.**39, **8.**40
  molding practice, **17.**2–**17.**6

Octave band, **14.**7
Operating pitch cylinders, **1.**12
  worm gear drives, **1.**21
Operating pitch surfaces, hypoid gear drive, **1.**23
Oscilloscope patterns, **13.**10
Out-of-mesh oil jet, **15.**4
Overhung mounting, **10.**28, **10.**29
Overload factors, **11.**13

Parallel axis gears, **3.**7
Parametric representation of curved surfaces, **1.**26–**1.**30
  normal, **1.**27, **1.**28
  plane curve, **1.**27
  tangent, **1.**27
Performance of gears, **12.**36
Phenolic laminates, **8.**40
Phosphor bronze, **8.**30
Pin measurement, **9.**7
Pin size, **24.**2
Pitch circle, **1.**2
Pitch cone, **1.**14
Pitch cylinder, **1.**9
  worm gears, **1.**21
  worm operating, **1.**21
Pitch diameter:
  operating, **4.**10
  standard, **4.**9
Pitch measurement, **23.**20–**23.**22

Pitch plane hypoid gear, **1**.24
Pitch relationships, tables, **24**.26, **24**.27
Pitch surfaces (axodes):
 bevel gears, **1**.14, **1**.16
 crossed helical gears, **1**.19
 helical gears, **1**.13
Pitting, **12**.49, **15**.16, **15**.20
Planar gearing, **1**.32
Planetary gears, arrangements, **3**.14–**3**.18
 bearing reactions, **10**.17, **10**.18
 efficiency, **12**.31
 (*See also* Epicyclic gears)
Planocentric gear, **3**.31, **3**.33
Planoid gears:
 description, **12**.16
 efficiency, **12**.16, **12**.17
Pointed teeth, **4**.13–**4**.16
 calculation, **4**.16
 diameter, **4**.27
Powdered metal gears (*see* Sintered gears)
Power consideration for noise, **14**.10
Power drives, **4**.19
Power formula, **11**.4, **12**.30
Power loss, **12**.37
Power, maximum gear capacity, **3**.4
 (*See also* Rating)
Preshave cutting, **18**.8
Pressure angle, **1**.3, **1**.6
Profile contact ratio m_p, **5**.4
Profile measurements, **23**.2, **23**.14, **23**.15
 by portable machines, **23**.47
 by projection, **23**.46
Profile modification, design recommendations, **4**.23–**4**.25
Propfan gears, **3**.23
Proportions (*see* Tooth proportions)
Proximity probes, **13**.5, **13**.13
PVT equation, **15**.20

Quality classes:
 AGMA classification, **7**.15–**7**.24
 AGMA fine pitch, **7**.16–**7**.29
 levels of quality, **7**.15

Rack, helical, **1**.10
Radial oil jet, **15**.5
Radioactive gear test, **12**.68
Rake angles (clearance angle), **22**.12
Rating, **11**.40
 bending fatigue limit, **11**.40
  adaptation factor A_F, **11**.45
  bending stress, **11**.47

Rating (*Cont.*):
 conventional fatigue limit, **11**.40
 geometry factor J calculation, **11**.40
  internal gears, **11**.44
  table external gears, **11**.42
 service factor, **11**.47
 size factor, **11**.46
consideration, **11**.1
coplanar gears, **11**.4
 bevel gear rating, **11**.7, **11**.8
 involute parallel and bevel, **11**.4–**11**.8
 life curves, **11**.3–**11**.5
  bending fatigue, **11**.3
  surface fatigue, **11**.5
 synthetic factor K, **11**.6, **11**.8
 tangent load, **11**.4
 torque, **11**.4
 unit bending load factor U_L, **11**.6, **11**.8
crossed helical gears, **11**.55
 capacity tables, **11**.56
cylindrical worm gearing, **11**.57
 capacity tables, **11**.58
detailed analysis, conventional surface and bending fatigue limits and service factors, coplanar gears, **11**.13–**11**.21
 application factor K_a, **11**.14
 dynamic factor K_v AGMA, **11**.14, **11**.15
 dynamic factor K_v ISO, **11**.15
 dynamic resonance, **11**.16–**11**.19
 dynamic vibration, **11**.16–**11**.19
 load distribution factor C_m, K_m, **11**.19
  table, **11**.26
 overload derating factors, **11**.13
 power sharing factor K_{sh}, **11**.14
hypoid gear load rating, **11**.56
life analysis coplanar gears, **11**.47
 life curves definition, **11**.48, **11**.49
 load cycles, **11**.51
 loading factors, **11**.48
 reliability, **11**.53
 tooth damage, **11**.53
 yielding, **11**.51, **11**.52
nomenclature, **11**.2
scoring and wear prevention, **11**.54
 progressive wear, **11**.54
 scoring and scuffing equation, **11**.55
simplified estimates and design criteria, **11**.8–**11**.13
 design procedure, **11**.9–**11**.13
state of standards, **11**.59

Rating (*Cont.*):
  surface fatigue limit of K_{lim} factors
    AGMA and ISO, **11.21–11.40**
    adaptation factors A_H, **11.30, 11.31**
    bevel gears, **11.24**
    contact ratio, **11.21–11.23**
    curvature coefficient X_B, **11.23**
    elastic coefficient C_p, **11.29**
    geometry factor table surface durability, **11.25, 11.26**
    Hertzian stress calculation, **11.33**
    ISO surface durability, **11.25**
    lubrication factor Z_L, **11.32**
    material, surface and bending fatigue limit, **11.30**
    power capacity tables, **11.34–11.39**
      helical gears, **11.37**
      spiral bevel, **11.39**
      spur gears, **11.35**
      straight bevel, **11.38**
    preliminary geometry calculation, **11.21**
    roughness factor Z_r, **11.32**
    service factor C_{SF}, **11.33, 11.34**
    unified geometry factor, G_H spur and helical, **11.24, 11.25**
  Unified geometry charts, **11.28**
    comments and comparisons, **11.27**
Ratio formula, **1.2**
  how to obtain, **3.3**
Rattle, **14.25**
Reactions:
  basic bearing, **10.4, 10.5**
  bevel gear bearings, **10.34–10.40**
  helical gears, **10.18–10.22**
  hypoid gear bearings, **10.34–10.40**
  internal gear bearings, **10.15**
  mechanics of, **10.1–10.3**
  planetary gears, **10.17, 10.18**
  Spiroid gears, **10.52, 10.53**
  spur gear bearings, **10.10–10.14**
  trains of gears, **10.16–10.18**
  worms and wormgears, **10.46–10.49**
Regimes of lubrication, **15.9, 15.18, 15.19**
Region of mesh, **4.11–4.14**
Reliability, **11.53**
Resonance of gear blanks, **14.22**
Revacycle:
  cutter, **20.18**
  machine, **20.18**
Reverter gear, **3.9**
Ridging, **12.48**

Right angle gear, **3.29**
Rockwell hardness, **24.28, 24.29**
Roll angle *e*, calculation, **5.4**
Roll size, **24.1**
Roll finishing, **18.18–18.27**
  data on roll finished gears, **18.25**
  double die gear rolling, **18.23**
  rolling dies, **18.22**
  seaming, **18.21**
  single-die gear rolling, **18.26, 18.27**
  tooth flow pattern, **18.20**
  vertical roll machine, **18.24**
Rolling, general information, **17.16, 17.17**
  worm, **21.21–21.22**
Root fillet, design recommendations, **4.25**
Roughness factor, **11.32**
Run-in of worm gears, **10.51**
Ryder gear tests, **15.27**

Scoring, criterion, **11.55**
Scoring index, **15.21**
Scroll gears, **2.28**
Scuffing/scoring, **15.14**
Sector gears, **2.29**
Sensing devices, **13.4–13.6**
Service factors, classes of service, **11.33, 11.34**
Shallow-hardening, **8.24**
Shaper cutters:
  application to CNC, **22.13**
  clearance angle, **22.12**
  clearance groove, **22.16**
  common types, **22.4–22.19**
  general principles, **22.11, 22.12**
  for helical gears, **22.15, 22.16**
  mounting, **22.14, 22.15**
  sharpening practice, **22.20, 22.21**
  tool clearance, **22.16**
  tooth modification, **22.17, 22.18**
  wafer cutter, **22.18, 22.19**
Shaping application, **16.27**
  clearance groove required, **16.27, 16.30, 22.16**
  CNC, **16.30**
  cutting speeds, **16.35–16.38**
  general information, **16.27–16.31**
  machines for, **16.28, 16.30**
  theory, **16.27**
Shaving:
  angular traverse, **18.5–18.11**
  axial traverse, **18.7, 18.8**
  automation, **18.14**

Shaving (*Cont.*):
  CNC, **18**.14, **18**.15
  contact area, **18**.5
  crossed-axes method, **18**.4–**18**.11
  crown method, **18**.9
  cutter, **18**.12, **18**.14
  cutting vs. guiding action, **18**.6
  diagonal, **18**.7–**18**.10, **22**.26, **22**.27
  feeds and speeds, **18**.16–**18**.18
  general information, **18**.1–**18**.3
  of internal gears, **18**.11–**18**.14
  machines, **18**.14
    machine setup, **18**.15–**18**.18
  methods, **18**.7
  mounting cutter, **18**.16
  mounting gear, **18**.15
  plunge, **18**.10, **22**.28
  preshave form, **18**.6, **18**.7
  production practice, **18**.15–**18**.18
  protuberance for, **18**.6, **18**.7
  right-angle traverse, **18**.5
  rotary crossed axes, **18**.4–**18**.12
  stock allowance, **18**.7, **18**.8, **22**.31
  tangential, **18**.10
  traverse, **18**.8–**18**.10, **22**.26
  undercut, **18**.7
  underpass, **18**.10, **22**.26, **22**.27
Shaving cutters:
  direction of feed, **22**.28, **22**.29
  general information, **18**.12, **18**.13, **22**.25
  general principles, **22**.25–**22**.30
  pivot point, **22**.28
  preshave gear, **22**.31
  resharpening, **18**.13, **22**.32, **22**.33
  tool geometry, **22**.30
Shear cutting application, **16**.38
  advantages, **16**.44
  general information, **16**.38–**16**.41
  machines for, **16**.40, **16**.43
  principles, **16**.39
  tools, **16**.41
  typical production rates, **16**.44
Silicon bronze, **8**.30, **8**.32
Simple gear train, **3**.8
Sintered gears, general information, **17**.11
Size, gear, relative, **3**.6
  tolerance control, **9**.24–**9**.27
  (*See also* Rating)
Size factor K_S, **11**.46
Sliding velocity:
  of bevel gears, **12**.3

Sliding velocity (*Cont.*):
  of crossed-axes helical gears, **12**.4
  of hypoid gears, **12**.4
  of spur gears, **12**.2
  of wormgears, **12**.3
Solar gears, **3**.17
  (*See also* Epicyclic gears)
Spacing measurement, definitions, **23**.3, **23**.19
Spacing variation:
  analysis of index check, **23**.20, **23**.29
  analysis of pitch check, **23**.20
  index checking, **23**.3, **23**.6
  pitch checking machines, **23**.21–**23**.22
Spectrum analysis, **13**.8
Speed-increasing drives, **4**.18
Spiral bevel gears:
  description, **2**.13
  sample drawings, **9**.16, **9**.17
Spiral bevel gear manufacture, **20**.11–**20**.37
Spiral bevel systems (other):
  cyclo-palloid system, **20**.42–**20**.45
  grinding, **20**.46
  hard finishing, **20**.46
  KNC machine, **20**.47, **20**.49
  longitudinal crowning, **20**.44
  tooling system, **20**.45
Spiroid gears, calculation of bearing loads, **10**.52–**10**.54
  description, **2**.22, **2**.26
  design, **4**.67–**4**.72
  drafting practice, **9**.22, **9**.23
  efficiency, **12**.14
  formulas, **2**.49
  nomenclature, **2**.48
  proportions, **4**.68–**4**.71
Splash lubrication, **15**.1
Spline failure, **12**.59
Split torque, **3**.27
Spur gears:
  bearing reactions, **10**.12–**10**.15
  broaching, **16**.45–**16**.53
  composite tolerances, **7**.18, **7**.22
  description, **2**.2, **2**.3
  drawing practice, **9**.2–**9**.6
  efficiency, **12**.5–**12**.8
  fillet coordinates, **6**.3–**6**.7
  form grinding, **19**.16–**19**.19
  generating grinding, **19**.12–**19**.16
  hobbing, **16**.6–**16**.26
  honing, **18**.27–**18**.33

Spur gears:
  made by die processes, **17.1–17.11**
  milling, **16.1–16.6**
  quality classes, **7.15**
  scoring criterion, **15.22**
  shaping, **16.26–16.38**
  shaving, **18.4–18.11**
  shear cutting, **16.38–16.45**
  strength rating **11.40–11.47**
  surface durability rating, **11.21–11.40**
  threaded wheel grinding, **19.12, 19.13**
  tolerances, **7.16–7.23**
    examples, **7.6, 7.7**
Square gears, **2.27**
Stamping:
  design data, **17.17–17.21**
  general information, **17.17**
  process, **17.19–17.21**
Standard system:
  for bevel gears, **4.52–4.58**
  for face gears, **4.63**
  general considerations, **4.28**
  for helical gears, **4.43**
  for Helicon gears, **4.67**
  for internal gears. **4.49**
  for Spiroid gears, **4.67**
  for spur gears, **4.29–4.38**
  for worm gears, **4.58**
Star gears, **3.17**
(*See also* Epicyclic gears)
Steels:
  alloy type, **8.12–8.17**
  carbon type, **8.12–8.17**
  carburizing, **8.18–8.21**
  choice of, **8.13**
  composition, **8.13**
  general properties, **8.2**
  hardenability, **8.7**
  heat treatment, **8.3–8.7**
  induction hardening, **8.22–8.26**
  localized hardening, **8.17**
  nitriding, **8.22**
  physical properties, **8.2**
  processing, **8.15**
  shallow hardening, **8.24**
Stock allowance for bevel gears, **20.33**
Straddle mounting, **10.27–10.29**
Straight bevel gears:
  description of, **2.11, 2.19**
  manufacturing practice, **20.11–20.19**
Straight bevel gear machines, **20.12**
  interlocking cutter machines, **20.16**

Straight bevel gear machines, interlocking cutter machines (*Cont.*):
  Coniflex provision, **20.16**
  cutters, **20.17**
  machine cycles, **20.17**
  required setup data, **20.17**
  revacycle machines, **20.18**
    automatic deburring, **20.19**
    cutter and machine cycle, **20.18**
    cutter maintenance, **20.20**
    required setup data, **20.19**
    two-cut operation, **20.19**
  two-tool generators and roughers, **20.14**
    coniflex, **20.15**
    finishing cycle, **20.16**
    required setup data, **20.14**
    roughing, **20.14**
    tools, **20.14**
Strength of steel gears, **8.2**
Strength rating, allowable fatigue stress, **11.40–11.47**
Sungear vibration, **13.10**
Surface action, **1.31**
Surface distress, **15.12, 15.13**
Surface durability:
  allowable contact stress, **11.21–11.40**
  bevel gears, **11.38, 11.39**
  cylindrical wormgears, **11.57**
  general formula, **11.21–11.40**
Surface fatigue life, **15.32**
Surface finish:
  definition of, **7.40**
  vs. dimensional tolerances, **7.40–7.43**
  of profile, **7.42**
  standard values, **7.41**
Surface principal directions, **1.37, 1.38**
Surface texture, **7.40**
Symbols:
  allocation of letters, **2.31, 2.32**
  Greek letters, **2.32**
  (*See also* Nomenclature)
Synthetic service factor, **11.6.11.8**

Taki, **15.17**
Tangent load, **11.4**
Tangential hobbing, **21.12**
Teeth (*see* Tooth...)
Temperature, effect on center distance, **4.90–4.92**
Tensile strength of steel, **8.2, 8.3**
Terminology:
  of bevel gears, **2.10**

Terminology (*Cont.*):
of internal gears, **2.8**
of spur gears, **2.2**
(*See also* Nomenclature)
Testing:
of accessory boxes, **12.63**
for efficiency, **12.65, 12.66**
by four-square torque loop, **12.61**
of lubricants, **12.68**
by power absorption, **12.60**
Testing gear units, **13.16**
Tests of power packages, **13.18**
Thermal stability, **15.15**
Thickness of teeth (*see* Tooth thickness)
Thread grinding:
worms, **21.18–21.20**
production rates, **21.20**
(*See also* Grinding)
Thread hobbing:
production time, **21.7**
speed and feed, **21.5**
worms, **21.4–21.7**
(*see also* Hobbing)
Thread milling, worms:
general information, **21.2, 21.3**
on hobber, **21.4**
production time, **21.6, 21.7**
(*See also* Milling)
Thread rolling (*see* Rolling)
Threaded wheel grinding, general principles, **19.12, 19.13**
Time domain modeling, **14.24**
Time synchronous averaging, **14.27**
Tolerances:
AGMA quality classes, **7.15–7.28**
allowance vs., **7.5**
by application, **7.34**
for bevel gears, **7.24**
on center distance, **7.44**
choice of, **7.11, 7.12**
comparison AGMA, DIN, JIS, ISO, **7.28–7.33**
composite, **7.18, 7.19**
definition, **7.3, 7.5**
drawing practices, **7.13**
effects of, **7.8**
error vs. deviation, **7.5**
examples of, **7.6**
by fabrication method, **7.29–7.31, 7.37**
factors affecting choice, **7.11**
for helical gears, **7.14–7.23**
magnitudes, **7.6**

Tolerances (*Cont.*):
by materials, **7.36**
mounting effects, **7.14–7.45**
by pitch line speed, **7.35**
position (tangential variation), **7.33**
size control, **7.31, 7.32**
for spur gears, **7.14–7.23**
standards, **7.15**
terminology, **7.1–7.4**
types, **7.9–7.11**
for wormgears, **7.14–7.26**
for various gear types, **7.12**
(*See also* Quality classes)
Tooth breakage, **12.55, 12.56**
Tooth numbers (*see* Number of teeth)
Tooth proportions:
basic considerations, **4.5, 4.6**
for bevel gears, **4.52**
definition of elements, **4.2–4.4**
for face gears, **4.63–4.66**
for helical gears, **4.43, 4.44**
for Helicon gears, **4.67–4.72**
for internal gears, **4.49**
low-friction drives, **4.19, 4.20**
obsolete systems, **4.42**
power drives, **4.19**
speed-increasing drives, **4.18, 4.19**
for Spiroid gears, **4.67–4.72**
for spur gears, **4.29–4.43**
standard data, **4.40**
for worm gears, **4.58–4.63**
(*See also* Standard systems)
Tooth thickness:
calculation, **5.6**
chordal (*see* Chordal thickness)
design considerations, **4.21–4.28**
factor, **5.11**
measurement:
at any diameter, **5.9, 5.10, 7.32**
at any radius, **5.6**
by dimension over pins, **23.45**
over rolls or balls, **5.9–5.13**
by span method, **23.45**
on spiral bevel gears, **23.28, 23.30**
tables of dimensions over wires, **24.3–24.20**
by tooth calipers, **23.45**
by vernier calipers, **23.30**
Top land, width, **4.26**
Toprem feature, **20.25**
Torque, **11.4**
Total contact ratio, **5.6**

Transmissions:
  automatic, **3.**10, **3.**24, **3.**32
  gear shift, **3.**10, **3.**13, **3.**31
  reverse gear units, **3.**26
Transmission error, **1.**40–**1.**42, **14.**11–**14.**19
  accuracy effects:
    eccentricity, **14.**18
    other, **14.**19
    tooth profile, **14.**18
  bevel and hypoid, **14.**17, **14.**18
  effect of face contact ratio, **14.**24
  effect of misalignment, **14.**25
  effect of profile modification, **14.**13–**14.**15
  effect of tip relief, **14.**25
  factors contributing to, **14.**11
  high contact ratio gears, **14.**15, **14.**16
  loaded, **14.**11, **14.**13
  measurement, **14.**12
  spur gears, **14.**11
Transverse shaving, **22.**26, **22.**27
Triangle gears, **2.**27
Trigonometric functions, **24.**23

Undercut:
  design table, **4.**33
  general discussion, **4.**16, **4.**17
  in spiral bevel gears, **4.**57
  in straight bevel gears, **4.**56
Undercutting, **5.**13–**5.**15
Underpass shaving, **22.**26, **22.**27
Unitool method, **20.**33
V and H check, **23.**32–**23.**36
Velocity limits, **13.**11, **13.**16
Velocity plots, **13.**9
Velocity polygon:
  crossed helical, **1.**20,
  hypoid, **1.**25
  worm, **1.**22
Velocity transducers, **13.**6
Versacut process, **20.**34
Vibration of gears, **12.**38, **13.**1–**13.**20
  approximate vibration limits, **13.**10–**13.**15
    acceleration limits, **13.**12, **13.**17
    displacements limits, **13.**13, **13.**15
    general vibration tendencies, **13.**13, **13.**14
    velocity limits, **13.**11, **13.**16
    vibration trade standard, **13.**14
  control of vibration, **13.**15
    field tests, **13.**18
    testing gear units, **13.**16
    tests of power packages, **13.**18

Vibration of gears   (*Cont.*):
  example, **13.**7–**13.**10
    oscilloscope patterns, **13.**10
    spectrum analysis, **13.**8
    sun gear vibration, **13.**10
    velocity plot, **13.**9
  fundamentals, **13.**1–**13.**3
  measurements, **13.**3–**13.**7
    displacement probes, **13.**5
    measurement problems, **13.**6
    proximity probes, **13.**5, **13.**13
    sensing devices, **13.**4–**13.**6
    velocity transducer, **13.**6
  vibration analysis techniques, **13.**19
    table of analysis methods, **13.**20

Waterfall plots, **14.**26, **14.**27
Wear, **11.**54, **12.**41–**12.**44, **15.**13, **23.**42–**23.**44
  from inadequate lubrication, **15.**15
Weight of gears, relative, **3.**5
Windage, **15.**3
Windage losses, **12.**24–**12.**28
Winter and Michaelis, **15.**24
Wire size:
  selection, **24.**2
  table, **24.**2
Workholding devices, **20.**13
Worm gears:
  blank proportions, **10.**49
  calculation of bearing loads, **10.**46–**10.**49
  calculation procedure, **4.**60–**4.**63
  cutter diameter, **4.**63
  cutting, double-enveloping, **21.**15
  cylindrical, design, **4.**60–**4.**63
  description, **2.**19–**2.**22
  drafting practice, **9.**18, **9.**21
  efficiency, **12.**11–**12.**13
  estimating power capacity, **11.**58
  formulas, **2.**44, **2.**46
  hobbing, **21.**4–**21.**13
    chip load, **21.**4, **21.**5
    fly cutting, **20.**13
    general information, **21.**4–**21.**6
    infeed method, **21,**10–**21.**12
    production time, **21.**15, **21.**18, **21.**20
    speed and feed, **21.**5, **21.**14
    tangential method, **21.**12, **21.**13
  inspection, contact check, **23.**38
    of double-enveloping gears, **23.**40–**23.**42
    worm checking, **23.**40

Worm gears (*Cont.*):
  lead angle, **4**.62
  nomenclature, **2**.43, **2**.45
  normal pitches, **4**.62
  operating pitch cylinder, **1**.21
  powder metal compacts, **21**.23–**21**.30
  rating, **11**.57, **11**.58
  time required, **21**.29
  tolerances, **7**.24, **7**.25
  tooth forms, **4**.59
Worms:
  carburizing, **21**.23
  cutter for, **21**.12, **21**.13
  cutting throated worm, **21**.16–**21**.18
  definition, **2**.40, **2**.43
  grinding threads, **21**.18–**21**.20
  hobbing threads, **21**.3–**21**.6

Worms (*Cont.*):
  inspection of, **23**.38–**23**.42
  milling threads, **21**.2, **21**.3
  optional finishing, **21**.23
  rolled threads, **17**.16, **17**.17
  rolling, **21**.21, **21**.22
  sample drawing, **9**.20, **9**.21
  tolerances, **7**.14

Yielding, **11**.51, **11**.52

Zerol bevel gears:
  description, **2**.11, **2**.13
  formulas, **2**.42
  nomenclature, **2**.40
Zinc-base die castings, **8**.33, **8**.34
Zinc die castings, **8**.33, **8**.34

About the Editor

Dennis P. Townsend is Senior Research Engineer and Manager of Gear Research at NASA's Lewis Research Center. He holds the B.S. degree in Mechanical Engineering from the University of West Virginia and, prior to joining NASA in 1962, worked for the Department of Defense and General Electric Company. Mr. Townsend has authored or coauthored over 85 papers in the gear and bearing research area. He serves as the resident gear consultant for NASA, several military groups, and numerous industrial companies. He has served with the Design Engineering Division of ASME for 16 years and was chairman of that division from 1989 to 1990 and chairman of ASME's Power Transmission and Gearing Committee from 1978 to 1983.